FOOD AND BEVERAGE CONSUMPTION AND HEALTH

GRAPES

PRODUCTION, PHENOLIC COMPOSITION AND POTENTIAL BIOMEDICAL EFFECTS

Food and Beverage Consumption and Health

Additional books in this series can be found on Nova's website under the Series tab.

Additional e-books in this series can be found on Nova's website under the e-book tab.

FOOD AND BEVERAGE CONSUMPTION AND HEALTH

GRAPES

PRODUCTION, PHENOLIC COMPOSITION AND POTENTIAL BIOMEDICAL EFFECTS

JOSÉ S. CÂMARA
EDITOR

New York

For permission to use material from this book please contact us:
Telephone 631-231-7269; Fax 631-231-8175
Web Site: http://www.novapublishers.com

Library of Congress Cataloging-in-Publication Data

ISBN: 978-1-63321-402-6

Library of Congress Control Number: 2014943357

Published by Nova Science Publishers, Inc. † New York

Contents

Preface ix

About the Editor xi

Contributors xiii

Section 1: Phenolic Compounds in Grapes and Factors Influencing Their Biosynthesis 1

Chapter 1 Influence of Genotype, Pedoclimatic Conditions, Viticultural Practices and Ripening on the Phenolic Composition of Grapes: A Review 1
Antonietta Baiano

Chapter 2 Grape-Berries Yeast Populations: Influence of Vineyard Management 27
Gustavo Cordero-Bueso, Teresa Arroyo and Eva Valero

Chapter 3 Phenolic Compounds in Grapes and Wines: Chemical and Biochemical Characteristics and Technological Quality 47
Ellen Silva Lago Vanzela, Milla Alves Baffi, Maurício Bonatto Machado de Castilhos, Marcos Roberto Moacir Ribeiro Pinto, Vanildo Luiz Del Bianchi, Afonso Mota Ramos, Paulo César Stringheta, Isidro Hermosín-Gutiérrez and Roberto Da-Silva

Chapter 4 Factors Affecting Antioxidant Activity of Grape Tissues 107
Francesca Cecchini

Chapter 5 Screening the Phenolic Composition of the Wide Diversity of Autochthonous Grape Cultivars from La Mancha (Spain) 135
Isidro Hermosín-Gutiérrez, Sergio Gómez-Alonso, Noelia Castillo-Muñoz, Mónica Fernández-González, Jesús Martínez-Gascueña and Esteban García-Romero

Chapter 6 Grape Seed Extract Antioxidant and Antimicrobial Properties: Use in Active Packaging 165
Javiera F. Rubilar and Rui M. S. Cruz

Chapter 7	Grape Phenols and Their Healthy Properties *A. Peralbo-Molina and M. D. Luque de Castro*	**181**
Chapter 8	Productivity of Stilbene Compounds by Wild Grapes Native to East Asia: *Vitis davidii*, *Vitis kiusiana*, *Vitis ficifolia* Var. *Lobata*, and *Vitis ficifolia* Var. *Ganebu* *Shuji Shiozaki, Taiji Nakamura and Tsuneo Ogata*	**211**
Chapter 9	Obtaining Polyphenolic Extracts from Wine By-Products *G. Davidov-Pardo, M. Navarro, I. Arozarena and M. R. Marín-Arroyo*	**225**
Chapter 10	Grape Stalks Based Sorbents: Towards a Sustainable Alternative for the Detoxification of Divalent Heavy Metal Polluted Effluents *Carlos Escudero-Oñate, Núria Fiol, Jordi Poch and Isabel Villaescusa*	**245**
Chapter 11	Thermal Stability of Major Classes of Polyphenols in Skins, Seeds and Stems of Grape Pomace *Jianmei Yu*	**273**
Chapter 12	Supercritical Fluid Extraction of Oil and Bioactive Compounds from Grape Residues: Experimental Optimization, Modeling and Economic Evaluation *Marcelo M. R. de Melo, Armando J. D. Silvestre and Carlos M. Silva*	**287**
Section 2: Health Benefits of Grape Poylyphenols		**303**
Chapter 13	Healthy Effects of Bioactive Metabolites from *Vitis vinifera* L. Grapes: A Review *R. Perestrelo, C. Silva, J. Pereira and José S. Câmara*	**305**
Chapter 14	Extraction, Characterization and Potential Health Benefits of Grape Seed Procyanidins *Joana Afonso, Cláudia P. Passos, Manuel A. Coimbra, Patrício Soares-da-Silva and Carlos M. Silva*	**339**
Chapter 15	Cardioprotective Effects of Grapes and Phenolic Constituents *Bolanle C. Akinwumi and Hope D. Anderson*	**365**
Chapter 16	*Vitis vinifera* Seed Extract Reduces Adenosine Deaminase Activity in High Glucose-Treated Human Lymphocytes *Luziane P. Bellé, Paula Eliete R. Bitencourt, Gustavo Scola, Mirian Salvador, Claúdia Funchal and Maria B. Moretto*	**387**
Chapter 17	Phenolic Compounds of Grapes (*Vitis vinifera*): Diversity of Molecules, Content and Bioavailability *Mariane Lutz*	**395**

Chapter 18 Potential Benefits of Phenolic Compounds Extracted from Winemaking Wastes **411**
Milla Alves Baffi, Ellen Silva Lago Vanzela, Eleni Gomes, and Roberto Da-Silva

Chapter 19 Health Benefits of Grape Seed Extract **429**
Snehal S. Joshi and Doris H. D'Souza

Index **447**

Preface

Grapes (Vitis spp.) are economically the most important fruit species in the world. Most cultivated vines belong to the European type (Vitis vinefera), the American bunch type (V. labrusca and its derivatives) or Muscadine type (V. rotundifolia). Grapes contain many of the most valuable elements necessary for life. In addition, it is known to have commendable medicinal qualities/properties, attributed, maynly to polyphenols. They are famous as antioxidants, hormones, constituents of essential oils, natural neurotransmitters, and as having many other biological activities. Their antioxidant ability is known to confer many health benefits such as reducing the risk of cardiovascular disease and cancer. They also provide antimicrobial activity for the plant's own defense against invading pathogens. The diversity of structure and activity of phenolic compounds has resulted in a multiplicity of research areas such as chemistry, biotechnology, ecology, physiology, nutrition, medicine, and cosmetics

In the past 30 years the view of plant phenolic metabolites as one of nature's meaningless waste products has been replaced by the current opinion that plant phenolics playcrucial roles in plant ecology and plant physiology. In addition, biomedical research has revealed that dietary phenolics, because of their antioxidant and free radical scavenging properties, play important roles in the prevention of many of the major contemporary chronic

The understanding of grape and wine phenolics is an increasingly important requirement for managing wine styles efficiently and an important way to enhance their healthy benefit effects. The aim of this book is to describe grape production, its phenolic composition and the potential biomedical effects which are currently being researched. The areas of most interest at present and the subjects in which this interest is likely to continue or to increase in the following years have been selected.

The book comprises 19 chapters, each written by authorities in the field with 59 contributing authors in total. Chapters are well referenced with regard to previous work, have appropriate chapter introductory sections accommodating non-specialists, students and experts in the field, and most end with a summary of conclusions and, in some cases, recommendations for future work or new standards, protocols or procedures.

It is helpfully arranged in two different sections and these are discussed separately below. Section I, entitled ''Phenolic compounds in grapes and factors influencing their biosynthesis'', comprises twelve chapters, concerning the phenolic and polyphenolic compounds in grapes and the most important aspects influencing their biosynthesis, the biochemical and physiological changes that occur during fruit development and how they influence the wine quality and its sensorial aspects and physiological activity. Most of chapters are focused on factors affecting the antioxidant and antimicrobial activity of grapes, and on the effect of traditional and novel processing methods of grapes on anthocyanin and polyphenol content. The thermal stability of major classes of polyphenols in skins, seeds and stems, the effect of traditional and novel processing methods of grapes and the screening the phenolic composition of the wide diversity of autochthonous grape cultivars from La Mancha, were also reviewed, as well as the productivity of stilbene compounds by wild grapes, and the supercritical fluid extraction of oil and bioactive compounds from grape residues.

Section II, entitled ''Health benefits of grape poylyphenols'', contains seven chapters. The potential health benefits of grape seed procyanidins; the cardioprotective effects of grapes and phenolic constituents; the influence of *Vitis vinifera* seed extract on the adenosine deaminase activity in high glucose-treated human lymphocytes; the bioavailability of phenolic compounds; and the potential benefits of phenolic compounds extracted from winemaking wastes, were outlined.

"Grapes: production, phenolic composition, and potential biomedical effects" is scientifically written including current trends but also in a style that is easy and clear to understand. It is hoped that it will serve as a most useful text and reference source for grape producers and winemakers as well as for other professionals of the sector, and students of oenology, food technology and similar disciplines.

The editor would like to express their thanks to Nova Science Publishers and all the authors who contributed their expertise and know-how to the success of this book.

I found this book very interesting with a high information content, and, as well as oenologists, physiologists and biologists, I would add to the list chemists, microbiologists, biomedical scientists and others (postgraduate or postdoctoral), who might be interested in reading and learning about some fascinating areas of research and discovery of biomedical effects of grapes.

About the Editor

José S. Câmara, has obtained his Ph.D. in Chemistry, Analytical Chemistry, at the University of Madeira (Portugal) on 16th July, 2004. His thesis entitled "Chemical Characterization of grapes used in Madeira wine production" focused on the establishment of the volatile signature of *Vitis vinifera* L. grapes from different varieties – Boal, Malvazia, Verdelho and Sercial, used in Madeira wine production. Then, from July 2004 until now, he was appointed assistant professor in Chemistry, Analytical Chemistry at the University of Madeira. In 2006-2008, he was awarded by UMa/Banco Santander Totta as the best Younger Scientist at University of Madeira, and in 2009 with "Prémio Zarco 2009" by Banif/Diário de Notícias. Currently, he is in charge with the Food Chemistry and Analytical Characterization of Nanomaterials teaching of the Biochemistry Bachelor's, Applied Biochemistry and Nanochemistry and Nanomaterials Master degree respectively.

He was an elected member of the General Council (2008), and President of the Group of Chromatography of the Portuguese Chemical Society (2007-2011). He is a member of Scientific Societies such as the Portuguese Chemical Society, American Chemical Society (ACS) and International Union of Pure and Applied Chemistry (IUPAC). He is also Associate Editor of the Journal of Integrated OMICS, and member of Editorial Board of International Journal of Spectroscopy (Hindawi Publishing Corporation). He is regular referee of some scientific journals. He was expert-reviewer of the scientific projects/proposals and member of the Chemistry evaluation panel, in the contest of individual fellowships for PhD and post-doctorate 2013 (Portuguese Foundation for Science and Technology).

Currently, he is the Principal Coordinator of the Human Cancer Volatome project (New Indigo, ERA-NET, FP 7). He was Principal Coordinator of VinSaudeMAC and AntiVinMAC projects (Interreg III, MAC), and also participate as investigator in several other R&D projects. In the Madeira Chemistry Research Centre (CQM) Unit (RRR), José S. Câmara manages a research group within the CQM team (Natural Products chemistry group) where he supervises post-doctorants, Ph.D. and Master students. The mains research activities of the group are to establish the volatile composition and chemical characterisation food or food-related matrices and evaluate the biological effect of polyphenols from different sources in oxidative stress and cardiovascular diseases. The development of analytical approaches (based on non-invasive sampling strategies - exhaled breath and urine) provides biochemical basis to understand the metabolic pathways related with several diseases and will be used as a diagnostic tool. The volatile metabolomics in several fields of application, including cardiovascular diseases, cancer and asthma, has emerged as a new frontier in medical diagnois and becomes a field of intensive and atractive studies.

Analytical Chemistry; Food chemistry; Medicinal chemistry; Natural Products Chemistry; Metabolomics and urine analysis; Food bioactive components; Multivariate processing of the instrumental signals, are its maism specialization domains.

In the last years José S. Câmara has authored or coauthored over 55 original papers and comprehensive reviews on Food Chemistry, Analytical Chemistry and Medicinal Chemistry, with more than 960 citations (by 692 documents) (h index =19). He is also co-author of 2 books (ISBN 978-989-97667-0-9 and ISBN 978-972-789-302-7), contributed to several book chapters and over 240 communications in International and Nacional scientific conferences.

He has been actively involved in the organization of several scientific conferences (on the Analytical Chemistry field) namely the VI ENC2009 (Funchal, 2009) (http://www.spq.pt/eventos/cromatografia09/), VII ENC2012 (Porto, January, 2012), XIX ENC2013 (Covilhã, 2013), and 16th ExTech, International Symposium on Advances in Extraction Technologies (Grece 2014).

He integrates the list of the World Top Authors in Experts in Solid Phase Microextraction (BiomedExperts), a solventless sample preparation method, invented in 1989, that uses a fused silica fiber which is coated with a stationary phase.

Contributors

Afonso Mota Ramos
Universidade Estadual Paulista - UNESP,
Cristovão Colombo, São Paulo
Brazil.

Ángela Peralbo Molina, Ph.D. student
Department of Analytical Chemistry,
University of Córdoba,
Spain.
Maimónides Institute of Biomedical Research (IMIBIC),
Reina Sofía University Hospital,
University of Córdoba,
Spain.

Antonietta Baiano, Ph.D.
Department of Agricultural Sciences, Food and Environment,
University of Foggia
Foggia. Italy.

Armando J.D. Silvestre, Ph.D.
Department of Chemistry, CICECO,
University of Aveiro,
Aveiro, 3810-193, Portugal

Bolanle Akinwumi, M.D.
Faculty of Pharmacy, University of Manitoba

Carlos Escudero-Oñate, Ph.D.
Norwegian Institute for Water Research (NIVA),
Oslo.
Chemical Engineering Department, Escola Politècnica Superior,
Universitat de Girona,
Spain.

Carlos M. Silva, Ph.D.
Department of Chemistry, CICECO,
University of Aveiro,
Aveiro, 3810-193, Portugal

Catarina Silva, M.D.
CQM/UMa, Centro de Química da Madeira,
Universidade da Madeira,
Portugal.

Claúdia Funchal, Ph.D.
Postgraduate Program in Biosciences and Rehabilitation,
Methodist University Center, Porto Alegre,
Brazil

Cláudia P. Passos, Ph.D.
Department of Chemistry, CICECO,
University of Aveiro, Campus Universitário de Santiago,
3810-193 Aveiro, Portugal

Derrick Brian Amoako
Department of Nutrition and Food Science,
Texas A&M University,
TX 77843-2472, U.S.A

Doris H. D'Souza
Department of Food Science and Technology,
University of Tennessee-Knoxville,
2600 River Drive, U.S.A.

Eleni Gomes, Ph.D.
Universidade Estadual Paulista - UNESP,
Cristovão Colombo, São Paulo
Brazil.

Ellen Silva Lago Vanzela, Ph.D.
Universidade Estadual Paulista - UNESP,
Cristovão Colombo, São Paulo
Brazil.

Esteban García-Romero, Ph.D.
Instituto de la Vid y el Vino de Castilla-La Mancha,
Spain.

Eva Valero, Ph.D.
Departamento de Biología Molecular e Ingeniería Bioquímica,
Universidad Pablo de Olavide.
Sevilla, Spain.

Francesca Cecchini, Ph.D.
Consiglio per la Ricerca e la Sperimentazione in Agricoltura (CRA).
Research Unit for Enology in Central Italy (ENC)
Italy

G. Davidov-Pardo, Ph.D.
Public University of Navarre.
Food Technology Department
Spain.

Gustavo Cordero-Bueso, Ph.D.
Departamento de Biología Molecular e Ingeniería Bioquímica,
Universidad Pablo de Olavide.
Sevilla, Spain.

Gustavo Scola, Ph.D.
Institute of Biotechnology,
University of Caxias do Sul,
Brazil

Hope D. Anderson, Ph.D.
Faculty of Pharmacy, University of Manitoba
Canadian Centre for Agri-Food Research in Health and Medicine,
St. Boniface Hospital Research Centre,
Canada.

I. Arozarena, Ph.D.
Public University of Navarre.
Food Technology Department
Spain.

Isabel Villaescusa, Ph.D.
Norwegian Institute for Water Research (NIVA),
Oslo.

Isidro Hermosín-Gutiérrez, Ph.D.
Instituto Regional de Investigación Científica Aplicada
Universidad de Castilla-La Mancha
Spain.

Javiera F. Rubilar, Ph.D., M.D.
Departamento de Ingeniería Química y Bioprocesos,
Pontificia Universidad Católica de Chile,
Chile.

Jesús Martínez-Gascueña, Bachelor
Instituto de la Vid y el Vino de Castilla-La Mancha,
Spain.

Jianmei Yu, Ph.D.
Food and Nutritional Sciences Program,
Department of Family and Consumer Science,
North Carolina A&T State University,
USA

Joana Afonso, M.D.
Department of Pharmacology and Therapeutics,
Faculty of Medicine, University of Porto,
4200-319 Porto, Portugal

José S. Câmara, Ph.D., M.D.
CQM/UMa, Centro de Química da Madeira,
Universidade da Madeira,
Portugal.
Centro de Ciências Exatas e de Engenharia
Universidade da Madeira,
Portugal

Jordi Poch, Ph.D.
Applied Mathematics Department, Escola Politècnica Superior,
Universitat de Girona,
Spain

Jorge Pereira, Ph.D.
CQM/UMa, Centro de Química da Madeira,
Universidade da Madeira,
Portugal.

Luziane Potrich Bellé, M.D.
Postgraduate Program in Pharmacology Health Sciences Center,
Federal University of Santa Maria (UFSM).
Brazil.

M. Dolores Luque de Castro, Ph.D.
Department of Analytical Chemistry,
University of Córdoba,
Spain.
Maimónides Institute of Biomedical Research (IMIBIC),
Reina Sofía University Hospital,
University of Córdoba,
Spain.

Marcelo M.R. de Melo, M.D.
Department of Chemistry, CICECO,
University of Aveiro,
Aveiro, 3810-193, Portugal

M. Navarro, Ph.D.
Public University of Navarre.
Food Technology Department
Spain.

Maria Beatriz Moretto, Ph.D.
Postgraduate Program in Pharmacology Health Sciences Center,
Federal University of Santa Maria (UFSM).
Brazil.
Postgraduate Program in Pharmaceutical Sciences Health Sciences Center,
Federal University of Santa Maria (UFSM).
Brazil.

Maria R. Marín-Arroyo, Ph.D.
Public University of Navarre.
Food Technology Department
Spain.

Marcos Roberto Moacir Ribeiro Pinto
Universidade Federal de Viçosa – UFV
Campus Universitário, Minas Gerais
Brazil

Mariane Lutz, Ph.D., M.D.
Centro de Investigación y Desarrollo de Alimentos Funcionales,
Facultad de Farmacia, Universidad de Valparaíso,
Chile.

Manuel A. Coimbra, Ph.D.
Department of Chemistry, QOPNA, University of Aveiro,
Campus Universitário de Santiago,
3810-193 Aveiro, Portugal

Maurício Bonatto Machado de Castilhos
Universidade Estadual Paulista - UNESP,
Cristovão Colombo, São Paulo
Brazil.

Milla Alves Baffi, Ph.D.
Universidade Federal de Uberlândia – UFU, Amazonas
Uberlândia, Minas Gerais, Brazil

Miriam Salvador, Ph.D.
Institute of Biotechnology,
University of Caxias do Sul,
Brazil

Mónica Fernández-González, Ph.D.
Instituto Regional de Investigación Científica Aplicada,
Universidad de Castilla-La Mancha
Spain.
Fundación Parque Científico y Tecnológico de Albacete
Spain.

Noelia Castillo-Muñoz, Ph.D.
Instituto Regional de Investigación Científica Aplicada,
Universidad de Castilla-La Mancha
Spain.

Núria Fiol, Ph.D.
Norwegian Institute for Water Research (NIVA)
Oslo.

Patrício Soares-da-Silva, Ph.D.
Department of Chemistry, CICECO,
University of Aveiro, Campus Universitário de Santiago,
3810-193 Aveiro, Portugal

Paula Eliete Rodrigues Bitencourt, Ph.D.
Postgraduate Program in Pharmacology Health Sciences Center,
Federal University of Santa Maria (UFSM).
Brazil.

Paulo César Stringheta
Universidade Estadual Paulista - UNESP,
Cristovão Colombo, São Paulo
Brazil.

Roberto Da-Silva, Ph.D.
Universidade Estadual Paulista - UNESP,
Cristovão Colombo, São Paulo
Brazil.

Rosa. Perestrelo, Ph.D.
CQM/UMa, Centro de Química da Madeira,
Universidade da Madeira,
Portugal.

Rui M.S. Cruz, Ph.D.
Departamento de Engenharia Alimentar,
Instituto Superior de Engenharia,
Universidade do Algarve,
Portugal.
CIQA-Centro de Investigação em Química do Algarve,
Departamento de Química e Farmácia,
Faculdade de Ciências e Tecnologia,
Universidade do Algarve,
Portugal.

Sergio Gómez-Alonso, Ph.D.
Instituto Regional de Investigación Científica Aplicada,
Universidad de Castilla-La Mancha
Spain.
Fundación Parque Científico y Tecnológico de Albacete
Spain.

Shuji Shiozaki, Ph.D.
Graduate School of Life & Environmental Sciences,
Osaka Prefecture University,
Japan

Snehal S. Joshi
Department of Food Science and Technology,
University of Tennessee-Knoxville,
2600 River Drive, U.S.A.

Taiji Nakamura, Bachelor
Graduate School of Life & Environmental Sciences,
Osaka Prefecture University,
Japan

Teresa Arroyo, Ph.D.
Departamento de Agroalimentación,
Instituto Madrileño de Investigación y Desarrollo Rural Agrario y Alimentario.
Madrid, Spain.

Tsuneo Ogata, Ph.D.
Faculty of Agriculture,
Kochi University, Otsu
Japan.

Vanildo Luiz Del Bianchi
Universidade Estadual Paulista - UNESP,
Cristovão Colombo, São Paulo
Brazil.

Section 1: Phenolic Compounds in Grapes and Factors Influencing Their Biosynthesis

In: Grapes
Editor: José S. Câmara
ISBN: 978-1-63321-402-6

Chapter 1

Influence of Genotype, Pedoclimatic Conditions, Viticultural Practices and Ripening on the Phenolic Composition of Grapes: A Review

Antonietta Baiano*
Department of Agricultural Sciences, Food and Environment, University of Foggia, Via Napoli, Foggia. Italy

Abstract

Phenolic compounds are the main phytochemicals in table and wine grapes. In table grapes, they strongly affect the sensory characteristics of pulps, skins, and seeds. In red wine, phenolics are the main responsible for color, mouth-feel and taste, while in white wines they are generally undesired due to their high susceptibility to oxidation. The importance of the phenolic composition is also related to the antioxidant activity of grapes and derivatives and to their use as functional foods.

The phenolic composition of grapes is affected by several factors: cultivars; pedoclimatic conditions (soil composition, climate, exposure to sunlight, rainfall, altitude and latitude); viticultural practices (irrigation/water stress, balancing growth-yield, type of farming, vine training systems, soil management, use of fertilizers); ripening time. All these parameters are strictly correlated and cannot be individually considered.

The aim of this work was to provide an overview of the most recent studies performed on the phenolic composition of grapes and to supply the readers with an objective analysis of the experimental data.

Keywords: Grapes; phenolic composition, genotype, climatic conditions, viticultutral practices

---

* Corresponding author: Email: antonietta.baiano@unifg.it.

Introduction

According to the Oxford Dictionary (2013), 'grape' is a berry (typically green, purple, or black) growing in clusters on a grapevine. It is a non-climacteric fruit that grows on the perennial and deciduous woody vines of the genus Vitis (Du et al., 2012). There are about 60 species of Vitis, which are mainly located in the temperate zones of the Northern hemisphere and almost equally distributed between America and Asia (Mullins et al., 1992), although they are also diffused in the South of the world.

From a commercial point of view, grapes are usually classified as either table or wine grapes, based on their method of consumption: table grapes are intended for consumption as fresh or dried fruit while wine grapes are used in making wine. On the other hand, as stated by the Council Regulation (EC) N. 479/2008, wine can only be made from grape varieties listed as allowed, and only those vine varieties may be planted for commercial purposes. Each EU member state draws up lists of Vitis vinifera varieties and certain crosses between V. vinifera and other species of the Vitis genus, while American vines, such as Vitis labrusca, may not be used for wine and are not allowed in EU vineyards. Regardless almost all table and wine grapes belong to Vitis vinifera, they are significantly different due to selective breeding operations. Table grape cultivars have large and sometimes seedless berries with relatively thin skin and medium sugar content, while wine grapes have smaller and usually seeded berries with relatively thick skins (from which many aroma precursors and all the pigments of wine comes from) and high sugar content (Donald and Dibbern, 2013). Approximately 71% of the world grape production is used for wine, 27% as fresh fruit, and 2% as dried fruit (Conde et al., 2007).

The quality of grapes, especially of wine grapes, has historically been loosely related to their sugar:acid balance. Thanks to the advances in analytical chemistry over the last half century, the number of chemical attributes associated with grape quality is grown. Some fruit quality indicators have been defined and quantified in order to understand the connection between quality attributes and the sensory properties of grape and wine (Tarara and Lee, 2013). In addition to sugar and acid contents, phenolics and (varietal) aroma compounds are the most important chemical indicators of grape quality.

Phenolic compounds are the main phytochemicals in grapes. They are secondary metabolites of plants, therefore they are only indirectly involved in physiological functions such as growth and reproduction (Harborne, 1994). They arise from the shikimate-phenylpropanoids-flavonoids pathways. According to their molecular structure, the phenolic compounds are divided into four classes: cinnamic and benzoic acids, with one phenolic ring; stilbenes, with two phenolic rings; anthocyanins, flavonols, and flavan-3-ols (also called flavonoid compounds), with three rings; and ellagic acids, with complex rings (Hartle et al., 2005). The remaining compounds are known as non-flavonoids. All phenolic compounds exhibit absorption in the UV region of the spectrum and the colored ones also absorb in the visible region. Each class of phenolic compounds has characteristic absorption ranges: phenols and phenolic acids, 250-290 nm; cinnamic acid derivatives, 290-330 nm; flavones and flavonols, 250 and 350 nm; chalcones and aurones, above 350 nm and a much less intense band at 250 nm; anthocyanins and betacyanins, 475-560 nm and 535-545 nm, respectively, and a subsidiary peak at about 270-275 nm (Harborne, 1964; Mabry et al., 1970; Lattanzio et al., 2006). Phenolics play an important role in pigmentation, growth, reproduction, resistance to pathogens, and sensory properties of berries and wines. Moreover, they exhibit antimicrobial and antioxidant activity and they may protect the human body against potential diseases (Harborne and Williams, 2000).

Grapes, in particular the red ones, are rich in phenolic compounds (Fuleki and Ricardo-Da-Silva, 2003) distributed among seeds (62%), skins (33%), and juices-pulps (5%) (Zoecklein et al., 1990). Anthocyanins are the main phenolic compounds in purple grapes whereas flavan-3-ols (i.e. catechins) are the more abundant phenolics in white varieties. The total phenolic content is higher in purple varieties than in the white ones, due almost entirely to the presence of anthocyanins in the first (Cantos et al., 2002). Table grapes have lower concentrations of total phenolics, total flavonoids, and total anthocyanins than wine grapes.

The specific expression of each cultivar is the dominating factor affecting the grape phenolic composition. On the other hand, concentration and distribution of the different phenolics are significantly influenced by environmental, season, time of harvesting, weather, agricultural practices, and diseases (Guidoni et al., 2008), i.e. all those factors that, in the case of wine grapes, are known with the name of terroir.

The following paragraphs were intended to supply the readers with information concerning the effects of the most important biological, environmental, and technical factors on the phenolic content and distribution of table and wine grapes.

Influence of Genotype on Phenolic Composition of the Grapes

It is not easy to study the influence of the cultivar on content and distribution of phenolics in grape, since it would require grapes growth in the same geographical region, at the same time, and under the same viticultural practices. Due to the difficult of maintain homogeneous experimental conditions, most of the researches on the influence of genotype also include other variables.

A research of Giovanelli and Brenna (2007) investigated the evolution of some phenolic components, carotenoids, and chlorophylls during ripening of a typical Italian white grape cultivar (Erbaluce) and two red ones (Barbera and Nebbiolo), all grown in the same vineyard. Very similar profiles of catechins (as the sum of (+)catechin and (−)epicatechin) and flavonols (as the sum of rutin, quercetin, and quercetin-3-glucoside) were observed in the three cultivars, thus suggesting that climatic conditions and sun exposure may play an important role in the evolution of these compounds..

According to Masa ad Vilanova (2008), a total of 16 flavonoids (five dihydroflavonols, seven quercetin derivatives, and four kaempferol derivatives) were found in Albarín blanco, a white Vitis vinifera L. cultivar native of Asturias (North of Spain). The results were particularly interesting, since dihydroflavonols were found to be the most abundant flavonoids in Albarín blanco grapes, while this class of phenolic compounds was always considered as very scarce in grape skins. Navarro et al. (2008) characterized during ripening two Spanish autochthonous grapevines, Bobal and Crujidera, in comparison with the well-known cultivars Tempranillo and Cabernet Sauvignon. The authors found higher values of the total phenolic index in Tempranillo, Cabernet Sauvignon, and Bobal. Although the berries of the four varieties accumulated high concentrations of anthocyanins at the harvest time, the ability of the grape to synthesize these compounds was greatly dependent upon the cultivar. Anthocyanin concentrations were very high in Cabernet and Bobal, lower in Tempranillo, and halved in the skins of Crujidera grapes. The biosynthetic potential of resveratrol production largely depended on the grape variety, with a remarkably high content found in Bobal berry skins.

To assess the genotypic variability in two consecutive years, Jin et al. (2009) investigated the phenolic profiles of the berry skins of nine red Vitis vinifera varieties. Cultivars such as Syrah, Cabernet Sauvignon, Cabernet Gernischt and Merlot contained more types of anthocyanins, flavonols, flavan-3-ols, stilbenes, and phenolic acids than Gamay, Yan73, Pinot Noir, Zinfandel, and Мускат Розовый, while Yan 73 and Pinot Noir showed high concentrations of anthocyanins, but only few types non-anthocyanin phenolic compounds. Gamay, Zinfandel and Мускат Розовый contained only a few anthocyanins and flavonols. The results also demonstrated that, except for Syrah and Yan 73, the phenolic profiles of the tested grape cultivars had no significant year-to-year variations.

In order to analyze the differences in the phenolic composition of red wine grapes with diverse genetic backgrounds, Zhu et al. (2012) estimated the skin phenolic composition of the following 21 cultivars: Cabernet Sauvignon and Merlot (Vitis vinifera L.); the East-Asian Changbaijiu, Shuanghong, Shuangyou, Zuoshanyi, Zuoshaner (V. amurensis), Black Pearl (V. dividii), Mao (V. quinquangularis), Mi (V. xunyangensis), Sangye (V. ficifolia); the Euro-Asian hybrids Hasang, Zuohongyi, and Zuoyouhong (V. amurensis x V. vinifera); the North American grapes Black Spanish (V. aestivalis), Catawba, Concord, Niagara Rosada (V. labrusca); the Euro-American hybrids Marechal Foch (V. riparia x V. rupestris x V. vinifera), Chambourcin, St. Croix (V. rupestris x V. lincecumii x V. vinifera); and 2 varieties of muscadine grapes, Alachua and Noble, (V. rotundifolia).. The researchers found that the skin phenolic compositions varied widely among the studied grape cultivars, as a function of genetic background and origin. Forty-five anthocyanins, 28 flavonols, 8 flavan-3-ols, 9 cinnamic acids, 5 benzoic acids, 5 ellagic acids, and 2 stilbenes were detected in all the samples. Monoglucoside anthocyanins were detected in V. vinifera, diglucoside anthocyanins were found in V. rotundifolia grapes, while the other grapes contaned both types. .In the Euro-Asian grapes, 3',4',5'-substituted and methylated anthocyanins were dominant, while the proportions of total anthocyanins were relatively low in grapes originating from North America. The content of acetyl anthocyanins were higher in V. vinifera than in other grapes. Quercetin-3-glucuronide and quercetin-3-glucoside were common in the skins of Euvitis species but were not detected in muscadine grapes. The muscadine grape skins possessed more non-anthocyanin phenolics than other grapes due to their richness in flavonols, flavan-3-ol monomers, and ellagic acids. The skins of the East-Asian grapes and hybrids showed low concentrations of flavan-3-ols. According to the principal component analysis, the grapes could be divided into three groups: V. rotundifolia grapes, East-Asian grapes and hybrids, and the other grapes.

In a study performed by Breska III et al. (2010), 16 raisin grape cultivars (Diamond Muscat, DOVine, Fiesta, Selma Pete, Summer Muscat, A50-33, A50-39, A56-66, A95-15, A95-27, B53-122, B58-41, B64-108, C65-81, C90-100, and Thompson Seedless) were analyzed for antioxidant activity (ABTS assay) and for total and individual phenolic compounds. The cultivar A95-27 had a total phenolic concentration more than three times higher than Thompson Seedless grapes, which is the most widely used grape variety in raisin production. The varieties A56-66, A95-15, and A95-27 had antioxidant activity values more than three times higher than Thompson Seedless grapes. Trans-caftaric acid was the predominant compound in all the samples. The lowest and highest concentrations of this acid were found in A95-15 and Fiesta, respectively. The selections A56-66, A95-15, and A95-27 had much higher levels of catechin and epicatechin than the other samples. Xu et al. (2010) analysed the phenolic compounds and the antioxidant properties of skins and seeds of 18 grape cultivars belonging to oriental and north American Vitis species/hybrids and to Vitis vinifera grown in China. In particular, five oriental Vitis species, Vitis vinifera, three Euro-Asian hybrids, one Euro-American hybrid, and muscadine grape (Vitis rotundifolia) were studied. Authors found a great variability of phenolic contents and antioxidant properties among the cultivars investigated. Seeds of V. vinifera Cabernet Sauvignon had the highest phenolic concentrations and antioxidant activity values, followed by Muscadines, while the lowest values were found in the oriental Vitis species. The Euro-Asian or Euro-American hybrids fell between the parents. Far less variation was observed in the phenolic contents and antioxidant activity values of the grape skins. Furthermore, also in the case of cultivars with similar total phenolic contents, the distribution of phenolic compounds in seeds and skins can vary greatly.

Gómez Gallego et al. (2012) compared young wines made from Bobal and Moravia Agria, which are two red grape varieties in danger of extinction in Castilla-La Mancha, to those made from Cencibel, the reference red grape cultivar of that Spanish region. Their phenolic composition, chromatic characteristics, and antioxidant activity were studied. The results highlighted that the phenolic contents were affected by both grape variety and vintage, whereas the phenolic profiles (especially those of anthocyanins and flavonols) were mainly affected by grape cultivar, thus allowing their differentiation. Bobal and Moravia Agria wines showed higher amounts of resveratrol than Cencibel wines although it occurred mainly as cis-isomers. Similar effects of variety and vintage were observed by Gómez Gallego et al. (2013) in a research performed on three minority red grape varieties (Moravia Dulce, Rojal, and Tortosí) cultivated in the same Spanish region. Du et al. (2012) compared the total phenolic contents and the total antioxidant activities of four wine grapes (Cabernet Sauvignon, Cabenet Franc, Merlot, Cabernet Gernischt) and four table grapes (Muscat, Red Globe, Vitis labruscana Kyoho, Milk grape). The results indicated that Cabernet Gernischt had the highest total phenolic (257.0±8.9 mg/100 g), total flavonoid (607.7±24.3 mg/100 g), and total anthocyanin (164.2±6.1 mg/100 g) contents, while Merlot had the highest resveratrol content (11.7±0.61μg/100 g). The table grape cultivars have the lowest total phenolic, total flavonoid, total anthocyanin contents, and consequently the lowest antioxidant activity values. Obreque-Slier et al. (2012) compared the phenolic composition of Carménère, Cabernet Sauvignon, Merlot, and Cabernet Franc grape seeds at four different ripening stages. Significant differences between Carménère and Cabernet Franc were observed throughout veraison to harvest in a number of parameters, including total tannins and polymeric flavanol fraction. According to the results, grape seeds present marked differences in phenolic composition during ripening. In particular, Merlot seeds presented high concentrations of (+)-catechin, (-)-epicatechin, epicatechin-3-O-gallate, proanthocyanidin trimers B1, B3 and B4, while Carménère seeds had a characteristic high concentration of other proanthocyanidins and Cabernet Franc showed the highest concentrations of gallic acid and proanthocyanidin gallates.

Pedoclimatic Conditions and Phenolic Profiles of Grapes

Soil Composition

The vine is a species resistant to environmental stress. Historically, fertile land was preferentially used for annual crops, while shallow or stony soils and steep slopes were used for vineyards (van Leeuwen and Seguin, 2006). In these conditions, yields tend to be from moderate to low, while potential for wine making is generally high (van Leeuwen et al., 2007). European vine growers have established how best to match the various wine-grape varieties to the soil types since they found that planting grape vines in the "right" soil can make the difference between a great and an ordinary wine. This vine–soil relationship is a part of the concept of terroir as highlighted by Mackenzie and Christi (2005), who performed a study in order to establish the influence of soil mineralogical and/or chemical composition on the composition and quality of wine grapes.

The effect of the availability of soil nitrogen on vine behavior can best be observed when vine water status is not limited. In 1997, which was a rainy vintage in Bordeaux, a clear effect of soil nitrogen supply was observed on vegetative development, yield and must composition of Cabernet Sauvignon vines located on various soils of a Grand Cru Classé of Saint-Julien (Bordeaux). The wines with the highest phenolic contents were produced from the plot where vine nitrogen status was the lowest and close to deficiency. The positive effect of low nitrogen status on grape quality for red wine production is the result of factors such as reduced vigor, limited berry size, and stimulated phenolic compound synthesis in berries (van Leeuwen et al., 2000). For white grapes (and particularly varieties containing volatile thiols like Sauvignon Blanc), a low vine nitrogen status is not desirable, because it limits the synthesis of aroma precursor (Choné et al., 2006; Peyrot des Gachons et al., 2005).

The effects of soil type on the phenolic composition can be used, together with other parameters, in a metabolic profiling based on proton nuclear magnetic resonance (1H-NMR) in order to study the multifactorial effects of the vine environment on intricate grape quality traits. In a study of Elias Pereira et al. (2006), the partial least square analysis separated clearly the wines from the 3 soil types (gravely, clayey, and sandy) independently on vintage (2002 and 2003) and cultivar (Merlot, Cabernet Sauvignon, and Cabernet franc).

In order to clarify the relationships between cultivar/clone and environment, a trial was carried by Santini et al. (2009) in order to investigate the performances of three clones of Nebbiolo in two different environments: the south-east Piedmont (in La Morra and Neive areas, which are hilly and characterized by a loamy and alkaline soil) and the north-east Piedmont (in Lessona area, which is plain and characterized by a sandy and acidic soil). The phenolic contents were quantitatively higher in the northern location than in the southern sites, although the anthocyanic profile was poorer in malvidin-3-glucoside and richer in peonidin-2-glucoside. These aspects were of particular relevance in winemaking because the tri-substituted anthocyanins are more determinant for the intensity and the stability of wine color than the di-substituted anthocyanins, which undergo enzymatic degradation at the end of fermentation. Environment did not much influence the amount of grape resveratrol.

Iron deficiency is an important constraint in grape growing in Mediterranean climate areas since it is responsible for the decrease of chlorophyll and carotenoid concentration in leaves (chlorosis), and for the reduction of photosynthesis, thus depressing the yield and vigor of vineyards and leading to poor wine grape quality, with reduced accumulation of sugar and anthocyanin (Castino et al., 1987; Meggio et al., 2010; Veliksar et al., 2005).

Climate

Many climatic factors influence the grape berry composition. It is often difficult, however, to separate effects that are combined each other, such as sun exposure and temperature. Obviously, the increasing exposure of berries to the sun led to their heating. In a study on V. vinifera L. cv. Merlot (Cohen, 2013), berry temperatures where manipulated independent on sun exposure, in order to study the effects of modulating daytime and nighttime temperatures as well as damping the diurnal temperature range. In order to assess the effects of treatments during two phases of the berry development characterized by accumulation of distinct phenolic metabolites, the research was divided into two steps: from berry set to veraison (phase I) and from veraison to commercial harvest (phase II). According to the results of this study, proanthocyanidin accumulation at veraison was linearly related to the heat summation over the developmental period with nighttime heating yielding the highest concentration and daytime cooling yielding the lowest one. Furthermore, damping the diurnal temperature fluctuation reduced proanthocyanidins, while day-cooling resulted in an increase of the concentration of flavonols at the end of the phase I and in a decrease at the end of the phase II.

van Leeuwen et al. (2004) studied the simultaneous effects of the three main components of terroir, i.e. soil (gravelly soil, a soil with a heavy clay subsoil, and a sandy soil with a water table within the reach of the roots), cultivar (Merlot, Cabernet franc, and Cabernet Sauvignon), and climate (with year-to-year variations of maximum and minimum temperatures, degree days (base of 10°C), sunshine hours, evapotranspiration, rainfall, and water balance for the period 1996 to 2000). The effect of climate was the greatest on most parameters, followed by soil and cultivar. The effects of climate on vine development and grape composition can be explained in large part by its influence on the vine water status. Vintage influences vine water status varying amounts of summer rain, thus confirming the studies of van Leeuwen and Seguin (1994) and of Ojeda et al. (2002) that indicate the positive impact of moderate water deficit stress on phenolic compound synthesis and grape quality. Hence, the optimum quality was reached in vintages where low summer rainfall led to a water deficit stress.

Climatic region had a significant effect on the content of several phenolic compounds of Pinotage wines (de Beer et al., 2006). In fact, as a general rule, the concentration of anthocyanin monoglucosides, flavonols, flavan-3-ols, and tartaric acid esters of hydroxycinnamic acids generally increased as the climatic region becomes cooler, while concentrations of acylated derivatives and free hydroxycinnamic acids decreased.

Nicholas et al. (2011) conducted a three-year study to collect and analyze grapes from Pinot noir vineyards in the Carneros and Sonoma Valley (American Viticultural Areas of California), and correlate anthocyanins, tannins, and total phenolics with climate statistics derived from hourly temperature measures at each vineyard site. According to their results, cool conditions following harvest the year before maturity, warm temperatures from budburst to bloom, and cool temperatures from bloom to veraison were positively correlated with concentrations of all three classes of phenolics. Anthocyanins were positively and significantly correlated with temperatures between 16 and 22 °C from veraison to harvest. Tannins were significantly increased by warm nights preceding budburst and warm days from budburst to bloom. Increased light interception was significantly correlated with lower levels of phenolic compounds in this study. Furthermore, warm temperatures from budburst to bloom seem to increase phenolic concentrations.

Ferrer-Gallego et al. (2012) studied the influence of two different climatic conditions (in terms of average temperature, accumulated rainfall and accumulated solar radiation in Atlantic and Mediterranean climate) during vegetative stage on the phenolic composition of grape skins and seeds of *Vitis vinifera* L. cv. Graciano, an autochthonous cultivar from Rioja and Navarra regions (Spain). They found that the samples with the highest phenolic concentrations in the grape seeds at harvest presented the lowest content of phenolic compounds (especially of flavonols and phenolic acids) in the grape skins. Accumulated solar radiation presented a direct relationship with anthocyanin and flavonol concentrations. On the other hand, accumulated rainfall presented an inverse relationship with the contents of antocyanins and flavonols.

Exposure to Sunlight

For the many vineyards with rows oriented north-south, the consequence of the daily solar cycle is that berries on the east side of the vine may be above air temperature in the morning and berries on the west side may be above air temperature in the afternoon, which is during the hours when the daily maximum air temperature also occurs. However, over the course of one day, the same total amount of solar radiation strikes the fruit on the two sides of the vine. In the case of rows oriented east-west, north-facing fruit is shaded for most of the day, whereas south-facing fruit is in the sun for most of the day (Tarara and Lee, 2013).

The biosynthetic pathways involved in flavonoid production in plant tissue are greatly influenced by sunlight. An experiment concerning the effects of sunlight exposure and temperature on the contents of quercetin, myceritin, and kaempferol revealed that berries from sun exposed cluster might contain as much ten times the content found in sample obtained from shaded cluster (Varache-Lembege et al., 1996). It is known that some phenolics require light to be synthesized (flavonols) and that the fruit accumulates higher concentrations of these compounds (anthocyanins, for example) in the presence of light. However, it is unclear whether the resulting phenolic accumulation was due to the light itself or to a combination of that and of the daytime radiation load driving berry temperature. Nevertheless, consistent with literature, flavonol accumulation was apparently insensitive to temperature but responsive to solar radiation, specifically UV. Instead, controlling solar radiation, anthocyanins decreased as berry temperatures rose while, controlling berry temperature, shaded fruit developed less anthocyanins than the sunlit one (Tarara and Lee, 2013).

Dokoozlian and Kliewer (1996) studied potted Cabernet Sauvignon and Pinot noir grapevines (Vitis vinifera L.) grown in a sunlit phytotron in order to understand the influence of cluster light exposure on berry growth and composition during various stages of the fruit development. The experiment was divided into three stages: fruit development stages I and II refer to the period between berry set (2 to 3 mm in diameter) and the initiation of berry softening, while stage III refers to the period between fruit softening and maturity. Four treatments were carried out: clusters exposed to light during stages I, II, and III (control); clusters grown without light during stages I and II, and exposed to light during stage III; clusters exposed to light during stages I and II, and grown without light during stage III; clusters grown without light during stages I, II, and III. The temperature of both light-exposed and non-exposed fruit was similar. Control berries had greater skin anthocyanins and phenolics compared to the other treatments. Berries grown without light during stages I and II, or stage III, were greater in anthocyanins and phenolics than fruit grown without light during stages I, II, and III.

UV radiation, as a natural component of sunlight, has several regulatory effects on grapevine physiology and biochemistry (Jug and Rusjan, 2012). According to Lafontaine et al. (2005), berries exposed to UV-B radiation increased both the concentration of total bound glycosidic secondary metabolites and phenolics. The accumulation of the UV-B absorbing compounds under enhanced UV-B radiation and the increase in antioxidant enzymes activities constitute the main mechanisms of grapevine adaptation since they decrease UV-penetration through the epidermis (Kolb et al., 2001; 2003). The UV light stimulates the biosynthesis of secondary metabolites through the regulation of some key enzymes involved in the phenylpropanoid pathway (Pontin et al., 2010). Keller et al. (2000) found that only a growing callus of grapevine irradiated with UV light was capable of producing stilbenes, including trans-resveratrol. A similar behavior was observed in ripening grape berries, which gradually lose their potential for synthesizing stilbenes as they reach full-ripe maturity (Pan et al., 2006). According to Gil et al. (2012), their profile seems to be dependent upon intensity and duration of UV-B.

Altitude

High altitude means low temperatures, high UV radiation, high light intensity and photosynthesis rate, low amounts of oxygen and carbon dioxide, and wide swing in diurnal temperature differences distinguished by lower night-time (cooling) temperatures. Grapes from vineyards grown at high altitude have high levels of tannins and anthocyanins (meaning a more stable color) and low levels of monomeric tannins (i.e. a more rounded mouthfeel). Obviously, the growth at high altitude also implicates some potential challenges such as: increased risk of frost and damages due to strong winds and heavy rains; low yields; high costs of management (Rieger, 2008).

Matheus et al. (2001) investigated the effects of altitude and its related climatic conditions on the polyphenolic composition (catechin monomers such as (+)-catechin, (-)-epicatechin, (-)-epicatechin O-gallate); procyanidin oligomers (B1 to B8, B2-3''-O-gallate and trimer C1), and total extractable proanthocyanidins in grape skin and seeds; anthocyanidin monoglucosides in grape skin). of grapes and wines from Touriga Nacional and Touriga Francesa red varieties. Higher levels of anthocyanins in grape skin were found at higher altitude, whereas procyanidins were synthetized in larger amounts at low altitude in both skin and seeds.

Another interesting study was aimed to understand the influence of different altitudes (slope land, 1280 m, and valley land, 909 m) on phenolic compounds and antioxidant activities of Cabernet Sauvignon grape berries grown in Loess Plateau region (Jiang et al.,

2011). The content of total flavanols in valley land was higher than that of slope land, but the content of anthocyanins was lower. Antioxidant activities of grape berries in valley land were higher than in slope land.

Effects of Viticultural Practices on Phenolics in Grapes

Irrigation/Water Stress

The vine is a species resistant to environmental stress (van Leeuwen et al., 2007). Water deficit is the most frequent form of environmental stress to which vines are subjected. In Mediterranean regions, the high temperatures and the low rainfall can affect vine water status and compromise vine growth, yield, and grape quality (Nadal and Vernet, 2010). Nevertheless, a limitation in water uptake generally increases berry anthocyanin and tannin contents (Koundouras et al., 2006).

In a work of Nadal and Vernet (2010), the effects of two irrigation trials, namely moderated and high water deficit, were tested on a four year-old Vitis vinifera cv. Syrah vineyard grafted on 41B and 140 Ruggeri rootstocks in the Spanish region of Catalunia. The authors found that moderate irrigation had a greater effect on rootstock 41B than on rootstock 140Ru. The anthocyanin content of 41B was higher under drought conditions as a consequence of the shrinkage of berries. Nevertheless, an irrigation extended after veraison is recommended for Syrah vines in order to improve grape composition and quality.

In another paper (Deis et al., 2011, the effects of combined in-field treatments such as water stress and exogenous abscisic acid were evaluated on phenolic accumulation in grape and wine. The abscisic acid regulated the content of phenols on the base of plant water-status. The effects of water-stress and exogenous abscisic acid were similar in magnitude, but affected different metabolic pathways. The abscisic acid supplied increased catechin and malvidin synthesis and reduced the quercetin content independently on water status while the resveratrol was enhanced only in water-stress conditions. The response of grape phenolic composition to variation in grapevine water status was studied in Vitis vinifera cv. Merlot (Bindon et al., 2011). The results showed increases of the concentration of grape anthocyanins and flavonols in response to water deficit due to the decrease of fresh weight of berries. Skin-derived tannin concentration was not affected by the irrigation treatments.

Intrigliolo and Castel (2010) studied the response of grapevine cv. Tempranillo to timing and amount of irrigation. The results showed that water application after veraison was beneficial to fruit ripening. In order to increase the concentration of phenolic substances in wines, water stress should be applied during the preveraison period. Santesteban et al. (2011) investigated the effects of the regulated deficit irrigation on growth, yield, grape quality, and anthocyanin composition in Vitis vinifera L. cv. Tempranillo. The most relevant effect of regulated deficit irrigation strategies was an increase in anthocyanin and phenolic content due to an enhanced anthocyanin synthesis but also to an indirect consequence of smaller berry size.

Romero et al. (2013) studied the effects of regulated deficit irrigation based upon optimum water status in Monastrell grapes. As a result of this research, the regulated deficit irrigation strategy, consisting in the application of a mild water stress during the early season and a moderate water stress during pre and post-veraison, was the most effective irrigation regime for increasing the anthocyanin content.

Balancing Growth-Yield

On a given trellis, there are some indications in training the vines, which are referred to as "canopy management". The grape grower's challenge is to configure the vine canopy to maximize solar radiation interception for photosynthesis and to create an optimum fruit microclimate for ripening (Tarara and Lee, 2013). In viticulture, the canopy of a grapevine includes the parts of the vine visible aboveground: trunk, cordon, stems, leaves, flowers, and fruit. The canopy plays a key-role in light energy capture via photosynthesis, transpiration, and microclimate of ripening grapes (Smart and Robinson, 1991). Canopy management is an important aspect of viticulture due to its influence on grape yields, quality, vigor, and on the prevention of grape diseases.

Vine balance implies constant quality and on-time yield. An accepted definition of this equilibrium between crop load and vegetative growth is the leaf area required for maximizing the yield and ripeness. This equilibrium has been reported to range from 7 cm^2 g^{-1} to 14 cm^2 g^{-1} (Howell, 2001) and between 8 cm2 g-1 and 12 cm2 g-1 (Kliewer and Dokoozlian, 2005). Smart and Robinson (1991) indicated that the balanced Ravaz index (fruit yield/pruning weight) is between five and ten. An excessive vegetative growth allowing shaped clusters as well as poor induction and development of buds. This unbalance negatively affects the fruit composition (Zoecklein et al., 2001). In order to avoid this problems, canopy management can be applied by practices such as pruning, cluster thinning, leaf and shoot removal.

Manipulation of canopy can affect gene expression behavior inside the phenylpropanoid biosynthetic pathway and thus the concentration of important berry compounds. Sternad Lemut et al. (2011) applied the leaf removal on vines of Pinot noir grown in Vipava Valley (Slovenia) in order to manipulate microclimate conditions within the grape area. In order to investigate the effects of leaf removal timing, defoliation was manually applied, by removing the basal 4-6 leaves of all shoots alternatively at berry set and at veraison. Untreated vines were used as control. According to the results, depending upon timing of leaf removal, different microclimates can be achieved, with different light exposure for the specific plant tissue and different temperatures on the berry surface. The concentration of hydroxycinnamic acids were affected only slightly, mostly at the beginning of the maturation period. Anthocyanins were significantly affected, while flavonols were the most affected by the practice and their concentration was clearly related to modification of the light within the canopy. Jogaiah et al. (2013) studied the effect of canopy management practices on the berry composition of red (Cabernet Sauvignon) and white (Sauvignon blanc) grape cultivars grown in Pune, a semi-arid region of India. Four-years old Cabernet Sauvignon and Sauvignon Blanc grapes grafted on 110R rootstock were selected for this study. The row orientation was in the direction north – south. The vines were trained to double cordon small T system. Shoot thinning was done 45 days after pruning, wherein approximately 32 shoots (eight shoots on either side of double cordon) per vine were retained. Cluster thinning was done after fruit set stage (3 to 4 mm stage) to maintain 40 basal one or two clusters per vine. Leaf removal was performed during version stage by removing two leaves above and two leaves below the cluster to expose bunches. Combination treatment of leaf removal either with shoot thinning or cluster thinning exhibited the highest anthocyanin content while the vines with the three treatments together and the control vines recorded the lowest anthocyanin concentration and phenolic compounds, thus indicating that excessive light exposure or shade to clusters are not congenial for producing better quality fruits. King et al. (2012) studied the effect of severity of leaf and cluster removal on juice and wine composition of Merlot grown in the cool-climate regions of Hawke's Bay in New Zealand. The following leaf removal (LF) and cluster thinning (CT) treatments were combined for a total of 9 treatments: no LR; moderate LR = remove 2 basal leaves; severe LR = remove all leaves up to 2 nodes above the second cluster; no CT = 40 clusters or 45 clusters; moderate CT = thin to 30 or 31 clusters; severe CT = thin to 20 or 18 clusters. In summary: leaf removal enhanced fruit exposure to light of heavily shaded VSP-trained Merlot and it increased anthocyanin and quercetin levels; cluster thinning had no effect on anthocyanin concentrations, but significantly increased the total phenols in the wine. The practices of leaf removal and cluster thinning are recommended in cool areas where "green" characters associated with suboptimal maturity are often present in the wines. The effects of fruit zone leaf removal were also studied on the concentrations of phenolic acids in Istrian Malvasia grape juice and wine (Bubola et al., 2012). Fruit zone leaf removal was applied at three different phenological stages: before blooming, at berry set and at veraison. In particular, at veraison, two intensities of leaf removal were applied. The leaf removal before blooming significantly lowered the concentration of all the hydroxycinnamic and hydroxybenzoic acids in grape juice in comparison with the other investigated treatments. The concentration of hydroxybenzoic acids in grape juice was lower in veraison treatments in comparison with berry set and control treatments. The effectiveness of pre-bloom leaf removal was investigated on a high-yielding Italian red cultivar (Vitis vinifera L., cultivar Ciliegiolo) (Palliotti et al., 2012). The pre-bloom, long-lasting source limitation, which led to the elimination of 75–80% of the leaves in the canopy improved the concentration of phenolics and anthocyanins.

Guidoni et al. (2002) investigated the effect of cluster thinning (removal of 50% of the cluster one month after bloom) on berry skin anthocyanin composition of Vitis vinifera cv. Nebbiolo in a vineyard located in the southern Piedmont (Italy). The concentrations of anthocyanins and flavonoids in berry skins were higher in berries from cluster-thinned plants. Cluster thinning increased the concentrations of cyanidin-3-glucoside, peonidin-3-glucoside, and, to a lesser extent, petunidin-3-glucoside, whereas did not affect the concentrations of malvidin-3-glucoside and of acylated anthocyanins. A research concerning the effects of cluster thinning in grapes of Cabernet Sauvignon planted in a vineyard in Mile (Yunnan, China) was published in 2006 (Zhao et al.). Five treatments were studied: basal cluster kept, distal cluster removed at anthesis (T1); distal cluster kept, basal cluster removed at anthesis (T2); basal cluster kept, distal cluster removed at veraison (T3); distal cluster kept, basal cluster removed at veraison (T4); basal cluster and distal cluster kept, no cluster removal (CK). From three weeks after veraison to ripeness, concentration of catechin, epicatechin and epigallocatechin decreased in each of the 4 treatments that included the removal of clusters and was larger than that in CK, especially when the distal cluster (T1, T3) was removed. This illustrates that cluster thinning promoted berry development. Vieira Da Mota et al. (2010) performed a work in order to evaluate the biochemical and agronomical responses to cluster thinning of two Vitis vinifera cultivars (Merlot and Cabernet Sauvignon) grown in Caldas (Minas Gerais, Brazil), a new winegrape region. Grapevines were submitted to cluster thinning (0%, 50% and 75% of cluster removal) and trimming treatments (trimmed and untrimmed), thus altering the source-sink ratio. In both cultivars, the anthocyanins and the total phenols extracted from berry skins were more affected by cluster thinning than shoot trimming. These compounds were significantly higher in berries from the vines with the lowest crop yield. Merlot berries with the highest yield (no cluster removal treatment) showed higher anthocyanin and total phenolic contents in the skins of untrimmed vines, whereas in 50% and 75% cluster thinning, the trimming treatment increased anthocyanins and total phenols levels. In Cabernet Sauvignon there was no significant effect of trimming on anthocyanins and phenols, while the cluster reduction increased the level of these phenolic compounds. Fanzone et al. (2011) evaluated the effect of cluster thinning (removal of 50% of clusters at 40 days (T1), 80 days (T2), and 100 days after flowering (T3)) on the phenolic composition of Malbec (Vitis vinifera L.) grape skins and seeds in a vineyard located in the southern Mendoza (Argentina). The effects were different depending upon vintage. In 2008, T1 encouraged the biosynthesis of individual anthocyanins in skins, as compared to the control, whereas in seeds, T1 and T2 mainly changed the concentrations of (+)-catechin, epicatechin-3-gallate, procyanidin B4, dimer gallate 1, trimer gallate 2, and tetramer. Conversely in 2009, T1 significantly affected the content of flavanols and flavonols in skins, whereas in seeds, T1 and T2 modified the level of (+)-catechin, procyanidins B4 and B6, and trimer gallate 2.

Kamiloğlu (2011) conducted a study to determine the effects of different cultural practices (cluster thinning, boron, topping, cluster thinning + boron, cluster thinning + topping, boron + topping, cluster thinning + boron + topping) on total anthocyanins and individual anthocyanins in 'Horoz Karası' table grape. The topping practice was made by leaving five leaves above the last cluster on summer shoots. Cluster thinning was made by pruning 1/3 of clusters on the vine. Boron at the rate of 100 mg l^{-1} as boric acid was applied as foliar. It was determined that cluster thinning practice increased the amounts of total anthocyanins and of individual anthocyanins significantly more than the other practices.

Beslic et al. (2010) investigated the effect of pruning severity on quercetin and catechin content in berry skins of cv. Blaufränkisch (Vitis vinifera L.). Different crop levels and canopy structures were obtained by retaining 8, 16, and 24 nodes per vine at pruning. Canopy density directly affects the intensity of photosynthetically active radiation. The quercetin content was highly dependent upon the light exposure of the berries in which it accumulates. An increase in node number linearly decreased the skin catechin, thus suggesting that the decrease was caused by an increased yield per vine.

Conventional, Organic, and Biodynamic Systems

Organic agriculture is a production management system that promotes and enhances biodiversity, biological cycles, and soil biological activity (Research Institute of Organic Farming, 2000; Alternative Farming Systems Information Center, 2005). Biodynamics is an ecological approach to agriculture thought by the Austrian philosopher Steiner (1993). Biodynamic farms operate as close to self-sufficient systems as possible with all fertilizers and animal feeds produced on the farm (Pfeiffer, 1983). To promote efficient cycling of nutrients, a series of eight fermented herbal and manure based preparations numbered 500 through 507 (Koepf, 1989) are applied regularly, in small quantities to the soil, compost and crops. These preparations help to stimulate the vitality of the soil and plant and to improve the health of the animals (Pfeiffer, 1983). Some biodynamic farmers plant their crops according to specific phases of the moon, maintaining as the moon causes tides in the oceans, sap flow in a plant is also influenced by the moon and planets (Koepf, 1989; Reeve, 2003).

In a rigorous 8-year study (Reeve et al., 2005), the authors compared two vineyards, one certified organic and one certified biodynamics, both located in Ukiah (California). The differences observed were too small to be of any significance and it is unlikely that biodynamic preparations contribute to grape quality.

Dani et al. (2007) investigated phenolic content and antioxidant activity of juices produced from Niagara (white) and Bordo (red) grapes obtained both from conventional and organic systems in Brazil. According to the results, organic juices presented higher total phenols, resveratrol, and anthocyanin content than conventional juices, whereas there were no differences in in-vitro antioxidant activities.

Mulero et al. [90] compared antioxidant activity and phenolic composition of organic and conventional Monastrell variety grapes. The phenolic content and antioxidant activity measured a month before harvesting were higher in the organic grapes than in the conventional ones, although these differences disappear at harvesting.

In a study of Corrales et al. (2010), grape skin extracts of Riesling Vitis vinifera L. from conventionally or organically managed cultivars were compared on the basis of their phenolic content and antioxidant capacity. The antioxidant capacity of conventional extracts was significantly higher, according to the higher content in catechin, epicatechin, and procyanidin B.

Vine Training Systems

The architecture of the plants influences the interception of the light in the canopy and the ratio between surface area of leaves and fruits, thus affecting the composition ad ripening of the grapes. This means that the training and the pruning systems affect these aspects in an important way (González-Neves et al., 2004).

Peterlunger et al. (2002) analyzed the effects of different training systems (simple Guyot, double Guyot, horizontal spurred cordon, and vertical spurred cordon) on the grape composition of Pinot noir grown in the Friuli hills (northeastern Italy). The training systems showed little or no impact on grape phenolics.

In a research on own-rooted Shiraz, grown in the Barossa Valley (South Australia), the effects of five training systems on fruit composition and crop yield were compared (Wolf et al., 2003). The training systems were the following: low single wire, in which vines were trained to bi-lateral cordons, non-shoot-positioned; high single wire, a higher cordon version of low single wire; vertically shoot positioned; Scott Henry, where canopies of alternate vines were cordon-trained and shoot-positioned upwards or downwards; and minimally pruned, the same as low single wire, except vines were not annually spur-pruned. The berry total anthocyanins and the total phenolics concentrations at harvest were not greatly affected by the training system.

In a study on anthocyanic composition of Tannat grapes produced in the south of Uruguay, the effects of two training systems (vertical trellis and lyre) and two pruning methods (Guyot and Royat cordon) in the lyre were considered (González-Neves et al., 2004).

According to the results, the total potential in anthocyanins in the lyre with cordon pruning was significantly higher than the others, whereas the grapes from the cane pruning presented the highest extractable anthocyanins contents. The phenolic content of the grapes from the lyre (independently on pruning systems) was significantly higher than in the vertical trellis. Grapes from the vertical trellis system had significantly higher proportions of malvidin, coumarylic and acetyl anthocyanins, while grapes from the lyre with cordon pruning had significantly higher proportions of delphinidin, cyanidin, petunidin, peonidin, and non-acylated glucosides were detected.

Río Segade et al. (2009) studied the phenolic composition of red wines elaborated from Sousón and Brancellao grapes (Vitis vinifera L.), whose vines were trained in cordon and Guyot systems. Sousón cultivar trained in Guyot system and Brancellao cultivar trained in cordon system showed the highest total anthocyanin concentration. However, greater ionization index and flavanols reagent with dimethyl aminocinemaldehyde were associated with Guyot and cordon systems, respectively, independently on cultivar. Gentisic acid, malvidin-3-glucoside, petunidin-3-acetylglucoside, and malvidin-3-caffeoylglucoside permitted to differentiate the training systems in both Sousón and Brancellao cultivars.

Bavougian et al. (2013) evaluated four trellis/training systems (Geneva double curtain (GDC), high cordon (HC), Smart-Dyson (SD), vertical shoot (VSP)) of ‘Frontenac’ grapes (Vitis sp. MN 1047) grown in southeastern Nebraska also in terms of phenolic and flavonoid concentrations. Flavonoid and total phenolic concentrations of the bound seed and bound skin extracts did not differ among trellises. Within the free extracts, VSP had higher total phenolic concentration than SD (GDC and HC were intermediate) and there were no differences in flavonoid concentration.

Soil Management

Vineyard floor management has multiple goals that encompass improving weed management and soil conservation, reducing soil resource availability to control vine vigor, and influencing desirable aspects in wine quality. Cover crops increased anthocyanins, and other phenolic components. They were associated with red wines judged superior to those issued from non-cover-cropped vines. (Guerra and Steenwerth, 2012).

Skinkis (2013) evaluated the effects of clean cultivated, alternate row tillage, and solid cover of an established perennial grass cover crop applied to two commercial Pinot noir vineyards. Berry skin anthocyanin concentration and total phenols was the highest in solid cover treatment in only one of the two vineyards. In a similar study, Qian (2013) analyzed the effects of three cover crop management regimes, including clean cultivated, alternate row tillage, and solid cover of an established grass mix cover crop, on two commercial Pinot noir vineyards. Berry skin anthocyanin concentration was the lowest in the clean cultivated treatments. However, the clean cultivated and the solid cover treatments in one of the two vineyards yielded higher total berry skin polyphenols than the alternately tilled. In the other vineyard, solid cover and alternate tilled treatments allowed higher total berry polyphenols than clean cultivated which may be due to the higher vine vigor at that site.

In another study, Muganu et al. (2013) evaluate the influence of soil management (complete bare soil, by mechanical tillage, and natural permanent vegetation) on vine growth and grape quality of Canaiolo nero and Trebbiano giallo varieties. Differences referable to the soil management type were detected on vine growth, bud break time, soluble solids, and anthocyanin content.

Another study (Xi et al., 2010) was conducted to determine the effect of cover crop inter-row in a Cabernet Sauvignon vineyard on the main mono-phenol contents of grape berry and wine. Three cover crops, included two perennial legumes (white clover and alfalfa) and a perennial gramineous grass (tall fescue), were sown in vineyard. Compared to the soil tillage, the sward treatments showed a higher content of the main mono-phenols and total non-anthocyanin phenolics in grapes and wines. The wine from tall fescue cover crop had the highest gallic acid, caffeic acid, and catechin concentrations. The cover crop system increased the total non-anthocyanin phenolics of grapes and wines, in the following decreasing order: tall fescue, white clover, alfalfa, and soil tillage (control). Cover crop in vineyard increased the total phenols of grape berry and wine, and thus evidently improved the quality of wine.

Use of Fertilizers

Grapes are deep-rooted perennial crops, and thus may not respond to fertilizer as predicted by a soil sample taken only 7 inches deep. Grape producers should monitor vine nutrient status with periodic soil testing and annual analysis of tissues, leaves in particular, because they provide a better evaluation of all nutrients with respect to petiole. As a general rule, applications of fertilizers should be in a 3 to 5 ft. wide band, parallel to the row, in established vineyards. The band must be kept 18" to 24" away from the base of the vines.

Grapes do not have high nitrogen requirement. In fact, a high nitrogen supplementation is often detrimental to the quality of grapes. Hilbert et al. (2003) evaluated the effects of nitrogen supply on anthocyanin accumulation in berries of cv. Merlot. The applied three nitrogen treatments from fruit set to leaf fall: were 1.4 mM N, 3.6 mM N and 7.2 mM N, respectively denominated N1, N2, N3 and considered as limited, mean and excessive nitrogen levels. Nitrogen was supplied as ammonium nitrate, potassium nitrate, ammonium phosphate, calcium nitrate and ammonium sulfate. Authors found that a high nitrogen supply delayed fruit maturation and gave berries with a higher arginine content, a lower anthocyanin content, and relatively more acylated anthocyanins than the berries of vines supplied with low nitrogen level. During maturation, the anthocyanin content in the skin of the berries decreased in a greater extent in high-N vines. In conclusions, high nitrogen supply delayed the quantitative and qualitative biosynthesis and enhanced their degradation during the final steps of berry maturation.

Grapes do not have a great demand for phosphorous. However, a crop grown on an acid soil has a less efficient use of the phosphorous in the soil, and should benefit from an adequate rate of P_2O_5. Also, most grapes are grown with a cover crop, which needs an addition of P_2O_5 to maintain a healthy, dense stand. Potassium is one of the most important fertilizers because vines are not particularly efficient at obtaining K from the soil, and thus they have a relatively high demand for it. Potassium can be added as K_2O. Calcium can be added as $CaSO_4$, although it generally is unnecessary. Magnesium stress in grapes can occur when grapes are grown on acid soils, low in Mg. Furthermore, the high K_2O rates that are typically needed can reduce Mg uptake. When the soil K to Mg ratio is higher than 1.5/1, Mg uptake decreases. Boron deficiency prevents the normal development of pollen tubes and reduces fruit set, causing the formation of much fewer berries, which are small and deformed.

Iron and manganese are most likely to become deficient where the soil pH is exceptionally high, such as pH 7.5 to 8.0.

Boonterm et al. (2010) studied the effects of nitrogen, potassium fertilizers, and clusters per vine on yield and anthocyanin content of Cabernet Sauvignon grape grown in a humid subtropical area located in the South of China. Three different levels of N-K were applied: 0-0, 100-20, and 200-60 g/plant. At the 0-0 and 100-20 levels of N-K, the phenolic contents in the grape were lower than in the case of the other treatment.

Evolution of Phenolics during the Ripening Time

The time course of berry growth in V. vinifera includes modifications of size, composition, color, texture, flavor, and pathogen susceptibility (Conde et al., 2007). Grape berries exhibit a double sigmoid growth pattern since it first occurs mostly by cell division and later by cell expansion (Coombe, 1992). The biosynthesis of soluble phenolics begins with the aromatic amino acid phenylalanine, a product of the shikimate pathway whose early precursors are erythrose-4-phosphate and phosphoenol pyruvate. The first enzyme responsible for the phenolic synthesis is the phenyl ammonia lysase, which catalyze the conversion of phenylalanine into cinnamic acid. Through a series of transformations, this compound gives rise to the formation of precursors of phenolic acids, lignin, etc. The phenylpropanoid pathway (Dias, 2003) generates complex phenolic compounds, like flavonoids and stilbenes. Skin anthocyanins have a little turnover once formed. The other phenolic classes accumulate and decline during ripening, suggesting their degradation and utilization for the biosynthesis of other compounds. Skin tannins are synthesized in a very early of the berry development, and change very little on a per berry basis from veraison to harvest, although their concentration declines along berry growth. The phenolic maturity corresponds to minimum seed tannin and maximum anthocyanin concentrations (Fredes et al., 2010). Qualitative changes, such as the increase of the polymerization degree, can also take place from veraison to harvest (Kennedy et al., 2001). Generally, there is a decline in seed tannins during ripening that accompanies seed browning, possibly due to tannin oxidation. There are two phases for anthocyanin pathway gene expression in grape berry skins. At early stages, most of the genes are expressed and then repressed during the lag phase of the development while, after veraison, a coordinate induction of all the genes of this pathway coincides with the accumulation of anthocyanins in the skin (Boss et al., 2001). On a per berry basis, total hydroxycinnamates increase in mesocarp tissues prior to véraison and then decline, leading to a constant amount (per berry) as the fruit ripens. In ripe fruit, the most abundant hydroxycinnamates are caftaric and coutaric acids (Romeyer et al., 1983). The expression of several enzymes of the phenylpropanoid pathway may be controlled by sugars since they enhance the anthocyanin production (Larronde et al., 1998; Vitrac et al., 2000).

Río Segade et al. (2008) studied the influence of ripeness grade on accumulation and extractability of grape skin anthocyanins in six red cultivars (Sousón, Brancellao, Ferrol, Merenzao, Mouratón, and Mencía) grown the Ribeiro region, which is one of the most traditional Spanish vine zones. They found that the concentrations of total anthocyanins and extractable anthocyanins usually increased during the ripening step until reaching a maximum value, and then began to diminish.

Oberholster et al. (2010) investigated the changes occurred in the chemical and phenolic composition of Cabernet-Sauvignon grapes during ripening in order to identify the optimal harvest time in several South African wine growing regions. The study was performed over a three year-period and grapes were sampled weekly from just after veraison until two weeks after commercial harvest. Beginning with veraison, anthocyanins accumulated in the grape berry, and correlated with sugar accumulation. Anthocyanin concentrations reached a maximum at full maturity, but are broken down if the grapes become overripe. Wine color or color density (A420 nm + A520 nm) correlated with the grape anthocyanin concentration. The latter also well correlated with the wine quality.

Ali et al. (2011) applied the 1H nuclear magnetic resonance (NMR) to monitor biochemical changes during grape berry development in three Portuguese cultivars (Trincadeira, Aragonês, and Touriga Nacional) at four developmental stages (green, veraison, ripe, harvest). They found that the accumulation of anthocyanins started at veraison. Condensed tannins were synthesized very early and very little change in tannin level has been observed from veraison to harvest. The total hydroxycinnamates showed the maximum levels prior to veraison and then a decline in the later stages due to: catabolic reactions, utilization in the synthesis of other compounds (anthocyanins and/or flavonoids), or dilution due to berry enlargement. The quercetin glucosides were highly accumulated in berries during early stages and showed a sharp decline after veraison due to the utilization of its precursors (dihydrokaempferol and/or dihydroquercetin) in the production of anthocyanins.

Another study compared the phenolic composition of seeds and skins of Carménére and Cabernet Sauvignon grape varieties (Vitis vinifera L.) during ripening (Obreque-Slier et al., 2010). The samples were harvested monthly from veraison (20th February) to the harvest time (22nd May). In Carménére and Cabernet Sauvignon skins, the extractable total phenols, total tannins, and total anthocyanins, as well as the color intensity, decreased during the study period, in particular between 20th March and 20th April, without further changes observed at the fourth sampling date (22nd May). The monomeric fraction was the more invariant and less abundant extractable fraction in the skins of both the varieties. In Carménére, extractable oligomers were also poorly represented, except for a sharp 3-fold increase occurred at the second sampling date. In the skin of Cabernet Sauvignon grapes, the levels of extractable oligomers were significant at the start of the study and then decreased continuously. The extractable polymeric fraction was far predominant over the monomeric and oligomeric fractions in the skins of both grape varieties at the first sampling date and then suffered a marked and continuous decrease. In Carménére and Cabernet Sauvignon seeds, total phenols and total tannins decreased between the first and the last sampling. Concerning flavanols, monomers were the less abundant fraction in the seeds of both the varieties. In Carménére, monomers increased up to a 10-fold peak at the third sampling date and returned to the basal levels at the last sampling date. Oligomers increased up to a 30-fold peak at the third sampling date. Polymers displayed a sharp 3-fold increase at the third sampling date. In the seeds of Cabernet Sauvignon, the contents of monomers, oligomers, and polymers displayed similar temporal progression profiles.

Conclusion

The phenolic composition of grapes is affected by a number of factors. Some of them cannot be modified, while others can be opportunely managed in order to improve fruit composition and quality. The effects of some parameters depend on the grape variety, while other practices exert the same influence independently on cultivar. The viticultural practices cannot be considered singularly because they interact with each other, enhancing or reducing their effects.

References

Ali, K., Maltese, F., Fortes, A. M., Pais, M. S., Choi, Y. H. & Verpoorte, R. (2011). Monitoring biochemical changes during grape berry development in Portuguese cultivars by NMR spectroscopy. *Food Chem.*, *124*, 1760–1769.

Alternative Farming Systems Information Center. (2005). Organic Agricultural Products: Marketing and Trade Resources. NAL, USDA.

Bavougian, Read, P. E., Schlegel, V. L. & Hanford, K. J. (2013). Canopy light effects in multiple training systems on yield, soluble solids, acidity, phenol and flavonoid concentration of 'Frontenac' grapes. *Hort Technol.*, *23*, 86-92.

Beslic, Z. S., Todic, S. R., Tesevic, V. V., Jadranin, M. B., Novakovic, M. M. & Tesic, D. (2010). Pruning effect on content of quercetin and catechin in berry skins of cv. Blaufränkisch (Vitis vinifera L.). *Turk. J. Agric. For.*, *34*, 461-466.

Bindon, K., Myburgh, P., Oberholster, A., Roux, K. & Du Toit, C. (2011). Response of grape and wine phenolic composition in Vitis vinifera L. cv. Merlot to variation in grapevine water status. *S. Afr. J. Enol. Vitic.*, *32*, 71-88.

Boonterm, V., Silapapun, A. & Boonkerd, N. (2010). Effects of nitrogen, potassium fertilizers and clusters per vine on yield and anthocyanin content in Cabernet Sauvignon grape. *Suranaree J. Sci. Technol.*, *17*, 155-163.

Boss, P. K. & Davies, C. (2001). Molecular biology of sugar and anthocyanin accumulation in grape berries. In: *Roubelakis-Angelakis KA (Eds) Molecular Biology and Biotechnology of the Grapevine*, Kluwer Academic Publishers, Dordrecht, The Netherlands, 1-33.

Breksa III, A. P., Takeoka, G. R., Hidalgo, M. B., Vilches, A., Vasse, J. & Ramming, D. W. (2010). Antioxidant activity and phenolic content of 16 raisin grape (Vitis vinifera L.) cultivars and selections. *Food Chem.*, *121*, 740–745.

Bubola, M., Persuric, D., Kovacevic Ganic, K., Karoglan, M. & Kozina, B. (2012). Effect of leaf removal on grape juice and wine phenolics. *Food Technol. Biotechnol.*, *50*, 159–166.

Cantos, E., Espín, J. C. & Tomás-Barberán, F. A. (2002). Varietal differences among the polyphenol profiles of seven table grape cultivars studied by LC-DAD-MS-MS. *J. Agric. Food Chem.*, *50*, 5691–5696.

Castino, M., Ubigli, M., Corino, L., Luzzati, A., Siragusa, N. & Nappi, P. (1987). Oenological effects of nutrients deficiencies on the grape variety Barbera cultivated in Piedmont vineyards. *Vignevini*, *14*, 37−54.

Choné, X., Lavigne-Cruège, V., Tominaga, T., van Leeuwen, C., Castagnède, C., Saucier, C. & Dubourdieu, D. (2006). Effect of vine nitrogen status on grape aromatic potential: flavor precursors (S-ysteine conjugates), glutathione and phenolic content in Vitis vinifera L. cv. Sauvignon Blanc grape juice. *J. Int. Sci. Vigne Vin*, *40*, 1–6.

Cohen, S. D. Assessing the impact of temperature on grape phenolic metabolism. Available at http://ir.library.oregonstate.edu/xmlui/bitstream/handle/1957/6661/cohen_thesis.pdf?sequence=1. Retrieved on 5th August 2013.

Conde, C., Silva, P., Fontes, N., Dias, A. C. P., Tavares, R. M., Sousa, M. G., Agasse, A., Delrot, S. & Gerós, H. (2007). Biochemical changes throughout grape berry development and fruit and wine quality. *Food*, *1*, 1-22.

Coombe, B. G. (1992). Research on development and ripening of the grape berry. *Am. J. Enol. Vitic.*, *43*, 101-110.

Corrales, M, Fernandez, A., Vizoso Pinto, M. G., Butz, P., Franz, C. M. A. P., Schuele, E. & Tauscher, B. (2010). Characterization of phenolic content, in vitro biological activity, and pesticide loads of extracts from white grape skins from organic and conventional cultivars. *Food Chem. Toxicol.*, *48*, 3471–3476.

Council Regulation (EC) N. 479/2008. of 29 April 2008 on the common organisation of the market in wine, amending Regulations (EC) No 1493/1999, (EC) No 1782/2003, (EC) No 1290/2005, (EC) No 3/2008 and repealing Regulations (EEC) No 2392/86 and (EC) No 1493/1999, *Official Journal of the European Union*, L148, 1-61.

Dani, D., Oliboni, L., Vanderlinde, R., Bonatto, D., Salvador, M. & Henriques J. (2007). Phenolic content and antioxidant activities of white and purple juices manufactured with organically- or conventionally-produced grapes *Food Chem. Toxicol.*, *45*, 2574-2580.

de Beer, D., Joubert, E., Marais, J., van Schalkwyk, D. & Manley, M. (2006). Climatic region and vine structure: effect on Pinotage wine phenolic composition, total antioxidant capacity and colour. *S. Afr. J. Enol. Vitic.*, *27*, 151-166.

Deis, L., Cavagnaro, B., Bottini, R., Wuilloud, R. & Silva, M. F. (2011). Water deficit and exogenous ABA significantly affect grape and wine phenolic composition under in field and in-vitro conditions. *Plant Growth Regul.*, *65*, 11–21.

Dias, A. C. P. (2003). The potential of in vitro cultures of Hypericum perforatum and Hypericum androsaemum to produce interesting pharmaceutical compounds. In: *Ernest E (Ed) The Genus Hypericum*, Harwood Academic Publishers, UK, 137-155.

Dokoozlian, N. K. & Kliewer, W. M. (1996). Influence of light on grape berry growth and composition varies during fruit development. *J. Amer. Soc. Hort. Sci.*, *121*, 869–874.

Donald, A. & Dibbern, Jr. Wine grapes and grape-y wines. http://www.wineloverspage.com/dibbern/grapetaste07.phtml. Retrieved on 31st July 2013.

Du, B., He, B. J., Shi, P. B., Li, F. Y., Li, J. & Zhu, F. M. (2012). Phenolic content and antioxidant activity of wine grapes and table grapes. *J. Med. Plants Res.*, *6*, 3381-3387.

Du, B., He, B. J., Shi, P. B., Li, F. Y., Li, J. & Zhu, F. M. (2012). Phenolic content and antioxidant activity of wine grapes and table grapes. *J. Med. Plants Res.*, *6*, 3381-3387.

Elias Pereira, G., Gaudillere, J.-P., Van Leeuwen, C, Hilbert, G., Maucourt, M., Deborde, C., Moing, A. & Rolin, D. (2006). 1H-NMR metabolic profiling of wines from three cultivars, three soil types and two contrasting vintages. In: Congrès International des Terroirs Viticoles, 6., International Terroir Congress, 6., 2006, Bordeaux/Montpellier. Comptes rendus... *Bordeaux: Enita*, *OIV*, 2006, v. *1*, 402-406.

Fanzone, M., Zamora, F., Jofré, V., Assof, M. & Peña-Neira, Á. (2011). Phenolic composition of Malbec grape skins and seeds from Valle de Uco (Mendoza, Argentina) during ripening. Effect of cluster thinning. *J. Agric. Food Chem.*, *59*, 6120-6136.

Ferrer-Gallego, R., Hernández-Hierro, J. M., Rivas-Gonzalo, J. C. & Escribano-Bailón, M. T. (2012). Influence of climatic conditions on the phenolic composition of Vitis vinifera L. cv. Graciano. *Anal. Chim. Acta*, *732*, 73–77.

Fredes, C., Von Bennewitz, E., Holzapfel, E. & Saavedra, F. (2010). Relation between seed appearance and phenolic maturity: a case study using grapes cv. Carménère. *Chil. J. Agr. L. Res.*, *70*, 381-389.

Fuleki, T. & Ricardo-Da-Silva, M. J. (2003). Effects of cultivar and processing method on the contents of catechins and procyanidins in grape juice. *J. Agric. Food Chem.*, *51*, 640-646.

Gil, M., Pontin, M., Berli, F., Bottini, R. & Piccoli, P. (2012). Metabolism of terpenes in the response of grape (Vitis vinifera L.) leaf tissues to UV-B radiation. *Phytochem.*, *77*, 89-98.

Giovanelli, G. & Brenna, O. V. (2007). Evolution of some phenolic components, carotenoids and chlorophylls during ripening of three Italian grape varieties. *Eur. Food Res. Technol.*, *225*, 145–150.

Gómez Gallego, M. A., Gómez García-Carpintero, E., Sánchez-Palomo, E., González Viñas, M. A. & Hermosín-Gutiérrez, I. (2012). Oenological potential, phenolic composition, chromatic characteristics and antioxidant activity of red single-cultivar wines from Castilla-La Mancha. *Food Res. Int.*, *48*, 7–15.

Gómez Gallego, M. A., Gómez García-Carpintero, E., Sánchez-Palomo, E., Hermosín-Gutiérrez, I. & González Viñas, M. A. (2013). Polyphenolic composition of Spanish red wines made from Spanish Vitis vinifera L. red grape varieties in danger of extinction. *Eur. Food Res. Technol.*, *236*, 647–658.

González-Neves, G., Barreiro, L., Gila, G., Franco, J., Ferrer, M., Moutounet, M. & Carbonneau, A. (2004). Anthocyanic composition of Tannat grapes from the south region of Uruguay. *Anal. Chim. Acta*, *513*, 197–202.

Guerra, B. & Steenwerth, K. (2012). Influence of floor management technique on grapevine growth, disease pressure, and juice and wine composition: a review. *Am. J. Enol. Vitic.*, *63*, 149-164.

Guidoni, S., Allara, P. & Schubert, A. (2002). Effect of cluster thinning on berry skin anthocyanin composition of Vitis vinifera cv. Nebbiolo. *Am. J. Enol. Vitic.*, *53*, 224-226.

Guidoni, S., Ferrandino, A. & Novello, V. (2008). Effect of seasonal and agronomical practices on skin anthocyanin profile of Nebbiolo grapes. *Am. J. Enol. Vitic.*, *59*, 22-29.

Harborne, J. B. (1964). *Methods in Polyphenol Chemistry*. Pridham, J.B., editor. Oxford: Pergamon Press.

Harborne, J. B. (1994). The flavonoid, advances in research since 1986. London: Chapman and Hall.

Harborne, J. B. & Williams, C. A. (2000). Advances in flavonoid research since 1992. *Phytochem.*, *55*, 481–504.

Hartle, D. K., Greenspan, P. & Hargrove, J. L. (2005). *Muscadine Medicine*, 1st ed., Blue Heron Nutraceuticals: Fullerton, CA, USA, 47–96.

Hilbert, G., Soyer, J. P., Molot, C., Giraudon, J., Milin, S. & Gaudillere, J. P. (2003). Effects of nitrogen supply on must quality and anthocyanin accumulation in berries of cv. Merlot. *Vitis*, *42*, 69–76.

Howell, G. S. (2001). Sustainable grape productivity and the growth-yield relationship: a review. *Am. J. Enol. Vitic.*, *52*, 165-174.

Intrigliolo, D. S. & Castel, J. M. (2010). Response of grapevine cv. 'Tempranillo' to timing and amount of irrigation: water relations, vine growth, yield and berry and wine composition. *Irrig. Sci.*, *28*, 113–125.

Jiang, B., Zhang, Z.-W, Zhang, X.-Z. & Tong, P. (2011). Influence of altitudes on phenolic compounds content and antioxidant activities of Cabernet Sauvignon berries in Loess Plateau region. *Journal of Northwest A&F University (Nat. Sci. Ed.)*, *39*, 197-202.

Jin, Z. M., He, J. J., Bi, H. q., Cui, X. Y. & Duan, C. Q. (2009). Phenolic compound profiles in berry skins from nine red wine grape cultivars in northwest China. *Molecules*, *14*, 4922-4935.

Jogaiah, S., Oulkar, D. P.: Vijapure, A. N., Maske, S. R., Kumar Sharma, A. & Somkuwar, R. G. (2013). Influence of canopy management practices on fruit composition of wine grape cultivars grown in semi-arid tropical region of India. *Afr. J. Agric. Res.*, *8*, 3462-3472.

Jug, T. & Rusjan, D. (2012). Advantages and disadvantages of UV-B radiations on grapevine (Vitis sp.). *Emir. J. Food Agric.*, *24*, 576-585.

Kamiloğlu, O. (2011). Influence of some cultural practices on yield, fruit quality and individual anthocyanins of table grape cv. 'Horoz Karasi'. *J. Anim. Plant Sci.*, *21*, 240-245.

Keller, M., Steel, C. C. & Creasy, G. L. (2000). Stilbene accumulation in grapevine tissues: Developmental and environmental effects. *Acta Hort.*, *514*, 275-286.

Kennedy, J. A., Hayasaka, Y., Vidal, S., Waters, E. J. & Jones, G. P. (2001). Composition of grape skin proanthocyanidins at different stages of berry development. *J. Agric. Food Chem.*, *49*, 5348-5355.

King, P., McClellan, D. & Smart, R. (2012). Effect of severity of leaf and cluster removal on juice and wine composition of Merlot. *Am. J. Enol. Vit.*, *63*, 500-507.

Kliewer, M. & Dokoozlian, N. (2005). Leaf area/crop weight ratios of grapevines: influence on fruit composition and wine quality. *Am. J. Enol Vitic.*, *56*, 170-181.

Koepf, H. H. The Biodynamic Farm. Anthroposophic Press, Hudson. New York.

Kolb, C. A., Käser, M. A., Kopecký, J., Zotz, G., Riederer, M. & Pfündel, E. E. (2001). Effects of natural intensities of visible and ultraviolet radiation on epidermal ultraviolet screening and photosynthesis in grape leaves. *Plant Physiol.*, *127*, 863-875.

Kolb, C. A., Kopecký, J., Riederer, M. & Pfündel, E. E. (2003). UV screening by phenolics in berries of grapevine (Vitis vinifera). *Funct. Plant Biol.*, *30*, 1177-1186.

Koundouras, S., Marinos, V., Gkoulioti, A., Kotseridis, Y. & van Leeuwen, C. (2006). Influence of vineyard location and vine water status on fruit maturation of non-irrigated cv Agiorgitiko (Vitis vinifera L.). Effects on wine phenolic and aroma components. *J. Agric. Food Chem.*, *54*, 5077–5086.

Lafontaine, M., Schultz, H., Lopes, C., Bálo, B. & Varadi, G. (2005). Leaf and fruit responses of Riesling grapevines to UV-radiation in the field. *Acta Hort.*, *689*, 125-132.

Larronde, F., Krisa, S., Decendit, A., Cheze, C., Deffieux, G. & Mérillon, J. M. (1998). Regulation of polyphenol production in Vitis vinifera cell suspension cultures by sugars. *Plant Cell Reports*, *17*, 946-950.

Lattanzio, V., Lattanzio, V. M. T. & Cardinali, A. (2006). Role of phenolics in the resistance mechanisms of plants against fungal pathogens and insects. In: Imperato F, editor. Phytochemistry: Advances in Research. Kerala, India: Research Signpost, 23-67.

Mabry, T. J., Markham, K. R. & Thomas, M. B. (1970). *The Systematic Identification of Flavonoids*. New York: Springer Verlag.

Mackenzie, D. E. & Christy, A. G. (2005). The role of soil chemistry in wine grape quality and sustainable soil management in vineyards. *Water Sci. Technol.*, *51*, 27–37.

Masa, A. & Vilanova, M. (2008). Flavonoid and aromatic characterisation of cv. Albarín blanco (Vitis vinifera L.). *Food Chem.*, *107*, 273–281.

Mateus, N., Proença, S., Ribeiro, P., Machado, J. M. & De Freitas, V. (2001). Grape and wine polyphenolic composition of red Vitis vinifera varieties concerning vineyard altitude. *Cienc. Tecnol. Aliment.*, *3*, 102-110.

Meggio, F., Zarco-Tejada, P. J., Núñez, L. C., Sepulcre-Cantó, G., González, M. R. & Martín, P. (2010). Grape quality assessment in vineyards affected by iron deficiency chlorosis using narrow-band physiological remote sensing indices. *Remote Sens. Environ.*, *114*, 1968–1986.

Muganu, M., Paolocci, M., Gnisci, D., Barnaba, F. E., Bellincontro, A., Mencarelli, F. & Grosu, I. (2013). Effect of different soil management practices on grapevine growth and on berry quality assessed by NIR-AOTF spectroscopy. *Acta Hort. (ISHS)*, *978*, 117-125.

Mulero, J., Pardo, F. & Zafrilla, P. (2010). Antioxidant activity and phenolic composition of organic and conventional grapes and wines. *J. Food Comp. Anal.*, *23*, 569–574.

Mullins, M. G., Bouquet, A. & Williams, L. (1992). *Biology of the grapevines*. New York, USA: Cambridge University Press.

Nadal, M. & Vernet, J. M. (2010). Deficit irrigation and rootstock effects on water stress, growth, and grape composition in a Mediterranean climate. *Contributions to Science*, *6*, 115–119.

Navarro, S., León, M., Roca-Pérez, L., Boluda, R., García-Ferriz, L., Pérez-Bermúdez, P. & Gavidia, I. (2008). Characterisation of Bobal and Crujidera grape cultivars, in comparison with Tempranillo and Cabernet Sauvignon: Evolution of leaf macronutrients and berry composition during grape ripening. *Food Chem.*, *108*, 182–190.

Nicholas, K. A., Matthews, M. A., Lobell, D. B., Willits, N. H. & Field, C. B. (2011). Effect of vineyard-scale climate variability on Pinot noir phenolic composition. *Agric. Forest Meteorol.*, *151*, 1556–1567.

Oberholster, A., Botes, M.-P. & Lambrechts, M. (2010). Phenolic composition of Cabernet-Sauvignon (Vitis vinifera) grapes during ripening in four South African winegrowing regions. *J. Int. Sci. Vigne Vin, special issue Macrowine*, 33-40.

Obreque-Slier, E., López-Solís, R., Castro-Ulloa, L., Romero-Díaz, C. & Peña-Neira, Á. (2012). Phenolic composition and physicochemical parameters of Carménère, Cabernet Sauvignon, Merlot and Cabernet Franc grape seeds (Vitis vinifera L.) during ripening. *LWT - Food Sci. Technol.*, *48*, 134-141.

Obreque-Slier, E., Peña-Neira, Á., López-Solís, R., Zamora-Marín, F., Ricardo-Da Silva, J. M. & Laureano, O. (2010). Comparative study of the phenolic composition of seeds and skins from Carménére and Cabernet Sauvignon grape varieties (Vitis vinifera L.) during ripening. *J. Agric. Food Chem.*, *58*, 3591–3599.

Ojeda, H., Andary, C., Kraeva, E., Carbonneau, A. & Deloire, A. (2002). Influence of pre- and postveraison water deficits on synthesis and concentration of skin phenolic compounds during berry growth of Vitis vinifera cv. Shiraz. *Am. J. Enol. Vitic.*, *53*, 261-267.

Oxford Dictionary. http://oxforddictionaries.com/definition/english/grape. Retrieved on 31st July 2013.

Palliotti, A., Gardia, T., Berriosb, J. G., Civardic, S. & Poni, S. (2012). Early source limitation as a tool for yield control and wine quality improvement in a high-yielding red Vitis vinifera L. cultivar. Sci. *Horticult.*, *145*, 10–16.

Pan, Q.-H., Wang, L. & Li, J.-M. (2009). Amounts and subcellular localization of stilbene synthase in response of grape berries to UV irradiation. *Plant Sci.*, *176*, 360–366.

Peterlunger, E., Celotti, E., Da Dalt, G., Stefanelli, S., Gollino, G. & Zironi, R. (2002). Effect of training system on Pinot noir grape and wine composition. *Am. J. Enol. Vitic.*, *53*, 14-18.

Peyrot des Gachons, C., van Leeuwen, C., Tominaga, T., Soyer, J.-P., Gaudillère, J.-P. & Dubourdieu, D. (2005). The influence of water and nitrogen deficit on fruit ripening and aroma potential of Vitis vinifera L. cv Sauvignon blanc in field conditions. *J. Sci. Food Agric.*, *85*, 73–85.

Pfeiffer, E. (1983). *Soil Fertility, Renewal and Preservation*. The Lanthorn Press, East Grinstead. UK.

Pontin, M. A., Piccoli, P. N., Francisco, R., Bottini, R., Martinez-Zapater, J. M. & Lijavetzky, D. (2010). Transcriptome changes in grapevine (Vitis vinifera L.) cv. Malbec leaves induced by ultraviolet-B radiation. *BMC Plant Biol.*, *10*, 224-237.

Qian, M. Effect of vineyard cover crop management on grape and wine quality II-grape composition and wine aroma. Available at: http://owri.oregonstate.edu/sites/owri.org/files/documents/Cover%20Crop%20Qian%202008.pdf. Retrieved on 6th August 2013.

Reeve, J., Carpenter-Boggs, L., Reganold, J., York, A., McGourty, G. & McCloskey, L. (2005). Soil and winegrape quality in biodynamically and organically managed vineyards. *Am. J. Enol. Vitic.*, *56*, 367-376.

Reeve, J. R. (2003). Effects of biodynamic preparations on soil, winegrape, and compost quality on a california vineyard. *A thesis submitted for the degree of Master of Science*, Washington State University, Department of Crop and Soil Sciences.

Research Institute of Organic Farming. (2000). Organic Farming Enhances Soil Fertility and Biodiversity: Results from a 21 Year Old Field Trial. FiBL, Frick, Switzerland.

Rieger, T. (2008). Exploring high altitude viticulture. Part One. *Vineyard & Winery Management*, 86-90.

Río Segade, S., Soto Vázquez, E. & Díaz Losada, E. (2008). Influence of ripeness grade on accumulation and extractability of grape skin anthocyanins in different cultivars. *J. Food Comp. Anal.*, *21*, 599-607.

Río Segade, S., Soto Vázquez, E., Vázquez Rodríguez, E. I. & Rego Martínez, J. F. (2009). Influence of training system on chromatic characteristics and phenolic composition in red wines. *Eur. Food Res. Technol.*, *229*, 763–770.

Romero, P., Gil-Muñoz, R., del Amor, Valdés, F. M., E., Fernández, J. I. & Martinez-Cutillas, A. (2013). Regulated Deficit Irrigation based upon optimum water status improves phenolic composition in Monastrell grapes and wines. *Agric. Water Manag.*, *121*, 85– 101.

Romeyer, F. M., Macheix, J. J., Goiffon, J. P., Reminiac, C. C. & Sapis, J. C. (1983). The browning capacity of grapes. 3. Changes and importance of hydroxycinnamic acid-tartaric acid esters during development and maturation of the fruit. *J. Agric. Food Chem.*, *31*, 346-349.

Santesteban, L. G., Miranda, C. & Royo, J. B. (2011). Regulated deficit irrigation effects on growth, yield, grape quality and individual anthocyanin composition in Vitis vinifera L. cv. 'Tempranillo'. *Agric. Water Manag.*, *98*, 1171–1179.

Santini, D., Cascio, P., Mazza, G. & Mannini, F. (2010). Influence of different environments on grape phenolic and aromatic composition of three clones of 'Nebbiolo' (Vitis Vinifera L.). *Quad. Vitic. Enol. Univ. Torino*, 2009-2010, *31*, 103-106.

Skinkis, P. A. Vineyard cover crop management in the North Willamette Valley: I. Effects of cover crop management on vine vigor and fruit quality of Pinot Noir. Available at: http://s3.amazonaws.com/zanran_storage/oregonwine.org/ContentPages/106805706.pdf. Retrieved on 6th August 2013.

Smart, R. & Robinson, M. (1991). *Sunlight into wine, a handbook for wine grape canopy arrangement.* Ed. Winetitles, Adelaide.

Steiner, R. (1993). *Agriculture: Spiritual foundations for the renewal of agriculture.* Anthroposophic Press, Hudson, New York.

Sternad Lemut, M., Trost, K., Sivilotti, P. & Vrhovsek, U. (2011). Pinot Noir grape colour related phenolics as affected by leaf removal treatments in the Vipava Valley. *J. Food Comp. Anal.*, *24*, 777–784.

Tarara, J. M. & Lee, J. An Introduction to environmental influences on ripening in grapes: focus on wine grapes and phenolics. http://www.extension.org/pages/33025/an-introduction-to-environmental-influences-on-ripening-in-grapes:-focus-on-wine-grapes-and-phenolics. Retrieved on 30th July 2013.

van Leeuwen, C., Friant, P., Choné, X., Tregoat, O., Koundouras, S. & Dubourdieu, D. (2004). Influence of climate, soil, and cultivar on terroir. *Am. J. Enol. Vitic.*, *55*, 207-217.

van Leeuwen, C., Friant, P., Soyer, J.-P., Molot, C., Choné, X. & Dubourdieu, D. (2000). L'intérêt du dosage de l'azote total et l'azote assimilable dans le moût comme indicateur de la nutrition azotée de la vigne. *J. Int. Sci. Vigne Vin*, *34*, 75–82.

van Leeuwen, C. & Seguin, G. (1994). Incidences de l'alimentation en eau de la vigne, appréciée par l'état hydrique du feuillage, sur le développement de l'appareil végétatif et la maturation du raisin (Vitis vinifera variété Cabernet franc, Saint-Emilion, 1990). *J. Int. Sci. Vigne Vin*, *28*, 81-110.

van Leeuwen, C. & Seguin, G. (2006). The concept of terroir in viticulture. *J. Wine Res.*, *17*, 1-10.

van Leeuwen, C., Trégoat, O., Choné, X., Gaudillère, J.-P. & Pernet, D. (2007). Different environmental conditions, different results: the role of controlled environmental stress on grape quality potential and the way to monitor it. *Proceedings of the ThirteenthAustralian Wine Industry Technical Conference*, Adelaide, South Australia, 28 July - 2 August 2007.

Varache-Lembege, M., Waffo-Teguo, P., Decendit, A., Devaux, G., Deffieux, G. & Mérillon, J. M. (1996). Polyhydroxystilbenes from Vitis vinifera L. cells: inhibitory effect on human platelet aggregation and molecular modeling, In: Vercauteren J, Cheze C, Dumon MC, Weber JF Eds. Polyhenols Communications 96, 18th International Conference on Polyphenols 1996, July 15-18.

Veliksar, S. G., Toma, S. I. & Kreidman, J. (2005). Effect of Fe-containing compounds on the chlorosis manifestation and grape quality. *Proceedings International Workshop on Advances in Grapevine and Wine Research*, Venosa, Italy, vol. *122*, 2005.

Vieira Da Mota, R., De Souza, C. R., Pinheiro Carvalho Silva, C., De Faria Freitas, G., Misuzu Shiga, T., Purgatto, E., Lajolo, F. M. & De Albuquerque Regina, M. (2010).

Biochemical and agronomical responses of grapevines to alteration of source-sink ratio by cluster thinning and shoot trimming. *Bragantia, Campinas* 69, 17-25.

Vitrac, X., Larronde, F., Krisa, S., Decendit, A., Deffieux, G. & Merillon, J. M. (2000). Sugar sensing and Ca2+-calmodulin requirement in Vitis vinifera cells producing anthocyanins. *Phytochem., 53*, 659-665.

Wolf, T. K., Dry, P. R., Iland, P. G., Botting, D., Dick, J., Kennedy, U. & Ristic, R. (2003). Response of Shiraz grapevines to five different training systems in the Barossa Valley, Australia. *Austr. J. Grape Wine Res., 9*, 82–95.

Xi, Z.-M., Zhang, Z.-W., Cheng, Y.-F. & Li, H. (2010). The effect of vineyard cover crop on main monomeric phenols of grape berry and wine in Vitis vinifera L. cv. Cabernet Sauvignon. *Agricultural Sciences in China, 9*, 440-448.

Xu, C., Zhang, Y., Cao, L. & Lu, J. (2010). Phenolic compounds and antioxidant properties of different grape cultivars grown in China. *Food Chem., 119*, 1557–1565.

Zhao, X. J., Wang, X., Liu, B., Li, J., Sun, Y. X. & Shu, H. R. (2006). Effect of cluster thinning on catechins in berries of Vitis vinifera cv. Cabernet Sauvignon. *Vitis, 45*, 103–104.

Zhu, L., Zhang, Y. & Lu, J. (2012). Phenolic contents and compositions in skins of red wine grape cultivars among various genetic backgrounds and originations. *Int. J. Mol. Sci., 13*, 3492-3510.

Zoecklein, B., Kenneth, F., Gump, B. & Nury, F. (2001). *Análisis y producción de vino.*, Editorial Acribia, 2001, Zaragoza, España. 613 p.

Zoecklein, B. W., Fugelsang, K. C., Gump, B. H. & Nury, F. S. (1990). *Production Wine Analysis*. New York: Van Nostrand Reinhold.

In: Grapes
Editor: José S. Câmara
ISBN: 978-1-63321-402-6

Chapter 2

Grape-Berries Yeast Populations: Influence of Vineyard Management

***Gustavo Cordero-Bueso*[1], *Teresa Arroyo*[2] *and Eva Valero*[1,*]**
[1]Departamento de Biología Molecular e Ingeniería Bioquímica,
Universidad Pablo de Olavide. Sevilla, Spain
[2]Departamento de Agroalimentación, Instituto Madrileño de Investigación y Desarrollo
Rural Agrario y Alimentario (IMIDRA). Madrid, Spain

Abstract

Wine production in most countries is based on the use of commercial strains leading to the colonisation of the wineries and vineyards by these strains with the consequent reduction of autochthonous biodiversity. This implies that wine styles could become standardised. Exploiting autochthonous yeasts present in grape must is re-emerging as a commercial option worldwide in winegrowing regions under pressure from market competition. Thus, vineyards could be an important source of native yeasts of oenological interest. This chapter reviews the influence of different agronomic practices on the biodiversity of yeast population in the vineyard, such as floor management strategies, the farming system (conventional and organic farming), the grape variety selected and the use of pesticides.

Keywords: Biodiversity, farming system, floor management, fungicide, grape variety

Introduction

Traditionally, wine fermentation is carried out in a spontaneous way by indigenous yeast communities present on the grapes when harvested from vineyards or by those present in the winery equipment.

* Corresponding author address Email: evalero@upo.es.

Thus, vineyards and grape berry surfaces are a primary source of natural yeasts in wine production. Generally, indigenous species present on the grape, such as *Hanseniaspora* (anamorph *Kloeckera*), *Lachancea*, *Candida*, *Meyerozyma*, *Pichia*, *Wickerhamomyces* and *Metschnikowia*, initiate the alcoholic fermentation of the grape juice (Cordero-Bueso et al., 2013; Barata et al. 2012; Tello et al. 2012; Fleet, 2008). During their growth these yeast populations intervene in the fermentation process through the synthesis of secondary metabolites (acids, alcohols, esters, polyols, aldehydes, ketones, volatile sulphur compounds, etc. (Fleet, 2003) and enzymes (esterases, glycosidases, lipases, β-glucosidases, proteases, cellulases, etc. (Strauss et al, 2001) which may affect the final quality of the wine. These species are able to grow at low concentrations of ethanol, but as its concentration exceeds 5 - 7% and the abundance of fermenting sugars decreases, they begin to decline and die off (Fleet, 2008).

At this time, and although several species of *Brettanomyces*, *Kluyveromyces*, *Schizosaccharomyces*, *Torulaspora* and *Zygosaccharomyces* may be found (Pretorius, 2000), alcohol-tolerant strains of *Saccharomyces cerevisiae* dominate the process until its completion (Martini and Vaughan-Martini, 1990). In addition to the gradual growth of different yeast species throughout the course of fermentation, there is an underlying successional development of strains within each species (Fleet, 2008).

Otherwise, the use of selected strains of *S. cerevisiae* has improved the fermentative processes and the quality of wines, but their continued use has led to a colonization and consequent elimination of the native microorganisms present in the wineries, including the nearest vineyards (Blanco et al. 2012; Cordero-Bueso et al. 2011a; Schuller et al. 2005; Valero et al. 2005, 2007).

This not only happen within different wineries, but also within different winemaking regions, resulting in wines with similar characteristics produced in different areas. So, given that commercial strains do not always enhance the sensory traits, they may be enhanced by yeasts isolated from specific geographical areas (Schuller et al. 2012; Tello et al. 2012; Rainieri and Pretorius, 2000).

Hence, although the presence of fermentative yeasts in the vineyard is low, (Mortimer and Polsinelli, 1999), the latter may be the main reservoir of native yeasts of oenological interest. It is therefore necessary to preserve, and even encourage, the presence of fermentative species in the vineyard. In fact, the use of native yeast strains could be useful since they are better acclimatized to the environmental conditions and assure the maintenance of the typical sensory properties of the wines of a given region (Cordero-Bueso et al. 2013; Callejón et al. 2010, Esteve-Zarzoso et al., 2000).

Numerous studies have evaluated the different species present in the wine ecosystem and demonstrated the impact of grape conditions on yeast populations and other external factors such as climatic conditions, the geographical location of the vineyard, the ripeness of the grape berries, the age of the vineyard, the application of agrochemicals, etc. (Milanović et al. 2013; Čadež et al. 2010; Comitini and Ciani, 2008; Raspor et al. 2006; Fleet et al. 2002; Pretorius, 2000). Winemakers recognize that some indigenous yeast species may enhance or damage the quality of wine, but a deeper knowledge of the effects of these factors on the yeast ecology of grapes is required.

In this chapter we will attempt to update the current awareness of yeast ecology on grape surfaces describing species diversity and the agronomic practices affecting the balance of these species.

The Vineyard Farming System Influencing Yeast Diversity

Nowadays wine has lost its function as a mere dietary component: its consumption is also linked with pleasure, conviviality, psychological satisfaction, cultural interest and health (Fotopoulos et al, 2003). In this sense, consumers' concern about health and protection of the environment has grown, increasing the demand for safe and/or organic food (Kallas et al. 2010). Among other organic products, organic wines are being demanded by consumers more and more frequently. Wheeler and Crisp (2009) suggest that organic grapes result in higher quality in the final wine than conventional grapes, for red but not white varieties. As a result, the number of new organically based wineries is increasing, and old traditional wineries are adding the production of organic wines to their production lines, including vineyard management (Tello et al. 2012; Callejón et al, 2010). Studies tend to report that, compared with conventional farming; the organic system is able to achieve a higher and richer abundance of species across a number of different areas. It has been shown that organic farming has higher microbial activity, biomass and biodiversity (Hole et al. 2005). Thus, depending on the agronomic practices (conventional and organic) there are differences in the distribution of the population of yeasts.

According to published research (Setati et al. 2012; Tello et al. 2012; Cordero-Bueso et al. 2011b) the farming system of vineyards has been shown to have an important influence on the yeast populations associated with the vineyard, not only in the different species but also on a strain level. Given that the use of different agrochemical products has a negative effect on grape yeasts, the organic vineyard looks like the best option as a natural reservoir where developing the tasks necessary for obtaining yeasts of interest for the wine-making process (Setati et al. 2012; Cordero-Bueso et al. 2011b; Čadež et al. 2010; Comitini and Ciani, 2008). These studies showed that a large proportion of non-*Saccharomyces* strains were found, representing more than 80% of all the yeasts isolated, and fewer than 20% of the isolations turned out to be *Saccharomyces* strains. A wide range of yeast species were found throughout the different vineyards sampled, corresponding to *Candida sorbosa* (anamorphic *Issatchenkia occidentalis*), *Candida stellata*, *Hanseniaspora guilliermondii*, *Lachancea thermotolerans* (formerly *Kluyveromyces thermotolerans*), *Metschnikowia pulcherrima*, *Wickerhamomyces anomalus* (formerly *Pichia anomala*), *Pichia toletana*, *Rhodotorula mucilaginosa*, *Saccharomyces cerevisiae*, *Torulaspora delbrueckii and Lachancea fermentati* (formerly *Zygosaccharomyces fermentanti*). The presence of these species was previously described in other surveys of yeast microbiota in wine (Francesca et al. 2010; Li et al. 2010; Chavan et al. 2009; Tofalo et al. 2009; Romancino et al. 2008; Zott et al. 2008; Di Maro et al. 2007; ; Ribéreau-Gayon et al. 2006; Mills et al. 2002; Pretorius, 2000;). Moreover, strains identified as *S. cerevisiae* by Tello et al. (2012) and Cordero-Bueso et al. (2011b) were genotyped by microsatellite multiplex PCR analysis (Vaudano and García-Moruno, 2008) using SC8132X, YOR267C and SCPTSY7 primers obtaining twenty-nine different electrophoretic patterns in total.

A field and laboratory study carried out by Cordero-Bueso et al. (2011b) shown that an important amount of non-*Saccharomyces* yeasts and *Saccharomyces* strains were isolated in musts from the organic vineyard in comparison with the isolates from the conventional one. A total of 876 non-*Saccharomyces* were isolated, 541 (62%) for the organic and 335 (38%) for the conventional vineyard. On the other hand, 204 *Saccharomyces* were found, 178 (87%) for the organic and 26 (13%) for the conventional vineyard. A great diversity of yeast species was observed in the musts obtained from the grapes harvested from the organic vineyard. *L. thermotolerans* was the most abundant, comprising half of the isolated yeast (50.2%), followed by *S. cerevisiae* (12.5%), *C. stellata, M. pulcherrima* and *H. guilliermondii*. With respect to fermentative yeasts, all genotypes were found within the *S. cerevisiae* analyzed in the organic vineyard. This demonstrates the great diversity of fermentative strains in this vineyard.

L. thermotolerans was the predominant non-*Saccharomyces* species found in the musts obtained from both vineyards. This yeast was previously found in several wine-producing regions (Kapsopoulou et al. 2007; Nurgel et al. 2005; Torija et al. 2001). It is common in warm and dry areas, such as the South of Europe, South Africa, Australia and New Zealand. It is characterized by the high production of L-lactic acid, the low production of volatile acidity, moderate alcohol productivity and the absence of off-flavour production (Ribèreau-Gayon et al. 2006). It has also been reported that this yeast shows a reduced vitality in mixed cultures with *S. cerevisiae* (Kapsopoulou et al. 2007). *H. guilliermondii* was the second most frequent species of non-*Saccharomyces* found. Apiculate yeasts, like *H. guilliermondii*, are known to be high producers of esters and glycerol (Viana et al. 2008). Moreover, they have the ability to secrete some important enzymes, such as β-glucosidases and proteases, which interact with some precursor compounds (glycosidically bound terpenes) derived from the grapes. This contributes to the subsequent expression of the varietal aroma, and, thereby, the wine-making process (Arroyo et al. 2010; Li et al. 2010; Zott et al. 2008; Jolly et al. 2006). Other species such as *C. stellata*, *T. delbrueckii* and *P. anomala* were also found in the vineyard with both farming systems. *C. stellata* is frequently associated with over-ripened and botrytized grapes and is able to survive in the fermenting must until the completion of the wine-making process (Csoma and Sipiczki, 2008). With respect to *T. delbrueckii*, it was more commonly isolated in the musts obtained from the grapes harvested from the conventional vineyard. This species has a strong fermentative capacity, being alcohol-tolerant (up to 10% v/v) and resistant to antiseptics. This yeast is commonly found in spontaneous fermenting musts, independently of the presence or absence of SO_2 (Renault, 2009). *P. anomala* is able to use a broad range of nitrogen and phosphor sources of different organic residues generated from some agricultural techniques (Passoth et al. 2006). The proportion of other species was relatively low. *M. pulcherrima* was found only in the organic vineyard grapes (Cordero-Bueso et al. 2011b). It is characterized for being a significantly proteolytic yeast, as well as being able to provide some important amino acids for the growth of *S. cerevisiae* by mid-fermentation (Fleet, 2003; Ganga and Martínez, 2004) and for displaying β-glucosidase activity (Fernández et al. 2000). A very low percentage (<5%) of *C. sorbosa* and *P. toletana* was isolated from musts obtained from the conventional vineyard (Cordero-Bueso et al. 2011b).

An analysis of the diversity and distribution of wine yeasts isolated from organically and conventionally grown grapes, and during the subsequent fermentation with or without starter cultures in six different commercial wineries in the Madrid vine-growing region, was carried out by Cordero-Bueso et al. (2011b) and Tello et al. (2012). The work concludes that organic farming management is more beneficial in terms of diversity and abundance than its conventional counterpart. Induced fermentation generated a great replacement of native yeasts. Although winery-resident yeasts were predominant in the process, some non commercial strains originally in the vineyard were found in the final stages of fermentation, confirming that autochthonous strains of *S. cerevisiae* are capable of conducting the fermentation process up to its end.

Setati et al. (2012) have shown that farming systems (organic, biodynamic and conventional) have a significant impact on fungal and yeast diversity but more importantly that there is significant species heterogeneity between samples in the same vineyard. Cultivation-based methods confirmed that while the same oxidative yeast species dominated in all vineyards sampled, the least treated vineyard displayed significantly higher species richness. The yeast-like fungus *A. pullulans* was found to be the dominant yeast inhabiting the grape berry surface in the conventional, integrated and biodynamic vineyards in a similar density. In contrast other researchers have reported higher levels of this yeast-like fungus in organic vineyards than in their conventional counterparts (Schmid et al. 2011). It is interesting to point out that a higher number of yeasts with biocontrol potential including *Rhodosporidium diobovatum*, *Meira geulakoningii* and *Cryptococcus laurentii* were isolated from the biodynamic vineyard.

Thus, the results of all these studies showed that the phytosanitary treatment affected the grape microbiota negatively, reducing the abundance of yeasts isolated and the diversity, especially of fermentative yeast strains of *S. cerevisiae*. This last aspect is very interesting, given that phytosanitary treatments affect not only the grape richness of species in general but also have a specific effect on fermentative strains. This is particularly relevant for winemakers, given that a larger proportion of fermentative yeasts favour the development of alcoholic fermentation, principally for those winemakers who produce wines with spontaneous fermentations, particularly those classified as organic, integrated, biodynamic or natural wines.

Influence of the Grape Variety on Yeast Ecology

The composition and properties of different grape varieties have been extensively investigated (Cordero-Bueso et al. 2013; Holt et al. 2013; Francesca et al. 2010; Chavan et al. 2009; Bauza et al. 2007; Raspor et al. 2006; Clemente-Jimenez et al. 2005). But wine quality is influenced, partially, by the composition of the grape juice and by the microorganisms present in the fermentation process (Blanco et al. 2012; Callejón et al. 2010). Varietal factors such as thickness of the grape skins can play an important role on the yeast microbiota present on grapes (Li et al. 2010). Thus, yeast populations and species may vary according to the berry development stage. As an example, Fleet (2003) reported an increased incidence of apiculate yeasts in mature grapes. Moreover, according to Renouf et al. (2005), when a large berry surface was available for the adhesion of yeasts and non agrochemical treatments were used, the microbial and yeast population was higher. Regarding *S. cerevisiae* strains, Mortimer and Polsinelli (1999) discussed the low occurrence of *S. cerevisiae* in grapes and musts. On the other hand, Nurgel et al. (2005) reported the high presence of this fermentative species in musts obtained from white and black grapes grown in Anatolia, where there was an excessive use of sulfite in the vineyard.

Species of non-*Saccharomyces* of the genera *Candida*, *Hanseniaspora*, *Hansenula*, *Issatchenkia*, *Lachancea*, *Kluyveromyces*, *Metschnikowia*, *Pichia*, *Torulaspora* and

Zygosaccharomyces are known to be present on the surface of grape berries (Cordero Bueso et al., 20011a, 2011b, 2013, Tello et al. 2012; Francesca et al. 2010; Li et al. 2010; Chavan et al. 2009; González et al. 2007; Renouf et al. 2005). According to Joly et al. 2006 grape berry surfaces of the varieties Chardonnay, Cabernet sauvignon and Merlot are dominated by species of the genera *Hanseniaspora* and *Metschnikowia* in warm regions like Spain, South Africa or Australia, while the yeast-like fungus *Aureobasidium pullulans* is mostly present in cooler regions of central Europe (Cadez et al. 2010; Raspor et al. 2006; Renouf et al. 2005). According to Sabaté et al. (2002), yeast populations on unripe grapes are very similar to those on other plant substrates, but maturation in grapes causes weakness in the peel and some juice might escape. The presence of a rich sugary medium on the grape surface would increase the must yeast proportion of some species such as *Hanseniaspora*, *Candida* and *Pichia*. The population density and diversity of indigenous yeasts on grape berries are intricately linked to other numerous factors such as the climatic conditions, the geographical location of the vineyard, the age of the vineyard, the soil type, the application of antifungals and the technique used to harvest (Setati et al. 2012; Li et al. 2010; Chavan et al. 2009; Nisiotou and Nychas, 2007; Valero et al. 2007; Raspor et al. 2006; Combina et al. 2005; Valero et al. 2005; Fleet et al. 2002; Pretorius, 2000). However, grape variety and ripeness has a strong influence: not always the same species or strains are present among different varieties (Csoma et al. 2010), but some studies demonstrated that there are minor differences between yeast communities on the same variety cultivated in different vine-growing regions (Shuangshil et al. 2011; Csoma et al. 2010).

Syrah, Grenache, Cabernet sauvignon, Chardonnay and Barbera (red varieties) are known to be easy-growing grapevines, as well as being resistant to the most common vineyard diseases. These varieties have a special interest, due to their easy adaptation to the soil, climatic conditions and other parameters which characterise the areas where they grow. Furthermore, these grapevines among others are used to obtain some of the wines produced in the wineries of different vine-growing regions such as Madrid (Cordero-Bueso et al. 2013; Cordero-Bueso et al. 2011b), "Serranía de Ronda", Málaga, southern Spain (Clavijo et al. 2010), La Mancha, Spain (Fernández-González and Briones, 2013), Hungary (Csoma et al. 2010), Dolenjska, Slovenia (Raspor et al. 2006), Italy (Capece et al. 2011), China (Li et al. 2010), India (Chavan et al. 2009), New Zealand (Gayevskiy and Goddard, 2012). In the case of the published study by Cordero-Bueso et al. (2011b), the yeast ecology of three vine varieties (Syrah, Grenache and Barbera) was analysed during three years. The amount and distribution of all yeast strains identified in this study differed among the vine varieties. Nine different yeast species were found in the Syrah grape variety, while five were isolated for both Grenache and Barbera, in the global data. Non-*Saccharomyces* yeasts were the most abundant species found in all the varieties (Syrah, Grenache and Barbera). *L. thermotolerans* was the predominant species isolated for the Syrah and Grenache. It was the third most abundant in Barbera must. *C. stellata* and a notable percentage of *H. guilliermondii* were found in all the varieties studied. *W. anomalus* was isolated from Syrah and Grenache, but not in the Barbera must. *T. delbrueckii* was detected principally in Barbera musts, only one isolate was found in Syrah must and none in Grenache must. A low percentage of the species *C. sorbosa*, *M. pulcherrima* and *P. toletana* were found only in the musts obtained from Syrah grapes. This study also showed the presence of *S. cerevisiae* species in all of the different varieties studied. With respect to the different genotypes of *S. cerevisiae* identified by microsatellite multiplex PCR, Barbera and Grenache grapevines were the most important reservoirs of fermentative yeast strains of *S. cerevisiae*, both in quantity and diversity.

The Barbera grapevine was the best variety in terms of the abundance and diversity of *Saccharomyces* yeast strains. This rich biodiversity found is very important for the food industry, especially for the oenological sector. Each identified genotype could influence the final quality of the wine and its properties, such as the capacity to form aroma. This capacity depends not only on the yeast species but also on the particular strain of the individual species (Torrens et al. 2008). Different strains of *Saccharomyces* can produce significant flavour differences when fermenting the same must. This is a consequence of both the differential ability of wine yeast strains to release varietal volatile precursors and the different ability to synthesise the novo yeast-derived volatile compounds (Callejón et al. 2010; Swiegers et al. 2006; Wondra et al. 2001). Therefore, the selection of the proper yeast strain may be critical for the development of the desired wine style (Callejón et al. 2010; Molina et al. 2009).

As has been already mentioned, non-*Saccharomyces* species represent the highest number of isolated yeasts in all vine varieties. Clavijo et al. (2010) have shown that most of the isolates (84%) corresponded to thirteen different non-*Saccharomyces* species with *L. thermotolerans*, *H. guilliermondii*, *H. uvarum* and *Issatchenkia orientalis* accounting for 42.7% of the total. Mitochondrial DNA restriction analysis from the *S. cerevisiae* isolates revealed a low diversity since only eleven different profiles were found, nine of them corresponding to local strains (Ronda, Málaga, Spain) and two to commercial ones that had been used in different campaigns and that were very likely disseminated from the winery to the adjacent vineyard. So, a different distribution of strains was found in the three grape varieties studied. The diversity of yeast species related with the grape variety Touriga Nacional was evaluated by Duarte et al. in 2011, in four distinct vineyards of three Portuguese wine regions. A total of twenty yeast species were found. As expected, higher species diversity was obtained for the initial sampling time. The most representative species was *H. uvarum*, which was detected in grape must from all the vineyards, followed by *C. zemplinina*. Some species like *P. membranifaciens*, *P. kluyveri var kluyveri*, *C. railenensis*, *Saccharomycopsis vini*, *C. diversa*, among others, were only detected in grape must from one vineyard. None of the isolates belonged to the species *S. cerevisiae*. Chavan et al. (2009) evaluated six grape varieties from India; Bangalore blue, Zinfandel, Cabernet, Chenin blanc, Sauvignon blanc and Shiraz. The yeast diversity was studied on the basis of morphological, colony, physiological characteristics and 5.8S-ITS sequencing of rDNA of the isolates. Eleven different species belonging to seven genera were identified as: *C. azyma*, *C. quercitrusa*, *D. hansenii*, *H. guilliermondii*, *H. viniae*, *H. uvarum*, *I. orientalis*, *I. terricola*, *P. membranifaciens*, *S. cerevisiae* and *Zygoascus steatolyticus*. This study has shown that *H. guilliermondii* was the predominant species while *S. cerevisiae* was observed occasionally in the six vine varieties.

Several studies have also tried to show that the geographic location and microclimatological conditions have a great impact on density and diversity of yeasts remaining on the grape surfaces. In 2013, Lederer et al. isolated yeast species from two interspecific grape varieties and one hybrid from three different vineyards grown in Denmark. A total of 200 isolates were grouped according to their rep-PCR profiles and representative isolates identified by sequencing of the 26S rRNA D1/D2 region and selected phenotypic tests. The yeast microbiota on the grapes was also dominated by non-*Saccharomyces* yeasts. Twelve different yeast species belonging to six genera were detected, including *Rhodotorula fujisanensis, Rhodotorula glutinis, Rhodotorula nothofagi, Cryptococcus carnescens, Cryptococcus festucosus, Cryptococcus* spp., *Cryptococcus victoriae, Cryptococcus wieringae, H. uvarum, Metschnikowia fructicola, Kazachstania servazzii* and *Sporobolomyces roseus* and the yeast-like ascomycetous fungi, *Aureobasidium pullulans.* On the other hand, Raspor et al. (2006) evaluate the impact of geographical locations in the Dolenjska vine-growing region (Slovenia) isolating yeasts from three different grape varieties. The predominant species were *R. glutinis*, *H. uvarum* and *A. pullulans*. The distribution and density was very similar to those isolated in Denmark by Lederer et al. (2013). Thus, the difference between yeast species isolated from Northern (cool climate) grape varieties with Southern varieties (warm climate) is noteworthy. Yeast microbiota present on grape surfaces differs within vine varieties as well as between different locations.

Effect of Floor Management Strategies on Yeast Species Diversity

The most common soil management used by growers is bare soil by tillage, but it should be pointed out that tillage is not a natural way of management of the vineyard soil, due to the breakdown of particles and destruction of the natural vegetation cover, leading to a higher loss of soil (Pastor et al. 2001). The use of chemical herbicides to prevent the growth of weeds is an increasing viticultural practice to reduce the use of tillage. Some advantages of herbicides are their effectiveness when correctly chosen, their low cost, and the ease of use. The main disadvantages include the risk of developing herbicide-resistant weeds, the risk of toxicity, both to the vines and operator, and the potential for herbicide residues leaching to waterways (Guerra and Steenwerth, 2012). Glyphosate (N-phosphonomethyl glycine) is a commonly used herbicide known for its effective control against competing vegetation, its rapid inactivation in soil, and the low mammalian toxicity (Busse et al. 2001; Franz et al. 1997). According to Roslycky (1982), low concentrations of glyphosate have little effect on total populations of soil actinomycetes, bacteria and fungi, while high concentrations initially increase actinomycetes and bacterial population in soil. Johal & Huber (2008) observed several ascomycetes, as yeasts are classified, in soils treated with herbicides. One of the consequences of the use of glyphosate is the change in the microbiological structure of soil, probably due to the result of the stimulation of the growth of fungi. This effect could be a consequence of direct or indirect interaction with other microorganisms (Araújo et al. 2003; Krzysko-Lupicka and Sudol, 2008).

On the other hand, the use of cover crops has been described as a sustainable alternative means of floor management with many advantages. Cover crops can increase the presence of organic matter and nutrients in the soil, due to aerial degradation and underground biomass. This improves some physical properties of the soil, like its porosity, its structure and the stability of the aggregates (Tésic et al. 2007). Furthermore, it has been stated that cover crops increase the moisture retention capacity, and the cation exchange capacity of the soil, as well as being able to reduce the water runoff, prevent erosion and increase soil biological activity (Frye and Blevins, 1989). As regards the vineyard, some other additional benefits are that it contributes to the control of the weed population, and it is a means of controlling some species of nematodes that damage the vineyard (Aballay and Insunza, 2002). The great disadvantage of ground cover is that it may compete for water with the vineyard (Marques et al. 2010; Tésic et al. 2007; Pastor et al. 2001). The effect of water and nutrient competition between sward resulting from the reduced use of herbicides and vines has also been studied, the relationship being well established in cooler and more humid areas. Nevertheless, it is poorly understood in warm and dry areas (Tésic et al. 2007). The permanent natural vegetation present in two Australian vineyard sites in Wagga Wagga (warm climate) and Tumbarumba (cool climate) increased the populations of parasitic beneficial nematodes several fold, and decreased the populations of plant parasitic nematodes after three years (Rahman et al. 2009). Cover crops also reduced soil water content and decreased *Botrytis* incidence by opening up the vine canopy (decreasing the leaf layer number and percentages of internal clusters and leaves and increasing the percentage of gaps) (Guerra and Steenwerth, 2012; Tésic et al. 2007). In addition, several experimental works have shown that inappropriate management affects the physical and chemical characteristics of the soil, affecting the productivity of the vineyard and grape composition (Marques et al. 2010; Murisier et al. 1999; Ovalle et al. 2007).

Cordero-Bueso et al. (2011c) carried out a study which involved the assessment of the influence of different floor management strategies of the vineyard on native yeast populations associated with grape-berries over a period of 3 consecutive years. Syrah grapes were collected from a conventional vineyard, which was treated using three floor management strategies: bare soil by tillage, bare soil maintained with herbicides and soil maintained with cover crop. A large proportion of non-*Saccharomyces* strains were found after fermentation, representing 84% of the total yeast isolated against 16% of *Saccharomyces* strains. The species *L. thermotolerans* and *P. anomala* and *S. cerevisiae* were the most abundant in the vineyards. *M. pulcherrima* and *P. guilliermondii* and *P. toletana* and *C. sorbosa* were found in minor percentages (Cordero-Bueso et al. 2011c). All strains identified as *S. cerevisiae* were genotyped by microsatellite multiplex PCR analysis and three different electrophoretic patterns were obtained. The species *L. thermotolerans and W. anomalus* were the predominant non-*Saccharomyces* in grape must obtained from the vineyard managed by bare soil by tillage. These yeasts were also found in several wine-producing regions and it is common in warm and dry areas (Cordero-Bueso et al. 2011b; Kapsopoulou et al. 2007; Nurgel et al. 2005). *S. cerevisiae* represented 28% of the total of isolates from the vineyard treated by tillage in the first year, but no *Saccharomyces* species were isolated in the other two years of study. Several factors could have affected the populations of fermentative species, such as the climatic conditions, including temperature and rainfall, the sunshine, the tillage intensity, the use of different antifungal chemicals, the grape ripeness and the sugar content.

In the bare soil maintained with the herbicide glyphosate, 150 yeast colonies were isolated of which 28% belonged to the *Saccharomyces* and 72% to the non-*Saccharomyces* genus. *L. thermotolerans*, *P. guilliermondii, C. apicola* and *W. anomalus* were the most isolated species. Contrary to what was expected, the number of *S. cerevisiae* strains was higher in comparison with those isolated in the vineyard with bare soil by tillage, previously commented. Three different strains were found within the *Saccharomyces* analyzed. Unwanted side effects on non-target organisms are an environmental concern with the use of xenobiotic compounds. In the case of glyphosate, its potential disruption of soil microbial communities and their processes has attracted interest because of the mode of action of those compounds (Carlisle and Trevors, 1988). This herbicide inhibits the 5-enolpyruvylshikimic acid-3-phosphate synthase, an intermediate enzyme in the aromatic amino acid synthesis via the shikimic acid pathway (Franz et al. 1997). Most living organisms, excluding plants, lack this pathway, so they are directly unaffected by the presence of glyphosate. However, the shikimic acid pathway is ubiquitous in microorganisms (Bentley, 1990). Reports of harmful effects to microorganisms are numerous in laboratory studies (Busse et al. 2001; Krzysko-Lupicka and Sudol, 2008; Krzysko-Lupicka and Orlik, 1997; Santos and Flores, 1995). Contrary to laboratory results, most agricultural studies have shown that glyphosate do not affect (or even encourage) the presence of soil microorganisms (Guerra and Steenwerth, 2012). An increment in cultivable bacteria and fungi (Busse et al. 2001; Johal and Huber, 2008; Roslycky, 1982), soil respiration (Busse et al. 2001; Haney et al. 2000), nitrogen mineralisation (Haney et al. 2000) and microbial biomass (Busse et al. 2001) have been reported. Some authors (Krzysko-Lupicka and Orlik, 1997) demonstrated that fungi can use this glyphosate as a nutrient and as an energetic substrate. Busse et al. (2001) showed that the use of glyphosate as an available carbon substrate increased microbial activity. These discrepancies between the results obtained in laboratory and field studies can be partially explained by the high concentrations of herbicide used in many of these laboratory studies, and by the herbicide chemistry itself (Busse et al. 2001; Wardle, 1995). The findings of Cordero-Bueso et al. (2011c) suggest that the rate of glyphosate used (Roundup 37% Monsanto), a value ranged within the safety intervals, has little or no repercussion on grape yeast communities. Besides, the use of this herbicide could have stimulated the occurrence of yeasts populations (including fermentative strains) as they might have used this glyphosate as a nutrient, as well as an energetic substrate.

In the floor managed with cover crop all isolates corresponded to non-*Saccharomyces* species such as *C. apicola*, *M. pulcherrima*, *W. anomalus* and *L. thermotolerans*. The *Saccharomyces* genus was not represented. This fact may be due to many intrinsic and extrinsic factors, such as the physical and chemical characteristics of the soil, the weather conditions, the grape ripeness, the vine vigor and the sward-vine competition. The presence of permanent vineyard cover crops or swards has been examined on a variety of soils and climates across Europe, including light-textured soils in dry regions. A reduction in vine vigor was observed in Bordeaux vineyards when comparing the presence of different swards with traditional soil cultivation (Coulon, 2002), although the reduced vigor was not necessarily associated with a reduced yield in Spain due to the greater exposure to sunshine. Sward-vine competition reduced vine vigor and yield in cool and dry areas of Germany so drastically that, in the 1950s, cover crops were abandoned, being considered as unsustainable (Schultz and Löhnertz, 2003). Marques et al. (2010) observed a reduction of 50% in the yield of a vineyard beneath a grass cover, due to the lower degree of soil moisture. This vineyard was also located in the Madrid region. Moreover, Uliarte et al. (2009) showed that Malbec vines were able to grow on poorly structured soils with low organic matter in Argentina with a permanent cover crop. Nevertheless, the growth and yields of the grapevine were reduced, modifying the vigor/production balance, and, subsequently, the composition of the grapes and the resulting wine. Tésic et al. (2007) observed that the use of different treatments resulted in strong differences in vegetative growth, canopy structure, and the yield of berry weight and composition.

The reaction of vines to floor covers is less pronounced under mild weather than under hot and dry conditions. However, floor covers have little effect on the available nutrients present in the soil for the plant. Thus, Tésic et al. (2007) reported a decrease in petiole nutrient status, an effect that was most pronounced for nitrogen, magnesium and, in a minor way, for low potassium concentrations. These results indicate that the low nutrient status of vines growing in competition with floor covers was mostly due to low water availability and, consequently, decreased nutrient uptake. The presence of a permanent grass sward decreases the amino acid (especially arginine) concentration in Riesling juice compared with cultivated soil. This was attributed to the reduced soil moisture and the limited availability of nitrate (Löhnertz et al. 2000). Besides, under hot and dry conditions, the competition for the little available water and nutrients (especially if it occurs at sensitive stages such as blooming and berry setting), may lead to a substantial decrease in the yield generated and the vine capacity.

The presence of floor cover crop may decrease water and nutrient availability in warm and dry vine-growing regions. This reduction in the vineyard soil moisture could be associated with a considerably decreased biodiversity of yeast populations, affecting the *Saccharomyces* genus considerably. Cover crops reduced vine vigor and yield. Additionally, due to the high humidity caused by the presence of the cover crop, the fungi population increased in this floor. These communities may have a competitive effect with grape yeasts populations. The bare soil by tillage could be a sustainable alternative for managing the soil, due to the reduced use of agrochemicals and the resulting high yeasts biodiversity (Cordero-Bueso et al. 2011c). Moreover, it is a low cost technique due to the absence of herbicides which generates less hillside erosion. As stated before, it would not reduce the amount of yeasts present in the vineyard, as well as not interfering in the vine and grape yield in warm climate areas. Nevertheless, the presence of herbicides in the vineyard has a minor impact on the diversity of yeast communities associated with grapes, but the use of some herbicides (such as glyphosate) could have increased the yeast populations, including fermentative strains as a result of its use as a nutrient and as an energetic substrate. Thus, from the fermentative yeasts (like *Saccharomyces*) point of view, in hot and arid environments the best recommendation would be the maintenance of bare soil with the use of herbicides or by tillage. Cover crop in the vineyard increased the fungi population on the floor, which may have a competitive effect with grape yeasts populations, reducing their quantity and biodiversity in the vineyard (Cordero-Bueso et al. 2011c).

Influence of Fungicide Treaments on Yeast Population Associate with Grape Berries

What happens in the vineyard with respect to the use of fungicides? Can these strategies have a significant effect on the natural yeast communities or on the final product? Developments in agricultural technology have continued to keep up with the burgeoning world population, with the proper application of pesticides to control weeds, pests and diseases as important elements in this technology (Lee, 1990).

Fungal obligate parasites are able to penetrate the intact grape skin by their own biochemical and mechanical activities (Barata et al., 2012; González and Tello, 2010). Grapevine oidium or powdery mildew (*Erisiphe necator*), grey mould (*Botrytiotinia fuckeliana* "anamorph *Botrytis cinerea*") and downy mildew (*Plasmopara viticola*) affect grape berry yield and quality worldwide, and it is recognized as a major disease that results in economic losses in the grape and wine industry (Chellemy and Marois, 1992; Crisp et al.

2006; Magarey et al., 1997; Ough and Berg, 1979). In order to control this disease, vine growers of conventional vineyards rely mainly on synthetic fungicides, which have become much more diverse in recent years as a result of demand for new products that specifically target constitutive organisms. But these products are generally more expensive than pesticides with broad-spectrum effects, and their cost needs to be justified in terms of positive or null environmental and human health consequences (Thomson and Hoffmann, 2007). Viticulturists involved in organic and biodynamic production do not use synthetic chemicals, the disease is controlled mainly by regular applications of sulfur and copper sulfate, as the sole fungicides allowed in organic production (Comitini and Ciani, 2008; Milanović et al. 2013).

Most fungicides act directly on essential fungal functions such as cell division, respiration and sterol biosynthesis (Leroux, 2003). Penconazole (1 (4-chlorobenzyl)-1-cyclopentyl-3-phenylurea) is a systematic fungicide against powdery mildew which is considered stable and tends to accumulate in soils and grape berries (Ribeiro et al. 2000; Singh, 2005). The extensive use of this antifungal in viticulture has resulted in contamination of surfaces and groundwater (Komárek et al. 2010). This synthetic fungicide acts as an inhibitor of ergosterol biosynthesis present in fungi membranes (Ribeiro et al. 2000), thus it could have negative effects on yeast population. On the other hand, synthetic organic fungicides are banned in European organic viticulture. Many pesticides are non-specific in their mode of action, and little is known about their impact on the yeast community in the grape berries. Some agrochemicals can inhibit the growth of individual bacteria, yeasts, and fungi and/or whole populations of non-target microorganisms (Walter et al., 2007; Zhang et al, 2009). Copper-based products, such as $CuCl_2$, $Cu(OH)_2$, $CuSO_4$ and Cu_2O, and sulfur-based antifungals are allowed and are indispensable for organic vine cultivation at limited doses according to the Commission Directive 2006/60/EC of the European Community. Sulfur is less persistent in the soil environment than penconazole and also has a lower toxicity to beneficial insects and aquatic species and mammals, including humans (Smith, 2008). It is one of the pesticides permitted as part of an organic pest management strategy.

Numerous studies in the last years have centered on yeast belonging to the genus *Saccharomyces* that are responsible for alcoholic fermentation (Fleet, 2008; Valero et al. 2007). Nonetheless, evidence exists that non-*Saccharomyces* strains may influence the unique enological characteristics of wines (Callejón et al. 2010; Francesca et al. 2010; Esteve-Zarzoso et al. 1998) and the presence of some fungicides on grapevines can affect yeast metabolic activity (Čadež et al. 2010; Calhelha et al. 2006; Ganga and Martínez, 2004). These studies are based on analysis after fungicide treatments and do not exclude the influence of other factors.

The application of organic fungicides on wine grapes during ripening caused a drastic reduction in the yeast microbiota and a shift in the yeast population towards *Aureobasidium pullulans*, whereas *H. uvarum* and *M. pulcherrima* were dominant on untreated grapes (Comitini and Ciani, 2008). Čadež et al. (2010) attributed a higher yeast population in grape berries treated with an anti-*Botrityotinia* fungicide (iprodione). Ganga and Martínez (2004) showed that non-*Saccharomyces* diversity was lower when fungicides were applied against grey mould, however *S. cerevisiae* species were not reduced. A recent qualitative and quantitative evaluation of the yeast communities associated with grape berry surfaces from vineyards organically treated with copper and sulphur formulations carried out by Milanović et al. (2013), showed that yeast population was lower on the berry surfaces from an organic vineyard in comparison with the samples obtained from the conventional one. On the contrary, a similar evaluation was carried out from 2006 to 2008 in our laboratory and we found that sulphur formulations at low doses had a good effectiveness against powdery mildew but maintained the yeast population on berry surfaces (data not published). In conclusion, the presence of fungicides clearly exerts effects over the abundance and diversity of the majority of the yeast species.

This chapter has reviewed the influence of some agronomic parameters on the richness and diversity of yeasts associated with grape berry surfaces. But knowledge of yeast ecology in the vineyard environment has a lot of gaps and these deserve to be studied. Other agronomic practices which are used by vinegrowers daily have been extensively studied in order to enhance the grape and vine yield, although the way in which these practices can affect microorganisms residing in the vineyard remains to be studied. This would be very useful in order to preserve the yeast microbiota of enological or biotechnological interest present in the vineyard.

References

Aballay, E., Insunza, V., 2002. Evaluación de plantas con propiedades nematicidas en el control de *Xiphinema index* en uva de mesa cv. Thompson Seedless en la zona central de Chile. *Agricult. Tecn. Chile* 62, 357-365.

Araújo, A.S.F., Monteiro, R.T.R., Abarkeli, R.B., 2003. Effect of glyphosate the microbial activity of two Brazilians soils. *Chemosphere* 52, 799-804.

Arroyo, T., Cordero-Bueso, G., Serrano, A., Valero E., 2010. β-glucosidase production by non-*Saccharomyces* yeast isolated from vineyards. In: Blank, I., Wüst, M., Yeretzian, C. (Ed.), *Expression of Multidisciplinary Flavour Science*, Winterthur (Switzerland), pp. 359-362.

Barata, A., Malfeito-Ferreira, M., Loureiro, V. 2012. The microbial ecology of wine grape berries. *Int. J. Food Microbiol.* 153, 243-259.

Bauza, T., Kelly, M.T., Blaise, A., 2007. Study of polyamines and their precursor amino acids in Grenache noir and Shiraz grapes and wine of the Rhone Valley. *Food Chem.* 105, 405-413.

Bentley, R., 1990. The shikimate pathway a metabolic tree with many braches. *Critic. Rew. Biochem. Molec. Bio.* 25, 307-384.

Blanco, P., Orriols, I. and Losada, A. (2011) Survival of commercial yeasts in the winery environment and their prevalence during spontaneous fermentations. *J. Ind. Microbiol. Biotechnol.* 38, 235-239.

Busse, M.D., Ratcliff, A.W., Shestak, C.J., Powers, R.F., 2001. Glysophate toxicity and the effects of long-term vegetation control on soil microbial communities". *Soil Bio. Biochem.* 33, 1777-1789.

Čadež, N., Zupan, J., Raspor, P., 2010. The effect of fungicides on yeast communities associated with grape berries. *FEMS Yeast Res.* 10, 619-630.

Calhelha, R.C., Andrade, J.V., Ferreira, I.C., Estevinho, L.M., 2006. Toxicity effects of fungicide residues on the wine-producing process. *Food Microbiol.* 23, 393-398.

Callejón, R.M., Clavijo, A., Ortigueira, P., Troncoso, A.M., Paneque, P., Morales, M.L., 2010. Volatile and sensory profile of organic red wines produced by different selected autochthonous and commercial *Saccharomyces cerevisiae* strains. *Anal. Chim. Acta* 660, 68-75.

Chavan, P., Mane, S., Kulkarni, G., Shaikh, S., Ghormade, V., Nerkar, D.P., Shouche, Y. Deshpande, M.V., 2009. Natural yeast flora of different varieties of grapes used for wine making in India. *Food Microbiol.* 26, 801-808.

Clemente-Jimenez, J.M., Mingorance-Cazorla, L., Martinez-Rodriguez, S., Las Heras-Vazquez, F.J., Rodriguez-Vico, F., 2005. Molecular characterization and oenological properties of wine yeasts isolated during spontaneous fermentation of six varieties of grape must. *Food Microbiol.* 21, 149-155.

Combina, M., Mercado, L., Borgo, P., Elia, A,. Jofré, V., Ganga, A., Martinez, C., Catanis, C., 2005. Yeasts associated to Malbec grape berries from Mendoza, Argentina. *Int. J. Food Microbiol.* 98, 1055-1061.

Comitini, F., Ciani, M., 2006. Survival of inoculated *Saccharomyces cerevisiae* strain on wine grapes during two vintages. *Lett. Appl. Microbiol.* 42, 248-253.

Comitini, F., Ciani, M., 2008. Influence of fungicide treatments on the occurrence of yeast flora associated with wine grapes. *Annals Microbiol.* 58, 489-493.

Coulon, T., 2002. Effect of a permament cover crop on vine physiology in Bordeaux vineyards. *Le progress agricole et viticole* 120, 154-161.

Commission Directive 2006/60/EC of the European Communities. *Official Journal of the European Union.* 27th of July 2006. L206, 1-11.

Cordero-Bueso, G., Esteve-Zarzoso, B., Cabellos, J.M., Gil-Díaz, M. Arroyo, T., 2013. Biotechnological potential of non-*Saccharomyces* yeasts isolated during spontaneous fermentations of Malvar (Vitis vinifera cv. L.). *Eur. Food Res. Technol.* 236, 193-207.

Cordero-Bueso, G., Arroyo, T., Serrano, A., Valero, E., 2011a. Remanence and survival of commercial yeast in different ecological niches of the vineyard. *FEMS Microbiol. Ecol.* 77, 429-437.

Cordero-Bueso, G., Arroyo, T., Serrano, A., Tello, J., Aporta, I., Vélez, M.D., Valero, E., 2011b. Influence of the farming system and vine variety on yeast communities associated with grape-berries. *Int. J. Food Microbiol.* 45, 132-139.

Cordero-Bueso, G., Arroyo, T., Serrano, A., Valero, E., 2011c. Influence of different floor management strategies of the vineyard on the natural yeast population associated with grape berries. *Int. J. of Food Microbiol.* 148, 23-29.

Crisp, P., Wicks, T.J., Lorimer, M., Scott, E.S., 2006. An evaluation of biological and abiotic controls for grapevine powdery mildew: Greenhouse studies. *Aus. J. Grape Wine Res.* 12, 192-202.

Csoma, H., Sipiczki, M., 2008. Taxonomic reclassification of *Candida stellata* strains reveals frequent occurrence of *Candida zemplinina* in wine fermentation. *FEMS Yeast Res.* 8, 328-336.

Di Maro, E. Ercolini, D., Coppola, S., 2007. Yeast Dinamic during spontaneous wine fermentation of Catalanesca grape. *Int. J. Food Microbiol.* 117, 201-210.

Duarte, F.L., Teixeira, A., Costa, A., Ramos, P., Schuller, D., Gomes, A.C., Alemão, M.F. 2011. *Yeast diversity related with Touriga Nacional grape variety*. Avaliable at http://repositorium.sdum.uminho.pt/bitstream/1822/13518/1/2011_OIV_Touriga_paper.pdf.

Esteve-Zarzoso, B., Manzanares, P., Ramón, D., Querol, A., 1998. The role of non-*Saccharomyces* yeasts in industrial winemaking. *Int. Microbiol.* 1, 143-148.

Esteve - Zarzoso, B., Belloch, C., Uruburu, F., Querol, A., 1999. Identification of yeast by RFLP analysis of the 5.8 rRNA and the two ribosomal internal transcribed spacers. *Int. J. Sys. Bacteriol.* 49, 329-337.

European Union pesticide database, 2010. Available at http://ec.europa.eu/sanco_pesticides/public/index.cfm.

Fernández-González, Briones, A.I., 2013. Study of yeast population and their enological properties in Guijoso Appellation of Origin (Spain). *Annals Microbiol.* 63, 371-379.

Fernández-Espinar, M.T., Esteve – Zarzoso, B., Querol, A., Barrio, E., 2000. RFLP analysis of the ribosomal transcribed spacers and the 5.8S rRNA gene region of the genus *Saccharomyces*: a fast method species identification and the differentiation of flor yeasts. *Ant. Van Leeuwen.* 78, 87 - 97.

Fleet, G.H., 2003. Yeast interactions and wine flavour. *Int. J.Food Microbiol.* 86, 11-22.

Fleet, G.H., 2008. Wine yeast for the future. *FEMS Yeast Res.* 8, 979-995.

Fleet, G.H., Prakitchaiwattana, C.J., Beh A.L. and Heard, G.M., 2002. The yeast ecology of wine grapes. *Biodiversity and Biotechnology of wine Yeasts* (Ciani, M. Ed), pp. 1-17.

Fotopoulos, C., Krystallis, A., Ness, M. 2003. Wine produced by organic grapes in Greece: using means-end chains analysis to reveal organic buyers' purchasing motives in comparison to the non-buyers. *Food qual. Pref.* 14(7), 549-566.

Francesca, N., Chiurazzi, M., Romano, R., Aponte, M., Settani, L., Moschetti, G., 2010. Indigenous yeast communities in the enviroment of "Rovello bianco" grape variety and their use in commercial white wine fermentation. *World J. Microbiol. Biotechnol.* 26, 337-351.

Franz, J.E., Mao, M.K., Sikorski, J.A., 1997. Glyphosate: a unique global herbicide. American Chemical Society Monograph 189. *Am. Chem. Soc.*, Washington D.C.

Frye, W.W., Blevins, R.L., 1989. Economically sustainable crop production with legume cover crops and conservation tillage. *J. Soil Water Cons.* 44, 57- 60.

Ganga, M.A., Martínez, C., 2004. Effect of wine yeast monoculture practice on the biodiversity of non-*Saccharomyces* yeasts. *J. Appl. Microbiol* 96, 76-83.

Gayevskiy, V., Goddard, M.R., 2012. Geographic delineations of yeast communities and populations associated with vines and wines in New Zealand. *ISME J.* 6:7, 1281-1290.

González, S.S., Barrio, E., Querol, A., 2007. Molecular identification and characterization of wine yeasts isolated from Tenerife (Canary Island, Spain). *J. Appl. Microbiol* 102, 1018-1025.

Guerra, B., Steenwerth, K., 2012. Influence of floor management technique on grapevine growth, disease pressure, and juice and wine composition: A review. *Am. J. Enol. Vitic.* 63, 149-164.

Haney, R.L., Senseman, S.A., Hons, F.M., Zuberer, D.A., 2000. Effect of glyphosate on soil microbial activity and biomass. *Weed Sci.* 48, 89-93.

Hole, D., Perkins, A., Wilson, J., Alexander, I., Grice, P., Evans, E., 2005. Does organic farming benefit biodiversity?. *Biological Conservation* 122, 113-130.

Johal, G.S. and Huber, D.M., 2009. Glyphosate effects on diseases of plants. *Eur. J. Agr.* 31, 144-152.

Jolly, N.P., Augustyn, O.P.H., Pretorius, I.S., 2006. The role and use of non-*Saccharomyces* yeasts in wine production. *S. Afr. J. Enol. Viticul.* 27, 15-39.

Kapsopoulou, K., Mourtzini, A., Anthoulas, M., Nerantzis, E., 2007. Biological acidification during grape must fermentation using mixed cultures of *Kluyveromyces thermotolerans* and *Saccharomyces cerevisiae*. *World J. Microbiol. Biotechnol.* 23, 735-739.

Kliewer, W.M., 1991. Methods for determining the nitrogenous status of vineyards. p. 133-147. *Proc. Int. Symp.* in Nitrogen in Grapes and Wines, Seattle. 18-19 June. *Am. Soc. Enol.Viticul. (ASEV)*, Seattle, Washington, USA.

Komárek, M., Čadková, E., Chrastný, V., Bordas, F. Bollinger, J.C., 2010. Contamination of vineyard soils with fungicides: A review of environmental and toxicological aspects. *Env. Int.* 36: 138-151.

Krzysko-Lupicka, T., Orlik, A., 1997. The use of glyphosate as the sole source of phosphorus or carbon for the selection of soil-borne fungal strains capable to degrade this herbicide. *Chemosphere* 34, 2601-2605.

Krzysko-Lupicka, T., Sudol, T., 2008. Interactions between glyphosate and autochthonous soil fungi surviving in aqueous solution of glyphosate. *Chemosphere* 71, 1386-1391.

Lederer, M.A., Nielsen, D.S., Toldam-Andersen, T.B., Herrmann, J.V., Arneborg, N., 2013. Yeast species associated with different wine grape varieties in Denmark. *Acta Agricult. Scandiv. B, soil plant sci.* 63, 89-96.

Lee, T.H., 1990. Pesticides and grapes and wine: present and future obligation, an update. Spring vineyard seminar 19 September 1990; Richmond Grove Winery. Hunter Valley Vineyard. *Association and Department of Agriculture and Fisheries*, 5 p.

Leroux, P., 2003. Modes d'action des produits phytosanitaires sur les organisms pathogens des plantes: mode of action of agrochemical towards plant pathogens. *Cahiers Recher. Biol.* 326, 9-21.

Li, S.S., Cheng, C., Li, Z., Chen, J.Y., Yan, B., Han B.Z., Reeves, M., 2010. Yeast species associated with wine grapes in China. *Int. J. Food Microbiol.* 138, 85-90.

Marques, M.J., García-Muñoz, S., Muñoz-Organero, G., Bienes, R., 2010. Soil conservation beneath grass cover in hillside vineyards under mediterranean climatic conditions (Madrid, Spain). *Land Degr. Develop.* 21, 122-131.

Martini, A., Vaughan-Martini, A., 1990. Grape must fermentation: Past and present in Spencer, J. F. T. and Spencer, D. M. (Eds), *Yeast Technology*, Springer-Verlag. 105-123.

Milanović, V., Comitini, F., Ciani, M., 2013. Grape berry yeast communities: influence of fungicide treatments. *Int. J. Food Microbiol.* 161, 240-246.

Mortimer, R., Polsinelli, M., 1999. On the origins of wine yeast. *Res. Microbiol.* 150, 199-204.

Murisier, F., Maigre, D., Spring, J.L., 1999. Gestione del suolo nella viticoltura svizzera. Esperienze con varie tecniche di inerbimento e riflessi sulla qualità del vino. Notiziario Tecnico *CRPV* 58, 1-17.

Nisiotou, A.A., Nychas, G.J.E., 2007. Yeast populations residing on healthy or *Botrytis* infected grapes from a vineyard in Attica, Greece. *Appl. Env. Microbiol.* 73, 2765–2768.

Nurgel, C., Erten, H., Canbas, A., Cabaroglu, T., Selli, S., 2005. Yeast flora during the fermentation of wines from *Vitis vinifera* L. cv. Emir and Kalecik Karasi grown in Anatolia. *W. J. Microbiol. Biotechnol.* 21, 1187-1194.

Ovalle, C., Del Pozo A., Lavín, A., Hirzel, J., 2007. Cover crops in vineyards: performance of annual forage legume mixtures and effects on soil fertility. *Agr. Téc. Chile* 67, 265 - 277.

Pastor, M., Castro, J. Humanes, M.D., Muñoz, J., 2001. Sistemas de cultivo con cubiertas en olivar en Andalucía (I). *Vida Rural*, 58-62.

Passoth, V., Fredlund, E., Druvefors, U.A., Schnürer, J., 2006. Biotechnology, physiology and genetics of the yeast *Pichia anomala*. *FEMS Yeast Res.* 6, 3-13.

Pretorius, I. S., 2000. Tailoring wine yeast for the new millennium: novel approaches to the ancient art of winemaking. *Yeast* 16, 675-729.

Rahman, L., Whitelaw-Weckert, M.A., Hutton, R.J., Orchard, B. 2009. Impact of floor vegetation on the abundance of nematode trophic groups in vineyards. *Appl. Soil. Ecol.* 42, 96-106.

Raspor, P., Milek, D.M., Polanc, J., Monzina, S.S., Čadež, N., 2006. Yeast isolated from three varieties of grapes cultivated in different locations of the Dolenjska vine-growing region, Slovenia. *Int. J. Food Microbiol.* 109, 97-102.

Renault, P., Miot-Sertier, C., Marullo, P., Henández-Orte, P., Lagarigue, L., Lonvaud-Funel, A., Bely, M., 2009. Genetic characterization and phenotipic variability in *Torulaspora delbrueckii* species: Potential applications in the wine industry. *Int. J. Food Microbiol.* 134, 201-210.

Renouf, V., Claisse, O., Lonvaud-Funel, A. 2005. Understanding the microbial ecosystem on the grape berry surface through numeration and identification of yeast and bacteria. *Aus. J. Grape Wine Res.* 11, 316-327.

Ribeiro, I.C., Veríssimo, I., Moniz, L., Cardoso, H., Sousa, M.J., Soares, A.M., Leao, C., 2000. Yeasts as a model for assessing the toxicity of the fungicides Penconazol, Cymoxanil and Dichlofluanid. *Chemosphere* 41, 1637-1642.

Ribéreau-Gayon, P., Dubourdieu, D., Donèche, B., Lonvaud, A., 2006. The Microbiology of wine and vinifications. In: *Handbook of Enology*, vol 1. Pp. 1-51. John Wiley and Sons, West Sussex (England).

Romancino DP, Di Maio S, Muriella R & Oliva D (2008) Analysis on non-*Saccharomyces* yeast populations isolated from grape musts from Sicily (Italy). *J. Appl. Microbiol.*105, 2248-2254.

Sabaté J., Cano J., Esteve-Zarzoso, B., Guillamón, J.M., 2002. Isolation and identification of yeasts associated with vineyard and winery by RFLP analysis of ribosomal genes and mitochondrial. *Microbiol. Res.*157, 267-274.

Santos, A., Flores, M., 1995. Effects of glyphosate on nitrogen fixation of free-living heterotrophic bacteria. *Lett. Appl. Microbiol.* 20, 349-352.

Schuller, D., Cardoso, F., Sousa, S., Gomes, P., Gomes, A.C., Santos, M.A.S., Casal, M. 2012. Genetic diversity and population structure of *Saccharomyces cerevisiae* strains isolated from different grape varieties and winemaking regions. *Plos One* 7, e-32507, 1-12.

Schuller, D., Alves, H., Dequin, S., Casal, M. 2005. Ecological survey of *Saccharomyces* strains from vineyards in the Vinho Verde region of Portugal. *FEMS Microbiol. Ecol.,* 51, 167-177.

Schultz, H.R., Löhnertz, O., 2003. Cover crop use in Germany and possible effects on wine quality. *Le Progrès agricole et viticole* 120,175-178.

Setati, M.E., Jacobson, D., Andong, U.C. Bauer, F. 2012. The vineyard microbiome, a mixed model microbial map. *Plos One* 7, e52609, 1-11.

Shuangshil, L., Ningl, S., Hengyul, L., Jingyul, C., Beizhongl, H. 2011. Diversity and distribution of yeast associated with Cabernet Gernischet. *China Brew*. 2011-05.

Singh, N., 2005. Mobility of four triazole fungicides in two Indian soils. *Pest Man. Sci.* 65, 191-196.

Smith, J., 2008. Fungicides revisited: a critical look at copper- and sulfur-based fungicides in organic agriculture. Farmstart, California, USA. Available at http://www.farmstart.ca/news/wp-content/uploads/2008/06/smith-2008-fungicides-revisited.pdf.

Swiegers J.H., Francis, I.L., Herderich, M.J., Pretorius, I.S., 2006. Meeting consumer expectations through management in vineyard and winery: the choice of yeast for fermentation offers great potential to adjust the aroma of Sauvignon Blanc wine. *Aus. New Z. Wine Indust. J.* 21, 34-42.

Tello, J., Cordero-Bueso, G., Aporta, I., Cabellos, J. M., Arroyo, T., 2012. Genetic diversity in commercial wineries: effects of the farming system and vinification management on wine yeasts. *J. Appl. Microbiol.* 112, 316-328.

Tesic, D., Keller, m., Huntton, R. J., 2007. Influence of vineyard floor management practices on grapevine vegetative growth, yield and fruit composition. *Am. J. Enol. Viticul.* 58, 1-11.

Thomson, L.J., Hoffmann, A.A., 2007. Ecologically sustainable chemical recommendations for agricultural pest control?. *J. Econ. Entomol.* 100, 1741-1750.

Torija M.J., Rozès N., Poblet M., Guillamón J.M., Mas A., 2001. Yeast population dynamics in spontaneous fermentations: comparison between two different wine-producing areas over a period of three years. *A. Van Leeuwen.* 79, 345 - 352.

Uliarte, E.M., Monte, R.F., Parera, C.A., Catania, C.D., Monte, M.S.A., 2009. Influence of soil management through established cover crops in the vegetative development, production and wines characteristics of vineyards under surface irrigation (cv. Malbec). *Bulletin de l'OIV*, Vol. 82 No. 938/939/940 pp. 205-227

Torrens, J., Urpi, P., Riu-Aumatell, M., Vichi, S., López-Tamames, E., Buxaderas, S., 2008. Different commercial yeast strains affecting the volatile and sensory profile of cava base wine, *Int. J. Food Microbiol.*124, 48-57.

Valero, E., Schuller, D., Cambon, B., Casal, M., Dequin, S. 2005. Dissemination and survival of comercial wine yeast in the vineyard: A large scale, three-year study. *FEMS Yeast Res.* 5, 959 - 969.

Valero, E., Cambon, B., Schuller, D., Casal, M., Dequin, S., 2007. Biodiversity of *Saccharomyces* yeast strains from grape berries of wine-producing areas using starter commercial yeast. *FEMS Yeast Res.* 7, 317-329.

Vaudano, E., Garcia-Moruno, E., 2008. Discrimination of *Saccharomyces cerevisiae* wine strains using microsatellite multiplex PCR and band pattern analysis. *Food Microbiol.* 25, 56-64.

Viana F., Gil J.V., Genovés S., Vallés S., Manzanares P., 2008. Rational selection of non-*Saccharomyces* wine yeasts for mixed starters based on ester formation and enological traits. *Food Microbiol.* 25, 778-785.

Wardle, D.A., 1995. Impacts of disturbance on detritus food webs in agro-ecosystems of contrasting tillage and weed management practices. *Adv. Eco. Res.* 26, 105-185.

Wheeler, S.A., and Crisp, P., 2009. Evaluating a range of the benefits and costs of organic and conventional production in a Clare Valley Vineyard in South Australia. Available at www.adelaide.edu.au/cies/research/ wine/pubs/Wheeler_WCO210.pdf.

Wondra, M., Boveric, M., 2001. Analyses of aroma components of Chardonnay wine fermented by different yeast strains. *Food Tech. Biotechnol.* 39, 141-148.

Zott, K., Miot-Sertier, C., Claisse, O., Lonvaud-Funel, A., Masneuf-Pomarede, I., 2008. Dynamics and diversity of non-*Saccharomyces* yeast during the early stages in winemaking. *Int. J. Food Microbiol*. 125, 197-203.

In: Grapes
Editor: José S. Câmara
ISBN: 978-1-63321-402-6

Chapter 3

Phenolic Compounds in Grapes and Wines: Chemical and Biochemical Characteristics and Technological Quality

Ellen Silva Lago Vanzela[1,*], *Milla Alves Baffi*[2], *Maurício Bonatto Machado de Castilhos*[1], *Marcos Roberto Moacir Ribeiro Pinto*[3], *Vanildo Luiz Del Bianchi*[1], *Afonso Mota Ramos*[3], *Paulo César Stringheta*[3], *Isidro Hermosín-Gutiérrez*[4] *and Roberto Da-Silva*[1]

[1]Universidade Estadual Paulista - UNESP, Cristovão Colombo, 2265, Jardim Nazareth, São José do Rio Preto, São Paulo, Brazil

[2]Universidade Federal de Uberlândia – UFU, Amazonas s/n, Umuarama, Uberlândia, Minas Gerais, Brazil

[3]Universidade Federal de Viçosa – UFV, Peter Henry Rolfs, s/n, Campus Universitário, Viçosa, Minas Gerais, Brazil

[4]Instituto Regional de Investigación Científica Aplicada, Escuela de Ingenieros Agrónomos de Ciudad Real, Universidad de Castilla-La Mancha, Ciudad Real, Spain

Abstract

The grape is one of the most widely consumed fruits in the world, both in natural and in processed forms and the economic importance generated by the sale of this products is notorious. It is considered one of the greatest sources of phenolic compounds when compared to other fruits and vegetables, but there is a great diversity among cultivars resulting in grapes with different characteristics, which is directly linked to its polyphenol profile. The phenolic compounds are secondary metabolites of plants, playing important physiological activities, and their concentration is subjected to great variation. They can

[*] Corresponding autor: Ellen Silva Lago Vanzela. E-mail address: ellen@ibilce.unesp.br.

be classified into two groups: non-flavonoids and flavonoid. In the grape, the non-flavonoids are mainly phenolic acids and the stilbenes and flavonoids are anthocyanins, flavonols and flavan-3-ols (monomers, dimers, oligomers and polymers, the latter also known as tannins). Epidemiological, clinical and in vitro studies have demonstrated some of the biological effects associated with dietary phenolic compounds such as antioxidant, anti-inflammatory, antimicrobial and anti-carcinogenic activities. In grapes and their products, phenolic compounds are directly related to sensory characteristics such as color and flavor.

However, grape juices and, mainly, red wines present phenolic composition more complex than that which is found in the grapes used for their preparation. The qualitative and quantitative differences are largely due to numerous chemical and biochemical oxidative processes. These oxidative processes may lead to changes in the color and the nutritional value of grapes and its products. In contrast, the deliberate addition of specific enzyme preparations in key stages in the manufacturing process plays a crucial role in improving the quality of products such as juices and, especially, wines. This is because the grape skins are limiting barriers that prevent the release of phenolics to the must. The extraction of these compounds during the maceration requires that the middle lamella and cell walls are broken to allow the extraction of constituents present within the cell vacuoles.

Furthermore, wines usually need to be stabilized to get rid of unstable compounds such as potassium bitartrate. That's why this chapter summarizes the current knowledge on the chemical characteristics of phenolic compounds in grapes and their products and highlights some aspects related to their impact on color, flavour and health properties, with examples taken mostly from research on grapes and wine. Furthermore, presents a discussion of the state of the art of the main natural and induced biochemical processes that occur in the grape during processing which can positively or negatively affect the phenolic compounds and, hence, the sensory and nutritional quality of any derived products.

Keywords: Phenolic compounds, grape, wine, enzyme, oxidative processes

Introduction

Epidemiological and intervention studies have pointed to an inverse relation between the ingestion of fruits and vegetables and the incidence of degenerative diseases such as cancer and cardiovascular diseases. There is strong evidence showing that the synergistic action of several bioactive compounds present in these foods, such as phenolic compounds, vitamin C, vitamin E and carotenoids, is responsible for these benefits to human health. It is well established and broadly accepted that the consumption of grapes and their derivative products constitutes one of the most important forms of incorporating phenolic compounds into a human diet. The demand for products which aggregate, beyond the basic nutritional exigencies, beneficial factors to health is an important tendency which has been orienting strategies in the food industry.

The anthocyanins, flavan-3-ols, flavonols and stilbenes found in grapes and derivatives are the phenolic compounds most investigated due to their functional properties (antioxidant, anti-inflammatory, antimicrobial and anti-carcinogenic activities). However, the chemical structure of many of these substances which propitiates high antioxidant activity, at the same time, can show susceptibility to oxidative processes during the fruit processing.

Oxidative processes of enzymatic origin can lead to damage to the color, aroma and nutritional value of grapes and their products. On the other hand, the addition of enzymatic preparations in the key steps of industrial process plays a fundamental role in the improvement of the quality of the products. Additionally, countless oxidative processes of chemical origin can lead to the progressive reduction of some original phenolic constituents from grapes on behalf of the production of polymeric compounds during the evolution of the color of juices and, primarily, wines. From this perspective, the understanding of the main chemical and biochemical processes which occur in grapes and their derived products and which can affect these substances positively or negatively and, consequently, the sensorial and nutritional quality, are crucial for both quality control and development of this sector. This chapter presents a brief description of grapes, followed by a profile of the composition of chemical and biochemical reactions in grapes and wines. Towards a better understanding the chapter has been separated into parts: Part 1, Part 2 and Part 3.

The Chemical Characteristics of Grapes

General Aspects of Grapes

Botanically, the grapes are small round berries which grow in bunches in a woody climbing plant of the *Vitis* genus. They are native to European and Mediterranean regions but, at the present time, they are widely cultivated throughout the world. The structure of a grape corresponds to a semi-translucent pulp, covered by a smooth and thin skin. Some varieties contain seeds while others are apirenic (seedless). The color of the fruit can vary from light green to dark red or purple. These latter ones are rich in anthocyanins while the former contain tannins, especially catechin.

A grape bunch consists of an herbaceous part known as the stem or rachis (inflorescence axis), whose extremity is called the pedicel (a transport channel for nutritious elements) and the other fleshy part is called the berry which is bound to the stem by the pedicel. The support that is fastened to the pedicel when it is separated from the berry is called the brush (Figure 1) [1].

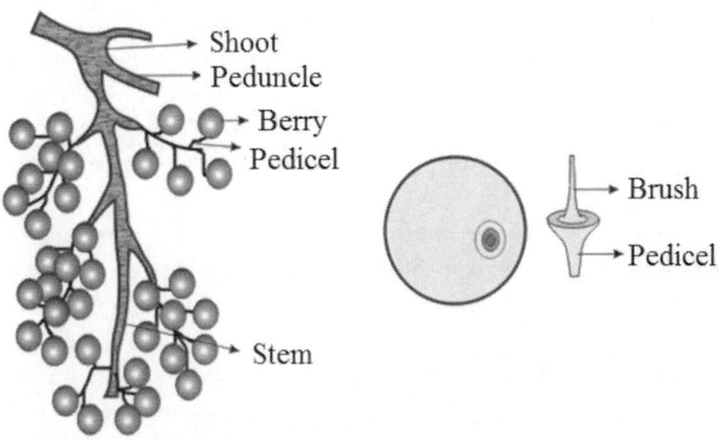

Figure 1. Structural organization of a grape bunch.

The stem is rich in resins, minerals and astringent tannins. During winemaking, it is normally separated from the berry before the processing of the fruit to avoid an excess of bitterness which gives an unpleasant taste to the final product. Although the proportion of the components of the berry can vary depending on the grape variety, maturation degree and plant productivity, in general, it presents the following composition (% in wet weight): 70 to 85 of water; 12 to 15 in reducing sugars; 0.4 to 1.3 of organic acids; 0.3 to 0.5 of minerals; 0.03 to 0.17 in nitrogen compounds and other components which appear in minimal quantities. In grapes with seeds, the seeds usually represent of 4 to 13% of the total weight of the berry and, depending on the fertilization, they are found in variable number (1 to 4). The pulp represents, on average, 74 to 88% of the total wet weight of the berry, contributing significantly to the yield of the must, while the skin corresponds to 8 to 13% of the total wet weight [1].

Consequently, the chemical composition of the pulp can sometimes be similar to that found in the must. The skin is the protective covering which is composed of three overlying layers: the external film (cuticle or epidermis), the intermediate epidermis, with one or two layers and the internal film, the hypodermis (Figure 2).

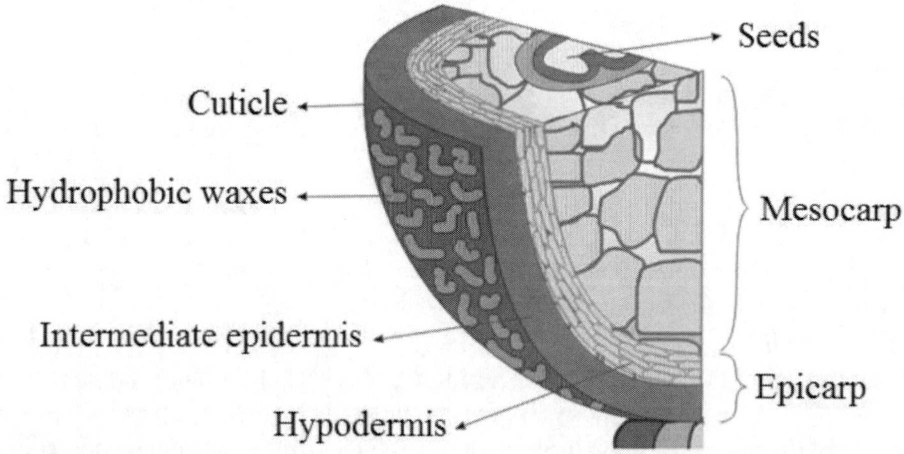

Figure 2. Structural organization of a grape berry.

The epidermis is composed of a layer of hydroxylated fatty acids (cutin), which, in turn, are overlaid by a thin film of hydrophobic waxes, known as the bloom. The bloom layer has the role of protecting the berry cells against the effects of warmth and humidity, as well as avoiding the entry of microorganisms into the internal part of the berry. The hypodermis is the skin region wherein the majority of phenolic compounds of red grapes and some aroma substances of oenological interest are found [2, 3].

The synthesis and distribution of phenolic compounds (secondary metabolites) of each plant species, which are largely responsible for the color of the berries, are determined by intrinsic genetically- controlled systems of enzymes [4, 5]. Once the enzymes are expressed under the genetic information of each species, the gene responses can be induced or inhibited according to the adaptive needs of the plant [6]. This is due to the fact that the secondary metabolites have an important role in the adaptation of plants to their environment, that is, they increase the survival probability of a species, since they are responsible for different biological activities. Some compounds can act as antibiotics, antifungals and antivirals with the purpose of protecting the plants against pathogens, as well as can present antigerminative or toxic activities in other plants. Moreover, some of these metabolites constitute important substances which can absorb ultraviolet light, thus preventing damage to the leaves [7]. The phenolic content is strongly influenced by extrinsic factors such as season of the year (climate), ultraviolet radiation incidence and soil composition, among others. It occurs because the integration of the environment with the physiological mechanisms of plants can influence the secondary metabolism, which gives rise to a large variation in the production of the secondary metabolites intra and interspecies [6]. That is, the synthesis of such compounds is subject to the interaction of both genetic (intrinsic) and environmental (extrinsic) factors.

The main secondary metabolites are distributed in three groups in accordance to their biosynthetic route: terpenes, phenolic compounds and nitrogen compounds (alkaloids) [8]. The following are detailed phenolic compounds present in grapes.

Phenolic Compounds in Grapes

The phenolic compounds found in grapes can be divided in two big groups:

i. *flavonoids*: found mainly in the skin and in the seeds, but rarely in the pulp, among them the anthocyanins, flavonols and flavan-3-ols are highlighted;
ii. *non-flavonoids:* found mainly in the pulp, constituting the phenolic acids, of which the benzoic and cinnamic acids are spotlighted, as well as other derivatives such as stilbenes, with emphasis on resveratrol, that mostly occurs in the skin [9].

The phenolic compounds are responsible for the color of the products derived from grapes, such as juices and wines. Nevertheless, the phenolic profile of these products is dependent on countless factors which include the grape species, grape variety, plantation location, culture system, climate, soil, extraction method of the phenolic compounds and the kind of process used. Furthermore, these profiles are also affected by chemical and enzymatic reactions that begin with the grape crushing and continue throughout the process, as well as the maturation or aging [10, 11]. Hence, the juices, and especially the wines, present a phenolic composition more complex than that found in the grapes which are used for their elaboration.

The main phenolic compounds from grapes are presented below:

Flavonoids

The structure of flavonoids is based on the skeleton of 15 carbons, characterized by two aromatic rings bounded by a chain composed of three atoms of carbon, forming an oxygenated heterocycle (C6-C3-C6). Depending on the oxidation state of the central ring C, the flavonoids can be 2-phenylbenzopyran (Figure 3A) or 2- phenylbenzopyrone (Figure 3B) [12].

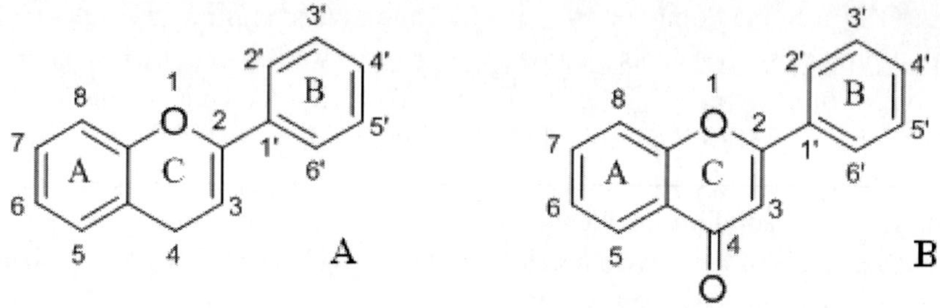

Figure 3. Basic structure of flavonoids: 2-phenylbenzopyran (A) and 2- phenylbenzopyrone (B).

In acid conditions, two bands of absorption are typical of flavonoids: the former (Band I) with maximum of absorption among 300 and 550 nm, apparently from the ring B of the molecule and, the second (Band II) with maximum of absorption among 240 and 285, supposedly from the ring A [13].

A description of the main flavonoids is described as it follows:

I) Anthocyanins

The anthocyanins are found primarily in the skins and in the third or fourth layers of the hypodermic cells, in intracellular scale in the cell vacuole of red grapes [3, 14, 15, 16] and, in a lesser proportion, in the pulp of some varieties [17, 18]. The basic structure of these compounds is the aglycon, named anthocyanidin, constituting of two phenolic rings (A and B) and a heterocyclic pyran ring (C) (Table 1) [19].

Table 1. Anthocyanidins (*flavilium* cation form) most common in grapes

Basic structure of aglycon	Anthocyanidins	Substituents groups
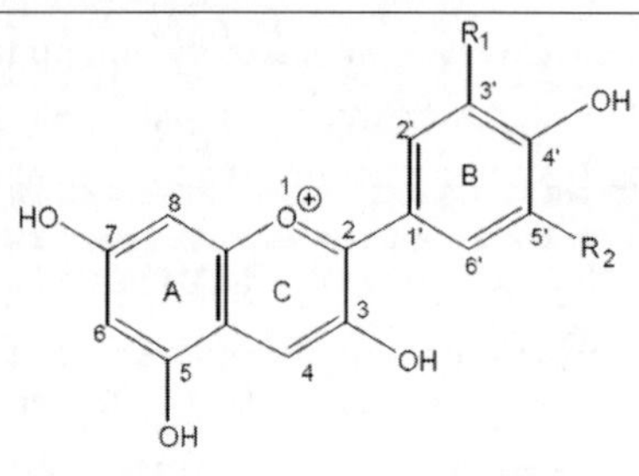	Pelargonidin (Pg)	$R_1 = R_2 = H$
	Cyanidin (Cy)	$R_1 = OH; R_2 = H$
	Peonidin (Pn)	$R_1 = OCH_3; R_2 = H$
	Delphinidin (Dp)	$R_1 = R_2 = OH$
	Petunidin (Pt)	$R_1 = OH; R_2 = OCH_3$
	Malvidin (Mv)	$R_1 = OCH_3; R_2 = OCH_3$

This structure contains chromophores of light absorption which give grapes their attractive pigmentation, with nuances between red and blue colors [19]. In literature, 23 anthocyanidins have already been mentioned, which differ among themselves by the number and position of the hydroxy and/or metoxy groups. However, only six of them are often found in foods: cyanidin, peonidin, delphinidin, petunidin, malvidin and pelargonidin [20]. In grapes, there is a predominance of the first five listed ones, however, in some varieties of non-vinifera grapes [21] and in the Garnacha Tintorera variety (*Vitis Vinifera*) [17] small quantities of anthocyanins, derived from pelargonidin, have been also found. Although this anthocyanin is not predominant in grapes, it is regularly present in other fruits such as mangosteen [22], strawberry [23], açai and acerola [24].

The electronic deficiency of the flavilium cation produces an elevated reactivity of the free aglycons, to such an extent that, in nature, they are observed bound to one or two sugar molecules (for example glucose, arabinose, rhamnose, galactose, or di- and trisaccharides constituted by these sugars) through hemiacetalic unions [14]. According to the sugar number bound to the anthocyanidin, the formed anthocyanins can be classified as mono-, di- or triglycosides [25]. Glucose is the most common sugar bound to anthocyanidins from grapes and the most common positions of binding among these molecules are the C-3 of ring C and C-5 of ring A [17] and, with rare exceptions, they can be found bound to C-7 position of the ring A (Figure 4) [26].

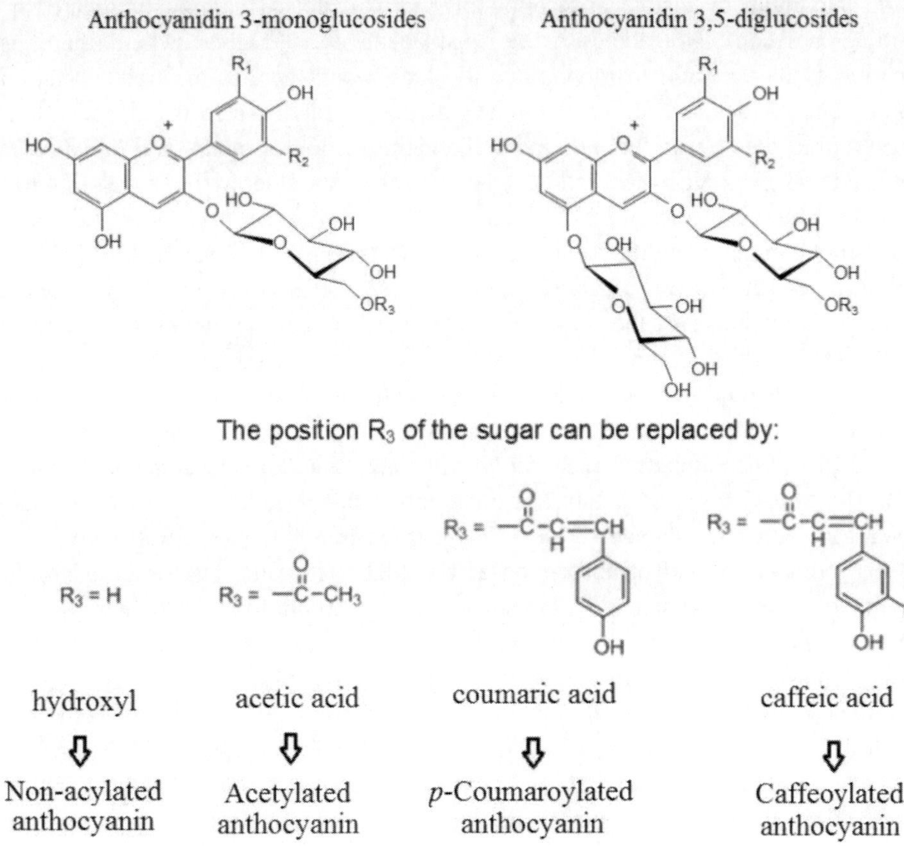

Figure 4. Structures of anthocyanins from grape.

In anthocyanins, the wavelength of absorption at visible light spectra is altered by the number of substitutions which are present in ring B and, in a small proportion, by the kind of substitutions. This way, as the number of substituents of ring B increases, there is a shifting of Band I to higher wavelengths around 8 to 12 nm. For example, the trisubstituted anthocyanidins (delphinidin, petunidin and malvidin) originate anthocyanins with maximum absorption at visible spectrum, at values of around 10 nm higher than the corresponding anthocyanins with disubstituted anthocyanidins (cyanidin and peonidin) [26].

These last ones, on the other hand, absorb more than the monosubstituted (pelargonidin) that presents the shorter wavelength (502 nm) (Figure 5A). These changes in wavelengths explain the color modifications of the molecule. In general, the increasing of hydroxylation induces a displacement from a red to a blue color (pelargonidin → cyanidin → delphinidin), which is shown by a bathochromic shift at the maximum wavelength of absorption of the spectrum of the pigment. Conversely, the increase in the degree of methoxylation (peonidin → petunidin → malvidin) reverses this trend, leading to pigmentations tending towards red or purple colors [27].

The degree and position of the glycosylation of the anthocyanidin also display profound influence in the pigmentation of anthocyanins. The color is modified due to a hypsochromic shift of the maximum of visible absorption of approximately 10 nm with regard to the corresponding aglycon. Nonetheless, this displacement is higher when there is a transformation of the aglycon in monoglucoside than when there is the transformation of mono- to di-glucoside and so on. The position of glycosylation also affects other zones of the visible spectrum in such a way that it helps in the identification of mono- and diglycosylated anthocyanins: the 3-monoglucoside anthocyanins present a shoulder at the region of 440 nm, a characteristic which is absent in 3,5-diglucosides (Figure 5B). On the other hand, the kind of sugar does not habitually interfere in the UV-Vis spectrum of anthocyanins [26] [28].

In many cases, the sugars of anthocyanins are sterified to organic acids, which are, normally bound to carbon-6 of the sugar molecule [14]. Among the organic acids, the most common are acetic acid, *p*-coumaric acid, caffeic acid, amongst others. Thus, non-sterified anthocyanins are called non-acylated anthocyanins, while the sterified ones with one or more phenolic acids are named monoacylated and polyacylated anthocyanins, respectively, which can present different denominations according to the acid bound to the sugar molecule [12] (Figure 4). The esterification of sugars also markedly influences the color of anthocyanins. Delgado-Vargas; Jiménez; Paredes-López [29] demonstrated that in pHs above 4, the acylation can produce a band of absorption at UV higher than the one observed for non-acylated anthocyanins, at the same pH, between 310 and 320 nm [30]. *p*-Coumaroylated and caffeoylated anthocyanins can present a new shoulder or maximum of absorbance correspondent to residue of coumaroyl or caffeoyl (Figure 5B) [31].

It should be highlighted that different colors exhibited by grapes depend not only on the structural differences but also on other factors such as the pH of vacuolar fluid of vegetal cell [32] or on the medium in which they are present, the concentration of anthocyanins, the presence of other pigments, the temperature, the light and the presence of co-pigments, metallic ions, enzymes, ascorbic acid and sugars [33]. Some studies suggest that the copigmentation of anthocyanins with other compounds (copigments) may be one of the main mechanisms of stabilization of color in plants, including fruits [34]. Several studies focused on the determination of the composition and content of anthocyanins from grapes [15, 16, 17, 18, 35, 36, 37, 38, 39, 40] and wines [17, 41, 42, 43] have demonstrated striking differences between anthocyanins from *Vitis vinifera* and American or hybrid grapes. In *Vitis vinifera* grapes and in their respective wines, monoglycosylated anthocyanins are predominantly identified, while the presence of diglycosylated anthocyanins is a dominant characteristic of the American and hybrid species. Based on these data and by the use of advanced equipment, it was possible to detect adulteration (using American or hybrid varieties) in European wines, which are fully produced by *Vitis vinifera.* It is possible because the pigments of grapes are partially transferred to the wines during their elaboration and, consequently, all the varietal wine presents a typical and characteristic anthocyanin profile resulting from the original fruit, that becomes its 'fingerprint" [31, 41, 44, 45, 46, 47].

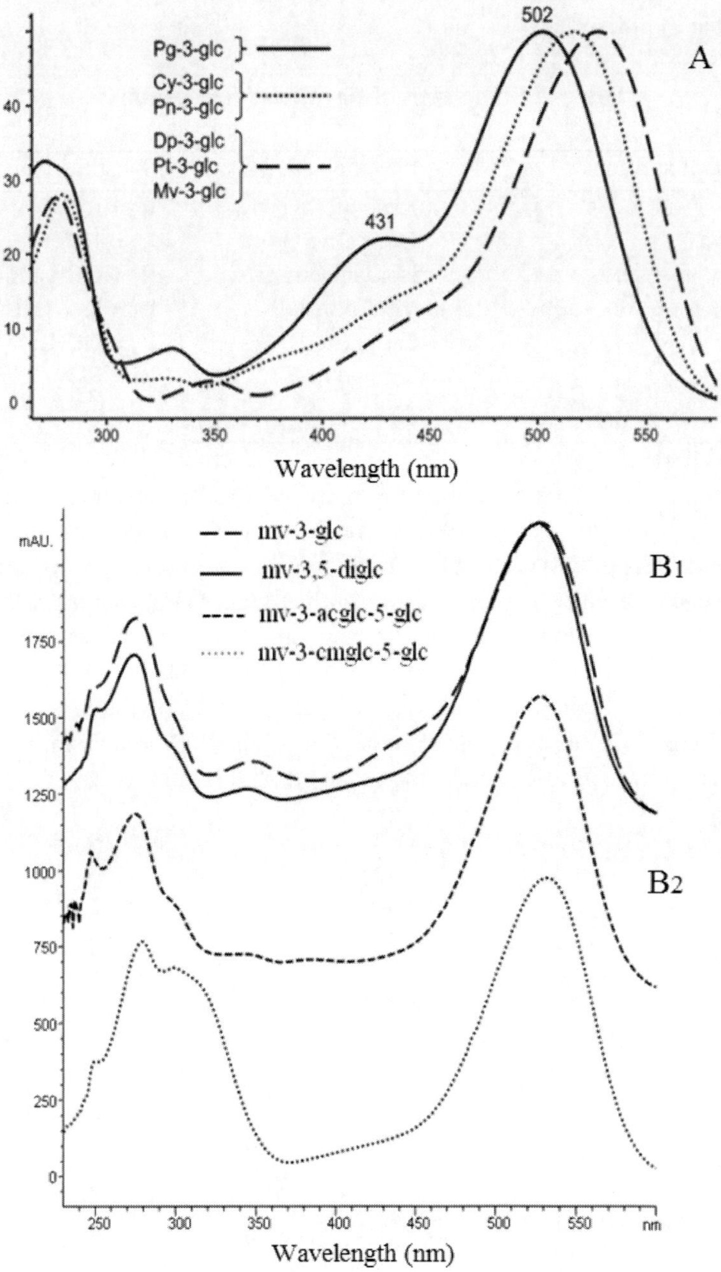

Figure 5. Absorption spectrum of different anthocyanins in acid pH (pH < 2): (A) 3-monoglucosides (the increasing of substitutions of ring B increases the wavelength of absorption); (B_1) monoglucoside x diglucoside: 3-monoglucoside (big dashed line) and 3,5-diglucoside (full line); (B_2) acetylated diglucosides x coumaroylated diglucoside: malvidin-3-acetyl glucoside-5-glucoside (small dashed line) and malvidin-3-(6"-*p*-coumaroyl)-glucoside-5-glucoside (dotted line).

II) Flavonols

The basic structure of these compounds is similar to that of the anthocyanidins (C6-C3-C6), although, with the presence of carbonyl and hydroxyl groups at the C-ring. It has been demonstrated that the occurrence of flavonols in the cultivars of red varieties is represented by six glycosylated structures (kaempferol, quercetin, isorhamnetin, myricetin, laricitrin and syringetin) (Table 2) [48, 49].

Table 2. Structures of flavonols from grapes

Basic structure of aglycon	Flavonol aglycons	Substituent groups
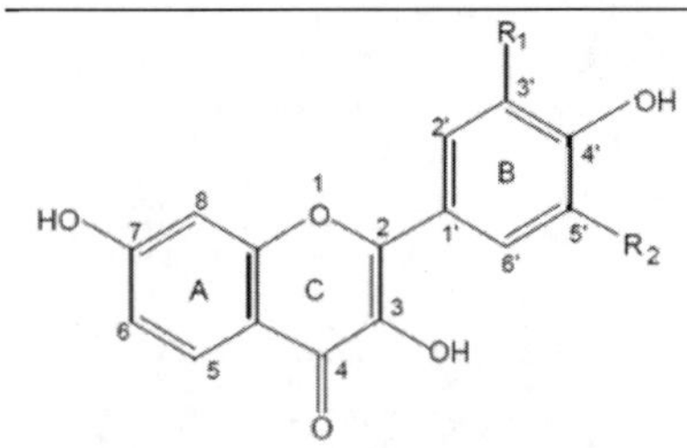	Kaempferol (K)	$R_1 = R_2 = H$
	Quercetin (Q)	$R_1 = OH; R_2 = H$
	Isorhamnetin (I)	$R_1 = OCH_3; R_2 = H$
	Myricetin (M)	$R_1 = R_2 = OH$
	Laricitrin (L)	$R_1 = OCH_3; R_2 = OH$
	Syringetin (S)	$R_1 = R_2 = OCH_3$

Although some works describe that the presence of the flavonols isorhamnetin and myricetin is unique of red grapes, the occurrence of such compounds has been already described in white grapes [50, 51, 52]. Furthermore, some studies have demonstrated the presence of myricetin in cultivars of non-vinifera white grapes (*Vitis rotundifolia*) and in their corresponding elaborated wines [53, 54]. In general, the predominant flavonols in *Vitis vinifera* cultivars are quercetin and myricetin and in *Vitis labrusca* cultivars only quercetin predominates [9, 15]. Flavonols, colorless or of pale yellow coloration, accumulate at the grape skins and comprise the function of protection against UV radiation, because of their strong absorbance at this specific wavelength range [48, 55] (Figure 6).

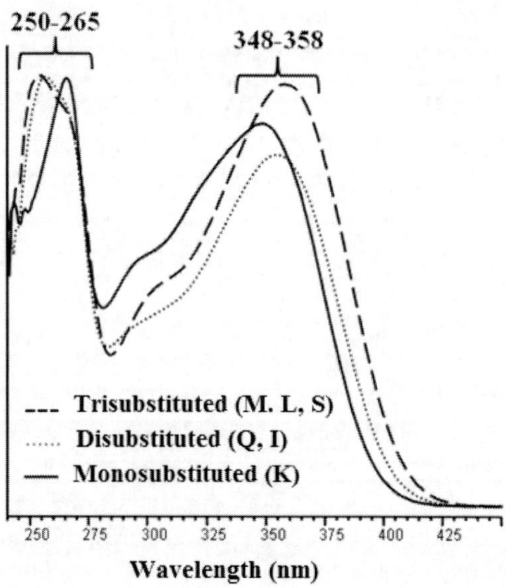

Figure 6. Spectra of flavonol glycosides according to B-ring substitution pattern.

Since its synthesis is sometimes stimulated by the light, the composition difference in the same grape variety is due to variations on the time of exposition to the light during the crop [55].

These compounds are found in grapes only in the 3-*O*-glucoside form, whereas the free corresponding aglycons can be found in wines, as a result of the acid hydrolysis that occurs during the elaboration and aging of wines. The degree of hydrolysis is dependent on the flavonol structure and also on the kind of sugar bound to the aglycon [48]. Glucose is the most common sugar linked to C-3 position of the C-ring of aglycone, but galactose and glucuronic acid have also been reported [48, 56]. Moreover, the occurrence of quercetin in grapes has been suggested as 3-*O*-(6″-rhamnosyl)-glucoside (named rutin), 3-*O*-glucosilgalactose and 3-*O*-glucosilxyloside [57].

Recently, it was demonstrated that the glycosilated forms of flavonols of *Vitis vinifera* grapes correspond to the complete series of 3-*O*-glucosides, 3-*O*-galactosides and 3-*O*-glucuronides of kaempferol, quercetin, isorhamnetin, myricetin, laricitrin and syringetin for red varieties [54], and just kaempferol, quercetin and isorhamnetin for white grapes [49] (Figure 7).

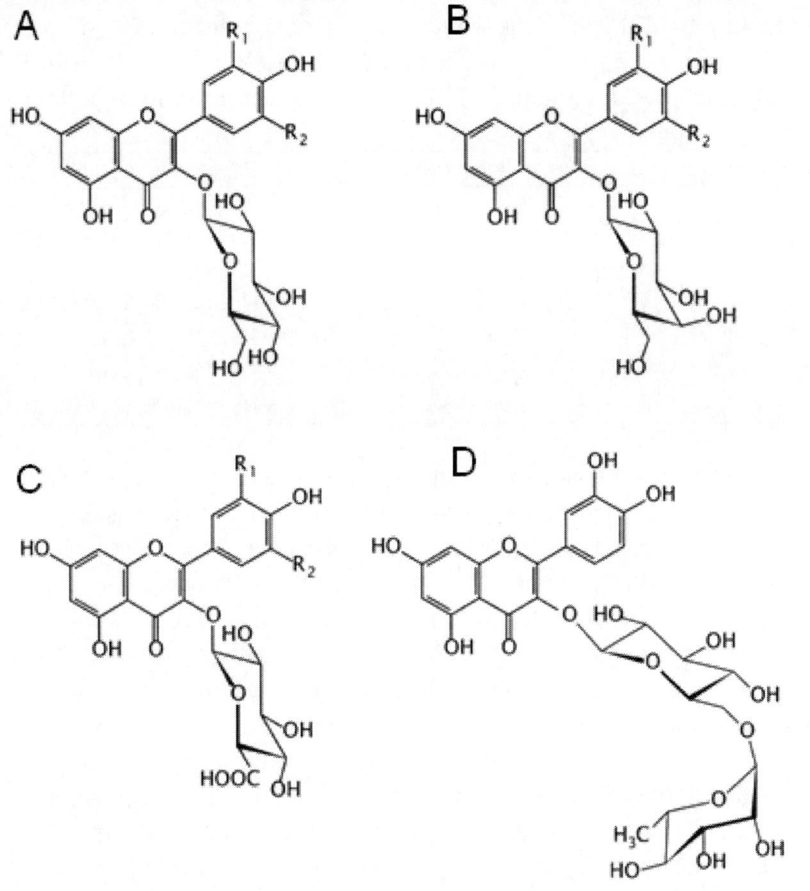

Figure 7. Structures of flavonols from *Vitis vinifera* grapes: (A) Flavonol-3-*O*-glucoside, (B) Flavonol-3-*O*-galactoside, (C) Flavonol-*O*-glucuronide, (D) Quercetin-3-*O*-(6-rhanmosil)-glucoside (rutin).

In both instances, rutin, a flavonol glucoside which has already been suggested as present in the majority of grapes and wines, is indeed, a minority flavonol in grapes which, in addition, shows a strong tendency to be hydrolysed, bringing about free quercetin in wines [58]. The flavonols directly contribute to the color of white wines and, in red wines, even though their coloration is dressed up by anthocyanins, they are robustly involved in copigmentation phenomena [59] [60]. Moreover, they contribute to the antioxidant capacity of grapes and their products [61]. Some very recent studies have contributed significantly to the knowledge about these grape compounds [15, 16, 18, 26, 62, 63].

III) Flavan-3-ols

The flavan-3-ols are compounds found in large amounts in the seeds and stems of grapes and present, as their fundamental units, the monomeric structures of 2-phenylbenzopyran (C6-C3-C6). The (+)-catechin and its isomer (-)-epicatechin are the main monomeric flavan-3-ols in grapes [64], and can, in part, be sterified to gallic acid ($C_7H_6O_5$), usually at carbon 3 (Table 3)[65, 66]. These compounds are mostly responsible for the flavour and astringency of wines and juices [41, 67, 68].

Table 3. Examples of flavan-3-ols from grapes: C: (+)-catechin; E: (-)-epicatechin; ECG: (-)-epicatechin 3-gallate; GC: (+)-gallocatechin; EGC: (-)-epigallocatechin; GCG: (+)-gallacatechin 3-gallate; EGCG: (-)-epigallocatechin-3-gallate

Basic structure of aglycon	Flavanol Aglycons	Substituent Group
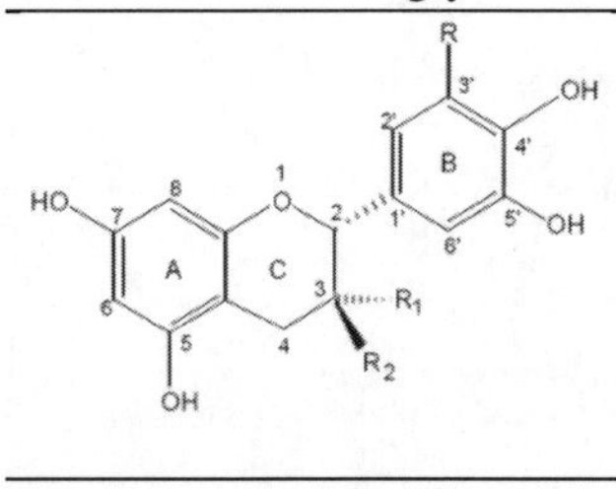	C	R = H; R_1 = OH; R_2 = H
	E	R = H; R_1 = H; R_2 = OH
	ECG	R = H; R_1 = H; R_2 = O-Galloyl
	GC	R = OH; R_1 = OH; R_2 = H
	EGC	R = OH; R_1 = H; R_2 = OH
	GCG	R = OH; R_1 = H; R_2 = O-Galloyl
	EGCG	R = OH; R_1 = O-Galloyl; R_2 = H

The association of several monomeric units of catechins or epicatechins produces compounds called proanthocyanidins or condensed tannins. Depending on the number of times this basic unit is repeated, the proanthocyanidins can be dimer, trimer, oligomeric (up to 5 units) or polymeric (above 5 units). The structural variations occur due to the number of linked monomers, the position of the occurrence of the bindings, the kind of oxygenation of the A and B rings of the unit and the stereochemistry of the substituents of C-ring [69].

The proanthocyanidins can liberate anthocyanidins when they are warmed in strongly acid and alchoolic medium, under the rupture of the linkage among the monomeric units [70, 71]. Depending on the liberated anthocyanidins, cyanidin and, in a lesser proportion, delphinidin, in grapes specially, the precursor molecules are called procyanidins (constituted of (+)-catechin and (-)-epicatechin) and prodelphinidins (constituted of (+)-gallocatechin and (-)-epigallocatechin) respectively. The procyanidins are the predominant flavan-3-ols in grape seeds [72], whereas both the procyanidins and the prodelphinidins are found in grape skins [73].

The dimeric procyanidins are classified according to their kind of interflavanolic linkage and are divided in two groups: A and B. The A-type procyanidins ($C_{30}H_{24}O_{12}$) possess C4-C8 or C4-C6 interflavan unions and a ether linking between carbons C-5 or C-7 of terminal unit (A-ring) with the carbon C-2 of the superior unit (C-ring) (Figure 8). The B-type procyanidins ($C_{30}H_{26}O_{12}$) are resultant from the condensation of the flavan-3-ol units through a linking among the carbon 4 of the superior monomer (C-ring) with the carbon 6 or 8 from the inferior unit (A-ring). The resulting compounds by C4-C8 bindings form a series of procyanidins B1, B2, B3 and B4, while the ones composed by C4-C6 linkings constitute the series of B5, B6, B7 and B8 procyanidins (Figure 8). As for the trimeric procyanidins, they are classified as type C (resultant from interflavan unions corresponding to the type B of dimers) (Figure 8) and type D (which have an interflavan binding of type B and another of type A) [70].

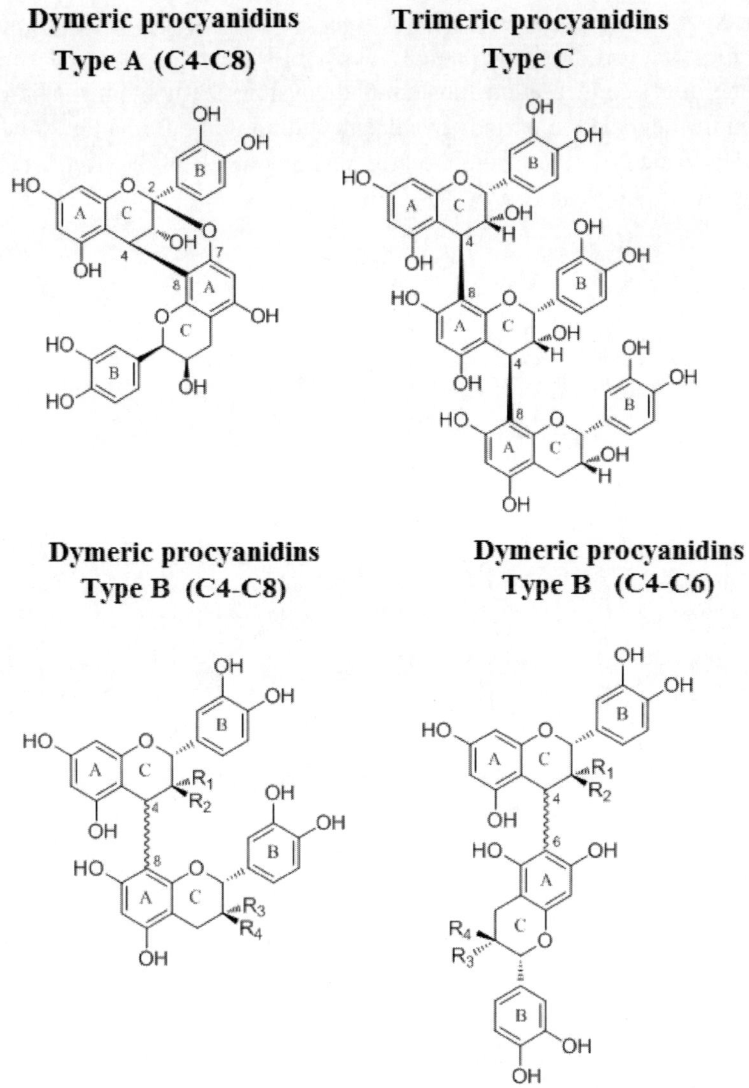

Figure 8. Examples of dymeric (Type A and Type B) and trimeric (Type C) procyanidins. Type B (Dymer C4-C8): B1: R_1=OH; R_2=H; R_3=H; R_4=OH; B2: R_1=OH; R_2=H; R_3=OH; R_4=H; B3: R_1=H; R_2=OH; R_3=H; R_4=OH; B4: R_1=H; R_2=OH; R_3=OH; R_4=H. Type B (Dymer C4-C6): B5: R_1=OH; R_2=H; R_3=OH; R_4=H; B6: R_1=H; R_2=OH; R_3=H; R_4=OH; B7: R_1=OH; R_2=H; R_3=H; R_4=OH; B8: R_1=H; R_2=OH; R_3=OH; R_4=H.

The polymeric forms (Figure 9) are the main flavan-3-ols in grapes and are constituted of an elevated number of monomeric units, joined either at the terminal position or at the superior position. The quantity, structure and degree of polymerization of proanthocyanidins in grapes differ according to their localization in the different parts of grapes [74]. Tannins are ordinarily associated to the evolution of the color and the organoleptic quality of juices and wines, mainly with respect to astringency and bitterness [75]. The ability to interact with salivary proteins, forming stable complexes which lead to a reduction of the lubrification of the palate, causes the sensation of dryness and constriction, known as astringency [76, 77].

Many chemical reactions occur among the extracted anthocyanins and tannins during the maceration, stabilization and evolution of wines, forming a great number of colorless and colorful compounds. The most important reactions are the indirect condensation flavanol-anthocyanin, the indirect polymerization flavanol-flavanol, both intermediated by acetaldehyde, the direct condensation flavanol-anthocyanin, the non-enzymatic oxidation of flavan-3-ols and the degradation of anthocyanins, resulting in the formation of new pigments with different chromatic properties from the original compounds of the fruit, which markedly influence the quality of juices and wines [78, 79].

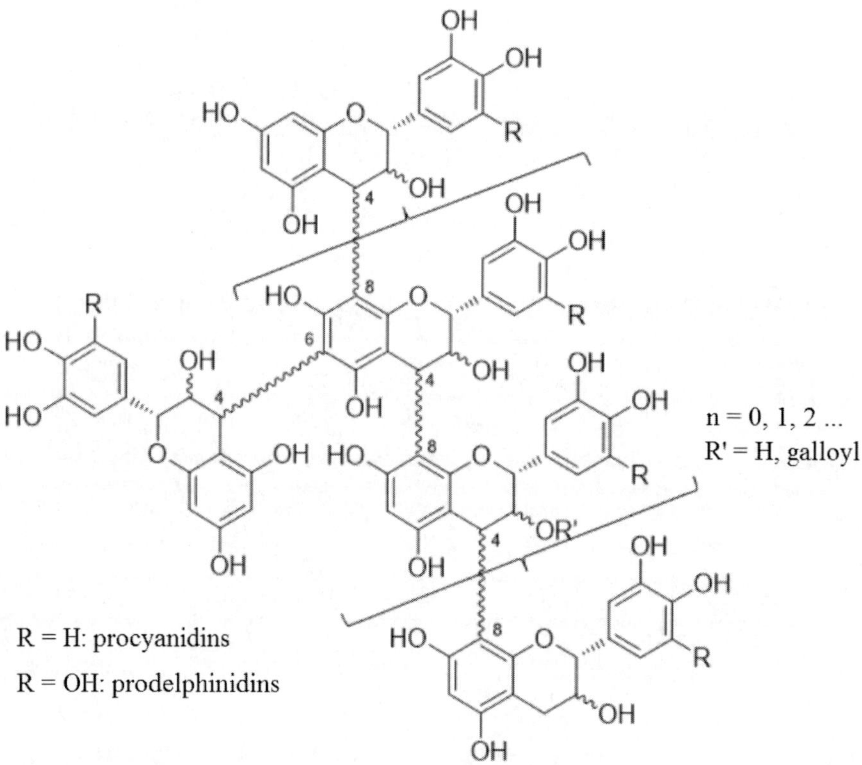

Figure 9. Polymeric proanthocyanidins in grapes.

Non-Flavonoids

The main non-flavonoids compounds present in grapes and wines are the phenolic acids (hydroxybenzoic and hydroxycinnamic), in free or sterified forms, and the stilbenes. In wines, some phenolic acids and other related derivative products can be found.

Phenolic Acids

The phenolic acids present in grapes are the hydroxybenzoic (C6-C1) and the hydroxycinnamic (C6-C3) acids, the former being found in skin and seeds and represented by gallic acid, producing esters (gallates) with the flavan-3-ols; and the second are predominantly found in the tartaric ester forms, such as caftaric and coutaric acid, in the skin and pulp (Table 4) [9, 80]. The hydroxycinnamic acids are considered the major constituents of the pulp [71] and, although they are present in low concentrations, they represent one of the main phenolic compounds in white grapes [81]. Among the hydroxybenzoic acids, only the gallic acid is present in grapes, being found in the pulp, seeds and skins in the free or flavanol ((-)-epicatechin-3-gallate) ester form. Nonetheless, it is possible to find other hydroxybenzoic acids [71].

Table 4. Substitution pattern phenolic acids

A

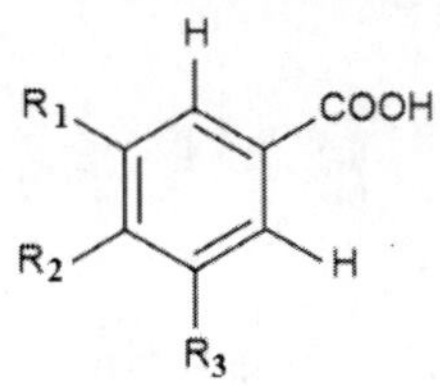

B

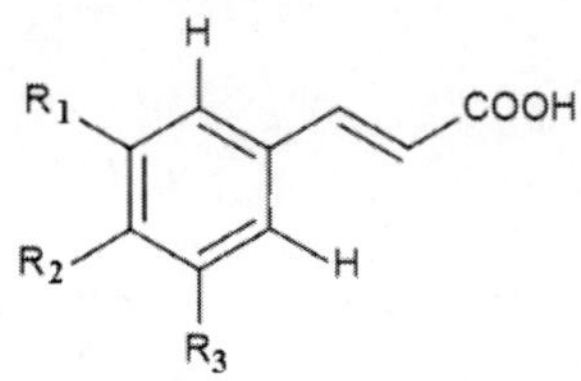

Compounds	Substituent		
A) Hydroxybenzoic acid (C6-C1)	R_1	R_2	R_3
gallic acid	OH	OH	OH
B) Hydroxycinnamic acid (C6-C3)	R_1	R_2	R_3
p-coumaric	H	OH	H
caffeic acid	OH	OH	H
ferulic acid	$O-CH_3$	OH	H
sinapic acid	$O-CH_3$	OH	$O-CH_3$

During winemaking, tartaric esters from hydroxycinnamic acids can undergo hydrolysis, releasing hydroxycinnamic acids in such a way that both the free and combined forms are found in wines. Thus, the content of these compounds will depend both on the type of grape and also on the hydrolysis process of these esters [82]. In principle, these compounds do not confer any flavor and smell to the wine. However, they are precursors of volatile phenolic compounds which affect these features in wines. During the fermentation, in particular, the carboxylase enzymatic action of some yeasts present in grapes can convert the *p*-coumaric and ferulic acids into 4-vinylphenol and 4-vinilguayacol, respectively [83, 84]. These compounds, in turn, can be reduced to ethyl derivatives (ethylphenol and ethylguayacol) by vinylreductases produced by certain yeasts like *Brettanomyces* [85] or by some lactic bacteria from the Lactobacillus genus. The presence of ethylphenols in wines is responsible for a great economic loss since these compounds can confer an unpleasant aroma, known by specialists as leather or animal aroma or horse sweat [86]. On the other hand, hydroxycinnamic acids, especially the *p*-coumaric and caffeic acids, are involved in the acylation process of the glycosidic substituents of anthocyanins [71] and, along with their 4-vinylphenol derivatives, they can react to anthocyanins, producing compounds of the pyranoanthocyanin type. These contain unsaturated polarizable ligations, that is, with substituents of different polarities in both margins of the double ligation, causing these reactions. The first identified structure of pyranoanthocyanin was vitisin A (Figure 10A) resulting from the reaction between malvidin-3-glucoside and pyruvic acid (a secondary metabolite of yeasts) through the enolic form of the latter and, after that, descarboxylation [87].

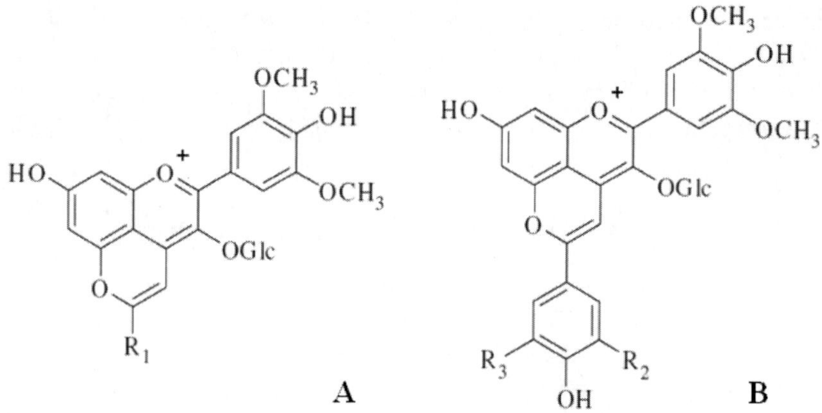

Figure 10. Chemical structures of wine pyranoanthocyanins derived from malvidin-3-glucoside: (A) pyranoanthocyanins of vitisin type (R_1 = COOH, vitisin A; R_1 = H, vitisin B); (B) pyranoanthocyanins of hydroxyphenyl type ($R_2 = R_3$ = H, malvidin-3-glucoside-4-vinylphenol; R_2 = H and R_3 = OH, malvidin-3-glucoside-4-vinylcatechol or pinotin A; R_2= H and R_3 = OCH_3, malvidin-3- glucoside-4-vinylguayacol).

The structures of pyranoanthocyanins are diverse and several modes of creation have been proposed [87]. In Figure 10B, other relevant pyranoanthocyanins are illustrated, which are formed from the reaction of anthocyanins with the hydroxycinnamic acids (pyranoanthocyanins of hydroxyphenyl type). These reactions are involved in the process of color stabilization during the aging of wine [88].

Stilbenes

The stilbenes have a molecular structure essentially composed of two benzene rings separated by an ethylene binding (C6-C2-C6). These molecules are synthesized especially in the skins, seeds and roots in response to stress conditions, such as microbial infection, fungi, mechanical damage or ultraviolet light irradiation [89]. So, in this way, they can be found in grapes [90, 91], juices [92, 93, 94, 95, 96] and wines [97, 98, 99]. Among the main compounds belonging to the class of stilbenes, the resveratrol (*trans*-3,5,4'–trihydroxystilbene), the piceid (*trans*-3,5,4'-trihydroxystilbene-3-*O*-β-D-glucoside) and the astringin (*trans*-3,3',5,4'-tetrahydroxystilbene-3-*O*-β-D-glucoside) show up.

Resveratrol contains a skeleton based on the structure of 1,2-diphenylethylene [100]. This substance exists in its natural form as a *trans* isomer, though, it can be isomerized (under UV light) to its corresponding *cis* isomer. In grapes, the resveratrol is present linked to a glucose molecule at position 3. The resveratrol-3-*O*-glucoside is also named as piceid and polydatin, and can also be presented in the *cis* and *trans* forms (Figure 11) [101].

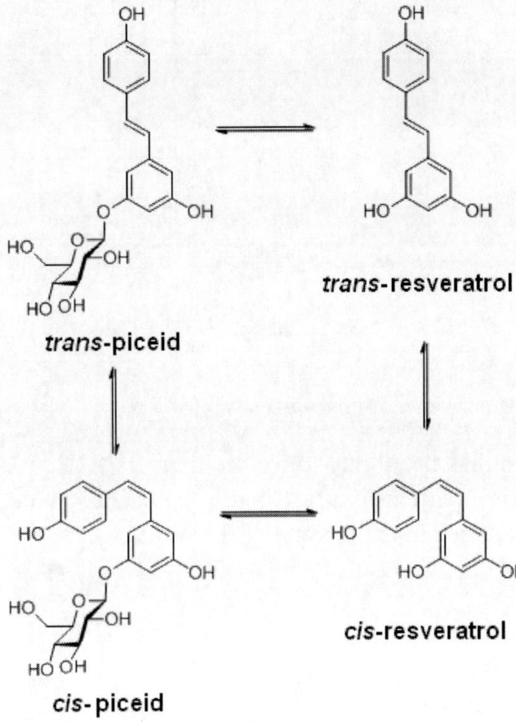

Figure 11. *Trans*-resveratrol and its derivatives.

There is also a compound called piceatannol, which is a superior homologue of resveratrol, with an additional OH group at position 3’ and its 3-*O*-glucoside is called astringin (Figure 12).

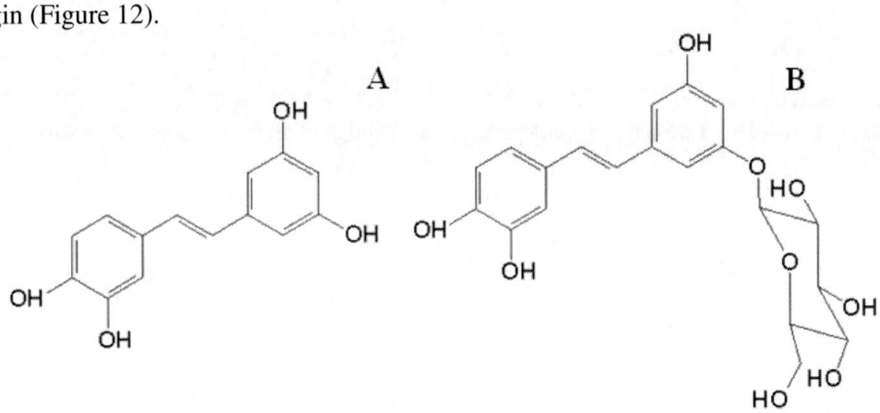

Figure 12. Homologues from resveratrol: piceatannol and its 3-*O*-glucoside (astringin).

The presence of an additional OH group means that these compounds (piceatannol and astringin) present UV absorption maximum at 302 and 321 nm, respectively [102], while resveratrol presents around 306 and 316 nm. Another glycosylated derivative from resveratrol found in Reisling wines is the *trans*-resveratrol-2-*C*-glucoside [103], that differenciates itself from piceid by the position of the glucoside at the C-2 position (Figure 13).

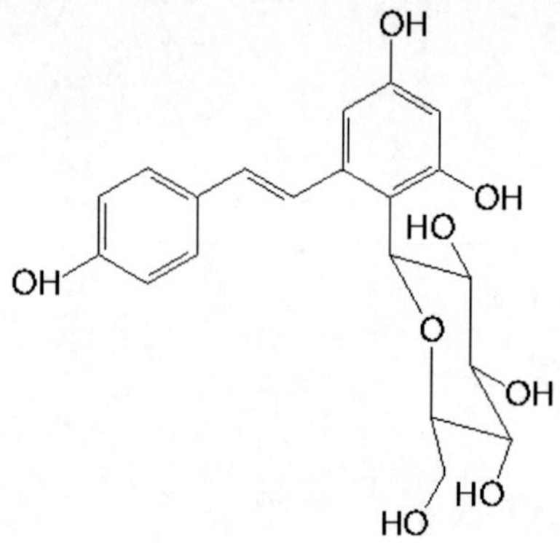

Figure 13. The 2-*C*-glucoside derivative from *trans*-resveratrol.

Finally, the resveratrol has the ability to form oligomers [102, 103, 104, 105], and some dimers (pallidol, viniferin, partenocissin) (Figure 14), and, more recently, a tetramer (hopeaphenol) has been reported in the literature (Figure 15).

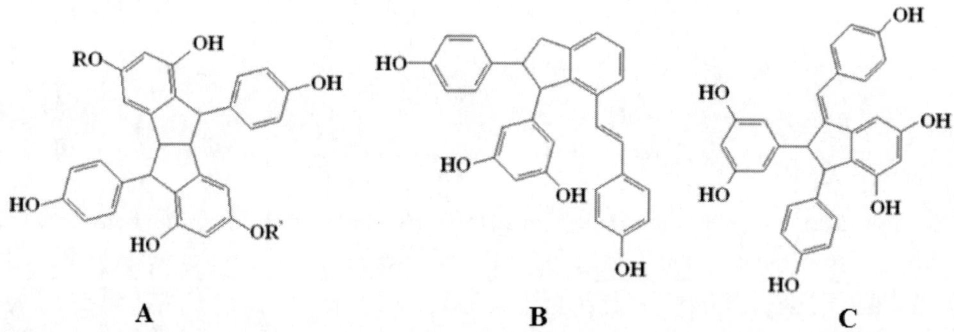

Figure 14. Dimers of resveratrol. A. Pallidol, B. *Trans*-ε-viniferin, C. Partenocissin. Specially for pallidol, the derivatives 3-glucoside (R= glycosyl; R' = H) and 3,3'-diglucoside (R = R' = glycosyl) are known.

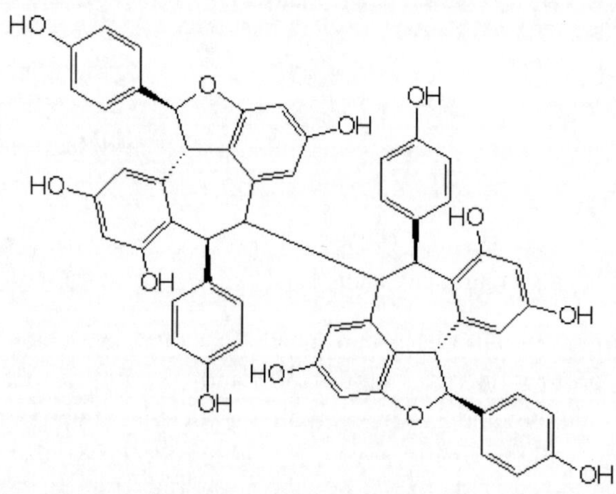

Figure 15. Resveratrol tetramer (hopeaphenol).

Some studies suggest that the winemaking techniques notably influence the content of resveratrol and its derivatives found in wines [106, 107], where there was greater content present in products which had been submitted to skin maceration for a longer time [108]. Conversely, operations such as clarification and filtration can lead to a decrease in the levels of resveratrol and piceid in juices and wines [110, 111].

In addition, some wines elaborated from botrytized grapes present lower levels of resveratrol when compared to ones which were elaborated from "normal" grapes. In this case, the same fungi (*Botrytis cin*erea) that attacks the grape is responsible by the development of differenciated aromas and tastes to the wines by the production of the laccase enzyme which oxidizes resveratrol [111, 112]. However, the major reason for interest by the researchers in resveratrol is associated to its benefits to human health [113, 114, 115, 116, 117].

There is increasing evidence that this compound can prevent or retard the emergence of cancer, cardiovascular diseases, diabetes, pathologic inflammations, viral infections, amongst others [115].

It should be considered that the studies about the bioactive properties of resveratrol have been basically carried out with the *trans* isomer, with the exception of a few works available in literature [118, 119]. So, it is relevant to perform more conclusive studies which demonstrate if these isomeric forms of resveratrol can present or not similar properties to those attributed to *trans*-resveratrol, or if they can show other different activities from those found in *trans*-resveratrol.

The Biochemical Characteristics of Grapes

Enzymatic Oxidation of Phenolic Compounds Present in Grapes

The phenolic compounds responsible for the color of juices and wines come from the grapes used for its production. However, the profile of phenolic compounds present in these products is dependent on numerous factors including the species and variety of grape, location of planting, cultivation system, climate, soil, form of extraction of phenolic compounds and the type of process employed. Moreover, they are also influenced by chemical and enzymatic reactions that begin with the crush and continue throughout the whole process, as well as during the maturation and aging of wines. Thus, the wines have a phenolic composition much more complex than that found in the original grapes used for its production [10]. The knowledge of these transformations is of undoubted importance to the quality control of the product and to the technological development of fruits and vegetables.

The enzymatic oxidative reaction of phenolic compounds, also known as enzymatic browning, is one of the main problems associated with post-harvest grapes and the processing of this fruit due to the formation of undesirable color and flavor and the loss of nutrients, and, therefore, causing considerable economic losses [120]. The enzymatic browning in plants is catalyzed by the group of enzymes collectively called polyphenol oxidases (PPOs). Another class of enzyme, oxidoreductases, also involved in the reactions of enzymatic browning of fruits and vegetables but with less importance in grapes, is the peroxidase or guaiacol peroxidase (POD) (EC. 1.11.1.7) [120, 121].

Despite numerous works on structure and characterization of PPOs, the classification and specificity of these enzymes remains an area of confusion [122, 123]. In this review, we are going to adopt the classification according to the Nomenclature Committee of the International Union of Biochemistry and Molecular Biology (NCIUBMB) [124]. So, they are catechol oxidase or o-diphenol:oxigen oxirreductase (EC 1.10.3.1), monophenol oxidase, cresolase or tyrosinase (EC.1.14.18.1) and laccases (benzenediol: oxygen oxidoreductase; EC 1.10.3.2). According to Aniszewski et al. [123], cresolase and tyrosinase are the same enzyme differentiated according to their organism specialization: cresolase occurs in plants but not in animals, and tyrosinase occurs in animals but not in plants. Catecholase and laccase differ from each other and from cresolase and tyrosinase. Although there is still not clear evidence at present to indicate that these proteins are fundamentally different in structure or that there are separate genes coding for these enzymes, there is clear evidence that these enzymes have different activities. These three different enzymes are widely distributed in nature and are important enzymes in plants, fungi, bacteria and animals, although their biological functions and mechanisms are not known in detail [123, 125].

Investigations to determine the biochemical characteristics of grape PPO and conditions that are more active have been conducted primarily with the enzymes of the species *Vitis vinifera* [126, 127, 128, 129, 130, 131]. Few studies have been conducted with other species, such as DeChaunac hybrid grapes [132] and seedless Crimson grapes [121].

The biochemical mechanism of phenol oxidation begins with the breakdown of the physical integrity of the grape berry caused by damage during harvest and transportation or the crushing/pressing of the fruit during processing, i.e, of juice and wine [133]. In this situation, there is the decompartmentalization of the organelles of plant cells, promoting the release of phenolics together with the PPOs, which are predominantly found in plastids [134].

Direct contact between the phenolic compounds (substrates) and proteins (enzymes) triggers numerous reactions in the presence of molecular oxygen (O_2), with consequent formation of melanin, the dark colored polymers [133, 135]. Besides the presence of molecular oxygen, the degree of browning depends on several factors such as the presence of reducing agents, metal ions, pH and temperature [126].

In the next section of this overview chapter we provide the current knowledge of advances in chemistry of enzymatic browning focusing on the group of enzymes that catalyze the oxidation of phenolic compounds, PPOs and peroxidases, including structure and function, mechanism, distribution and substrates involved in the formation of brown products.

Poliphenol Oxidases

Catechol Oxidase and Monophenol Oxidase

I) Structure and Function

Cathecol oxidase and monophenol oxidases (as well as laccase) belong to the group of multi-copper proteins oxidases. Proteins containing copper ions at their active site are usually involved as redox catalysts in a range of biological processes, such as electron transfer or oxidation of various organic substrates. These enzymes share similar active sites and physiological functions. Tyrosinases are activated *in vivo* by limited proteolytic cleavage, which might open up substrate access to the catalytic site. Monophenol oxidase and catechol oxidases are proteins containing a single type-3 center with two atoms of coppers, whereas laccases are proteins containing three different centers with four atoms of coppers.

In this review, in order to avoid unnecessary repetition, we use the terminology PPOs when referring to properties that are common to (shared by) monophenol oxidase and catechol oxidases typical of plants. In cases where it is necessary to include the laccase this will be indicated. The structure and mechanism of catalysis PPOs have been the subject of much research [123, 136, 137, 138, 139].

The PPOs of grapes are characterized as enzymes containing di-coppers in its active site, in which three histidines bind each of the two catalytic copper ions, and one of the histidines is engaged in a thioether linkage with a cysteine residue (Figure 16) [139]. The two copper ions are capable of binding molecular oxygen to form a complex dioxygen-dicopper (II). The PPO carries out the oxidation of phenols such as catechol forming benzoquinone. Hydrogens removed from catechol combine with oxygen to form water. This reaction produces the yellow compound benzoquinone and the benzoquinone is then oxidised by the air to form dark-brown melanin. This is a form of enzymatic browning exhibited in many plants and foods upon exposure to oxygen.

Catechol is present in the vacuoles of cells and catechol oxidase is present in the cytoplasm of cells. If the plant tissues are damaged, the catechol is released and the catechol oxidase converts the catechol to ortho-quinone, which is a natural antiseptic. Catechol oxidase, therefore, has a role in plant defense mechanisms, helping to protect damaged plants against both bacterial and fungal disease.

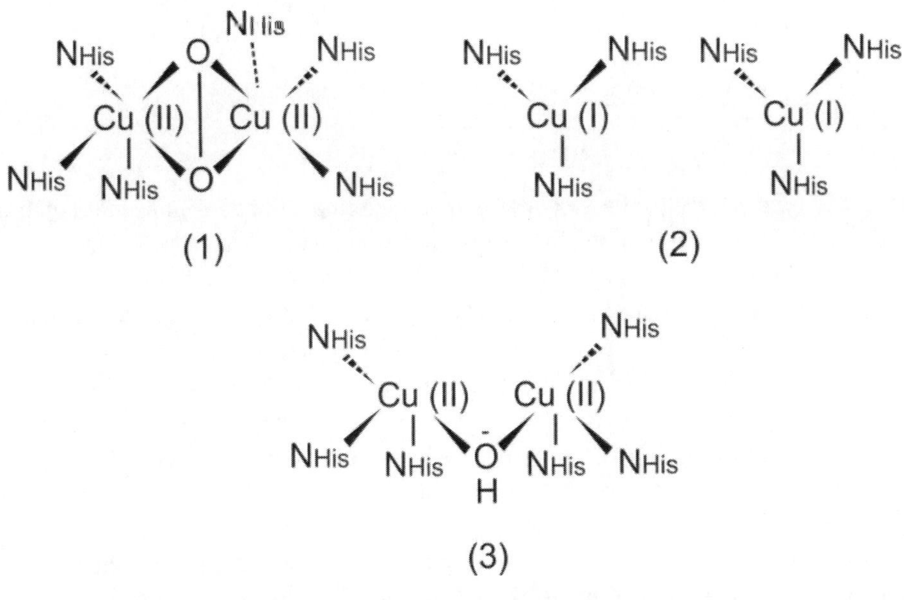

Figure 16. Intermediate states of the active site of PPOs (1) Oxytirosinase (Oxy); (2) Deoxytirosinase (Deoxy); (3) Mettirosinase (Met).

Although the active site of the enzyme is well preserved, the PPOs show considerable variability regarding the amino acid sequences, size, glycosylation and kinetics between species [139]. Some authors suggest that the presence of multiple forms may in some cases be generated by artifacts due to partial release of the enzyme-linked to the membranes by partial denaturation, proteolysis or fragmentation, as well as the activation of latent forms [140]. The molecular weight of PPOs varies, depending on the tissue and the organism from which it has been extracted. The molecular mass found ranges of 38 to 60 kDa. This difference is possibly due to C-terminal processing [141].

Monophenol oxidase or tyrosinase (EC.1.14.18.1) also contains two copper atoms. Each of them is coordinated by three histidine ion residues [142, 143]. But different of cathecol oxidase, it is a mixed-function oxidase that is, an enzyme which oxidizes two different substrates simultaneously. It is able to catalyze ortho-quinone formation by the oxidation of both catechols (oxidase activity) and phenols (monooxygenase activity).

During the monooxygenase cycle, one of the bound oxygen atoms is transferred to a monophenol (such as L-tyrosine), generating an o-diphenol intermediate (Figure 17), which is subsequently oxidized to an *o*-quinone and released, along with a water molecule. The enzyme remains in an inactive deoxy state, and is restored to the active oxy state by the binding of a new oxygen molecule. The enzyme is referred to as tyrosinase in animals, since L-tyrosine is the major monophenolic substrate. In plants, the enzyme is sometimes referred to as cresolase owing to the ability of the enzyme to utilize the monophenolic substrate, cresol.

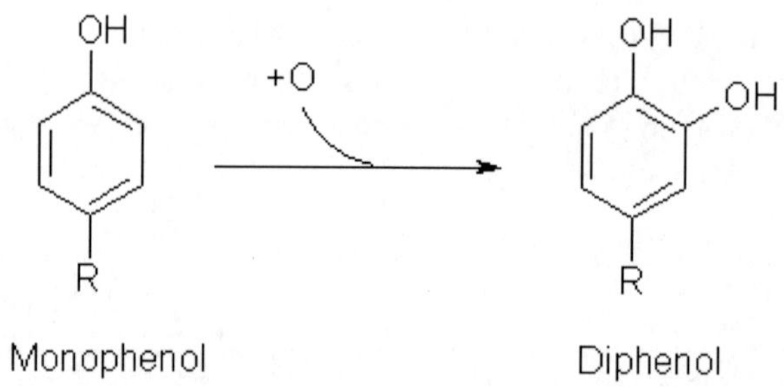

Figure 17. Monophenol oxidase pathway producing the diphenol.

During the oxidase cycle, the enzyme binds an external diphenol molecule (such as L-dopa) and oxidizes it to an *o*-quinone that is released along with a water molecule, leaving the enzyme in the intermediate met state. The enzyme then binds a second diphenol molecule and repeats the process, ending in a deoxy state. The second reaction is identical to that catalyzed by the related enzyme catechol oxidase or o-diphenol:oxygen oxireductase (EC 1.10.3.1). However, the latter cannot catalyze the hydroxylation or monooxygenation of monophenols.

Catechol oxidase (COs) is a type 3 copper protein that catalyzes exclusively the oxidation of catechols (i.e., o-diphenols) to the corresponding o-quinones. The latter highly-reactive compounds undergo an auto-polymerization leading to the formation of a brown polyphenolic pigment, i.e. melanin, a process thought to protect damaged tissue against pathogens or insects [144]. The enzyme also acts on a variety of substituted catechols. Cathecol oxidases are found in plant tissues and in crustaceans. It is different from tyrosinase, EC 1.14.18.1 (monophenol monooxygenase), which can catalyze both the monooxygenation of monophenols and the oxidation of catechols.

II) Mechanism

The monooxygenase cycle of the mono phenol oxidase. The catalytic cycle begins with the enzyme binding dioxygen to form oxy-tyrosinase. One of the oxygen atoms is transferred to a monophenol generating an o-diphenol intermediate which is subsequently oxidized to an o-quinone.

The catalytic cycle of cathecol oxidase begins with the oxidized met form of the enzyme (resting form of CO binding OH). The binding of the diphenolic substrate to CuB reduce the Cu(II)–Cu(II) form to the dicuprous state (Cu(I)–Cu(I) leading to the formation of quinone. In the process of the binding, a catechol proton has therefore been transferred to an external base. In the next step, the catechol loses its second proton to the bridging hydroxide and leaves as the first quinone product.

After formation of the first quinone, dioxygen binds to the dicuprous state Cu(I)–Cu(I) active site replacing the solvent molecule bonded to CuA in the reduced enzyme form. The copper dimer is oxidized again to Cu(II)–Cu(II) form. As in the first step, it is suggested that catechol loses a proton to an external base and form a monodentate binding to one of the coppers (a simultaneous binding of catechol and dioxygen is possible). After oxidation of the second substrate molecule and loss of the bound water, the dicopper center is in its met form again and ready to undergo another catalytic cycle [122, 145] (Figure 18).

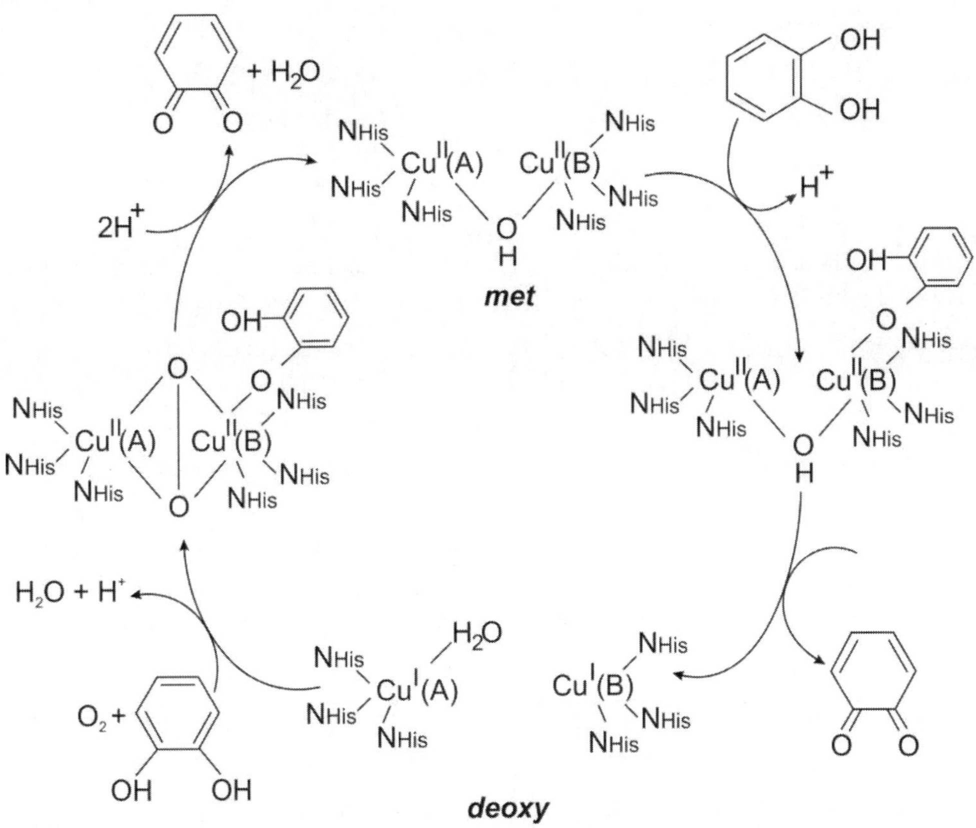

Figure 18. Catalytic cycle of catechol oxidase from *Ipomoea batatas* (Adapted by [122, 145]). Two catechol molecules (or *o*-diphenols) are oxidized, coupled with the reduction of molecular oxygen to water.

III) Distribution

Monophenol oxidase are type III copper proteins found in a broad variety of bacteria, fungi, plants, insects, crustaceans and mammals, which are involved in the synthesis of betalains and melanin. The catechol oxidase activity is particularly high in fruits and vegetables containing significant levels of phenolic compounds [146], including grapes [126, 127, 139] (Table 5).

Table 5. Characteristics of PPO extract from grapes

Source	PPO	Optimum pH	Optimum T (ºC)	Substrate	Km (mM)	Ref
Vitis vinifera						
Grenache	Active	-	-	Catechol/TBC	-	[139]
Crimson Seedless	Active	3.0	25	TBC	2.8	[121]
	Latent (activated with 0.2mM SDS)	6.0	25	TBC	12.7	
Napoleon	Active	3.0	25	TBC	8.2	[126]
	Latent (activated with 0.2% SDS)	6.0	25	TBC	3.1	
Dominga	Active	3.0	25	TBC	3.0	[171]
	Latent (activated with 0.2% SDS)	6.0	25	TBC	7.6	
De Chaunac	Active	6.0	20	Caffeic acid	16.0	[132]
	Active	6.0	20	4-methyl-catechol	25.0	
Monastrell	Active	3.5-5.0	30	4-methyl-catechol	9.0	[172]
Victoria	Active	5.0	25	Catechol	52.6	[127]
	Active	5.0	25	4-methyl-catechol	7.5	
	Active	5.0	25	(+) Catechin	4.3	
	Active	5.0	25	Chlorogenic acid	3.2	
	Active	5.0	25	D,L-DOPA	68.2	
Emir	Active	4.2	30	Catechol	25.1	[128]
Merlot	Active	5.0	30	4-methyl-catechol	7.5	[165]
	Active	5.0	30	(+)-Catechin	7.9	
	Active	5.0	30	Caffeic acid	5.8	
	Active	5.0	30	Caffeoyl-quinate (chlorogenic acid)	1.5	
	Active	5.0	30	Caffeoyl-tartrate	2.1	
	Active	5.0	30	*p*-Coumaric acid	2.0	
	Active	5.0	30	*p*-Coumaroyl-tartrate	1.1	
Mazruma	Active	6.5	20	4-methyl-catechol	6.8	[131]
	Active	6.0	15	Catechol	15.7	
	Active	7.0	15	(+)-Catechin	28.6	
	Active	7.5	25	Epicatechin	18.2	
	Active	6.0	20	Pyrogallol	13.3	
Sultaniye	Active	3.4	30	Catechol	44.5	[129]
Vitis labrusca						
Concord	Active	6.0	25	Catechol	67	[173]
Airén	Active	3.5-4.5	-	4-methyl-catechol	9.5	[174]
Hybrids						
Muscat Bailey A	Active	6.3	25-30	Catechol	18.4	[175]
BRS Clara	Active	6.0	25	Catechol	-	[149]
	Latent (activated with 0.25% SDS)	5.5	25	Catechol	-	
BRS Morena	Active	6.0	25	Catechol	-	[149]
	Latent (activated with 0.25% SDS)	5.5	25	Catechol	-	

* TBC - 4-tert-butylcatechol.

Especially in plants, most of the studies reported cresolase absence or low activity compared to those found catechol oxidase [147, 148]. Investigations to determine the biochemical characteristics of grape PPO and conditions that are more active have been conducted primarily with the enzymes of the species *Vitis vinifera* [126, 127, 128, 129, 130, 131, 139]. Limited studies have been conducted with other species, such as hybrid DeChaunac grapes [132] and Crimson Seedless grapes [121].

The latency property of PPO has been reported for some varieties of grapes [121, 126, 149]. Considering the difficulty of experiments in the fruit itself (in vivo), the proposed activation of these enzymes is based on the hypothesis that the activation properties in vitro are comparable to that *in vivo* activation [150]. This activation is mainly carried out by proteases [151] and non-ionic, anionic and cationic detergents [121, 126, 149, 152]. The use of the activating agent (SDS-ionic detergent sodium dodecyl sulfate) is particularly interesting because the PPO can be activated at concentrations of SDS for which most of the enzymes are denatured [126].

Lago-Vanzela et al. [149] reported that using catechol as a substrate, a positive effect on latent PPO activity was observed with the addition of SDS in the reaction mixture at a concentration of 0.25% for the BRS Clara grape, and 0.3 - 0.6% for the BRS Morena grape. SDS concentrations higher than those cited decreased the activity of PPO in all the grapes. Fortea et al. [121] showed that the detergent SDS showed an inhibitory effect for peroxidase (POD) when used in the same concentration that activated PPO enzymes. The activator effect and inhibitor effect of the detergent SDS for latent PPO and POD, has been reported by authors of several studies [153, 154, 155].

Detailed information about the phenomenon of PPOs latency at the molecular level and its physiological role require purification and characterization of latent enzymes without the presence of activated PPO enzymes [156]. Lago-Vanzela et al. [149] showed good results in extracting latent PPO from seedless grapes using aqueous two-phase detergent Triton-X-114.

Methods for purification of the PPO in its active and latent were performed and from the characterization results, it was found that the isoforms of PPO present molecular weights ranging from 30 to 72 kDa [139, 149, 157, 158, 159]. This range in molecular weight of enzyme is due to the different sources from which they were isolated and, moreover, it is the result of modification of the peptide chain of the enzyme during purification, or by proteolytic cleavage or chemical modification by the action of reactive quinones [156].

In some reaction models, it is described that hydrogen peroxide (H_2O_2) is released during the catalysis of catechol oxidase [136], which is subsequently used as a co-substrate for peroxidase (POD) in the oxidative process [160] (Figure 19). The action of POD and laccase will be discussed in due course.

IV) Substrates

Different PPOs have different specific substrates and the most tested are *p*-cresol, tyrosine and *p*-coumaric acid to determine monophenol oxidase (cresolase) activity, and catechol, 4-methylcatechol, L-DOPA (L-3,4-dihydroxyphenylalanine) (-)-epicatechin, (+)-catechin, chlorogenic acid and dopamine to determine catechol oxidase activity.

However, the in vivo presence of a number of substrates has not been demonstrated yet and their relevance in a physiological role remains uncertain [123]. The catechol and its derivatives are among the main physiological substrates used for the determination of catechol oxidase, including *Vitis vinifera* [127, 128].

The PPOs have a high affinity for hydroxycinnamic acid derivatives, especially tartaric esters such as *p*-caftaric acid and to a lesser extent, cutaric acid, oxidizing them into their respective o-quinones [161, 162, 163, 164, 165]. The *p*-caftaric acid, for example, after being oxidized to o-quinone-caffeoyl tartaric acid becomes a potent and highly reactive oxidant able to oxidize spontaneously other components of the grape that are not substrates for PPO in a reaction which is called coupled oxidation. Metals, amino acids and proteins when interact with quinone become unavailable and therefore the nutritional quality of the food is oxidized impaired. Although the polymerized compounds resulting from these reactions are colorless, the increase in the degree of condensation of these compounds leads to the formation of polymeric dyes with colors between yellow and brown (melanins) (Figure 19).

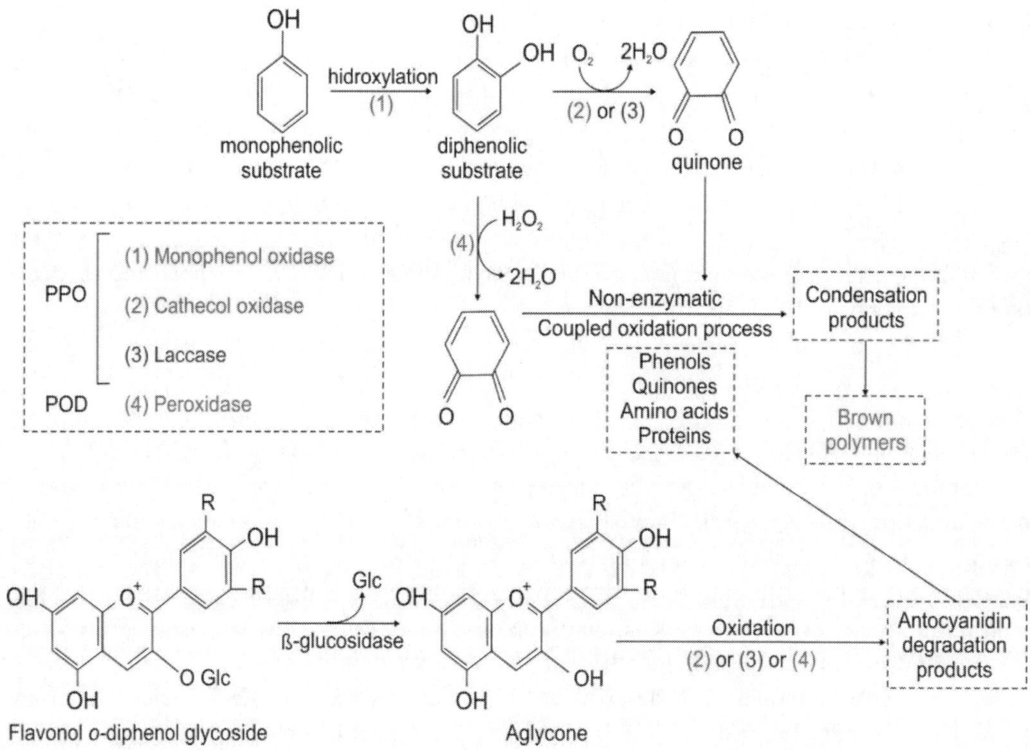

Figure 19. Mechanism of enzymatic oxidation of grape flavonoids (adapted by [166]).

The PPOs existing in grapes also have affinity, but to a lesser extent, to the monomeric flavan-3-ols, catechin, epicatechin [167, 168]. Among the substrates that are not directly oxidized by the action of PPO, but are oxidized by coupled oxidation mechanisms, there are reducing agents such as ascorbic acid and sulphites, *o*-diphenolic compounds, epicatechin gallate, procyanidins and *o*-diphenolic anthocyanins [168].

The flavonol *o*-diphenols glycosides (anthocyanins and flavonols) and procyanidins in particular, can be directly oxidized by the peroxidase or laccase [166]. Furthermore, *o*-diphenols glycosylated may be hydrolyzed by β-glycosidases (EC 3.2.1.21) to their respective aglycones, which are potential substrates for other enzymes such as catechol oxidase, peroxidase and/or laccase, accelerating the process of degradation of compounds (Figure 19) [169, 170].

Laccase

I) Structure and Function

Laccases (benzenediol: oxygen oxidoreductase; EC 1.10.3.2), like cathecol oxidase and monophenol oxidases, also belong to the group of multi-copper proteins oxidases. Usual other names of this enzyme are urishiol oxidase, urushiol oxidase and *p*-diphenol oxidase. Laccases are dimeric or tetrameric glycoproteins and their four Cu atoms are distributed at the three different copper centers, namely: Type-1 or blue copper center, Type-2 or normal copper and Type-3 or coupled binuclear copper center, differing in their characteristic electronic paramagnetic resonance (EPR) signals (Figure 20) [176, 177, 178].

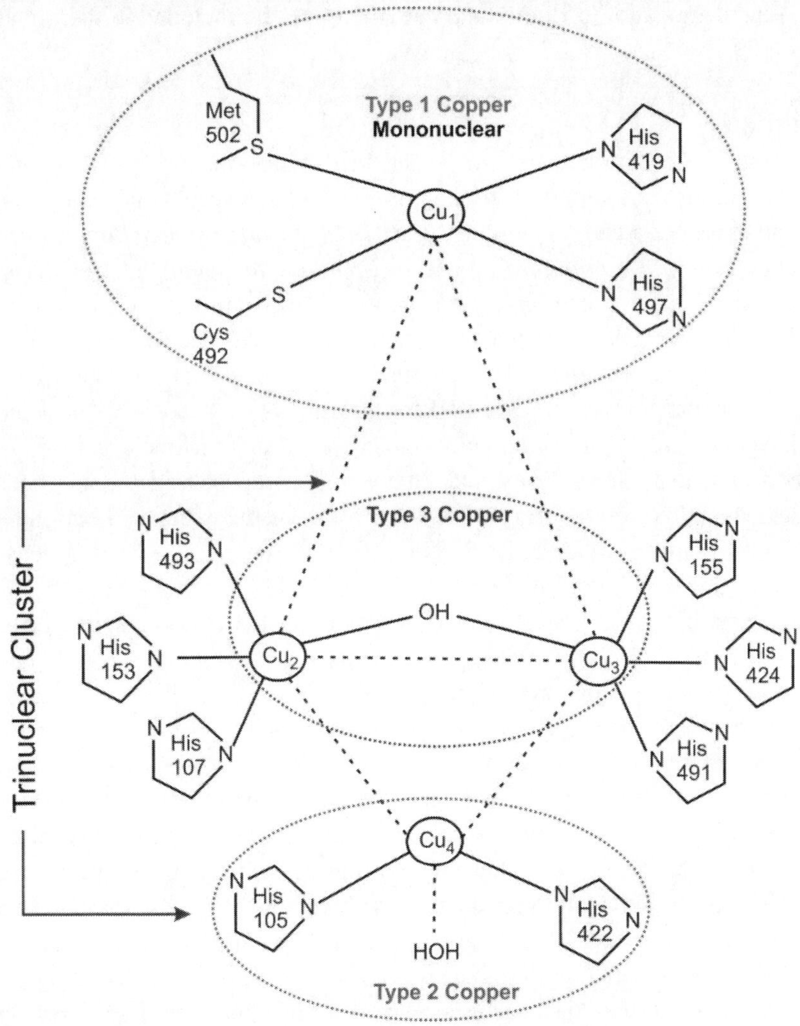

Figure 20. Schematic representation of copper centers of laccase from *B. subtilis* [176, 177, 178].

The functions exhibited by laccases can be divided into three categories: (a) cross-linking of monomers (synthesis of lignin in plants), (b) degradation of polymers and (c) ring cleavage of aromatic compounds. Thus, in plants, the functions carried out by laccases include lignification, wound healing and iron oxidation. In grapes and wine these enzymes are mostly derived from fungal contaminants, especially *Botrytis cinerea* responsible for noble rot (which may be beneficial to wine) and gray mold (bad for grapes, see more below).

Laccase is a multi-copper proteins oxidase. In plants, it acts on *o*- and *m*-diphenols and similar molecules by consuming O_2 and performing one-electron oxidations, which remain poorly defined. It is proposed that laccases play a role in the formation of lignin by promoting the oxidative coupling of monolignols, a family of naturally occurring phenols [142]. According to Dwivedi et al. [178], the catalytic cycle begins with Type-1 Cu reduction by reducing substrate, then there is an internal electron transfer from Type-1 Cu to Type-2 and Type-3 Cu tri-nuclear cluster, which is coupled to the reduction of oxygen (to water) at Type-2 and Type-3 Cu. The best known mechanisms of laccases are related to fungal ones, which play a role in the degradation of lignin and can, therefore, be included in the broad category of ligninases [179].

II) Distribution

Laccase was first identified in the Japanese lacquer tree, *Rhus vernicifera* [180]. Subsequently, laccases were reported from a variety of plants such as lacquer, mango, mung bean, peach, pine, prune, sycamore, loblolly pine (*Pinus taeda*), poplar (*Populus trichocarpa*) yellow-poplar (*Liriodendron tulipifera*), tobacco (*Nicotiana tabacum*) and Zea mays [181].

These enzymes are present in grape and wine, but they are mostly from contaminating fungi, especially *Botrytis cinerea* responsible for the gray mold (bad for grapes) and noble rot (which may be beneficial to wine as it is an enhancer of wine quality). The gray rot occurs when the still immature grapes are affected by the fungus. It becomes a pest that spreads quickly, disintegrates grapes, invading cells skins that rot and become a gray goo. Infection occurs, especially with high humidity and little aeration (dense vegetation). For the same fungus to cause the noble rot, the grapes must reach full maturity before being infected. The mature sugar rich grapes provide a food source for the fungus, the berries dehydrate and concentrate the sugars which become rich in flavors typical of botrytis resembling honey. The noble rot is favored by hot, dry sunny days with morning fog, for example when there is a lake nearby. The fight against gray mold is usually made with fungicides. But it is tricky since both fermentative yeasts as fungi are eukaryotes, killing one also kills the other. Therefore, in humid regions, the use of specially selected stronger cultured yeast, may be necessary. The production of laccase by the *B. cinerea* fungi oxidizes phenolic compounds in the grape juice, such as caftaric acid to its corresponding quinone, which can form brown polymeric products.

Other oxidation enzymes, like tyrosinases and cathecol oxidases (PPO) are also produced in grapes (by the plants themselves), but the substrate specificity of laccase is wider which makes it less inhibited by oxidation products than the others enzymes. Laccase is also more resistant to sulfur dioxide than PPO. The number of acetic acid bacteria can also increase drastically on *Botrytis* infected grapes, because of the rupture of the grape berry skin due to the infection, which allows the bacteria entry to the juice of the berry. Normally, *Gluconobacter* species occur in grapes but in rotten grapes *Acetobacter* species (which prefer ethanol as a carbon source) can also dominate, probably due to the production of small concentrations of ethanol by wild yeasts. Acetic acid bacteria are known to produce high levels of SO_2 binding substances from glucose and fructose, like gluconic and 2-diketogluconic acid among others. Lactic acid bacterial numbers can also increase in rotten grapes. This, together with the potential oxidation problem, can pose a serious problem for the winemaker [182].

III) Substrates

Laccases are a large group of oxidative enzymes that exhibit great diversity with regard to catalytic activity. They display low specificity acting on both *o*- and *p*-quinols, and often acting also on aminophenols and phenylenediamine, in the presence of molecular oxygen. The semiquinone may react further either enzymatically or non-enzymatically [183, 184]. Some studies have shown that the *o*-substituted compounds (guaiacol, pyrocatechol, caffeic acid, gallic acid, etc.) are better substrates for this enzyme than the *p*-substituted ones (hydroquinone and *p*-cresol) [185].

In grapes, there are different substrates for the action of laccase. The preferred for the action of laccase are the non-flavonoid compounds, such as caffeic acid, *p*-coumaric acid and esters of tartaric acid. To a lesser extent, it acts on the ferulic acid and flavonoid compounds like (+)-catechin and (-)- epicatechin [186]. The presence of the enzyme monophenol oxidase influences the action of laccase, because it transforms monophenolic to diphenolic substrates which can be rapidly oxidized by the laccases [187].

IV) Applications

Musts and wines are complex mixtures of different chemical compounds such as organic acids, tartrates, proteins, tannins and other phenolic compounds. These compounds can interact through cross-linking (mainly proteins and polyphenols) producing haze. These interactive products may, sooner or later, become so big that they reach colloidal size. When the protein-polyphenols reaction goes further, haze becomes visible in all conditions (permanent haze). Dead yeast cells, bacteria, cell wall debris, tartrates and other compounds can also contribute to insoluble matter suspended in the wine. Continued reactions may produce larger molecules and larger particles that will precipitate to the bottom of the storage vessel.

The soluble wine can then be siphoned from the compact solids into a new container [188]. But this process may take many months, or even years in order to produce a perfectly clear wine. Producers can accelerate the process by using clarification and stabilization, which are processes by which insoluble matter suspended in the wine is removed before bottling [189].

Clarification and stabilization may involve fining agents, filtration, centrifugation, flotation, refrigeration and physical chemical adsorbents, such as bentonite. The use of laccase as an alternative to the traditional treatment has been tested by different authors [188, 189]. Laccases catalyze the oxidation of polyphenols increasing earlier flocculation, and then preventing the phenomenon occurring in the bottled wine. This might be especially interesting in the case of white wines. However, this use of laccase in wine should be considered case by case since some winemakers deliberately leave more tartrates and phenolics in wines designed for long aging in the bottle so that they are able to develop the aromatic compounds that constitute bouquet [190, 191, 192]. There is another advantage of use of laccase in winemaking.

During storage, reaction of oxydation of various off-flavor precursors may occur due presence of oxygen, resulting in off-flavor compounds. Laccases could be added at the end of the process in order to remove the unwanted oxygen in the finished wine and, thereby, the storage life of wine can be enhanced [192]. Microbial laccases have also been applied to the degradation of complex natural polymers, such as lignin or humic acids [193]. The reactive radicals generated lead to the cleavage of covalent bonds and to the release of monomers. It was also found that if compounds such as 2,2-azinobis (3-ethylbenzothiazoline-6-sulfonic acid) (ABTS) were present, they act as a mediator and allow laccase to oxidize and cleave lignin substrates [194, 195].

Peroxidases

I) Structure and Function

Heme-containing peroxidases (donor: H_2O_2 oxidoreductases, EC 1.11.x) are versatile catalysts oxidizing diverse substrates. In plants, the most well-known enzyme is horseradish peroxidase or guaiacol peroxidase (EC. 1.11.1.7). Peroxidases belong to the group of heme proteins oxidases with histidine as the proximal ligand, usually involved as redox catalysts in a range of biological processes, such as electron transfer or oxidation of various organic substrates by hydrogen peroxide.

Plant peroxidases, e.g. horseradish peroxidase, and fungal peroxidases, e.g. yeast cytochrome c peroxidase, consist of about 300 amino acids and non-covalently bound heme, whereas mammalian peroxidases are much larger (576–738 amino acids) and the heme is covalently bound. On the basis of amino acid sequence homologies, three families of plant peroxidases have been proposed: class I – intracellular, organelle associated and bacterial peroxidases (e.g. cytochrome cperoxidase), class II – secretory fungal peroxidases including ligninolytic peroxidases [196]. This class is exclusively found in wood and leaf-litter decomposing *Basidiomycotina* (*mostlyagaricomycetes*) and comprise manganese, lignin and versatile per-oxidases (MnP, LiP and VP). They possess high redox potentials, which enable them to oxidize a large number of recalcitrant substrates including methoxylated aromatics present in woodand other lignocelluloses [197, 198], and class III – secreted plant peroxidases (e.g. horseradishperoxidase, HRP) [121]. Others names for this enzyme: lactoperoxidase; guaiacol peroxidase; plant peroxidase; Japanese radish peroxidase; horseradish peroxidase (HRP); soybean peroxidase (SBP); extensin peroxidase; heme peroxidase; oxyperoxidase; protoheme peroxidase; pyrocatechol peroxidase; scopoletin peroxidase, *Coprinus cinereus* peroxidase, *Arthromyces ramosus* peroxidase.

The POD has the remarkable feature of thermostability. For this reason, and also because of its ease of detection, this enzyme is often used as an index of effectiveness of bleaching fruits and vegetables, to prevent loss of quality during storage. However, if sufficiently high temperatures are not applied to effect the irreversible inactivation of POD, it can regenerate itself and shorten the shelf life of many processed vegetables. The regeneration of peroxidase activity has been reported in grapes [199].

The catalytic reactions of the oxidative enzymes POD and PPO, have been studied in fruits and vegetables for many years. Both enzymes have some common substrates, but each also has its own specific substrates [200, 201, 202, 203]. Their common diphenolic substrates lead to products with brown colors (Figure 21). Moreover, both enzymes have some common inhibitors and some specific inhibitors. Inhibition of POD and PPO activities can reduce the browning process. A binding of ligand and protein may result in the activation or the inhibition of the enzyme [203, 204].

The prevention of off-flavor development in frozen vegetables has been reported after inactivation of the POD enzyme [205]. Furthermore, the related activity of PPO and POD is due to the generation of hydrogen peroxide during the oxidation of phenolic compounds in PPO-catalyzed reactions [138, 206, 207].

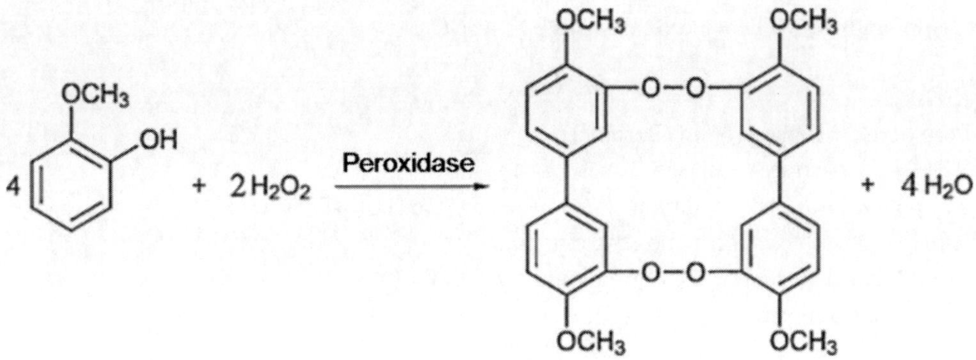

Figure 21. The assay showing peroxidase/H_2O_2 oxidation of guaiacol to the chromophore tetraguaiacol.

II) Mechanism

The catalytic cycle of heme peroxidases (Figure 22) is initiated by a rapid two electron transfer from the native POD ferric heme protein (PorFeIII) to H_2O_2, to give a higher oxidation state of the enzyme called compound I (•+PorFeIV= O), Equation (1).

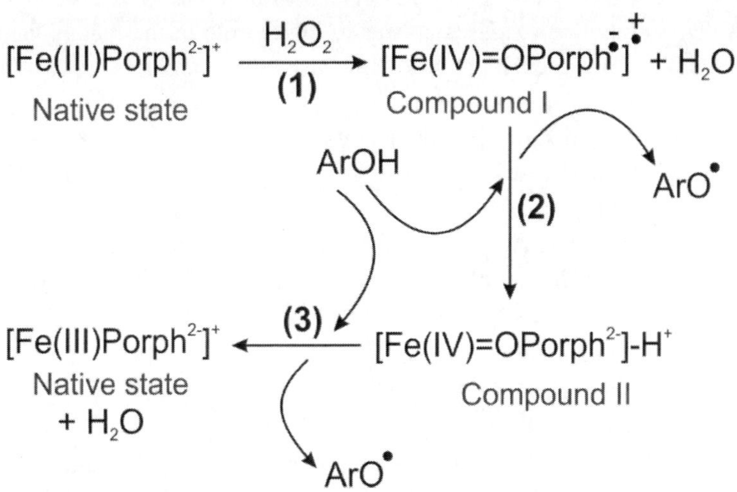

Figure 22. Schematic representation of oxidative cycle of peroxidase.

It is a ferryl π-cation radical with a formal oxidation state for the iron (FeV) and oxygen coupled to this iron by a double bond (Equation 1). In the peroxidative cycle, compound I is reduced by two consecutive one-electron reductions via compound II (PorFeIV= O) (Equation 2) to the native ferric enzyme (Equation 3). Independence of the redox potentials of compound I and II, numerous substrates can serve as electron donors in these reactions, e.g. phenolic and non-phenolic aromatics, metal ions or complex dye-stuff molecules [200].

Equations of the reactions:

$$\text{Peroxidase} + \text{ROOH} = \text{compound I} + \text{ROH} \tag{1}$$

$$\text{Compound I} + \text{XOH} = \text{Compound II} + \text{XO}^{\bullet} \tag{2}$$

$$\text{Compound II} + \text{XOH} = \text{Peroxidase} + \text{XO}^{\bullet} + H_2O \tag{3}$$

Key:
Peroxidase = heme protein (PorFeIII)
ROOH = peroxide, i.e., H_2O_2
compound I = higher oxidation state of the enzyme ($\text{Por}^{\bullet}$+FeIV = O)
XOH = substrates electron donors, e.g. guaiacol
Compound II = intermediary compound (PorFeIV= O)
$\text{XO}^{\bullet}$ = oxidized substrate

PPO activity can be monitored by oxygen consumption or spectrophotometrically using a variety of substrates such as pyrogallol, pyrocatechol, 4-methylcatechol, 3,4-dihydroxyphenylacetic acid, 4-tert-butylcatechol and chlorogenic acid [200].

III) Distribution

POD can be found in plants, animals and microbes. Currently, more than 40 peroxidase genes have been identified in Arabidopsis plants, which have five chromosomes and the smallest genome size (~100 Mb) of higher plants with few and short introns and little repetitive DNA. These plant peroxidases appear to play a role in the metabolism of auxin, a hormone that signals growth:development. They also play a role in extracellular defense against pathogens and stress, biosynthesis and degradation of lignin, intracellular removal of hydrogen peroxide and oxidation of toxic reductants.

Plant peroxidases are induced by stress, e.g. pathogen attack, wounding, heat, cold or draught, UV light. Plant:fungal peroxidases likely arose by gene duplication from an ancestral gene that encoded five helices, a His heme ligand, an Asp or Asn ligand hydrogen-bonded to His, and a Ca^{+2} binding site. Mammalian peroxidases likely arose from a different ancestral gene from plant peroxidases and play a role in thyroid hormone synthesis or defense against pathogens. Reviews of plant and heme peroxidases have recently appeared [196].

IV) Substrates

They also peroxidize non-phenolic substrates such as 3,3',5,5'-tetramethylbenzidine (TMB) and 2,2'-azinobis(3-ethylbenzthiazoline-6-sulfonic acid) (ABTS). Certain peroxidases (e.g. lactoperoxidase, SBP) oxidize bromide, iodide and thiocyanate.

The Chemical and Biochemical Characteristics and the Technological Quality of Grapes

Grape juices and wines are considered important sources of phenolic compounds for human diet and present a much more complex phenolic composition than that found in the grapes used for their elaboration. The observed qualitative and quantitative differences arise from a number of factors which include the grape species, grape variety, location of plantation, system of culture, climate, soil, extraction method of phenolic compounds and the kind of process used, as well as from the influence of oxidative processes of chemical and enzymatic origin, which initiate with the grape crushing and occur throughout the process of fermentation, maturation and aging of wines.

It is well known that, at room temperature and under normal conditions of pressure, the oxygen of the must and wine is dissolved during the main operations of winemaking, such as pumping, racking, filtration or bottling. Among wine components, phenolic compounds are the most promptly oxidable. During these oxidative processes, the content of some original phenolic constituents from grapes can gradually reduce due to the formation of polymeric compounds during the evolution of the color in juices and, mainly, in wines [208, 209, 210].The knowledge of these transformations is of undoubted importance for the control and improvement of quality in products derived from fruits.

Reactions of Enzymatic Origin

The enzymatic oxidation of phenolic compounds from red and white grapes occurs mainly during the initial steps of winemaking due to the natural presence of enzymes from the oxidoreductase class, previously discussed. Because of the oxidative reactions catalyzed by these enzymes, the oxygen consumption is very fast and they are largely correlated to the contents of caffeoyl tartaric (caftaric) and *p*-coumaroyl tartaric (coutaric) acids [211].

The oxidative reactions of enzymatic origin promote great alterations in the color tonality and intensity, depending on the oxidized compound and the type of reaction [208, 212].

In white grape musts, for example, the main phenolic compounds are *trans*-caftaric, *trans*-coutaric and *cis*-coutaric acids [213]. These compounds have been identified as the precursors for browning in white wines [214]. In the presence of oxygen, and especially, polyphenol oxidase, the compounds are quickly oxidized to *o*-quinones. Primary and secondary *o*-quinones can act like electrophiles and react to nucleophilic substances, with the formation of the 2-*S*-glutathionyl-caftaric acid as the key example, also called Grape Reaction Product or GRP, resulting from the binding between glutathione and the *o*-quinone of caftaric acid (Figure 23) [209, 214, 215].

It should be highlighted that the GRP is not oxidized by a PPO enzyme, which limits in part the oxidative browning [215, 216, 217]. Nonetheless, it can undergo an additional oxidation to GRP *o*-quinone if the laccase from *Botrytis cinerea* is present in grapes attacked by the fungus [218]. These compounds can continue reacting and result in the formation of 2,5-di-S-glutathionyl caftaric acid, in the presence of an excess of glutathione [219]. Considering that laccase persists in such wines, the GRP oxidation can be responsible for the browning observed in wines elaborated with botrytized grapes.

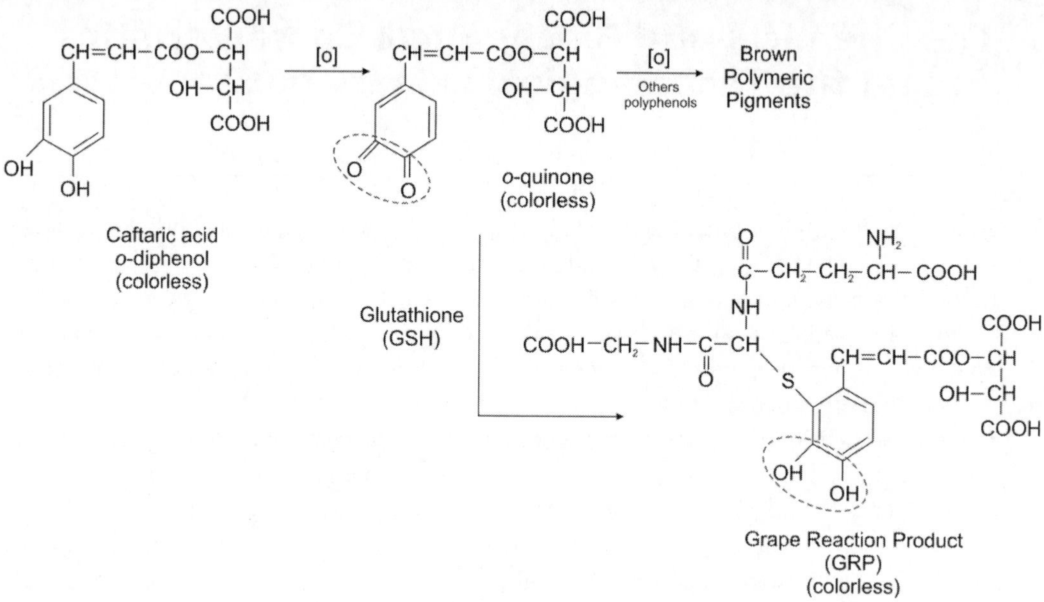

Figure 23. Scheme of enzymatic oxidation of caftaric acid in wines.

Sulfiting agents as SO_2 or sodium, potassium and calcium bisulphites, as well as inhibiting the enzymatic and non-enzymatic browning reactions during grape processing and storage of derived products, can also act in the inhibition of spoilage caused by detrimental bacteria, fungi and yeasts [220]. It is accepted that laccase is more resistant to the SO_2 action when compared to catechol oxidase, which is easily inactivated with a small quantity of sulfiting agent [221]. After the treatment of the must with SO_2, fermentation development and consequential ethanol production, the PPO activity becomes nearly null in wines [161, 221, 222]. More recently, the toxicity and allergenic effects of this additive [223, 224], along with the increase in knowledge about the oxidation mechanisms of wines, gave rise to new biotechnological approaches and production trends, that led to a significant reduction of sulfites in wine production. The use of natural antioxidants, such as glutathione (GSH) and ascorbic acid, to supplement the sulfite action has been a well employed alternative [210].

According to Cheynier; Souquet; Moutounet [225], the GSH concentrations in grapes and musts can vary from 56-372 μmol/kg (17-114 mg/kg) and 46-333 μM (14-102 mg/L), respectively. In musts with an overload of glutathione, that is an indication of the grape variety, the caftaric acid is totally transformed to GRP, maintaining stability and without browning. Whereas glutathione is available, the GRP production hinders the participation of *o*-quinones in the coupled reactions that lead to the production of brown pigments. This way, it can be concluded that the oxygen consumption in musts is related to the concentration of caftaric and coutaric acids, while the tendency to enzymatic browning is related to the reason among the molar concentration of caftaric acid and glutathione [217, 226].

Similar studies of the kinetics of must oxidation in model solutions containing the same proportions of caftaric acid and glutathione demonstrated that other components of the must can compete with the glutathione in the reaction of production of the caftaric acid *o*-quinone [227]. A lag stage in which the oxygen is consumed while the acid concentration is maintained constant can be attributed to the immediate reduction of quinone by reducing agents like the ascorbic acid [164]. According to Cheynier et al. [225], in musts which contain an excess of glutathione or have been submitted to the addition of ascorbic acid, the levels of flavan-3-ols are maintained during the winemaking process.

Interestingly, the enzymatic oxidation of white wines promotes the browning, but does not necessarily represent a problem for the quality of the final product, considering that a limited exposition to oxygen can result in a reduction of astringency and color stabilization [228]. One of the techniques employed with success for this purpose is the hyperoxygenation of the white grape must. This pre-fermentation technique is characterized by oxygen being added to a non-sulfited must up to saturation [229]. The oxygen addition enhances the enzymatic oxidation of some precursors of phenolic compounds from the must which could produce oxidized and brown polymeric compounds of high molecular weight. After the precipitation of these compounds, which are responsible for the bitterness, astringency and browning during the aging of wine, they can be removed before the alcoholic fermentation due to their high solubilization in ethanol. As a result, the hyperoxygenation method produces lighter white wines with respect to sensorial parameters, less susceptible to oxidative browning and more stable regarding the color when compared to wines produced by conventional techniques [221, 229, 230].

Moreover, the association of hyperoxygenation and the addition of glutathione to the must has resulted in a better sensorial acceptability of wines regarding the typical aroma of the fruit and stability of color, as well as a higher content of phenolic compounds [231].

In red wines, the enzymatic browning causes a limited impact on the color when properly sulfited [10, 232]. Under the sensory viewpoint, the oxidation can be advantageous to red wines because, when controlled, the oxidation reduces the astringency and stabilizes the color [233]. Some of the effects from the O_2 addition to red wines include the decrease of some phenolic compounds like (+)-catechin, (-)-epicatechin, quercetin, caffeic acid and anthocyanins, and the increase of polymeric pigments, improving the density of color. However, although it has been shown that this oxygenation can improve the evolution of red wines during aging, at the same time it reduces the quantity of some mono and olygomeric phenolic compounds that present nutritional benefits [234].

Hence, it can be concluded that oxidative processes of enzymatic origin can lead to the loss of color and nutritional value of grapes and products. From another aspect, the addition of specific enzymatic preparations (exogenous) in key steps of the industrial process can play a fundamental role in the improvement of the quality of the products. This is due to the fact that the grape skins are limiting barriers that prevent the liberation of phenolic compounds to the must. The extraction of these compounds during the crushing of grapes requires the degradation of the middle lamella rich in pectins to liberate the cells and that the cell walls be affected to permit the extraction of the constituents in the cellular vacuoles [235]. Mostly, the action of endogenous hydrolitic enzymes is not enough for the winemaking conditions. Thus, suitable pectinases are used, as well as some cellulases and hemicellulases, with primary and secondary enzymatic activities on the grape skin, with the aim to hydrolyze the polysaccharides from the cell wall and optimize the release of anthocyanins and other compounds of relevance to the must, like tannins [236, 237, 238, 239, 240]. Therefore, if done correctly, the use of enzymatic crushing can considerably influence the velocity of wine production, the maintenance of the chemical stability of certain polyphenolic molecules and, consequently, the sensorial characteristics and the potential protection of health [241].

The international organization with competence in the world viticultural sector, called Organisation Internationale de la Vigne et du Vin (OIV), has established that only enzymes produced by *Aspergillus niger* and *Trichoderma fungi* can be used in wine production [240]. The enzymatic commercial complexes used during the processing of grapes are Pectinex® BE3-L (pectin-esterase, pectin-lyase, polygalacturonase, hemicellulase and cellulase of *Aspergillus niger*), Vinozym® EC (pectinase and cellulase of *Aspergillus niger* and *Trichoderma longibrachiatum*), Vinozym® G (pectin-lyase, polygalacturonase, hemicellulase and cellulase of *Aspergillus niger*) [242], Vinozym® FCE G (polygalacturonase of *Aspergillus aculeatus* and *Aspergillus niger*), Celluclast® 1.5 L FG (cellulase of *Trichoderma reesei*) [243], among others. It is interesting to observe that on the basis of the different compositions and enzymatic activities and the heterogeneity of grapes and the winemaking process [244, 245], commercial enzymatic complexes presenting constrasting results regarding the advantages of the use of exogenous enzymes are found. While some studies show an increase in the content of anthocyanins, tannins [246] and on the density of color [247], other studies demonstrated the contrary effect [236].

Indeed, many enzymatic preparations present enzymes such as cinnamate-esterase, β-glucosidase, tannase and pectin methylesterase which can cause some collateral effects in musts and wines. The cinnamate-esterase liberates the caffeic and *p*-coumaric acids from these respective esters of tartaric acids [248] and their respective acylated anthocyanins [249], which often results in the loss of stability of anthocyanins. The β-glucosidase liberates the aglycons of flavonoids and anthocyanidins of their corresponding glucosides, resulting in turbidity and loss of color [250]. Some fungi such as *Aspergillus niger* and *Botrytis cinerea* present tannin acyl hydrolase (EC 3.1.1.20) that cleaves the galloyl esters of flavan-3-ols [251]. Pectin methylesterase hydrolyzes the pectin present in the grape berry, promoting an increase in the amount of methanol in juices and wines. The toxicity of methanol is low; however, formic aldehyde and formic acid are produced in the metabolic process.

In humans, the symptons of methanol intoxication are conditional to the ingested quantity and vary from headaches, nausea, vomiting up to blindness and death [252]. The liberation of methanol from methylated polygalacturonic acids during wine production, in particular, can be minimized by the use of commercial preparations of pectinases with high activity of pectin lyase and low activity of pectin methylesterase. Studies reveal that the methanol content, as well as the phenolic extraction, depends on the type and the quantity of added enzymes, grape species and time of contact of the enzymes with the skins [253, 254, 255, 256]. Consequently, efforts have been focused on the development of enzymatic preparations that satisfactorily execute their functions and in a safe form, under the specific conditions of the winemaking process [257, 258].

Reactions of Chemical Origin

The non-enzymatic oxidation, also known as chemical oxidation or non-enzymic browning, occurs in wines after the incorporation of oxygen during the steps of racking, by the surface, due to empty spaces in the storage tanks, which depend on the practices of topping up, through the pores of wood of the maturing vessels and during the bottling [259].

According to Waterhouse and Laurie [260], the oxygen does not react directly with phenolic compounds present in wines, in fact the oxidation mechanism occurs through reactive species of oxygen, being this process mediated by the catalytic action of metallic ions, mainly iron and copper [217, 260, 261, 262, 263]. These metallic ions are present in wines due to natural presence of low amounts in grapes or because of their contamination with the soil, as well as the contact with some equipment used for winemaking [264]. Thus, there is the formation of radical oxygen with emphasis on the hydroxyperoxyl radical ($HO_2^{\bullet}$) (Figure 24a) and the hydrogen peroxide (H_2O_2) (Figure 24b). The solubility of the oxygen in the wine will be higher the greater the alcoholic degree and the smaller the temperature [260]. In this context, it is important to emphasize the hydrogen peroxide as a powerful oxidant able to react with ferrous ion (Fe^{2+}) or cuprous ion (Cu^{2+}) (Fenton reaction) (Figure 24d), creating reactive species of oxygen, like the hydroxyl radical ($^{\bullet}OH$), which is extremely powerful to oxidate non-especifically almost all the organic compounds present in wines [265].

It is well established that monomeric anthocyanins, originally found in grapes and considered the main pigment which promotes the color in young wines, are fairly unstable and quickly react to other compounds in the must, as well as with other compounds from yeasts or produced during the storage in wooden barrels, producing new pigments with different chemical structures which alter the coloration and steadiness during the storage and aging of wine [266, 267]. The *o*-diphenols in grapes more susceptible to oxidation process are caffeic acid and esters and its esters, catechins, epicatechins, anthocyanins and their derivatives, and the gallic acid [163, 268, 269]. During winemaking, these *o*-diphenols can be oxidized to *o*-quinones and free radicals of semiquinone can be produced while the oxygen is reduced to hydrogen peroxide [260, 261]. The formation of semiquinones is a key step to the velocity of oxidation. The secondary *o*-quinones generated by coupled oxidation, as well as the primary *o*-quinones formed by enzymatic oxidation (Figure 24c), spontaneously react to the nucleophile molecules of the fermented must producing numerous products of condensation, which in turn can be concerned in new reactions of degradation, that lead to structural modifications of the compounds [166].

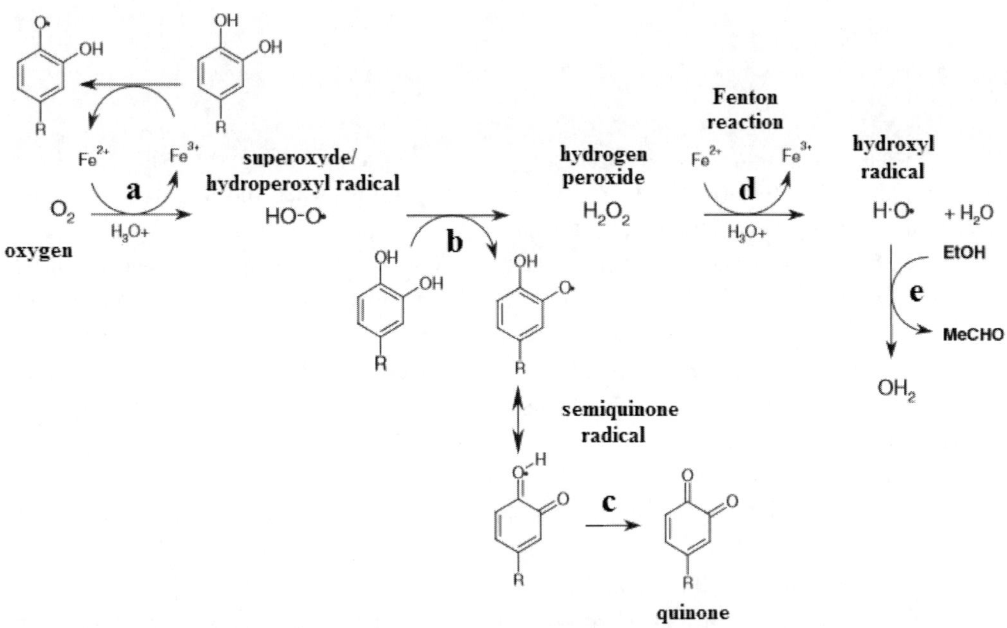

Figure 24. Oxidation mechanisms in wine (Adapted by [260]).

Several chemical reactions arising from the oxidation of these and others phenolic compounds and from the subsequent polymerization of the oxidized products are responsible for significant alterations in the color of wines [74, 208]. The produced dimers and polymers during the oxidation have redox potential smaller than the phenolic compounds that have originated them, which make them more likely to suffer additional oxidation, accelerating the reactions of polymerization of phenolic compounds [208].

In this process, the produced dimers and polymers can suffer new structural rearrangements through condensation reactions, which lead to the formation of brown pigments which are insoluble (melanin) [271]. For example, during the autoxidation of *o*-diphenol catalyzed by enzymes, hydrogen peroxide is produced. In consequence, the generated *o*-diquinone reacts to *o*-diphenol, eventually regenerating a dimer of *o*-diphenol (Figure 25), as well as two or more molecules of tannin can react and produce pigments with high degree of polymerization, which are more unstable and precipitate, being observed in aged wines [272, 273, 274, 275].

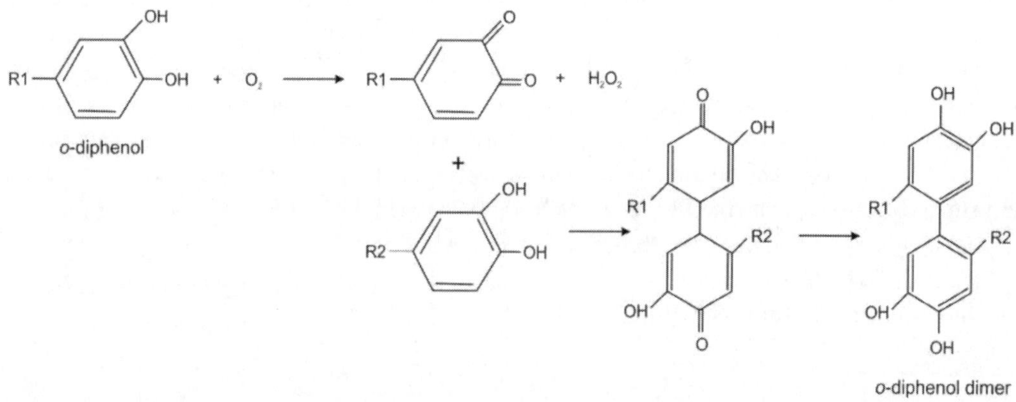

Figure 25. Production of *o*-diphenol polymers through the formation of *o*-diquinone by the autoxidation of an *o*-diphenol.

The formation of polymeric pigments allows explain the loss of the intensity of color as well as the change in the tonality of red wines during aging, since the initial cromophores (monomeric anthocyanins) practically disappear and cannot even form copigmentation complexes [266]. Generally, young red wines present a red staining with a tonality tending towards violet-blue due to the higher content of anthocyanins, while in the oldest ones, a red color with a brown-orange hue is observed, sometimes defined as tile red [276]. Meanwhile, more H_2O_2 is produced and more ethanol is oxidized to acetaldehyde (Figure 24e), at the presence of transition metals, and some phenolic compounds which are not easily oxidizable can suffer oxidation through one of these reactive molecules produced during this process [271].

Beyond ethanol, other organic compounds of the wine, such as tartaric acid, can also be oxidized [262, 277] by $^{\bullet}OH$ radical generating glyoxylic acid [274]. In consequence of these oxidative reactions, the glyoxylic acid can react to flavan-3-ol units to form compounds of yellow color. This acid can also condense to degradation products of anthocyanins (floroglucynol) and catechins, producing golden pigments [278]. Likewise, studies also reported the occurrence of other reactions, such as the direct condensation flavanol-anthocyanin that can lead to the formation of compounds of yellow or orange coloration and the non-enzymatic oxidation of flavan-3-ols, producing brown or colorless compounds. These reactions occur very slowly, but, can negatively influence the color and aroma of wines [228].

Conversely, the acetaldehyde, produced by oxidation or also during the alcoholic fermentation as a secondary metabolite from yeast, can produce a carbocation in acidic conditions which presents ability to participate in condensation reactions among anthocyanins and flavan-3-ols (tannins), which generate polymeric pigments. These pigments are not much stable and can be cleaved in small molecules (dimers, trimers and tetramers) of different molecular weights (as 8-vinylflavan-3-ols) which, in turn, can react to anthocyanins to give origin to pigments of orange color (absorption maximum among 498 - 512 nm in aqueous medium at pH < 2), named pyranoanthocyanins or vitisins (Figure 26) [279].

These pigments are less sensitive to pH changes and to SO_2 bleaching and, therefore, present an important role for the color stabilization of wine [279]. These compounds are also formed by different mechanisms during the processing and storage of wines from precursors of various origins.

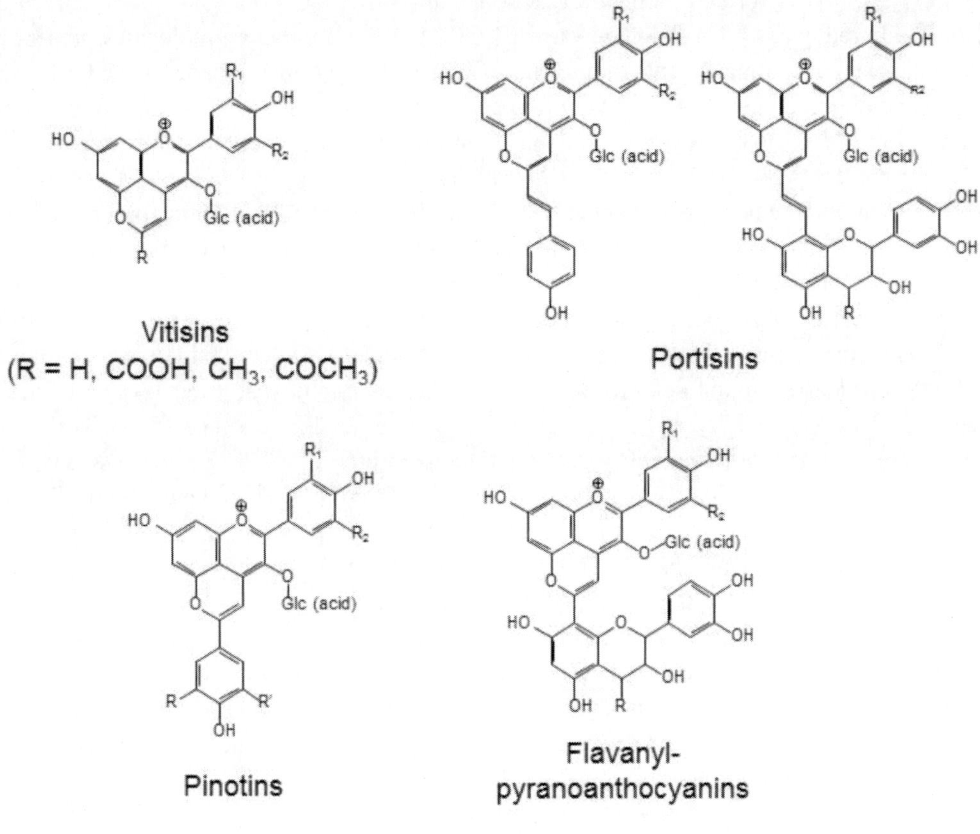

Figure 26. Different families of pyranoanthocyanins found in wines.

According to Hermosín-Gutiérrez [266], for the formation of a pyranoanthocyanin is necessary the reaction among a monomeric anthocyanin and a substance containing a polarized double bond, that is, with substituents of different polarities in both the extremes of the double bond (Figure 27).

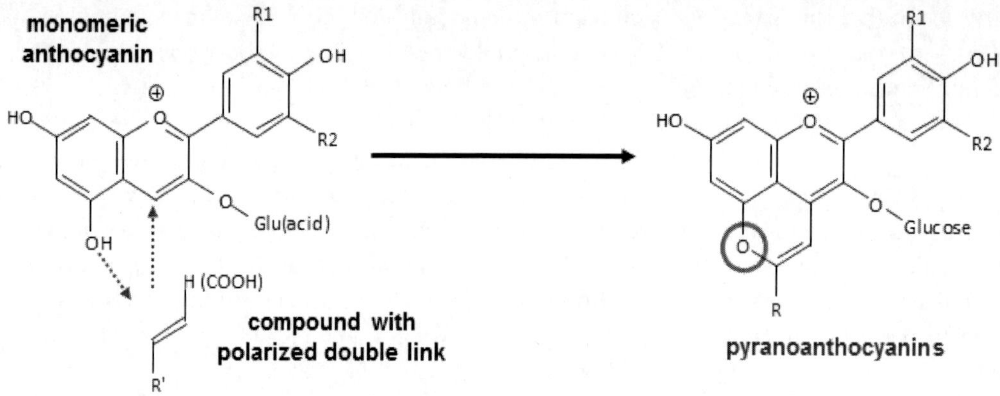

Figure 27. Development of pyranoanthocyanins.

Some examples are the vitisin A, which originates from the reaction between malvidin-3-glucoside and piruvic acid (a secondary metabolite from yeasts) [274] and the compound pinotin A, which is formed from the reaction with caffeic acid and has been isolated and identified in red wines from Pinotage variety [280]. Furthermore, these molecules present low molecular weight (similar to the anthocyanins involved in the reactions responsible for their origin) and do not demonstrate inclination to polymerize, thus being maintained in dissolution in aged wines. Pyranoanthocyanins play an important role for the color stabilization of wine [279].

Phenolic compounds are subjected to several chemical transformations due to reactions of oxidation, condensation, polymerization and complexation to other compounds. Many of the phenolic compounds in these reactions will be gradually transformed in more stable pigments and with different physico-chemical traits of their precursors, contributing to the alteration of the sensory characteristics and color of wines [74]. The oxidative reactions can be firstly influenced by the composition of phenolic compounds and their levels, as well as factors such as wine acidity, SO_2 and ascorbic acid contents, the presence of metals like Fe^{2+} (autoxidation), amount of available oxygen and temperature.

In conclusion, the oxidative reactions are a well defined process in winemaking. Enzymic browning almost entirely occurs in grape must, which has been thorough studied and can be controlled at present; non-enzymic browning prevails in fermented wine. The chemistry underpinning the technique is not clearly understood at this stage, but the process is certainly developing. The effect of oxidative reactions on a wine depends on the kind of wine (variety, origin, etc.), the vintage process and the differences in the phenolic composition between grape varieties used. Currently, the researches seek elucidate the structure of these molecules, their interactions at the molecular level, understand their functions and associate their effects on the sensory and nutritional quality of products.

Acknowledgments

The authors are grateful to CAPES, CNPq and FAPESP (Brazil) for financial support and for scholarships.

References

[1] Boulton R. B. Teoría y práctica de la elaboración del vino. *Viticultura para elaboradores de vino*. Zaragoza, España: Editorial Acribia, 1995. 15-65 pp.

[2] Lecas, M., & Brillouet, J. M. (1994). Cell wall composition of grape berry skins. *Phytochemistry, 35*(5), 1241-1243.

[3] Pinelo, M., Arnous, A., & Meyer, A. S. (2006). Upgrading of grape skins: significance of plant cell-wall structural components and extraction techniques for phenol release. *Trends in Food Science & Technology, 17*(11), 579-590.

[4] Lima, C. P., Cunico, M. M., Miyazaki, C. M. S., Miguel, O. G., Côcco, L. C., Yamamoto, C. I., & Miguel, M. D. (2012). Polyphenolic content and antioxidant capacity of fruits of Juçara (*Euterpe edulis Martius*) palm tree. *Revista Brasileira de Plantas Medicinais, 14*(2), 321-326.

[5] Kühnau, J. (1975). The flavonoids. A class of semi-essential food components: their role in human nutrition. *World Review of Nutrition and Dietetics, 24*, 117-191.

[6] Leite, J. P. V. *Fitoterapia: bases científicas e tecnológicas*. São Paulo: Atheneu. 328 p.

[7] Fumagali, E., Gonçalves, R. A. C., Machado, M. F. P. S., Vidoti, G. J., & Oliveira, A. J. B. D. (2008). Produção de metabólitos secundários em cultura de células e tecidos de plantas: O exemplo dos gêneros Tabernaemontana e Aspidosperma. *Revista Brasileira de Farmacogia, 18*, 627-641.

[8] Silva, M. L. C, Costa R.S, Santana, A. S., & Koblitz, M. G. B. (2010). Phenolic compounds, carotenoids and antioxidant activity in plant products. *Revista Semina: Ciências Agrárias, 31*(3), 669-682.

[9] Jackson R. S. *Wine science: principles and applications*. Academic P. San Diego; 1994. 475p.

[10] Cheynier, V., & Fulcrand, H. Oxidación de los polifenoles en los mostos y los vinos. In: Flanzy, C. (Coord.). *Enología: fundamentos científicos y tecnológicos*. Madrid: AMV ediciones, 2000. p. 369-376.

[11] Garrido, J., & Borges, F. (2013). Wine and grape polyphenols – A chemical perspective. *Food Research International, 54*(2), *1844-1858*.

[12] Andersen, O. M., & Markham, K. R. (Eds.). *Flavonoids: chemistry, biochemistry and applications*. Boca Raton, Florida, USA: CRC Press, 2010.

[13] Robards, K., & Antolovich, M. (1997). Analytical Chemistry of Fruit Bioflavonoids. A Review. *Analyst, 122*(2), 11R-34R.

[14] Brouillard, R. Chemical structure of anthocyanins. In: Markakis P, editor. *Anthocyanins as food colors.* New York: Academic P, 1982.

[15] Lago-Vanzela, E. S., Da-Silva, R., Gomes, E., García-Romero, E., & Hermosín-Gutiérrez, I. (2011). Phenolic composition of the edible parts (flesh and skin) of Bordô grape (*Vitis labrusca*) using HPLC-DAD-ESI-MS/MS. *Journal of Agricultural and Food Chemistry, 59*(24), 13136-13146.

[16] Rebello, L. P. G., Lago-Vanzela, E. S., Barcia, M. T., Ramos, A. M., Stringheta, P. C., Da-Silva, R., Castillo-Munhõz, N., Gómez-Alonso, S., & Hermosín-Gutiérrez, I. (2013). Phenolic composition of the berry parts of hybrid grape cultivar BRS Violeta (BRS Rubea × IAC 1398-21) using HPLC–DAD–ESI-MS/MS. *Food Research International, 54*(1), 354-366.

[17] Castillo-Muñoz, N., Fernández-González, M., Gómez-Alonso, S., García-Romero, E., & Hermosín-Gutiérrez, I. (2009). Red-color related phenolic composition of Garnacha Tintorera (*Vitis vinifera* L.) grapes and red wines. *Journal of Agricultural and Food Chemistry, 57* (17), 7883-7891.

[18] Lago-Vanzela, E. S., Da-Silva, R., Gomes, E., García-Romero, E., & Hermosín-Gutiérrez, I. (2011). Phenolic composition of the Brazilian seedless table grape varieties BRS Clara and BRS Morena. *Journal of Agricultural and Food Chemistry, 59*(15), 8314-8323.

[19] Harborne, J. B., & Williams, C. A. (1995). Anthocyanins and other flavonoids. *Natural Product Reports*, 12, 639 - 657.

[20] Kong, J. M., Chia, L. S., Goh, N. K., Chia, T. F., & Brouillard, R. (2003). Analysis and biological activities of anthocyanins. *Phytochemistry, 64*(5), 923-933.

[21] Wang, H., Race, E. J., & Shrikhande, A. J. (2003). Characterization of anthocyanins in grape juices by ion trap liquid chromatography-mass spectrometry. *Journal of Agricultural and Food Chemistry, 51*(7), 1839-1844.

[22] Zarena, A. S., & Sankar, U. (2012). Isolation and identification of pelargonidin 3-glucoside in mangosteen pericarp. *Food Chemistry, 130*(1), 665-670.

[23] Crespo, P., Bordonaba, J. G., Terry, L. A., & Carlen, C. (2010). Characterisation of major taste and health-related compounds of four strawberry genotypes grown at different Swiss production sites. *Food Chemistry, 122*(1), 16-24.

[24] De Rosso, V. V., Hillebrand, S., Cuevas Montilla, E., Bobbio, F. O., Winterhalter, P., & Mercadante, A. Z. (2008). Determination of anthocyanins from acerola (Malpighia emarginata DC.) and açai (*Euterpe oleracea* Mart.) by HPLC-PDA-MS/MS. *Journal of Food Composition and Analysis, 21*(4), 291-299.

[25] Mazza, G., & Miniati, E. *Anthocyanins in fruits, vegetables, and grains*. Boca Raton, Florida, USA: CRC press, 1993, 362p.

[26] Castillo-Muñoz, N., Winterhalter, P., Weber, F., Gómez, M. V., Gómez-Alonso, S., García-Romero, E., & Hermosín-Gutiérrez, I. (2010). Structure elucidation of peonidin 3, 7-O-β-diglucoside isolated from Garnacha Tintorera (*Vitis vinifera* L.) grapes. *Journal of Agricultural and Food Chemistry, 58*(20), 11105-11111.

[27] Carreño, J., Martínez, A., Almela, L., & Fernández-López, J. A. (1995). Proposal of an index for the objective evaluation of the colour of red table grapes. *Food Research International, 28*(4), 373-377.

[28] Durst, R. W., & Wrolstad, R. E. Separation and characterization of anthocyanins by HPLC. *Current protocols in food analytical chemistry*. Wrolstad, R. E., Ed., New York: John Wiley & Sons, 2001.

[29] Delgado-Vargas, F., Jiménez, A. R., & Paredes-López, O. (2000). Natural pigments: carotenoids, anthocyanins, and betalains—characteristics, biosynthesis, processing, and stability. *Critical Reviews in Food Science and Nutrition, 40*(3), 173-289.

[30] Baublis, A., Spomer, A. R. T., & Berber-Jiménez, M. D. (1994). Anthocyanin pigments: comparison of extract stability. *Journal of Food Science, 59*(6), 1219-1221.

[31] Nixdorf, S. L., & Hermosín-Gutiérrez, I. (2010). Brazilian red wines made from the hybrid grape cultivar Isabel: Phenolic composition and antioxidant capacity. *Analytica Chimica Acta, 659*(1), 208-215.

[32] Moskowitz, A. H., & Hrazdina, G. (1981). Vacuolar contents of fruit subepidermal cells from Vitis species. *Plant Physiology, 68*(3), 686-692.

[33] Delgado-Vargas, F., & Paredes-López, O. *Natural colorants for food and nutraceutical uses*. Boca Raton, Florida, USA: CRC Press, 2002.

[34] Castañeda-Ovando, A., Pacheco-Hernández, M. D. L., Páez-Hernández, M. E., Rodríguez, J. A., & Galán-Vidal, C. A. (2009). Chemical studies of anthocyanins: A review. *Food Chemistry*, *113*(4), 859-871.

[35] Gómez-Plaza, E., Gil-Muñoz, R., Hernández-Jiménez, A., López-Roca, J. M., Ortega-Regules, A., & Martínez-Cutillas, A. (2008). Studies on the anthocyanin profile of *Vitis Vinifera* intraspecific hybrids (Monastrell × Cabernet Sauvignon). *European Food Research and Technology*, *227*(2), 479-484.

[36] Guerrero, R. F., Liazid, A., Palma, M., Puertas, B., González-Barrio, R., Gil-Izquierdo, Á., Garcia-Barroso, C. & Cantos-Villar, E. (2009). Phenolic characterization of red grapes autochthonous to Andalusia. *Food Chemistry*, *112*(4), 949-955.

[37] Hermosín-Gutiérrez, I., & García-Romero, E. (2004). Anthocyanins of red wine grape cultivars grown in the Spanish region of La Mancha: characteristic cultivar patterns of grapes and single cultivar wines, and evolution during the ripening of the berry. *Alimentaria*, *41*, 127-139.

[38] Kıralp, S., & Toppare, L. (2006). Polyphenol content in selected Turkish wines, an alternative method of detection of phenolics. *Process Biochemistry*, *41*(1), 236-239.

[39] Liang, Z., Wu, B., Fan, P., Yang, C., Duan, W., Zheng, X., Liu, C. & Li, S. (2008). Anthocyanin composition and content in grape berry skin in *Vitis germplasm*. *Food Chemistry*, *111*(4), 837-844.

[40] Pomar, F., Novo, M., & Masa, A. (2005). Varietal differences among the anthocyanin profiles of 50 red table grape cultivars studied by high performance liquid chromatography. *Journal of Chromatography A*, *1094*(1), 34-41.

[41] Lago-Vanzela, E. S., Rebello, L. P. G., Ramos, A. M., Stringheta, P. C., Da-Silva, R., García-Romero, E., Gómez-Alonso, S., & Hermosín-Gutiérrez, I. (2013). Chromatic characteristics and color-related phenolic composition of Brazilian young red wines made from the hybrid grape cultivar BRS Violeta ("BRS Rúbea" × "IAC 1398-21"). *Food Research International*, *54*(1), 33-43, 2013.

[42] Hermosin-Gutierrez, I., Sanchez-Palomo Lorenzo, E., & Espinosa, A. V. (2005). Phenolic composition and magnitude of copigmentation in young and shortly aged red wines made from the cultivars, Cabernet Sauvignon, Cencibel, and Syrah. *Food Chemistry*, *92*(2), 269-283.

[43] Hrazdina, G. (1975). Anthocyanin composition of Concord grapes. *Lebensmittel-Wissenschaft & Technologia*, *8*, 111-113.

[44] Arozarena, I., Casp, A., Marín, R., & Navarro, M. (2000). Differentiation of some Spanish wines according to variety and region based on their anthocyanin composition. *European Food Research and Technology*, *212*(1), 108-112.

[45] Berente, B., De la Calle García, D., Reichenbächer, M., & Danzer, K. (2000). Method development for the determination of anthocyanins in red wines by high-performance liquid chromatography and classification of German red wines by means of multivariate statistical methods. *Journal of Chromatography A*, *871*(1), 95-103.

[46] García-Beneytez, E., Revilla, E., & Cabello, F. (2002). Anthocyanin pattern of several red grape cultivars and wines made from them. *European Food Research and Technology*, *215*(1), 32-37.

[47] Castilhos, M. B. M. do, Cattelan, M. G., Conti-Silva, A. C., & Del-Bianchi, V. L. (2013). Influence of two different vinification procedures on the physicochemical and sensory properties of Brazilian non-*Vitis vinifera* red wines. *Food Science and Technology*, *54* (2), 360-366.

[48] Castillo-Muñoz, N., Gómez-Alonso, S., García-Romero, E., & Hermosín-Gutiérrez, I. (2007). Flavonol profiles of *Vitis vinifera* red grapes and their single-cultivar wines. *Journal of Agricultural and Food Chemistry*, *55*(3), 992-1002.

[49] Mattivi, F.; Guzzon, R.; Vrhovsek, U.; Stefanini, M.; Velasco, R. (2006). Metabolite profiling of grape: Flavonols and anthocyanins. *Journal of Agricultural and Food Chemistry*, 54, 7692-7702.

[50] Castillo-Muñoz, N., Gomez-Alonso, S., Garcia-Romero, E., & Hermosin-Gutierrez, I. (2010). Flavonol profiles of *Vitis vinifera* white grape cultivars. *Journal of food composition and analysis*, *23*(7), 699-705.

[51] Hermosín-Gutiérrez, I., Castillo-Muñoz, N., Gómez-Alonso, S., & García-Romero, E. (2011). Flavonol profiles for grape and wine authentication. In *ACS symposium series. American Chemical Society, Washington, DC.*

[52] Rodriguez Montealegre, R., Romero Peces, R., Chacon Vozmediano, J. L., Gascuena, J. M., & Garcia Romero, E. (2006). Phenolic compounds in skins and seeds of ten grape *Vitis vinifera* varieties grown in a warm climate. *Journal of Food Composition and Analysis*, *19*(6-7), 687-693.

[53] Talcott, S. T., & Lee, J. H. (2002). Ellagic acid and flavonoid antioxidant content of muscadine wine and juice. *Journal of Agricultural and Food Chemistry*, *50*(11), 3186-3192.

[54] Pastrana-Bonilla, E., Akoh, C. C., Sellappan, S., & Krewer, G. (2003). Phenolic content and antioxidant capacity of muscadine grapes. *Journal of Agricultural and Food Chemistry*, *51*(18), 5497-5503.

[55] Price, S. F., Breen, P. J., Valladao, M., & Watson, B. T. (1995). Cluster sun exposure and quercetin in Pinot noir grapes and wine. *American Journal of Enology and Viticulture*, *46*(2), 187-194.

[56] Castillo-Muñoz, N., Gómez-Alonso, S., García-Romero, E., Gómez, M. V., Velders, A. H., & Hermosín-Gutiérrez, I. (2009). Flavonol 3-O-glycosides series of *Vitis vinifera* Cv. Petit Verdot red wine grapes. *Journal of Agricultural and Food Chemistry*, *57*(1), 209-219.

[57] Cheynier, V., Moutounet, M., & Sarni-Manchado, P. Los compuestos fenólicos. In Enología. *Fundamentos Científicos y Tecnologicos*, 2nd ed.; Flanzy, C., Ed.; AMV Ediciones and Ediciones MundiPrensa: Madrid, Spain, 2003, p. 120-121.

[58] Jeffery, D. W., Parker, M., & Smith, P. A. (2008). Flavonol composition of Australian red and white wines determined by high-performance liquid chromatography. *Australian Journal of Grape and Wine Research*, *14*(3), 153-161.

[59] Boulton, R. (2001). The copigmentation of anthocyanins and its role in the color of red wine: a critical review. *American Journal of Enology and Viticulture*, *52*(2), 67-87.

[60] Schwarz, M., Picazo-Bacete, J. J., Winterhalter, P., & Hermosín-Gutiérrez, I. (2005). Effect of copigments and grape cultivar on the color of red wines fermented after the addition of copigments. *Journal of Agricultural and Food Chemistry*, *53*(21), 8372-8381.

[61] Burda, S., & Oleszek, W. (2001). Antioxidant and antiradical activities of flavonoids. *Journal of Agricultural and Food Chemistry*, *49*(6), 2774-2779.

[62] Flamini, R., Mattivi, F., De Rosso, M., Arapitsas, P., & Bavaresco, L. (2013). Advanced knowledge of three important classes of grape phenolics: anthocyanins, stilbenes and flavonols. *International Journal of Molecular Sciences*, *14*, 19651-19669.

[63] Ledda, S., Sanna, G., Manca, G., Franco, M. A., & Porcu, A. (2010). Variability in flavonol content of grapes cultivated in two Mediterranean islands (Sardinia and Corsica). *Journal of Food Composition and Analysis*, *23*(6), 580-585.

[64] Haslam, E. (1980). In vino veritas: oligomeric procyanidins and the ageing of red wines. *Phytochemistry*, *19*(12), 2577-2582.

[65] Ricardo-da-Silva, J. M., Cheynier, V., Souquet, J. M., Moutounet, M., Cabanis, J. C., & Bourzeix, M. (1991). Interaction of grape seed procyanidins with various proteins in relation to wine fining. *Journal of the Science of Food and Agriculture*, *57*(1), 111-125.

[66] Liu, Y., Pan, Q., Yan, G., He, J., & Duan, C. (2010). Changes of flavan-3-ols with different degrees of polymerization in seeds of 'Shiraz', 'Cabernet Sauvignon' and 'Marselan' grapes after veraison. *Molecules*, *15*, 7763-7774.

[67] Lesschaeve, I., & Noble, A. C. (2005). Polyphenols: factors influencing their sensory properties and their effects on food and beverage preferences. *American Society for Clinical Nutrition*, *81*, 330S-335S.

[68] Perez-Trujillo, J. P., Hernandez, Z., Lopez-Bellido, F. J., & Hermosín-Gutierrez, I. (2001). Characteristic phenolic composition of single-cultivar red wines of the Canary Islands (Spain). *Journal of Agricultural and Food Chemistry*, *59*, 6150–6164.

[69] Terrier, N., Poncet-Legrand, C., & Cheynier, V. Flavanols, flavonols and dihydroflavonols. In *Wine chemistry and biochemistry.* New York: Springer, 2009. p. 463-507.

[70] Porter, L. J. Structure and chemical properties of the condensed tannins. In *Plant Polyphenols* New York: Springer, 1992. p. 245-258.

[71] Ribéreau-Gayon, P., Glories, Y., Maujean, A., & Dubourdieu, D. Tratado de Enología: Química del Vino, Estabilización y Tratamientos. v. 2. *Buenos Aires: Hemisfério Sur.*, 2003.

[72] Prieur, C., Rigaud, J., Cheynier, V., & Moutounet, M. (1994). Oligomeric and polymeric procyanidins from grape seeds. *Phytochemistry*, *36*(3), 781-784.

[73] Souquet, J. M., Cheynier, V., Brossaud, F., & Moutounet, M. (1996). Polymeric proanthocyanidins from grape skins. *Phytochemistry*, *43*(2), 509-512.

[74] Monagas, M., Bartolomé, B., & Gómez-Cordovés, C. (2005). Updated knowledge about the presence of phenolic compounds in wine. *Critical Reviews in Food Science and Nutrition*, *45*(2), 85-118.

[75] Bate-Smith, E. C. (1973). Haemanalysis of tannins: the concept of relative astringency. *Phytochemistry*, *12*(4), 907-912.

[76] Haslam, E., Lilley, T. H., & Butler, L. G. (1988). Natural astringency in foodstuffs - a molecular interpretation. *Critical Reviews in Food Science & Nutrition*, *27*(1), 1-40.

[77] McRae, J. M., & Kennedy, J. A. (2011). Wine and grape tannin interactions with salivary proteins and their impact on astringency: A review of current research. *Molecules*, *16*, 2348-2364.

[78] Dallas, C., Ricardo-da-Silva, J. M., & Laureano, O. (1996). Interactions of oligomeric procyanidins in model wine solutions containing malvidin-3-glucoside and acetaldehyde. *Journal of the Science of Food and Agriculture*, *70*(4), 493-500.

[79] Dallas, C., Ricardo-da-Silva, J. M., & Laureano, O. (1996). Products formed in model wine solutions involving anthocyanins, procyanidin B2, and acetaldehyde. *Journal of Agricultural and Food Chemistry*, *44*(8), 2402-2407.

[80] De Beer, D., Joubert, E., Gelderblom, W. C. A., & Manley, M. (2002). Phenolic compounds: A review of their possible role as *in vivo* antioxidants of wine. *South African Journal of Enology and Viticulture*, *23*(2), 48-61.

[81] Singleton, V. L. In: Hemingway, R. W. (ed.). *Plant polyphenols-synthesis, properties and significance.* Plenum Press, 1992.

[82] Ritchey, J. G., & Waterhouse, A. L. (1999). A standard red wine: monomeric phenolic analysis of commercial Cabernet Sauvignon wines. *American Journal of Enology and Viticulture*, *50*(1), 91-100.

[83] Edlin, D. A. N., Narbad, A., Gasson, M. J., Dickinson, J. R., & Lloyd, D. (1998). Purification and characterization of hydroxycinnamate decarboxylase from *Brettanomyces anomalus*. *Enzyme and Microbial Technology*, *22*(4), 232-239.

[84] Salameh, D., Brandam, C., Medawar, W., Lteif, R., & Strehaiano, P. (2008). Highlight on the problems generated by p-coumaric acid analysis in wine fermentations. *Food Chemistry*, *107*(4), 1661-1667.

[85] Dias, L., Pereira-Da-Silva, S., Tavares, M., Malfeito-Ferreira, M., & Loureiro, V. (2003). Factors affecting the production of 4-ethylphenol by the yeast Dekkera bruxellensis in enological conditions. *Food Microbiology*, *20*(4), 377-384.

[86] Chatonnet, P., Dubourdieu, D., & Boidron, J. N. (1995). The influence of Brettanomyces/Dekkera sp. yeasts and lactic acid bacteria on the ethylphenol content of red wines. *American Journal of Enology and Viticulture*, *46*(4), 463-468.

[87] Rentzsch, M., Schwarz, M., & Winterhalter, P. (2007). Pyranoanthocyanins–an overview on structures, occurrence, and pathways of formation. *Trends in Food Science & Technology*, *18*(10), 526-534.

[88] Rentzsch, M., Schwarz, M., Winterhalter, P., Blanco-Vega, D., & Hermosín-Gutiérrez, I. (2010). Survey on the content of vitisin A and hydroxyphenyl-pyranoanthocyanins in Tempranillo wines. *Food Chemistry*, *119*(4), 1426-1434.

[89] Bravo, A. E., 1996. Resveratrol in wine: contribution to potential cardiovascular protective activity. *Alimentaria*, 269: 71-72.

[90] Cantos, E., Espín, J. C., & Tomás-Barberán, F. A. (2001). Postharvest induction modeling method using UV irradiation pulses for obtaining resveratrol-enriched table grapes: A new "functional" fruit? *Journal of Agricultural and Food Chemistry*, *49*(10), 5052-5058.

[91] González-Barrio, R., Beltrán, D., Cantos, E., Gil, M. I., Espín, J. C., & Tomás-Barberán, F. A. (2006). Comparison of ozone and UV-C treatments on the postharvest stilbenoid monomer, dimer, and trimer induction in var.'Superior'white table grapes. *Journal of Agricultural and Food Chemistry*, *54*(12), 4222-4228.

[92] Romero-Pérez, A. I., Ibern-Gómez, M., Lamuela-Raventós, R. M., & de la Torre-Boronat, M. C. (1999). Piceid, the major resveratrol derivative in grape juices. *Journal of Agricultural and Food Chemistry*, *47*(4), 1533-1536.

[93] Sautter, C. K., Denardin, S., Alves, A. O., Mallmann, C. A., Penna, N. G., & Hecktheuer, L. H. (2005). Determinação de resveratrol em sucos de uva no Brasil. *Ciência e Tecnologia de Alimentos*, *25*(3), 437-442.

[94] Soleas, G. J., Goldberg, D. M., Diamandis, E. P., Karumanchiri, A., Yan, J., & Ng, E. (1995). A Derivatized Gas Chromatographic-Mass Spectrometric Method for the Analysis of Both Isomers of Resveratrol in Juice and Wine. *American Journal of Enology and Viticulture*, *46*(3), 346-352.

[95] Soleas, G. J., Goldberg, D. M., Karumanghiri, A., Diamandis, E. P., & Ng, E. (1995). Influences of viticultural and oenological factors on changes in cis-and trans-resveratrol in commercial wines. *Journal of Wine Research*, *6*(2), 107-121.

[96] Yasui, Y., Nakabayashi, T., Naito, A., Kawaguchi, M., Yunoki, K., Ohnishi, M., & Ito, S. (1997). Changes in concentration of resveratrol during fermentation of musts from grapes grown in Tokachi. *American journal of enology and viticulture*, *48*, 129.

[97] Siemann, E. H., & Creasy, L. L. (1992). Concentration of the phytoalexin resveratrol in wine. *American Journal of Enology and Viticulture*, *43*(1), 49-52.

[98] Moreno-Labanda, J. F., Mallavia, R., Pérez-Fons, L., Lizama, V., Saura, D., & Micol, V. (2004). Determination of piceid and resveratrol in Spanish wines deriving from Monastrell (*Vitis vinifera* L.) grape variety. *Journal of Agricultural and Food Chemistry*, *52*(17), 5396-5403.

[99] Perez-Trujillo, J. P., Hernández, Z., López-Bellido, F. J., & Hermosín-Gutiérrez, I. (2011). Characteristic phenolic composition of single-cultivar red wines of the Canary Islands (Spain). *Journal of Agricultural and Food Chemistry*, *59*, 6150-6164.

[100] Flanzy, C. Enologia: *Fundamentos científicos y tecnológicos*. Madrid: Ediciones Mundi-Prensa, 2000.1º Edición. 783p.

[101] de la Lastra, C. A., & Villegas, I. (2007). Resveratrol as an antioxidant and pro-oxidant agent: mechanisms and clinical implications. *Biochemical Society Transactions*, *35*(Pt 5), 1156-1160.

[102] Püssa, T., Floren, J., Kuldkepp, P., & Raal, A. (2006). Survey of grapevine *Vitis vinifera* stem polyphenols by liquid chromatography-diode array detection-tandem mass spectrometry. *Journal of Agricultural and Food Chemistry*, *54*(20), 7488-7494.

[103] Baderschneider, B., & Winterhalter, P. (2000). Isolation and characterization of novel stilbene derivatives from Riesling wine. *Journal of Agricultural and Food Chemistry*, *48*(7), 2681-2686.

[104] Amira-Guebailia, H., Valls, J., Richard, T., Vitrac, X., Monti, J. P., Delaunay, J. C., & Mérillon, J. M. (2009). Centrifugal partition chromatography followed by HPLC for the isolation of cis-ε-viniferin, a resveratrol dimer newly extracted from a red Algerian wine. *Food Chemistry*, *113*(1), 320-324.

[105] Vitrac, X., Castagnino, C., Waffo-Téguo, P., Delaunay, J. C., Vercauteren, J., Monti, J. P.,Deffieux, G. & Mérillon, J. M. (2001). Polyphenols newly extracted in red wine from southwestern France by centrifugal partition chromatography. *Journal of Agricultural and Food Chemistry*, *49*(12), 5934-5938.

[106] Bavaresco, L., Fregoni, C., Trevisan, M., & Fortunati, P. (2000). Effect of cluster stems on resveratrol content in wine. *Italian Journal of Food Science*, *12*(1), 103-108.

[107] Moreno-Labanda, J. F., Mallavia, R., Pérez-Fons, L., Lizama, V., Saura, D., & Micol, V. (2004). Determination of piceid and resveratrol in Spanish wines deriving from

monastrell (*Vitis vinifera* L.) grape variety. *Journal of Agricultural and Food Chemistry*, *52*(17), 5396-5403.

[108] Mattivi, F., Reniero, F., & Korhammer, S. (1995). Isolation, characterization, and evolution in red wine vinification of resveratrol monomers. *Journal of Agricultural and Food Chemistry*, *43*(7), 1820-1823.

[109] Lamuela-Raventos, R. M., Romero-Perez, A. I., Waterhouse, A. L., & De La Torre-Boronat, M. C. (1995). Direct HPLC analysis of cis-and trans-resveratrol and piceid isomers in Spanish red Vitis vinifera wines. *Journal of Agricultural and Food Chemistry*, *43*(2), 281-283.

[110] Vrhovsek, U., Wendelin, S., & Eder, R. (1997). Effects of various vinification techniques on the concentration of cis-and trans-resveratrol and resveratrol glucoside isomers in wine. *American Journal of Enology and Viticulture*, *48*(2), 214-219.

[111] Adrian, M., Rajaei, H., Jeandet, P., Veneau, J., & Bessis, R. (1998). Resveratrol oxidation in *Botrytis cinerea* conidia. *Phytopathology*, *88*(5), 472-476.

[112] Minussi, R. C., Rossi, M., Bologna, L., Rotilio, D., Pastore, G. M., Durán, N. (2007). Phenols removal in musts: strategy for wine stabilization by laccase. *Jounal of Molecular Catalysis B: Enzymatic*, *45*, 102-107.

[113] Yang, J., Martinson, T. E., & Liu, R. H. (2009). Phytochemical profiles and antioxidant activities of wine grapes. *Food Chemistry*, *116*, 332-339.

[114] Bertelli, A. A., & Das, D. K. (2009). Grapes, wines, resveratrol, and heart health. *Journal of Cardiovascular Pharmacology*, *54*(6), 468-476.

[115] Baur, J. A., & Sinclair, D. A. (2006). Therapeutic potential of resveratrol: the in vivo evidence. *Nature reviews Drug discovery*, *5*(6), 493-506.

[116] Espín, J. C., García-Conesa, M. T., & Tomás-Barberán, F. A. (2007). Nutraceuticals: facts and fiction. *Phytochemistry*, *68*(22), 2986-3008.

[117] Stivala, L. A., Savio, M., Carafoli, F., Perucca, P., Bianchi, L., Maga, G., Forti, L., Pagnoni, U. M., Prosperi, E., & Vannini, V. (2001). Specific structural determinants are responsible for the antioxidant activity and the cell cycle effects of resveratrol. *Journal of Biological Chemistry*, *276*(25), 22586-22594.

[118] Waffo-Téguo, P., Hawthorne, M. E., Cuendet, M., Mérillon, J. M., Kinghorn, A. D., Pezzuto, J. M., & Mehta, R. G. (2001). Potential cancer-chemopreventive activities of wine stilbenoids and flavans extracted from grape (*Vitis vinifera*) cell cultures. *Nutrition and Cancer*, *40*(2), 173-179.

[119] He, S., Yang, J., Wu, B., Pan, Y., Wan, H., Wang, Y., Du, Y., & Wang, S. (2010). Neuroprotective effect of parthenocissin A, a natural antioxidant and free radical scavenger, in focal cerebral ischemia of rats. *Phytotherapy Research*, *24*(S1), S63-S70.

[120] Lima, M. A. C., Alves, R. E., Assis, J. S., Filgueiras, H. A. C., Costa, J. T. A. (2002). Aparência, compostos fenólicos e enzimas oxidativas em uva Itália sob influência do cálcio e do armazenamento refrigerado. *Revista Brasileira de Fruticultura*, *24*(1), 39-43.

[121] Fortea, M. I., López-Miranda, S., Serrano-Martínez, A., Carreño, J., & Núñez-Delicado, E. (2009). Kinetic characterisation and thermal inactivation study of polyphenol oxidase and peroxidase from table grape (Crimson Seedless). *Food Chemistry*, *113*(4), 1008-1014.

[122] Eicken, C. C., Krebs, B. B., & Sacchettini, J. C. J. C. (1999). Cathecol oxidase – structure and activity. *Current Opinion in Structural Biology*, 9(6), 677-683.

[123] Aniszewski, T., Gulewicz, K., & Lieberei, R. (2008). Rresearch on catecholases, laccases and cresolases in plants. recent progress and future needs. *Acta Biologica Cracoviensia*, *50*(1), 7-18.

[124] Nomenclature Committee of the International Union of Biochemistry and Molecular Biology (NCIUBMB). Available: <http://www.chem.qmul.ac.uk/iubmb/enzyme/EC1/cont1bb.html>.

[125] Mayer, A. M. (2006). Polyphenol oxidases in plants and fungi: Going places? A review. *Phytochemistry*, *67*, 2318-2331.

[126] Núñez-Delicado, E., Serrano-Megías, M., Pérez-López, A. J., & López-Nicolás, J. M. (2007). Characterization of polyphenol oxidase from Napoleon grape. *Food Chemistry*, *100*(1), 108-114.

[127] Rapeanu, G., Van Loey, A., Smout, C., & Hendrickx, M. (2006). Biochemical characterization and process stability of polyphenoloxidase extracted from Victoria grape (*Vitis vinifera* ssp Sativa). *Food Chemistry*, *94*(2), 253-261.

[128] Ünal, M. Ü., & Şener, A. (2006). Determination of some biochemical properties of polyphenol oxidase from Emir grape (*Vitis vinifera* L. cv. Emir). *Journal of the Science of Food and Agriculture*, *86*(14), 2374-2379.

[129] Ünal, M., Şener, A., & Şen, K. (2007). Characterization of Sultaniye grape (*Vitis vinifera* L. cv. Sultana) polyphenol oxidase. *International Journal of Food Science & Technology*, *42*(9), 1123-1127.

[130] Weemaes, C. A., Ludikhuyze, L. R., Van den Broeck, I., Hendrickx, M. E., & Tobback, P. P. (1998). Activity, electrophoretic characteristics and heat inactivation of polyphenoloxidases from apples, avocados, grapes, pears and plums. *LWT-Food Science and Technology*, *31*(1), 44-49.

[131] Yilmaz, H., Sakiroglu, H., & Küfrevioglu, I. (2003). Polyphenol oxidase from Mazruma grape (*Vitis vinifera* L.). *Agrochimica*, *47*(1-2), 21-27.

[132] Lee, C. Y., Smith, N. L., & Pennesi, A. P. (1983). Polyphenoloxidase from DeChaunac grapes. *Journal of the Science of Food and Agriculture*, *34*(9), 987-991.

[133] Toivonen, P. M. A.; Brummell, D. A. (2008). Biochemical basesof appearance and texture changes in fresh-cut fruit and vegetables. *Postharvest Biology and Technology*, *48*(1), 1-14.

[134] Pinelo, M.; Arnous, A.; Meyer, A. S. 2006). Upgrading of grape skins: significance of plant cell-wall structural components and extraction techniques for phenol release. *Trends in Food Science & Technology*, *17*, 579-590.

[135] Oren-Shamir, M. (2009). Does anthocyanin degradation play a significant role in determining pigment concentration in plants? *Plant Science*, *177*(4), 310-316.

[136] Born, K., Comba, P., Daubinet, A., Fuchs, A., & Wadepohl, H. (2007). Catecholase activity of dicopper (II)-bispidine complexes: stabilities and structures of intermediates, kinetics and reaction mechanism. *JBIC Journal of Biological Inorganic Chemistry*, *12*(1), 36-48.

[137] Mukherjee, J., & Mukherjee, R. (2002). Catecholase activity of dinuclear copper (II) complexes with variable endogenous and exogenous bridge. *Inorganica Chimica Acta*, *337*, 429-438.

[138] Nokthai, P., Lee, V. S., & Shank, L. (2010). Molecular modeling of peroxidase and polyphenol oxidase: substrate specificity and active site comparison. *International Journal of Molecular Sciences*, *11*(9), 3266-3276.

[139] Virador, V. M., Reyes Grajeda, J. P., Blanco-Labra, A., Mendiola-Olaya, E., Smith, G. M., Moreno, A., & Whitaker, J. R. (2009). Cloning, sequencing, purification, and crystal structure of Grenache (*Vitis vinifera*) polyphenol oxidase. *Journal of Agricultural and Food Chemistry*, *58*(2), 1189-1201.

[140] Harel, E., Mayer, A. M., & Lehman, E. (1973). Multiple forms of *Vitis vinifera* catechol oxidase. *Phytochemistry*, *12*(11), 2649-2654.

[141] Gerdemann, C., Eicken, C., & Krebs, B. (2002). The crystal structure of catechol oxidase: new insight into the function of type-3 copper proteins. *Accounts of Chemical Research*, *35*(3), 183-191.

[142] Solomon, E. I., Sundaram, U. M., & Machonkin, T. E. (1996). Multicopper oxidases and oxygenases. *Chemical Reviews*, *96*(7), 2563-2606.

[143] Solomon, E. I., Baldwin, M. J., & Lowery, M. D. (1992). Electronic structure of active sites in copper proteins: contributions to reactivity. *Chemical Reviews*, *92*, 521-542.

[144] Dervall, B. J. (1961). Phenolase and pectin enzyme activity in the chocolate spot disease of beans. *Nature*, *189*, 311.

[145] Koval, I. A., Gamez, P., Belle, C., Selmeczi, K., & Reedijk, J. (2006). Synthetic models of the active site of catechol oxidase: mechanistic studies. *Chemical Society Reviews*, *35*, 814-840.

[146] Cantos, E., Tudela, J. A., Gil, M. I., & Espín, J. C. (2002). Phenolic compounds and related enzymes are not rate-limiting in browning development of fresh-cut potatoes. *Journal of Agricultural and Food Chemistry*, *50*(10), 3015-3023.

[147] Espín, J. C., Morales, M., Varón, R., Tudela, J., & García-Cánovas, F. (1995). A continuous spectrophotometric method for determining the monophenolase and diphenolase activities of apple polyphenol oxidase. *Analytical Biochemistry*, *231*(1), 237-246.

[148] Selinheimo, E., NiEidhin, D., Steffensen, C., Nielsen, J., Lomascolo, A., Halaouli, S.,Record, E., O'Beirne, D., Buchert, J.,& Kruus, K. (2007). Comparison of the characteristics of fungal and plant tyrosinases. *Journal of Biotechnology*, *130*(4), 471-480.

[149] Lago-Vanzela, E. S., Pavezzi, F. C., Martin, N., Gomes, E., & Da Silva, R. (2011). Isolation and characterization of latent and active polyphenoloxidase in BRS Clara (CNPUV 154-147× Centennial seedless) and BRS Morena (Marroo seedless× Centennial seedless) seedless table grapes. *Plant Physiology and Biochemistry*, *49*(11), 1251-1258.

[150] Gandía-Herrero, F., Jiménez-Atiénzar, M., Cabanes, J., García-Carmona, F., & Escribano, J. (2005). Evidence for a common regulation in the activation of a polyphenol oxidase by trypsin and sodium dodecyl sulfate. *Biological Chemistry*, *386*(6), 601-607.

[151] Laveda, F., Núñez-Delicado, E., García-Carmona, F., & Sánchez-Ferrer, A. (2001). Proteolytic activation of latent paraguaya peach PPO. Characterization of monophenolase activity. *Journal of Agricultural and Food Chemistry*, *49*(2), 1003-1008.

[152] Laveda, F., Núñez-Delicado, E., García-Carmona, F., & Sánchez-Ferrer, A. (2000). Reversible sodium dodecyl sulfate activation of latent peach polyphenol oxidase by cyclodextrins. *Archives of Biochemistry and Biophysics*, *379*(1), 1-6.

[153] Jiménez, M., & García-Carmona, F. (1996). The effect of sodium dodecyl sulphate on polyphenol oxidase. *Phytochemistry*, *42*(6), 1503-1509.

[154] Nazari, K., Mahmoudi, A., Shahrooz, M., Khodafarin, R., & Moosavi-Movahedi, A. A. (2005). Suicide-peroxide inactivation of horseradish peroxidase in the presence of Sodium n-Dodecyl Sulphate: A study of the enzyme deactivation kinetics. *Journal of Enzyme Inhibition and Medicinal Chemistry*, *20*(3), 285-292.

[155] Saeidian, S., Keyhani, E., & Keyhani, J. (2007). Effect of ionic detergents, nonionic detergents, and chaotropic agents on polyphenol oxidase activity from dormant saffron (*Crocus sativus* L.) corms. *Journal of Agricultural and Food Chemistry*, *55*(9), 3713-3719.

[156] Selles-Marchart, S., Casado-Vela, J., & Bru-Martinez, R. (2006). Isolation of a latent polyphenol oxidase from loquat fruit (*Eriobotrya japonica* Lind l.): Kinetic characterization and comparison with the active form. *Archives of Biochemistry and Biophysics*, *446*(2), 175-185.

[157] Flurkey, W. H. (1985). In vitro biosynthesis of Vicia faba polyphenoloxidase. *Plant Physiology*, *79*(2), 564-567.

[158] Robinson, S. P., & Dry, I. B. (1992). Broad bean leaf polyphenol oxidase is a 60-kilodalton protein susceptible to proteolytic cleavage. *Plant Physiology*, *99*(1), 317-323.

[159] Vaughn, K. C., Lax, A. R., & Duke, S. O. (1988). Polyphenol oxidase: the chloroplast oxidase with no established function. *Physiologia Plantarum*, *72*(3), 659-665.

[160] Tomás-Barberan, F. A., & Espín, J. C. (2001). Phenolic compounds and related enzymes as determinants of quality in fruits and vegetables. *Journal of the Science of Food and Agriculture*, *81*(9), 853-876.

[161] Cheynier, V., Basire, N., & Rigaud, J. (1989). Mechanism of *trans*-caffeoyltartaric acid and catechin oxidation in model solutions containing grape polyphenoloxidase. *Journal of Agricultural and Food Chemistry*, *37*(4), 1069-1071.

[162] Cheynier, V., Rigaud, J., Souquet, J. M., Duprat, F., & Moutounet, M. (1990). Must browning in relation to the behavior of phenolic compounds during oxidation. *American Journal of Enology and Viticulture*, *41*(4), 346-349.

[163] Oszmianski, J., Cheynier, V., & Moutounet, M. (1996). Iron-catalyzed oxidation of (+)-catechin in model systems. *Journal of Agricultural and Food Chemistry*, *44*(7), 1712-1715.

[164] Rigaud, J., Cheynier, V., Souquet, J. M., & Moutounet, M. (1991). Influence of must composition on phenolic oxidation kinetics. *Journal of the Science of Food and Agriculture*, *57*(1), 55-63.

[165] Gunata, Y. Z., & Moutounet, M. (1988). Activités de la polyphénoloxydase du raisin et de la laccase de *Botrytis cinerea* sur les dérivés hydroxycinnamiques comparées à d'autres substrats phénoliques. *Revue Française D'Oenologie*, *28*(113), 42-47.

[166] Pourcel, L., Routaboul, J. M., Cheynier, V., Lepiniec, L., & Debeaujon, I. (2007). Flavonoid oxidation in plants: from biochemical properties to physiological functions. *Trends in Plant Science*, *12*(1), 29-36.

[167] Harel, E., & Mayer, A. M. (1971). Partial purification and properties of catechol oxidases in grapes. *Phytochemistry*, *10*(1), 17-22.

[168] Cheynier, V., Schneider, R., Salmon, J. M., & Fulcrand, H. (2010). Chemistry of wine. *Comprehensive Natural Products II, Chemistry and Biology*, *3*, 1119-1172.

[169] Barbagallo, R. N., Palmeri, R., Fabiano, S., Rapisarda, P., & Spagna, G. (2007). Characteristic of β-glucosidase from Sicilian blood oranges in relation to anthocyanin degradation. *Enzyme and Microbial Technology*, *41*(5), 570-575.

[170] Oren-Shamir, M. (2009). Does anthocyanin degradation play a significant role in determining pigment concentration in plants? *Plant Science*, 177 (4), 310-316.

[171] Núñez-Delicado, E., Serrano-Megías, M., Pérez-López, A. J., López-Nicolás, J. M. (2005). Polyphenol oxidase from Dominga table grape. *Journal of Agricultural and Food Chemistry*, *27*(15), 6087-6093.

[172] Sánchez-Ferrer, A., Bru, R., Cabanes, J., & García-Carmona, F., (1988). Characterization of catecholase and cresolase activities of Monastrell grape polyphenoloxidase. *Phytochemistry*, *27*, 319-321.

[173] Cash, J.N., Sistrunk, W.A., & Stutte, C.A., (1976). Characteristics of Concord grape polyphenoloxidase involved in juice color loss. *Journal of Food Science, 41*, 1398-1402.

[174] Valero, E., Varón, R., & García-Carmona, F. (1988). Characterization of polyphenoloxidase from Airén grapes. *Journal of Food Science, 53*, 1482-1485.

[175] Okuda, T., Pue, A.G., Fujiyama, K., Yokotsuka, K., (1999). Purification and characterization of polyphenoloxidase from Muscat Bailey A grape juice. *American Journal of Enology and Viticulture, 50*, 137-143.

[176] Enguita, F. J., Martins, L. O., Henriques, A. O., & Carrondo, M. A. (2003). Crystal structure of a bacterial endospore coat component a laccase with enhanced thermostability properties. *Journal of Biological Chemistry*, *278*(21), 19416-19425.

[177] Claus, H. (2004). Laccases: structure, reactions, distribution. *Micron*, *35*(1), 93-96.

[178] Dwivedi, U. N., Singh, P., Pandey, V. P., & Kumar, A. (2011). Structure–function relationship among bacterial, fungal and plant laccases. *Journal of Molecular Catalysis B: Enzymatic*, *68*(2), 117-128.

[179] Cohen, R., Persky, L., & Hadar, Y. (2002). Biotechnological applications and potential of wood-degrading mushrooms of the genus Pleurotus. *Applied Microbiology and Biotechnology*, *58*(5), 582-594.

[180] Bertrand, M.G. (1894). Sur le latex de l'arbre laque. Comptes Rendus de l'Académie des Sciences, *118*, 1215-1218.

[181] Hoopes, J. T., & Dean, J. F. D. (2004). Ferroxidase activity in a laccase-like multicopper oxidase from *Liriodendron tulipifera*. *Plant Physiology and Biochemistry*, *42*, 27-33.

[182] Du Toit, W. J., & Pretorius, I. S. (2002). The occurrence, control and esoteric effect of acetic acid bacteria in winemaking. *Annals of Microbiology*, *52*, 155-179.

[183] Mayer, A. M., & Staples, R. C. (2002). Laccase: new functions for an old enzyme. *Phytochemistry*, *60*(6), 551-565.

[184] Ranocha, P., McDougall, G., Hawkins, S., Sterjiades, R., Borderies, G., Stewart, D., & Goffner, D. (1999). Biochemical characterization, molecular cloning and expression of laccases–a divergent gene family–in poplar. *European Journal of Biochemistry*, *259*(1-2), 485-495.

[185] López-Nicolas, J. M.; García-Carmona, F. Enzymatic and nonenzy degradation of polyphenols. In: Rosa, L. A.; Alvares-Parrilla, E.; González-Aguilar, G. A. (2010). *Fruit and vegetables phytochemistry*. Iowa: Wiley-Blackwell, 101-130.

[186] Ribéreau-Gayon, P., Duboudieu, D., Doné Che, B., Lonvaus, A. *Handbook of enology: the microbiology of wine and vinifications.* Chichester: John Wiley & Sons, v. 1, 2006.

[187] Steel, C. C., Blackman, J., & Schmidtke, L. M. (2013). Grapevine bunch rots: Impacts on wine composition, quality and potential procedures for the removal of wine faults. *Journal of Agricultural and Food Chemistry*, *61* (22), 5189-5206.

[188] Osma, J. F., Toca-Herrera, J. L., & Rodriguez-Couto, S. (2010). Uses of laccases in the food industry. *Enzyme Research*, 918761.

[189] Minussi, R. C., Pastore, G. M., & Duran, N. (2002). Potential applications of laccase in the food industry. *Trends in Food Science and Technology*, *13*(6-7), 205-216.

[190] Robinson (a), J. (ed). *The Oxford Companion to Wine*. Third Edition, Oxford University Press, 2006. pp. 173, 661-662.

[191] Robinson(b), J. (ed). *The Oxford Companion to Wine*. Third Edition, pp. 271-272. Oxford University Press, 2006.

[192] Dhillon, G. S, Kaur, S., Brar, S. K., & Verma, M. (2012). Flocculation and haze removal from crude beer using in-house produced laccase from trametes versicolor cultured on brewer's spent grain. *Journal of Agricultural and Food Chemistry*, *60*, 7895-7904.

[193] Claus, H. (2004). Laccases: structure, reactions, distribution. *Micron*, *35*, 93-96.

[194] Bourbonnais, R., Paice, M. G. (1990). Oxidation of non-phenolic substrates: an expanded role for laccase in lignin biodegradation. *FEBS Letters*, *267*, 99-102.

[195] Dwivedi, U. N., Singh, P., Pandey, V. P., & Kumar, A. (2011). Structure–function relationship among bacterial, fungal and plant laccases. *Journal of Molecular Catalysis B: Enzymatic*, *68*, 117-128.

[196] O'Brien, P. J. (200). Peroxidases. *Chemico-Biological Interactions*, *129*, 113-139.

[197] Calderón, A. A., Zapata, J. M., Muñoz, R., & Ros Barceló, A. (1993). Localization of peroxidase in grapes using nitrocellulose blotting of freezing/thawing fruits. *Hortscience*, *28*(1), 38-40.

[198] Falguera, V., Garza, S., Pagán, J., Garvín, A., & Ibarz, A. (2013). Effect of UV–Vis irradiation on enzymatic activities and physicochemical properties of four grape musts from different varieties. *Food Bioprocess Technology*, *6*, 2223-2229.

[199] Clemente, E. (1998). Purification and thermostability of isoperoxidase from oranges. *Phytochemistry*, 49(1), 29-36.

[200] Banci, L. (1997). Structural properties of peroxidases. *Journal of Biotechnology*, *53*, 253-263.

[201] Onsa, G. H., Saari, N. B., Selamat, J., & Bakar, J. (2000). Latent polyphenol oxidase from Sago Log (*Metroxylon sagu*): Partial purification, activation, and some properties. *Journal of Agricultural and Food Chemistry*, *48*, 5041-5045.

[202] Onsa, G. H., Saari, N. B., Selamat, J., & Bakar, J. (2004). Purification and characterization of membrane-bound peroxidases from *Metroxylon sagu*. *Food Chemistry*, *85*, 365-376.

[203] Furumo, N. C., & Furutani, S. (2008). A simple method for assaying total protein, polyphenol oxidase and peroxidase activity from 'Kaimana' Litchi chinensis Sonn. *Journal of Hawaiian and Pacific Agriculture*, *15*, 1-7.

[204] Altschul, S. F., Madden, T. L., Schaffer, A. A., Zhang, J., Zhang, Z., Miller, W., & Lipman, D. J. (1997). Gapped BLAST and PSI-BLAST: A new generation of protein database search programs. *Nucleic Acids Research*, *25*, 3389-3402.

[205] Yemenicioglu, A., Özkan, M., & Cemeroglu, B. (1999). Some characteristics of polyphenol oxidase and peroxidase from taro (*Colocasia antiquorum*). *Turkish Journal of Agriculture & Forestry*, *23*, 425-430.

[206] Francisco, A. T-B., Juan Carlos, E. (2001). Phenolic compounds and related enzymes as determinants of quality in fruits and vegetables. *Journal of the Science of Food and Agriculture*, *81*, 853-876.

[207] Khan, A. A., & Robinson, D. S. (1994). Hydrogen donor specificity of mango isoperoxidases. *Food Chemistry*, *49*, 407-410.

[208] Li, H., Guo, A., & Wang, H. (2008). Mechanisms of oxidative browning of wine. *Food Chemistry*, *108*(1), 1-13.

[209] Oliveira, C. M., Ferreira, A. C. S., De Freitas, V., & Silva, A. M. S. (2011). Oxidation mechanisms occurring in wines. *Food Research International*, 44, 1115-1126.

[210] Comuzzo, P., & Zironi, R. (2013). Biotechnological strategies for controlling wine oxidation. *Food Engineering Reviews*, *5*, 217-229.

[211] Oszmianski, J., Cheynier, V., & Moutounet, M. (1996). Iron-catalyzed oxidation of (+)-catechin in model systems. *Journal of Agricultural and Food Chemistry*, *44*(7), 1712-1715.

[212] Robards, K., Prenzler, P. D., Tucker, G., Swatsitang, P., & Glover, W. (1999). Phenolic compounds and their role in oxidative processes in fruits. *Food chemistry*, 66(4), 401-436.

[213] Betés-Saura, C., Andrés-Lacueva, C., & Lamuela-Raventós, R. M. (1996). Phenolics in white free run juices and wines from Penedès by high-performance liquid chromatography: changes during vinification. *Journal of Agricultural and Food Chemistry*, *44*(10), 3040-3046.

[214] Singleton, V. L., Zaya, J., Trousdale, E., & Salgues, M. (1984). Caftaric acid in grapes and conversion to a reaction product during processing. *Vitis*, *23*(2), 113-120.

[215] Singleton, V. L., Salgues, M., Zaya, J., & Trousdale, E. (1985). Caftaric acid disappearance and conversion to products of enzymic oxidation in grape must and wine. *American Journal of Enology and Viticulture*, *36*(1), 50-56.

[216] Cheynier, V. F., Trousdale, E. K., Singleton, V. L., Salgues, M. J., & Wylde, R. (1986). Characterization of 2-S-glutathionyl caftaric acid and its hydrolysis in relation to grape wines. *Journal of Agricultural and Food Chemistry*, *34*(2), 217-221.

[217] Du Toit, W., Pretorius, I. S., Marais, J., & Du Toit, M. (2006). Oxygen in must and wine: a review. *South African Journal for Enology and Viticulture*, *27*(1), 76-94.

[218] Salgues, M., Cheynier, V., Gunata, Z., & Wylde, R. (1986). Oxidation of grape juice 2-S-glutathionyl caffeoyl tartaric acid by *Botrytis cinérea* laccase and characterization of a new substance: 2,5-di-S-glutathionyl caffeoyl tartaric acid. *Journal of Food Science*, 51(5), 1191-1194.

[219] Cheynier, V., Van Hulst, M. W. (1988). Oxidation of trans-caftaric acid and 2-S-glutathionyl caftaric acid in model solutions. *Journal of Agricultural Food Chemistry*, *36*, 10-15.

[220] Taylor, S. L., Higley, N. A., & Bush, R. K. (1986). Sulfites in Foods: Uses, Analytical Methods, Residues, Fate, Exposure Assessment, Metabolism, Toxicity, and Hypersensitivity. *Advances in Food Research*, *30*, 1-76.

[221] Ribéreau-Gayon, P. *The chemistry of wine: stabilization and treatments.* Chichester, England: John Wiley and Sons Ltd. v. 2, 2006.

[222] Spagna, G., Barbagallo, R. N., & Pifferi, P. G. (2000). Fining treatments of white wines by means of polymeric adjuvants for their stabilization against browning. *Journal of Agricultural and Food Chemistry*, *48*(10), 4619-4627.

[223] Gao, Y. C., Zhang, G., Krentz, S., Darius, S., Power, J., & Lagarde, G. (2002). Inhibition of spoilage lactic acid bacteria by lysozyme during wine alcoholic fermentation. *Australian Journal of Grape and Wine Research*, *8*(1), 76-83.

[224] Garde-Cerdan, T., Marselles-Fontanet, A., Arias-Gil, M., Martin-Belloso, O., & Aanci-Azpilicueta, C. (2007). Influence of SO_2 on the consumption of nitrogen compounds through alcoholic fermentation of must sterilized by pulsed electric fields. *Food Chemistry*, *103*(3), 771-777.

[225] Cheynier, V., Rigaud, J., Souquet, J. M., Barillere, J. M., & Moutounet, M. (1989). Effect of pomace contact and hyperoxidation on the phenolic composition and quality of Grenache and Chardonnay wines. *American Journal of Enology and Viticulture*, *40*(1), 36-42.

[226] El Hosry, L., Auezova, L., Sakr, A., & Hajj-Moussa, E. (2009). Browning susceptibility of white wine and antioxidant effect of glutathione. *International Journal of Food Science & Technology*, *44*(12), 2459-2463.

[227] Cheynier, V., Rigaud, J., Souquet, J. M., Duprat, F., & Moutounet, M. (1990). Must browning in relation to the behavior of phenolic compounds during oxidation. *American Journal of Enology and Viticulture*, *41*(4), 346-349.

[228] Atanasova, V., Fulcrand, H., Cheynier, V., & Moutounet, M. (2002). Effect of oxygenation on polyphenol changes occurring in the course of wine-making. *Analytica Chimica Acta*, *458*(1), 15-27.

[229] Cejudo-Bastante, M. J., Hermosín-Gutiérrez, I., Castro-Vázquez, L. I., & Pérez-Coello, M. S. (2011). Hyperoxygenation and bottle storage of chardonnay white wines: effects on color-related phenolics, volatile composition, and sensory characteristics. *Journal of Agricultural and Food Chemistry*, *59*(8), 4171-4182.

[230] Schneider, V. (1998). Must hyperoxidation: A review. *American Journal of Enology and Viticulture*, *49*(1), 65-73.

[231] Vaimakis, V., & Roussis, I. G. (1996). Must oxygenation together with glutathione addition in the oxidation of white wine. *Food Chemistry*, *57*(3), 419-422.

[232] Cejudo-Bastante, M. J., Pérez-Coello, M. S., & Hermosín-Gutiérrez, I. (2010). Identification of new derivatives of 2-S-Glutathionylcaftaric acid in aged white wines by HPLC-DAD-ESI-MS n. *Journal of Agricultural and Food Chemistry*, *58*(21), 11483-11492.

[233] Singleton, V. L. A. Survey of wine ageing reactions, especially with oxygen. In: *ASEV 50th ANNIVERSARY ANNUAL MEETING,* 2000, Davis. Proceedings… Davis: California, 2000, p. 323-336.

[234] Castellari, M., Matricardi, L., Arfelli, G., Galassi, S., & Amati A. (2000). Level of single bioactive phenolic compounds in red wine as a function of the oxygen supplied during storage. *Food Chemistry*, *69*, 61-67.

[235] Romero-Cascales, I., Ortega-Regules, A., López-Roca, J. M., Fernández-Fernández, J. I., & Gómez-Plaza, E. (2005). Differences in anthocyanin extractability from grapes to wines according to variety. *American Journal of Enology and Viticulture*, *56*(3), 212-219.

[236] Bautista-Ortín, A. B., Martínez-Cutillas, A., Ros-García, J. M., López-Roca, J. M., & Gómez-Plaza, E. (2005). Improving colour extraction and stability in red wines: the use of maceration enzymes and enological tannins. *International Journal of Food Science & Technology*, *40*(8), 867-878.

[237] Revilla, I., & José, G. S. (2003). Addition of pectolytic enzymes: an enological practice which improves the chromaticity and stability of red wines. *International Journal of Food Science & Technology*, *38*(1), 29-36.

[238] Revilla, I., & González-SanJosé, M. L. (2003). Compositional changes during the storage of red wines treated with pectolytic enzymes: Low molecular-weight phenols and flavan-3-ol derivative levels. *Food Chemistry*, *80*(2), 205-214.

[239] Rodriguez-Nogales, J. M., Ortega, N., Perez-Mateos, M., & Busto, M. D. (2008). Pectin hydrolysis in a free enzyme membrane reactor: An approach to the wine and juice clarification. *Food Chemistry*, *107*(1), 112-119.

[240] Romero-Cascales, I., Fernández-Fernández, J. I., Ros-García, J. M., López-Roca, J. M., & Gómez-Plaza, E. (2008). Characterisation of the main enzymatic activities present in six commercial macerating enzymes and their effects on extracting colour during winemaking of Monastrell grapes. *International Journal of Food Science & Technology*, *43*(7), 1295-1305.

[241] Bautista-Ortín, A. B., Fernández-Fernández, J. I., López-Roca, J. M., & Gómez-Plaza, E. (2007). The effects of enological practices in anthocyanins, phenolic compounds and wine colour and their dependence on grape characteristics. *Journal of Food Composition and Analysis*, *20*(7), 546-552.

[242] Muñoz, O., Sepulveda, M., & Schwartz, M. (2004). Effects of enzymatic treatment on anthocyanic pigments from grapes skin from Chilean wine. *Food Chemistry*, *87*(4), 487-490.

[243] Arnous, A., & Meyer, A. S. (2010). Discriminated release of phenolic substances from red wine grape skins (Vitis vinifera L.) by multicomponent enzymes treatment. *Biochemical Engineering Journal*, *49*(1), 68-77.

[244] Alvarez, I., Aleixandre, J. L., García, M. J., & Lizama, V. (2006). Impact of prefermentative maceration on the phenolic and volatile compounds in Monastrell red wines. *Analytica Chimica Acta*, *563*(1), 109-115.

[245] Zimman, A., Joslin, W. S., Lyon, M. L., Meier, J., & Waterhouse, A. L. (2002). Maceration variables affecting phenolic composition in commercial-scale Cabernet Sauvignon winemaking trials. *American Journal of Enology and Viticulture*, *53*(2), 93-98.

[246] Ducasse, M. A., Canal-Llauberes, R. M., De Lumley, M., Williams, P., Souquet, J. M., Fulcrand, H., Doco, T., & Cheynier, V. (2010). Effect of macerating enzyme treatment on the polyphenol and polysaccharide composition of red wines. *Food Chemistry*, *118*, 369-376.

[247] Bakker, J., Bellworthy, S. J., Reader, H. P., & Watkins, S. J. (1999). Effect of enzymes during vinification on color and sensory properties of Port wines. *American Journal of Enology and Viticulture*, *50*(3), 271-276.

[248] Günata, Z. Y., Sapis, J. C., & Moutounet, M. (1986). Substrates and aromatic carboxylic acid inhibitors of grape phenol oxidases. *Phytochemistry, 26*, 1-3.

[249] Dugelay, L., Günata, Y. Z., Sapis, J. C., Baumes, R., & Bayonove, C. (1993). Role of cinnamoyl esterase activities from enzyme preparations on formation of volatile

phenols during winemaking. *Journal of Agriculture and Food Chemistry*, *41*, 2092-2096.

[250] Somers, T. C., & Ziemelis, G. (1985). Flavonol haze in white wines. *Vitis*, *24*, 43-50.

[251] Okamura, S., Watanabe, M. (1982). Purification and properties of hydroxycinnamic acid hydrolase from *Aspergillus japonicum*. *Agricultural and Biological Chemistry*, *46*, 1839-1848.

[252] Biasoto, A. C. T. (2008). Vinhos tintos de mesa produzidos no Estado de São Paulo: caracterização do processo de fabricação, de parametros fisico-quimicos, do perfil sensorial e da aceitação. Dissertação (Mestre em Alimentos e Nutrição) – Faculdade de Engenharia de Alimentos da Universidade Estadual de Campinas –UNICAMP, Campinas.

[253] Da-Silva, R., Franco, C. M., & Gomes, E. (1997). Pectinases, hemicelulases e celulases, ação, produção e aplicação no processamento de alimentos: revisão. *Revista Brasileira de Ciência e Tecnologia de Alimentos*, *31*(2), 249-260.

[254] Gómez-Plaza, E., Gil-Muñoz, R., López-Roca, J. M., Martínez-Cutillas, A., & Fernández-Fernández, J. I. (2001). Phenolic compounds and color stability of red wines: effect of skin maceration time. *American Journal of Enology and Viticulture*, *52*(3), 266-270.

[255] Meyer, A. S., Jepsen, S. M., & Sørensen, N. S. (1998). Enzymatic release of antioxidants for human low-density lipoprotein from grape pomace. *Journal of Agricultural and Food Chemistry*, *46*(7), 2439-2446.

[256] Pardo, F., Salinas, M. R., Alonso, G. L., Navarro, G., & Huerta, M. D. (1999). Effect of diverse enzyme preparations on the extraction and evolution of phenolic compounds in red wines-source and amounts. *Food Chemistry*, *67*(2), 135-142.

[257] Clare, S., Skurray, G., & Theaud, L. (2002). Effect of a pectolytic enzyme on the colour of red wine. *The Australian & New Zealand Grapegrower & Winemaker*, 456, 29-35.

[258] Gil, J. V., & Vallés, S. (2001). Effect of macerating enzymes on red wine aroma at laboratory scale: exogenous addition or expression by transgenic wine yeasts. *Journal of Agricultural and Food Chemistry*, *49*(11), 5515.

[259] Schneider, V. (1998). Must hyperoxidation: A review. *American Journal of Enology and Viticulture*, *49*(1), 65-73.

[260] Waterhouse, A. L., & Laurie, V. F. (2006). Oxidation of wine phenolics: A critical evaluation and hypotheses. *American Journal of Enology and Viticulture*, *57*(3), 306-313.

[261] Danilewicz, J. C. (2003). Review of reaction mechanisms of oxygen and proposed intermediate reduction products in wine: Central role of iron and copper. *American Journal of Enology and Viticulture*, *54*(2), 73-85.

[262] Danilewicz, J. C. (2007). Interaction of sulfur dioxide, polyphenols, and oxygen in a wine-model system: Central role of iron and copper. *American Journal of Enology and Viticulture*, *58*(1), 53-60.

[263] Danilewicz, J. C., Seccombe, J. T., & Whelan, J. (2008). Mechanism of interaction of polyphenols, oxygen, and sulfur dioxide in model wine and wine. *American Journal of Enology and Viticulture*, *59*(2), 128-136.

[264] Lasanta, C., Caro, I., & Pérez, L. (2005). Theoretical model for ion exchange of iron (III) in chelating resins: Application to metal ion removal from wine. *Chemical Engineering Science*, *60*(13), 3477-3486.

[265] Laurie, V. F., & Waterhouse, A. L. (2006). Oxidation of glycerol in the presence of hydrogen peroxide and iron in model solutions and wine. Potential effects on wine color. *Journal of Agricultural and Food Chemistry*, *54*(13), 4668-4673.

[266] Hermosín-Gutierrez, I. (2007). Copigmentación y piranoantocianos. El papel de los flavonoles y los ácidos hidroxicinámicos en el color del vino tinto. *ACE: Revista de enología*, (81), 2. Disponível em:<http://www.acenologia.com/ciencia81_2.htm>. Acesso em: 01 out. 2013.

[267] Vivar-Quintana, A. M., Santos-Buelga, C., & Rivas-Gonzalo, J. C. (2002). Anthocyanin-derived pigments and colour of red wines. *Analytica Chimica Acta*, *458*(1), 147-155.

[268] Fernández-Zurbano, P., Ferreira, V., Escudero, A., & Cacho, J. (1998). Role of hydroxycinnamic acids and flavanols in the oxidation and browning of white wines. *Journal of Agricultural and Food Chemistry*, *46*(12), 4937-4944.

[269] Lopez-Toledano, A., Mayen, M., Merida, J., & Medina, M. (2002). Yeast-induced inhibition of (+)-catechin and (-)-epicatechin degradation in model solutions. *Journal of Agricultural and Food Chemistry*, *50*(6), 1631-1635.

[270] Monagas, M., Bartolome, B., & Gomez-Cordoves, C. (2005). Updated knowledge about the presence of phenolic compounds in wine. *Critical Reviews in Food Science and Nutrition*, *45*(2), 85-118.

[271] Karbowiak, T., Gougeon, R. D., Alinc, J., Brachais, L., Debeaufort, F., Voilley, A., & Chassagne, D. (2010). Wine oxidation and the role of cork. *Critical Reviews in Food Science and Nutrition*, 5020-5052.

[272] Es-Safi, N. E., Cheynier, V., & Moutounet, M. (2002). Role of aldehydic derivatives in the condensation of phenolic compounds with emphasis on the sensorial properties of fruit-derived foods. *Journal of Agricultural and Food Chemistry*, *50*(20), 5571-5585.

[273] Es-Safi, N. E., Cheynier, V., & Moutounet, M. (2003). Implication of phenolic reactions in food organoleptic properties. *Journal of Food Composition and Analysis*, *16*(5), 535-553.

[274] Fulcrand, H., Dueñas, M., Salas, E., & Cheynier, V. (2006). Phenolic reactions during winemaking and aging. *American Journal of Enology and Viticulture*, *57*(3), 289-297.

[275] Liu, S. Q., & Pilone, G. J. (2000). An overview of formation and roles of acetaldehyde in winemaking with emphasis on microbiological implications. *International Journal of Food Science & Technology*, *35*(1), 49-61.

[276] Hermosín-Gutierrez, I., Lorenzo, E. S. P., & Espinosa, A. V. (2005). Phenolic composition and magnitude of copigmentation in young and shortly aged red wines made from the cultivars, Cabernet Sauvignon, Cencibel, and Syrah. *Food Chemistry*, *92*(2), 269-283.

[277] Elias, R. J., Andersen, M. L., Skibsted, L. H., & Waterhouse, A. L. (2009). Identification of free radical intermediates in oxidized wine using electron paramagnetic resonance spin trapping. *Journal of Agricultural and Food Chemistry*, *57*(10), 4359-4365.

[278] Furtado, P., Figueiredo, P., Chaves das Neves, H., & Pina, F. (1993). Photochemical and thermal degradation of anthocyanidins. *Jounal of Photochemistry and Photobiology A: Chemistry*, *75*, 113-118.

[279] Sarni, P., Fulcrand, H., Souillol, V., Souquet, J. M., & Cheynier, V. (1995). Mechanisms of anthocyanin degradation in grape must-like model solutions. *Journal of the Science of Food and Agriculture*, *69*(3), 385-391.

[280] Schwarz, M., Jerz, G., & Winterhalter, P. (2003). Isolation and structure of Pinotin A, a new anthocyanin derivate from Pinotage wine. *Vitis: Journal of Grapevine Research*, *42*(2), 105-106.

In: Grapes
Editor: José S. Câmara
ISBN: 978-1-63321-402-6

Chapter 4

Factors Affecting Antioxidant Activity of Grape Tissues

Francesca Cecchini*
Consiglio per la Ricerca e la Sperimentazione in Agricoltura (CRA)
Research Unit for Enology in Central Italy (ENC),
Velletri (Rome) Italy

Abstract

The close relationship between pro-oxidant conditions and cancer development is an issue that has aroused much interest in recent years, as mutagenesis events downstream of oxidative stress have been widely indicated as one of the principal causes of tumor transformation. Many epidemiological studies have shown that consumption of a fruit and vegetable-based diet significantly reduces the risk of cancer, especially tumors of the digestive tract. For this reason an antioxidant-rich-diet, or supplementation with antioxidant compounds such as nutraceuticals as polyphenols, have been suggested to reduce risk of tumor onset. Polyphenols are secondary metabolites produced from plants. They have been proposed to exert beneficial effects in a multiple of disease states, including cardiovascular disease, anti-inflammatory, anti-allergic activities and neurodegenerative disorders. Polyphenols can be divided in two groups: non-flavonoid (hydroxybenzoic acid, hydroxicinnamic acids and stilbenes) and flavonoid compounds (anthocyanins, flavan-3-ols and flavonols). Those secondary compounds have been demonstrated *in vitro* experiments to protect against oxidative damage by inhibiting or quenching free radicals and reactive oxygen species (ROS). Grapes are a polyphenol-rich fruit and one of the earliest domesticated fruit crops. Grapes have been widely cultivated all over the world. Grapes are classified as table grapes, wine grapes and raisin grapes. People consume the various grape products, such as fruit, raisins, juice and wine. Polyphenols in grape are distributed in various grape tissues (skin, pulp and seeds). The qualitative and quantitative phenolic compounds in grapes depend on the variety of the grapevine and are influenced by viticultural and agronomic practices and environmental factors.

* E-mail: francesca.cecchini@entecra.it.

In this chapter, we will summarize research carried out in our laboratories regarding factors affecting antioxidant activities of different grape tissue extracts from several grape cultivars (red and white), through the analysis of phenolic composition, antioxidant activity and an anti-proliferative effect in cancer cells (the latter only for seed extracts). The results point out statistically significant differences in the content of total polyphenols, flavans, individual phenols and antiradical activity between different grape tissue extracts between different cultivars. Promising results have also been obtained by seed grape extracts on the anti-proliferative effects on cancer cells (HeLa).

Keywords: Antioxidant activity; grapes; phenolic composition

Introduction

The grape is a fruiting berry of the deciduous woody vines of the botanical genus *Vitis*. Grapes can be eaten raw or they can be used for making wine, jam, juice, jelly, grape seed extract, raisins, vinegar, and grape seed oil. Grapes are a non-climacteric type of fruit, generally occurring in clusters. *Vitis vinifera*, L. (common grape vine) is a species of *Vitis*, whereas the term subsp. *vinifera* is restricted to cultivated forms. Table 1 reports the botanical classification (Fregoni, 1999).

Grapes are a type of fruit that grow in clusters of 15 to 300, and can be black, dark blue, yellow, green, orange and pink. "White" grapes are usually green-yellow in color, and are evolutionarily derived from the purple grape. The anthocyanins, which are responsible for the color of the skin of purple grapes coincides with expression of the gene encoding the final step of anthocian biosynthesis, UDP-glucose: flavonoid 3-O-glucosyl transferase (UFGT) (Boss, Davies and Robinson 1996 a). In white grapes, the gene encoding UFGT is not expressed (Boss, Davies, and Robinson 1996 b).

Anthocyanins and other pigment chemicals of the larger family of polyphenols in purple grapes are responsible for the varying shades of purple in red wines. (Waterhouse, 2002) Grapes are typically a rounded or spherical ellipsoid shape.

Table 1. Vitis vinifera botanical classification

Kingdom:	Plantae
Division:	Magnoliophyta
Class:	Magnoliopsida
Order:	Rhamnales
Family:	Vitaceae
Genus:	*Vitis*
Species:	*V. vinifera*

The grape is a polyphenols rich-fruit and is one of the earliest domesticated fruit crops. It has been widely cultivated all over the world. According to the Organization International de la Vigne et du Vin (OIV 2013), world vineyards reached a total area surface of 7,528 mha (millions of hectares) in 2012. There are more than 10,000 different wine grape varieties in the world, all fall into two major families – *Vitis Vinifera*, which is prevalent in Europe, and *Vitis Labrusca*, which is native to Canada and the eastern United States. Global grape production in 2012 was 691 million quintals (OIV 2013). Approximately 71% of world grape production is used for wine, 27% as fresh fruit, and 2% as dried fruit.

Grapes have been also shown to be a rich source of high-added-value bioactive compounds, named phytochemicals. Phenolic compounds are the most important phytochemicals in grapes because they possess many biological activities and health-promoting benefits (Shrikhande, 2000; Xia, Deng, Guo, and Li 2010)

These compounds possess strong antioxidant properties that enable them to scavenge free radicals, donate hydrogen, chelate metals, break radical chain reactions, and quench singlet oxygen *in vitro* and *in vivo* (Rice-Evans, et al. 1997, Havsteen, 2002). All these properties enable them to act in the prevention of oxidative-stress-related diseases (God, Tate, Larcom, 2007; Jung, Wallig, Singletary, 2006). There is growing interest in using naturally occurring compounds as potential cancer chemo preventative agents in human populations. In this regard, a significant correlation between dietary intake and many types of cancer has been shown in epidemiological data generated throughout the world, while in animal experiments, many dietary substances have been documented to have anti-cancer properties (Aherne, Obrien, 2002; Yao, Jiang, Shi, Tomàs-Barberàn, Datta, Singanusong, Chen, 2004).

Polyphenols have shown scavenging activity against the reactive oxygen species (ROS) and the ability to prevent lipid oxidation. ROS, including free radicals, are a class of highly reactive molecules derived from the normal oxygen metabolism (Klaus & Heribert 2004). ROS induces oxidative damage to biomolecules, causing human diseases including cancer, cardiovascular atherosclerosis, cataracts, chronic inflammation, and neurodegenerative diseases, such as Alzheimer's or Parkinson's disease (Halliwell, Gutteridge, Cross, 1992; Gutteridge, 1993; Knight, 1995; Shanmuganayagam, Warner, Krueger, Reed, Folts, 2007). Furthermore, ROS are known to induce both apoptosis (Yeh, Chiang, Hsu, Lin, Wang, Guh, 2007; Chou, Kaur, Thompson, Agarwal, & Agarwal, 2010) and authophagic cell death (Chen, Debnath, 2009; Cecchini, Moretti, Mulinacci, Innocenzi, Pasqua, Ciriolo, Filomeni, 2012).

Several studies have documented the anticancer and chemopreventive efficacy of grape seed extract (GSE) against various malignant cancers such as prostate cancer (Agarwal, Veluri, Kaur, Chou, Thompson & Agarwal 2007; Chou et al. 2010), and colon cancer (Engelbrecht, Mattheyse, Ellis, Smith, Peters, Smith, Myburgh, 2007; Kaur, Mandair, Agarwal & Agarwal 2009). GSE have been shown to reduce the incidence of carcinogen-induced mammary tumors in rats and skin cancer in mice and to inhibit the growth of human cancer cells both *in vitro* and *in vivo*, after transplantation into animals (Singletary & Brandon, 2001; Eng, Ye, Williams, Phung, Moore, Young, Gruntmanis, Braunstein G, Chen, 2003).

In this chapter, we will summarize research carried out in our laboratories about factors affecting the antioxidant activity of different grape tissue extracts from several grape cultivars (red and white), through the analysis of phenolic composition, antioxidant activity and anti-proliferative effects in cancer cells.

Grape Phenolic Compositon

Phenols, sometimes called phenolics, are a class of chemical compounds consisting of a hydroxyl group (-OH) bonded directly to an aromatic ring. They have a weak acidity due to the hydroxyl group bonded to an unsaturated ring; consequently, the acidity is due to stabilization of the conjugate base through resonance in the aromatic ring.

Phenolic compounds are classified as simple phenols, or polyphenols based on the number of phenol units in the molecule.

Phenolic acids are simple phenols that have one carboxylic acid functional group; they include hydroxybenzoic and hydroxycinnamic acids. The most common hydroxybenzoic acids in grape berry include *p*-hydroxybenzoic acid, vanillic acid, gallic acid, gentisic acid and salicylic acid. In terms of concentration, *p*-cumaric acids, caffeic and ferulic acids are the predominant simple phenols in grape, and they are the third most abundant class of soluble phenolics in grape berries, after proanthocyanidins and anthocyanins. These three hydroxycinnamic acids are present primarily as *trans*- isomers. They differ by type and number of substituents on the aromatic ring. When these hydroxycinnamic acids are esterified with tartaric acid, they are named coutaric acid (*trans-p*-coumaroyl-tartaric acids), caftaric acid (*trans*-caffeoyl-tartaric acid) and fertaric acid (*trans*-feruloly-tartaric acid) (Teixeira, Erias-Dias, Castellarin, & Gròs, 2013).

The term flavonoid refers to a broad class of structurally different compounds. They have a diphenylpropane ($C_6C_3C_6$) skeleton. The basic structure of the fused A and C rings, with the phenyl B ring attached through its 1′- position to the 2-position of the C ring (numbered from the pyran oxygen) is shown in Figure 1.

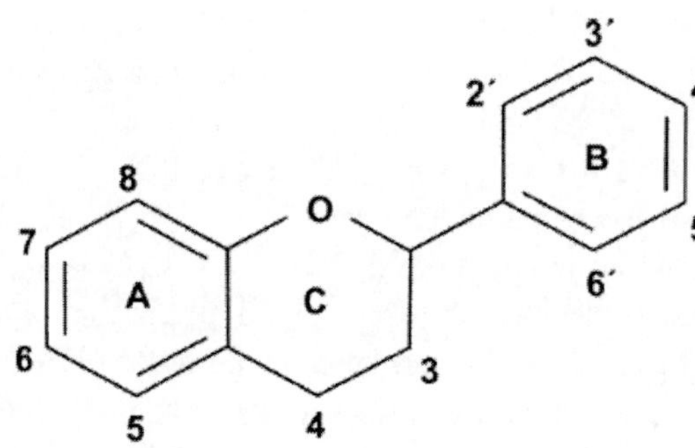

Figure 1. The basic structure of the flavonoids.

These flavonoids contain several phenolic hydroxyl functions attached to ring structures, designated as A, B and C. Structural variations within the rings subdivide the flavonoids into several families: monomeric flavanols, flavanones, anthocyanidins, flavones, flavonols and isoflavones. The flavonoid variants are all related by a common biosynthetic pathway, incorporating precursors from both the shikimate and the acetate-malonate pathway (Ric-Evans et al. 1997)

Further modification occurs at various stages, resulting in alteration in the extent of hydroxylation, methylation, isoprenylation, dimerization and glycosylation (Rice-Evans, Miller, Paganga, 1996). Quercetin, myricetin, kaempherol, laricitrin, syringetin isorhamnetin, and their derivatives are the most common flavonols found in grapes.

Flavanols or flavan-3-ols (also named flavans) are the most abundant class of phenolics in the grape berry. They have monomeric (catechins) or a polymeric structure known as proantocyanidis also called condensed tannins. The flavans in grape are (+)-catechin and (-)-epicatechin, (+)-gallocatechin and (-)-epigallocatechin. Monomeric flavans are characterized by the presence of a hydroxyl group at the position of the C ring (Teixeira et al. 2013). Proanthocyanidins are characterized by flavan units with C_4-C_6 and C_4-C_8 interflavan bonds, the polymerization degree ranging mainly between 3 and 11, up to 17 and more, with molecular mass above 30 kDa (Dreosti, 2000).

The proanthcyanidins dimers (named procyanidins B) and trimers (named procyanidins C), as well as (+)-catechin, are also present as esters linked to gallic acid at the 3-hydroxyl group in the C ring. The procyanidins B can also be esterified with two gallic acid molecules as 3-3′-O-digallate. The anthocyanins (Figure 2 (http://it.wikiwikipedia.org/wiki/antociani)) are common red-blue grape pigments, located within vacuoles of the skin cells of red grape berries. They are glycosides of polyhydroxy and polymethoxy derivatives of 2-phenylbenzopyrylium or flavylium salts (Mazza and Miniati, 1993). The differences between individual anthocyanins are the number of hydroxyl groups in the molecule, the degree of methylation of these hydroxyl groups, the nature and number of sugars attached to the molecule and the position of the attachment (Mazza and Maniti, 1993). The anthocyanidins are the sugar-free anthocyanins (aglicon form).

The color of anthocyanidins is dependent on pH. At a low pH (acidic conditions), colored anthocyanidins are present, whereas at a higher pH (basic conditions) the colorless chalcone forms are present. Of these anthocyanidins, five occur most frequently in grapes. They are cyanidin, peonidin, delphinidin, petunidin and malvidin. Resveratrol (3,5,4'-trihydroxy-*trans*-stilbene) is the main stilbenoid present in grape (Figure 3 http://it.wikiwikipedia.org/wiki/file: resveratrol.svg). In plants, resveratrol functions as a phytoalexin that protects against biotic and abiotic stresses. It is mainly synthesized in the berry skin of red grapes.

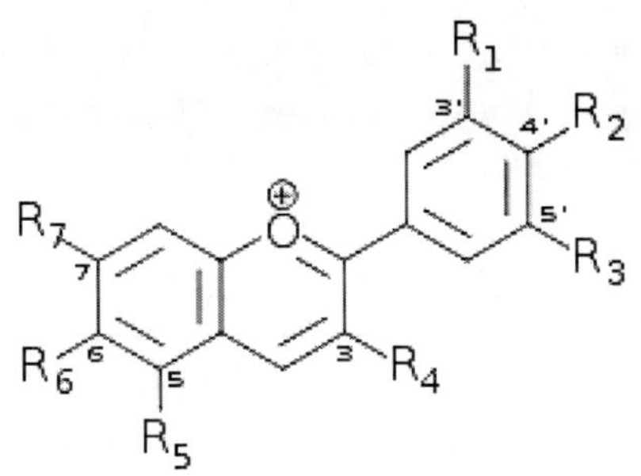

Figure 2. Basic structure of anthocyanins.

The principal components of ripe grapes are carbohydrates, glucose and fructose. These sugars, which are converted in alcohol during fermentation, are a major factor in the taste of table grapes, raisins, juices and other products. The organic acids, although present in small quantities, contribute significantly to the taste of grapes and grape products, and their concentration has a marked effect on the color intensity of anthocyanins. Phenols represent the third most abundant constituent in ripe grapes after carbohydrates and organic acids (Ribéreau-Gayon, Glories, Maujean, Ubourdìeu, 2003). They are broadly distributed inside grapes and are mainly located in the seeds and the skins. On the contrary, the juice contains a very small amount, 3 to 5% of total phenols. The total extractable phenolics in grapes are present at only 10% or less in pulp, 60-70% in the seeds and 28-35% in the skins.

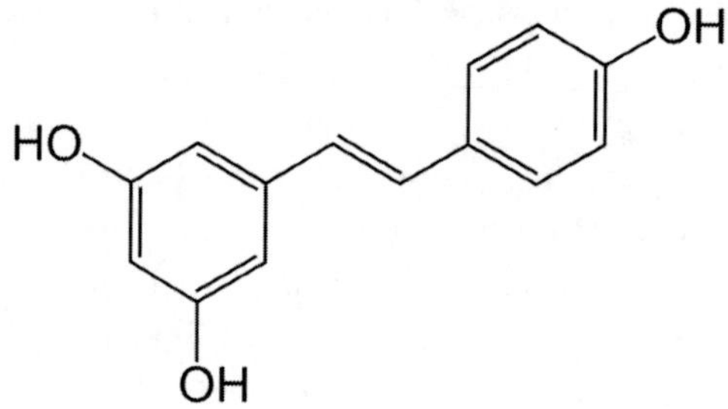

Figure 3. Basic structure of resveratrol.

Grape skins are rich source of anthocyanins (red grapes), hydroxcinnamic acids, flavanols and flavanol glycosides, whereas phenolic compounds in grape seeds are mainly flavonols, including gallic acid, the monomeric flavan-3-ols: (+)-catechin, (-)-epicatechin, gallocatechin, and epicathechin-3-O-gallate, procyanidin dimers, trimers, and more highly polymerized proanthocyanidins and they are known for their high antiradical power. In pulp the main phenols present are hroxycinnamic tartaric acids.

The functional properties of the flavonoids are influenced by their structural characteristics. For example the position of hydroxyl groups and other features in the chemical structure of flavonoids are important for their antioxidant and free radical scavenging activities. Flavonoids aglycons having 3-OH free are antioxidants more powerful than those substituted in the same group. The presence of a double bond between C2 and C3 increases antioxidant activity as well as the number of hydroxyl groups on the B ring (Rice-Evans et al., 1996; Teixeira et al. 2013).

The beneficial effects on human health of the grapes are mainly due to the antioxidant activity of flavonoids. The protective effects of these phytochemicals in biological systems are ascribed to their capacity to transfer electrons, free radicals, chelate metal catalysts, activate antioxidant enzymes, reduce α-tocopherol radicals, and inhibit oxidases (Heim, Tagliaferro, Bobilya, 2002).

The different antioxidant activity of the skin, pulp and seeds in red and white grapes depend on their different phenolic composition. The phenol composition and antioxidant activity of the grape is clearly affected by many environmental factors (Connor, Luby and Tong, 2002; Downey, Dokoozlian, and Kristic, 2005; Cecchini, et al. 2012 a; Cecchini, et al., 2012 b).

In the next paragraphs, we will discuss the factors affecting the antioxidant activity of grape tissues. We shall report some examples derived from studies performed in our laboratories.

Factors Affecting Antioxidant Activity

There are many factors reputed to affect polyphenols biosynthesis in grape, including grape variety, climate, temperature, light, maturation degree, ripeness, pathogen attacked, defoliation, and various developmental processes (Connor, et al., 2002, Downey, et al., 2005; Cecchini, & Giannini, 2010) Cecchini, et al. 2012 a; Cecchini, et al. 2012 b; Cecchini, Giannini, Serra 2013).

Considering the close correlation between polyphenols and antioxidant activity, it is clear that all the factors affect polyphenol biosynthesis since they affected the antioxidant activity. All the results reported in this chapter were performed using the following analytical methods: Phenols of skin and seeds were extracted by hydro-alcoholic solution (10% v/v) buffered pH 3.2, the pulp was centrifuged and supernatants were used for analysis (Checchini et al. 2012 a). Total polyphenols (TP) in the skin and seeds extracts were determined according to Folin-Ciocalteau method (Singleton, & Rossi, 1965), using catechin as a standard and results expressed as cathechin equivalents (CE) (mg CE /g of tissue).

The flavan-3-ols (FL) in skin, seed extracts and in pulp were determined by the vanillin assay (Di Stefano, & Cravero, 1991) using catechin as standard and the results were expressed as cathechin equivalents (CE) (mg CE /g of tissue)

The anthocyanins in skin extracts were determined according to Di Stefano, & Cravero, (1991) method using malvidin-O-glucoside as standard and results expressed as mg malvidin-O-glucoside/ g of tissue.

Antioxidant activity (AA) was measured by the free stable radical 2,2-diphenil-1-picryl-hydrazyl ($DPPH^{\cdot}$) spectrophotometric assay (Brand-Williams, Cuvellier, Berset, 1995, Cecchini, et al.2012 a) and was defined as mg of fresh or dry (DM) tissue necessary to reduce the initial $DPPH^{\cdot}$ concentration by 50%, which is a value generally defined as “Efficient Concentration” (EC_{50}): the higher the antiradical activity, the lower this value is. AA was expressed as ARP ($1/EC_{50}$). The higher the ARP, the antioxidant becomes more efficient.

The monomeric flavans, procyanidins and condensed proanthocyanidin were determined by HPLC/DAD/MS method (Mulinacci, Innocenzi, Pasqua, Santamaria, 2011). All the measurements were made in triplicate.

The statistical analysis was carried out with analysis of variance (ANOVA) and the means were compared with Least Significant Difference (LSD) test. Pearson’s correlation coefficient was used to determine the correlation among variables. For data analysis, the “Statistical package” (version 7.1, StatSoft., Italy) was used.

Although the genetic traits of the cultivar determine its phytochemical profile and related antioxidant power, the environment plays a key role in modulating this biosynthetic activity. Figure 4 summarizes the results of AA of two tissue extracts from Dolcetto (red grape cultivar native of North Italy) and Verdicchio (white grape cultivars native of Central Italy).

These cultivars were grown at the experimental vineyard of the Research Unit for Enology of Central Italy (CRA-ENC), located in Velletri (Rome) in 2009, and they were harvested at ripeness. The AA was significantly different ($p \leq 0.05$) between the two cultivars. For both cultivars the AA was higher in seeds and decreased in the skin and pulp, respectively. The significant differences observed between the cultivars can be attributed to their different polyphenols content that showed the same trend of the AA. Considering that these cultivars were grown in the same geographic area, harvested at the same maturation degree in the same year, with the same training system, and adopting the same agronomic practice, the differences observed can be attributed to different genotypic characteristics of the variety. In addition, different AA was also found between clones of the same variety. A clone is a genetically identical group of individuals derived originally from a single individual by asexual propagation (i.e. plant cuttings, or grafting) (Hartmann, Kester, & Davies, 1990).

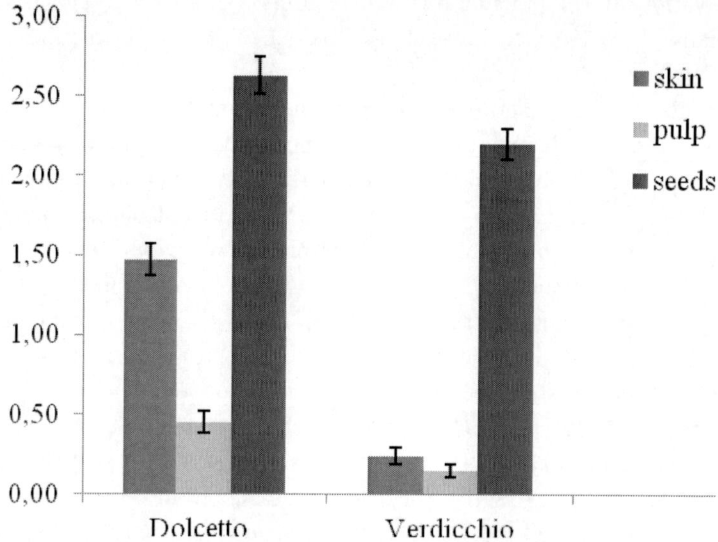

Figure 4. AA expressed as ARP ($1/EC_{50}$); EC_{50}= mg of DM that reduce of 50% the DPPH. The Values represent the mean of three replications ± standards deviation.

Some of the viticultural characteristics that allow us to distinguish the clones of one another include berry and cluster size, morphology, yield, fruit chemistry, color, phenolics, flavor, the time of budbreak and the ripening period, vine vigor, and others. All of these parameters help to determine the antiradical power of the clone. The following is a summary of data from our studies carried out on grapes coming from different clones of the same cultivar harvest in 2011. In these studies we have obtained interesting results concerning the characterization and comparison of seven clones (named A5, A8, A9, A10, A19, A20, A21) of the red cultivar Cesanese d'Affile (*Vitis vinifera* L.), a native red grape variety of Central Italy, appreciated for its high oenological quality and for its varietal specificity.

Clones were grown in the experimental vineyards located in Lazio region (Italy) under the same pedoclimatic conditions and with the same training system (cordon spur).

The grape of seven clones harvested at different times and at the same ripening (between 21 and 22 g/L of reducing sugars) showed differences in morphological and chemical characteristics and antiradical activity. Figure 5, shows the AA of seven clones of Cesanese d'Affile.

The AA was significantly different ($p \leq 0.05$) in both the skin and seed extracts of the seven clones of the Cesanese d'Affile cultivar.

A5 and A21 clones showed the lowest mean weight of berry and the lowest number of seeds *per* berry. In previous research by Garofolo , Gianni & Favale, (2006), carried out using the same clones during years 2002-2005, though with different values, the A5 clone confirmed the same trend, while clone A21, in 2002, showed a greater berry weight and a greater number of seeds *per* berry, when compared to all others. In particular, clone A21 showed a greater sensitivity to climatic variations. Significant differences were also found in anthocyanins content of the skins between the analyzed clones.

For A5 clone, polyphenols, as well as the sum of catechin, procyanidins B, and AA were always significantly lower ($p \leq 0.05$) than those of other clones.

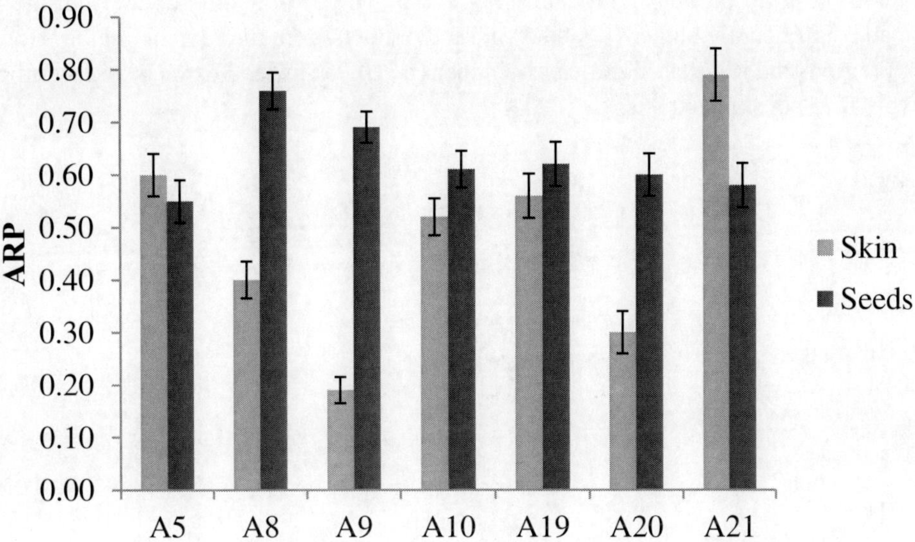

Figure 5. AA expressed as ARP ($1/EC_{50}$); EC_{50}= mg of fresh tissue that reduce of 50% the DPPH. Values represent the mean of three replications ± standards deviation.

These results were consistent with those reported by Garofolo et al. (2006) even if these authors calculated EC_{50} values as moles of TP (P) per Moles of DPPH (D) ($EC_{50P/D}$). Moreover, A5 and A21 clones, having the smallest berries, were found to be more efficient in terms of AA of the skins and less efficient in terms of AA of the seeds. However, the AA as well as the phenol content of the other clones was significantly different, both in skin and seed extracts. Also in this case the differences between the clones can be ascribed to a variability of the genetic characteristic. It follows that the contribution of the morphological, chemical and phytochemical characteristics of the clone affects the antiradical power.

In Figure 6 gives an example of the AA expressed of the white grape Verdicchio cultivar, an Italian variety, grown in two different regions of Italy, Lazio and Marche. This cultivar is native of Marche region, but it is cultivated in several Italian Regions. The Verdicchio grape from the Lazio Region came from a vineyard 14 years old of the institute of Enology, located in Velletri (Rome) in Lazio in Central-West Italy (41° 40′ 30″ N latitude, 08° 11′ 58″ E longitude) at 335 m above sea level. The Verdicchio grape from the Marche Region came from a vineyard 14 years old, located in Jesi (Ancona) Central-East Italy (43° 31′ 46″ N latitude, 13° 14′ 27″ E longitude) at 97 m above sea level. Climate, soil composition (clay-gravel, loam and sandy) and altitude are very different between these regions, accounting for the environmental impact on the grape quality. Grapes were harvested at the same degree of maturation. Data highlighted significant differences ($p \leq 0.05$) in grape berry morphology between the grapes from the two regions. The primary effect of the different geographic area was the difference in berry size. The main weight of the berry of the Verdicchio grape from Marche was 60% less than the average weight of the berry of the Verdicchio grape from Lazio (1.71 g and 2.82 g for berries from Marche and Lazio regions respectively). The number of seeds per berry was also significantly different ($p\leq 0.05$) between the two regions. The mean number of seeds per berry was 2.15 and 1.90 for berries from Marche and Lazio regions, respectively.

The AA (Figure 6) was significantly higher ($p \leq 0.05$) in skin extracts from the Lazio region, and significantly higher ($p \leq 0.05$) in seed extracts from the Marche region. However, for both regions the AA was significantly higher ($p \leq 0.05$) in seed extracts and significantly lower ($p \leq 0.05$) in skin extracts.

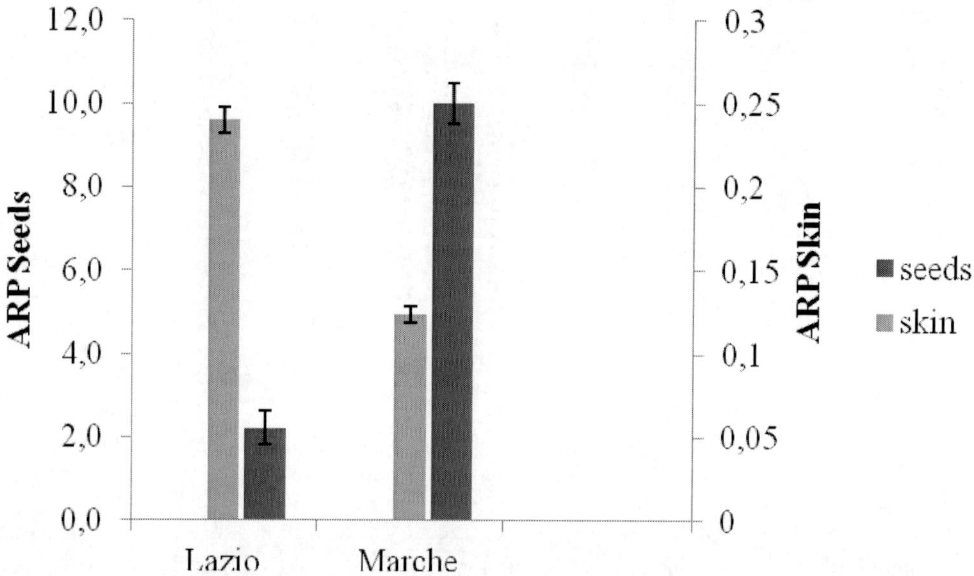

Figure 6. AA expressed in ARP ($1/EC_{50}$); EC_{50}= mg of DM tissues that reduce 50% of the DPPH. Values represent the mean of three replications. Bars indicate ± standards deviation.

Also in this case the AA followed the same trend of phenolic compounds. TP, FL, Catechins and Procyanidins B were significantly higher ($p \leq 0.05$) in skin extracts from the Lazio region and significantly higher ($p \leq 0.05$) seeds extracts from the Marche Region. The AA of the pulp for both regions was negligible.

Obtained results, in agreement with our previous research (Cecchini, et al. 2012 a), suggest that the different geographical origins of grape belonging to the same genotype strongly influence the morphology of the berry, its phenolic composition and AA. This in turn could have some effect on the antiradical capacity of this cultivar during the winemaking process.

Figure 7 and Figure 8 report an example of the AA of skin and seeds from the red grape cultivar Nero d'Avola (Italian red variety) and the white grape cultivar Verdicchio during ripening in 2011. Berry development is characterized by two successive sigmoid curves, separated by a lag phase corresponding to the third stage of development (*Coombe*, 1973). Stage I is characterized by the first period of berry growth, from bloom to approximately 60 days afterward. During this period, the pericarp and seed cell numbers increase and approach full size. Also during the first growth period, the berry expands in volume as solutes accumulate. There are several solutes that accumulate in the berry during this period, all of which reach an apparent maximum around véraison. Stage II is characterized by a change in berry size; seed embryos develop with a concomitant hardening of their coat (Coombe, 1960). Stage III, termed the vérasion, is characterized by berry softening and antocyanin accumulation for the red variety, while seeds are fully developed.

During grape ripening significant changes in polyphenol amount, composition and molecular weight occur.

Nero d'Avola and Verdicchio grapes were taken from pre-veraison to post-harvest. For both cultivars, even with different levels in skin extracts, TP, FL, cathehins and proanthocyanidin accumulation occurred from fruit set until véraison (sampling 4, 5, 6) and then declined rapidly from véraison to harvest (sampling 7 and 8, reducing sugars were 180 g L^{-1} 200 g L^{-1} respectively). Anthocyanins skin extracts from Nero d'Avola increased from véraison until harvest. Then, in both cultivars, PT, FL, Anthocyanins (only for skin of red grapes) decreased slowly until post-harvest (sampling 10, 270 g L^{-1} of reducing sugars). The main period of FL, and the proanthocyanidin accumulation in grape seeds, in both cultivars, occurred immediately after fruit-set with maximum levels observed at harvest (sampling 7), and then declined gradually. Seeds in post-harvest due, also to the seed lignification, TP, FL and the radical scavenging activity, sharply decrease. At harvest (sampling 7 and 8) the seed extracts showed the maximum concentration of galloilated forms (cathechins and Procyanydins esterefied with gallic acid) for both cultivars and a lower level of TP with respect to the skin extracts.

At harvest, 70% and 75% of extractable berry proanthocyanidin was in the seed extracts for Nero d'Avola and Verdiccho grapes respectively. Accumulation of proanthocyanidins in the seeds appears to be independent of that in the skins. However in both tissues, and for both cultivars, synthesis occurs early in berry development and maximum levels are reached around véraison for the skin and around harvest for the seeds. In skin, for both cultivars, the AA (Figure 7) followed the same trend of phenolic compounds; the highest AA was found at the end véraison (sampling 6, 140g/L reducing sugars) and then rapidly declined until harvest. Therefore there was a high correlation between TP and AA ($r= 0.92$, $p \leq 0.01$). On the other hand, the AA in seeds extracts for both cultivars (Figure 8) was higher at harvest, with respect the other previous sampling, even if the TP were lower.

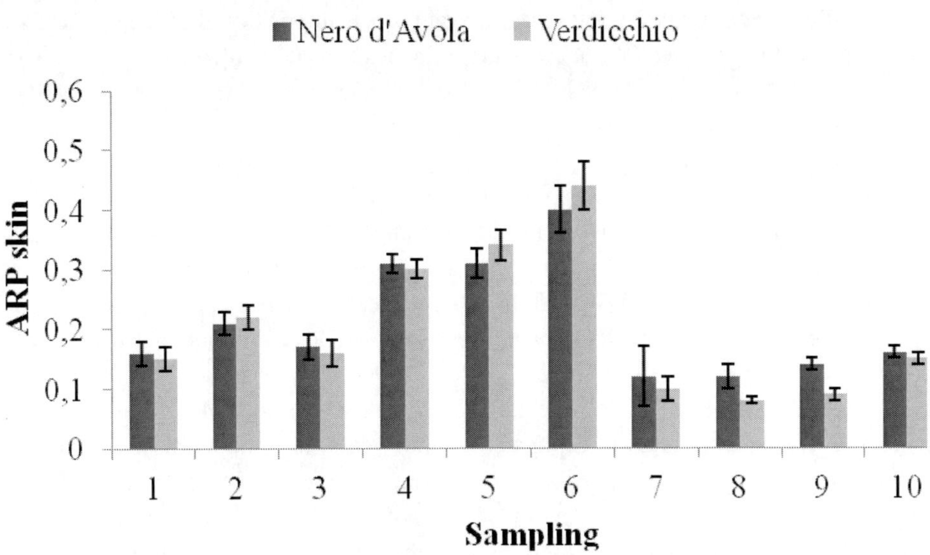

Figure 7. The AA expressed in ARP (1/EC50, where EC50 = mg/g DM of the skin extracts from Nero d'Avola and Verdicchio cultivars. Values represent the mean of three replications. Bars indicate ± standard deviation.

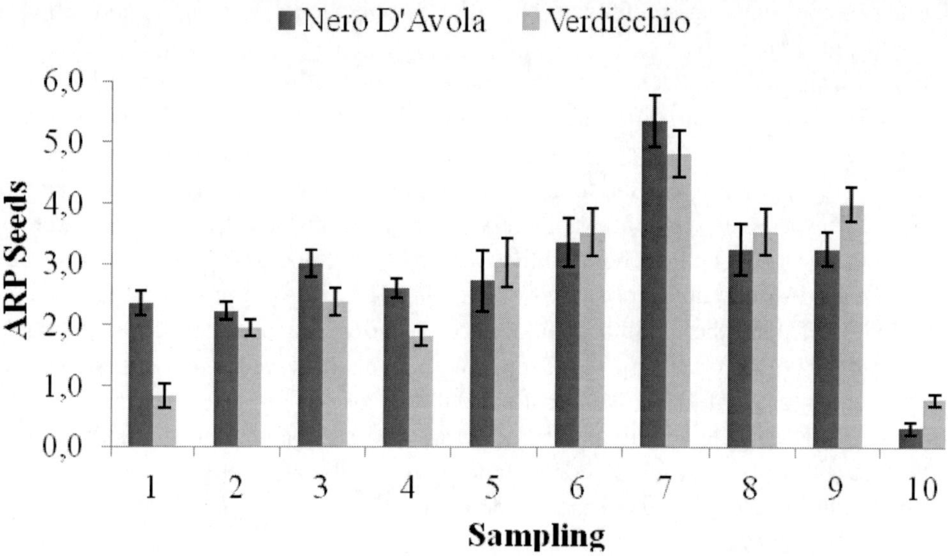

Figure 8. The AA expressed in ARP (1/EC50, where EC50 = mg/g DM of the seed extracts from Nero d'Avola and Verdicchio cultivars. Values represent the mean of three replications. Bars indicate ± standard deviation.

This fact is probably due to the higher concentration of the galloilated form that accounts for the most superior antioxidant activity due to enlargement of the number of -OH group, therefore at harvest the seeds extracts AA appear influenced by these molecules. These results agree with Kennedy, Mark, Matthews, and Waterhouse (2002), Govannelli & Brenna (2007) that found the same trend of individual phenolic during ripening of grapes.

Many studies show the effect of light exposure on specific phenolic compounds of the berries (Smart, 1987; Spayd, Tarara, Mee, &. Ferguson, 2002). These studies report on the importance of the effect of canopy microclimate on berry composition. In grapevine canopies, depending on canopy architecture, leaves and bunches can develop in conditions varying from heavily shaded to middle shaded to fully exposed canopies. The application of shading treatments has been shown to substantially alter temperature and humidity within grapevine canopies (Smart, 1987; Dokoozlian, & Kliewer, 1996). The relationship between sunlight exposure and temperature cluster is important to berry composition and metabolism (Spayd, et al. 2002). Generally, grape berries exposed to sunlight are higher in sugar, anthocyanins, and phenolics, and pH, and lower in titratable acidity, with increased berry phenolics. The following report the values of AA of skin and seed extracts from the red grape Dolcetto cultivar where three levels of bunch exposure were imposed (Cecchini, et al. 2013).

The study was carried out using the red grape Dolcetto cultivar grown in the experimental field of the Research Unit for Enology of Velletri in the Lazio region (Italy) during 2011 and 2012 vintages. Trials were made in triplicate. In both years the same cultural practices were found in vineyard. Three replicates of each treatment were applied to 12 vines. Vines were trained to the Single Cordon Spur with plant density of 2.60 × 2.00 m and pruned approximately 10 nodes per meter of cordon. Three levels of bunch exposure were imposed: heavily shaded; middle shaded; fully exposed. All of the treatments were made after winter vine pruning. Bunches on the cordon were used for all treatments.

Bunches on shoots where leaves had not been removed represent the heavily shaded treatment. Bunches on shoots where leaves had been removed represent the fully exposed treatment. Bunches on shoots where leaves had been partially removed represented the middle shaded treatment.

In the fully exposed treatment the leaves were removed from the shoot in the vicinity of bunches from the lower and upper cordon. In the middle shaded treatment the leaves were removed, only lower shoot and infertile shoots were removed. All the treatments were applied when berries were at the véraison stage.

In both years the grapes were harvested to maturation. There were no significant differences in the juice oenological parameters between the three treatments. In agreement with other authors (Crippen & Morrison 1986), the light had no effect on juice oenological parameters even if the fully exposed grape was harvested three and six days before in middle shaded and in heavily shaded, respectively.

Data of grape berry morphology showed significant differences ($p \leq 0.05$) in the morphological characteristic of the berry among heavily shaded, middle shaded and fully exposed and the significant differences ($p \leq 0.05$) between the same treatments in two years. Grapes developed on fully exposed bunches, for both years, showed a higher berry weight and a higher number of seeds per berry. The differences observed between the same treatment in 2011 and 2012 depended exclusively on different climatic conditions in two years.

Both skin and seeds differences were significant ($p \leq 0.05$) in phenolic content among the three treatments and between same treatments in two years were also found. The amount of Total Anthocyanis (TA), FL and TP was significantly higher in skin berries from heavily shaded bunches in both years, even if the differences were significant between two years. While TA, FL and TP were significantly higher in skin of berries from fully exposed bunches in 2012 compared to 2011. In seeds the higher concentrations of TP and FL were found in all three treatments in 2011 with the only exception being the FL, which was higher in berry seeds from fully exposed bunches in 2012.

The antioxidant activity in both skin and seeds, figure 9 and figure 10, expressed as ARP was significant higher ($p \leq 0.05$) in 2011 compared to 2012 for all the treatments and was significantly higher ($p \leq 0.01$) in seeds compared to the skin in agreement with other authors (BaliK, Kyselakova, Vrchotova, Triska, Kumsta, Veverka, Hic, Totusek, Lefnerova, 2009). In 2011 the skin from berries heavily shaded showed higher value of ARP was significantly different from middle shaded and fully exposed. In 2012 the ARP value was not significantly different between heavily shaded and middle shaded treatments that showed lower values.

In seed extract from 2011 the ARP value was significantly higher ($p \leq 0.05$) in fully exposed and significant lower ($p \leq 0.05$) in middle shaded. Also in 2012 the ARP was significantly higher in fully exposed and there was no significant differences between heavily shaded and middle shaded. From the results of three experiments, it appeared that the plant metabolism responded to change in light conditions. In conditions where bunches are heavily shaded the light should be a limiting factor in the accumulation of phenols during ripening, as reported by other authors (Haselgrove, Botting, van Heeswuck, Høj, Dry, Ford, Iland, 2008). Contrarily, in our results, the higher content of all phenolic compounds in both years, even if with different concentrations, were found in the skin and seeds of heavily shaded berries. In this research the heavily shaded berries represented a treatment where the leaves were not removed, so low lighting is due exclusively to the shade made from leaves. While in research of other authors, the heavily shaded treatment was made by covering the bunches completely with aluminum foil. Therefore, in this case, the high content of phenolic compounds in heavy shade is due to strong natural abiotic stress response in which the vine was exposed, while in the bunches were completely covered and so there is an inhibition in the accumulation of phenol compounds. The lower phenol content in fully exposed bunches among the three treatments was due to less stress by low lighting. The AA in both skin and seeds in 2012 from grape bunches from fully exposed treatment was higher compared to the other two treatments. Although the high correlation between polyphenols and antioxidant activity was amply demonstrated, the seeds extract from fully exposed grapes, in both years, showed the lower phenolic content and the higher antioxidant activity. While the skin extract from heavily shaded berries, for both years, showed the higher phenolic content, the antioxidant activity was higher only in 2011. This fact suggests that the light exposure of the bunches has an effect both on the quantity and on the quality of polyphenols. In addition, the differences observed both in phenol content and in AA, between same treatments in two years, were exclusively due to climatic differences.

The climatic analysis showed that climate conditions were different in the site analyzed in summertime. From véraison to ripening, in the 2012 GDD (Growing Degree Days) values, with daily and monthly mean temperatures, the sum of radiation was generally higher from late June to August 20. In addition, the year 2012 showed a positive anomaly of about +4°C in the maximum temperature and about +3°C in the minimum temperature with reference to the mean values in the period 1993 - 2010. The 2012 season was more rainy (1198.4 mm) than 2011 (1082.2 mm). During the period from April to September rainy days were 39 in 2011 and 52 in 2012.

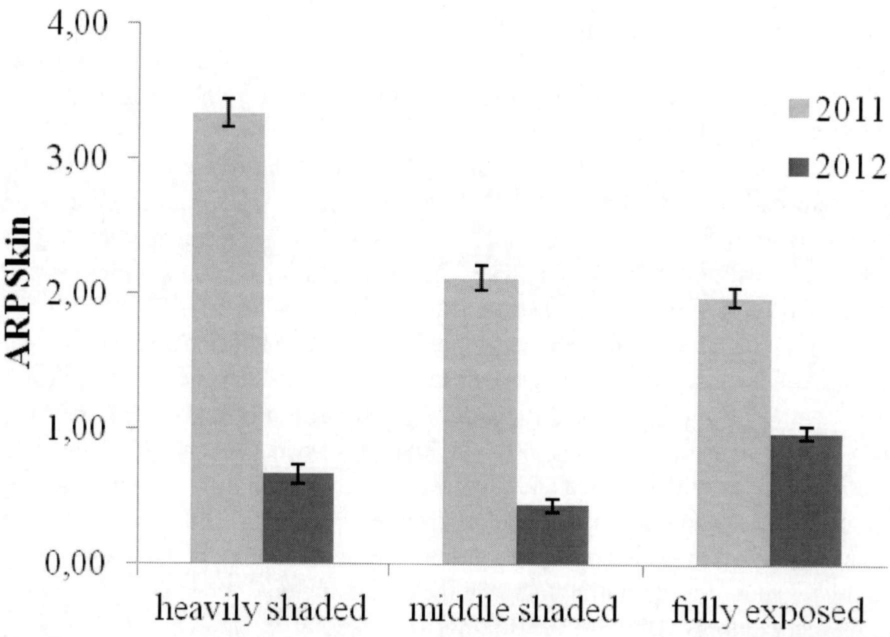

Figure 9. Antioxidant activity in grape skin extract from Dolcetto cv. expressed as ARP ($1/EC_{50}$) in the three treatments. Values represent the mean of three replications. Bars indicate ± standard deviation.

The interaction among climatic conditions and canopy treatments significantly affect the composition and concentration of phenolic compounds and consequentially, on the antioxidant activity.

The importance of polyphenols, which are present in many vegetables, is well-known and documented. Their main function is to protect the plants against biotic and abiotic stresses. Among biotic stress is the infection from *Uncinula necator,* known as *Oidium,* a fungus that causes powdery mildew of grape. It is a common pathogen of the Vitis species, including the wine grape, *Vitis vinifera*. *Uncinula necator* infects all green tissue on the grapevine, including the leaves and young berries. It can cause crop loss and poor wine quality if untreated. The sexual stage of this pathogen requires free moisture to release ascospores from its cleistothecia in the spring. Briefly, the following reports an example of AA in the skin and seeds extracts from an infected red grape Dolcetto cultivar *Oidium*. Results showed that during the *Oidium* attack the weight of the berry, skin and seeds significantly descreased, due to the precocious senescence induced from the pathogen infection.

During the pathogen attack the grape produces large quantities of ROS as a response of the biotic stress. ROS plays a critical role during the natural course of senescence and during oxidative stress (Bhattacharjee, 2005; Foyer, & Noctor 2005). Figure 11 reports the AA of skin and seed extracts from sound and *Oidium* infected Dolcetto grapes. The statistical analysis pointed out the significant differences ($p \leq 0.05$) between sound skin extracts and *Oidium* infected ones, and between sound seed extracts and *Oidium* infected ones. The AA was significantly higher ($p \leq 0.05$) in Skin and seed extracts from *Oiduim* infected grapes compared to the skin and seed extracts from sound grapes. The TP followed the same trend of the AA in both skin and seeds. The high value of phenolics was related to their high biochemical activity during the pathogen attack.

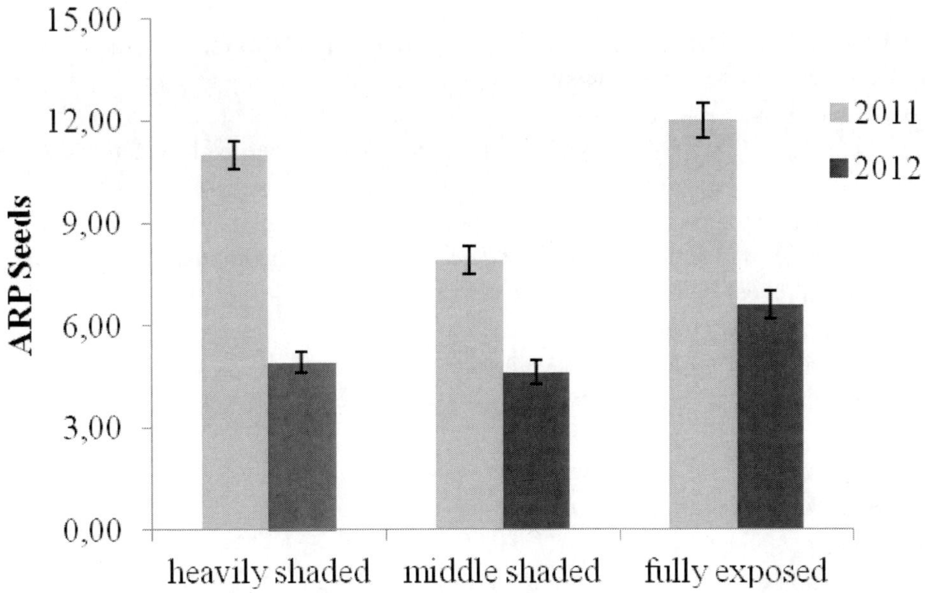

Figure 10. Antioxidant activity in grape seeds extract from Dolcetto cv. expressed as ARP ($1/EC_{50}$ DM) in the three treatments. Values represent the mean of three replications. Bars indicate ± standard deviation.

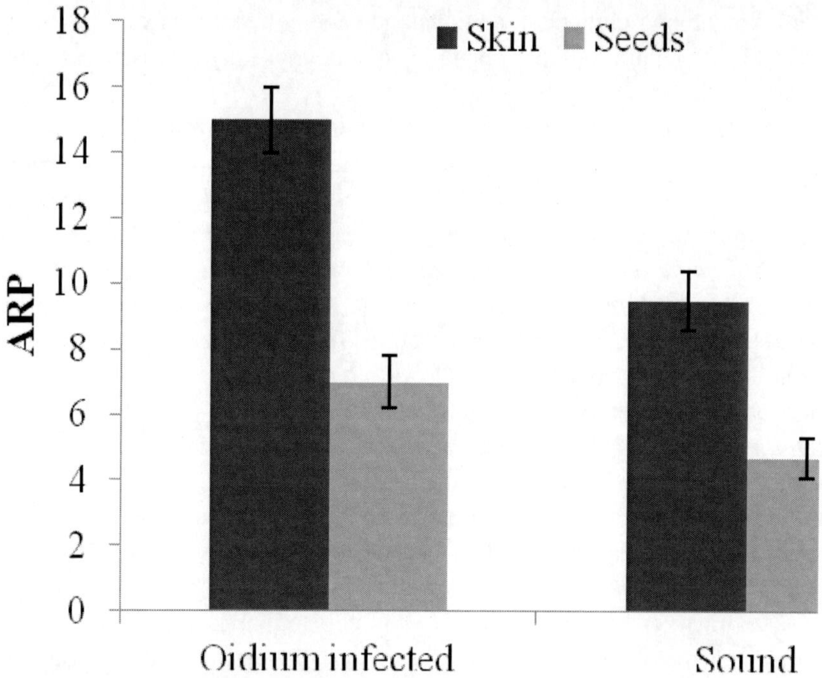

Figure 11. AA in skin and seed extracts both sound and *Oidium* infected Dolcetto grape ARP ($1/EC_{50}$). Values represent the mean of three replications. Bars indicate ± standard deviation.

Biological Activity of Flavanoids

Flavonoids have been proposed to exert beneficial effects in multiple of disease states. Many biological actions of flavonoids have been attributed to their antioxidant properties. The term antioxidant is commonly used in scientific literature but it can be defined in multiple ways according to the methods used to measure antioxidant activity (Carbone, Picchi, Giannini, Lo Scalzo, Cecchin, 2013). Therefore, Halliwell et al. (1992) proposed a definition of an antioxidant as "any substance that delays, prevents or removes oxidative damage to a target molecule". The physiological role of these compounds, as this definition suggests, is to prevent damage to cellular components arising as a consequence of chemical reactions involving free radicals.

As deduced by much research, flavonoids can prevent injury caused by free radicals by the following mechanisms (Prochàzkovà, Bošovà, I., Wilhelmovà, 2011):

- direct scavenging of reactive oxygen species (ROS);
- activation of antioxidant enzymes);
- metal chelating activity ;
- reduction of α-tocopheryl radicals (Hirano, Sasamoto, Matsumoto, Itakura, Igarashi, Kondo. 2001; Heim, et al., 2002);
- inhibition of oxidases (Heim, et al. 2002);

- increase in antioxidant properties of low molecular antioxidants (Yeh, Wang, Huang, Hu, 2005).

As attested by many authors, the antioxidant activity of flavonoids and their metabolites *in vitro* depends upon the arrangement of functional groups about the nuclear structure (Heim, et al. 2002). The B ring hydroxyl configuration is the most significant determinant of ROS scavenging (Burda, Oleszek, 2001), while the substitution of ring A and ring C has little impact on superoxide anion radical scavenging rate constants (Taubert, Breitenbach, Lazar, Censarek, Harlfinger, Berkels, Klaus, Roesen, 2003; Amić, Davidovič-Amić, Bešlo, Rastija, Lučić, Trinajstić, 2007). The *in vitro* antioxidant activity could be increased by polymerization of flavonoid monomers, e.g. proanthocyanidins, the polymers of catechins, are excellent *in vitro* antioxidants due to the high number of hydroxyl groups in their molecules. The antioxidant capacity of proanthocyanidins depends on their oligomer chain length and the type of ROS with which they react (Lotito, Actis-Goretta, Renart, Caligiur, Rein, D. Schmitz, et.al 2000). The glycosylation of flavonoids reduces their *in vitro* antioxidant activity when compared to the corresponding aglycons (Prochàzkovà, et al. 2011). The torsion angle of the B-ring with respect to the rest of the molecule strongly influences free radical scavenging ability. Flavonols and flavanols with a 3-OH are planar, while the flavones and flavanones, lacking this feature, are slightly twisted (van Acker, de Groot, van den BergTromp, M., Donnè-Op den Keldervan der Vijgh, Bast, 1996). Planarity permits conjugation, electron dislocation, and a corresponding increase in flavonoid phenoxyl radical stability (van Acker et al. 1996) Removal of a 3-OH abrogates coplanarity and conjugation, thereby compromising scavenging.

The flavonoids can also act, under certain conditions, as pro-oxidants and, hence, promote the oxidation of other compounds.

Flavonoids including flavones, isoflavones and flavonones, play two opposite roles: they act as antioxidant scavenging hydroxyl radicals and other reactive species produced by reaction among Cu^{2+}, H_2O_2 and O_2, and they act as prooxidants producing reactive species through direct reaction with Cu^{2+} and O^2.

Prooxidant activity is directly proportional to the total number of hydroxyl groups in a flavonoid molecule (Cao, Sofic, Prior, 1997). Series of mono and dihydroxyflavonoids demonstrated no detectable prooxidant activity, while multiple hydroxyl groups, especially in the B-ring, significantly increased production of hydroxyl radicals in Fenton reaction (Hanasaki, Ogawa, Fukui, 1994; Heim et al., 2002). The single hydroxyl substitution at position 5 provides no activity, while the conjugation between rings A and B is very important for copper-initiated prooxidant action of the flavonoids. The O-methylation of the hydroxyl substitutions inactivates both the antioxidant and prooxidant activities of the flavonoids (Cao et al, 1997).

It is possible that flavonoid prooxidant function could play a role in their other beneficial functions. For example, epigallocatechin gallate promotes apoptosis and has bactericidal activity, which is attributed to its ability to reduce O_2 to yield H_2O_2 (Arakawa, Maeda, Okubo, Shimamura, 2004; Nakagawa, Hasumi, Nagai, Wachi 2004; Maeta, Nomur, Takashume, Izawa, Inove, 2007.)

Flavonoid prooxidant properties seem to be concentration-dependent. They act as antioxidants at low concentration while high concentrations act as prooxidants (Yen, Duh, Tsai, Huang 2003).

Similarly, protection of human leucocytes against superoxide-induced oxidative DNA damage by quercetin was shown. Incubation concentrations of quercetin (1–50 μM) reduced levels of oxidative DNA damage, while at 100 μM the amount of damage increased. These results were supported by ESR-findings on quercetin in solution, also showing a prooxidant effect at 100 μM (Wilms, Kleinjans, Moonen, Briedé, 2008).

Some flavonoids have a mutagenic effect and can also interfere with essential biochemical pathways. Among proteins that interact with flavonoids, cytochromes P450 (CYPs) monooxygenases metabolize xenobiotics (e.g. drugs, carcinogens). Inhibition of CYPs involved in carcinogen activation and scavenging reactive species are formed from carcinogens by CYP-mediated reactions. Flavonoids induce the expression of several CYPs and modulate (inhibit, stimulate) their metabolic activity. Flavonoids enhance activation of carcinogens and/or influence the metabolism of drugs via induction of specific CYPs. (Hodek, Trefil, Stiborovà, 2002)

There has been an exponential increase in the number of studies investigating how different components of the diet interact at the molecular and cellular levels to determine and witness the fate of a cell.

Many epidemiologic investigations, and human clinical trials showed that polyphenols have important effects on cancer chemoprevention and chemotherapy. In many molecular mechanisms of action for prevention against cancer, flavonoids play a major role by interacting between different types of genes and enzymes. Many mechanisms of action have been identified, including carcinogen inactivation, antiproliferation, cell cycle arrest, induction of apoptosis, inhibition of angiogenesis, antioxidation, and reversal of multidrug resistance or a combination of these mechanisms. In addition, the mechanism of inhibition of polyamine biosynthesis can contribute to the antiproliferative activities of flavonoids. The process of carcinogenesis is generally considered a complex and multistep process and can be divided into three stages (Clere, Faure, Martinez, Andriantsitohaina, 2011):

1. Initiation is a rapid phase, comprised of exposure and interaction of cells, especially DNA, with a carcinogenic agent;
2. Promotion is relatively lengthier than the previous stage, abnormal cells persist, replicate and may originate a focus of preneoplastic cells;
3. Progression stage is the final phase of the tumorigenesis that involves the gradual conversion of premalignant cells to neoplastic ones with an increase in invasiveness, metastasis potential, and new blood vessel formation (angiogenesis).

Studies on animals and investigations using different cellular models suggest that some flavonoids could inhibit tumor initiation as well as tumor progression.

The effects of EGCG on tumor cell cycle arrest have been widely described. EGCG induces cell cycle arrest in many cancer cells without affecting normal cells (Ahmad, Feyes, Nieminen, Agarwal, Mukhtar, 1997).

Delphinidin inhibits cell growth and induces apoptosis of highly aggressive human prostate carcinoma PC-3 cells. Flavonoids can act as inhibitors of cyclin-dependent kinase from breast carcinoma cells. Quercitin has also been shown to mediate the down regulation of mutant p53 in a human breast cancer cell line and other studies indicate that the quercitin induced growth-inhibitory activity in ovarian cancer cells (Rice-Evans, et al., 1996).

Catechins, *in vitro* inhibit the invasion of mouse MO4 cells into embryonic chick heart fragments (Bracke, Vyncke, Opdenakker, Foidart, De Pestel, 1991) Mareel M. A. Quercetin and apigenin inhibited melanoma cell (B16-BL6) growth and metastatic potential in syngenetic mice, *in vitro* (Yin, Giuliano, Van Herle, 1999) They were found to significantly decrease the invasion of B16-BL6 cells. Also apigenin significantly decreased the invasion of lymphatic vessels of intestinal adenocarcinomas that are induced by azoxymethane and are of cancer peritoneal metastasis. The inhibitory effect of apigenin on cancer metastasis may be through the inhibition of phosphorylation of mitogen-activated protein kinase (MAPK) (Tatsuta, Iishi, Baba, Yano, Murata, Mukai, Akedo, 2000)

In Western countries, breast cancer is one of the most common causes of death in women and prostate cancer is the second most common cause of death in men.

Some flavanols (isoflavones) act as phytoestrogens that bind to the estrogen receptor (ER) and induce various estrogenic and antiestrogenic responses (Wall, 2004). Evidences support that estrogens are involved in mammary carcinomas. High concentrations of the isoflavones may reduce the risk of cancer, particularly in the breast and prostate.

Several studies have documented the anticancer and chemopreventive efficacy of grape seed extract (GSE) against various malignant tumors including prostate cancer (PCA). GSE is a complex of polyphenols including gallic acid monomeric flavans and dimeric, trimeric and plomeric procyanidins, some of which are esterefied with gallic acids (Agarwal 2007; Kaur, Mandair, Agarwal, and Agarwal, 2008; Kaur, Agarwal, and Agarwal, 2009; Shen-Chieh et al. 2010). These studies clearly show the important role of the gallate esterification on the anticancer activity.

The authors show that both procyanidin B2-3-O-gallate and procyanidin B2-3-3′-O-digallate have strong anti-cancer efficacy against human PCA LNCaP cells, and that the presence of a gallate moiety at 3′ rather than the 3 position is responsible for most of the biological activity of procyanidin B2-3-3′-O-digallate, while procyanidin B2 and gallic acid were ineffective despite their high antioxidant activity.

In our preliminary study we evaluated the anti-proliferative effect in cancer cells of grape seed extract (GSE) from white and red *vitis vinifera* L. grape cultivars and clones of the same cultivar, and autochthons of Lazio region (central Italy) in years 2006 and 2008. The cultivars were grown in the experimental field of the Research Unit for Enology of Velletri in the Lazio region (Italy) during vintage 2006 and 2008. The trials were made in triplicate. Grapes were harvested at maturation. The cultivars had training system to Cordon Spur. In both years the same cultural practices were adopted in the vineyard.

All the cultivar and clones from the same cultivar showed AA as significantly different ($p \leq 0.05$), and significant differences ($p \leq 0.05$) were also found between the same cultivars in two years (figure 12). TP and FL followed the same trend of AA. A highly significant correlation was found between the antiradical activity of GSE and its phytochemical content, especially with total polyphenols ($r = 0.84$; $p < 0.001$).

After the chemical analysis, the GSEs were used to test their anti-proliferative effect *in vitro* on the cervix adenocarcinoma cells (HeLa). The experiments were performed in both high-glucose (HG) and low-glucose (LG) supplemented media, which contain 25mM and 5 mM glucose respectively. The cells of different concentration of dried GSE 2006-2008 (from 0 to 2.0 mg dried extract mL^{-1}) were added. Results obtained demonstrated that only GSE from Abbuoto 2008 was able to decrease the cell proliferation rate at concentrations similar to those expected to be assumed by food intake (0.01-0.15 mg dried extracts mL^{-1}). The chemical characterization (by HPLC/DAD/MS) of Abbuoto GSE 2008 highlighted an amount of flavanols (monomeric, dimmers, trimers and polymeric and galloylated forms) significantly higher ($p \leq 0.05$) compared to the extracts obtained from other cultivars for both years. In particular, the polymeric procyanidins were significantly higher in 2006, while the gallate esters were significantly higher in 2008.

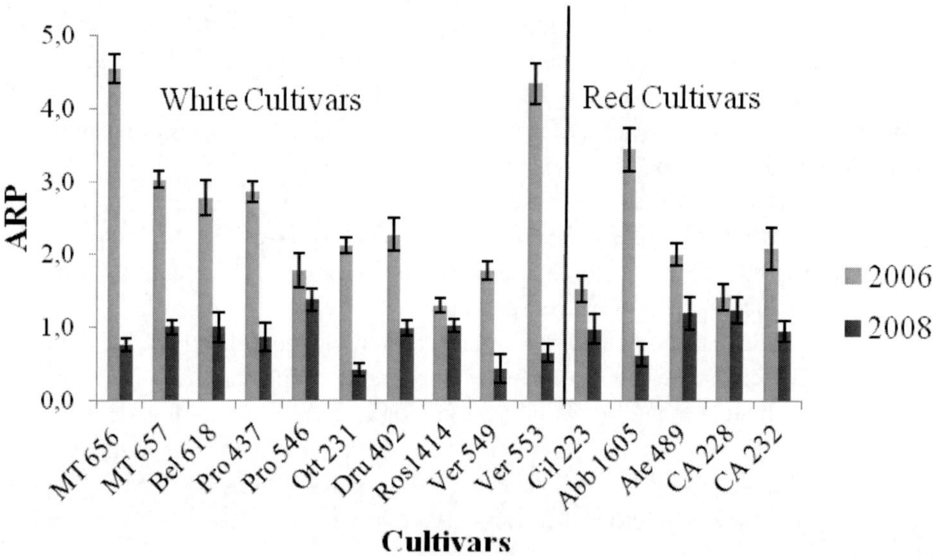

Figure 12. Antioxidant activity (ARP) of GSE from 15 cultivars: MT= Moscato di Terracina, Bel= Bellone, Pro= Procanico, Ott= Ottonese, Dru= Drupeggio; Ros= Rossetto, Ver = Verdicchop, Cil = Ciliegiolo, Abb= Abbuoto, Ale= Aleatico, CA= Cesanese d'Affile. The number indicates the type of clones. Values represent the mean of three replications. Bars indicate ± standard deviation.

In 2006 the polymeric procyanidins and gallate esters represent the 77.90% and 5.39% respectively with respect to the total flavanols; while in 2008, the polymeric procyanidins and gallate esters represent the 42.96% and the 16.73% respectively with respect to the total flavanols. GSE 2008 of Abbuoto 1065 showed the highest values of gallate esters (10.10 mg/g dried extract) compared to all other GSEs (where gallate esters ranging between 3.08 and 5. 13 mg/g of dried extract).

The growth inhibitory effect of GSE from Abbuoto 2008 was evident only in LG conditions. We provide evidence that GSE from Abbuoto 2008 induces authophagy, the major catabolic mechanism used by eukaryotic cells to degrade proteins and organelles, as demonstrated by measurement of acid compartment (e.g. lysosomes) upon acridine orange staining, by Western blot analyses of microtubule associated protein light chain (LC3) and p62, and by cellular distribution of green fluorescent protein/GFP)-conjugated LC3, which appears localized in discrete spots, reasonably standing for autophagosomes. In the LG medium the mitochondrial function and structure were completely collapsed, confirming that mitochondria represents specific targets of Abbuoto GSE 2008 (Cecchini et al. 2012).

All GSEs from other cultivars in 2008 were able to induce the autophagic response but only at higher concentrations, while All GSEs from other cultivars 2006 were ineffective even though they showed higher AA with respect to GSEs from 2008 cultivars. This suggests that higher antioxidant activity does not always correspond to greater anti-carcinogenesis activity.

Therefore, from a chemical point of view, the polymeric flavanols with several -OH substituents have higher antioxidant activity, and from a biological point of view the cancer chemopreventive effect is more molecular-structure dependent. In agreement with other authors (Agarwal et al.2007; Chou et al. 2010) the esterification of procyanidins with gallic acid increases the anti-cancer effect.

Considering that in two years, the grape cultivars were cultivated under the same conditions of microclimate and soil, and the agronomic practices were adopted in agreement with other authors (Lachman, Šulc, Faitovà, Pivec 2009, Cecchini et al. 2012 a, b); the significant difference found in phenolic compounds for the same cultivar in the two years was exclusively due to different meteorological conditions.

The climatic condition of 2006 seemed to affect the higher synthesis of polymeric procyanidin, while the climatic condition of 2008 seemed to affect the higher synthesis of gallate esters.

In 2006 the rainfall in spring was plentiful, but the ripening period was dry. The temperatures were mild with few fluctuations and average temperatures were not extreme, and about 2° higher than in 2008. In the spring and summer of 2008 there was considerable rain, especially during the ripening period. The average temperature was very low during the ripening period and mild in summer.

Considering that the polymeric procyanidins account for the most superior antioxidant and oxid-reduction capacity due to the enlargement of the number of reactive sites (OH groups), the higher antioxidant activity and higher TP content of grape seeds in 2006 appears to be largely influenced by the polymeric content, with other flavanols playing a minor role. From these results it is clear that all the phenomena that affect the physiology of the plant, especially the climatic conditions, lead to different metabolic pathways in the synthesis of the polyphenols.

Conclusion

In this chapter we have reported a summary of the results on the antioxidant activity of grape tissues obtained in our laboratories in the last few years. In defining the nutraceutical potential of these products, we have taken into account both their phytochemical profile, expressed in terms of both their content of polyphenols and flavanols, and their antiradical power. Reported data underline that the antioxidant activity is greatly dependent on the polyphenolic profile, and the latter is greatly dependent on genotype and environmental interactions (i.e. climate, temperature, light, maturation degree, ripeness, pathogen attacks, defoliation, and various developmental processes). The beneficial effects of polyphenols are mainly due to the antioxidant properties of flavonoids. These properties not only depend on the total polyphenol content, but also on the type and concentration of individual compounds. In particular, the flavonols esterefied with gallic acid seem to show a greater anticancer effect.

References

Agarwal, C., Veluri, R., Kaur, M., Chou S-C. Thompson, J.A. & Agarwal R. (2007). Fractionation of high molecular weight tannins in grape seed extract and identification of procyanidins B2-3,3′-diO-gallate as a major active constituent causing growth inhibition and apoptotic death of DU145 Human prostate carcinoma cells. *Carcinogenesis*. 28,1478-1484.

Aherne, A. S., Obrien, N.M. (2002). Dietary flavonoids: chemistry, food content and metabolism. *Nutrition* 18, 175-81.

Ahmad, N., Feyes, D.K., Nieminen, A.L., Agarwal, R. Mukhtar, H. (1997). Green tea constituent epegallocatechin-3-gallate and induction of apoptosis and cell cycle arrest in human carcinoma cells. *J. Natl. Cancer. Inst.* 89 (24), 1881-1886.

Amić, D., Davidovič-Amić, D., Bešlo, D., Rastija, V., Lučić, B., Trinajstić, N. (2007). SAR and QSAR of antioxidant activity of flavonoids. *Cur.r Med. Chem.* 14, 827-845.

Arakawa, H. Maeda, M. Okubo, S. Shimamura T. (2004). Role of hydrogen peroxide in bactericidal action of catechin. *Biol Pharm Bull*, 27, 277–281.

Balik J.; Kyselakova M.; Vrchotova N.; Triska J.; Kumsta M.;Veverka J.; Hic P.; Totusek J.;Lefnerova D.(2009) Relation between Polyphenols Content and Antioxidant Activity in Vine, Grapes and Leaves. Czech. *J. Food Sci.* 26,special Issue S 25-S32.

Bhattacharjee, S., (2005). Reactive oxygen species and oxidative burst: Roles in stress, senescence and signal transduction in plants. *Current Science*. 89-97.

Boss, P.K., Davies C., & Robinson, S.P. (1996 a). Expression of anthocyanin biosynthesis pathway genes in red and white grapes. *Plant. Mol. Biol*. 32, 652-663.

Boss, P.K., Davies, C., and Robinson, S.P. (1996 b). Analysis of the expression of anthocyanin pathway genes in developing *Vitis vinifera* L. cv. Shiraz grape berries and the implications for pathway regulation. *Plant Physiol*. 111, 1059-1066.

Bracke, M., Vyncke, B., Opdenakker, G., Foidart, J.M., De Pestel, G., Mareel, M. (1991). Effect of catechins and citrus flavonoids on invasion *in vitro*. *Clin. Exp. Metastasis*. 9, 13-25.

Brand-Williams, W., Cuvellerand, M. E. & Berset, C.(1995). Use of a free radical method to evaluate antioxidant activity. *Food Sci. Tech.,* 28, 25-30.

Burda S, Oleszek W. (2001). Antioxidant and antiradical activities of flavonoids. *J Agric Food Chem* 49, 2774–2779.

Cao, G., Sofic, E., Prior, R.L, 1997. Antioxidant and prooxidant behavior of flavonoids: structure-activity relationship. *Free Radic. Boil. Med.* 22,749-760.

Carbone K., Picchi, V. Giannini, B. Lo Scalzo, R., Cecchini F. 2013. On the importance of using different radical sources to investigate the scavenging properties of grape tissues and wines. *Int. J. Nut. Dietetics*. 1 (1) 1-20.

Cecchini, F., & Giannini, B. (2010). Comparison of the free radical-scavenging activity in Oidium and sound Dolcetto grape cultivars grown in a terroir of central Italy. *In Proceedings VIII International Terroir Congress*, 14-18 June 2010 Soave Italy pp 41-46.

Cecchini, F., Moretti, S. Mulinacci, N. Innocenzi, M., Pasqua, G. Ciriolo, M.R., Filomeni G. (2012). Anti-proliferative effects in cervical cancer effects (HeLa) of grape seeds extract from *Vitis vinifera* cultivar of central Italy. *In proceedings XXVIth International Conference on Polyphenols* Florence Italy 23-26 July 2012. pp 471-472.

Cecchini, F.,Giannini, B. Morassut, M., Moretti, S. (2012 a). Effect of the basic terroir on the nutraceutical proprieties in *Vitis vinifera* L. cv Dolcetto red grape variety. *Le Progrès agricole et Viticole*. 88-92.

Cecchini, F.,Giannini, B., Serra M.C., Moretti, S. (2013). Canopy microclimate: the effect of exposure on the nutraceutic properties of *Vitis vinifera* L cv. Dolcetto grape berries. (In press in *Infowine online J. Vtic. Enol*. 2013)

Cecchini, F.,Giannini, B.,Casadei, G., Serra M.C. (2012 b). Influenza del clima sul contenuto polifenolico e sull'attività antiossidante della cultivar Dolcetto. *In proceedings del del XV Convegno nazionale di Agrometereologia*. pp.55-56.

Chen, N., and Karantza-Wadsworth V. (2009) Role and regulation of autophagy in cancer. Biochem. Biophis. Acta. 1793,1516-1523.

Chou S-C., Kaur M., Thompson, J. A., Agatwal, R. & Agarwal, C. (2010). Influence of gallate esterification on the activity of procyanidin B2 in Androgen-dependent prostate carcinoma LNCaP cells. *Pharmaceutical Research*. 27, 619-627.

Clere, N., Faure, S., Martinez, M.C., Andriantsitohaina, R. (2011). Anticancer properties of flavonoids: roles in various stages of carcinogenesis. *Cardiovascular & Hematological Agents in Medicinal Chemistry*. 9, 1-16.

Connor, A. M., Luby, J.J., &Tong C.B.S. (2002). Genotypic and Environmental variation in antioxidant activity, total phenolic content, and anthocyanin content among blueberry cultivars. *J. Amer. Soc. Hort.* 127 (1), 89-97.

Coombe, B.G. (1960). *Plant Physiol*. 35 (2), 241-250.

Coombe, B.G. (1973). The regulation of set and development of the grape berry. *Acta Hortic*.34, 261-273.

Cos P, Ying L, Calomme M, Hu JP, Cimanga K, Van Poel B, et al. (1998). Structure–activity relationship and classification of flavonoids as inhibitors of xantine oxidase and superoxide scavengers. *J. Nat. Prod.* 61,71–6.

Crippen D.D., and Morrison J. C. (1986). The Effects of Sun Exposure on Phenolic Content of Cabernet Sauvignon Berries during Development. *Am. J. Enol. Vitic*. (37) 243.247.

Crozier, A. (2005). The influence of moderate red wine consumption on antioxidant status and indices of oxidative stress associated with CHD in healthy volunteers. *Br. J. Nutr.*, *93*, 233–240.

Di Stefano R., Cravero M.C., (1991). Metodi per lo Studio dei Polifenoli delle Uve. *Riv. Vitic. Enol.* 2,37-43.

Dokoozlian, N.K., Kliver, W. M. (1996). Influence of light on grape berry and composition varies during fruit development. *J. amer. Soc. Hort. SCI.*121 (5), 869-874.

Doweney, M.O., Dokoozlian, N. K., and Krstic, M. P.(2005). Cultural practice and environmental impacts on the flavonoids composition of grapes and wine: a review of recent research. *Am. J. Enol. Vitic.*57 (3), 257-267.

Doweney, M.O., Harvey, J. S., Robinson, S.P. (2003). Analysis of Tannins in Seeds and Skins of Shiraz Grapes throughout Berry Development. *Au. J. Grape and Wine Res.* 9 (1), 15-17.

Dreosti, I. E. 2000. Antioxidant polyphenols in tea, cocoa, and wine. *Nutrition* 16, 692-694.

Eng , E. T., Ye J. , Williams, D., Phung, S., Moore , R. E. , Young M.K , Gruntmanis, U. , Braunstein G , & Chen S. (2003). Suppression of Estrogen Biosynthesis by Procyanidin Dimers in Red Wine and Grape Seeds. *Cancer Res.* 63, 8516-8522.

Engelbrecht, A. M, Mattheyse, M., Ellis, B., Loos, B., Thomas, M., Smith, R., Peters, S., Smith, C. Myburgh K. (2007). Proanthocyanidin from grape seeds inactivates the PI3-kinase/PKB pathway and induces apoptosis in a colon cancer cell line. *Cancer letters*.144-153.

Foyer, C. H., & Noctor, G. (2005). Oxidant and signalling in plants: A re-evaluation of the concept of oxidative stress in a physiological contex. *Plant cell and Environ.* 28, 1056-1071.

Fregoni M. (1999) Viticultura di qualità (2 th ed.). Informatore Agrario Italy pp. 25-48

Garofolo, A., Giannini, B., and Favale, S., Savino, M. (2006). Studio del potere antiradicalico in relazione al profile polifenolico di uve e vini ottenuti da cloni della cv. Cesanese d'Affile. *Riv. Vitic. Enol.* 4, 35-57.

Giannini B, Cecchini F, Moretti S. (2011) Nutraceutic Property of Italian Red and White Grape Cultivar and Wine. In proceedings *of the Second Edition of the International Conference Series on Wine Active Compounds* 2011,1 321-322 Bourgogne France.

Giovannelli, G., Brenna, O.I.V. (2007). Evolution of some phenolic components, carotenoids and chlorophylls during ripening of three Italian grape varieties. *Eur Food Res Technol.* 225, 145-150.

God, J.M.; Tate, P.; Larcom, L.L. (2007). Anticancer effects of four varieties of muscadine grape. *J.*

Gutteridge, J.M. (1993). Free radicals in disease processes: A compilation of cause and consequence. *Free Rad. Res.*, 19, 141–158.

Halliwell, B., Gutteridge, J.M., Cross C.E. 1992. Free radicals, antioxidants, and human disease: where are we now? *J. Lab. Clin. Med.*119, 598–620.

Hanasaki, Y. Ogawa, S. Fukui, S. (1994). The correlation between active oxygens scavenging and antioxidative effects of flavonoids. *Free Radic. Biol. Med.* 16, 845-850.

Hartmann, H. T. Kester, D. E. & Davies, F. T. (1990). Plant Propagation Principles and Practices. Englewood Cliffs, New Jersey (1990).

Haselgrove L., D. Botting, R. van Heeswuck, P.B. Højј, P.R. Dry, C. Ford, and P.G. Iland. (2008). Canopy Microclimate and Berry Composition: The Effect of Bunch Exposure on the Phenolic Composition of vitis vinifera L cv. Shiraz Grape Berry. *Aus. J. Grape and Wine Res.h.* 6 (2), 141-149.

Havsteen, B.H., (2002) The biochemistry and medical significance of the flavonoids. *Pharmacology and Therapeutics*, 67–202.

Heim KE, Tagliaferro AR, Bobilya DJ. (2002). Flavonoid antioxidants: chemistry, metabolism and structure–activity relationships. *J. Nutr. Biochem.* 13, 572–84.

Heim, K. E., Tagliaferro, Bobilya, D. J. (2002). Flavonoid antioxidants: chemistry, metabolism and structure-activity relationships *J. Nutrition. Biochem.* 13, 572–584

Hirano R, Sasamoto W, Matsumoto A, Itakura H, Igarashi O, Kondo K. (2001). Antioxidant ability of various flavonoids against DPPH radicals and LDL oxidation. *J. Nutr. Sci. Vitaminol* (Tokyo).47:357–62.

Hodek, P., Trefil, P., Stiborovà, M. (2002). Flavonoids-potent and versatile biologically active compounds interacting with cytochromes P450. *Chemico-Biol. Interaract.* 139, 1-21.

Hvsteen, B. H. (2002). The biochemistry and medical significance of the flavonoids. *Pharmacology & Therapeutics* 96, 67-202.

Jung, K., Wallig, M., Singletary, K. (2006). Purple grape juice inhibits 7,12-dimethylbenz-anthracene (DMBA)-induced rat mammary tumorigenesis and *in vivo* DMBA-DNA adduct formation, *Cancer Lett.* 233, 279–288. R

Kaur M, Mandair, R, Agarwal R & Agarwal C. (2008). Grape Seed Extract Induces Cell Cycle Arrest and Apoptosis in Human Colon Carcinoma Cells. *Nutrition and Cancer.* 2-11.

Kaur, M Agarwal,C and Agarwal R. (2009). Rew. Anticancer and Cancer Chemopreventive Potential of Grape Seed Extract and Other Grape-Based. *J. Nutr.* 1806-1812.

Kennedy, J. A., Mark A., Matthews M.A., Waterhouse A. L. (2002). Effect of Maturity and Vine Water Status on Grape Skin and Wine Flavonoids. *Am. J. Enol. Vitic.*, 53 (4),268-274.

Klaus, A., Heribert, H. (2004). Reactive oxygen species: metabolism, oxidative stress, and signal transduction. *Annu. Rev. Bio.*55, 373-399.

Knight, J.A. (1995). Diseases related to oxygen-derived free radicals. *Annals of Clinical and Laboratory Sciences.* 25, 111-121.

Lachman, J., Šulc, M., Faitovà, K., Pivec, V. (2009). A Review. Major factors influencing antioxidant contents and antioxidant activity in grapes and wines. *Int. J. Wine Res.* 1, 101-121

Lotito, S.B., Actis-Goretta, Renart, M.L., Caligiur, M., Rein, D. Schmitz, H.H.et.al (2000). A Review. Influence of oligomer chain length on the antioxidant activity of procyanidins. *Biochem. Biophys. Res. Commun.* 276, 945-951.

Maeta, K., Nomur, W., Takashume, Y., Izawa S., Inove Y. (2007). Green tea polyphenols function as prooxidants to active oxidative-stress-responsive transcription factors in yeasts. *Appl. Environ. Microbiol.*, 73,, 572–580.

Mazza, G., Miniati, E. (1993). Antocyanins in fruits and vegetables, and grains. CRC Press Inc. USA *Med. Food. 10*, 54-59.

Mulinacci, N., Innocenzi, M., Pasqua, G., Santamaria, A. (2011). Antiossidanti dei semi dell'uva. *In proceedings del Seminario CRA* Velletri (Roma) Italy 30 Marzo 2011 CD.

Nakagawa, H.K. Hasumi, J.T. Nagai, K. Wachi M. (2004). Generation of hydroxide peroxide primarily contributes to the induction of Fe (II)-dependent apoptosis in Jurkat cells by (−)-epigallocatechin-3-gallate. *Carcinogenesis.* 25, 1567-1574.

Organization International de la Vigne et du Vin. (OIV) (2013). Statistical report on world vitiviniculture 2013. http://www.oiv.int/oiv/cms/index?lang=it

Prochàzkovà, D., Bošovà, I., Wilhelmovà, N. (2011). A Revew: Antioxidant and prooxidant properties of flavonoids. *Fitoterapia* . 82, 513-523.

Ribéreau-Gayon,P., Glories, Y., Maujean, A., Ubourdieu, D. (2003). Trattato di enologia II. (2° ed.) Edagricole Editor.

Rice-Evans, C. A. Miller, N.J. Paganga, G. (1997). Reviews Antioxidant properties of phenolic compounds. *Trend in Plant Science.* 2, 152-159.

Rice-Evans, C. A. Miller, N.J. Paganga, G. 1996. Review Structure-antioxidant activity relationships of flavonoids and phenolic acids. *Free Radic. Biol. Medic.* 20, 933-956.

Shanmuganayagam, D.; Warner, T.F., Krueger, C.G., Reed, J.D., Folts, J.D. (2007). Concord grape juice attenuates platelet aggregation, serum cholesterol and development of atheroma in serum cholesterol and development of atheroma in hypercholesterolemic rabbits. *Atherosclerosis.* 190(1), 135-42.

Shrikhande, A.J. (2000). Wine by-products with health benefits. *Food Res. Internat. 33*, 469–474.

Singletary, K. W. & Brandon, M. (2001). Effect of Grape Seed Proanthocyanidins on Colon Aberrant Crypts and Breast Tumors in a Rat Dual-Organ Tumor Model. *Nutrition and Cancer*, 2, 252–258.

Singleton V. L., Joseph A. Rossi, Jr.(1965). Colorimetry of Total Phenolics with Phosphomolybdic-Phosphotungstic Acid Reagents *Am. J. Enol. Vitic.* 16,144-158.

Smart R. E., (1987). A Review. Canopy Management to Improve Yield, Fruit Composition and Mechanization. *In Proceedings of the 6th Australian wine industry Technical Conference*, Adelaide 205-211.

Spayd S. E., J.M., Tarara, D.L. Mee, and J. C. Ferguson (2002). Separation of Sunlight and Temperature Effects on the Composition of vitis vinifers cv. Merlot Berries. *Am. J. Enol. Vitic.*171-182.

Tatsuta, M., Iishi, H., Baba, M., Yano, H., Murata, K., Mukai, M., Akedo, H. (2000). Suppression by apigenin of peritoneal metastasis of intestinal adenocarcinomas induced by azoxymethane in Wistar rats. *Clin. Exp. Metastasis*. 18, 657–62.

Taubert, D., Breitenbach, T., Lazar, A., Censarek, P., Harlfinger, S., Berkels, R., Klaus, W., Roesen, R. (2003). Reaction rate constants of superoxide scavenging by plant antioxidants. *Free Radic. Biol. Med.* 35, 1599-1607

Teixeira, A., Erias-Dias, J., Castellarin, S. D., and Geròs E. (2013). A Review. Berry phenolics of grapevine under challenging environments. *Int. J. Mol. Sci.*14, 18711-18739.

van Acker, S. A., de Groot, M. J., van den Berg, D.-J., Tromp, M. N., Donnè-Op den Kelder, G., van der Vijgh, W. J., Bast, A. (1996). A quantum chemical explanation of the antioxidant activity of flavonoids. Chem. Res. Toxicol., 9(8), 1305-1312.

Wall, T. (2004). Serial Review: flavonoids and isoflavones (phytoestrogens): Absorption, metabolism, and bioactivity. *Free rad. Biol. Med. 36 (7), 829-837.*

Waterhouse, A. L. (2002). Wine phenolics. *Annals of the New York Academy of Sciences* 957, 21-36.

Wilms, L.C. Kleinjans, J.C. Moonen, E.J. Briedé J.J. (2008). Discriminative protection against hydroxyl and superoxide anion radicals by quercetin in human leucocytes *in vitro. Toxicol In Vitro*. 22, 301–307.

Xia, En-Q. Deng, G-F., Guo, Ya-J. and Li H-B. (2010). Biological activities of polyphenols from grapes. *Int. J. Mol Sci.* 11(2), 622–646.

Yao, l.H., Jiang, Y. M.Shi, Tomàs-Barberàn, F.A., Datta, N., Singanusong, R. & Chen, S.S. (2004). Flavonoids in food and their Health Benefits. *Plant Foods for Human Nutrition* 59, 113-122.

Yeh T-C., Chiang, P-C., Li T-K., Hsu, J-L, Lin, C-J, Wang, S-W., Peng, C-Y. Guh, J-H. (2007). Genistein induces apoptosis in human hepatocellular carcinomas via interaction of endoplasmic reticulum stress and mitochondrial insult. *Biochem.pharmaco...* 73, 782-792.

Yeh, S.L., Wang, W.Y., Huang, C.H., Hu, M.L.(2005). Pro-oxidative effect of β-carotene and the interaction with flavonoids on UVA-induced DNA strand breaks in mouse fibroblast C3H10T1/2 cells. *.J Nut.r Biochem.* 16, 729–735.

Yen, G.C. Duh, P.D. Tsai, H.L. Huang S.L. (2003). Pro-oxidative properties of flavonoids in human lymphocytes. *Biosci. Biotechnol. Biochem.* 67, 1215–1222.

Yin, F., Giuliano, A.E., Van Herle, A.J. (1999). Growth inhibitory effects of flavonoids in human thyroid cancer cell lines. *Thyroid*. 9, 369-76.

In: Grapes
Editor: José S. Câmara
ISBN: 978-1-63321-402-6

Chapter 5

Screening the Phenolic Composition of the Wide Diversity of Autochthonous Grape Cultivars from La Mancha (Spain)

Isidro Hermosín-Gutiérrez[1*], *Sergio Gómez-Alonso*[1,2], *Noelia Castillo-Muñoz*[1], *Mónica Fernández-González*[1,2], *Jesús Martínez-Gascueña*[3] *and Esteban García-Romero*[3]
[1]Instituto Regional de Investigación Científica Aplicada, Universidad de Castilla-La Mancha, Campus Universitario s/n, Ciudad Real, Spain
[2]Fundación Parque Científico y Tecnológico de Albacete, Paseo de la Innovación, Albacete, Spain
[3]Instituto de la Vid y el Vino de Castilla-La Mancha, Carretera Toledo-Albacete s/n, Tomelloso, Spain

Abstract

The Spanish region of La Mancha is the largest grape-growing and winemaking region in the world and has an ancient wine tradition. It possesses a wide diversity of grape cultivars but many of the autochthonous ones are being replaced by more well-known and more widely grown grape cultivars. The Collection of Vine Germplasm from the La Mancha region currently includes 44 new *Vitis vinifera* grape genotypes that have never been described before. In this study, we have performed a screening of the phenolic composition of some of these interesting grape cultivars, covering those phenolic compounds influencing wine quality and related with healthy properties: anthocyanins (only for red grape cultivars), flavonols, hydroxycinnamic acid derivatives, resveratrol, and flavan-3-ol monomers and dimers. An analysis of qualitative (profiles of phenolic compounds) and quantitative data revealed that phenolic composition is a useful chemo-taxonomical tool for cultivar differentiation and also allows making suggestions about the suitability of a specific grape cultivar for making a specific type of wine.

[*] Corresponding Author address: Email: isidro.hermosin@uclm.es.

Keywords: Grape; phenolic composition; anthocyanin; flavonol; cultivar differentiation; wine

Introduction

Grape diversity is really very high. Around 6154 principal vine cultivar names have been recorded by the OIV from data submitted by the 35 nations involved in this international organization; after removing the records that are common to different countries, it results in 4020 different vine cultivar names (Organisation Internationale de la Vigne et du Vin, 2013). This compilation contains the names and synonyms of cultivars from all destinations, without classification, whether for fresh consumption (table grapes), drying (dried grapes), making wine (wine grapes) or grafting (rootstocks). However, this figure must not be considered as the final number of existing *Vitis* cultivars. In contrast, the number of grape cultivars that are widely cultivated in the world is rather reduced. Focusing on wine grapes, this number is around several dozens including the very well known red cultivars like Cabernet Sauvignon, Merlot, Pinot Noir and Syrah and the common white cultivars like Chardonnay and Sauvignon Blanc. In Spain, the red grape Tempranillo, Garnacha and Bobal cultivars together with the white grape Airén and Macabeo cultivars are some of the most cultivated. Even in the region of La Mancha, Cencibel (a synonym of Tempranillo) and Airén still dominate the vineyards. This panorama makes it difficult to be successful in the very competitive wine market and some lucrative selling strategies include the improvement of grape and wine quality as well as the use of grape cultivars which are very well adapted to local or regional cultivation zones and thus give rise to high quality wines. On the basis of the latter strategy, some winemaking regions (usually identified with countries) have associated the name of one grape cultivar with the origin of its quality wine. Some examples are the French cultivars that have been widely spread around the world during the last few decades, but also the relatively recent Shiraz wines from Australia, Sauvignon Blanc and Pinotage wines from South Africa, Carménère wines from Chile, or Malbec wines from Argentina.

The production of wine in the Iberian Peninsula is an ancient tradition which had already been initiated before the Romans conquered the Iberian Peninsula and has reached its maximum expression in the region of La Mancha. The Spanish region of La Mancha (central-southern Spain) is currently the largest grape-growing (ca. 0.5 million ha of vineyards) and wine-making (ca. 20 million hL of wine production) region in the world. This long term cultivation of wine grape cultivars in La Mancha has led to a wide diversity of grape cultivars, many of which are autochthonous. However, in current times, many of these ancient grape cultivars are being replaced by more well-known grape cultivars because the former probably offered less crop yields or were less resistant to vine diseases. Since the year 2004, a great effort has been made by the Instituto de la Viña y el Vino de Castilla-La Mancha (IVICAM; Institute for Vine and Wine of Castilla-La Mancha, an institution belonging to the regional government) to create a Collection of Vine Germplasm from Castilla-La Mancha. This collection has been constantly expanded with new materials and it currently includes 137 different genotypes of *Vitis vinifera* grapes, 44 of them being genotypes that have never been described before. In this study, we have performed a screening of the phenolic composition of some of these interesting grape cultivars, including both white and red grapes. The study was focused on a very complete list of types of phenolic compounds, including those which mostly determine wine quality and related with health properties associated with the moderate consumption of wine: anthocyanins (only for red grape cultivars), flavonols, hydroxycinnamic acid derivatives, resveratrol, and flavan-3-ol monomers and dimers. An analysis of qualitative (profiles of phenolic compounds) and quantitative data revealed that phenolic composition is a useful chemo-taxonomical tool for cultivar differentiation and also allows making suggestions about the suitability of a specific grape cultivar for making a specific type of wine.

Materials and Methods

Chemicals

All solvents were of HPLC quality, and all chemicals were analytical grade (> 99%). Water was of Milli-Q quality. The following commercial standards from Phytolab (Vestenbergsgreuth, Germany) were used: malvidin 3-glucoside, caffeic acid, *p*-coumaric acid, *trans*-caftaric acid, *trans*-piceid, (−)-epigallocatechin, and (−)-gallocatechin. These commercial standards from Extrasynthese (Genay, France) were used: cyanidin 3-glucoside, procyanidins B1 and B2, kaempferol, quercetin, isorhamnetin, myricetin, syringetin, and the 3-glucosides of kaempferol, quercetin, isorhamnetin, and syringetin. The following commercial standards from Sigma (Tres Cantos, Madrid, Spain) were used: gallic acid, *trans*-resveratrol, (+)-catechin, (−)-epicatechin, (−)-epicatechin 3-gallate, and (−)-gallocatechin 3-gallate. Some other non-commercial flavonol standards (myricetin 3-glucoside, quercetin 3-glucuronide, and laricitrin 3-glucoside) were previously isolated from Petit Verdot grape skins (Castillo-Muñoz et al., 2009b). Finally, a sample of procyanidin B4 was kindly supplied by Prof. Fernando Zamora (Department of Biochemistry and Biotechnology, Universitat Rovira i Virgili, Spain).

All standards were used for identification. Quantitation (in mg per kg of grape) was done as equivalents of the most representative compounds for each family of phenolic compounds: malvidin 3-glucoside was used for anthocyanins; caffeic acid for hydroxycinnamic acid derivatives; quercetin 3-glucoside for flavonol 3-glycosides and their free aglycones; individual flavan-3-ol monomers and dimers for their corresponding standards, but their sum total was taken as (+)-catechin equivalents.

Grapes and Sample Preparation

During the 2010 harvest season, healthy grapes were collected at optimum ripeness for harvesting in the Collection of Vine Germplasm located at IVICAM (Tomelloso, Spain). The grape samples corresponded to the following cultivars (NG means new found genotype): a) 7 of red grapes, including Azargón (NG), Churriago (NG), Gallera Negra (NG), Garnacha Gris, Granadera, Pintailla (NG), and Sinsó; b) 7 of white grapes, including Albillo Mayor, Castellana Blanca, Garnacha Blanca, Jarrosuelto (NG), Moscatel Serrano (NG), Moscatel Grano Menudo, and Planta Fina. All the grape cultivars represented in this vine collection are cultivated in the same vineyard (geographical coordinates 39° 10’ 34.18” N, 3° 0’ 0.94” W) and under the same conditions. The vines were grafted into 110-R rootstock and grown on trellises at a distance of 3 × 1.5 m (row by vine spacing). The plants were trained to a double cordon Royat canopy, with 3 spurs of 2 buds on each arm. The orientation was 120° E-SE / 300° W-NW.

The vineyard was irrigated by a drip system with two drippers per vine-stock. Three batches of 200 grape berries were sampled following a zigzag path between two marked rows of ten vines, picking grape berries from both sun-shaded and sun-exposed cluster sides and from the top, central, and bottom parts of each cluster. Two subsamples of 100 grape berries each were randomly separated from each batch. The berries of one of the subsamples were manually finger-pressed to separate the skins and then softly pressed with filter paper to remove excess juice.

The corresponding grape seeds were manually separated, rinsed with water and dried by softly pressing with filter paper. The skins and seeds were extracted separately: 8 g of skins and 1 g of seeds were immersed in 50 mL of methanol–water–formic acid (50:48.5:1.5) and homogenized (Heidolph Silent crusher M; 20,000 rpm, 3 min). After centrifugation, the resulting pellets were extracted a second time (50 mL) and the combined supernantants were stored at −18 °C until use. The berries of the second subsample from each grape batch were entirely homogenized, the first time using methanol–formic acid (97:3) and the second time, after centrifugation, with methanol–water–formic acid (50:48.5:1.5). Then the combined supernantants were stored at −18°C until use. Extracts of entire berries were only used for analyzing the total content of hydroxycinnamic acid derivatives, whereas skin extracts were used for analyzing the total content of anthocyanins (only for red grape cultivars), flavonols and stilbenes, together with the skin contents of flavan-3-ol monomers and dimers. Seed extracts were used for analyzing their content in flavan-3-ol monomers and dimers. The analysis of anthocyanins present in red grape skins was performed using the corresponding skin extracts. However, it was necessary to obtain anthocyanin-free fractions from the red grape skin extracts to analyze their non-anthocyanin phenolic compounds. ECX SPE cartridges (40 µm, 500 mg, 6 mL; Scharlab, Sentmenat, Barcelona, Spain) allowed the isolation of non-anthocyanin phenolic compounds from the red grape skin extracts, and these anthocyanin-free fractions were used to analyze flavonols and stilbenes.

To carry out this step, 3 mL of grape skin extracts were diluted with 9 mL of 0.1 N HCl and the prepared samples were then passed through the SPE cartridges that had been previously conditioned with 5 mL of methanol and 5 mL of water. After the cartridges were washed with 5 mL of 0.1 N HCl and 5 mL of water, the anthocyanin-free fraction was eluted with 3 × 5 mL of methanol. This eluate was dried in a rotary evaporator (35 °C) and re-dissolved in 1 mL of 20% methanol in water and directly injected into the HPLC equipment. In the case of white grape skin extracts, the analysis of their non-anthocyanin phenolic compounds did not need their previous isolation by ECX SPE cartridges.

Flavan-3-ols (monomers and B-type procyanidin dimers) were isolated from the grape skin extracts by SPE on C18 cartridges (Sep-pak Plus C18, Waters Corp., Mildford, MA; cartridges filled with 820 mg of adsorbent). A mixture of 2 mL of each extract with 12 mL of water was then passed through the C18 cartridge which had been previously conditioned with methanol (5 mL) and water (5 mL).

After the cartridge was dried under reduced pressure, methanol (15 mL) and ethyl acetate (5 mL) were added to recover the adsorbed phenolic compounds. Once the solvent was evaporated in a rotary evaporator (35 °C), the residue was dissolved in methanol (2 mL) and stored at −18 °C until use.

HPLC-DAD-ESI-MS Identification and Quantitation of Phenolic Compounds

Anthocyanins and non-anthocyanin phenolic compounds (flavonols and hydroxycinnamic acid derivatives) were separately analyzed after the previously described methods (Castillo-Muñoz et al., 2007 and 2009b) were adapted to the use chromatography columns of narrow bore and smaller particle size. In the case of grape skin anthocyanins, 10 μL of the diluted extracts (1:4 with 0.1 N HCl) and 20 μL of anthocyanin-free extract fraction in the case of the analysis of non-anthocyanin phenolics, were injected after filtration (0.20 μm, polyester membrane, Chromafil PET 20/25, Machery-Nagel, Düren, Germany) into a reversed-phase column Zorbax Eclipse XDB-C18 (2.1 × 150 mm; 3.5 μm particle; Agilent, Germany), and thermostated at 40 ºC. The flow rate was 0.19 mL/min. The solvents were: solvent A (acetonitrile–water–formic acid, 3:88.5:8.5, v/v/v), solvent B (acetonitrile–water–formic acid, 50:41.5:8.5, v/v/v), and solvent C (methanol–water–formic acid, 90:1.5:8.5, v/v/v). The linear solvents gradient for anthocyanin analysis was: zero min, 94% A and 6% B; 10 min, 70% A and 30% B; 30 min, 50% A and 50% B; 34 min, 100% B; 36 min, 100% B; 42 min, 96% A and 4% B; 50 min, 96% A and 4% B. The linear solvents gradient for non-anthocyanin analysis was: zero min, 98% A and 2% B; 8 min, 96% A and 4% B; 37 min, 70% A, 17% B and 13% C; 51 min, 50% A, 30% B and 20% C; 51.5 min, 30% A, 40% B and 30% C; 56 min, 50% B and 50% C; 57 min, 50% B and 50% C; 64 min, 96% A and 4% B. For identification, ESI-MS/MS was used in both positive (anthocyanins) and negative (flavonols and hydroxycinnamic acid derivatives) ion modes, setting the following parameters: dry gas, N_2, 8 L/min; drying temperature, 325 ºC; nebulizer, N_2, 50 psi. Ionization and fragmentation parameters were optimized by direct infusion of appropriate standard solutions (malvidin-3-glucoside in positive ionization mode; quercetin-3-glucoside and caftaric acid in negative ionization mode). The scan range was 50−1200 *m/z*. Identification was mainly based on the spectroscopic data (UV-vis and MS/MS) obtained from authentic standards or previously reported (Castillo-Muñoz et al., 2009b; Gómez Gallego et al., 2012; Lago-Vanzela et al., 2011a and 2011b). For quantitation, DAD-chromatograms were extracted at 520 (anthocyanins), 360 (flavonols) and 320 nm (hydroxycinnamic acid derivatives). The analyses were performed in triplicate.

In the analysis of stilbenes and flavan-3-ol monomers and B-type dimer procyanidins, 0.25 mL of the SPE-C18 extract was diluted with 4.75 mL of water–formic acid (98.5:1.5) in a chromatographic vial that was sealed, and the solution was then injected. The analysis was performed using an Agilent 1200 series system equipped with DAD (Agilent, Germany) and coupled to an AB Sciex 3200 Q TRAP (Applied Biosystems) electrospray ionization mass spectrometry system (ESI-MS/MS). The chromatographic system was managed by the Agilent Chem Station (version B.01.03) data-processing station. The mass spectra data was processed with the Analyst MSD software (Applied Biosystems, version 1.5). The samples were injected (10 μL) into a reversed-phase column Agilent Eclipse XDB-C18 (2.1 × 150 mm; 3.5 μm particle; Agilent, Germany), and thermostated at 16 °C. The solvents used were water–methanol–formic acid (97:2:1, v/v/v, solvent A) and methanol (solvent B), and the flow rate was 0.1 mL/min. The linear gradient for solvent B was as follows: 0 min, 5%; 2

min, 5%; 25 min, 30%; 40 min, 55%; 50 min, 65%; 55 min, 95%; 65 min, 95%; 70 min, 5%; 80 min, 5%. Two MS scan types were used: enhanced MS (EMS) for compound identification and multiple reaction monitoring (MRM) for quantitation (Lago-Vanzela et al., 2011a and 2011b). MS conditions for both scan types were as follows: ion spray voltage, −4000 V; ion source temperature, 500 °C; collision gas, high; curtain gas, 15 psi; ion source gas 1, 50 (arbitrary units); ion source gas 2, 50 (arbitrary units); declustering potential, −35 V; entrance potential, −10 V; collision energy, −30 V; collision cell exit potential, −3 V. For the identification and quantitation of diverse flavan-3-ols we used standards of the monomers (+)-catechin, (−)-epicatechin, (−)-epigallocatechin, (−)-gallocatechin, and (−)-epicatechin 3-gallate and the dimers procyanidins B1, B2 and B4.

Statistical Treatment of Data

The phenolic composition data of the grapes (red and white grapes were separately studied) was subjected to ANOVA (Student-Newman-Keuls test, $\alpha = 0.05$) looking for diffreences associated with the grape cultivar. Additionaly, this data was subjected to a Principal Component analysis to obtain the most representative phenolic compounds contributing to the grouping or separation of grape cultivars.

Results and Discussion

The study of the content of phenolic compounds in grapes was based on the implications of such compounds in interesting properties closely related to wine quality, namely: a) sensory properties like red and white wine color, bitterness and astringency (especially in the case of red wines), but also undesired coloration developed by the browning of white wine; b) technological properties like the ability for aging in oak barrels or being subjected to micro-oxygenation in vats; c) functional properties, not only those associated with wine, but also the use of winemaking waste as a source of bioactive compounds.

Anthocyanins of Red Grape Cultivars

Anthocyanins were found exclusively in the skins of the red grape cultivars, as none of these cultivars was a *teinturier*. All anthocyanins found were those usually described for *Vitis vinifera* red grapes, namely nonacylated and acylated anthocyanidin 3-glucosides [6]. All the anthocyanin structures derived from the five anthocyanidins expected to be found in *Vitis vinifera*, namely the B-ring disubstituted cyanidin (cy) and peonidin (pn), and the B-ring trisubstituted delphinidin (dp), petunidin (pt), and malvidin (mv). We could also confirm the absence of anthocyanins based on B-ring monosubstituted pelargonidin, which has rarely been found as minor structure in some *Vitis vinifera* grape cultivars (Castillo-Muñoz et al., 2009a). The glucose moiety was found as itself (nonacylated anthocyanins, 3glc) or acylated in position six of glucose with acetyl (3acglc), *p*-coumaroyl (3cmglc, mainly as the *trans* isomer, with some minor proportions of the corresponding *cis* isomer sometimes observed) and, to a lesser extent, caffeoyl substituents (3cfglc, only found for the major malvidin derivatives). Figure 1 and Table 1 summarize all the anthocyanins found for the seven red grape cultivars under study.

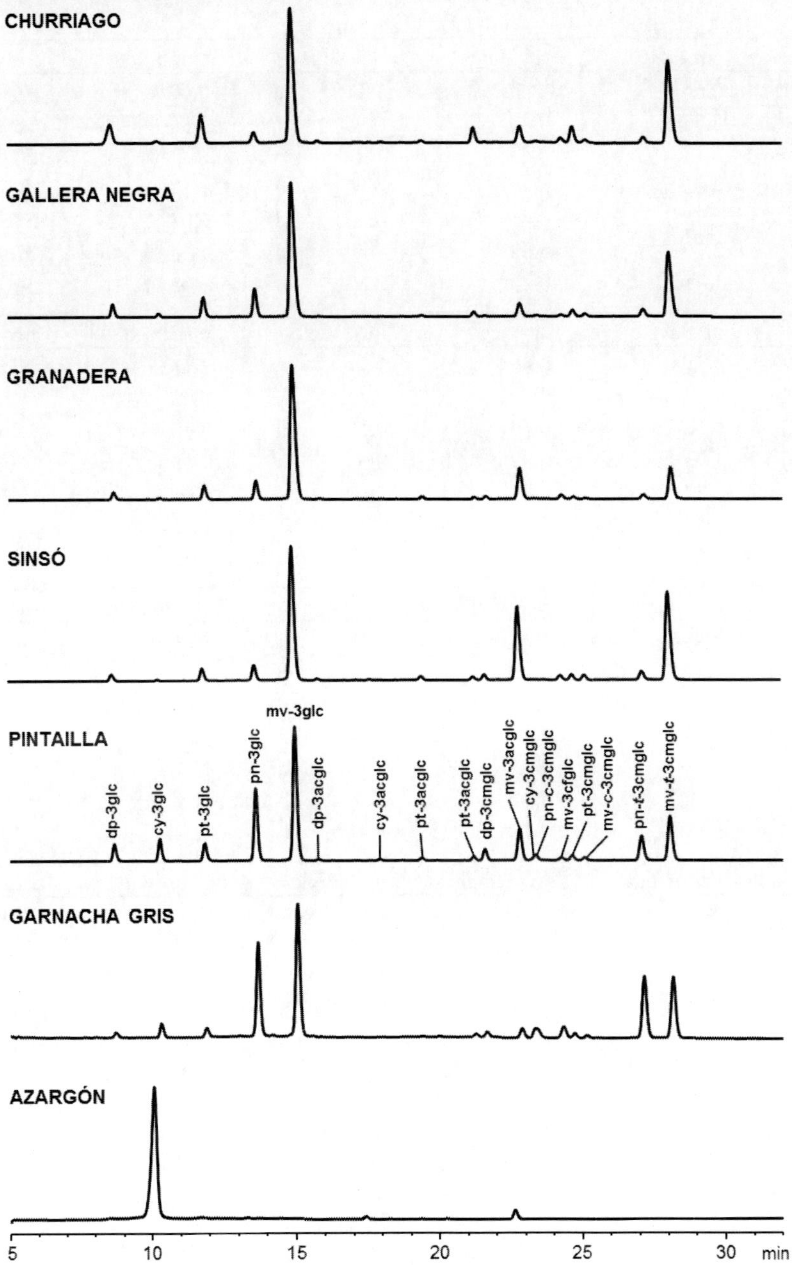

Figure 1. Anthocyanin profiles (DAD-chromatograms at 520 nm) of red grape cultivars. Abbreviations: dp, delphinidin; cy, cyanidin; pt, petunidin; pn, peonidin; mv, malvidin; 3glc, 3-glucoside; 3acglc, 3-(6"-acetyl)-glucoside; 3cmglc, 3-(6"-*p*-coumaroyl)-glucoside; 3cfglc, 3-(6"-caffeoyl)-glucoside; *t*, *trans*; *c*, *cis*.

Table 1. Anthocyanin profiles and total anthocyanin contents (mean value; malvidin 3-glucoside equivalents) in the red grape cultivars. Abbreviations: see Figure 1. Different letters in the same column mean significant differences according to ANOVA (α = 0.05). ND, non-detected

Red cultivar	dp-3glc		cy-3glc		pt-3glc		pn-3glc		mv-3glc	
Azargón	0.66	a	92.32	d	1.11	a	0.62	a	0.4	a
Churriago	5.53	f	0.49	a	6.92	f	2.86	b	38.7	d
Garnacha Gris	1.19	b	3.22	b	1.99	b	21.59	e	28.6	b
Gallera Negra	3.01	d	0.75	a	4.76	e	7.17	c	45	e
Granadera	2.45	c	0.41	a	4.25	d	6.01	c	54.3	f
Pintailla	3.76	e	5.62	c	4.3	d	17.5	d	32.9	c
Sinsó	1.45	b	0.25	a	2.82	c	3.63	b	34.3	c

Red cultivar	dp-3acglc		cy-3acglc		pt-3acglc		pn-3acglc		mv-3acglc	
Azargón	ND		0.96	b	ND		ND		ND	
Churriago	1.14	e	0.59	a	0.85	bc	0.42	a	4.2	c
Garnacha Gris	0.16	a	0.19	a	0.17	a	0.76	c	1.26	a
Gallera Negra	1.03	de	0.99	b	0.63	b	0.58	b	3.9	c
Granadera	0.77	cd	0.47	a	1.10	c	0.91	d	10.7	d
Pintailla	0.43	b	0.46	a	0.75	b	1.01	e	2.89	b
Sinsó	0.63	bc	0.48	a	1.43	d	1.58	f	19.4	e

Red cultivar	dp-3cmglc		cy-3cmglc		pt-3cmglc		*c*-pn-3cmglc		*t*-pn-3cmglc	
Azargón	ND		3.9	c	ND		ND		ND	
Churriago	4.71	d	0.75	a	4.41	e	ND		1.61	b
Garnacha Gris	1.48	b	1.72	b	2.18	d	1.02	b	0.64	a
Gallera Negra	2.06	c	0.5	a	2.14	c	ND		2.12	c
Granadera	0.93	a	0.55	a	0.87	a	ND		1.61	b
Pintailla	1.76	b	0.87	a	1.26	b	ND		0.76	a
Sinsó	1.11	a	0.45	a	1.44	b	0.14	a	2.49	d

Red cultivar	*c*-mv-3cmglc		*t*-mv-3cmglc		mv-3cfglc		disubstituted		trisubstituted	
Azargón	ND		ND		ND		97.81	f	2.18	a
Churriago	0.83	bc	24.4	c	1.4	b	6.74	a	93.25	f
Garnacha Gris	15.9	e	14.94	b	2.25	c	29.17	e	70.82	b
Gallera Negra	0.87	bc	23.66	c	0.71	a	12.13	c	87.86	d
Granadera	0.45	ab	12.81	a	1.33	b	9.98	bc	90.01	de
Pintailla	6.24	d	11.61	a	7.84	d	26.24	d	73.75	c
Sinsó	1.45	c	25.57	c	1.44	b	8.91	ab	91.08	ef

Red cultivar	nonacylated		acetylated		coumaroylated		caffeloylated		total mg/kg	
Azargón	95.13	f	0.96	a	3.9	a	ND		11.83	a
Churriago	54.56	b	7.27	d	36.8	e	1.40	b	482.10	d
Garnacha Gris	56.61	b	2.56	b	38.6	e	2.25	c	7.21	a
Gallera Negra	60.76	c	7.15	d	31.4	d	0.71	a	671.09	e
Granadera	67.46	e	13.95	e	17.2	b	1.33	b	379.45	c
Pintailla	64.07	d	5.56	c	22.5	c	7.84	d	222.76	b
Sinsó	42.47	a	23.52	e	32.6	d	1.44	b	337.72	c

Table 1 also reports on the structural features of the characteristic anthocyanin profiles found for each red grape cultivar, together with the total content of anthocyanins (mg of mv3glc equivalents per kg of grape). Some of the studied cultivars showed a slightly colored skin (e.g., Azargón and Garnacha Gris) and, thus, they contained the lowest amounts of anthocyanins (7.2–11.8 mg/kg). These grape cultivars are unable to produce red wines and so they are candidates for making rosé wines or even white wines (blanc des noirs, in French).

The rest of the red grape cultivars, with the exception of Gallera Negra, accounted for a medium content of anthocyanins, within the range of 223–482 mg/kg, as compared to the literature data (Fragoso et al., 2008): Cabernet Sauvignon (range, 686–883 mg/kg; average, 784 ± 58 mg/kg); Merlot (range, 843–1296 mg/kg; average, 1021 ± 139 mg/kg); Syrah (range, 826–1316 mg/kg; average, 1024 ± 142 mg/kg); Tempranillo (range, 651–1002 mg/kg; average, 782 ± 103 mg/kg); Garnacha (range, 348–482 mg/kg; average, 411 ± 41 mg/kg); and Cariñena (range, 485–759 mg/kg; average, 639 ± 70 mg/kg). The Gallera Negra cultivar raised an anthocyanin content similar to those previously mentioned (671 mg/kg) and close to those reported for the emblematic red grape cultivar of this region, namely, Cencibel (average of 690 ± 30 mg/kg, and range of 210–1500 mg/kg) (Hermosín-Gutiérrez et al., 2007). Some of the latter grape cultivars actually accounted for anthocyanin contents which are insufficient for making red wines as the color potential is low, and only Gallera Negra could be considered suitable for red wine production.

In general, the grape phenolic content, particularly the anthocyanin content, is affected by many factors, including the ripening stage, the viticulture and the edafo-climatic conditions (Mazza and Miniati, 1993). However, all the grape cultivars considered in our study were grown under the same conditions, thus suggesting that the differences found in the anthocyanin contents could be considered a cultivar characteristic or, at least, a cultivar trend. In contrast, it is commonly assumed that the phenolic profiles (molar proportions in which individual phenolic compounds account for the whole content of a family of phenolic compounds, like anthocyanins) is a cultivar characteristic and phenolic profiles have been used as chemo-taxonomical tools for grape cultivar identification and/or differentiation (Gómez Gallego et al., 2012; Hermosín-Gutiérrez and García-Romero, 2004). In fact, the anthocyanin profiles shown in Figure 1 already showed visually appreciable differences among grape cultivars, and detailed differentiation was revealed by ANOVA when applied to the data represented in Table 1. Thus, the Churriago and Gallera Negra cultivars showed anthocyanin profiles dominated by nonacylated derivatives, mainly mv-3glc, but also the corresponding *p*-coumaroylated derivatives were important contributors (mainly *t*-mv-3-cmglc). In the case of the Granadera and Sinsó cultivars, the proportion of acetylated derivatives increases in importance up to values similar to those of the *p*-coumaroylated type, always dominated by malvidin structures, and moreover the total sum of acylated anthocyanins was higher than that of nonacylated in Sinsó (45.0% vs. 34.3%, respectively for mv-based anthocyanins). The Garnacha Gris cultivar accounted for higher proportions of B-ring disubstituted anthocyanins, especially derived from peonidin, together with an important contribution of *p*-coumaroylated derivatives and an almost total absence of acetylated ones. Finally, Azargón was a unique cultivar, as it mostly accounted for one single anthocyanin, cy3glc, which is B-ring disubstituted and nonacylated.

The Principal Component (PC) analysis of the anthocyanin profiles data allowed obtaining a clearer picture of the differences among red grape cultivars associated to anthocyanins.

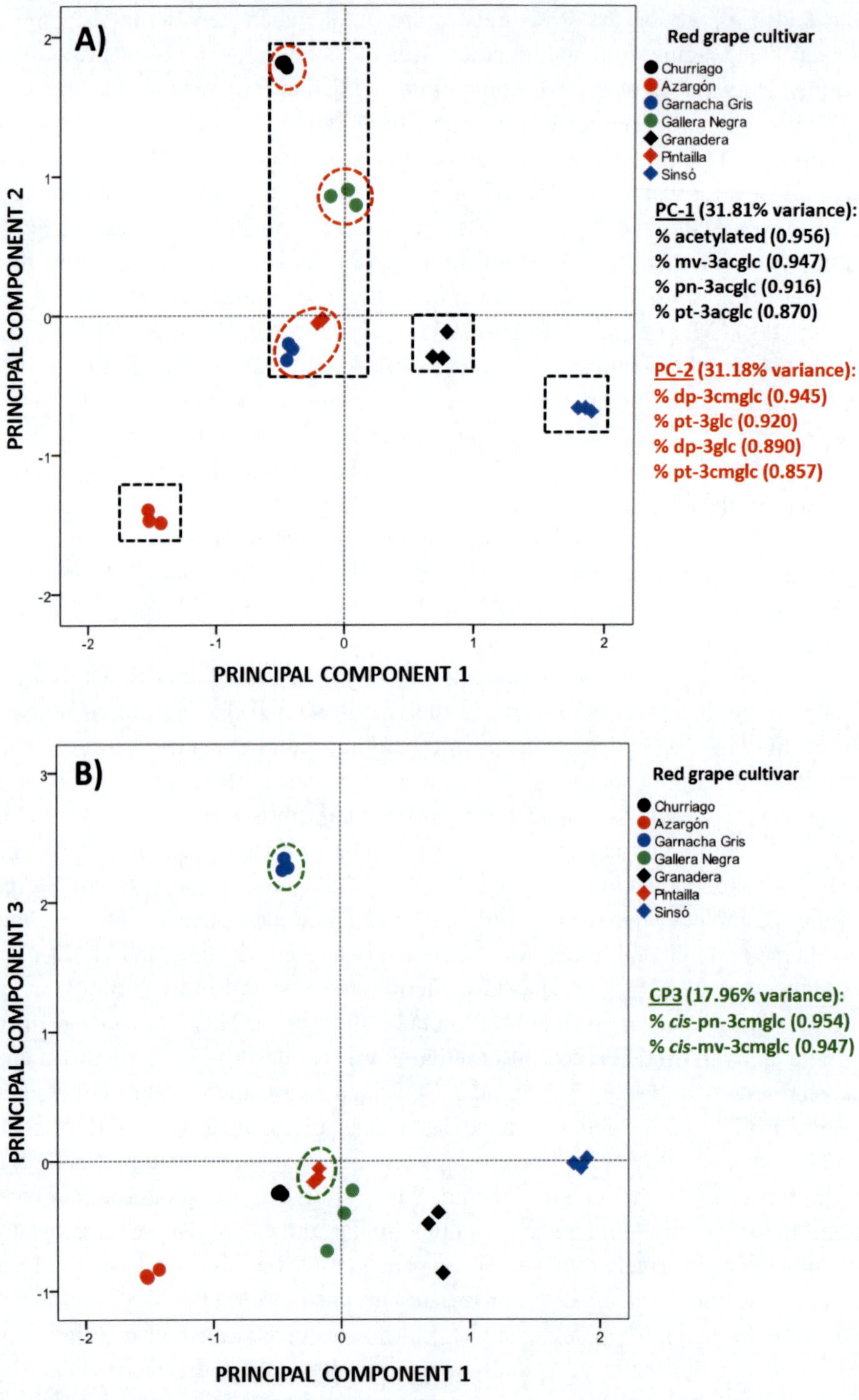

Figure 2. Principal Component (PC) analysis applied to anthocyanin data of red grape cultivars. For each PC is indicated the percentage of total explained variance and the most correlated variables with their respective loading values. Grape samples are represented in the graphs: A) PC-1 vs. PC-2; B) PC-1 vs. PC-3.

According to PC-1, the red grape cultivars were separated into 4 groups with 3 of them containing only one cultivar (Azargón, Granadera and Sinsó, respectively), mainly due to differences in the molar proportions of the acetylated anthocyanins (Figure 2A). The differences found in the molar proportions of the minor B-ring trisubstituted anthocyanins (based on delphinidin and petunidin) allowed the separation along the PC-2 of 2 out of 4 of the cultivars not separated by PC-1 (Gallera Negra and Churriago, repectively). The differentiation of the two remaining non-separated cultivars (Garnacha Gris and Pintailla) was only possible on the basis of the very minor *cis* isomers of *p*-coumaroylated anthocyanins along the PC-3 (Figure 2B).

Flavonols

Flavonols are a type of phenolic compound mainly found in grape skins as anthocyanins. They are yellow pigments and their role in white wine color can be relevant in some cases, especially in wines produced after pre-fermentative cold maceration. The usual white wine elaboration process (quick extraction of must with limited maceration with skins) leads to low contents of wine flavonols, within the range of 0.1–33.0 mg/L (as equivalents of quercetin 3-glucoside) (De Pablos Muñoz, 2010). In contrast, red wines are elaborated with skin maceration and their content in flavonols is higher, within the range of 18–220 mg/L (Hermosín-Gutiérrez et al., 2013), although their role as pigments in red wines is always shadowed by higher amounts of anthocyanins. However, flavonols have an important impact on red wine color due to the fact of their participation in the physical phenomenon of copigmentation, which helps enhance the extraction of anthocyanins during winemaking, increases the color intensity and gives purple nuances to the color of young red wines, and also enhances anthocyanin stability during the storage and aging of red wines (Castillo-Muñoz et al., 2007; Schwarz et al., 2005).

In all the grape cultivars considered in this study, the flavonols identified were those expected for *Vitis vinifera* and non-vinifera cultivars (Castillo-Muñoz et al., 2007, 2009b, and 2010; Hermosín-Gutiérrez et al., 2011; Lago-Vanzela et al., 2011a and 2011b): 3-glycosides (at least the 3-glucoside; in addition 3-galactoside, 3-glucuronide and even 3-rutinoside derivatives can be found) of six flavonoid structures (the B-ring monosubstituted kaempferol; the B-ring disubstituted quercetin and isorhamnetin; and the B-ring trisubstituted myricetin, laricitrin and syringetin). Flavonols derived from B-ring trisubstituted structures were absent in the white grape cultivars.

The flavonol profiles of grapes have been used for cultivar characterization and differentiation (Hermosín-Gutiérrez et al., 2011). In contrast, the total content of grape flavonols is influenced by particular growing conditions of grape clusters, like the degree of light exposure for example (Downey et al., 2004; Haselgrove et al., 2000; Price et al., 2005).

Flavonols of White Grape Cultivars

A simple visual inspection of chromatographic profiles (Figure 3), combined with the ANOVA analysis of the data (Table 2), highlighted interesting differences in the flavonol profiles of white grape cultivars.

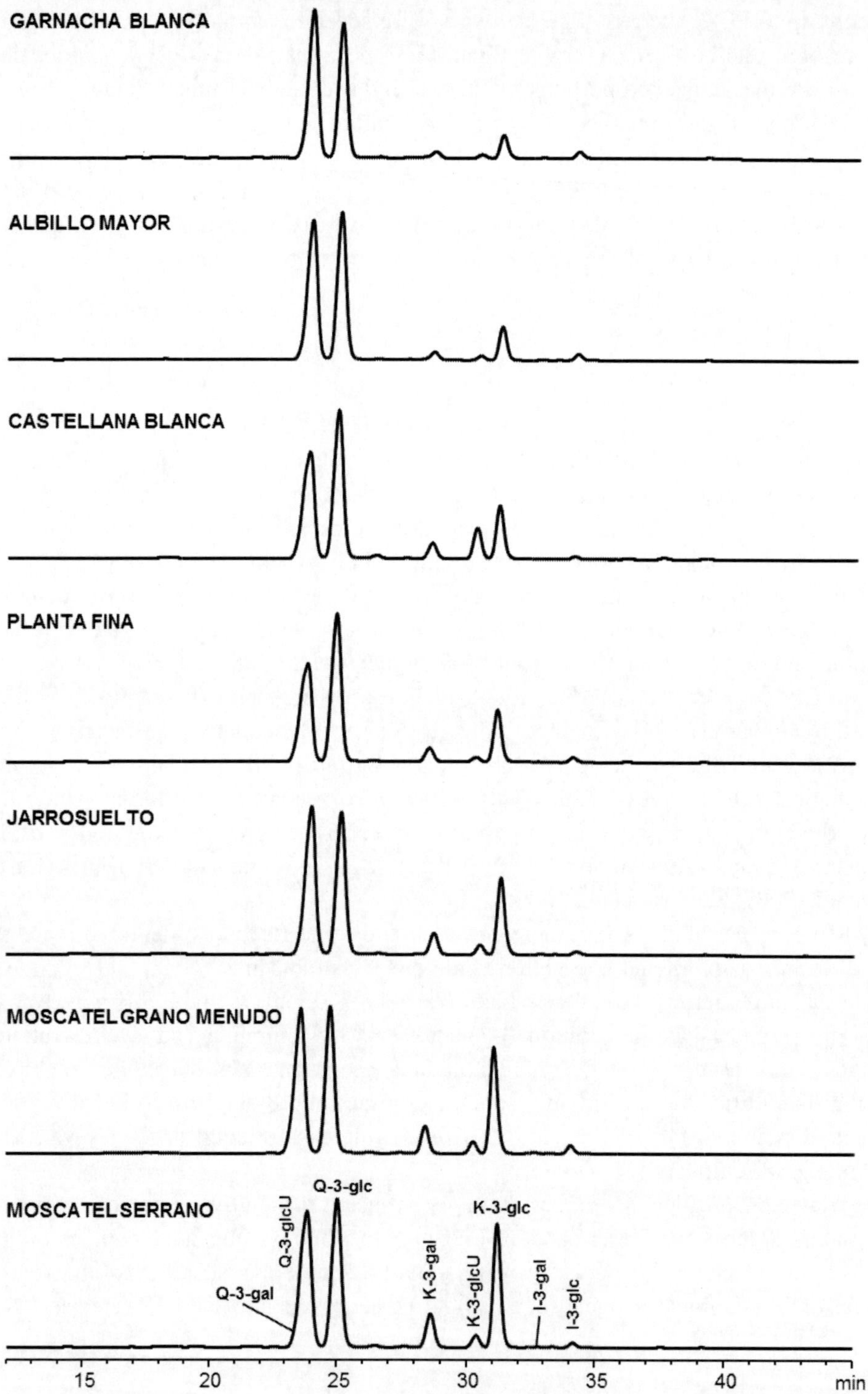

Figure 3. Flavonol profiles (DAD-chromatograms at 360 nm) of white grape cultivars. Abbreviations: : M, myricetin; Q, quercetin; L, laricitrin; K, kaempferol; I, isorhamnetin; S, syringetin; gal, galactoside; glcU, glucuronide; glc, glucoside.

Table 2. Flavonol profiles and total flavonol contents (mean value; quercetin 3-glucoside equivalents) in the white grape cultivars. Abbreviations: see Figure 2. Different letters in the same column mean significant differences according to ANOVA (α = 0.05). ND, non-detected

White grape cultivar	Q-gal		Q-glcU		Q-glc	
Albillo Mayor	17.47		26.49	b	42.28	d
Castellana Blanca	16.76		22.25	ab	37.5	c
Garnacha Blanca	19.19		32.37	c	38.44	c
Jarrosuelto	18.46		26.19	b	35.93	bc
Moscatel Grano Menudo	14.75		21.84	ab	33.37	ab
Moscatel Serrano	16.71		21.77	ab	31.65	a
Planta Fina	16.74		15.56	a	45.95	e

White grape cultivar	K-gal		K-glcU		K-glc	
Albillo Mayor	2.30	a	1.11	a	8.94	ab
Castellana Blanca	3.51	b	7.99	b	11.33	bc
Jarrosuelto	1.64	a	0.92	a	5.82	a
Garnacha Blanca	3.47	b	1.87	a	12.9	cd
Moscatel Grano Menudo	5.35	c	1.94	a	20.87	e
Moscatel Serrano	5.97	c	2.20	a	20.42	e
Planta Fina	4.04	b	1.00	a	15.25	d

White grape cultivar	I-gal		I-glc		methoxylated	
Albillo Mayor	0.08		1.31	bc	1.39	bc
Castellana Blanca	0.09		0.54	a	0.63	a
Jarrosuelto	0.17		0.97	b	1.14	b
Garnacha Blanca	0.19		1.39	bc	1.58	bc
Moscatel Grano Menudo	0.24		1.61	c	1.85	c
Moscatel Serrano	0.19		1.06	b	1.26	b
Planta Fina	0.13		1.29	bc	1.42	bc

White grape cultivar	monosubstituted		disubstituted		total mg/kg	
Albillo Mayor	12.37	b	87.64	c	90.59	b
Castellana Blanca	22.83	c	77.16	b	23.97	a
Jarrosuelto	18.25	c	81.74	b	34.39	a
Garnacha Blanca	8.39	a	91.60	d	83.42	b
Moscatel Grano Menudo	28.16	d	71.83	a	106.62	b
Moscatel Serrano	28.59	d	71.40	a	170.44	c
Planta Fina	20.30	c	79.69	b	85.39	b

Quercetin derivatives were predominant (69.98–90.02%), followed by kaempferol derivatives (8.39–28.59%), with isorhamnetin derivatives being minor flavonols (0.63–1.85%). 3-Glucosides were the major derivatives within kaempferol and isorhamnetin flavonols, followed by 3-galactosides; the 3-glucuronide of kaempferol was very minor, with the exception of the Castellana Blanca cultivar in which it was the second most important derivative of the kaempferol series (7.99%), after the corresponding 3-glucoside (11.33%). In contrast, the 3-glucuronide of isorhamnetin was not detected in any of the grape cultivars.

Regarding quercetin derivatives, the 3-glucuronide accounted for slightly lower proportions (15.56–32.37%) than those of the 3-glucoside (31.65–45.95%), and the 3-galactoside (partially co-eluting with the 3-glucuronide) usually accounted for lower proportions (16.71–19.19%).

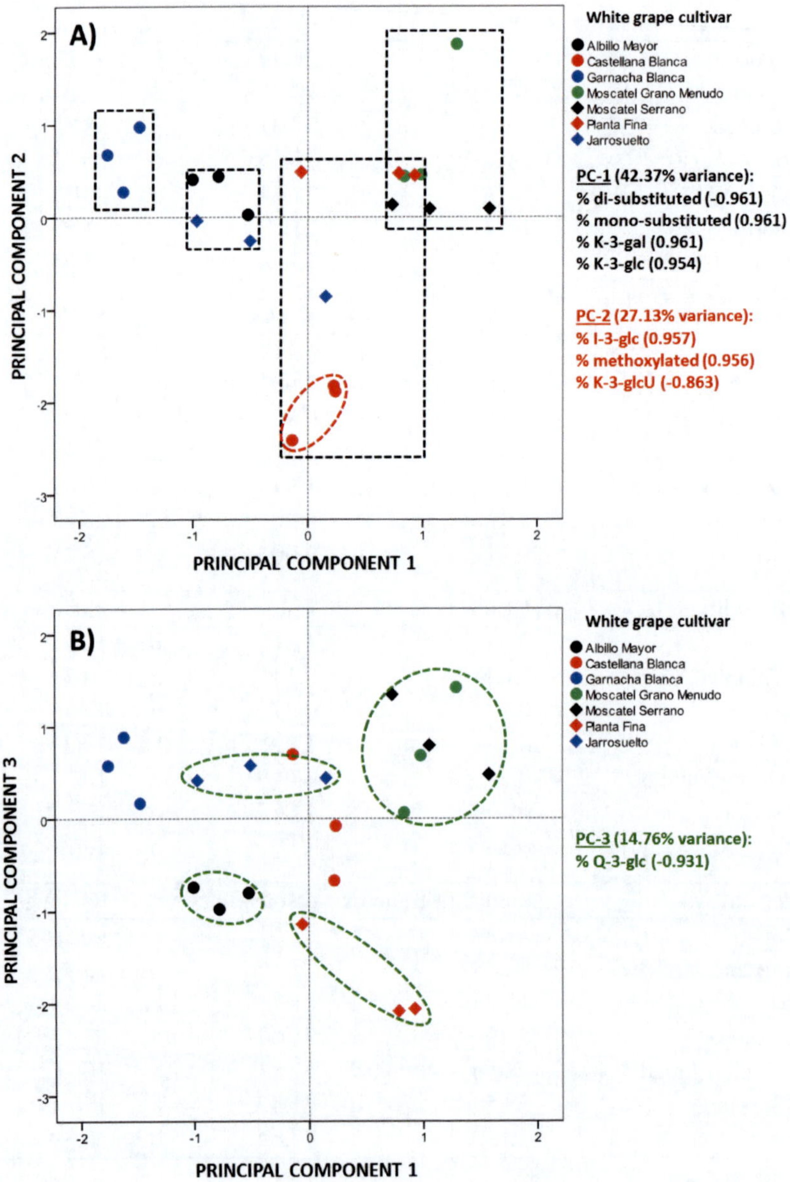

Figure 4. Principal Component (PC) analysis applied to flavonol data of white grape cultivars. For each PC is indicated the percentage of total explained variance and the most correlated variables with their respective loading values. Grape samples are represented in the graphs: A) PC-1 vs. PC-2; B) PC-1 vs. PC-3.

Similar to anthocyanins, the total content of flavonols (mg of quercetin 3-glucoside equivalents per kg of grape) suggested some trends associated with the grape cultivars. The Castellana Blanca and Jarrosuelto cultivars showed the lowest contents (24-34 mg/kg) and the highest content was found in the Moscatel Serrano cultivar (170 mg/kg), with the rest of the white cultivars ranging within 83–107 mg/kg. Reference values found in literature (Castillo-Muñoz et al., 2010; Ledda et al., 2006; Makris et al., 2006; Mattivi et al., 2006; Rodríguez Montealegre et al., 2006) were of similar magnitude, within the range 2.1–170 mg/kg. Therefore, we can suggest that the winemaking waste (unfermented solid parts of the grapes) of the Albillo Mayor, Garnacha Blanca, Moscatel Grano Menudo, Pintailla and especially Moscatel Serrano cultivars could be used as a source of bioactive flavonols.

The Principal Component (PC) analysis of flavonol data showed that kaempferol derivatives contributed to the cultivar differentiation of white grape cultivars into 4 groups along PC-1 (Figure 4A) but only the Garnacha Blanca cultivar was separated from the rest. Isorhamnetin derivatives, especially its 3-glucoside, and the 3-glucuronide of kaempferol contributed to differentiating the Castellana Blanca and Planta Fina cultivars along PC-2.

Finally, the proportion of the main individual flavonol (quercetin 3-glucoside) allowed a better differentiation of Planta Fina from both Moscatel cultivars, and also Albillo Mayor from Jarrosuelto. Only the two Moscatel cultivars were not completely differentiated by means of their respective flavonol profiles.

Flavonols of Red Grape Cultivars

The analysis of the flavonol profiles of red grapes also revealed interesting cultivar differences (Figure 5 and Table 3).

Firstly, a great similitude was found between the flavonol profiles of the lighter colored red grape cultivars (Azargón, Garnacha Gris, and even Pintailla) and those already described for white grapes: B-ring trisubstituted flavonols accounted for very low proportions (< 10% and even almost undetectable, only 0.46% of myricetin 3-glucoside for Azargón), whereas the disubstituted ones predominated (70–90%), mainly the quercetin derivatives, especially the 3-glucoside (37–45%).

In contrast, the rest of the red cultivars showed high proportions of B-ring trisubstituted flavonols (33–45%), the three possible structures contributing in similar proportions, especially their main 3-glucoside derivatives (myricetin 3-glucoside, 11–18%; laricitrin 3-glucoside, 6–11%; syringetin 3-glucoside, 5–14%), with the Granadera cultivar showing the highest values.

However, in all cases the B-ring disubstituted flavonols (mainly querectin derivatives) accounted for the highest proportions (46–60%), with the two predominant individual flavonols contributing with similar proportions (quercetin 3-glucuronide, 17–28%; quercetin 3-glucoside, 17–25%).

In all the latter cultivars, an important increase in the contribution of methoxylated flavonols was observed, up to 16–31%. For all the red grape cultivars, the contribution of B-ring monosubstituted flavonols (mainly kaempferol 3-glucoside) was usually lower than 10%, with the exceptions of the Pintailla and Sinsó cultivars (11.02 and 15.25%, respectively).

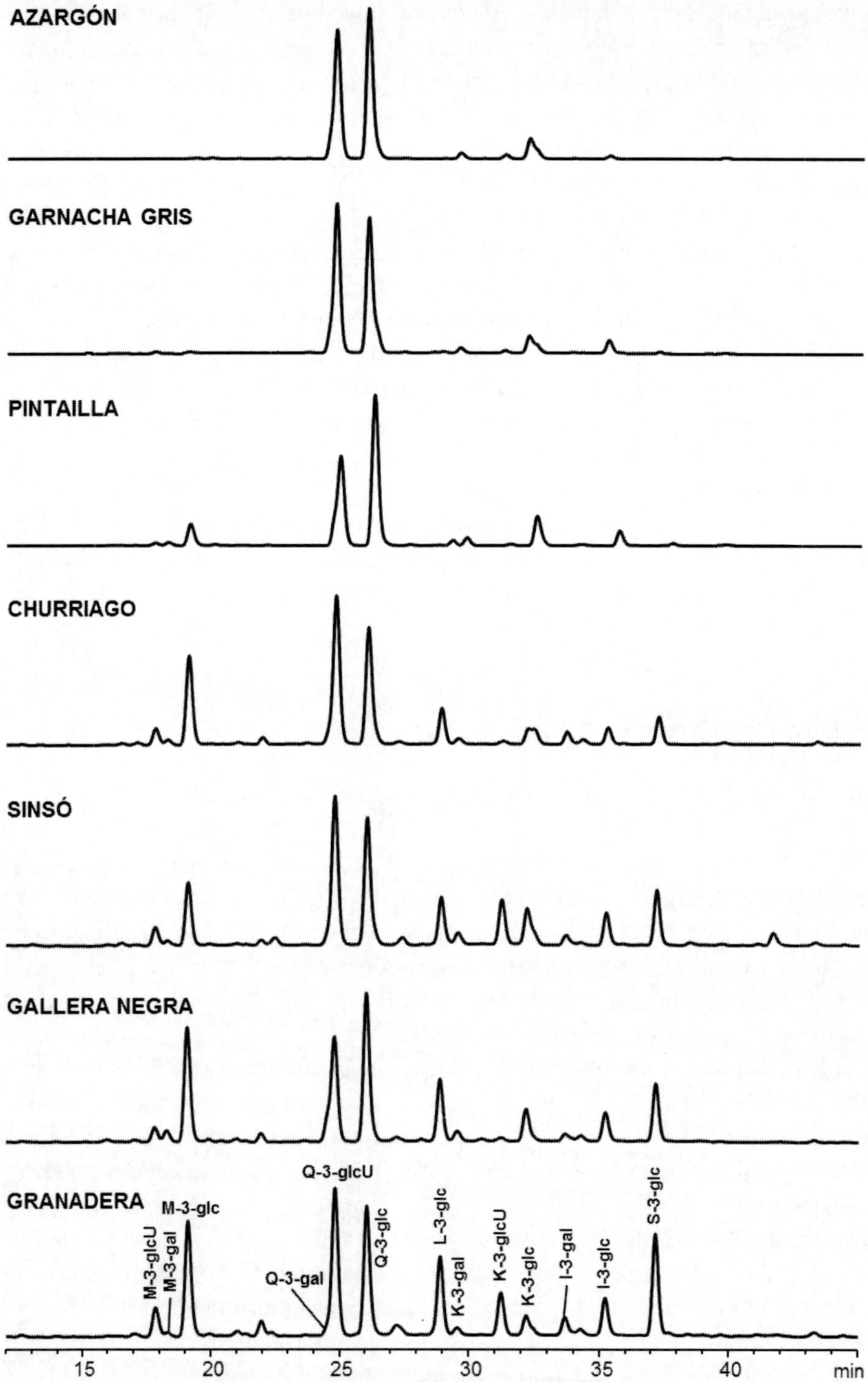

Figure 5. Flavonol profiles (DAD-chromatograms at 360 nm) of red grape cultivars. Abbreviations: M, myricetin; Q, quercetin; L, laricitrin; K, kaempferol; I, isorhamnetin; S, syringetin; gal, galactoside; glcU, glucuronide; glc, glucoside.

Table 3. Flavonol profiles and total flavonol contents (mean value; quercetin 3-glucoside equivalents) in the red grape cultivars. Abbreviations: see Figure 3. Different letters in the same column mean significant differences according to ANOVA (α = 0.05). ND, non-detected

Red cultivar	M-glcU		M-gal		M-glc	
Azargón	ND		ND		0.46	a
Churriago	3.2	d	1.18	d	18.13	e
Garnacha Gris	0.45	a	0.02	a	0.87	a
Gallera Negra	2.55	c	1.9	e	19.19	e
Granadera	3.9	e	0.63	b	15.41	d
Pintailla	0.82	b	0.99	c	6.07	b
Sinsó	2.93	d	0.96	c	11.25	c
Red cultivar	Q-gal		Q-glcU		Q-glc	
Azargón	5.07	b	38.68	c	45.41	d
Churriago	2.91	a	27.63	b	25.34	b
Garnacha Gris	4.97	b	43.83	d	37.21	c
Gallera Negra	3.64	a	17.38	a	23.55	b
Granadera	2.75	a	20.27	a	16.94	a
Pintailla	5.34	b	24.85	b	43.93	d
Sinsó	2.83	a	20.25	a	22.25	b
Red cultivar	K-gal		K-glcU		K-glc	
Azargón	1.84	cd	1.29	a	6.07	c
Churriago	1.23	b	0.78	a	3.1	a
Garnacha Gris	1.72	c	0.90	a	5.15	b
Gallera Negra	1.61	c	0.70	a	4.53	b
Granadera	0.90	a	5.81	b	2.35	a
Pintailla	2.26	e	0.56	a	8.20	d
Sinsó	2.14	de	6.48	b	6.62	c
Red cultivar	I-gal		I-glc		L-glc	
Azargón	ND		1.08	a	0.05	a
Churriago	1.02	a	3.11	b	7.15	c
Garnacha Gris	0.04	a	3.84	c	10.17	ab
Gallera Negra	0.96	c	4.39	c	10.88	e
Granadera	0.83	c	5.20	d	1.37	e
Pintailla	0.60	b	4.33	c	8.46	b
Sinsó	0.51	b	4.85	e	5.52	d
Red cultivar	S-glc		monosubstituted		disubstituted	
Azargón	ND		9.21	c	90.26	e
Churriago	4.98	b	5.12	a	60.02	c
Garnacha Gris	0.37	a	7.78	bc	89.91	e
Gallera Negra	9.38	c	6.85	b	49.93	b
Granadera	14.06	d	9.07	c	46.02	b
Pintailla	0.62	a	11.02	d	79.07	d
Sinsó	9.40	c	15.25	e	51.71	b

Table 3. (Continued)

Red cultivar	trisubstituted		methoxylated		total mg/kg	
Azargón	0.52	a	1.13	a	130.72	c
Churriago	34.84	c	16.27	c	36.38	a
Garnacha Gris	2.30	a	4.82	b	64.47	b
Gallera Negra	43.21	d	24.91	d	68.35	b
Granadera	44.90	d	30.99	e	28.33	a
Pintailla	9.90	b	6.94	b	134.88	c
Sinsó	33.02	c	24.23	d	48.74	ab

Three main groups were obtained after the PC analysis of flavonol data along the PC-1 (Figure 6A), mostly made up of the proportions of two of the main individual flavonols (quercetin 3-glucuronide and myricetin 3-glucoside, Q-3-glcU and M-3-glc respectively). On the one hand, the Gallera Negra cultivar was separated because of its high proportion of M-3-glc and low proportion of Q-3-glcU. On the other hand, the Azargón and Garnacha Gris cultivars showed the opposite situation (low proportions of M-3-glc and high proportions of Q-3-glcU) and the rest of the red grape cultivars accounted for intermediate proportions of both flavonols. The combined kaempferol derivatives allowed the separation along the PC-2 into three subgroups of the central group found along the PC-1: Pintailla; Sinsó; and Churriago and Granadera. Finally, the minor kampferol 3-glucuronide contributed to the differentiation of the Churriago and Granadera cultivars and accentuated the separation between the Pintailla and Sinsó cultivars along the PC-3 (Figure 6B).

Regarding total flavonol content, it was possible to consider three groups, thus suggesting some trends associated with the grape cultivar: low content (28–49 mg/kg), Churriago, Granadera and Sinsó; medium content (64–68 mg/kg), Garnacha Gris and Gallera Negra; high content (131–135 mg/kg), Azargón and Pintailla. All these values are within the range 5.8–160.9 mg/kg as found in the literature (Castillo-Muñoz et al., 2007; Makris et al., 2006; Mattivi et al., 2006); a range quite similar to that also found for the white grape cultivars (see latter section).

No relationship was found between the total content of flavonols and the corresponding anthocyanins for the grape cultivars considered in this study. This result could be predictable, even considering that flavonol and anthocyanin biosyntheses share the same pathway (Castillo-Muñoz et al., 2009b), because the factors influencing the biosynthesis of these two kinds of flavonoids are not necessarily the same (Downey et al., 2004; Haselgrove et al., 2000; Price et al., 2005). Thus, the two grape cultivars with the highest flavonol contents, Azargón and Pintailla, showed flavonol profiles closer to those of white cultivars and their total anthocyanin contents were lowest for Azargón and low-medium for Pintailla. In contrast, the grape cultivar with the highest content in anthocyanins, Gallera Negra, accounted for a medium content of flavonols.

On the basis of the aforementioned results and combined with the results found for anthocyanins, the elaboration of white wines (blanc de noirs) in the case of Azargón, Garnacha Gris and Pintailla cultivars would be of interest, together with the use of their unfermented winemaking by-products (solid waste) as a source of bioactive flavonols, especially in the cases of Azargón and Pintailla.

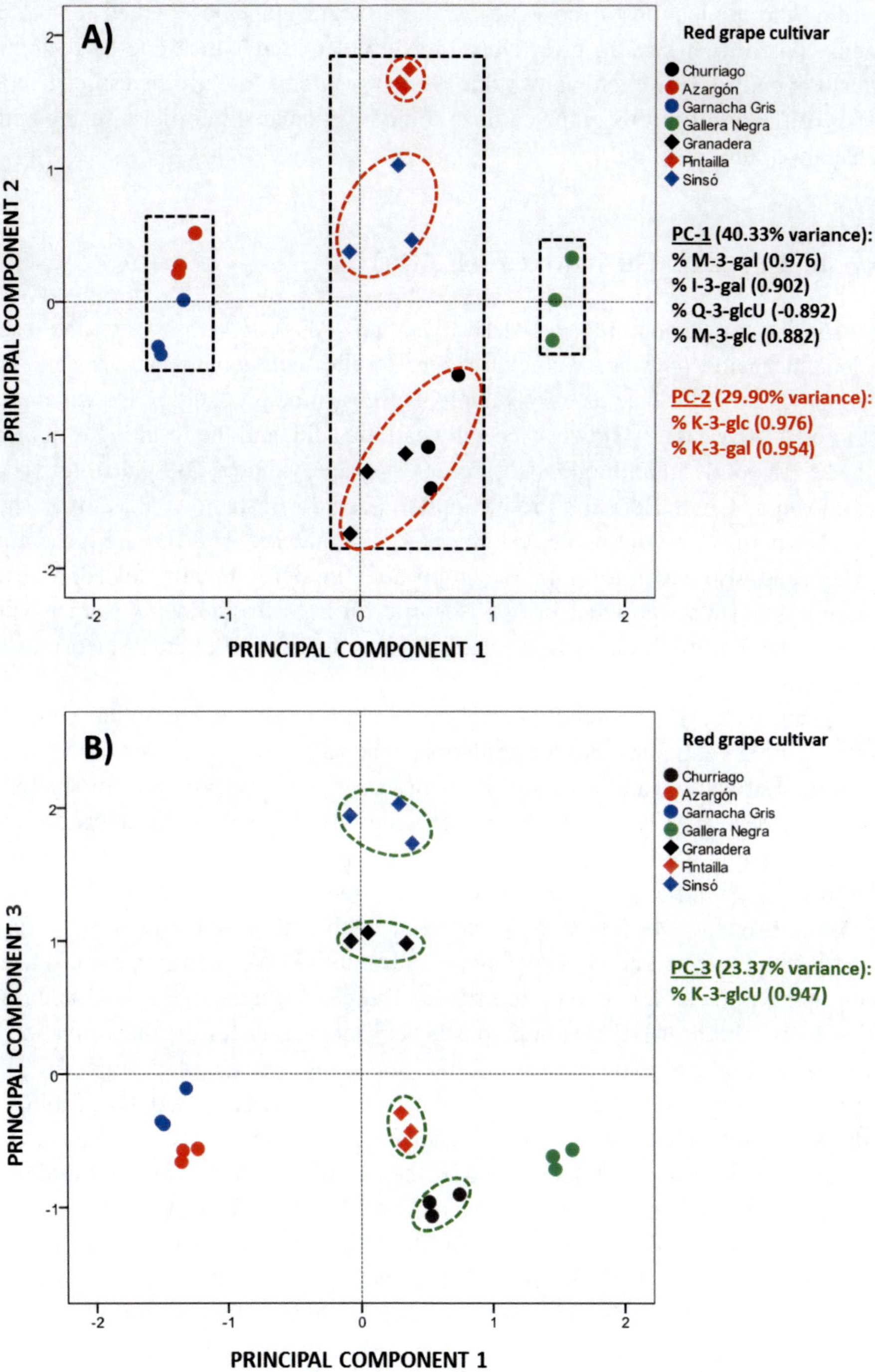

Figure 6. Principal Component (PC) analysis applied to flavonol data of red grape cultivars. For each PC is indicated the percentage of total explained variance and the most correlated variables with their respective loading values. Grape samples are represented in the graphs: A) PC-1 vs. PC-2; B) PC-1 vs. PC-3.

Regarding the production of red wines, Gallera Negra grapes accounted for the highest anthocyanin content that can be effectively transferred to red wines due to their medium content in flavonols. However, it would be very interesting to investigate how the biosynthesis of flavonols in this grape cultivar could be enhanced in order to make better use of its high anthocyanin potential.

Hydroxycinnamic acid Derivatives (HCADs)

The main hydroxycinnamic acid derivatives (HCADs) in grapes are the esters of hydroxycinnamic acids (caffeic, *p*-coumaric and ferulic acids) with tartaric acid, namely caftaric, coutaric and fertaric acids, respectively. During grape preparation for winemaking or analysis those HCADs easily oxidize, especially caftaric acid, and the resulting *o*-quinone can react with the tripeptide glutathione also present in grape yielding 2-*S*-glutathionyl-caftaric acid (also known as Grape Reaction Product, GRP) (Cejudo-Bastante et al., 2010). The GRP can be found in both wines and processed grapes in our analyses. HCADs are present mainly in grape flesh and also account for minor proportions in skins. Finally, all HCADs have a double olefinic bond that can exist in two isomeric forms, with the *trans* isomer being the naturally occurring form, although it is relatively easy for it to be transformed into the *cis* isomer by UV light.

The content of HCADs in grapes is important for the quality of the future wine. On the one hand, they are a good substrate for oxidation, especially the derivatives based on caffeic acid (*o*-diphenol structure), and they are the main compounds responsible for the oxidative browning which decreases the quality of white wines. Hydroxycinnamic acids are also the origin of the formation 4-ethylphenols, a kind of volatile compound responsible for the off-flavor known as "brett". On the other hand, free hydroxycinnamic acid or their decarboxylation products, 4-vinylphenols, can react with anthocyanins giving rise to a new class of red wine pigments known as pyranoanthocyanins. These pigments are considered to largely contribute to the color of aged red wines. Therefore, the study of the qualitative and quantitative composition of HCADs in grapes is very interesting for the determination of the enological potential of any grape cultivar.

The grape cultivars which accounted for the highest contents of HACDs (Table 4) were three of the white cultivars (Albillo Mayor, Castellana Blanca and Garnacha Blanca; 110–121 mg/kg, as caftaric acid equivalents), followed by the two red cultivars which showed very low contents of anthocyanins (Azargón and Garnacha Gris; 71–86 mg/kg).

The rest of the white cultivars showed HACDs contents within 39–68 mg/kg, whereas the red cultivars accounted for 29–52 mg/kg. Caftaric acid predominated for all grape cultivars (including GRP: 50–78% in the white cultivars and 51–70% in the red ones), with the only exception being Garnacha Gris in which coutaric acid predominated (54.67%). Coutaric acid was the second most important HACD and fertaric acid only contributed by 4–13%. The presence of the artifact GRP was observed in all the samples, measuring within the range of 2–31%, although it was usually lower than 18%.

Regarding red grape cultivars, HACDs play a similar role to flavonols in the formation of copigmentation complexes that enhance anthocyanin extractability during winemaking as well as the color intensity and stability of the resulting red wines (Schwarz et al., 2005).

Table 4. Hydroxycinnamic acid derivatives (HCADs) profiles and total HACDs contents (mean value; caftaric acid equivalents) in the white and red grape cultivars. Abbreviations: GRP, Grape Reaction Product. Different letters in the same column mean significant differences according to ANOVA (α = 0.05)

White grape	GRP		caftaric acid		GRP + caftaric	
Albillo Mayor	7.13	a	59.95	de	67.08	b
Castellana Blanca	1.76	a	7305	f	74.82	bc
Garnacha Blanca	18.5	b	55.37	cd	73.88	bc
Jarrosuelto	4.39	a	65.35	e	69.74	bc
Moscatel Grano Menudo	3.73	a	49.85	bc	53.59	a
Moscatel Serrano	17.69	b	32.53	a	50.23	a
Planta Fina	32.41	c	45.53	b	77.95	c
White grape	**coutaric acid**		**fertaric acid**		**total mg/kg**	
Albillo Mayor	26.70	b	6.20	ab	121.06	b
Castellana Blanca	17.68	a	7.49	b	109.77	b
Garnacha Blanca	22.15	ab	3.96	ab	114.44	b
Jarrosuelto	16.93	a	13.31	c	47.03	a
Moscatel Grano Menudo	41.30	c	5.10	ab	68.31	a
Moscatel Serrano	47.20	d	2.55	a	63.34	a
Planta Fina	17.34	a	4.69	ab	39.48	a

Red grape	GRP		caftaric acid		GRP + caftaric	
Azargón	3.29	a	65.94	d	69.24	d
Churriago	7.49	bc	62.34	d	69.83	d
Garnacha Gris	10.32	c	29.32	a	39.64	a
Gallera Negra	17.03	d	52.41	c	69.45	d
Granadera	15.45	d	45.58	b	61.03	c
Pintailla	5.5	ab	48.42	bc	53.92	b
Sinsó	1.97	a	48.66	bc	50.63	b
Red grape	**coutaric acid**		**fertaric acid**		**total mg/kg**	
Azargón	22.48	a	8.27	ab	71.64	c
Churriago	20.06	a	10.09	bc	28.98	a
Garnacha Gris	54.67	d	5.67	a	86.08	d
Gallera Negra	22.84	a	7.69	ab	52.46	b
Granadera	31.00	b	7.95	ab	40.92	ab
Pintailla	33.41	b	12.66	c	52.28	b
Sinsó	41.15	c	8.20	ab	51.75	b

In addition, the release of free hydroxycinnamic acids into the wine after hydrolysis of their respective hydroxycinnamoyl-tartaric esters leads to the formation of hydroxyphenyl-pyranoanthocyanins which are more stable red pigments than grape anthocyanins (Rentzsch et al., 2007). Therefore, a high content of HACDs would be desirable for red grape cultivars. However, the best candidates for red winemaking under this condition, the Azargón and Garnacha Gris cultivars, accounted for very low contents of anthocyanins which means that red wine production is not possible. In contrast, HACDs are implicated in the undesirable oxidative browning of white wines and the high content of HCADs found in the white Albillo Mayor, Castellana Blanca and Garnacha Blanca cultivars could be considered a handicap for their use in making white wines.

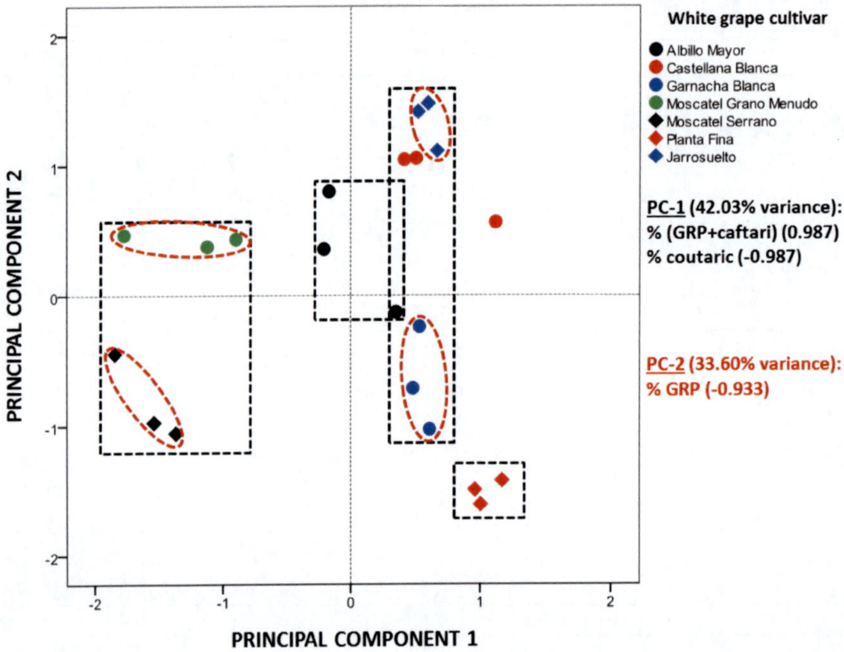

Figure 7. Principal Component (PC) analysis applied to hydroxycinnamic acid derivatives data of white grape cultivars. For each PC is indicated the percentage of total explained variance and the most correlated variables with their respective loading values. Grape samples are represented in the graphs: A) PC-1 vs. PC-2; B) PC-1 vs. PC-3.

The ability of HACDs profiles for cultivar characterization and differentiation was also studied by PC analysis. Because of the low number of individual hydroxycinnamic acid derivatives and the prevalence of caffeic-type structures, this analysis did not allow good separation according to the grape cultivar as previously obtained using the profiles of anthocyanins (only for red cultivars) and flavonols (for both red and white cultivars). Perhaps the most remarkable finding was the differentiation between the two Moscatel white cultivars (Figure 7) although it was only possible because of the formation of the artifact GRP which was seen in higher proportions in Moscatel Serrano (PC-2), something which can not be considered a cultivar characteristic. However, neither white cultivar was separated when the proportion of total caftaric acid, including GRP, was considered (PC-1).

Stilbenes

The only stilbene found in the grape samples analyzed was resveratrol 3-glucoside which can occur as the *trans* and *cis* isomers and only appeared in the skins. The total content of resveratrol 3-glucoside was within 0.27–1.02 µmol/kg in whole grape (3.02–9.78 µmol/kg referring only to skin weight) in the white cultivars (Table 5).

Table 5. Resveratrol 3-glucoside isomers profiles and total contents (mean values) in the white and red grape cultivars. Different letters in the same column mean significant differences according to ANOVA (α = 0.05). ND, non-detected

White cultivar	*trans*-resveratrol 3-glc	*cis*-resveratrol 3-glc	total µmol/kg
Albillo Mayor	51.63 ab	48.36 ab	1.02 c
Castellana Blanca	35.48 a	64.52 b	0.28 ab
Garnacha Blanca	50.77 ab	49.22 ab	0.38 ab
Jarrosuelto	ND	ND	ND
Moscatel Grano Menudo	43.40 a	56.59 b	0.60 ab
Moscatel Serrano	68.89 b	31.10 a	0.57 ab
Planta Fina	53.15 ab	46.84 ab	0.27 ab
Red cultivar	***trans*-resveratrol 3-glc**	***cis*-resveratrol 3-glc**	**total µmol/kg**
Azargón	53.86 a	46.14 c	3.65 a
Churriago	73.19 c	26.81 a	0.61 a
Garnacha Gris	58.60 ab	41.40 bc	0.37 a
Gallera Negra	73.82 c	26.18 a	19.89 b
Granadera	76.08 c	23.91 a	7.23 a
Pintailla	75.13 c	24.87 a	17.38 b
Sinsó	69.02 bc	30.97 ab	2.28 a

As resveratrol equivalents, these values mean 0.062–0.233 mg/kg in the whole grape and 0.689–2.232 mg/kg in the skin. The Albillo Mayor cultivar significantly doubled the content of resveratrol 3-glucoside compared to the others, whereas no form of resveratrol was detected in the Jarrosuelto cultivar. In contrast, the red cultivars accounted for a wider range of resveratrol 3-glucoside content: 0.37–19.89 µmol/kg in the whole grape (4.47–244.61 µmol/kg in the skin), or the resveratrol equivalent content within 0.084–4.540 mg/kg in the grape (1.020–55.832 mg/kg in the skin).

The Gallera Negra and Pintailla cultivars showed similar high contents of resveratrol 3-glucoside that significantly multiplied by a factor of 3 to 54 times those of the other cultivars. The *trans* isomer of resveratrol 3-glucoside predominated over the *cis* isomer, especially in the red grape cultivars (54–76%; 35–69% for the white cultivars) and only two of the white cultivars showed the predominance of the *cis* isomer (Castellana Blanca and Moscatel Grano Menudo, 35 and 43% respectively).

The minimum content of resveratrol in the whole grape reached similar values for both white and red grape cultivars. In contrast, the maximum content found in some red cultivars reached values (Gallera Negra and Pintailla, 4.540 and 3.967 mg/kg in the whole grape respectively) of up to 20 times higher than the highest value found in the white cultivars (Albillo Mayor, 0.233 mg/kg). However, the resveratrol content of the all these cultivars can be considered generally low, as could be expected for grapes growing in a warm, dry climate that contributes to the production of healthy grapes, thus limiting the need for the vine to produce resveratrol, a phytoalexin. It has been proposed that the grape cultivars are classified into two groups according to their intrinsic potential to produce resveratrol (average values) (Gatto et al., 2008): high resveratrol producers (*trans*-resveratrol, 2.37 mg/kg; *cis*-resveratrol 3-glucoside, 4.19 mg/kg; *trans*-resveratrol 3-glucoside, 4.18 mg/kg; total resveratrol, 10.74

mg/kg); and low resveratrol producers, having one degree of magnitude less than the high producers (*trans*-resveratrol, 0.53 mg/kg; *cis*-resveratrol 3-glucoside, 0.29 mg/kg; *trans*-resveratrol 3-glucoside, 0.48 mg/kg; total resveratrol, 1.30 mg/kg).

According to the latter classification, most of the grape cultivars studied can be considered low producers of resveratrol, and only the Gallera Negra and Pintailla red cultivars (total resveratrol content around 4 mg/kg) could be considered high resveratrol producing grape cultivars.

Flavan-3-ol Monomers and Dimers

Quantitatively, the flavan-3-ols (including monomers, dimers and polymers, the latter also known as tannins) are the main phenolic compounds in grapes. They are located in both the skins and the seeds and some traces of monomers and dimers have been also found in the flesh.

The seeds always account for higher flavan-3-ol content and there are some structural differences among the tannins from seeds and from skins: skin tannins contain epigallocatechin units not found in seed tannins, and they show a higher mean degree of polymerization and a lower proportion of galloylated units in comparison to seed tannins (Cheynier et al., 2003). The relevance of these compounds with regard to wine is that they contribute to the evolution and stabilization of red wine color during aging and they also are responsible for the red wine sensory attributes of bitterness and astringency. In addition, there is increasing interest in the potential health effects associated with the consumption of tannins (Yilmaz and Toledo, 2004).

The study of molar profiles of flavan-3-ol monomers and dimers (Tables 6 and 7) showed that the main contributor was, in most cases, the dimer procyanidin B1 (PB1) in both skins (53.3–65.4% in the white cultivars; 50.3–67.4% in the red cultivars) and seeds (31.8–49.3% in the white cultivars; 22.4–48.3% in the red cultivars). Only the seeds of two red cultivars, Pintailla and Sinsó, showed a different trend and procyanidin B2 (PB2) dominated their profiles over PB1 (PB2, 27.7% and 35.0% respectively, against 22.4% and 34.4% of PB1).

In the seeds, the second most important contributor in most cases was (+)-catechin, accounting for molar proportions of 24.6–35.7% in the white cultivars and 11.8–29.4% in the red ones. However, in both Pintailla and Sinsó cultivars, the second most important contributor was PB1, followed by (+)-catechin for Pintailla and (–)-epicatechin for Sinsó, with the latter cultivar being the only case in which (–)-epicatechin accounted for a higher proportion than (+)-catechin.

The flavan-3-ol profiles of the grape skins were more diverse. For most of the white cultivars, (+)-catechin was the second most important compound, except for Albillo Mayor and Castellana Blanca cultivars in which it was (–)-gallocatechin. For most of the red cultivars, (+)-catechin and (–)-gallocatechin accounted for similar proportions, with a slight prevalence of the latter compound. In contrast, the red Churriago and Sinsó cultivars showed a predominance of (+)-catechin over (–)-gallocatechin.

Table 6. Flavan-3-ol monomer and dimer profiles in the skins (SK) and the seeds (SD) of the white cultivars and total contents (as mg/kg of catechin equivalents; referred to grape weight and also to separated skin/seed weight). Abbreviations: C, catechin; EC, epicatechin; GC, gallocatechin; EGC, epigallocatechin; ECG, epicatechin 3-gallate; CG, catechin 3-gallate; PB1, procyanidin B1; PB2, procyanidin B2; PB4, procyanidin B4. Different letters in the same column mean significant differences according to ANOVA ($\alpha = 0.05$). ND, non-detected

White cultivar	C-SK	EC-SK	GC-SK		EGC-SK		ECG-SK		CG-SK	
Albillo Mayor	14.02	1.37	17.62	d	1.75	abc	0.44		ND	
Castellana Blanca	17.31	3.11	11.99	bc	2.78	bc	1.58		ND	
Garnacha Blanca	11.81	0.95	15.47	cd	0.94	ab	1.02		ND	
Jarrosuelto	13.97	2.80	4.41	a	1.63	abc	2.15		0.63	
Moscatel GM	15.80	2.33	10.10	b	3.47	c	1.95		ND	
Moscatel Serrano	18.58	1.51	8.99	b	1.11	ab	0.74		ND	
Planta Fina	20.81	2.60	9.92	b	ND	a	1.05		ND	
White cultivar	C-SD	EC-SD	GC-SD		EGC-SD		ECG-SD		CG-SD	
Albillo Mayor	35.70	3.53	0.10		ND		8.90	ab	0.40	ab
Castellana Blanca	24.56	6.26	0.13		ND		13.56	c	0.20	a
Garnacha Blanca	31.20	7.63	ND		ND		8.46	ab	0.20	a
Jarrosuelto	28.66	7.30	ND		ND		13.70	c	0.33	ab
Moscatel GM	30.63	5.50	ND		ND		11.86	bc	1.16	c
Moscatel Serrano	27.56	4.06	0.10		ND		6.86	a	0.16	a
Planta Fina	26.30	8.36	0.10		ND		15.56	c	0.50	b

White cultivar	PB1-SK		PB4-SK		PB2-SK		µmol/kg grape-SK		µmol/kg skin	
Albillo Mayor	59.40		1.91		3.48		246.70	b	2548.72	c
Castellana Blanca	56.31		3.02		3.85		21.03	a	581.21	a
Garnacha Blanca	64.67		1.96		3.15		218.00	b	2202.22	bc
Jarrosuelto	65.44		4.06		4.89		34.70	a	331.54	a
Moscatel GM	58.68		3.33		4.31		85.96	a	718.97	a
Moscatel Serrano	61.84		2.60		4.60		137.04	ab	1161.16	ab
Planta Fina	56.47		3.98		5.15		52.37	a	588.56	a
White cultivar	PB1-SD		PB4-SD		PB2-SD		µmol/kg grape-SD		µmol/kg seed	
Albillo Mayor	44.36	ab	1.43	a	5.56	a	5.14	ab	174.66	
Castellana Blanca	38.23	ab	4.63	b	12.36	b	1.07	a	61.30	
Garnacha Blanca	35.96	a	3.83	ab	12.70	b	3.56	ab	134.83	
Jarrosuelto	34.93	a	2.30	ab	12.70	ab	3.18	ab	124.66	
Moscatel GM	38.90	ab	2.70	ab	9.26	ab	2.85	ab	99.93	
Moscatel Serrano	49.26	b	2.06	a	9.86	b	6.39	b	119.83	
Planta Fina	31.80	a	3.73	ab	13.66	b	4.25	ab	114.36	

Table 7. Flavan-3-ol monomer and dimer profiles in the skin (SK) and the seeds (SD) of the red cultivars and total contents (as mg/kg of catechin equivalents; referred to grape weight and also to separated skin/seed weight). Abbreviations: C, catechin; EC, epicatechin; GC, gallocatechin; EGC, epigallocatechin; ECG, epicatechin 3-gallate; CG, catechin 3-gallate; PB1, procyanidin B1; PB2, procyanidin B2; PB4, procyanidin B4. Different letters in the same column mean significant differences according to ANOVA ($\alpha = 0.05$). ND, non-detected

Red cultivar	C-SK	EC-SK		GC-SK		EGC-SK		ECG-SK		GC-SK
Azargón	10.48	4.79	c	12.00	ab	ND		2.64	b	ND
Churriago	13.84	1.54	a	5.02	a	1.97	ab	1.47	a	0.11
Garnacha Gris	13.23	1.26	a	16.83	ab	1.10	a	1.09	a	ND
Gallera Negra	14.29	1.29	a	14.17	b	2.12	ab	0.53	a	ND
Granadera	12.67	3.31	b	16.12	ab	5.61	c	1.39	a	ND
Pintailla	10.44	2.95	ab	10.79	ab	4.85	c	1.45	a	ND
Sinsó	12.41	1.29	a	9.62	ab	3.73	bc	1.05	a	ND

Red cultivar	C-SD		EC-SD		GC-SD		EGC-SD	ECG-SD	CG-SD
Azargón	20.90	a	4.88	a	0.17	ab	ND	10.62	0.19
Churriago	28.89	b	10.41	a	0.22	b	ND	6.88	0.31
Garnacha Gris	29.42	b	7.33	a	ND		ND	9.04	0.18
Gallera Negra	18.84	a	9.14	a	ND		ND	11.66	0.23
Granadera	20.05	a	10.33	a	ND		ND	13.16	0.30
Pintailla	16.80	a	11.09	a	ND		ND	7.21	0.11
Sinsó	11.77	a	17.50	b	ND		ND	9.57	0.29

Red cultivar	PB1-SK		PB4-SK		PB2-SK		µmol/kg grape-SK		µmol/kg skin	
Azargón	56.50	ab	7.35	c	6.21	b	14.60	a	232.50	a
Churriago	67.35	b	4.76	b	3.92	a	56.45	ab	932.00	bc
Garnacha Gris	59.88	ab	2.27	a	4.31	a	109.76	c	1282.66	c
Gallera Negra	59.59	ab	4.35	ab	3.65	a	155.00	d	1910.00	d
Granadera	50.27	a	6.73	c	3.86	a	32.40	ab	407.00	ab
Pintailla	56.30	ab	9.76	d	3.44	a	76.25	bc	969.00	bc
Sinsó	64.05	b	4.30	ab	3.51	a	71.26	ab	968.50	bc

Red cultivar	PB1-SD		PB4-SD		PB2-SD		µmol/kg grape-SD		µmol/kg seed	
Azargón	48.27	c	2.41	a	12.52	ab	1.57	b	56.16	b
Churriago	29.76	ab	5.66	c	17.84	ab	2.67	c	75.70	c
Garnacha Gris	35.96	b	4.64	bc	9.79	a	1.89	b	92.93	c
Gallera Negra	31.53	b	4.69	bc	19.46	ab	0.69	a	25.86	a
Granadera	33.44	ab	3.66	ab	20.95	ab	0.77	a	30.30	a
Pintailla	22.41	b	3.67	ab	27.67	bc	1.59	b	47.13	ab
Sinsó	34.42	a	3.41	ab	35.03	c	0.88	a	44.50	ab

As described for tannins (Cheynier et al., 2003), we found that flavan-3-ol monomer composition differed between the grape skins and seeds. On the one hand, skin flavan-3-ol monomers were characterized by a higher degree of hydroxylation and considerable proportions of (–)-gallocatechin (GC; 4.4–17.6% in the white cultivars; 5.0–16.8% in the red cultivars) but lower proportions of (–)-epigallocatechin (EGC; 0.9–3.5% in the white cultivars; 0.0–5.6% in the red cultivars) were found. In contrast, GC and EGC were undetectable or accounted for trace amounts in the seeds of all cultivars. On the other hand, the seed flavan-3-ol monomers were characterized by a higher degree of galloylation (3-gallic acid esters of flavan-3-ol monomers), accounting for considerable proportions of (–)-epicatechin 3-gallate (ECG; 6.9–15.6% in the white cultivars; 6.9–13.2% in the red cultivars) and lower proportions of (+)-catechin 3-gallate (CG; 0.16–1.16% in the white cultivars; 0.11–0.31% in the red cultivars).

For all grape cultivars, the total content of flavan-3-ol monomers and dimers (μmol per kg of grape) was mainly provided by their seeds, ranging within 690–5140 μmol/kg over only 15–246 μmol/kg from the skins. According to the total content of flavan-3-ol monomers and dimers found in both skins and seeds, the grape cultivars were grouped as follows:

1) Skins of the white cultivars:
   - High content: Albillo Mayor (246 μmol/kg) and Garnacha Blanca (218 μmol/kg).
   - Medium content: Moscatel Serrano (137 μmol/kg).
   - Low content: Castellana Blanca, Jarrosuelto, Planta Fina and Moscatel Grano Menudo (21–86 μmol/kg).
2) Seeds of the white cultivars:
   - High content: Moscatel Serrano (6390 μmol/kg).
   - Medium content: Moscatel de Grano Menudo, Jarrosuelto, Garnacha Blanca, Planta Fina and Albillo Mayor (2850–5140 μmol/kg).
   - Low content: Castellana Blanca (1070 μmol/kg).
3) Skins of the red cultivars:
   - High content: Gallera Negra (155 μmol/kg).
   - Medium-high content: Garnacha Gris (110 μmol/kg).
   - Medium-low content: Pintailla, Sinsó, Churriago and Granadera (32–76 μmol/kg).
   - Low content: Azargón (14.6 μmol/kg).
4) Seeds of the red cultivars:
   - High content: Churriago (2670 μmol/kg).
   - Medium content: Azargón, Pintailla and Garnacha Gris (1570–1890 μmol/kg).
   - Low content: Gallera Negra, Granadera and Sinsó (690–880 μmol/kg).

Conclusion

To conclude, in this work we have screened the phenolic potential of some autochthonous grape cultivars from the Spanish region of La Mancha, some of them being new genotypes (NG) which had never been described before. Among the red cultivars, the new genotype Gallera Negra stands out for its higher content in anthocyanins, medium content in flavonols, and hydroxycinnamic acid derivatives, thus suggesting that this cultivar could be used for making interesting red wines. The latter cultivar also has a high content of flavan-3-ols in its skins and seeds, which could be considered an advantage for making long maceration red wines destined for long aging. In contrast, three red cultivars, Azargón (NG), Pintailla (NG), and Garnacha Gris, accounted for high contents of flavonols and hydroxycinnamic acid derivatives; however, their content in anthocyanins was low, especially for Azargón and

Garnacha Gris, and their suggested use is for the production of rosé wines or even white wines. In fact, the red Azargón and Garnacha Gris cultivars can be considered very close to white cultivars, including flavonol profiles reminiscent of those typical of white grapes, with an almost total absence of B-ring trisubstituted structures. Regarding white grape cultivars, the high content of flavonols and flavan-3-ols shown by the new genotype Moscatel Serrano led to suggesting them as a good candidate for use of their winemaking waste (unfermented solid parts of the grape) as a profitable source of such bioactive compounds. From a phenolic point of view, the high content of hydroxycinnamic acid derivatives found in the white Albillo Mayor, Castellana Blanca and Garnacha Blanca cultivars suggested that the elaboration of their corresponding white wines should be carefully controlled to avoid oxidative browning. Most of the studied grape cultivars can be considered low producers of resveratrol, and only the new red Gallera Negra and Pintailla genotypes (total resveratrol content around 4 mg/kg) could be considered high resveratrol producing grape cultivars.

Finally, the phenolic profiles of these autochthonous grapes were suitable for cultivar characterization and differentiation. Anthocyanin and flavonol profiles were the most distinguishing factors for the red cultivars, whereas only flavonol profiles were for the white ones. The hydroxycinnamic acid derivative profiles helped to differentiate two white cultivars which were not well differentiated by the flavonol profiles, although this did not prove to be very useful.

References

Castillo-Muñoz N., Fernández-González M., Gómez-Alonso S., García-Romero E., Hermosín-Gutiérrez, I., (2009b). Red-color related phenolic composition of Garnacha Tintorera (*Vitis vinifera* L.) grapes and red wines. *J. Agric. Food Chem.*, 57, 7883-7891.

Castillo-Muñoz, N., Gómez-Alonso, S., García-Romero, E., Gómez, M. V., Velders, A., Hermosín-Gutiérrez, I., (2009a). Flavonol 3-*O*-glycoside series of *Vitis vinifera* cv. Petit Verdot red wine grapes. *J. Agric. Food Chem.* 57, 209-219.

Castillo-Muñoz, N., Gómez-Alonso, S., García-Romero, E., Hermosín-Gutiérrez, I., (2007). Flavonol profiles of *Vitis vinifera* red grapes and their single-cultivar wines. *J. Agric. Food Chem.* 55, 992-1002.

Castillo-Muñoz, N., Gómez-Alonso, S., García-Romero, E., Hermosín-Gutiérrez, I., (2010). Flavonol Profiles of *Vitis vinifera* White Grape Cultivars. *J. Food Compos. Anal.*, 23, 699-705.

Cejudo-Bastante, M. J., Pérez-Coello, M. S., Hermosín-Gutiérrez, I., (2010). Identification of new derivatives of 2-*S*-glutathionylcaftaric acid in aged white wines by HPLC-DAD-ESI-MS^n. *J. Agric. Food Chem.*, 58, 11483-11492.

Cheynier, V., Moutounet, M., Sarni-Manchado, P. Los Compuestos Fenólicos. In *Enología. Fundamentos Científicos y Tecnológicos*, 2nd ed.; Flanzy, C., Ed.; AMV Ediciones and Ediciones Mundi-Prensa: Madrid, Spain, 2003.

De Pablos Muñoz, L., (2010). *Estudio del contenido fenólico de vinos blancos y sus efectos en el vino*, Proyecto Fin de Carrera, Escuela Universitaria de Ingeniería Técnica Agrícola, Universidad de Castilla-La Mancha, Ciudad Real.

Downey, M. A., Harvey, J. S., Robinson, S. P., (2004). The effect of bunch shading on berry development and flavonoid accumulation in Shiraz grapes. *Aust. J. Grape Wine Res.*, 10, 55-73.

Fragoso, S., Aceña, L., Guasch, J., Busto, O., Mestres, M., (2011). Application of FT-MIR spectroscopy for fast control of red grape phenolic ripening. *J. Agric. Food Chem.*, 59, 2175-2183.

Gatto, P., Vrhovsek, U., Muth, J., Segala, C., Romualdi, C., Fontana, P., Pruefer, D., Stefanini, M., Moser, C., Mattivi, F., Velasco, R., (2008). Ripening and genotype control stilbene accumulation in healthy grapes. *J. Agric. Food Chem.*, 56, 11773-11785.

Gómez Gallego, M. A., Gómez García-Carpintero, E., Sánchez-Palomo, E., Hermosín-Gutiérrez, I., González Viñas, M. A., (2012). Study of phenolic composition and sensory properties of red grape varieties in danger of extinction from the Spanish region of Castilla La Mancha. *Eur. Food Res. Technol.*, 234, 295-303.

Haselgrove, L., Botting, D., van Heeswijck, R., Høj, P. B., Dry, P. R., Ford, C., Illand, P., (2000). Canopy microclimate and berry composition: The effect of bunch exposure on the phenolic composition of *Vitis vinifera* L. cv. Shiraz grape berries. *Aust. J. Grape Wine Res.*, 6, 141-149.

Hermosín-Gutiérrez, I., Castillo-Muñoz, N., Gómez-Alonso, S., García-Romero, E., (2011). Chapter 8. Flavonol profiles for grape and wine authentication. In *Progress in Authentication of Food and Wine*; Ebeler, S. E., Takeoka, G. R., Winterhalter, P.; ACS Symposium Series; American Chemical Society: Washington, DC, 2011, pp. 113-129.

Hermosín-Gutiérrez, I., García-Romero, E., (2004). Antocianos de variedades tintas cultivadas en La Mancha: perfiles varietales característicos de la uva y de los vinos monovarietales, y evolución durante la maduración de la baya. *Alimentaria*, 41, 127-139.

Hermosín-Gutiérrez, I., Verdejo Fernández, M. C., Alcaide Quintana, J., Alía Robledo, J. M., Montero Riquelme, F., Brasa Ramos, A., (2007). Contribución al estudio para la zonificación de la Denominación de Origen Valdepeñas: análisis de la uva de la vendimia del 2004. *Enólogos*, 45, 58-66.

Lago-Vanzela, E. S., Da-Silva, R., Gómes, E., García-Romero, E., Hermosín-Gutiérrez, I., (2011a). Phenolic composition of the Brazilian seedless table grape varieties BRS Clara and BRS Morena. *J. Agric. Food Chem.*, 59, 8314-8323.

Lago-Vanzela, E. S., Da-Silva, R., Gómes, E., García-Romero, E., Hermosín-Gutiérrez, I., (2011b). Phenolic composition of the edible parts (flesh and skin) of Bordô grape (*V. labrusca*) using HPLC-DAD-ESI-MS/MS. *J. Agric. Food Chem.*, 59, 13136-13146.

Ledda, S., Sanna, G., Manca, G., Franco, M. A., Porcu, A., (2010). Variability in flavonol content of grapes cultivated in two Mediterranean islands (Sardinia and Corsica). *J. Food Compos. Anal.*, 23, 580-585.

Makris, D. P., Kallithraka, S., Kefalas, P., (2006). Flavonols in grapes, grape products and wines: Burden, profile and influential parameters. *J. Food Comp. Anal.*, 19, 396-404.

Mattivi, F., Guzzon, R., Vrhovsek, U., Stefanini, M., Velasco, R., (2006). Metabolite profiling of grape: Flavonols and anthocyanins. *J. Agric Food Chem.*, 54, 7692-7702.

Mazza, G., Miniati, E. *Anthocyanins in fruits, vegetables, and grains*; CRC Press: Boca Raton, FL, 1993.

Organisation Internationale de la Vigne et du Vin. *Liste internationale des variétés de vigne et de leurs synonymes*. Edition 2013; Editions OIV: Paris, 2013.

Price, S. F., Breen, P. J., Valladao, M., Watson, B. T., (1995). Cluster sun exposure and quercetin in Pinot Noir grapes and wine. *Am. J. Enol. Viticult.*, 46, 187-194.

Rentzsch, M., Schwarz, M., Winterhalter, P., Hermosín-Gutiérrez, I., (2007). Formation of Hydroxyphenyl-Pyranoantocianins in Grenache Wines: Precursor Levels and Evolution during Aging. *J. Agric. Food Chem.*, 55, 4883-4888.

Rodríguez Montealegre, R., Romero Peces, R., Chacón Vozmediano, J. L., Martínez Gascueña, J., García Romero, E., (2006). Phenolic compounds in skins and seeds of ten grape *Vitis vinifera* varieties grown in a warm climate. *J. Food Compos. Anal.*, 19, 687-693.

Schwarz, M., Picazo-Bacete, J. J., Winterhalter, P., Hermosín-Gutiérrez, I., (2005). Effect of copigments and grape cultivar on the color of red wines fermented after the addition of copigments. *J. Agric. Food Chem.*, 53, 8372-8381.

Yilmaz, Y., Toledo, R. T., (2004). Health aspects of functional grape seed constituents. *Trends Food Sci. Technol.*, 15(9), 422-433.

In: Grapes
Editor: José S. Câmara
ISBN: 978-1-63321-402-6

Chapter 6

Grape Seed Extract Antioxidant and Antimicrobial Properties: Use in Active Packaging

***Javiera F. Rubilar**[*1] **and Rui M. S. Cruz**[2,3]*

[1]Departamento de Ingeniería Química y Bioprocesos, Pontificia Universidad Católica de Chile, Santiago, Chile

[2]Departamento de Engenharia Alimentar, Instituto Superior de Engenharia, Universidade do Algarve, Faro, Portugal

[3]CIQA-Centro de Investigação em Química do Algarve, Departamento de Química e Farmácia, Faculdade de Ciências e Tecnologia, Universidade do Algarve, Faro, Portugal

Abstract

The high cost of some naturally occurring and biologically active agents constitutes a major limitation on the use of this type of compounds in food production, as a result, efforts have been focused on less expensive sources such as agricultural wastes rich in polyphenols. Grape seed extract (GSE) is an industrial derivative from whole grape seeds, and is a rich source of monomeric phenolic compounds such as (+)- catechins, (-)-epicatechin, (-)-epicatechin-3-o-gallate, gallic acid and dimeric, trimeric and tetrameric procyanidins, which have antioxidant (AOX) and antimicrobial (AM) properties. GSE AM activity is dependent according to the flesh, whole fruit grape extracts, fermented fruit, fermented pomace, skin, leaves and seeds. GSE can be applied to edible films and coatings and is one of the most promising coating materials for different food matrices. This chapter will focus on the GSE AOX and AM properties, and on the application of this compound in different edible films or active food packaging towards shelf-life extension.

Keywords: Grape seed extract, active packaging, antimicrobial, antioxidant.

[*] Corresponding author: Dr. Javiera F. Rubilar Parra e-mail: jrubilar@ing.puc.cl.

Introduction

The increasing demand, by consumers, for safe, high quality, minimally processed, and extended shelf-life food has led to the development of new packaging systems (Guarda et al. 2011). As a result of this requirement, the area of food packaging had to make a major change in the direction of development and innovation. Thus, it had to shift from the traditional concept of packaging aimed at protecting the product with minimum package/product interaction, to a new concept thet focuses on the idea that some active interactions between the package and the product may have positive effects (Suppakul et al. 2003c). Active packaging (AP) is one of the most innovative food packaging concepts that was introduced as a response to the consumers' demands. AP was defined as a system in which the conditions of the packaged products are changed in order to better preserve the properties, improve safety and/or sensory quality and extend the shelf-life while maintaining the quality of foods (European FAIR-Project CT 98-4170) (Mauriello et al. 2005; Miltz and Passy 1995; Rooney 1995). Antimicrobial (AM) packaging is one of the most interesting and promising component of AP, which potentially allows industry to combine the preservative functions of AM agents with protective functions for foods (Mauriello et al. 2004). In most packaged foods microbial proliferation occurs primarily on the surface; therefore AM activity should be focused on solid or semi-solid surfaces (Cagri et al. 2004). This activity could be achieved by either indirect contact between the AM package and the product using volatiles AM releasing systems or by direct contact beteen the AM package and the food using non-migratory AM systems (Guarda et al. 2011). Different preservatives, such as silver substituted zeolites, bacteriocins, organics acids, salts and essential oils (EO) from various plant extract have been incorporated into packaging materials to provide antimicrobial activity (Appendini and Hotchkiss 2002; Coma et al. 2001; Dawson et al. 2005; Guarda et al. 2011; Hofmann et al. 2001; Pranoto et al. 2005; Rubilar et al. 2013a; Suppakul et al. 2008). The increase of bacterial resistance to antibiotics and the efforts to use natural preservatives in food manufacturing and its storage has increased the interest in possible application of natural antimicrobial agents (Guarda et al. 2011). Also, these agents possess the GRAS (Generally Recognized as Safe) status as defined by FDA (US Food an Drug Administration).

Utilization of plant as an alternative to chemical or synthetic antimicrobial and antioxidants to combat the foodborne pathogens, inhibiting lipid oxidation and thus extending the shelf-life is an increasing trend in the food insdustry (Perumalla and Hettiarachchy 2011). Major classes of antimicrobial and/or antioxidant compounds from plant extracts include simple phenols, phenolic acids, quinones, flavonoids, flavones, flavonols, tannins, terpenoids, essential oils, alkaloids, lectins and polypeptides and polyacetylenes (Cowan 1999). On the other hand, lipid oxidation is one of the major deteriorative chamical changes that would decrease the shelf-life of the food product and thus decrease its overal acceptability (Cortinas et al. 2005). Oxidation of labile double bonds in polyunsaturated fatty acids (PUFA) produces secondary oxidative compounds such as hexanal, pentanal, heptanal, and octanal that is responsible for quality deterioration, warmed over flavors (WOF) and present health risks (Grun et al. 2006).

Grape seed extract (GSE), a well known AM agent, is a rich source of polyphenolic compounds, mainly monomeric catechins and epicatechin (Figure 1), gallic acid, and polymeric and oligomeric procyanidins (Monagas et al. 2003; Rubilar et al. 2013b). These polyphenolic compounds showed various biological effects, including AM capacity (Baydar et al. 2004; Baydar et al. 2006; Furiga et al. 2009; Jayaprakasha et al. 2003; Rubilar et al. 2013b). The use of natural AM agents like grape seed extract that contain different polyphenolic compounds may increase the spectrum of AM activity, reduce the resistance of microorganisms and may also produce synergistic interaction exhibiting AM activity greater than that of simple sum of AM agents (Choi et al. 2009; Rubilar et al. 2012; Rubilar et al. 2013b) and may also contribute to improve the physico-mechanical properties of the films (Rubilar et al. 2013a).

1.1. Antimicrobial Active Packaging

AP is a system that changes the package structure and condition to improve the preservation of food properties, improve safety, enhance sensory quality and further extend shelf-life (Cruz et al. 2010). On the other hand, intelligent packaging solutions may include an external or internal indicator providing information on the product history or food quality (Robertson 2005). Such intelligent packaging systems can sense and record changes in the external or internal environment of the product and inform the consumer on the effect of these changes upon safety and quality of the food product (Cruz et al. 2010; de Kruijf et al. 2002; Huff 2008).

AP can be classified into scavenging (absorbers) and release systems (emitters). Scavenging systems can remove oxygen, excessive moisture, ethylene, carbon dioxide and other specific undesirable compounds. Release systems emit into packaged food compounds such as carbon dioxide, water, antioxidants (AOX) and antimicrobials (AM) (Cruz et al. 2010; de Kruijf et al. 2002).

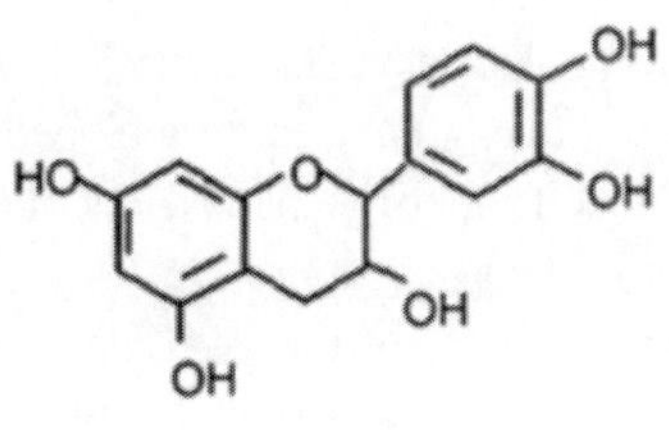

(-)-Epicatechin

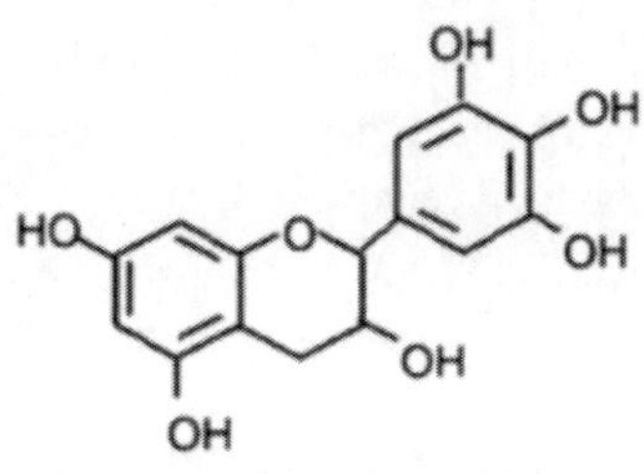

(-)-Epigallocatechin

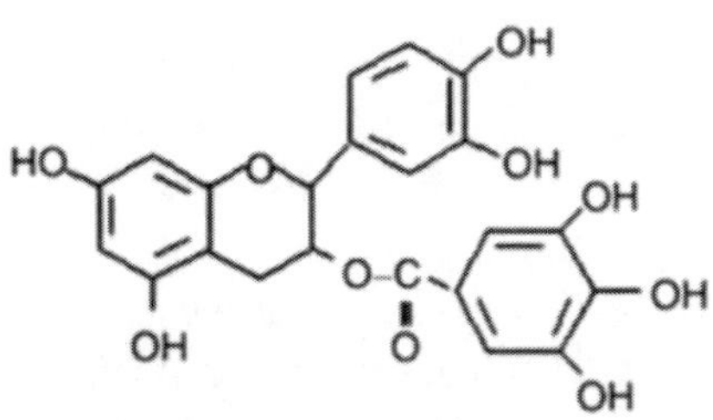

(-)-Epicatechin-3-gallate

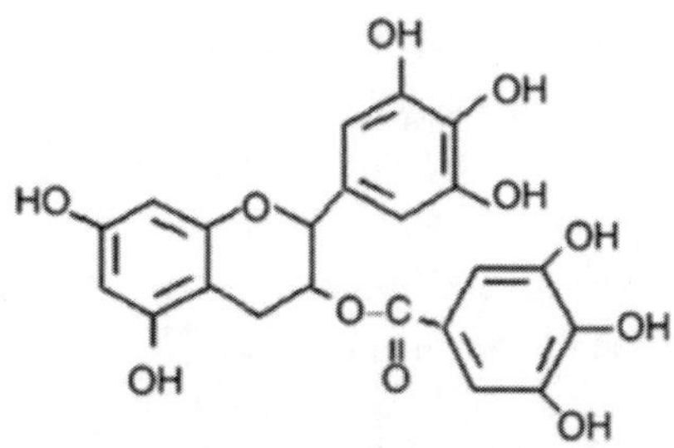

(-)-Epigallocatechin-3-gallate

Figure 1. Several polyphenolic compounds from grape seed extract.

AM food packaging is a packaging system that can reduce, inhibit and/or stop the growth of microorganisms in food. AM packaging can minimize food spoilage caused by microorganisms contaminating the food surface (Gill 1979; Maxcy 1981; Rubilar et al. 2013a). By releasing preservatives from the package (rather than mixing preservatives into the bulk of the food), a smaller amount of AM compound is needed to prevent the growth of microorganisms on food surfaces (Guarda et al. 2011; Han 2005; Yalpani et al. 1992).

For the consumer, it seems safer when active agents are indirectly integrated in the food package and released into the food product thereafter. Moreover, consumers tend to accept products to which naturally occurring substances have been added rather than those containing synthetic agents (Han 2005). This idea was first developed in the early 1980's when Ghosh et al. (Ghosh et al. 1973) developed a cheese wrapping paper containing sorbates as antifungal agents, using carboxy methyl cellulose (CMC) as a binder. In the last decade, AM packaging has become more interesting as shown by the increasing number of publications on this topic (Suppakul et al. 2003b). Several types of packaging materials, in which various polymeric matrices are mixed with different AM agents, have been described. This technology has resulted in advancements in food packaging and has opened up new options for AM packaging systems. Such packaging systems not only protect food from post-contamination but also inhibit growth of pre-contaminating microorganisms (Appendini and Hotchkiss 2002; Guarda et al. 2011; Rubilar et al. 2012; Rubilar et al. 2013a; Suppakul et al. 2003a; Yalpani et al. 1992).

1.1.1. Plant extracts - Emphasis in Grape Seed Extract

The extension of shelf-life has been an increasing trend in the food industry. For this reason, plant extracts have been used as an alternative to chemical or synthetic AM and AOX against foodborne pathogens and as inhibitors of lipid oxidation (Perumalla and Hettiarachchy 2011). Major groups of chemicals present in plant extracts include polyphenols, quinines, flavanols/flavanoids, alkaloids, and lectins (Cowan 1999). Phenolic extracts prepared from sage, rosemary, thyme, hops, coriander, green tea, grape seed, cloves, and basil are known to have antimicrobial effects against foodborne pathogens (Perumalla and Hettiarachchy 2011).

Natural plant extracts in conjunction with other hurdles like low storage temperature, low pH, anaerobic conditions, organic acids, bacteriocins, and irradiation showed synergistic AM action in various food systems (Beuchat et al. 1994; Gadang et al. 2008; Over et al. 2009).

1.1.1.1. Grape Seed Extract Like Antimicrobial Agent

The core structures with 3, 4, 5-trihydroxyphenyl groups found in epigallocatechin, epigallocatechin-3-Ogallate, castalagin and prodelphinidin might be important for antibacterial activity (Taguri et al. 2004). As GSE is a rich source of polymers of flavan-3-ols like (+)-catechin and (-)epicatechin, itsantimicrobial properties can be attributed to the general mechanisms of phenolics. The outer cell membrane or cytoplasmic membrane of a bacterium is essentially composed of a phospholipid bilayer and proteins ans is the major site of interaction with antimicrobial compounds (Perumalla and Hettiarachchy 2011). Damage to this vital membrane can result in death of the bacterium and can occur in the following ways (i) physical disruption of membrane (Shimamura et al. 2007); (ii) dissipation of the proton motive force (PMF) (Juven et al. 1994) and (iii) inhibition of membrane-associated enzyme activity (Perumalla and Hettiarachchy 2011).

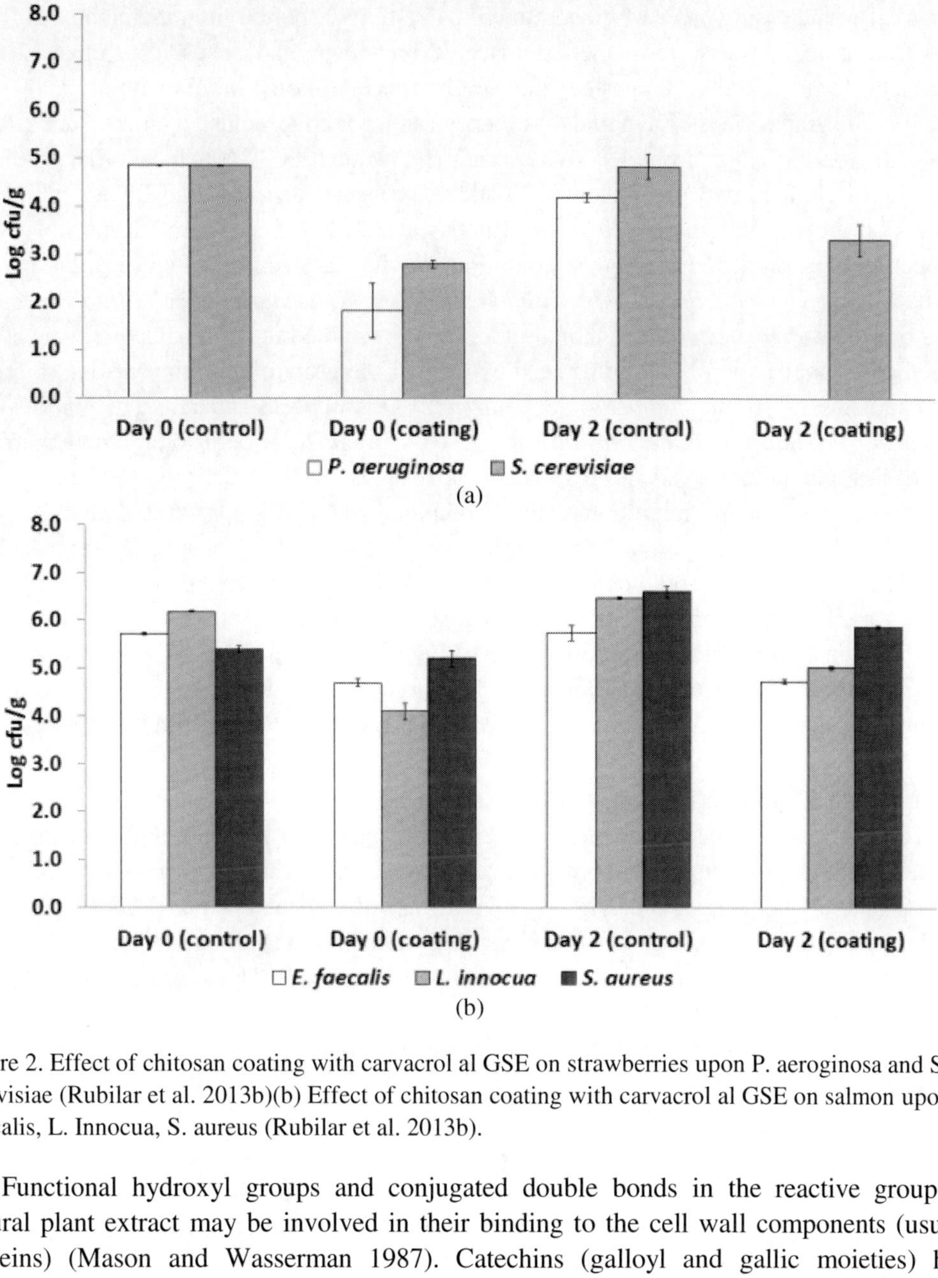

Figure 2. Effect of chitosan coating with carvacrol al GSE on strawberries upon P. aeroginosa and S. cerevisiae (Rubilar et al. 2013b)(b) Effect of chitosan coating with carvacrol al GSE on salmon upon E. Faecalis, L. Innocua, S. aureus (Rubilar et al. 2013b).

Functional hydroxyl groups and conjugated double bonds in the reactive groups of natural plant extract may be involved in their binding to the cell wall components (usually proteins) (Mason and Wasserman 1987). Catechins (galloyl and gallic moieties) have deteriorating effect on the lipid bilayer membrane that results in the loss of cell structure and function eventually leading to cell death (Cox et al. 2001; Ikigai et al. 1993; Tsuchiya et al. 1996). Presence of gallic acid esters in (-)-Epicatechin, (-)-Epigallocatechin-3-gallate are responsible for their high affinity for lipid bilayers, and effect the membrane structure (Hashimoto et al. 1999). Major phenolic constituents like epicatechin, caffeic acid, benzoic acid, and syringic acid may alter the cell morphology by influencing the osmotic presure of the cell, thus disrupting the cytoplasmic membrane and causing leakage of cel constituents

(Davidson and Naidu 2000; Sivarooban et al. 2008). The red pigmented polymeric phenolics from juice and skin showed H-dependent antilisterial activity, while the unpigmented polymeric phenolics demostrated antilisterial activity that was independent of pH (Rhodes et al. 2006). The increasing order of the antimicrobial activity reported in grape plant was flesh, whole fruit grape estracts, fermented pomace, skin, leaves and seeds (Xia et al. 2010). Anastasiadi et al. (2009) suggested that high concentrations of flavonoids and their derivatives in grape seeds, flavonoids, stilbenes, and phenolic acids in grape stems were responsible for the antimicrobial activity. Vaquero et al. (2007) concluded that the non-flavonoid caffeic acid and the flavonoids such as rutin and quecetin had higher inhibitory activities on growth of *L. monocytogenes*. Rhodes et al. (2006) demostrated that polymeric phenolic fractions produced the highes inhibition activity for all *Listerial* species, but not for other bacteria, such as *Bacillus cereus*, *Salmonella menston*, *Escherichia coli*, *Staphylococcus aureus* or *Yersenia enterocolitica*. Rubilar et al. (2013b) studied the optimal synergistic effect between chitosan, carvacrol and grape seed extract for the growth inhibition of five different microorganisms (*Listeria innocua*, *S. aureus*, *Enterococcus faecalis*, *Pseudomonas aeruginosa*, *Saccharomyces cerevisiae*), and test its usage in two different food matrices (strawberries and salmon) and the results of the response surface analysis showed several synergistic effects for all microorganisms. Combinations of 60 ppm-400 ppm-1.2% w/v (carvacrol-GSE-chitosan; optimal AM combination 1, OAMC-1); 9.6 ppm-684 ppm-1.25% w/v (OAMC-2); 90 ppm-160 ppm-1.24% w/v (OAMC-3) were found to be the optimal mixtures for all microoganisms. The best radical scavenging activity (OAMC-3) was applied as a coating in two matrices mentiones above. For strawberries, *P. aeruginosa* was more sensitive to the action of OAMC-3 than *S. cerevisiae*. For salmon, *S. aureus* was more resistant to the action of OAMC-3 than *E. faecalis* and *L. innocua* as showed the Figure 2.

1.1.1.2. Grape Seed Extract like antioxidant agent

Lipid oxidation is one of the major deteriorative chemical changes that decreases the shelf-life of food products and could decrease their overall acceptability (Cortinas et al. 2005). Oxidation of labile double bonds in polyunsaturated fatty acids (PUFA) produces secondary oxidative compounds such as hexanal, pentanal, heptanal, and octanal, which are responsible for quality deterioration and warmed over flavors (WOF), also presenting health risks (Grun et al. 2006).

The reaction-chain mechanism involved in lipid oxidation in foods can be described by three steps.

Step 1. Initiation: the free radicals (R•) are formed by loss of hydrogen radical (H•) from an unsaturated fatty acid (RH).

$$RH \xrightarrow{\text{Initiator}} R\bullet + H\bullet$$

Step 2. Propagation: the free radicals react with oxygen (O_2) in presence of light, heat or trace metals to form peroxyl radicals (ROO•). Next, peroxyl radicals react with more unsaturated fatty acids to form lipid hydroperoxides (ROOH) and more free radicals (R•), which in turn react with oxygen feeding the reaction chain. This propagation cycle will continue until step 3 takes place.

$$R\bullet + O_2 \longrightarrow ROO\bullet$$
$$ROO\bullet + RH \longrightarrow ROOH + R\bullet$$

Step 3. Termination: the chain oxidation process terminates when two free radicals react to produce a non-radical species (ROOR or RR).

$$ROO\bullet + ROO\bullet \longrightarrow ROOR + O_2$$
$$ROO\bullet + R\bullet \longrightarrow ROOR$$
$$R\bullet + R\bullet \longrightarrow RR$$

The AOX activity refers to the act of delaying or inhibiting the oxidation of lipids or other molecules by inhibiting partially or totally the initiation or propagation step of the oxidative chain reactions or forming stable radicals (A•) that are either unreactive or form non-radical products (Huang et al. 2005).

$$ROO\bullet + AH \longrightarrow ROOR + A\bullet$$

For example, polyphenolic compounds (mainly flavonoids) present in grape seed extract have demonstrated potential antioxidant properties due to their redox potential; this enables them to act in various ways, such as hydrogen donors, reducing agents, reactive oxygen quenchers, and chelating metal ions in numerous food applications (Gramza et al. 2006; Perumalla and Hettiarachchy 2011). The active hydroxyl groups present in the molecular structure of polyphenols ((-)-epicatechin, (-)-epigallocatechin, (-)-epicatechin-3-gallate, (-)-epigallocatechin-3-gallate) are the active components of green tea extract that can interact with the free radicals to inhibit lipid oxidation (Mitsumoto et al. 2005). Furthermore, tea polyphenols can exhibit scavenging activity against free radicals (Rice-Evans et al. 1997), superoxide radicals, peroxynitrite, and can chelate copper and iron, preventing metal catalyzed free radical formation (Rice-Evans et al. 1997).

Flavonoids present in plant extracts terminate the radical chain reactions that occur during the oxidation of triglycerides in food systems (fats, oils, and emulsions) and thus can act as free radical scavengers (Turkoglu et al. 2007).

Winery and grape juice industries by-products such as grape seeds are waste products that can be separated, extracted, dried and purified into grape seed extract (GSE) which contains proanthocyanidins and phenolic compounds (Lau and King 2003).

Grape seed polyphenols contain, aside from phenolic acid precursors (gallic acid), the monomeric flavan-3-ols (catechin, epicatechin, gallocatechin, epigallocatechin and epicatechin 3-O-gallate) and procyanidin dimers, trimers, and highly polymerised procyanidins (Chedea et al. 2011). The benefits of using GSE include antioxidant (Chedea et al. 2010), cardioprotective (Schewe et al. 2001), hepatoprotective, neuroprotective, anti-inflammatory, antidiabetic, anti-carcinogenic, and anti-ageing effects (Nassiri-Asl and Hosseinzadeh 2009; Xia et al. 2010). Moreover, the antioxidant potential of GSE is twenty and fifty fold greater than vitamins E and C, respectively (Shi et al. 2003). The red color and astringency taste of the GSE can be attributed to polyphenol rich compounds especially proanthocyanidins which may affect the color and sensory characteristics of the product when used at higher concentrations (Monteleone et al. 2004; Weber et al. 2007).

The mechanism of antioxidative activity is expressed in its ability of radical scavenging (superoxide, hydroxyl, and 1,1-diphenyl-2-picrylhydrazyl (DPPH)), metal chelating properties, reduction of hydroperoxide formation and their effects on cell signaling pathways, gene expression (Sato et al. 1996; Soobrattee et al. 2005) and synergism with other antioxidants (Lu and Foo 2001).

The antioxidant capacity is determined by the existence of a functional group -OH in the structure and its position on the ring of the flavanoid molecule (Arora et al. 1998). Addition of -OH groups to the flavonoid nucleus will enhance the antioxidant activity, while substitution by -OCH3 groups reduces the antioxidant activity (Majo et al. 2008). Degree of polymerization of the procyanidins may also determine the antioxidant activity, as the higher the degree of polymerization, the higher the antioxidant activity (Spranger et al. 2008). Among the different parts of grape plant, grape seeds exhibit highest antioxidant activity followed by the skin and the flesh (Pastrana-Bonilla et al. 2003).

GSE has been evaluated for its antioxidative effect on meat products and has been reported to improve the oxidative stability of cooked beef (Ahn et al. 2002), beef patties (Banon et al. 2007), turkey thigh meat (Lau and King 2003) and ground chicken breast (Brannan 2009). For example, GSE added at 10 and 20 g/kg to turkey thigh meat decreased TBARS values nearly ten-fold as compared to the control (Lau and King 2003). It was suggested that the optimal level of GSE supplemented to poultry meat was between 1 and 10 g/kg meat.

In cooked pork, the minimum concentration level of GSE required to produce an antioxidant effect was 400 μg/g and 0.1% (w/w) in ground chicken to reduce the TBARS (Lau and King 2003). Rojas and Brewer (2007) also reported that 0.01-0.02% of GSE (89% proanthocyanidins) were highly effective at maintaining low TBARS in beef. Supplementation of GSE prior to cooking significantly improved oxidative stability of minced turkey meat during heat treatment and storage (Mielnik et al. 2006).

Kulkarni et al. (2011) reported that GSE used at concentrations of 100, 300, and 500 ppm in precooked beef sausages cooked at 70 °C and stored for 4 months (-18 °C), lowered rancid odor and flavor values and maintained their fresh cooked beef odor and flavor longer than the control. Additionally, TBARS (0.2 mg MDA/kg) of the samples treated with the lowest GSE concentrations (100 and 300 ppm) remained similar or decreased slightly during storage.

In a study reported by Lorenzo et al. (2013) GSE showed to be a more effective antioxidant against lipid oxidation of sausage "chorizo" compared to chestnut extracts and buthylatedhydroxytoluene (BHT). However, in a study reported by Shirahigue et al. (2010) Isabel grape extract (IGE) and Niagara grape extract (NGE) significantly increased the oxidative stability of processed and cooked chicken meat during storage time. Results similar to BHT were observed applying the extracts at concentrations of 40 and 60 mg of gallic acid equivalent (GAE).

Regarding fish products, Yerlikaya and Gokoglu (2012) reported that the most effective GSE concentration in preserving oxidation of bonito fillets was 1.0 g/100 g.

Addition of GSE (≥1000 μg/g) may also result in an increase of redness ('color a value') in the surface of raw meat and retention (presence of anthocyanins in GSE) in cooked meat, which may have a negative impact on consumer acceptability based on the color of the meat product (Ahn et al. 2002; Carpenter et al. 2007).

The application of GSE in food formulations or food packaging may reduce the oxidation processes that may occur during shelf-life and thus, contributing for a high quality food product.

Conclusion

The response to the increasing demand, by consumers, for safe, high quality, minimally processed, and exended shelf-life food has led to the employ of natural and safety alternatives, several plant extracts are being used in the food industry. Grape seed extract is a natural source of antimicrobial and antioxidant compound that have the potential to improve the overall quality and extend the shelf life.

Acknowledgments

The author Javiera F. Rubilar would like to thank to the financial support of the FONDECYT Postdoctoral Project #3140349, as also to DIPEI of the Pontificia Universidad Católica de Chile. The author Rui M. S. Cruz would like to thank Fundação para a Ciência e Tecnologia (grant SFRH/BPD/ 70036/2010).

Abbreviations

A•	stable radicals
AM	antimicrobial
AOX	antioxidants
AP	active packaging
BHT	buthylatedhydroxytoluene
CMC	carboxy methyl cellulose
DPPH	1,1-diphenyl-2-picrylhydrazyl
FDA	US Food an Drug Administration
GAE	gallic acid equivalent
GRAS	Generally Recognized as Safe
GSE	grape seed extract
IGE	Isabel grape extract
MDA	malondialdehyde
NGE	Niagara grape extract
O2	oxygen
OAMC	optimal AM combination
PMF	proton motive force
PUFA	polyunsaturated fatty acids
R•	free radicals
RH	unsaturated fatty acid
ROO•	peroxyl radicals

ROOH	lipid hydroperoxides
TBARS	Thiobarbituric acid reactive substances
WOF	warmed over flavors

References

Ahn, J., Grün, I. U. & Fernando, L. N. (2002). Antioxidant properties of natural plant extracts containing polyphenolic compounds in cooked ground beef, *Journal of Food Science*, *67*, 1364-69.

Anastasiadi, M., Chorianopoulos, N. G., Nychas, G. J. & Haroutounian, S. A. (2009). Antilisterial activities of polyphenol-rich extracts of grapes and vinification byproducts, *Journal of Agricultural and Food Chemistry*, *57*, 457-563.

Appendini, P. & Hotchkiss, J. H. (2002). Review of antimicrobial food packaging, *Innovative Food Science and Emerging Tehcnologies*, *3*, 113-26.

Arora, A., Nair, M. G. & Strasburg, G. M. (1998). Structure-activity relationships for antioxidant activities of a series of flavonoids in a liposomal system, *Free Radical Biology and Medicine*, *24* (9). 1355-63.

Banon, S., Diaz, P., Rodriguez, M., Garrido, M. D. & Price, A. (2007). Ascorbate, green tea and grape seed extracts increase the shelf life of low sulphite beef patties, *Meat Science*, *77* (4), 626-33.

Baydar, N. G., Özkan, G. & Sağdiç, O. (2004). Total phenolic contents and antibacterial activities of grape (Vitis vinifera L.) extracts, *Food Control*, *15* (5), 335-39.

Baydar, N. G., Sagdic, O., Ozcan, G. & Cetin, S. (2006). Determination of antibacterial effects and total phenolic contents of grape (Vitis vinifera L.) seed extracts, *International Journal of Food Science and Technology*, *41*, (799-804).

Beuchat, L. R., Brackette, R. E. & Doyle, M. P. (1994). Lethality of carrot juice to Listeria monocytogenes as affected by pH, sodium chloride, and temperature, *Journal of Food Protection*, *57*, 470-74.

Brannan, R. G. (2009). Effect of grape seed extract on descriptive sensory analysis of ground chicken during refrigerated storage, *Meat Science*, *81*, 589-95.

Cagri, A., Ustunol, Z. & Ryser, E. (2004). Antimicrobial Edible Films and Coatings-Review, *Journal of Food Protection*, *67* (4), 833-48.

Carpenter, R., O'Grady, M. N., O'Callaghan, Y. C., O'Brien, N. M. & Kerry, J. P. (2007). Evaluation of the antioxidant potential of grape seed and bearberry extracts in raw and cooked pork, *Meat Science*, *76* (4), 604-10.

Coma, V., Sebti, I., Pardon, P., Deschamps, A. & Pichavant, F. H. (2001). Antimicrobial edible packaging based on cellulosic ethers, fatty acids, and nisin incorporation to inhibit *Listeria innocua* and *Staphylococcus aureus*, *Journal of Food Protection*, *64*, 470-75.

Cortinas, L., Barroeta, A., Villaverde, C., Galobart, J., Guardiola, F. & Baucells, M. D. (2005). Influence of the dietary polyunsaturation level on chicken meat quality: Lipid oxidation, *Poultry Science*, *84*, 48-55.

Cowan, M. M. (1999). Plant products as antimicrobial agents, *Clinical Microbiology Reviews*, *12*, 564-82.

Cox, S. D., Mann, C. M. & Markham, J. L. (2001). Interactions between components of essential oil of *Melaleuca alternifolia*, *Journal of Applied Microbiology*, *91*, 492-97.

Cruz, R. M. S., Rubilar, J. F., Ulloa, P., Torres, J. A. & Vieira, M. C. (2010). New Food Processing Technologies: development and impact on the consumer acceptability, in F. Columbus (ed.), *Food Quality: Control, Analysis and Consumer Concerns* (In Press); New York, USA: Nova Science.

Chedea, V. S., Braicu, C. & Socaciu, C. (2010). Antioxidant/pro-oxidant activity of a polyphenolic grape seed extract, *Food Chemistry*, *121*, 132-39.

Chedea, V. S., Echim, C., Braicu, C., Andjelkovic, M., Verhe, R. & Socaciu, C. (2011). Composition in polyphenols and stability of the aqueous grape seed extract from the Romanian variety ''Merlot Recas'', *Journal of Food Biochemistry*, *35*, 92-108.

Choi, J. G., Kang, O. H., Lee, Y. S., Oh, Y. C., Chae, H. S., Jang, H. J. & Kwon, D. Y. (2009). Antibacterial activity of methyl gallate isolated from Galla Rhois or carvacrol combined with nalidixic acid against nalidixic acid resistant bacteria, *Molecules*, 14 (5), 1773-80.

Davidson, P. M. & Naidu, A. S. (2000). Phyto-phenols, in A. S. Naidu (ed.). *Natural food antimicrobial systems* (BOca Raton, Fl, USA: CRC Press LLC), 265-94.

Dawson, P. L., Harmon, L., Sotthibandhu, A. & Han, I. Y. (2005). Antimicrobial activity of nisin-adsorbed silica and corn starch powders, *Food Microbiology*, *22* (1), 93-99.

de Kruijf, N., van Beest, M., Rijk, R., Sipiläinen-Malm, T., Paseiro Losada, P. & De Meulenaer, B. (2002). Active and intelligent packaging: applications and regulatory aspects, *Food Additives and Contaminants*, *19*, (144-162).

Furiga, A., Lonvaud-Funel, A. & Badet, C. (2009). In vitro study of antioxidant capacity and antibacterial activity on oral anaerobes of a grape seed extract, *Food Chemistry*, *113*, 1037-40.

Gadang, V. P., Hettiarachchy, N. S., Johnson, M. G. & Owens, C. M. (2008). Evaluation of antibacterial activity of whey protein isolate coating incorporated with nisin, grape seed extract, malic acid, and EDTA on a turkey frankfurter system, *Journal of Food Science: Food Microbiology and Safety*, *73* (8), M389-M94.

Ghosh, K. G., Srivatsava, A. N., Nirmala, N. & Sharma, T. R. (1973). Development and application of fungistatic wrappers in food presearvation. Part I. Wrappers obtained by impregnation method, *Journal of Food Science and Technology*, *10* (4), 105-10.

Gill, C. O. (1979). A review. Intrinsic bacteria in meat, *Journal of Applied Bacteriology*, *47*, 367-78.

Gramza, A., Khokhar, S., Yoko, S., Gliszczynska-Swiglo, A., Hes, M. & Korczak, J. (2006). Antioxidant activity of tea extracts in lipids and correlation with polyphenol content, *European Journal of Lipid Science and Technology*, *108*, 351-62.

Grun, I. U., Ahn, J., Clarke, A. D. & Lorenzen, C. L. (2006). Reducing oxidation of meat, *Food Technology*, *60*, 36-43.

Guarda, A., Rubilar, J. F., Miltz, J. & Galotto, M. J. (2011). The antimicrobial activity of microencapsulated thymol and carvacrol, *International Journal of Food Microbiology*, *146*, 144-50.

Han, J. H. (2005). Antimicrobial packaging systems, in J.H. Han (ed.). *Innovation in Food Packaging* (London: Elsevier, Academic Press). 517.

Hashimoto, T., Kumazawa, S., Nanjo, F., Hara, Y. & Nakayama, T. (1999). Interaction of tea catechins with lipid bilayers investigated with liposome systems, *Bioscience, Biotechnology and Biochemistry*, *63*, 2252-55.

Hofmann, K. L., JHan, I. Y. & Dawson, P. L. (2001). Antimicrobial effect of corn zein films impregnated with nisin, lauric acid, and EDTA, *Journal of Food Protection*, 64, 885-89.

Huang, D., Ou, B. & Ronald, L. (2005). The chemistry behind antioxidant capacity assays, *Journal of Agricultural and Food Chemistry*, *53* (6), 1841-56.

Huff, K. (2008). Active and Intelligent Packaging: Innovations for the Future. Department of Food Science and Technology, 1-13.

Ikigai, H., Nakae, T., Hara, Y. & Shimamura, T. (1993). Bactericidal catechins damage the lipid bilayer, *Biochimica et Biophysica Acta (BBA)-Biomembranes*, *1147* (1), 132-36.

Jayaprakasha, G. K., Selvi, T. & Sakariah, K. K. (2003). Antibacterial and antioxidant activities of grape (Vitis vinifera) seed extracts, *Food Research International*, *36* (2), 117-22.

Juven, B. J., Kanner, J., Schved, F. & Weisslowicz, H. (1994). Factors that interact with the antibacterial action of thyme essential oil and its active constituents, *The Journal of Applied Bacteriology*, *76*, 626-31.

Kulkarni, S., DeSantos, F. A., Kattamuri, S., Rossi, S. J. & Brewer, M. S. (2011). Effect of grape seed extract on oxidative, color and sensory stability of pre-cooked, frozen, re-heated beef sausage model system, *Meat Science*, *88*, 139-44.

Lau, D. W. & King, A. J. (2003). Pre- and post-mortem use of grape seed extract in dark poultry meat to inhibit development of thiobarbituric acid reactive substances, *Journal of Agricultural and Food Chemistry*, *51* (6), 1602-07.

Lorenzo, J. M., González-Rodríguez, R. M., Sánchez, M., Amado, I. R. & Franco, D. (2013). 'Effects of natural (grape seed and chestnut extract) and synthetic antioxidants (buthylatedhydroxytoluene, BHT) on the physical, chemical, microbiological and sensory characteristics of dry cure sausage "chorizo"', *Food Research International*, *54*, 611-20.

Lu, Y., Foo, Y. L. (2001). Antioxidant activities of polyphenols from sage (Salvia officinalis). *Food Chemistry*, 75, 197-202.

Majo, D. D., Guardia, M. L., Giammanco, S., Neve, L. L. & Giammanco, M. (2008). The antioxidant capacity of red wine in relationship with its polyphenolic constituents, *Food Chemistry*, *111* (1), 45-49.

Mason, T. L. & Wasserman, B. P. (1987). Inactivation of red beet betaglucan synthase by native and oxidative phenolic compounds, *Phytochemistry*, *26*, 2197-202.

Mauriello, G., Ercolini, D., La Storia, A., Casaburi, A. & Villani, F. (2004). Development of polythene films for food packaging activated with an antilisterial bacteriocin from Lactobacillus curvatus 32Y, *Journal of Applied Microbioloogy*, *97* (2), 314-22.

Mauriello, G., De Luca, E., La Storia, A., Villani, F. & Ercolini, D. (2005). Antimicrobial activity of a nisin-activated plastic film for food packaging, *Letters in Applied Microbiology*, *41* (6), 464-9.

Maxcy, R. B. (1981). Surface microenvironment and penetration of bacteria into meat, *Journal of Food Protection*, *44*, 550-52.

Mielnik, M. B., Olsen, E., Vogt, G., Adeline, D. & Skrede, G. (2006). Grape seed extract as antioxidant in cooked, cold stored turkey meat, *LWT-Food Science and Technology*, *39* (3), 191-98.

Miltz, J. & Passy, N. (1995). Shelf life grapefruit segments in multilayer Plastic Trays, *LWT - Food Science and Technology*, *28*, 442-47.

Mitsumoto, M., O'Grady, M. N., Kerry, J. P. & Buckley, D. J. (2005). Addition of tea catechins and vitamin C on sensory evaluation, colour and lipid stability during chilled storage in cooked or raw beef and chicken patties, *Meat Science*, *69*, 773-79.

Monagas, M., Gómez-Cordovés, C., Bartolomé, B., Laureano, O. & da Silva, J. M. R. (2003). Monomeric, oligomeric, and polymeric flavan-3-ol composition of wines and grapes from Vitis vinifera L. cv. Graciano, Tempranillo, and Cabernet Sauvignon, *Journal of Agricultural and Food Chemistry*, *51* (22), 6475-81.

Monteleone, E., Condelli, N., Dinnella, C. & Bertuccioli, M. (2004). Prediction of perceived astringency induced by phenolic compounds, *Food quality and preference*, *15* (7), 761-69.

Nassiri-Asl, M. & Hosseinzadeh, H. (2009). Review of the pharmacological effects of Vitis vinifera (Grape) and its bioactive compounds, *Phytother Research: PTR*, *23* (9), 1197-204.

Over, K. F., Hettiarachchy, N., Johnson, M. G. & Davis, B. (2009). Effect of organic acids and plant extracts on Escherichia coli O157: H7, Listeria monocytogenes, and Salmonella Typhimurium in broth culture model and chicken meat systems, *Journal of Food Science*, *74* (9), M515-M21.

Pastrana-Bonilla, E., Akoh, C. C., Sellappan, S. & Krewer, G. (2003). Phenolic content and antioxidant capacity of muscadine grapes, *Journal of Agricultural and Food Chemistry*, *51* (18), 5497-503.

Perumalla, A. V. S. & Hettiarachchy, Navam, S. (2011). Green tea and grape seed extracts — Potential applications in food safety and quality, *Food Research International*, *44* (4), 827-39.

Pranoto, Y., Rakshit, S. K. & Salokhe, V. M. (2005). Enhancing antimicrobial activity of chitosan films by incorporating garlic oil, potassium sorbate and nisin, *LWT-Food Science and Technology*, *38* (8), 859-65.

Rhodes, P. L., Mitchell, J. W., Wilson, M. W. & Melton, L. D. (2006). Antilisterial activity of grape juice and grape extracts derived from Vitis vinifera variety Ribier, *International Journal of Food Microbiology*, *107* (3), 281-6.

Rice-Evans, C., Miller, N. & Paganga, G. (1997). Antioxidant properties of phenolic compounds, *Trends in plant science*, *2* (4), 152-59.

Robertson, G. L. (2005). Active and intelligent packaging, in G. L. Robertson (ed.), *Food packaging: principles and practice*, *6*, Fl, USA CRC Press, Boca Raton.

Rojas, M. C. & Brewer, M. S. (2007). Effect of natural antioxidants on oxidative stability of cooked, refrigerated beef and pork, *Journal of Food Science*, *72* (4), S282-S88.

Rooney, M. L. (1995). *Active food packaging* (Glasgow, UK: Blackie Academic & Professional) 255.

Rubilar, J. F., Cruz, R. M. S., Khmelinskii, I. & Vieira, M. C. (2013a). Chitosan films with antioxidant and antimicrobial properties as active packaging, in F. Pedreschi and Z. Ciesarova (eds.), *Chemical Food Safety and Health* (Nova Publishers).

Rubilar, J. F., Cruz, R. M. S., Khmelinskii, I. & Vieira, M. C. (2013b). Effect of antioxidant and optimal antimicrobial mixtures of carvacrol, grape seed extract and chitosan on different spoilage microorganisms and their application as coatings on different food matrices, *International Journal of Food Studies*, *2*, 22-38.

Rubilar, J. F., Cruz, R. M. S., Khmelinskii, I. & Vieira, M. C. (2012). Antioxidant and optimal antimicrobial mixtures of carvacrol, grape seed extract and chitosan on different spoilage microorganisms and their applications in different food matrices, *International Journal of Food Studies*.

Rubilar, J. F., Cruz, Rui M. S., Silva, H. D., Vicente, A. A., Khmelinskii, I. & Vieira, Margarida C. (2013c). Physico-mechanical properties of chitosan films with carvacrol and grape seed extract, *Journal of Food Engineering*, *115* (4), 466-74.

Sato, M., Ramarathnam, N., Suzuki, Y., Ohkubo, T., Takeuchi, M. & Ochi, H. (1996). Varietal differences in the phenolic content and superoxide radical scavenging potential of wines from different sources, *Journal of Agricultural and Food Chemistry*, *44* (1), 37-41.

Schewe, T., Sadik, C., Klotz, L. -O., Yoshimoto, T., Kühn, H. & Sies, H. (2001). Polyphenols of cocoa: Inhibition of mammalian 15-lipoxygenase, *Biological Chemistry*, *382*, 1687-96.

Shi, J., Yu, J., Pohorly, J. E. & Kakuda, Y. (2003). Polyphenolics in grape seeds-biochemistry and functionality, *Journal of Medicinal Food*, *6* (4), 291-99.

Shimamura, T., Zhao, W. H. & Hu, Z. (2007). Mechanism of action and potential for use of tea catechin as an anti-infective agent, *Anti-infective Agents in Medicinal Chemistry*, *6*, (57-62).

Shirahigue, L. D., Plata-Oviedo, M., Alencar, S. M., Regitano-D'Arce, M. A. B., Vieira, T. M. F. S., Oldoni, T. L. C. & Contreras-Castillo, C. J. (2010). Wine industry residue as antioxidant in cooked chicken meat, *International Journal of Food Science and Technology*, *45* (5), 863-70.

Sivarooban, T., Hettiarachchy, N. S. & Johnson, M. G. (2008). Transmission electron microscopy study of *Listeria monocytogenes* treated with nisin in combination with either grape seed or green tea extract, *Journal of Food Protection*, *71*, 2105-09.

Soobrattee, M. A., Neergheen, V. S., Luximon-Ramma, A., Aruoma, O. I. & Bahorun, T. (2005). Phenolics as potential antioxidant therapeutic agents: mechanism and actions, *Mutation Research — Fundamental and Molecular Mechanisms of Mutagenesis*, *57*, 200-13.

Spranger, I., Sun, B., Mateus, A. M., Freitas, V. & Ricardo-da-Silva, J. M. (2008). Chemical characterization and antioxidant activities of oligomeric and polymeric procyanidin fractions from grape seeds, *Food Chemistry*, *108* (2), 519-32.

Suppakul, P., Miltz, J., Sonneveld, K. . & Bigger, S. W. (2003a). Antimicrobial properties of basil and its possible application in food packaging, *Journal of Agricultural and Food Chemistry*, *51*, 3197-207.

Suppakul, P., Miltz, J., Sonneveld, K. & Bigger, S. W. (2003b). Active packaging technologies with an emphasis on antimicrobial packaging and its applications, *Journal of Food Science and Technology*, *68* (2), 408-20.

Suppakul, P., Sonneveld, K., Bigger, S. W. & Miltz, J. (2008). Efficacy of polyethylene-based antimicrobial films containing principal constituents of basil, *LWT - Food Science and Technology*, *41* (5), 779-88.

Taguri, T., Tanaka, T. & Kouno, I. (2004). Antimicrobial activity of 10 different plant polyphenols against bacteria causing food-borne diseas, *Biological and Pharmaceutical Bulletin*, *27* (12), 1965-69.

Tsuchiya, H., Sato, M., Miyazaki, T., Fujiwara, S., Tanigaki, S., Ohyama, M. & Linuma, M. (1996). Comparative study on the antibacterial activity of phytochemical flavanones against methicillin resistant *Staphylococcus aureus*, *Journal of Ethnopharmacology*, *50* (1), 27-34.

Turkoglu, A., Duru, M. E., Mercan, N., Kivrak, I. & Gezer, K. (2007). Antioxidant and antimicrobial activities of Laetiporus sulphureus (Bull.) Murrill, *Food Chemistry*, *101* (1), 267-73.

Vaquero, M. J. Rodríguez, A., de Nadra, M. R. & Manca, M. C. (2007). Antibacterial effect of phenolic compounds from different wines, *Food Control*, *18* (2), 93-101.

Weber, H. A., Hodges, A. E., Guthrie, J. R., O'Brien, B. M., Robaugh, D., Clark, A. P. & Smith, C. S. (2007). Comparison of proanthocyanidins in commercial antioxidants: grape seed and pine bark extracts, *Journal of Agricultural and Food Chemistry*, *55* (1), 148-56.

Xia, E. Q., Deng, G. F., Guo, Y. J. & Li, H. B. (2010). Biological activities of polyphenols from grapes, *International Journal of Molecular Sciences*, *11* (2), 622-46.

Yalpani, M., Johnson, F., Robinson, LE, Brine, CJ, Sandford, PA & Zikakis, JP. (1992). *Advance in chitin and chitosan* (Elsevier Applied Science, London).

Yerlikaya, P. & Gokoglu, N. (2012). Protective effect of grape seed extract treatment before and after freezing on lipids of bonito, *CIGR-International Conference of Agricultural Engineering* (Valencia, Spain).

In: Grapes
Editor: José S. Câmara
ISBN: 978-1-63321-402-6

Chapter 7

Grape Phenols and Their Healthy Properties

A. Peralbo-Molina[1–3] and M. D. Luque de Castro[1–3]*

[1]Department of Analytical Chemistry, Annex Marie Curie Building, Campus of Rabanales, University of Córdoba, Córdoba, Spain
[2]University of Córdoba Agrifood Excellence Campus, ceiA3
[3]Maimónides Institute of Biomedical Research (IMIBIC), Reina Sofía University Hospital, University of Córdoba, Córdoba, Spain

Abstract

This chapter deals with the properties of grapes in general and the phenols they contain in particular. Thus, it is concerned with grape production (viz., world figures and major producing countries) and uses, and also with the phenol families, classes and subclasses present in berries (particularly as regards formulae and chemical properties). The main parts of grapes, and their absolute and relative contents of phenols, are described, and analytical information about specific compounds is discussed in order to provide a critical overview of (*a*) the different types of residues (raw materials) from which phenols can be obtained; (*b*) the extraction and leaching techniques typically used to obtain phenols from grapes, and their efficiency, expeditiousness and performance; (*c*) available methods for extracting analytical information about target phenols or phenol groups; and (*d*) the specific technologies currently used to identify and quantify each type of phenol.

The proven healthy properties of grape phenols (e.g. improvement and prevention of neurodegenerative diseases, cardiovascular disorders and tumors), which have been widely demonstrated but not comprehensively assessed to date in various clinical areas are discussed and potential metabolomic research lines for these compounds proposed.

Finally, predictable trends in grape phenol technology are outlined, with special emphasis on major goals such as the integral exploitation of all grape parts and residues by grape derivative industries.

* Corresponding author: qa1lucam@uco.es.

Keywords: Grape, phenols, integral exploitation, grape residues

Introduction

Epidemiological studies have contributed convincing evidence that diets rich in fruits and vegetables prevent or delay chronic degenerative diseases such as cancer, cardiovascular disease, atherosclerosis and type 2 diabetes (Spormann et al., 2008).

The Mediterranean diet is widely known by its well-documented health benefits (Kushi et al., 1995), which include the previous protective effects (Biesalski et al., 2004; Choi, 2005; Dryden et al., 2006; Giacosa et al., 2013; Kok et al., 2004). Grapes constitute one of the traditional crops of the Mediterranean region. By comparing diets between countries, researchers found that, notwithstanding the French tendency to eat abundant animal fat, heart disease had a lower incidence in France than in other countries. This phenomenon was called the "French paradox" and is thought to result from the protective effects of regular consumption of red wine. Studies on the healthful effects of wine promoted by this discovery led to its antioxidant and free radical scavenging properties being ascribed to polyphenols — which are even more influential on health than is the mere consumption of wine (Keys, 1970; Feinleib 1981).

A pressing need currently exists to obtain natural dyes as substitutes for artificial dyes. The need was revealed by a study of 2009 of the University of Southampton involving 6 artificial colorants that were found to cause hyperactivity and allergy in children. Subsequently, the European Food Safety Authority (EFSA) conducted a toxicological study on five azo colorants (viz., E102, tartrazine yellow; E110, orange sunset yellow; E122, carmoisine; E124, Ponceau 4R; and E129, red Allura AC) and one quinoline colorant (E104, quinoline yellow); based on the results, EFSA ruled that the target colorants had no hyperactive, allergenic or other toxic effects. However, the Directorate General for Health and Consumers (DG SANCO) of the European Parliament decided, as a precautionary measure, that the labeling of all foods containing these colorants should state that they can produce actions and negative effects on children's activity. This measure was set in Annex V of Regulation (EC) 1333/2008 (food additives) [www.cen-online.org January 21 (2008) 33; www.cen-online.org June 9 (2008) 32]. The situation has boosted the search for, and increasing use of, natural colorants as replacements for synthetic colorants [www.cen-online.org December 15 (2008) 18; www.cen-online.org March 24 (2008) 28]. In response, grapes have been the targets of a number of studies aimed at fulfilling this emerging need since colorants present in berries and their residues are subjected to no legal delay in their use (Wada et al. 2007).

A significant fraction of substances present in grapes consists of phenols, which constitute are a vast, varied family of compounds many of which occur naturally in an array of food plants (particularly fruits, seeds and leaves). In fact, grapes contain more than 8000 known compounds ranging from simple molecules such as phenol itself to complex substances of variable composition such as tannins (Bravo, 1998; Harborne, 1995). In wine jargon, the word "polyphenols" is used to refer collectively to both simple and complex phenols with mixed chemical structures that are often imprecise even to researchers (Tsao, 2010). To the authors' minds, "phenols" is a more appropriate term in this context as it comprises both simple phenols and polyphenols. Chemically, phenols encompass a wide variety of biomolecules (e.g., flavonoids, stilbenes, lignans) containing several hydroxyl groups (D'Archivio et al. 2007; Tsao, 2010). All types of phenols are believed to contribute to the antioxidant properties of grapes, other fruits and wine, and hence to be largely responsible for their health benefits (Moini et al. 2000).

This chapter provides an overview of grapes, their production and uses, and describes the classes of phenols they contain and their concentrations. Also, deals with the extraction techniques typically used to isolate grape phenols, the healthy properties of these compounds and their contributions to improvement in various clinical areas. Finally, it outlines some predictable trends in grape phenol research and exploitation.

Grape Production

Grape production worldwide amounts to about 70 million tons, 80% of which is used for winemaking and virtually all other for grape juice (Moini et al. 2000). In fact, only 2% is used as fresh or dried fruit. Most grapes are of the *Vitis vinifera* variety, which is native from the Mediterranean basin and Central Asia. Other, minor contributors include American and Asian species such as *Vitis labrusca*, *Vitis ripalia* and *Vitis rotundifolia*. According to the Food and Agriculture Organization (FAO), the Mediterranean basin and Central Asia produced 16.7 million ton of wine in 2010 (64% of the global wine production that year). Also, the Mediterranean region spans 63% of the total vine cultivated area in the world, with 4.5 million hectares (FAO, 2013). Mediterranean countries have the largest total area of vineyards and include the three greatest wine producers (viz., France, Italy and Spain, which jointly account for 48 % of the global wine production). However, wine production in this region has decreased by 18% over the past two decades owing to the emergence of new prominent wine producers such as Australia, South Africa and the United States (FAO, 2013).

Vineyards generate vast amounts of agricultural residues (particularly vine shoots and leaves). According to Jiménez et al. (2006) an estimated 11.2–16.0 million tons of vine shoots are produced each year. Also, grape pomace —the solid waste remaining after grape pressing— is the most abundant residue from the winemaking process, with about 10.5–13.1 million ton in the world each year (Kammerer et al. 2005). A growing demand currently exists for exploitation of agricultural residues, which have proved a key source of phenols (Peralbo-Molina and Luque de Castro, 2013).

Climate change is having a strong, growing influence on vine phenomenology and grape composition in terms of vinification, and of wine microbiology, chemistry and sensory properties. According to Mira de Orduña (2010), there is evidence that grape quality is being affected by climate change, which can be expected to influence wine quality (especially as regards expression of varietal grape aromas, microbiological and chemical stability and sensory balance) in the future. Phenols in plant foods can vary considerably in quantity and quality depending on various internal and external factors such as plant genetics and cultivar, soil composition, growing conditions, ripeness and postharvest conditions (Faller and Fialho, 2010). These issues should be studied in depth in order to acquire basic scientific knowledge on how climate change may affect wine consumption by humans.

Grape Phenols

Phenols contribute to color and sensory properties such as bitterness and astringency in fruits, vegetables, nuts, seeds, flowers and even some herbs. These compounds constitute an integral part of human diet that takes benefit from their antioxidant properties —the main reason for the growing interest in these compounds (Caillet et al., 2006).

Classification of Phenols

Phenols, which are among the most abundant and widely distributed natural compounds in the plant kingdom, are highly diverse. Most phenols in plants are present as glycosides containing various sugar units and acylated sugars at different positions of the phenol backbone. Phenols are classified mainly in terms of origin, biological function or the chemical structure of the corresponding aglycon. The last classification is used in this chapter and illustrated with a chart showing phenol classes and subclasses in terms of substituents, and typical examples of each class and subclass.

The antioxidant compounds present in grapes have been identified as phenolic acids (benzoic and hydroxycinnamic acids), stilbene derivatives (resveratrol), flavan-3-ols (catechin, epicatechin), flavonols (kaempferol, quercetin, myricetin) and anthocyanins; all have been classified as flavonoids or nonflavonoids (Caillet et al., 2006).

Nonflavonoids

Flavonoids are phenols possessing a C6-C3-C6 backbone with two benzene rings connected via a heterocyclic pyrane or pyrane ring, and nonflavonoids all others not possessing this chemical structure (e.g., phenolic acids, stilbenes, lignans).

Phenolic Acids

Phenolic acids in grapes, which are usually present as esters, can also be found in free form in grape pomace and leaf extracts as a result of partial hydrolysis of esters during extraction (Luque-Rodríguez et al., 2008). As can be seen in Figure 1, there are two classes of phenolic acids depending on the type of aromatic ring from which they originate, namely: benzoic acids and cinnamic acids and their derivatives (Manach et al. 2004).

Hydroxybenzoic acids occur at very low levels in plants; however, they are components of complex structures such as hydrolyzable tannins, which have gallic acid as their precursor (Garrido and Borges, 2013). Several types of hydroxybenzoic acids have been identified in both grapes and wine. The most abundant acids in this class are *p*-hydroxybenzoic, protocatechuic, vanillic, gallic and syringic. Because of their scant presence in human edible plants, hydroxybenzoic acids have neither been extensively studied nor assigned nutritional value (Manach et al. 2004).

Hydroxycinnamic acids, which are also rarely found in free form, are located mainly in grape skin, leaves and stems. The most abundant are present either as tartaric esters or as parts of acylated anthocyanins (Luque-Rodríguez et al., 2008). These acids are usually found at low concentrations in grapes (Morata et al. 2007); however, they are present in large amounts in some varieties use of which may trigger the formation of phenolic pigments.

Hydroxycinnamic acids can form a stack group with grape anthocyanins, thus increasing and stabilizing the color of anthocyanins, and that of wine as a result —a copigmentation effect of caffeic acid on red wine color has been reported (Darias-Martín et al., 2001). In addition, these compounds contribute to other sensory attributes such as bitterness, astringency and flavor (Andrade et al. 2001), and their antioxidant properties have been ascribed some beneficial physiological effects on human health (Monagas et al., 2005). The presence of hydroxycinnamic acids is an indicator of wine quality (Gambelli et al. 2004). Mulero et al. (2010) found the most abundant derivatives of hydroxycinnamic acids in grapes to be *trans-p*-coumaroyltartaric acid and *trans*-caffeoyltartaric acid.

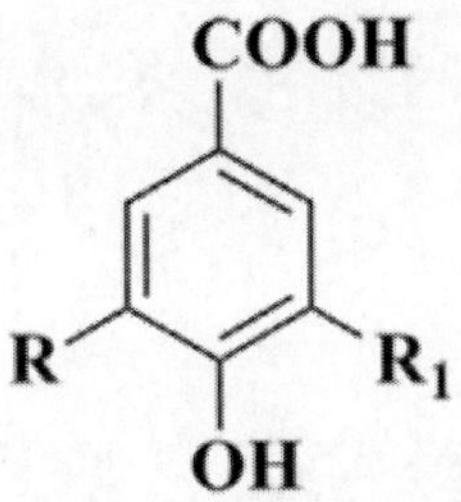

Hydroxybenzoic acids	R	R1
Hydroxybenzoic acid	H	H
Protocatechuic acid	OH	H
Vanillic acid	OCH_3	H
Gallic acid	OH	OH
Syringic acid	OCH_3	OCH_3

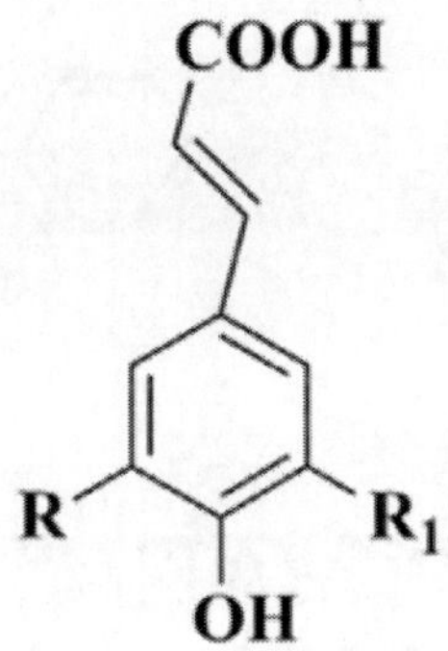

Hydroxycinnamic acids	R	R1
p-Coumaric acid	H	H
Caffeic acid	OH	H
Ferulic acid	OCH_3	H
Sinapic acid	OCH_3	OCH_3

Figure 1. Chemical structure of hydroxybenzoic and hydroxycinnamic acids.

Stilbenes

Stilbenes are nonflavonoid phenols present in a number of plant families which constitute their main sources in human diet. These phenols, which possess two aromatic rings linked by an ethane bridge (see Figure 2), are constituents of woody organs (roots, vine shoots and stems) and substances of induced production (in leaves and berries) acting as phytoalexins in the mechanisms of grape resistance to certain pathogens such as fungi (Luque-Rodríguez et al., 2008).

Stilbenes are also present in grapes and, after processing, in wine (Zamora-Ros et al., 2008). The main stilbenes in grapes are *trans*- and *cis*-resveratrol, their glucosides (known as "piceids"), picetannol (Cantos et al., 2002), astringin, pterostilbene and various oligomers (viniferins and hopeaphenol) (Guerrero et al., 2009); all have been ascribed a number of beneficial (antioxidant, antibacterial, antifungal, cardioprotective, neuroprotective, antiaging and anticancer) effects (Guerrero et al., 2009). Also, *trans*-resveratrol has been found to influence stress resistance and lifespan in some organisms (Gruber et al. 2009). Resveratrol, which is usually present at higher concentrations than other stilbenes in grapes, is also found as glucosylated derivatives or oligomeric forms called “viniferins” (Luque-Rodríguez et al. 2008).

Sun et al. (2006) examined the stilbenoid composition of grapes and wine, and obtained interesting results including that *cis*-piceid is the most abundant stilbene in grape skin (150.64 mg kg^{-1}), followed by *trans*-piceid (72.19 mg kg^{-1}) and *trans*-resveratrol (66.67 mg kg^{-1}). By contrast, *cis*-resveratrol has been found in wine but not in grape skin (Roggero and García-Parrilla, 1995; Romero-Pérez et al., 2001). Also, *cis*- and *trans*-astringin have not been detected in grape skin or red wines —some authors, however, have found the *trans* isomer in red wines (Ribeiro de Lima et al., 1999).

Stilbenes	R1	R2
trans-resveratrol	H	H
trans-piceid	H	Gluc
trans-pterostilbene	CH_3	CH_3

Stilbenes	R1	R2
cis-resveratrol	H	H
cis-piceid	H	Gluc

Figure 2. Chemical structure of stilbenes and derivatives.

Lignin

Lignin, which serves key biological functions in plants (e.g., increasing the mechanical strength of cell walls), accounts for about 17–28% dry weight of grapevines —20% on average in vine shoots (Luque-Rodríguez et al., 2008). Its three-dimensional structure (Figure 3) is a result of polymerization of phenolic alcohols units such as coniferyl or sinapyl alcohol (Garrido and Borges, 2013).

Flavonoids

Flavonoids are molecules with two aromatic rings linked by three carbon atoms usually arranged as an oxygenated heterocycle (C6-C3-C6). Flavonoids are classified according to the degree of oxidation of the three-carbon ring into anthocyans, flavanols (also named “catechins”), flavonols and flavanonols. Most flavonoids —except catechins— are present in nature as *O*-glycosides and other conjugates, which increases their complexity and number — more than 5000 different compounds have so far been identified (Anhê et al., 2013).

Anthocyanins

Anthocyanins account for most phenols remaining in skin from red pomace (3-glycosides, 3-acetylglycosides and 3-*p*-coumaroylglycosides of malvidin, peonidin, delphinidin, petunidin and cyanidin, mainly) (Luque-Rodríguez et al., 2007). However, tartaric esters of hydroxycinnamic acids, monomeric and dimeric flavanols, flavonols and stilbenes are also present in grape skin and seeds (Rodríguez-Montealegre et al., 2006). The main and nearly exclusive difference between red and white grape varieties is in these compounds, which are virtually absent from the skin of white grapes (Luque-Rodríguez et al. 2008).

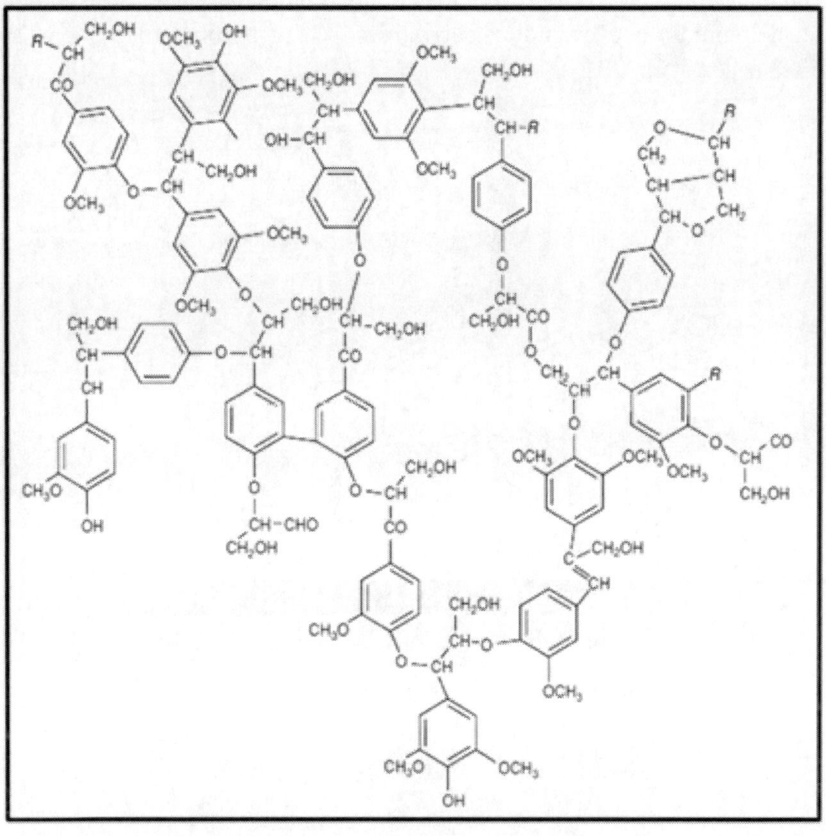

Figure 3. Lignin structure.

Figure 4 shows the basic structural backbone of anthocyanins. Chemically, these compounds are heterosides of an aglycone unit: anthocyanidin. Anthocyanins differ mainly in the number of hydroxylated groups in the anthocyanidin backbone, the nature and number of bonded sugars in their structure, the aliphatic or aromatic carboxyl derivatives bonded to the sugar in the molecule, and the position of the bonds (Valls et al., 2009).

These compounds have been ascribed potential health benefits derived from anticarcinogenic (Zhao et al. 1999), angioprotective (Vennat et al. 1988), antiinflammatory (Pietta et al. 2003) and antibacterial activity (Fukai et al. 1991). In fact, anthocyanins are powerful antioxidants and free radical scavengers (Woodman et al. 2005) that reduce the risk of coronary heart disease (Hung et al. 2000), and of some types of tumors (Williamson, 1990) and chronic diseases (Dryden et al. 2006).

Anthocyanins are extracted into must during winemaking, but not completely. As a result, the skin from grape pomace provides an abundant, inexpensive source of these phenols (Luque-Rodríguez et al., 2007), which are also present in other parts of the grapevine. Thus, leaves are rich in anthocyanins when they turn red in autumn (Luque-Rodríguez et al., 2008).

Anthocyanin contents have been determined in ripe grapes of some varieties including Gran Negro (1720 mg kg^{-1} berry) and Mouraton (1000 mg kg^{-1} berry), malvidin 3-*O*-glucoside being the most abundant anthocyanin (797 mg kg^{-1} berry) and malvidin-3-*O*-(6-*O*-*p*-coumaryl)-glucoside the second (119 mg kg^{-1} berry). Malvidin derivatives account for 52% of the total anthocyanin content, followed by peonidin compounds with 38% and others such as petunidin, delphinidin and cyaniding derivatives, which account for a mere 2–5% each (Figuereido-González et al., 2013).

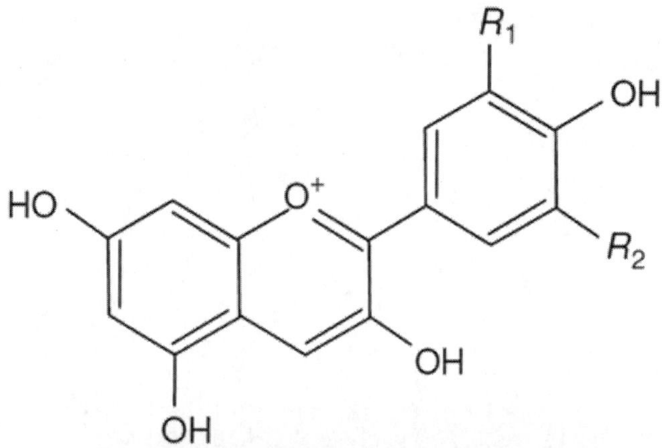

Anthocyanins	R1	R2
Cyanidin	OH	H
Delphinidin	OH	OH
Peonidin	OCH_3	H
Petunidin	OCH_3	OH
Malvidin	OCH_3	OCH_3

Figure 4. Anthocyanins frequently present in grapes.

Flavanols

Flavanols occur mainly in the outer coating of seeds, but are also present in some skin cells, stems, shoots and leaves (Luque-Rodríguez et al., 2008). The most abundant flavanols in grapes are flavan-3-ols, which are a subclass of flavonoids with a variable number of hydroxyl groups at positions 5 and/or 7 on the A-ring and at positions 3′, 4′ and 5′ on the B-ring (Figure 5). Position 3 on the C-ring is commonly occupied by a hydroxyl group or esterified by gallic acid. Catechin, epicatechin, catechin gallate, epicatechin gallate, gallocatechin, epigallocatechin, gallocatechin gallate and epigallocatechin gallate are the most common flavan-3-ol monomers. On the other hand, oligomers and polymers are mainly procyanidins, which include (+)-catechin, (–)-epicatechin and epicatechin-3-*O*-gallate in seeds and skins; partially galloylated units in shoots; and ungalloylated prodelphinidin derivatives in leaves (Luque-Rodríguez et al., 2008; Ali et al., 2010).

Figuereido-González et al. (2013) determined the contents in flavan-3-ol monomers of fresh (24 mg kg^{-1}) and dried grapes (42 mg kg^{-1}); the contents were consistent with those previously found in dried grapes by other authors (Márquez et al., 2012; Peinado et al., 2009; Serratosa et al., 2008a, 2008b, 2011). (+)-Catechin is the most concentrated flavan-3-ol monomer, followed by (–)-epicatechin and epicatechin-3-*O*-gallate.

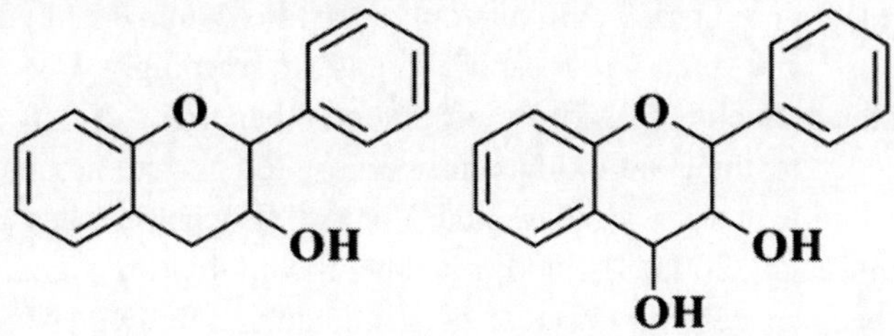

(A) (B)

Figure 5. Chemical structure of flavan-3-ols (A) and flavan-3,4-diols (B).

Flavonols

Flavonols are flavonoids containing a carbonyl group at position 4 and a double bond between C2 and C3 (Figure 6) which abounds in the skin of fresh grapes. These yellow pigments contribute directly to color in white wines, but are masked by anthocyanins in red wines. However, flavonols also affect the color of red wine —mainly through copigmentation (Schwarz et al. 2005).

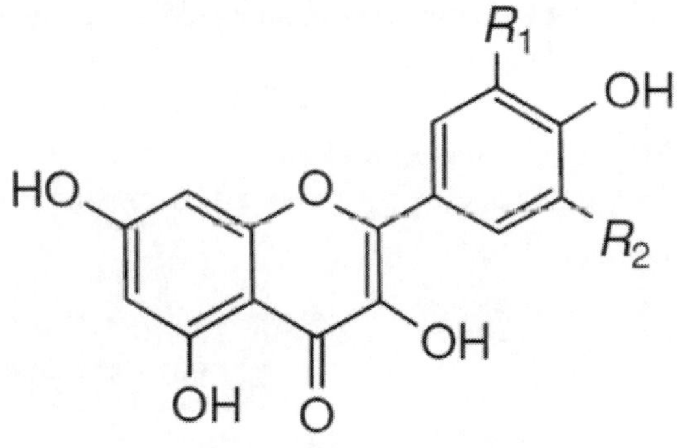

Flavanols	R1	R2
Kaempferol	H	H
Quercetin	OH	H
Myricetin	OH	OH
Isorhamnetin	OCH_3	H

Figure 6. Flavonols frequently present in grapes.

Castillo-Muñoz et al. (2007) recently demonstrated that flavonols in red grape cultivars belong to three glycosylated series (3-*O*-glucosyl, 3-*O*-galactosyl, and 3-*O*-glucuronyl) among the six possible flavonoid structures, namely: kaempferol, quercetin, isorhamnetin, myricetin, laricitrin and syringetin. The presence of isorhamnetin-type flavonols in the skin of white grapes has been suggested on the basis of hydrolysis and subsequent detection of the free aglycone isorhamnetin (Mattivi et al. 2006), and also of the coincidence of its UV–Vis spectrum and chromatographic retention time with those for a true standard of isorhamnetin 3-*O*-glucoside (Rodríguez et al., 2006). However, isorhamnetin- and myricetin-type flavonols are believed to be present in red grape cultivars only (Castillo-Muñoz et al., 2007).

Although the most widespread role of flavonols is seemingly UV protection, these compounds have attracted much interest for their potential beneficial effects on human health, which are linked mainly to their antioxidant properties. Recent studies have highlighted a possible role in preventing cardiovascular, neurological (López-Sánchez et al. 2007) and tumor diseases (Boots et al. 2008). In addition, the flavonol profile and the quercetin to myricetin ratio have been successfully used to discriminate varietal wines (Castillo-Muñoz et al., 2007).

Figueiredo-González et al. (2013) determined the total content of flavonols in grapes to be 44 mg kg^{-1} berry, which is similar to the value previously reported by Castillo-Muñoz et al. (2007) for the same grape variety: 45 mg kg^{-1} berry. The most abundant flavonols in fresh grapes are quercetin derivatives (46%), followed by myricetin compounds (22%), glucoside derivatives being present at higher concentrations than other derivatives (glucuronides, galactosides and rutinosides) (Figuereido-González et al., 2013).

Flavanonols and Flavones

Flavanonols are 2,3-dihydroflavonols, of which astilbin and engeletin in white grape skin (Trousdale and Singleton, 1983) and stems (Souquet et al. 2000) are the best known. Taxifolin, astilbin and dihydromyricetin 3-*O*-rhamnoside have also been often detected (Garrido and Borges, 2013). Flavones are structurally similar to flavonols but lack the hydroxyl group at C3 (Figure 7). These compounds, the most important of which are apigenin-7-glucoside and luteolin-7-glucoside in vine grape leaves (Hmamouchi et al. 1996), are known to have significant pharmacological activity (Garrido and Borges, 2013).

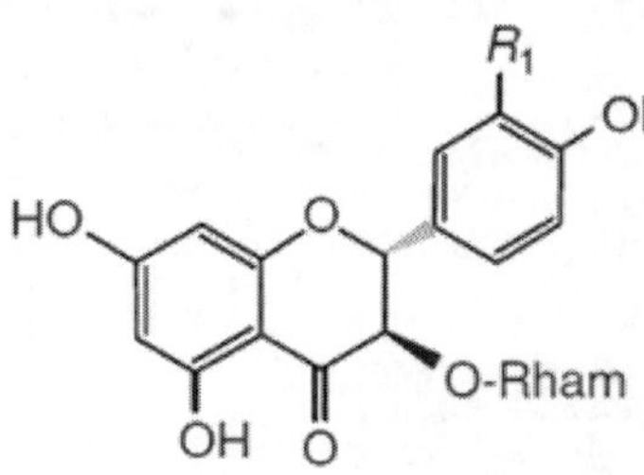

Flavanonol	R1
Astilbin	OH
Engeletin	H

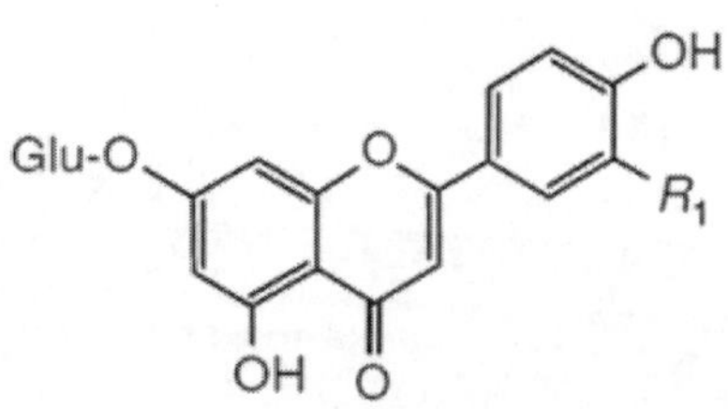

Flavone	R1
Luteolin-7-*O*-glucoside	OH
Apigenin-7-*O*-glucoside	H

Figure 7. Flavanonols and flavones found in grapes.

Analytical Research on Grape Phenols

Grape phenols have been extensively studied (Flamini et al., 2003; Jaitz et al., 2010; Modun et al., 2008; Serafini et al. 2000) on account of their antioxidant and bactericidal effects, which are beneficial to human health. The composition of grapes strongly affects the properties and sensory quality of wine (Bonghi et al. 2012). The effects of wine ageing in barrels have been examined in depth and tannins from wood found to provide an uptake route (Jaitz et al., 2010). The metabolism and gene expression of phenols in the skin of wine grape berries subjected to partial postharvest dehydration have also been studied. This process seems to selectively affect specific pathways in skins (e.g., altering berry composition —and hence the sensory characteristics of the ensuing wines—, and affecting health-promoting properties of grapes) (Bonghi et al., 2012).

As noted earlier, grapes are among the world's largest fruit crops (Moini et al., 2000); also, they are very much appreciated for their high contents in phenols such as gallic acid, catechin and resveratrol in addition to a wide variety of procyanidins (Xu et al., 2010). Although the nutritional qualities of grapes are affected by environmental, cultural and postharvest conditions, genotype is the determinant factor leading to variation (Prior et al., 1998; Proteggente et al., 2002).

Grapes processing for wine and juice globally leads to an annual output of at least 10 million ton of press residues (Maier et al., 2009). Wine orujo has become popular as a nutritional supplement —grape skin and seeds accumulate high proportions of phenolic compounds— in recent years. The winemaking process provides high added value products which, however, have remained largely underexploited. A distinction is made here between skin, seeds and lees in this respect.

The characteristics of grape skin depend on grape variety and color, red skin being more appreciated ever since several studies showed that the number and concentration of colorants it contains makes it a profitable raw material for colorant production from fresh grapes —which is costly— or after winemaking of red grapes —a more inexpensive source (Corrales et al., 2009).

In addition to colorants, aromas and colorless phenols are present at higher concentrations in skin and seeds than in pulp, which facilitates their obtainment prior to (aromas) and simultaneously with or subsequently to (colorless phenols) colorants (Peralbo-Molina and Luque de Castro, 2013). Luque-Rodríguez et al. (2007) developed a superheated liquid extraction method for the inexpensive removal of compounds of interest from this raw material.

Between 38 and 52% of grape is constituted by seeds; this gives an overall idea of the amount of residue produced by each harvest (Maier et al., 2009). The composition of grape seed is basically 40% (w/w) fiber, 16% essential oil, 11% protein, 7% complex phenolic compounds (e.g., tannins) and the remainder sugars, minerals and other substances (de Campos et al., 2008). The lipid fraction of grape seeds has a high commercial value thanks to its high contents of unsaturated fatty acids (e.g., linoleic acid, with 72–76% w/w) and tocopherols —which are also interesting because they possess antioxidant properties.

Phenolic compounds are present at concentrations of 5–8% in grapes depending on the particular cultivar (mainly as simple phenols such as gallic acid, flavonoids including monomeric flavan-3-ols catechin, epicatechin, gallocatechin, epigallocatechin, and epicatechin 3-*O*-gallate, and procyanidin dimers, trimers and more polymerized procyanidins). These compounds have been successfully extracted with methods using superheated liquids to reduce extraction costs (Luque-Rodríguez et al., 2007).

Vinification lees, which accumulate in the bottom of wine containers after fermentation, during storage or after authorized treatment, play a key role in winemaking; thus, they remove undesirable compounds or interact with the volatile fraction of wine. Lees constitute an underexploited residue; in fact, their use has been virtually limited to the recovery of tartaric acid (Pérez-Serradilla and Luque de Castro, 2011) even though they have been shown to provide highly valuable oils for cosmetic and nutritional purposes, phenols with high antioxidant capacity, and useful amino acids for enriching foods in addition to mannoproteins and colorants (Pérez-Serradilla and Luque de Castro, 2008, 2011).

The residue remaining after ethanol distillation from wine pomace (i.e., a residue from a previous residue) has proved an effective source of flavonoids (particularly anthocyanins, flavonols 3-glucoside derivatives and simple flavonols; phenolic compounds such as acids, esters, methoxy and ethoxy derivatives; and stilbenes) (Peralbo-Molina et al., 2012, 2013). Despite the degradation caused by the high temperatures used in the distillation process, the resulting extracts contain a variety of compounds potentially useful as food additives, and also for cosmetics and nutraceuticals.

Techniques and Methods for Extracting Valuable Phenols from Vineyard and Wine Residues

Green extraction processes (viz., extraction with subcritical water, or subcritical or supercritical carbon dioxide) are the most promising choices for the industrial extraction of grape phenols. However, prior development of methods at the laboratory scale is required for optimal exploitation. Extraction methods based on the previous extractants have proved highly efficient in isolating anthocyanins and other phenols from grape processing residues (Khanal et al., 2010).

Supercritical Fluid Extraction (SFE)

Using supercritical CO_2 as extractant has the advantage that it is nontoxic, nonflammable and noncorrosive; also, it can be easily removed after depressurization, which allows the target compounds to be extracted in powder form without the need for drying (Shi et al., 2005). However, the relatively high purchasing costs of SFE equipment and the need for highly pure supercritical carbon dioxide are two major constraints for wider use of this technique. In any case, SFE has been successfully used to extract phenols from grape pomace, usually by use of CO_2 modified with methanol (Casas et al., 2010; Mantell et al. 2001, 2003) or ethanol as cosolvent (Chafer et al., 2005) in order to increase its polarity.

Superheated Liquid Extraction (SHLE)

SHLE is a more affordable technique than SFE. Also, it has two crucial advantages over conventional extraction, namely: (*a*) raising the temperature above the boiling point of the solvent increases the diffusion rate, solubility and mass transfer of the target compounds while decreasing the viscosity and surface tension of the solvent, thereby leading to improved contact of the target compounds with the solvent and hence to faster extraction and reduced solvent consumption; and (*b*) the absence of light and air significantly reduces degradation and oxidation of the target compounds during extraction (Santos-Buelga and Williamson, 2003).

SHLE has so far been almost exclusively used as a preliminary step in methods for analyzing phenols from fresh grapes. Thus, Piñeiro et al. (2006) determined *trans*-resveratrol in grapes by using a method involving a triple discrete SHLE step with methanol. Also, Ju and Howard (2003, 2005) examined the effect of various solvents and temperature conditions on the SHLE of anthocyanins from fresh grape skin obtained from a highly pigmented advanced breeding line of wine grape and concluded that either superheated water (110 ºC) or sulfured water provide extraction efficiencies similar to those of plain hot water or a 60:40 methanol–water mixture but with dramatically shortened extraction times. Few SHLE methods for grape pomace have to date been reported (particularly for red grape pomace because skin phenols from this material are extracted during the winemaking process so the pomace contains strongly retained phenols only). González-Rodríguez et al. (2003) and Luque-Rodríguez et al. (2007) optimized the extraction of anthocyanins and other phenols by using either superheated ethanol–water mixtures or a batch mode for grape pomace and a continuous mode for skin from grape pomace. Although SHLE yields exceeded those obtained with conventional solvent extraction, the total amounts extracted were obviously smaller than those from fresh grapes.

Forms of energy other than temperature and pressure have also been used to accelerate extraction of phenols from solid samples. The most widely used among such alternative sources of energy are microwaves (MW) and ultrasound (US), which have given rise to microwave assisted extraction (MAE) and ultrasound assisted extraction (UAE), respectively.

Microwave Assisted Extraction (MAE)

The MAE technique is recognized as an outstanding means for promoting efficient removal of target compounds from solid samples, cutting down extraction times, increasing yields and, frequently, boosting extract quality. Also, MAE is considered to be expeditious, ecocompatible —it requires little solvent—, reproducible and highly effective (Cravotto et al., 2008). This technique has been used to develop methods for extracting organic compounds from matrices such as soils (Serrano and Gallego, 2006), sediments (Morales et al., 2005), seeds (Prado-Rosales et al., 2009) and foods (Robards, 2003; Wang and Sun, 2003). Microwaves have been found to expedite extraction from a variety of matrices relative to conventional extraction methods. Thus, MAE is seemingly well suited to the extraction of phenolic compounds unaltered from grapes at relatively high temperatures (110–150 ºC) —a critical feature with a view to handling antioxidants without degradation (Grigonis et al., 2005). MAE has also proved suitable for phenol removal without altering the antioxidant potential of the extract and largely surpassing conventional extraction methods in performance (Spigno et al., 2009). Peng et al. (2001) found the extractant to greately expedite the extraction of phenols from grape seeds; in fact, the extractant was the most influential variable —even more so than microwave power and extraction time by effect of microwaves not interacting with nonpolar extractants. Finally, a method for extracting anthocyanins from grapes involving the simultaneous processing of several samples in order to increase throughput was recently developed by Liazid et al. (2011).

Ultrasound Assisted Extraction (UAE)

UAE has shown a number of advantages over classical leaching methods including shorter processing times, less drastic conditions —especially temperature—, limited use of hazardous reagents and less intervention of the analyst resulting from automatability.

One of the greatest disadvantages of classical extraction methods such as Soxhlet extraction for analytical purposes or solvent extraction for analytical or industrial purposes is the long solid sample–extractant contact times typically needed. Albu et al. (2004) extracted antioxidants from rosemary by using an ultrasonic bath, an ultrasonic probe and conventional stirring in combination with different extractants at variable temperatures and found ultrasound to dramatically reduce extraction times. Similar results were obtained by Cho et al. (2006) in the extraction of resveratrol from grapes. The favorable effects of US on extraction, which are due mainly to the phenomenon of "cavitation" (viz., the creation of highly effective tiny temperatures and pressures) efficiently operating at the interface between a sonicated solution and a solid, may be offset by denaturation of phenols as a result (Kammerer et al., 2005).

Enzyme Assisted Extraction

Assistance of extraction by enzymes is one other potentially effective alternative to conventional extraction methods for phenols as shown by studies at the laboratory and pilot plant scales. Thus, Kammerer et al. (2005) found preextracting pomace with hot water, followed by treatment with cell wall degrading enzymes, to provide extraction rates comparable to those obtained by extracting grape pomace with sulfite.

Analytical Information from Phenols Extracts

Phenol extracts can be mined to a variable degree for information ranging from overall determinations to identification and/or quantitation of individual compounds depending on the sophistication of the particular analytical equipment used.

Overall Analytical Information from Phenol Extracts

The type of overall information that can be obtained from phenol extracts depends on the particular analytical method used and types of phenols contained in the extracts.

Total Contents in Target Compounds Families

This overall information can be obtained by using general reagents acting on a specific group in a molecule. A generic calibration curve is run by using a representative compound of the target family; as a consequence, the results are always semiquantitative in nature and relative to the particular standard used.

One of these methods, due to Folin–Ciocalteu (Singleton and Rossi, 1965), is typically used for the overall determination of phenols. Thus, all compounds containing a phenol group in their molecule react with the Folin–Ciocalteu reagent to form a colored compound that can be monitored at 765 nm for overall determination purposes. This method is the most widely used for determining total phenolic contents (Corrales et al. 2010; Negro et al. 2003), which, however, can also be accomplished with the enzymatic method of Stevanato et al. (2004) and that of Peinado et al. (2009). The latter involves supplying the extract with 0.1 M potassium phosphate buffer at pH 8 containing 3 mM horseradish peroxidase and monitoring the reaction at 510 nm after 10 min of rest. The results are expressed as catechin equivalents.

A number of assays for specific classes of phenols have been reported including one for total flavonoids and anthocyans based on absorbance measurements at 280 and 540 nm; the results are expressed as catechin and/or malvidin equivalents (Di Stefano et al., 1989). Also, the total content in free anthocyans can be determined by monitoring the absorbance at 540 nm after purification by solid phase extraction with C18 as sorbent (Di Stefano et al., 1989). Orak (2007) reported an alternative method for total anthocyans based on one previously developed by Wrolstad (1976) involving monitoring at 520 and 700 nm of grape extracts obtained at pH 1 and 4.5. The total anthocyanin content was calculated from a mathematical formula and expressed in milligrams of malvidin-3-*O*-glucoside per 1000 g fruit.

Total flavonoids can be determined according to Makris et al. (2007) by diluting the grape extract with MeOH, mixing with distilled water containing 5% $NaNO_2$ and allowing the mixture to react for 5 min. This is followed by addition of $AlCl_3$ and standing for a further 5 min. Finally, 1 M Na_2CO_3 and distilled water are added to the reaction mixture, and its absorbance at 510 nm against a blank prepared in a similar manner by replacing the extract with distilled water is measured. The total flavonoid content is calculated from a calibration curve run from a catechin standard and expressed in milligrams of catechin per 100 g fresh grape.

Nagel and Glories (1991) successfully determined total flavanols by using an optimized methodology after derivatization with *p*-DMACA. Grape extract was appropriately diluted with methanol, centrifuged in a microtube and supplied with 0.24 N HCl and 0.2% DMACA, both in methanol. The mixture was allowed to react at room temperature for 5 min and the absorbance monitored at 640 nm. A control sample obtained by replacing the sample with methanol was also measured. The results were expressed in milligrams of catechin per 100 g fresh grape.

Procyanidins can be determined as a whole by using the Habertson–Adams method as modified by Skogerson et al. (2007). To this end, tannins are precipitated with serum bovine albumin, and the precipitate is washed and redissolved to determine phenols by reaction with $FeCl_3$. The method uses catechin as standard and the results are expressed as catechin equivalents.

Chemical Information About the Grape Extract

Free radical scavenging activity is a frequently demanded chemical property of phenols extracts. Peralbo-Molina et al. (2012, 2013) succeeded in extracting this type of information by measuring total antioxidant activity in terms of ferric reducing antioxidant power (FRAP). The assay involves reducing the ferric tripyridyltriazin [Fe(III)–TPTZ] complex to ferrous tripyridyltriazin [Fe(II)–TPTZ] at low pH by effect of the action of antioxidants in the target solution. The ferrous complex [Fe(II)–TPTZ] has an intense blue color that can be monitored at 593 nm. The assay response was standardized against the antioxidant standard Trolox. The method was automated by Benzie et al. (1996, 1999), using a Cobas Fara centrifugal analyzer as follows: freshly prepared FRAP reagent was warmed to 37 ºC and a reagent blank reading taken at 593 nm, after which the sample was supplied with water. Absorbance readings were taken after 0.5 s and every 15 s thereafter during the monitoring period. The absorbance change between the selected final reading and the blank reading was calculated for each sample and interpolated in a calibration curve run from a Fe(II) standard solution tested in parallel. Halvorsen et al. (2002) introduced minor modifications to miniaturize the method by using a 96-well microtiter plate.

Other choices for measuring antioxidant activity include the 6-hydroxy-2,5,7,8-tetramethylchroman-2-carboxylic acid (Trolox) equivalent antioxidant capacity (TEAC) assay of Miller et al. (1993), and the oxygen radical absorbance capacity (ORAC) assay of DeLange and Glazer (1989). Both are based on the ability of the antioxidant to react with or neutralize free radicals generated in the assay.

Only FRAP among the previous assays affords direct measurement of antioxidants or reductants in a sample. This is a result of its using antioxidants as reductants in a redox linked colorimetric reaction; on the other hand, the other assays are indirect as they measure the inhibition of reactive species (free radicals) produced in the reaction mixture, so their results depend strongly on the type of reactive species used.

Quantitative Information about Individual Phenols in Grape Extracts

Extracting this type of information requires prior separation of the target phenols by using the technique best suited to their chemical characteristics (viz., liquid or gas chromatography, or even capillary electrophoresis, followed appropriate detection depending on the level of selectivity sought and on whether total or only partial separation is required). Thus, molecular absorption detection (usually by use a diode array detector, DAD) uses the retention times and spectral features of the analytes together with standards for accurate quantitation. Similar performance, but greater sensitivity, is provided by a fluorescence detector —preferably with laser excitation—; however, fluorescence detection requires the analyte to possess intrinsic emission or be derivatized to an emissive product. Comparably sensitive, but much more expensive to purchase and maintain, are mass detectors.

Ling et al. (2009) have reported examples of determinations of specific phenol families by use of a fast, sensitive method for quantifying four anthocyanins by liquid chromatography in combination with mass spectrometry (LC–MS).

One alternative method for anthocyanins is that used by Oh et al. (2008) to characterize and quantify this phenol family in grape cultivars. The procedure involves partial purification through XAD-7 cartridges, followed by C-18 LC–DAD, LC–MS, and LC–MSMS analysis.

Comprehensive Characterization of Phenols

Liquid chromatography with hybrid quadrupole time-of-flight detection (LC–QqTOF) is one of the most sensitive and accurate approaches to the comprehensive characterization of phenols. LC–QqTOF has become an unarguably powerful tool for the separation and characterization of phenols. The detector affords accurate mass determination of components in complex mixtures, after which the phenol composition of a grape extract can be accurately estimated from other chemical information such as neutral loss and product ions. LC–QqTOF has proved highly useful for characterizing most phenols in complex matrices; also, it provides crucial information about unknown compounds and isomers that is often pivotal for characterizing natural products from plants.

Perestrelo et al. (2012) conducted the first phytochemical research on phenolic compounds in Sercial and Tinta Negra *Vitis vinifera* L. grapes. The phenol profile of grape skins (white and red varieties, respectively, at different ripening stages) was analyzed by using LC–DAD–ESI-MS^n. A total of 40 phenols including 3 hydroxybenzoic acids, 8 hydroxycinnamic acids, 4 flavanols, 5 flavanones, 8 flavonols, 4 stilbenes, and 8 anthocyanins were identified in terms of their MS^n fragmentation profiles. Such valuable information can be useful for the oenological management and valorization of these grape varieties as sources of bioactive compounds.

Di Lecce et al. (2014) recently reported the first comprehensive characterization of phenols in different anatomical parts of Albariño grape. Complementary equipment (LC–QqTOF and LC coupled to a triple quadrupole mass spectrometer, LC–QQQ) was used to identify and quantify phenols in this grape variety. Also, a more complete phenolic profile was obtained from product ion and precursor ion scans, and a neutral loss scan at 152 mass units enabled fast screening of procyanidin dimers, trimers and their galloylated derivatives. Confirmation via accurate mass measurements in the TOF/MS and TOF-MS/MS modes at a high resolution provided good fits for all investigated ions, with errors ranging from 0.2 to 4.5 mDa.

Potential Biomedical Effects of Phenols

Phenols are secondary metabolites with key roles in plants such as pigmentation to attract pollinators and seed dispersers, and defense of the whole plant against pathogenic microorganisms and predators (Petti et al. 2009).

Concerning human health, a number of studies support the protective effect of phenols against oxidative damage associated to development of most major age-related degenerative diseases; thus, phenols are known to decrease the risk for cancer, cardiovascular diseases and neurodegenerative disorders (Manach et al., 2004; Petti et al., 2009).

The effects of phenols depend on the amount and periodicity of the intake. According to Manach et al. (2004), the mean daily intake of phenols by adults in the world is 863 mg, with phenolic acids accounting for 75%, proanthocyanidins for 14%, and anthocyanidins and other flavonoids jointly for 10%. The main sources for intake of phenols are coffee, cereals, fruits and vegetables (Petti et al. 2009).

Although phenols are commonly present in human diets, they are not necessarily active in the body, either because they possess limited activity or because they are poorly absorbed from the intestine, highly metabolized or rapidly eliminated (Manach et al. 2004). Little is known about phenol absorption, distribution and metabolism even though some authors have proposed a likely common pathway (Petti et al., 2009). As a rule, phenol metabolites are rapidly eliminated from plasma, which indicates that maintaining their required levels in blood requires consumption of plant products on a daily basis (Graefe et al., 2001; Hollman et al., 1997; Manach et al., 1998).

Petti et al. (2009) reviewed studies published from 1993 to 2008 with the aim of finding a relationship between phenols and dental diseases. They found that the antioxidant ability of phenols potentially accounts for their preventive effects against diseases of the oral cavity (Halliwell et al. 2000). Also, catechins, methoxylated flavonoids and proanthocyanidins have shown potential preventive activity against oral squamous cell carcinoma (Ho et al. 2007; Hsu et al. 2001, Khafif et al. 1998). In addition, antioxidant-rich diets seemingly inhibit development and progression of periodontal diseases, particularly in subjects exposed to environmental and dietary sources of oxidative stress (Battino et al., 1999; Ritchie et al., 2003; Sculley et al., 2002).

Research into the effects of phenols on Alzheimer's disease concluded that these compounds may exert a variety of modulatory actions with positive influence on the pathogenesis of the disease (Jayasena et al. 2013). Phenols also seemingly exert antidiabetic effects, either directly (dependent on absorption) or indirectly (dependent on gut microbiota modulation and pancreatic lipase inhibition) (Anhê et al., 2013).

In short, phenols seem to have favorable effects on human health thanks, among others, to their antioxidants properties.

Other Effects of Phenols

Other interesting properties making phenols especially useful are related to their uses by the cosmetic and food industries. Al Bittar et al. (2013) reported a study where fresh grape berries were pressed and the residue from the juice was extracted for phenols with a green technique (microwave hydrodiffusion and gravity). Then, the extract was added to the total volume of natural juice in order to obtain juice enriched with phenols. The supplemented juice had a more attractive color and acceptable organoleptic characteristics. No studies on the effect of this juice as compared to conventional juice have to date been reported, however.

A study investigating the potential relationship between antimicrobial activity and phenol content was conducted on extracts from grape seeds and bagasse obtained with two different solvent extraction methods. The study concluded that the grape seed extracts had high total phenolics relative to the bagasse extracts —the latter failed to inhibit all bacteria tested. Shoko et al. (1999) confirmed that phenols were the most important compounds with antibacterial activity. Also, they identified gallic acid as the most active compound for inhibition of bacteria. Their results suggest that grape seed extracts may be effective against bacterial deterioration of stored foods by bacteria provided a reduction in sensory quality is accepted.

Osmotic treatment (OT) is widely used to modify the composition of solid foods by partially removing water and adding solutes. Recently, Rozek et al. (2007) used OT a unit operation to introduce controlled quantities of solutes with antioxidant properties such as grape phenols into a food matrix. By using concentrated red grape must as a source of phenols, they found the phenolic content in osmotreated solid food to be similar to that of some phenol-rich fruits. Then, they examined other commercial sources of phenols such as grape seed and skin extracts, which they used to infuse a model food with grape phenols (Rozek et al., 2008, 2009). Also, they studied the influence of the source of grape phenols and the phenol concentration of the osmotic solution on the properties of osmotreated food (Rozet et al., 2010a, 2010b).

Winemakers' interest in the phenol content of grapes has grown in parallel to evidence that such a content influences the color, bitterness, astringency, "mouth feel" and "ageability" of wines. In fact, the content in phenols is one of the main factors dictating the quality of grapes and wine. The phenolic composition of wine depends primarily on the phenol content of the grapes and secondarily on the winemaking techniques used. In a recent study, grapes of three autochthonous varieties from Andalusia (viz., Jaén tinto, Palomino negro and Tintilla de Rota), and two others of the most commonly used in Spain and worldwide (Cabernet Sauvignon and Tempranillo), were analyzed by LC–DAD–MS and compared in order to determine and quantify their phenol composition. The results were found to be useful with a view to selecting effective winemaking techniques for improving the quality of the final wine (Guerrero et al., 2009). Schwartz et al. (2009) examined the relationship between antioxidant activity and phenolic content, and their correlation with wine age, in commercial and experimental sherry brandies. Although they identified a significant relationship, some nonphenol compounds were also found to be potential contributors to antioxidant activity. The results revealed certain differences possibly due to the addition of caramel coloring between commercial and experimental sherry brandies. Also, the antioxidant power of Brandy de Jerez was found to exceed that of other commercial aged distillates of different geographical origin.

The volatile fraction of wine comprises several hundreds of chemically different compounds. The presence or absence of some compounds in a particular wine depends on various factors such as the environment (climate and soil), grape ripeness and variety, winemaking conditions, and degree of aging. Some phenols are present in grapes, whereas others are formed during wine fermentation and aging. Ruiz-Bejarano et al. (2013) studied the evolution under different aging conditions of the main volatiles in sweet Sherry wines obtained from Muscat grapes dried in chambers and fermented under variable enological conditions. Two factors, and their potential influence on the volatile profile of wine, were thus considered simultaneously: aging and vinification conditions. These authors found the three factors considered (viz., fermentation method, aging method and aging time) to affect the volatile composition and sensory characteristics of the wines, with aging method as the individual factor most markedly influencing their volatile content.

The relationship between the contents in phenols and furan compounds of commercial Sherry brandies and other aged distillates of different geographical origin was examined in order to explain the contribution of the relative contents of these families to the highly specific character of Brandy de Jerez. The aging method was confirmed to have a strong influence on the analytical profile of aged distillates (discrimination up to 100%). Differences between commercial brandies and experimentally aged beverages of an equivalent average age were also confirmed by the large differences between Solera Gran Reserva Brandies de Jerez and the other distillates (discrimination up to 80%) —an indication of their highly specific character (Rodríguez et al. 2010).

Conclusion

Despite the many uses of grape byproducts and byproduct residues (i.e., "residues from residues"), their integral exploitation remains an unaccomplished goal. In fact, there has been little research focusing on this goal even though there is evidence that some residues might be useful sources for compounds of great interest. For example, Maier et al. (2009) assessed the press residues of grape seed oil production as phenol sources and found them to constitute a phenol-rich byproduct with antioxidant activity.

Peralbo-Molina et al. (2012, 2013) developed an extraction method with interesting analytical characteristics for isolating priced compounds from a raw material of very little value, namely: residues from distillery industries using the byproduct after alcohol distillation from red and white grape pomace. Also, the fact that the extracts contain interesting compounds suggests that, despite the degradation caused by the drastic conditions of the distillation process, this industrial waste is a useful raw material for uses other than simply heating. The large variety of compounds identified in the extracts makes them useful as food additives (colorants, flavor modifiers, antioxidants), and also in cosmetics and nutraceuticals.

Farías Campomanes et al. (2013) demonstrated the economic feasibility of large-scale operations of supercritical fluid extraction for recovery of phenolic compounds in grape bagasse from Pisco residues.

References

Al Bittar S, Perino-Issartier S, Dangles O, Chemat F. An innovative grape juice enriched in polyphenols by microwave-assisted extraction. *Food Chem*. 2013; 141(3):3268-72.

Albu S, Joyce E, Paniwnyk L, Lorimer JP, Mason TJ. Potential for the use of ultrasound in the extraction of antioxidants from *Rosmarinus officinalis* for the food and pharmaceutical industry. *Ultrason Sonochem*. 2004; 11(3-4):261–75.

Ali K, Maltese F, Choi YH, Verpoorte R. Metabolic constituents of grapevine and grape-derived products. *Phytochem Rev*. 2010; 9(3):357–78.

Andrade PB, Oliveira BM, Seabra RM, Ferreira MA, Ferreres F, García-Viguera C. Analysis of phenolic compounds in Spanish Albrariño and Portuguese Alvarinho and Loureiro wines by capillary zone electrophoresis and high-performance liquid chromatography. *Electrophoresis*. 2001; 22(8):1568–72.

Anhê FF, Desjardins Y, Pilon G, Dudonné S, Genovese MI, Lajolo FM, et al. Polyphenols and type 2 diabetes: a prospective review. *Pharma Nutrition.* (DOI: 10.1016/j.phanu.2013.07.004).

Battino M, Bullon P, Wilson M, Newman H. Oxidative injury and inflammatory periodontal diseases: the challenge of anti-oxidants to free radicals and reactive oxygen species. *Crit Rev Oral Biol Med.* 1999; 10(4):458–76.

Benzie IFF, Strain JJ. The ferric reducing ability of plasma (FRAP) as a measure of "antioxidant power": the FRAP assay. *Anal Biochem*. 1996; 239(1):70–6.

Benzie IF, Chung W, Strain JJ. "Antioxidant" (reducing) efficiency of ascorbate in plasma is not affected by concentration. *J Nutr Biochem*. 1999; 10(3):146–50.

Biesalski HK. Diabetes preventive components in the Mediterranean diet. *Eur J Nutr*. 2004; 43(1):26–30.

Bonghi C, Rizzini FM, Gambuti A, Moio L, Chkaiban L, Tonutti P. Phenol compound metabolism and gene expression in the skin of wine grape (*Vitis vinifera* L.) berries subjected to partial postharvest dehydration. *Postharvest Biol Technol.* 2012; 67(0):102–9.

Boots AW, Wilms LC, Swennen ELR, Kleinjans JCS, Bast A, Haenen GRMM. *In vitro* and *ex vivo* anti-inflammatory activity of quercetin in healthy volunteers. *Nutrition.* 2008; 24(7–8):703–10.

Bravo L. Polyphenols: chemistry, dietary sources, metabolism, and nutritional significance. *Nutr Rev.* 1998; 56(11):317–33.

Caillet S, Salmiéri S, Lacroix M. Evaluation of free radical-scavenging properties of commercial grape phenol extracts by a fast colorimetric method. *Food Chem.* 2006; 95(1):1–8.

Cantos E, Espín JC, Tomás-Barberán FA. Postharvest stilbene-enrichment of red and white table grape varieties using UV-C irradiation pulses. *J Agric Food Chem.* 2002; 50(22):6322–9.

Casas L, Mantell C, Rodríguez M, Ossa EJMdl, Roldán A, Ory ID, et al. Extraction of resveratrol from the pomace of Palomino fino grapes by supercritical carbon dioxide. *J Food Eng.* 2010; 96(2):304–8.

Castillo-Muñoz N, Gómez-Alonso S, García-Romero E, Hermosín-Gutiérrez I. Flavonol profiles of *Vitis vinifera* red grapes and their single-cultivar wines. *J Agric Food Chem.* 2007; 55(3):992–1002.

Chafer A, Pascual-Marti MC, Salvador A, Berna A. Supercritical fluid extraction and HPLC determination of relevant polyphenolic compounds in grape skin. *J Sep Sci.* 2005; 28(16):2050–6.

Cho Y-J, Hong J-Y, Chun HS, Lee SK, Min H-Y. Ultrasonication-assisted extraction of resveratrol from grapes. *J Food Eng.* 2006; 77(3):725–30.

Choi HK. Dietary risk factors for rheumatic diseases. *Curr Opin Rheumatol.* 2005; 17(2):141–6.

Corrales M, Fernández A, Vizoso Pinto MG, Butz P, Franz CM, Schuele E, et al. Characterization of phenolic content, *in vitro* biological activity, and pesticide loads of extracts from white grape skins from organic and conventional cultivars. *Food Chem Toxicol.* 2010; 48(12):3471–6.

Corrales M, García AF, Butz P, Tauscher B. Extraction of anthocyanins from grape skins assisted by high hydrostatic pressure. *J Food Eng.* 2009; 90(4):415–21.

Cravotto G, Boffa L, Mantegna S, Perego P, Avogadro M, Cintas P. Improved extraction of vegetable oils under high-intensity ultrasound and/or microwaves. *Ultrason Sonochem.* 2008; 15(5):898–902.

D'Archivio M, Filesi C, Di Benedetto R, Gargiulo R, Giovannini C, Masella R. Polyphenols, dietary sources and bioavailability. *Ann Ist Super Sanita.* 2007; 43(4):348–61.

Darias-Martín J, Carrillo M, Díaz E, Boulton RB. Enhancement of red wine colour by pre-fermentation addition of copigments. *Food Chem.* 2001; 73(2):217-20.

de Campos LM, Leimann FV, Pedrosa RC, Ferreira SRS. Free radical scavenging of grape pomace extracts from Cabernet sauvingnon (*Vitis vinifera*). *Biores Technol.* 2008; 99(17):8413–20.

DeLange RJ, Glazer AN. Phycoerythrin fluorescence-based assay for peroxy radicals: a screen for biologically relevant protective agents. *Anal Biochem.* 1989; 177(2):300–6.

Di Lecce G, Arranz S, Jáuregui O, Tresserra-Rimbau A, Quifer-Rada P, Lamuela-Raventós RM. Phenolic profiling of the skin, pulp and seeds of Albariño grapes using hybrid quadrupole time-of-flight and triple-quadrupole mass spectrometry. *Food Chem.* 2014; 145:874–82.

Di Stefano R, Cravero MC, Gentilizi M. Metodi per lo studio dei polifenoli dei vini. *L'Enotecnico.* 1989; May:83–9.

Dryden GW, Song M, McClain C. Polyphenols and gastrointestinal diseases. *Curr Opin Gastroenterol.* 2006; 22(2):165–70.

Faller ALK, Fialho E. Polyphenol content and antioxidant capacity in organic and conventional plant foods. *J Food Comp Anal*. 2010; 23(6):561–8.

Farías-Campomanes AM, Rostagno MA, Meireles MAA. Production of polyphenol extracts from grape bagasse using supercritical fluids: yield, extract composition and economic evaluation. *J Supercrit Fluids*. 2013; 77(0):70–8.

Feinleib M. Seven countries: A multivariate analysis of death and coronary heart disease. *JAMA*. 1981; 245(5):511–2.

Figueiredo-González M, Cancho-Grande B, Simal-Gándara J. Evolution of colour and phenolic compounds during Garnacha Tintorera grape raisining. *Food Chem*. 2013; 141(3):3230–40.

Flamini R. Mass spectrometry in grape and wine chemistry. Part I: Polyphenols. *Mass Spectrom Rev*. 2003; 22(4):218–50.

Food and Agriculture Organization of the United Nations. 2013.

Fukai K, Ishigami T, Kara Y. Antibacterial activity of tea polyphenols against phytopathogenic bacteria. *Agric Biol Chem*. 1991; 55(7):1895–7.

Gambelli L, Santaroni GP. Polyphenols content in some Italian red wines of different geographical origins. *J Food Comp Anal*. 2004; 17(5):613–8.

Garrido J, Borges F. *Wine and grape polyphenols – A chemical perspective. Food Res Internat*. (DOI: 10.1016/j.foodres.2013.08.002).

Giacosa A, Barale R, Bavaresco L, Gatenby P, Gerbi V, Janssens J, et al. Cancer prevention in Europe: the Mediterranean diet as a protective choice. *Eur J Cancer Prev*. 2013; 22(1):90–5.

González-Rodríguez J, Pérez-Juan P, Luque de Castro MD. Superheated liquids for extraction of solid residues from winemaking processes. *Anal Bioanal Chem*. 2003; 377(7-8):1190–5.

Graefe EU, Wittig J, Mueller S, Riethling AK, Uehleke B, Drewelow B, et al. Pharmacokinetics and bioavailability of quercetin glycosides in humans. *J Clin Pharmacol.* 2001; 41(5):492–9.

Grigonis D, Venskutonis PR, Sivik B, Sandahl M, Eskilsson CS. Comparison of different extraction techniques for isolation of antioxidants from sweet grass (*Hierochloë odorata*). *J Supercrit Fluids*. 2005; 33(3):223–33.

Gruber JAN, Tang SY, Halliwell B. Evidence for a trade-off between survival and fitness caused by resveratrol treatment of *Caenorhabditis elegans*. *Ann New York Academy Sci.* 2007; 1100(1):530–42.

Guerrero RF, García-Parrilla MC, Puertas B, Cantos-Villar E. Wine, resveratrol and health: a review. *Nat Prod Commun*. 2009; 4(5):635–58.

Guerrero RF, Liazid A, Palma M, Puertas B, González-Barrio R, Gil-Izquierdo A, et al. Phenolic characterisation of red grapes autochthonous to Andalusia. *Food Chem*. 2009; 112(4):949–55.

Halliwell B, Zhao K, Whiteman M. The gastrointestinal tract: a major site of antioxidant action. *Free Radic Res.* 2000; 33(6):819–30.

Halvorsen BL, Holte K, Myhrstad MCW, Barikmo I, Hvattum E, Remberg SF, et al. A systematic screening of total antioxidants in dietary plants. *J Nutrition.* 2002; 132(3):461–71.

Harborne, J.B. The flavonoids: advances in research since 1986. *J Chem Educ.* 1995; 72(3):A73.

Hmamouchi M, Es-Safi N, Lahrichi M, Fruchier A, Essassi EM. Flavones and flavonols in leaves of some Moroccan *Vitis vinifera* cultivars. *Am J Enol Viticulture.* 1996; 47(2):186–92.

Ho YC, Yang SF, Peng CY, Chou MY, Chang YC. Epigallocatechin-3-gallate inhibits the invasion of human oral cancer cells and decreases the productions of matrix metalloproteinases and urokinase-plasminogen activator. *J Oral Pathol Med.* 2007; 36(10):588–93.

Hollman PC, van Trijp JM, Buysman MN, van der Gaag MS, Mengelers MJ, de Vries JH, et al. Relative bioavailability of the antioxidant flavonoid quercetin from various foods in man. *FEBS Lett.* 1997; 418(1-2):152–6.

Hsu S, Lewis JB, Borke JL, Singh B, Dickinson DP, Caughman GB, et al. Chemopreventive effects of green tea polyphenols correlate with reversible induction of p57 expression. *Anticancer Res.* 2001; 21(6A):3743–8.

Hung LM, Chen JK, Huang SS, Lee RS, Su MJ. Cardioprotective effect of resveratrol, a natural antioxidant derived from grapes. *Cardiovasc Res.* 2000; 47(3):549–55.

Jaitz L, Siegl K, Eder R, Rak G, Abranko L, Koellensperger G, et al. LC–MS/MS analysis of phenols for classification of red wine according to geographic origin, grape variety and vintage. *Food Chem.* 2010; 122(1):366–72.

Jayasena T, Poljak A, Smythe G, Braidy N, Munch G, Sachdev P. The role of polyphenols in the modulation of sirtuins and other pathways involved in Alzheimer's disease. Ageing Res Rev. 2013; 12(4):867–83.

Jiménez L, Angulo V, Ramos E, de la Torre MJ, Ferrer JL. Comparison of various pulping processes for producing pulp from vine shoots. *Industrial Crops Products.* 2006; 23(2):122–30.

Ju ZY, Howard LR. Effects of solvent and temperature on pressurized liquid extraction of anthocyanins and total phenolics from dried red grape skin. *J Agric Food Chem.* 2003; 51(18):5207–13.

Kammerer D, Claus A, Schieber A, Carle R. A novel process for the recovery of polyphenols from grape (*Vitis vinifera* L.) pomace. *J Food Sci.* 2005; 70(2):C157–C63.

Kammerer D, Gajdos Kljusuric J, Carle R, Schieber A. Recovery of anthocyanins from grape pomace extracts (*Vitis vinifera* L. cv. Cabernet Mitos) using a polymeric adsorber resin. *Eur Food Res Technol.* 2005; 220(3-4):431–7.

Khafif A, Schantz SP, al-Rawi M, Edelstein D, Sacks PG. Green tea regulates cell cycle progression in oral leukoplakia. *Head Neck.* 1998; 20(6):528–34.

Khanal RC, Howard LR, Prior RL. Effect of heating on the stability of grape and blueberry pomace procyanidins and total anthocyanins. *Food Res Internat.* 2010; 43(5):1464–9.

Keys, A. Coronary heart disease in seven countries. Summary. *Circulation.* 1970; 41(4 Suppl):186–95.

Kok FJ, Kromhout D. Atherosclerosis-epidemiological studies on the health effects of a Mediterranean diet. *Eur J Nutr*. 2004; 43(1):2–5.

Kushi LH, Lenart EB, Willett WC. Health implications of Mediterranean diets in light of contemporary knowledge. 1. Plant foods and dairy products. *Am J Clin Nutr*. 1995; 61(6 Suppl):1407S–15S.

Liazid A, Guerrero RF, Cantos E, Palma M, García-Barroso C. Microwave assisted extraction of anthocyanins from grape skins. *Food Chem*. 2011; 124(3):1238–43.

Ling Y, Ren C, Mallery SR, Ugalde CM, Pei P, Saradhi UVRV, et al. A rapid and sensitive LC–MS/MS method for quantification of four anthocyanins and its application in a clinical pharmacology study of a bioadhesive black raspberry gel. *J Chromatogr B*. 2009; 877(31):4027–34.

López-Sánchez C, Martín-Romero FJ, Sun F, Luis L, Samhan-Arias AK, García-Martí V, et al. Blood micromolar concentrations of kaempferol afford protection against ischemia/reperfusion-induced damage in rat brain. *Brain Res*. 2007; 28:123–37.

Luque-Rodríguez JM, Luque de Castro MD, Pérez-Juan P. Dynamic superheated liquid extraction of anthocyanins and other phenolics from red grape skins of winemaking residues. *Bioresource Technol*. 2007; 98(14):2705–13.

Luque-Rodríguez JL, Luque de Castro MD, Japón-Luján R. *Exploitation of Residues from Vineyards, Olive Groves, and Wine and Oil Production to Obtain Phenolic Compounds of High-Added Value in Methods of Analysis for Functional Foods and Nutraceuticals,* Second Edition: CRC Press, Boca Raton, FL, USA; 2008.

Maier T, Schieber A, Kammerer DR, Carle R. Residues of grape (*Vitis vinifera* L.) seed oil production as a valuable source of phenolic antioxidants. *Food Chem*. 2009; 112(3):551–9.

Makris DP, Boskou G, Andrikopoulos NK. Polyphenolic content and in vitro antioxidant characteristics of wine industry and other agri-food solid waste extracts. *J Food Comp Anal*. 2007; 20(2):125–32.

Manach C, Morand C, Crespy V, Demigne C, Texier O, Regerat F, et al. Quercetin is recovered in human plasma as conjugated derivatives which retain antioxidant properties. *FEBS Lett*. 1998; 426(3):331–6.

Manach C, Scalbert A, Morand C, Remesy C, Jiménez L. Polyphenols: food sources and bioavailability. *Am J Clin Nutr*. 2004; 79(5):727–47.

Mantell C, de la Ossa EM, Rodríguez M. Supercritical fluid extraction of anthocyanins from grape pomace. In: Hidalgo M, Valiente M, editors. *Solvent Extraction for the 21st Century Proceedings of ISEC'99. Barcelona, Spain: Society of Chemical Industry*; 2001.

Mantell C, Rodríguez M, de la Ossa EM. Measurement of the diffusion coefficient of a model food dye (malvidin 3,5-diglucoside) in a high pressure CO_2+methanol system by the chromatographic peak-broadening technique. *J Supercrit Fluids*. 2003; 25(1):57–68.

Márquez A, Serratosa MP, López-Toledano A, Mérida J. Colour and phenolic compounds in sweet red wines from Merlot and Tempranillo grapes chamber-dried under controlled conditions. *Food Chem*. 2012; 130(1):111–20.

Mattivi F, Guzzon R, Vrhovsek U, Stefanini M, Velasco R. Metabolite profiling of grape: flavonols and anthocyanins. *J Agric Food Chem*. 2006; 54(20):7692–702.

Miller NJ, Rice-Evans C, Davies MJ, Gopinathan V, Milner A. A novel method for measuring antioxidant capacity and its application to monitoring the antioxidant status in premature neonates. *Clin Sci*. 1993; 84(4):407–12.

Mira de Orduña R. Climate change associated effects on grape and wine quality and production. *Food Res Internat*. 2010; 43(7):1844–55.

Modun D, Music I, Vukovic J, Brizic I, Katalinic V, Obad A, et al. The increase in human plasma antioxidant capacity after red wine consumption is due to both plasma urate and wine polyphenols. *Atherosclerosis*. 2008; 197(1):250–6.

Moini H, Guo Q, Packer L. Enzyme inhibition and protein-binding action of the procyanidin-rich French maritime pine bark extract, pycnogenol: effect on xanthine oxidase. *J Agric Food Chem*. 2000; 48(11):5630-9.

Monagas M, Bartolomé B, Gómez-Cordovés C. Evolution of polyphenols in red wines from *Vitis vinifera* L. during aging in the bottle. *Eur Food Res Technol*. 2005; 220(3-4):331–40.

Morales S, Canosa P, Rodríguez I, Rubi E, Cela R. Microwave assisted extraction followed by gas chromatography with tandem mass spectrometry for the determination of triclosan and two related chlorophenols in sludge and sediments. *J Chromatogr A*. 2005; 5(2):128–35.

Morata A, González C, Suárez-Lepe JA. Formation of vinylphenolic pyranoanthocyanins by selected yeasts fermenting red grape musts supplemented with hydroxycinnamic acids. *Internat J Food Microbiol*. 2007; 116(1):144–52.

Mulero J, Pardo F, Zafrilla P. Antioxidant activity and phenolic composition of organic and conventional grapes and wines. *J Food Comp Anal*. 2010; 23(6):569–74.

Nagel CW, Glories Y. Use of a modified dimethylaminocinnamaldehyde reagent for analysis of flavanols. *Am J Enol Viticulture*. 1991; 42(4):364–6.

Negro C, Tommasi L, Miceli A. Phenolic compounds and antioxidant activity from red grape marc extracts. *Bioresource Technol*. 2003; 87(1):41–4.

Oh YS, Lee JH, Yoon SH, Oh CH, Choi DS, Choe E, et al. Characterization and quantification of anthocyanins in grape juices obtained from the grapes cultivated in Korea by HPLC/DAD, HPLC/MS, and HPLC/MS/MS. *J Food Sci*. 2008; 73(5):C378–C89.

Orak HH. Total antioxidant activities, phenolics, anthocyanins, polyphenoloxidase activities of selected red grape cultivars and their correlations. *Sci Hortic*. 2007; 111(3):235–41.

Peinado J, López de Lerma N, Moreno J, Peinado RA. Antioxidant activity of different phenolics fractions isolated in must from Pedro Ximenez grapes at different stages of the off-vine drying process. *Food Chem*. 2009; 114(3):1050–5.

Peng Z, Hayasaka Y, Iland PG, Sefton M, Hoj P, Waters EJ. Quantitative analysis of polymeric procyanidins (Tannins) from grape (Vitis vinifera) seeds by reverse phase high-performance liquid chromatography. *J Agric Food Chem*. 2001; 49(1):26–31.

Peralbo-Molina A, Priego-Capote F, Luque de Castro MD. Comparison of extraction methods for exploitation of grape skin residues from ethanol distillation. *Talanta*. 2012; 101(0):292–8.

Peralbo-Molina A, Priego-Capote F, Luque de Castro MD. Characterization of grape seed residues from the ethanol-distillation industry. *Anal Methods*. 2013; 5(8):1922–9.

Peralbo-Molina A, Luque de Castro MD. Potential of residues from the Mediterranean agriculture and agrifood industry. *Trends Food Sci Technol*. 2013; 32(1):16–24.

Perestrelo R, Lu Y, Santos SAO, Silvestre AJD, Neto CP, Câmara JS, et al. Phenolic profile of Sercial and Tinta Negra *Vitis vinifera* L. grape skins by HPLC–DAD–ESI-MSn: novel phenolic compounds in *Vitis vinifera* L. grape. *Food Chem*. 2012; 135(1):94–104.

Pérez-Serradilla JA, Luque de Castro MD. Role of lees in wine production: A review. *Food Chem.* 2008; 111(2):447–56.

Pérez-Serradilla JA, Luque de Castro MD. Microwave-assisted extraction of phenolic compounds from wine lees and spray-drying of the extract. *Food Chem*. 2011; 124(4):1652–9.

Petti S, Scully C. Polyphenols, oral health and disease: a review. *J Dent*. 2009; 37(6):413–23.

Prados-Rosales RC, Luque García JL, Luque de Castro MD. Rapid analytical method for the determination of pesticide residues in sunflower seeds based on focused microwave-assisted Soxhlet extraction prior to gas chromatography–tandem mass spectrometry. *J Chromatogr A*. 2003; 993(1-2):121–9.

Pietta P, Minoggio M, Bramati L. Plant Polyphenols: structure, occurrence and bioactivity. In: Atta-ur R, editor. *Studies in Natural Products Chemistry*: Elsevier; 2003. pp 257–312.

Piñeiro Z, Palma M, García-Barroso C. Determination of trans-resveratrol in grapes by pressurised liquid extraction and fast high-performance liquid chromatography. *J Chromatogr A*. 2006; 1110(1–2):61–5.

Prados-Rosales RC, Luque García JL, Luque de Castro MD. Rapid analytical method for the determination of pesticide residues in sunflower seeds based on focused microwave-assisted Soxhlet extraction prior to gas chromatography–tandem mass spectrometry. *J Chromatogr A*. 2003; 993(1-2):121–9.

Prior RL, Cao G, Martin A, Sofic E, McEwen J, O'Brien C, et al. Antioxidant capacity as influenced by total phenolic and anthocyanin content, maturity, and variety of Vaccinium species. *J Agric Food Chem*. 1998; 46(7):2686–93.

Proteggente AR, Pannala AS, Paganga G, Van Buren L, Wagner E, Wiseman S, et al. The antioxidant activity of regularly consumed fruit and vegetables reflects their phenolic and vitamin C composition. *Free Radic Res*. 2002; 36(2):217–33.

Ribeiro De Lima MT, Waffo-Téguo P, Teissedre PL, Pujolas A, Vercauteren J, Cabanis JC, et al. Determination of stilbenes (*trans*-astringin, *cis*- and *trans*-piceid, and *cis*- and *trans*-resveratrol) in Portuguese wines. *J Agric Food Chem*. 1999; 47(7):2666–70.

Ritchie CS, Kinane DF. Nutrition, inflammation, and periodontal disease. *Nutrition*. 2003; 19(5):475-6.

Robards K. Strategies for the determination of bioactive phenols in plants, fruit and vegetables. *J Chromatogr A*. 2003; 1000(1–2):657–91.

Rodríguez Dodero MC, Guillén Sánchez DA, Rodríguez MS, García-Barroso C. Phenolic compounds and furanic derivatives in the characterization and quality control of Brandy de Jerez. *J Agric Food Chem*. 2010; 58(2):990–7.

Rodríguez Montealegre R, Romero Peces R, Chacón Vozmediano JL, Martínez Gascueña J, García Romero E. Phenolic compounds in skins and seeds of ten grape *Vitis vinifera* varieties grown in a warm climate. *J Food Comp Anal*. 2006; 19(6–7):687–93.

Roggero JP, García-Parrilla C. Effects of ultraviolet irradiation on resveratrol and changes in resveratrol and various of its derivatives in the skins of ripening grapes. *Sci Aliments.* 1995; 15(5):411–22.

Romero-Pérez AI, Lamuela-Raventós RM, Andrés-Lacueva C, de la Torre-Boronat MC. Method for the quantitative extraction of resveratrol and piceid isomers in grape berry skins. Effect of powdery mildew on the stilbene content. *J Agric Food Chem.* 2001; 49(1):210–5.

Rozek A, Achaerandio I, Almajano MP, Guell C, López F, Ferrando M. Solid foodstuff supplemented with phenolics from grape: antioxidant properties and correlation with phenolic profiles. *J Agric Food Chem.* 2007; 55(13):5147–55.

Rózek A, Achaerandio I, Güell C, López F, Ferrando M. Direct formulation of a solid foodstuff with phenolic-rich multicomponent solutions from grape seed: effects on composition and antioxidant properties. *J Agric Food Chem.* 2008; 56(12):4564–76.

Rózek A, Achaerandio I, Güell C, López F, Ferrando M. Grape phenolic impregnation by osmotic treatment: influence of osmotic agent on mass transfer and product characteristics. *J Food Eng.* 2009; 94(1):59–68.

Rózek A, Achaerandio I, Güell C, López F, Ferrando M. Use of commercial grape phenolic extracts to supplement solid foodstuff. *LWT - Food Sci Technol.* 2010; 43(4):623–31.

Rózek A, García-Pérez JV, López F, Güell C, Ferrando M. Infusion of grape phenolics into fruits and vegetables by osmotic treatment: phenolic stability during air drying. *J Food Eng.* 2010; 99(2):142–50.

Ruiz-Bejarano M, Castro-Mejías R, Rodríguez-Dodero MD, García-Barroso C. Study of the content in volatile compounds during the aging of sweet Sherry wines obtained from grapes cv. Muscat and fermented under different conditions. *Eur Food Res Technol.* 2013:1–18.

Santos-Buelga C, Williamson G, Chemistry RS. *Methods in Polyphenol Analysis: Royal Society of Chemistry*; 2003.

Schwarz M, Rodríguez M, Martínez C, Bosquet V, Guillén D, García-Barroso C. Antioxidant activity of Brandy de Jerez and other aged distillates, and correlation with their polyphenolic content. *Food Chem.* 2009; 116(1):29–33.

Schwarz M, Picazo-Bacete JJ, Winterhalter P, Hermosin-Gutiérrez I. Effect of copigments and grape cultivar on the color of red wines fermented after the addition of copigments. *J Agric Food Chem.* 2005; 53(21):8372–81.

Sculley DV, Langley-Evans SC. Salivary antioxidants and periodontal disease status. *Proc Nutr Soc.* 2002; 61(1):137–43.

Serafini M, Laranjinha JAN, Almeida LM, Maiani G. Inhibition of human LDL lipid peroxidation by phenol-rich beverages and their impact on plasma total antioxidant capacity in humans. *J Nutritional Biochem.* 2000; 11(11–12):585–90.

Serrano A, Gallego M. Continuous microwave-assisted extraction coupled on-line with liquid–liquid extraction: determination of aliphatic hydrocarbons in soil and sediments. *J Chromatogr A.* 2006; 1104(1–2):323–30.

Serratosa MP, López-Toledano A, Mérida J, Medina M. Changes in color and phenolic compounds during the raisining of grape cv. Pedro Ximenez. *J Agric Food Chem.* 2008; 56(8):2810–6.

Serratosa MP, Lopez-Toledano A, Medina M, Merida J. Drying of Pedro Ximenez grapes in chamber at controlled temperature and with dipping pretreatments. Changes in the color fraction. *J Agric Food Chem.* 2008; 56(22):10739-46.

Serratosa MP, Márquez A, López-Toledano A, Medina M, Mérida J. Changes in hydrophilic and lipophilic antioxidant activity in relation to their phenolic composition during the chamber drying of red grapes at a controlled temperature. *J Agric Food Chem.* 2011; 59(5):1882–92.

Shi J, Nawaz H, Pohorly J, Mittal G, Kakuda Y, Jiang Y. Extraction of polyphenolics from plant material for functional foods – engineering and technology. *Food Rev Internat.* 2005; 21(1):139–66.

Singleton VL, Rossi JA. Colorimetry of total phenolics with phosphomolybdic–phosphotungstic acid reagents. *Am J Enol Viticulture*. 1965; 16(3):144–58.

Skogerson K, Downey M, Mazza M, Boulton R. Rapid determination of phenolic components in red wines from UV–visible spectra and the method of partial least squares. *Am J Enol Viticulture*. 2007; 58(3):318–25.

Souquet JM, Labarbe B, Le Guerneve C, Cheynier V, Moutounet M. Phenolic composition of grape stems. *J Agric Food Chem*. 2000; 48(4):1076–80.

Spigno G, De Faveri DM. Microwave-assisted extraction of tea phenols: a phenomenological study. *J Food Eng*. 2009; 93(2):210-7.

Spormann TM, Albert FW, Rath T, Dietrich H, Will F, Stockis JP, et al. Anthocyanin/polyphenolic-rich fruit juice reduces oxidative cell damage in an intervention study with patients on hemodialysis. *Cancer Epidemiol Biomarkers Prev.* 2008; 17(12):3372–80.

Stevanato R, Fabris S, Momo F. New enzymatic method for the determination of total phenolic content in tea and wine. *J Agric Food C*hem. 2004; 52(20):6287–93.

Sun B, Ribes AM, Leandro MC, Belchior AP, Spranger MI. Stilbenes: quantitative extraction from grape skins, contribution of grape solids to wine and variation during wine maturation. *Anal Chim Acta*. 2006; 563(1–2):382–90.

Trousdale Vernon L, Singleton EK. Astilbin engeletin in grapes and wine. *Phytochem*. 1983; 22(2):619–20.

Tsao R. Chemistry and biochemistry of dietary polyphenols. *Nutrients*. 2010; 2(12):1231–46.

Valls J, Millán S, Martí MP, Borràs E, Arola L. Advanced separation methods of food anthocyanins, isoflavones and flavanols. *J Chromatogr A*. 2009; 1216(43):7143–72.

Vennat B, Pourrat A, Pourrat H, Gross D, Bastide P, Bastide J. Procyanidins from the roots of *Fragaria vesca*: characterization and pharmacological approach. *Chem Pharm Bull.* 1988; 36(2):828–33.

Wada M, Kido H, Ohyama K, Ichibangase T, Kishikawa N, Ohba Y, et al. Chemiluminescent screening of quenching effects of natural colorants against reactive oxygen species: Evaluation of grape seed, monascus, gardenia and red radish extracts as multi-functional food additives. *Food Chem.* 2007; 101(3):980–6.

Wang LJ, Sun DW. Recent developments in numerical modeling of heating and cooling processes in the food industry: a review. *Trends Food Sci Technol*. 2003; 14:408–23.

Williamson EM. Selected bibliography. *Phytother Res.* 1990; 4(4):5-10.

Woodman OL, Meeker WF, Boujaoude M. Vasorelaxant and antioxidant activity of flavonols and flavones: structure–activity relationships. *J Cardiovasc Pharmacol*. 2005; 46(3):302–9.

Wrolstad RE. *Color and Pigment Analyses in Fruit Products*: Oregon State University; 1976.

www.cen-online.org. June 9 (2008) 32.

www.cen-online.org. January 21 (2008) 33.

www.cen-online.org. March 24 (2008) 28.

www.cen-online.org. December 15 (2008) 18.

Xu C, Zhang Y, Cao L, Lu J. Phenolic compounds and antioxidant properties of different grape cultivars grown in China. *Food Chem.* 2010; 119(4):1557–65.

Zamora-Ros R, Andrés-Lacueva C, Lamuela-Raventós RM, Berenguer T, Jakszyn P, Martínez C, et al. Concentrations of resveratrol and derivatives in foods and estimation of dietary intake in a Spanish population: European Prospective Investigation into Cancer and Nutrition (EPIC)-Spain cohort. *Br J Nutr*. 2008; 100(1):188–96.

Zhao J, Wang J, Chen Y, Agarwal R. Anti-tumor-promoting activity of a polyphenolic fraction isolated from grape seeds in the mouse skin two-stage initiation–promotion protocol and identification of procyanidin B5-3′-gallate as the most effective antioxidant constituent. *Carcinogenesis*. 1999; 20(9):1737-45.

In: Grapes
Editor: José S. Câmara
ISBN: 978-1-63321-402-6

Chapter 8

Productivity of Stilbene Compounds by Wild Grapes Native to East Asia: *Vitis davidii*, *Vitis kiusiana*, *Vitis ficifolia* Var. *Lobata*, and *Vitis ficifolia* Var. *Ganebu*

Shuji Shiozaki[1*], Taiji Nakamura[1] and Tsuneo Ogata[2]
[1]Graduate School of Life & Environmental Sciences, Osaka Prefecture University, Sakai, Osaka, Japan
[2]Faculty of Agriculture, Kochi University, Otsu 200, Monobe, Nankoku, Kochi, Japan

Abstract

Productivity of stilbene compounds, such as resveratrol, piceatannol and piceid, were estimated in the leaves and berries of *Vitis davidii* native to China, *V. kiusiana*, *V. ficifolia* var. *lobata* and *V. ficifolia* var. *ganebu,* which are native to Japan, during 2005 and 2006. Ultraviolet-C (UV-C) irradiation at 420 μW· cm^{-1} for 15 min was used to stimulate stilbene production. Productivity of resveratrol in the leaf disc of all species was determined in 2005. Resveratrol level in the nonirradiated leaf discs (Ø 12 mm) was the highest in *V. ficifolia* var. *ganebu.* UV-C irradiation significantly increased the levels of resveratrol in each species except *V. davidii*; the most remarkable increase was observed in *V. ficifolia* var. *ganebu.* Productivity of resveratrol and piceattanol in the berries were compared between *V. ficifolia* var. *lobata* and *V. ficifolia* var. *ganebu* in 2005. Resveratrol levels in the nonirradiated berries differed little between *V. ficifolia* var. lobata and *V. ficifolia* var. *ganebu.* The resveratrol levels in immature berries of both varieties increased significantly 24 hr after UV-C irradiation. Resveratrol production in *V. ficifolia* var. *lobata* declined during berry development and maturation, whereas that of *V. ficifolia* var. *ganebu* declined until veraison before it increased to almost the same level as that during the most immature stage at harvest. Piceatannol levels in the nonirradiated berries in both varieties were less than 20% of resveratrol. The piceatannol

[*] Corresponding author: Email: ssgvo67@plant.osakafu-u.ac.jp.

levels in most immature berries after UV-C irradiation significantly increased in *V. ficifolia* var. *lobata* but not in *V. ficifolia* var. *ganebu.* UV-C irradiation did not affect the piceatannol levels at mature stage in both varieties. Productivity of resveratrol in the immature and mature berries was compared between *Vitis davidii*, *V. kiusiana*, *V. ficifolia* var. *lobata* and *V. ficifolia* var. *ganebu*, and piceid levels compared between *V. ficifolia* var. *lobata* and *V. ficifolia* var. *ganebu,* in 2006. No significant increase in resveratrol after UV-C irradiation was found in the berries of *V. davidii*, irrespective of the developmental stage of berries and time after UV-C irradiation. The resveratrol levels in the immature berries of *V. kiusiana*, *V. ficifolia* var. *lobata,* and *V. ficifolia* var. *ganebu* were significantly increased 24 hr after UV-C irradiation. The increased level of resveratrol was highest for *V. ficifolia* var. *ganebu* and lowest for *V. kiusiana*, and declined 48 hr after UV-C irradiation in *V. ficifolia* var. *lobata* and *V. ficifolia* var. *ganebu* but in *V. kiusiana.* The resveratrol levels were slightly increased by UV-C irradiation in the mature berries of *V. kiusiana* and *V. ficifolia* var. *lobata.* Increased resveratrol in the mature berries of *V. ficifolia* var. *ganebu* was also detected 48 hr after UV-C irradiation; the levels were higher in the mature berries than that in the immature berries. UV-C irradiation had no effect on the piceid level of either variety. *V. ficifolia* var. *ganebu* is a wild grape with a distinctive stilbene productivity, especially of resveratrol in the berry.

Keywords: Grape berry, leaf disc, piceatannol, piceid, resveratrol, wild grapes

Introduction

Three areas of geographical speciation are found in the grape genus (*Vitis*): European–West Asian species, North American species, and East Asian species, which comprise one (*Vitis sylvestris*), 28 (e.g., *V. rupestris, V. californica, V. labrusca*), and >40 species, respectively. *V. vinifera* was developed from domesticated *V. sylvestris* and is used worldwide as a wine grape. North American species and intraspecific hybrids are indispensable as phylloxera-resistant rootstocks, while some interspecific hybrids with *V. vinifera* are used as table grapes. East Asian species, such as *V. amurensis* Rupr., *V. davidii* (Roman.) Foëx, *V. coignetiae* Pulliat, *V. kiusiana* Momiyama, *V. flexuosa* Thunb., and *V. ficifolia* Bunge var. *lobata* (Regel) Nakai, have limited use compared with the European-West Asian species and North American species.

The wild grapes investigated in this chapter are found in the following habitats. *V. davidii* inhabits a region south of the Yellow River in China. The habitat of *V. kiusiana* is in the southern region of Kyushu, Japan. *V. ficifolia* var. *lobata* is endemic to low-altitude areas throughout Japan, except for the northern part of Hokkaido and Okinawa. *V. ficifolia* Bunge var. *ganebu* Hatusima is endemic to coastal areas of the southwest islands of Japan.

Grapes contain useful bioactive compounds such as flavonoids, including anthocyanins and nonflavonoids. The nonflavonoids found in grapes are stilbenes, including resveratrol (3,5,4′-trihydroxy stilbene), piceid (3-β-glucoside of resveratrol), and piceatannol (3′,4′,3,5-tetrahydroxy *trans*-stilbene) (Bavaresco and Fregoni 2001, Bavaresco et al. 2002, Cantos et al. 2003) (Figure 1). Resveratrol acts as a phytoalexin during the response to fungal infections in grapes. In addition to biotic stresses, resveratrol synthesis is highly stimulated by ultraviolet (UV) irradiation (Bavaresco and Fregoni 2001). Resveratrol, piceid, and piceatannol have antioxidant properties (Fauconneau et al. 1997, Fremont 2000, Su et al.

2013). A growing number of studies have demonstrated the beneficial health effects of resveratrol, including its antiatherogenic (Ramprasath and Jones 2010), anticancer (Aggarwal et al. 2004), antiplatelet aggregation (Wang et al. 2002), and anti-inflammatory effects (Fremont 2000). Both piceid and resveratrol can inhibit platelet aggregation and eicosanoid synthesis (Kimura et al. 1985, Shan et al. 1990). Piceid is effective on its own and after its hydrolysis by β-glucosidase in the intestine to produce resveratrol (Hackett 1986). Piceatannol has antileukemia activity (Mannila et al. 1993) and inhibitory effects on protein-tyrosine kinases, which are the targets of anticancer drugs (Geahlen and McLaughlin 1989). The important physiological and biological activities of resveratrol and its analogues have stimulated the development of new products containing resveratrol, such as food, nutraceuticals, and cosmetics, and also the consumer interest.

Resveratrol production in the leaves and berry skins in response to UV irradiation has been investigated in many white and red wine grape cultivars, including North American species (Bonomelli et al. 2004, Borie et al. 2004, Jeandet et al. 1991, Okuda and Yokotsuka 1996, Takayanagi et al. 2004). However, little is known about the resveratrol production levels in the leaves and berries of wild grapes native to East Asia. Thus, the aim of this chapter was to evaluate the resveratrol productivity of the wild grape varieties native to East Asia, i.e., *V. davidii*, *V. kiusiana*, *V. ficifolia* var. *lobata*, and *V. ficifolia* var. *ganebu*. We investigated their resveratrol, piceatannol, and piceid (Figure 1) levels and their potential for production in response to UV-C irradiation of their leaves and berries.

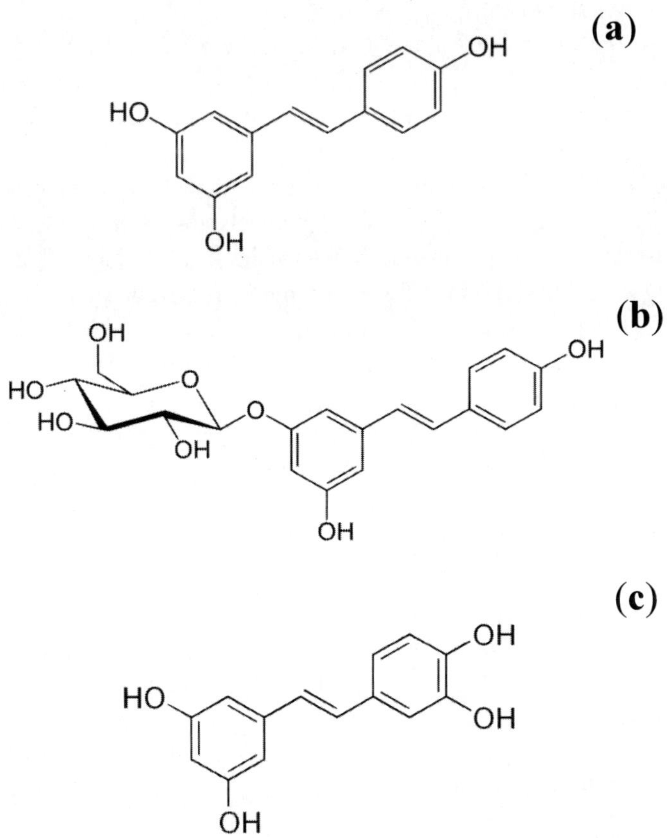

Figure 1. Chemical structures: (a) *trans*-resveratrol, (b) piceid, and (c) piceatannol.

Methods

Wild Grapes

The experiments were carried out in 2005 and 2006 with *Vitis davidii*, *V. kiusiana*, *V. ficifolia* var. *lobata*, and *V. ficifolia* var. *ganebu* grown in a research field at Osaka Prefecture University. Morphology of the leaves, canes, and berries of these grapes were observed.

Resveratrol Analysis with Leaf Discs

Resveratrol analysis of leaf discs of wild grapes was conducted in 2005. Leaves with no visual defects were sampled from the fifth to the tenth nodes, counting upward from the base of two-month-old shoots. The abaxial sides of the leaf discs (∅ 12 mm) were irradiated with UV-C light (260 nm) at a distance of 13 cm (420 $\mu W/cm^2$) for 15 min at 25°C. Samples were then incubated in the dark at 25°C for 24 h or 48 h. Resveratrol was extracted from a 0.5 g sample of leaf powder using aqueous methanol (MeOH). After purification by solvent partitioning and Sep-Pak C18-ENV column chromatography, the resveratrol was analyzed using a high performance liquid chromatography (HPLC) equipped with an UV detector (306 nm). Resveratrol was identified based on the retention time compared with an authentic standard.

Stilbene Analysis with Berries

Analysis of resveratrol and piceatannol in *V. ficifolia* var. *lobata* and *V. ficifolia* var. *ganebu* berries was conducted during berry development in 2005. Grape berries with pedicels were placed in a plastic container on wetted river sand, where the pedicel was inserted into the sand. The samples were irradiated with UV-C in the same manner as the leaves. The berries were sampled 24 h after UV-C irradiation. The stilbene analysis was conducted using the whole berries (ca 3 g) after excision of the pedicels for each species. Resveratrol and piceatannol were homogenized in aqueous MeOH using a bio-mixer. After purification by solvent partitioning and Sep-Pak C18-ENV column chromatography, the resveratrol and piceatannol were analyzed using a HPLC equipped with an UV detector (306 nm). Véraison of *V. ficifolia* var. *lobata* and *V. ficifolia* var. *ganebu* berries occurred about 50 and 60 days after full bloom, respectively.

The resveratrol levels in the berries of *V. davidii*, *V. kiusiana*, *V. ficifolia* var. *lobata*, and *V. ficifolia* var. *ganebu* were compared in UV-C-nonirradiated and-irradiated berries, which were sampled at the immature green stage (40 DAB) and the mature stage (80 or 120 DAB, which was before the full ripening stage, except in the case of *V. ficifolia* var. *lobata*) in 2006. The experimental methods were the same as those described above.

Comparison of the piceid levels was conducted in *V. ficifolia* var. *lobata*, and *V. ficifolia* var. *ganebu* berries in 2006. The piceid standard was prepared from a MeOH extract of *Polygonum cuspidatum* Sieb. et Zucc. (Kojo-kon) dried roots. The piceid extract was partially purified by C-18 reverse phase column chromatography and fractionated by reverse phase HPLC. Piceid was identified by comparing the retention time of a hydrolysate of the fractionated sample with that of authentic resveratrol, and by HPLC-mass spectrometry analysis.

After extracting resveratrol by solvent partitioning using ethyl acetate, the aqueous phase was used for piceid extraction. The aqueous sample was centrifuged and the supernatant, which contained piceid, was purified using a Sep-Pak C18-ENV cartridge. The cartridge was then washed with phosphate buffer (pH 7.0) and 0.1% HCl, before the residual aqueous solution was eliminated from the cartridge by suction. Piceid was then eluted using 16% acetonitrile containing 0.1% HCl.

Morphological Characteristics of Wild Grapes

Figures 2-4 shows the morphology of the leaves, canes, and berries of the wild grapes used in this chapter. The leaf blade of *V. davidii* is very broad and that of *V. ficifolia* var. *lobata* is the smallest (Figure 2).

The abaxial side of the leaf blades is pubescent, except in *V. davidii*. *V. kiusiana* and *V. davidii* have glandular hairs and aculei on their canes, respectively (Figure 3). The canes of *V. ficifolia* var. *lobata* and *V. ficifolia* var. *ganebu* are also pubescent. All of the berries accumulate anthocyanin in their skin at maturity (Figure 4). The mature berry of *V. kiusiana* is ca 11 mm in diameter, which is relatively large for the species of wild grapes native to Japan. The berry of *V. davidii* is larger (ca 16 mm) than that of *V. kiusiana*. *V. ficifolia* var. *ganebu* lacks endodormancy and can bear fruit continuously.

Figure 2. Leaf blades of *V. davidii*, *V. kiusiana*, *V. ficifolia* var. *lobata*, and *V. ficifolia* var. *ganebu*. Left, adaxial side. Right, abaxial side.

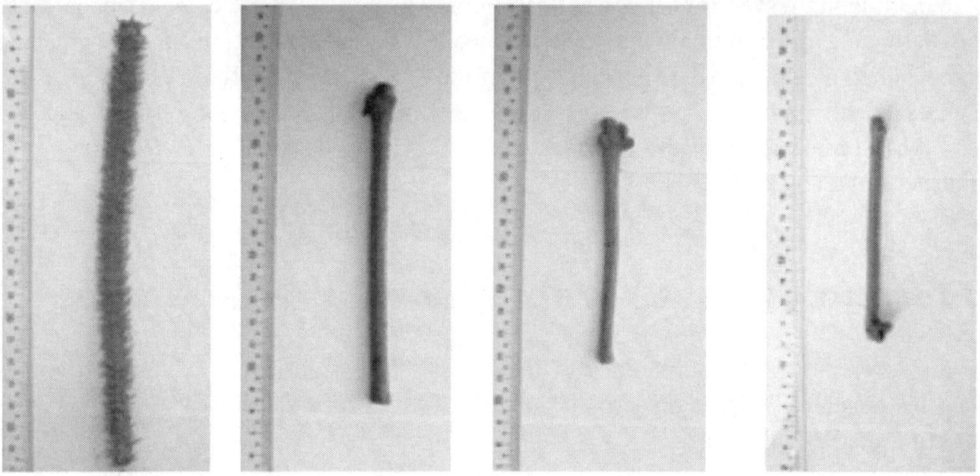

V. davidii ***V. kiusiana*** ***V. ficifolia* var. *lobata*** ***V. ficifolia* var. *ganebu***

Figure 3. Canes of *V. davidii, V. kiusiana, V. ficifolia* var. *lobata*, and *V. ficifolia* var. *ganebu*.

The berries of *V. ficifolia* var. *lobata* and *V. ficifolia* var. *ganebu* are very small (ca 8–10 mm). The berries of *V. kiusiana* have very little waxy bloom compared with the other three species. The berries of *V. ficifolia* var. *lobata* mature most rapidly, followed by *V. ficifolia* var. *ganebu*, *V. kiusiana*, and *V. davidii*, in that order.

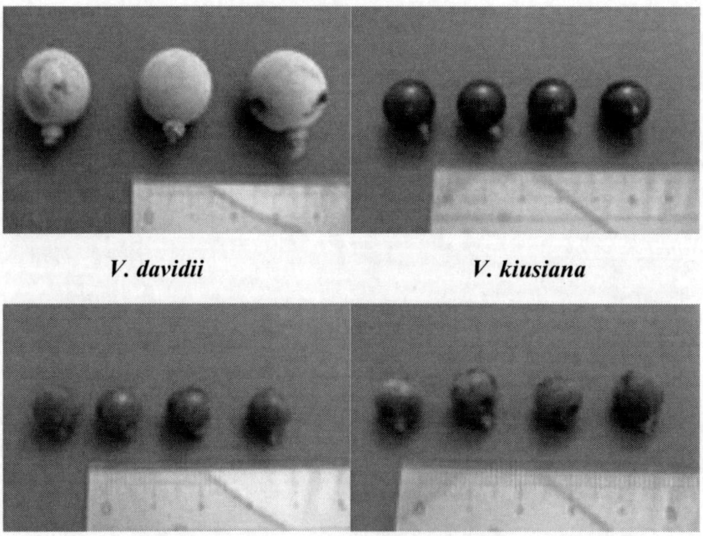

V. davidii ***V. kiusiana***

V. ficifolia* var. *lobata ***V. ficifolia* var. *ganebu***

Figure 4. Mature berries of *V. davidii, V. kiusiana, V. ficifolia* var. *lobata*, and *V. ficifolia* var. *ganebu*.

Resveratrol Accumulation in Leaf Discs after UV-C Irradiation

The *trans*-resveratrol content of the nonirradiated control leaf discs was highest in *V. ficifolia* var. *ganebu* and lowest in *V. ficifolia* var. *lobata* (Table 1). There was no significant difference in the resveratrol content of the nonirradiated control and the UV-C-irradiated sample in *V. davidii*. However, after UV-C irradiation, the *trans*-resveratrol contents of *V. kiusiana* and *V. ficifolia* var. *lobata* leaf discs were almost 11 times and 24 times that of the control, respectively, while that of *V. ficifolia* var. *ganebu* leaf discs was almost 29 times the control level. The *trans*-resveratrol content of UV-C-irradiated *V. ficifolia* var. *ganebu* leaf discs was 96.3 μg/gfw, which was 4.4 times greater than that of *V. ficifolia* var. *lobata*.

Pool et al. (1981) reported that only the abaxial side of grape leaves responded strongly to UV light, which resulted in a significant accumulation of resveratrol. *V. kiusiana, V. ficifolia* var. *lobata*, and *V. ficifolia* var. *ganebu* possess a pubescence on their abaxial surfaces of their leaves. Therefore, the UV-C responses of their leaves might be weaker compared with other *Vitis* species that lack a pubescence on the abaxial surfaces of their leaves. A strong UV-C response has been reported in the leaves of *V. rupestris*, the abaxial surface of which has no pubescence, where the *trans*-resveratrol content increased to ca 800 times that of the nonirradiated control at 24 h after UV-C irradiation (Bonomelli et al. 2004, Borie et al. 2004). However, there was no UV-C irradiation response in the leaf discs of *V. davidii*, which lacked a pubescence. Thus, the resveratrol productivity of *V. davidii* leaves is considered to be very low.

Changes in Resveratrol and Piceatannol Accumulation during Berry Development

The *trans*-resveratrol levels in the nonirradiated control of each variety increased slightly during berry development (Figure 5). During the immature stage, *trans*-resveratrol accumulation was induced significantly by UV-C irradiation in each variety. At 20 days after full bloom (DAB), there was no significant difference in the *trans*-resveratrol levels in response to UV-C in the two varieties. The accumulation of resveratrol in UV-C-irradiated *V. ficifolia* var. *lobata* berries declined as the berries developed, until it reached an insignificant level at harvest (80 DAB). By contrast, the accumulation of *trans*-resveratrol in UV-C-irradiated *V. ficifolia* var. *ganebu* berries followed a distinct pattern during berry development. The content decreased at 60 DAB (at véraison = beginning of berry maturation), before it increased and reached almost the same level as that found at 20 DAB when harvested (100 DAB). At harvest, the *trans*-resveratrol level of the UV-C-irradiated *V. ficifolia* var. *ganebu* berries was about 17 times higher than that of the control.

The piceatannol levels of *V. ficifolia* var. *lobata* and *V. ficifolia* var. *ganebu* berries were much lower than that of resveratrol, i.e., 2–5% of the resveratrol levels in the nonirradiated controls for each variety (Figure 6). Similar to resveratrol, the levels of piceatannol increased slightly in the nonirradiated control of each variety during berry development. The accumulation of piceatannol after UV-C irradiation occurred only in the most immature berries of *V. ficifolia* var. *lobata*.

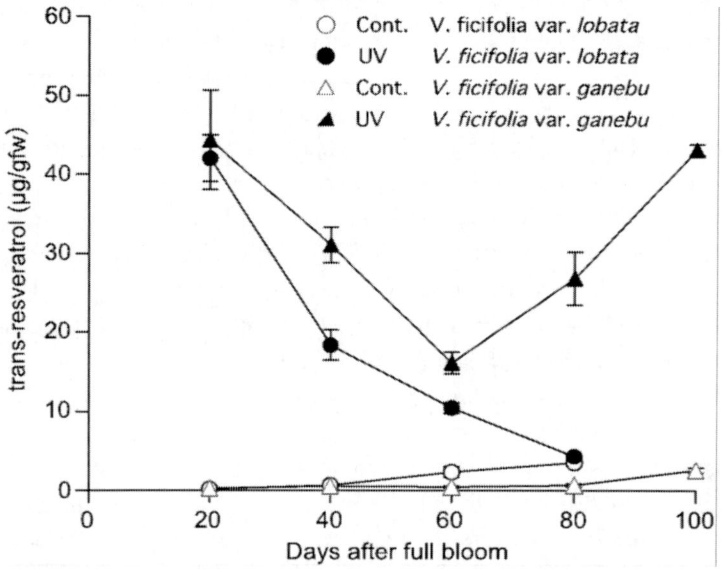

Figure 5. Changes in the *trans*-resveratrol levels in the UV-C-irradiated berries of *V. ficifolia* var. *lobata* and *V. ficifolia* var. *ganebu* throughout their development. Vertical bars indicate the standard error (SE) of the means (n = 3). When the data has no SE bar, the SE is smaller than the symbol. Shiozaki et al. 2013, Reprinted by permission of ASEV.

Table 1. *trans*-Resveratrol in the leaf discs of *V. davidii*, *V. kiusiana*, *V. ficifolia* var. *lobata*, and *V. ficifolia* var. *ganebu* 24 hr after UV-C irradiation

Wild grapes	Fw of leaf disc (mg)	*trans*-Resveratrol (µg/gfw) None		UV-C		Signif.[y]
V. davidii	23.1 ± 1.0	2.1 ± 0.02	ab[z]	10.3 ± 5.3	b	ns
V. kiusiana	18.7 ± 1.1	2.0 ± 0.2	ab	23.1 ± 0.7	b	**
V. ficifolia var. *lobata*	26.4 ± 3.6	0.9 ± 0.02	b	21.8 ± 2.2	b	*
V. ficifolia var. *ganebu*	30.1 ± 2.6	3.3 ± 0.2	a	96.3 ± 8.6	a	**

[z] Means with different letters in single rows are significantly different by Scheffé test 5%.
[y] ns, *, ** not significant, and significant at $P < 0.05$ and $P < 0.01$ by Student's *t*-test, respectively.

Resveratrol accumulation occurs in the skin and seeds of grape berries, but not in the flesh (Creasy and Coffee 1988, Bavaresco and Fregoni 2001). In general, the resveratrol productivity of grape berries was high at the immature stage and it declined during berry development and ripening, as reported by Jeandet et al. (1991). The change in resveratrol productivity in the whole berry of *V. ficifolia* var. *lobata* is similar to that found in other species in previous studies (Figure 5). However, the change in the resveratrol productivity of *V. ficifolia* var. *ganebu* during berry development and maturation was unique. The productivity decreased at véraison, before increasing during ripening until it reached a level equivalent to that found in the most immature stage (Figure 5). It has been hypothesized that the resveratrol synthesis pathway competes with the anthocyanin synthesis pathway for a common substrate (Jeandet et al. 1995). In contrast to other species or cultivars, resveratrol production in *V. ficifolia* var. *ganebu* berries might not be affected by anthocyanin synthesis in the skin.

Resveratrol production in the seeds should also be considered when evaluating the resveratrol productivity of the whole berry. The level of *trans*-resveratrol in grape seeds depends on the species or cultivar and ranges from 1 to 62 μg/g (Bavaresco and Fregoni 2001). No useful information is available on changes in the resveratrol levels in grape seeds during seed development. The differences in the resveratrol levels of *V. ficifolia* var. *lobata* and *V. ficifolia* var. *ganebu* did not appear to correlate with the number and size of seeds in the berries, because there were no significant differences in the seed numbers per berry or the average fresh seed weights of the two species (data not shown). In grapes, stilbenes are present in lignified organs, such as seeds, roots, and stems, where they may play a role in wood resistance to decay (Bavaresco and Fregoni 2001). The lignification of the grape seed testa occurs after the end of size enlargement, several days before véraison (Pratt 1971). Therefore, the high resveratrol level found only in the mature *V. ficifolia* var. *ganebu* berries was probably attributable to that produced in the skin rather than the seeds.

In a previous study, piceatannol was not detected in the mature berries of Cabernet Sauvignon, even when infected by *Plasmopara viticola* (van Zeller de Macedo Basto Gonçalves et al. 2011). Our data support the results reported for Cabernet Sauvignon, because the level of piceatannol was substantially lower than that of resveratrol and its accumulation did not appear to be stimulated by biotic or abiotic elicitors. In contrast to resveratrol, piceatannol might not be involved with disease resistance in grapes. Piceatannol is a naturally occurring molecule with one more hydroxyl group than resveratrol on C3′ of ring B (Figure 1). During stilbene metabolism, the resveratrol levels do not appear to be a rate limiting factor for piceatannol accumulation.

Resveratrol Accumulation in the Berries of Four Wild Grapes at 24 H and 48 H after UV-C Irradiation

Table 2 shows the levels of *trans*-resveratrol accumulation in the berries at 24 h and 48 h after UV-C irradiation for each variety. There were no significant differences in the *trans*-resveratrol levels of the controls for each variety at 24 h after UV-C irradiation on each sampling date. However, leaving the berries for 48 h reduced the levels of *trans*-resveratrol in the control berries of *V. ficifolia* var. *lobata* and *V. ficifolia* var. *ganebu*, irrespective of the berry development stage. By contrast, the resveratrol levels of *V. kiusiana* berries increased slightly after 48 h leaving compared with that after 24 h. No obvious changes were found in the resveratrol levels of *V. davidii* berries on each sampling date, even if after UV-C irradiation. The results for *V. davidii* berries agreed with the low resveratrol productivity of *V. davidii* leaves. Resveratrol accumulation in the berries was stimulated by UV-C irradiation in the immature green stage berries of *V. kiusiana, V. ficifolia* var. *lobata*, and *V. ficifolia* var. *ganebu.* The resveratrol level in the berries of *V. ficifolia* var. *ganebu* was the highest whereas that of *V. kiusiana* was the lowest at 24 h after UV-C irradiation. At the mature stage, however, resveratrol accumulation was detected in *V. kiusiana* and *V. ficifolia* var. *ganebu* at 24 h after UV-C irradiation. At 48 h after UV-C irradiation, there were significant increases in the resveratrol levels of *V. ficifolia* var. *lobata* and *V. ficifolia* var. *ganebu*, but not of *V. kiusiana*. The resveratrol level in *V. ficifolia* var. *ganebu* berries was about five times greater than that in *V. ficifolia* var. *lobata* berries. Interestingly, the resveratrol level at 48 h after UV-C irradiation was higher than that at 24 h after UV-C irradiation, whereas there was no increase in the level in the immature berries.

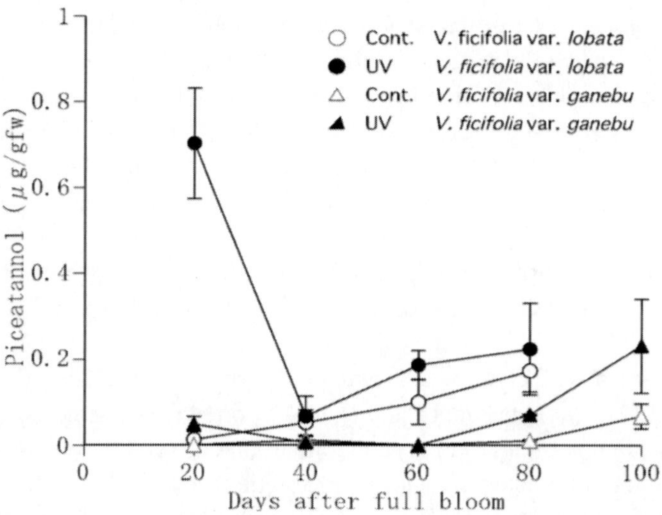

Figure 6. Changes in the piceatannol levels in the UV-C-irradiated berries of *V. ficifolia* var. *lobata* and *V. ficifolia* var. *ganebu* during development. Vertical bars indicate the standard error (SE) of the means (n = 3). When the data has no SE bar, the SE is smaller than the symbol.

Table 2. *trans*-Resveratrol accumulation in the berries of *V.davidii, V.Kiusiana, V. ficifolia* var. *lobata*, and *V. ficifolia* var. *ganebu* 24 hr and 48 hr after UV-C irradiation

		trans-Resveratrol (μg/gfw)					
		24 hr			48hr		
Wild grapes	DAB[z]	Cont	UV-C	Signif.[x]	Cont	UV-C	Signif.
V. davidii	40	0.04 ± 0.01 ns	9.5 ± 3.5 ab[y]	ns	0.04 ± 0.01 b	28.1 ± 12.1 ns	ns
V. kiusiana	40	0.20 ± 0.1	5.8 ± 1.0 b	*	0.70 ± 0.2 a	9.4 ± 1.9	*
V. ficifolia var. *lobata*	40	0.63 ± 0.03	31.3 ± 0.3 ab	**	0.45 ± 0.05 ab	27.3 ± 1.4	**
V. ficifolia var. *ganebu*	40	0.80 ± 0.3	43.8 ± 6.0 a	*	0.20 ± 0.1 ab	31.9 ± 2.0	**
V. davidii	120	0.03 ± 0.004 ns	1.6 ± 1.2 b	ns	0.01 ± 0.01 ns	1.2 ± 0.7 b	ns
V. kiusiana	80	1.10 ± 0.1	7.4 ± 0.6 b	**	3.30 ± 1.4	11.2 ± 2.3 b	ns
V. ficifolia var. *lobata*	80	3.70 ± 2.6	3.7 ± 0.5 b	ns	1.30 ± 0.5	9.4 ± 1.1 b	**
V. ficifolia var. *ganebu*	80	5.30 ± 2.6	23.8 ± 2.0 a	**	1.80 ± 0.3	49.2 ± 3.3 a	**

[z] days after full bloom.

[y] Means with different letters in single rows and same developmental stage are significantly by Scheffé test 5%.

[x] ns, *, ** not significant, and significant at $P < 0.05$ and $P < 0.01$ by student's *t*-test, respectively.

Bais et al. (2000) reported that the resveratrol levels only increased in the fully mature berry skins of specific cultivars of wine grapes (*V. vinifera*) given a sufficiently long delay after UV irradiation, although the levels were much lower than those in the immature grape skins.

Resveratrol productivity in the skins of fully mature table grapes was also reported in a series of studies by the CEBAS-CSIC research group (Spain), which aimed to produce grapes with enhanced health-promoting properties based on their high resveratrol content. Postharvest treatment of Napoleon grapes (*V. vinifera*) with UV-C light increased the resveratrol content of the skins by approximately threefold compared with the untreated control after refrigerated storage for 15 days (Cantos et al. 2000). In the same cultivar, Cantos et al. (2001) reported that the resveratrol content of the skins was up to 10-fold higher than that of the control after UV-C irradiation followed by storage at room temperature for 3 days. The resveratrol induction kinetics and the maximum content during storage at room temperature after UV-C irradiation depended on the grape cultivars, i.e., there was an early peak on the third day and a final peak on the seventh day, and the maximum levels in the skins were approximately 10–23 μg/gfw (Cantos et al. 2002). Jeandet et al. (1991) reported that the resveratrol level per unit of skins (fresh weight) was very low when expressed relative to the level per unit of berries (fresh weight). Therefore, the high resveratrol level in the whole berries after UV irradiation and the increase in this level up to 48 h after UV irradiation (by 49.2 μg/gfw in berries, which was an increase of over 27-fold compared with the nonirradiated control) suggests that mature *V. ficifolia* var. *ganebu* berries have a greater resveratrol production potential compared with other *Vitis* species/cultivars. The berries of *V. ficifolia* var. *ganebu* are unsuitable as a table grape because the berry is very small. However, the resveratrol content of the mature berries could be increased artificially and used to produce materials such as red wine or juice with high health-promoting effects, as well as various products with useful bioactivities for the nutraceutical and cosmetic industries. The optimal UV exposure conditions and requisite delay to obtain the maximum level of response to UV irradiation in fully mature *V. ficifolia* var. *ganebu* berries needs to be elucidated.

Piceid Accumulation in the Berries of v. Ficifolia var. Lobata and V. Ficifolia var. Ganebu at 24 h and 48 h after UV-C Irradiation

The piceid level (as *trans*-resveratrol equivalents) was one or two orders of magnitude lower than that of *trans*-resveratrol in the control (Table 3). UV-C irradiation did not induce piceid accumulation in *V. ficifolia* var. *lobata* berries at each developmental stage. At 40 DAB, no piceid was detected in the control or UV-C-irradiated *V. ficifolia* var. *ganebu* berries. However, piceid was detected in *V. ficifolia* var. *ganebu* berries at 80 DAB, although the increase in the piceid level in response to UV-C irradiation was negligible compared with the change in the *trans*-resveratrol level. Constantly low piceid levels have also been reported in grape leaves irradiated with UV-C (Douillet-Breuil et al. 1999). Thus, the turnover of piceid is probably not related to changes in the resveratrol levels in grapes, even after UV irradiation. Likewise, the resveratrol levels might not affect the piceid levels.

Table 3. Piceid levels in berries of V. ficifolia var. lobata and V. ficifolia var. ganebu 24 hr and 48 hr after UV-C irradiation. Shiozaki et al. 2013, Reprinted by permission of ASEV

		Piceid (*trans*-resveratrol equibvalent μg/gfw)					
		24 hr			48 hr		
Wild grapes	DAB[z]	Cont	UV-C	Signif.[x]	Cont	UV-C	Signif.
V. ficifolia var. *lobata*	40	0.02 ± 0.01	0.03 ± 0.01	ns	0.01 ± 0.006	0.02 ± 0.004	ns
V. ficifolia var. *ganebu*	40	nd[y]	nd	—	nd	nd	—
V. ficifolia var. *lobata*	80	0.17 ± 0.13	0.13 ± 0.04	ns	0.10 ± 0.05	0.04 ± 0.01	ns
V. ficifolia var. *ganebu*	80	0.07 ± 0.01	0.29 ± 0.18	ns	0.12 ± 0.03	0.37 ± 0.14	ns
Signif.		ns	ns		ns	ns	

[z] days after full bloom.
[y] not detected
[x] ns, not significant by Student's *t*-test.

Conclusion

V. kiusiana, *V. ficifolia* var. *lobata*, and *V. ficifolia* var. *ganebu*, which are wild grapes native to Japan, produced resveratrol in their leaves and berries in response to UV-C irradiation. By contrast, *V. davidii* grapes, which are native to China, had a very low resveratrol production potential. The piceatannol and piceid levels were much lower than that of resveratrol and their accumulation was stimulated little by UV-C irradiation. The highest accumulation of resveratrol after UV-C irradiation occurred in the leaves of *V. ficifolia* var. *ganebu*. A comparison of resveratrol accumulation in the UV-C-irradiated berries of *V. ficifolia* var. *lobata* and *V. ficifolia* var. *ganebu* showed that the resveratrol production potential of mature *V. ficifolia* var. *ganebu* berries was unique. In *V. ficifolia* var. *lobata*, the production potential decreased throughout development and maturation. By contrast, the production potential of *V. ficifolia* var. *ganebu* berries decreased at véraison, before increasing until maturation where the resveratrol production potential at maturation was equivalent to that during the immature green stage. Therefore, *V. ficifolia* var. *ganebu* may be a useful wild grape resource in terms of its resveratrol production potential, although the berries are very small.

References

Aggarwal, B. B., Bhardwaj, A., Aggarwal, R. S., Seeram, N. P., Shishodia, S. & Takada, Y. (2004). Role of resveratrol in prevention and therapy of cancer: preclinical and clinical studies. *Anticancer. Res.*, *24*, 2783-2840.

Bais, A. J., Murphy, P. J. & Dry, I. B. (2000). The molecular regulation of stilbene phytoalexin biosynthesis in *Vitis vinifera* during grape berry development. *Aust. J. Plant Physiol.*, *7*, 425-433.

Bavaresco, L. & Fregoni, B. (2001). Physiological role and molecular aspects of grapevine stilbenic compounds. In: Roubelakis-Angelakis K. editor. *Molecular Biology & Biotechnology of the Grapevine*, 153-182, Kluwer, Dordrecht.

Bavaresco, L., Fregoni, M., Trevisan, M., Mattivi, F., Vrhovsek, U. & Falchetti, R. (2002). The occurrence of the stilbene piceatannol in grapes. *Vitis.*, *41*, 133-136.

Bonomelli, A., Mercier, L., Franchel, J., Baillieul, F., Benizri, E. & Mauro. M. C. (2004). Response of grapevine defenses to UV-C exposure. *Amer. J. Enol. Vitic.*, *55*, 51-59.

Borie, B., Jeandet, P., Parize, A., Bessis, R. & Adrian, M. (2004). Resveratrol and stilbenesynthase mRNA production in grapevine leaves treated with biotic and abiotic phytoalexin elicitors. *Amer. J. Enol. Vitic.*, *55*, 60-64.

Cantos, E., Francisco, A., Baberan, T., Martinez, M. & Espin, J. C. (2003). Differential stilbene induction susceptibity of seven red wine grape varieties upon post-harvest UV-C irradiation. *Eur. J. Food Res. Tech.*, *217*, 253-258.

Cantos, E., Espin, J. C. & Tomas-Barberan, F. A. (2001). Postharvest induction modeling method using UV irradiation pulses for obtaining resveratrol-enriched table grapes: a new "functional" fruit? *J. Agr. Food Chem.*, *49*, 5052-5058.

Cantos, E., Espin, J. C. & Tomas-Barberan, F. A. (2002). Postharvest stilbene-enrichment of red and white table grape varieties using UV-C irradiation pulses. *J. Agr. Food Chem.*, *50*, 6322-6329.

Cantos, E., Garcia-Viguera, C., de Pascual-Teresa, S. & Tomas-Barberan, F. A. (2000). Effect of postharvest ultraviolet irradiation on resveratrol and other phenolics of cv. Napoleon table grapes. *J. Agr. Food Chem.*, *48*, 4606-4612.

Creasy, L. L. & Coffee, M. (1988). Phytoalexin production potential of grape berries. *J. Amer. Soc. Hort. Sci.*, 113, 230-234.

Douillet-Breuil, A-C., Jeandet, P., Adrian, M. & Bessis, R. (1999). Changes in the phytoalexin content of various *vitis* spp. in response to ultraviolet C elicitation. *J. Agri. and Food Chem.*, *47*, 4456-4461.

Fauonneau, B., Waffo-tegul, P., Huget, F., Barrier, L., Decendit, A. & Merillon, J-M. (1997). Comparative study of radical scavenger and antioxidant properties of phenolic compounds from *Vitis vinifera* cell cultures using *in vitro* tests. *Life Sci.*, *61*, 2103-2110.

Fremont, L. (2000). Biological effects of resveratrol. *Life Sci.*, *66*, 663-673.

Geahlen, R. L. & McLaughlin, J. L. (1989). Piceatannol (3,4,3',5'-tetrahydroxy-trans-stilbene) is a naturally occurring protein-tyrosine kinase inhibitor. *Biochem. Biophys. Res. Commun.*, *165*, 241-245.

Hacett, A. M. (1986). The metabolism of flavonoid compounds in mammals. In: Cody, V., Middleton, E.J., Harborne, J.B., editors. Plant Flavonoids in Biology and Medicine: Biochemical Pharmacological and Structure-Activity Relationships; *Prog. Clin. Biol. Res.*, *213*, 177-194, New York.

Jeandet, P., Bessis, R. & Gautheron, B. (1991). The production of resveratrol (3,5,4'-trihydroxystilbene) by grape berries in different developmental stages. *Am. J. Enol. Vitic.*, *42*, 41-46.

Jeandet, P., Bessis, R., Sbaghi, M. & Meunier, P. (1995). Production of the phytoalexin resveratrol by grapes as a response to *Botrytis* attack under natural conditions. *J. Phytopathol.*, *143*, 135-139.

Kimura, Y., Okuda, H. & Arichi, S. (1985). Effects of stilbene on arachidonate metabolism in leukocytes. *Biochim. Biophys. Acta.*, *834*, 275-278.

Mannila, E., Talvitie, A. & Kolehmainen, E. (1993). Anti-leukaemic compunds derived from stilbenes in *Picea abies* bark. *Phytochem.*, *33*, 813-816.

Okuda, T. & Yokotsuka, K. (1996). *Trans*-resveratrol concentrations in berry skins and wines from grapes grown in Japan. *Amer. J. Enol. Vitic.*, *47*, 93-99.

Pool. R. M., Creasy, L. L. & Frackelton, A. S. (1981). Resveratrol and the viniferins, their application to screening for disease resistance in grape breeding programs. *Vitis.*, *20*, 136-145.

Pratt, C. (1971). Reproductive anatomy in cultivated grapes – a review. *Amer. J. Enol. Vitic.*, *22*, 92-109.

Ramprasath, V. R. & Jones, P. J. H. (2010). Anti-atherogenic effects of resveratrol. *Eur. J. Clin. Nutr.*, *64*, 660-668.

Shan, C. W., Yang, S. Q., He, H. D., Shao, S. L. & Zhang, P. W. (1990). Influence of 3,4,5-trihydroxystilbene-3-*β*-mono-*D*-glucoside on rabbits' platelet aggregation and thromboxane B_2 production *in vitro*. *Acta Pharmacol. Sin.*, *11*, 527-530.

Shiozaki, S., Nakamura, T. & Ogata, T. (2013). Resveratrol productivity of wild grapes native to Japan: *Vitis ficifolia* var. *lobata* and *V. ficifolia* var. *ganebu*. *Amer. J. Enol. Vitic.*, *64*, 163-168.

Su, D., Cheng, Y., Liu, M., Liu, D., Cui, H., Zhang, B., Zhou, S., Yang, T. & Mei, Q. (2013). Comparison of piceid and resveratrol in antioxidation and antiproliferation activities *in vitro*. *PLOS ONE*, *8*, doi: 10.1371/journal.pone.0054505

Takyanagi, T., Okuda, T., Mine, Y. & Yokotsuka. K. (2004). Induction of resveratrol biosynthesis in skins of three grape cultivars by ultraviolet irradiation. *J. Jap. Soc. Hort. Sci.*, *73*, 193-199.

vanZeller de Macedo Basto Gonçalves, M. I., Bavaresco, L., Civardi, S. & Ferrari, F. (2011). Interactions between Plasmopara viticola infection and stilbene synthesis in leaves and berries of ten 'Cabernet Sauvignon' clones. *Vitis.*, *50*, 119-122.

Wang, Z., Zou, J., Huang, Y., Cao, K., Xu, Y. & Wu, J. M. (2002). Effect of resveratrol on platelet aggregation *in vivo* and *in vitro*. *Chin. Med. J.*, *115*, 378-380.

In: Grapes
Editor: José S. Câmara
ISBN: 978-1-63321-402-6

Chapter 9

Obtaining Polyphenolic Extracts from Wine By-Products

G. Davidov-Pardo, M. Navarro, I. Arozarena and M. R. Marín-Arroyo *
Public University of Navarre. Food Technology Department.
Campus Arrosadia s/n. Pamplona, Spain

Abstract

Among wine by-products, marcs represent 20 % of the total of 69 million tons of grapes intended for the production of wine a year, generating about 13-14 million tons of marcs, which unless used to make spirits, they are considered waste. Grapes have shown to be a rich source of polyphenols. Within the different wine waste, skins are rich in: anthocyanins, flavonols mainly 3-glucosides and 3-glucuronides of myricetin and quercetin, stilbenes mainly *trans*-resveratrol; hydroxycynamic acids mainly caftaric acid and condensed tannins. On the other hand, grape seeds are the part of the wine by-products that has the greatest amount of total phenolic compounds, being mainly flavan-3-ols (catechins and their polymers proanthocyanidins). It is known that polyphenols have a beneficial effect on human health and desirable technical properties such as antioxidant and anti-microbial activity. To use the polyphenols contained in wine by-products they must be extracted. Conventional batch solid-liquid extraction is the most commonly method used. There are a number of factors that can affect the performance and yield of the extraction and the relative phenolic composition and functional properties (i.e. antioxidant power) of the final extract. These factors have been extensively studied in a variety of scientific and technological papers: (a) pre-treatments of the raw material, (b) the choice of the solvent or the combination of solvents; (c) the solvent-to-solid ratio; (d) the pH of the solvent; (e) the particle size; (g) the temperature and (f) time of extraction; and (h) the post-extraction treatments of the extracts. Besides the traditional solid-liquid extraction method, there are novel techniques that can improve the yield and quality of the polyphenolic extracts. These novel techniques include: microwave and ultrasound-assisted extraction, enzyme-assisted extraction, pressurized-

* Corresponding author: Tel: +34 948 169108. Fax:+34 948 169893, E-mail: remedios.marin@unavarra.es.

liquid, supercritical fluid and subcritical water extractions and pulsed electric fields, high voltage electrical discharges and pulsed ohmic heating assisted extraction. This chapter will describe the phenolic profiles wine by-products and will focus on the main techniques and parameters that concern their extraction, and the subsequent separation in different fractions for different uses.

Keywords: Wine by-products, flavanols, anthocyanins, antioxidant capacity, conventional extraction, novel extraction techniques

Introduction

Among wine by-products, marcs represent 20 % of the total of 69 million tons of grapes intended for the production of wine a year, generating about 13-14 million tons of marcs, which unless used to make spirits, are considered waste. The high amount of wine by-products, represents a considerable pollution from the chemical demand of oxygen (COD) point of view, as well as, the biochemical demand of oxygen (BOD), due to their high content of sugars, tannins, polyphenols, polyhydric alcohols, pectins and lipid present in the marcs (Lafka et al., 2007). Using such by-products and giving them an added value, may represent a business opportunity and a decrease in pollution to soil and water in the wine regions of the world.

Marcs have shown to be a rich source of polyphenols and it is known that polyphenols have a beneficial effect on human health. The beneficial effects of polyphenols could be due to their free radical scavenger properties, blocking the deleterious action of these molecules on cells. In certain studies, it has been observed that during the process of aging, free radicals cause degenerative changes in the immune system, leading to the formation of cataracts, atherosclerotic plates, arthritis or even Parkinson's disease (Murcia et al., 2003). Flavonoids can be absorbed by cell membranes and protect them from the action of free radicals (Martínez-Flóres et al., 2002).

Considering that polyphenols are versatile bioactives rather than mere antioxidants, polyphenols can benefit human health in other metabolic ways (Hollman et al., 2011; Sies, 2010). They can inhibit certain enzymes that catalyze the release of histamine, which is responsible for causing inflammation and allergies (Shi et al., 2003). Another explanation of why polyphenols can function as anti-inflammatory agents is by their inhibitory action on the kappa B nuclear transcription factor, activator of many pro-inflammatory cytokines. They would also exert their anti-inflammatory action by inhibiting the enzymatic activities of the metabolism of the arachidonic acid by the 5-lipoxygenase ways (Martínez-Flóres et al., 2002).

Polyphenols inhibit the progression of atherosclerosis. This inhibition may be due to two main reasons. The first reason is by preventing oxidation of the lipoproteins of low-density in the arterial wall (Yamakoshi et al., 1999). The second reason could be because flavonoids and proanthocyanidins create a pseudo force of friction on the blood vessels, thus regulating the production of nitric oxide and prostacyclin that promote vessel dilation and suppress the production of endothelin-1 a substance vessel constrictor (Kar et al., 2006). The production of nitric oxide from nitrate in the stomach also induces a smooth muscle relaxation (Sies, 2010).

Proanthocyanidins can be used as potential drugs in the treatment of patients with diabetes mellitus that are not insulin dependent. Proanthocyanidins increase glucose intake and inhibit lipogenesis (Pereira et al., 2009). On the same path, a reduction of body weight gain as well as decrease in glycemia was evidenced after feeding rats with grape seed extracts in a dose of polyphenols adjusted to a human dealcoholised wine consumption of about two glasses of wine per meal (Décordé et al., 2009). In vitro and in vivo studies demonstrated beneficial effects of resveratrol in obesity managment. The results of the in vitro studies, indicate that resveratrol exerts the following effects: a decrease in lipid synthesis in adipocytes by PPAR-γ suppression, an increase in lipolysis, and a reduction in lipid accumulation in maturing preadipocytes. In primates, resveratrol reduced the body weight gain by increasing satiety and resting metabolic rate. In mice fed with a high-fat diet, resveratrol decreased food intake and down-regulated adipogenic and inflammatory processes (Alves, 2012).

Phenolic compounds including flavan-3-ols and proanthocyanidins, can possible inhibit the development and progression of cancer by intervening in various forms of intracellular signals that are responsible for the development and progression of cancer (Valenzuela, 2004).

Along with their nutraceutical properties, polyphenolic grape extracts (PGE) can be used as additives in the food industry. Several studies have been done to evaluate the possibility of using PGE to substitute synthetic antioxidants and antimicrobials. Grape seed extracts and PGE have been added to different food products such as fish, fresh vegetables, dairy products, meat, potato chips and cookies to reduce rancidity and microbial spoilage during processing and storage with positive results. Resveratrol has also been identified as an effective antibacterial against Gram-positive pathogens due to its bacteriostatic action (Yu and Ahmedna, 2013).

Wine By-Products Polyphenolic Profile

Stems are the first by-product generated during wine processing, representing the 3-7 % of the weight of the grape. Although stems appears to be very rich in polyphenols, not only in monomeric and oligomeric flavanols, but also in flavonols (Chira et al., 2008; Karvela et al., 2009), much more attention has been given to the recovery of polyphenols from the marcs or pomaces. These are the mixtures of grape seeds and peels that are produced in different moments depending of the type of vinification system, representing around 20 % of the grape weight. For white wines, the pomace is the result of the pressing of grapes, which always occurs prior to fermentation. Therefore, it may be assumed that the phenolic content and composition of the white-vinification by-products will be very similar to that observed in raw grapes. In the case of the red wine processing, the marc is obtained once the alcoholic fermentation has finished, when the solids are separated from the liquid and subsequently subject to pressing. Therefore, the red marcs will have a lower phenolic content, and a different relative composition than the raw grape. This is particularly true for the peels fraction, which shows a great extractability during fermentation in comparison to the seeds.

Seeds are the wine by-product that has the greatest amount of total phenolic compounds being mainly the monomeric flavan-3-ols (+)-catechin and (-)-epicatechin and their oligomers and polymers (condensed tannins or proanthocyanidins) (Chira et al., 2008; Makris et al., 2007; Negro et al., 2003).

Table 1 summarizes the phenolic composition and the antioxidant capacity of five commercial grape seed extracts (GSE) provided by five companies from different locations: China, France, Spain and USA (Davidov-Pardo et al., 2011). On average, the monomeric flavan-3-ols (+)-catechin and (-)-epicatechin jointly accounted the 68.5 % of the total individual compounds quantified by HPLC, followed by the sum of the dimers procyanidin B1 and B2 (21.5 %). The remaining compounds were the free gallic acid (1.2 %) and several esters between the gallic acid and the monomeric flavanols (8.8 %).

Table 1. Summary of the phenolic composition and the antioxidant capacity of five commercial grape seed extracts (GSE)

Compound	Mean (mg/g dw)	Min-Max (mg/g dw GSE)	RSD (%)
(+)-catechin (mg/g dw)	49.00	24.1-76.0	40
(-)-epicatechin (mg/g dw)	41.65	28.70-57.19	31
ProcyanidinB2 (mg/g dw)	16.00	11.17-19.41	23
Procyanidin B1 (mg/g dw)	12.52	6.00-16.87	34
(-)-gallo-catechin (mg/g dw)	4.53	nd-6.54	59
(-)-epigallocatechin (mg/g dw)	2.97	nd-6.67	58
(-)-epigallo-catechin-gallate (mg/g dw)	2.33	nd-4.58	72
(-)-epicatechin-gallate (mg/g dw)	1.23	nd-3.70	141
Gallic acid (mg/g dw)	1.60	0.69-4.60	106
Sum of individual compounds (mg/g dw)	132.39	79.35-184.68	32
Total phenolic content (mg GAE/g dw)	870	801-926	7
Antioxidant capacity (mmol trolox/gdw)	5.94	5.67-6.16	3

All the variables are expressed per gram of GSE dry weight. GAE: gallic acid equivalents. Source: Davidov-Pardo et al. (2011).

The total amount of the individual compounds in the five GSE varied from 79.35 to 184.68 mg/g GSE (dry weight). These values are far from the total phenolic contents obtained through the Folin-Ciocalteu method, which varied from 800 to 926 mg of gallic acid equivalents per gram of GSE. This suggests that a great part of the polyphenols in these commercial GSE corresponded to polymeric compounds that could not be individualized and quantified through the HPLC method applied. In addition, it must be noted that the relative standar deviation observed for the total phenolic content and the antioxidant capacity were much lower than those shown for the individual phenolics, which suggests that these polymeric polyphenols could play an important role with regard to the antiradical power of the extracts.

Peels from marcs have greater moisture and lower phenolic burden compared to seeds (Makris et al., 2007). However, peels from red marcs are a very interesting source of anthocyanins. They are present mainly in their monomeric form, as 3-monoglucosides of anthocyanidins, both acylated or not with acetic acid and several hydroxycynamic acids. During vinification the monomeric anthocyanins interact with different wine components in a wide type of chemical and bio-chemical reactions. So, the resultant anthocyanin-derived pigments will be also present in the marcs. Peels, those from red marcs in particular, also contain appreciable amounts of monomeric flavan-3-ols and proanthocyanidins. Besides anthocyanins and flavan-3-ols, lower quantities of other phenolic compounds are part of the phenolic composition of red and white peels. These include flavonols, mainly 3-glucosides and 3-glucuronides of myricetin and quercetin; stilbenes, mainly *trans*-resveratrol; and hydroxycynamic acids mainly caftaric acid (Castillo-Muñoz et al., 2007; Chira et al., 2008).

Main Parameters in the Conventional Solid-Liquid Extraction of Polyphenols

To be able to use the polyphenols contained in the grape by-products they must be extracted. Extraction is a delicate process and there is no single and standard extraction method. Conventional batch solid-liquid extraction is the most commonly method used. There are a number of factors that can affect the performance and yield of the extraction and the relative phenolic composition and functional properties (e.g. antioxidant power) of the final extract. These factors have been extensively studied in a variety of scientific and technological papers. Table 2 summarizes the conditions found in some conventional extractions of polyphenols from grape vinification by-products. The most relevant are: pre-treatment of the by-product, type of solvent/s, solvent-to-solid ratio, pH of the solvent, particle size of the solids, temperature, extraction time and post-extraction treatments.

(a) Pre-treatment of the by-product. Different operations can be applied to obtain a more stable and homogeneous raw material in order to assure a more reliable performance of the extraction.

The addition of sulphure dioxide to the grape pomace is very usual in order to prevent microbiological grow and enzymatic/chemical oxidation of polyphenols. Most of the times peels and seeds are separated prior to extraction, although sometimes all the marc is submitted to extraction as a whole.

Drying is common, in particular with regard to the peels, due to their high moisture content. Freeze-drying is sometimes applied at laboratory-scale extractions, but less expensive alternatives, such as air and sun-drying, are chosen for commercial-scale processes. Larrauri et al. (1997) reported that high air temperatures (100-140 ºC) affected negatively to the extractable polyphenols, antioxidant capacity and color of red pomace peels, while the peels air-dried at 60 ºC have shown very similar characteristics to those of the freeze-dried peels.

The seeds appear to be more resistant to heating than peels. Negative effects have been shown at very high temperatures, at 180 ºC for grape seed flours (Ross et al., 2011), while significant increments of the total phenolic content and radical scavenging activity have been observed on the extracts obtained from seeds previously heated 10 minutes at 100 ºC and 40 minutes at 150 ºC, in powder and whole form, respectively (Kim et al., 2006). In coherence with these findings, extracts from red and white grape wastes that were subjected to distillation, reaching temperatures of 100-130 ºC, shown higher antiradical power than those from grape wastes that came directly from pressing (Pinelo et al., 2005).

Cutting/crushing/milling followed by sieving are key operations to reduce and homogenize the particle size of the solid material. Finally, the seeds contain an appreciable amount of lipids, so defatting is sometimes applied in order to obtain a more pure polyphenolic extract.

Table 2. Description of several conventional liquid-solid methods for the extraction of polyphenols from grape vinification by-products

Raw material	Solvents	Solvent-to-solid ratio (ml/g)	Conditions	Total phenol content (g GAE/100 g dw)	Reference
WVBP (peels, seeds, stems)	Ethanol:water 28.5, 57 and 85.5 % (v/v) with/without 0.1 % HCl	5.5:1	Stirring (700 rpm) for 15 min (x 3 stages) Room temperature	Peels: 1.91-4.29 Stems: 3.1-7.5 Seeds: 8.0-13.7	Makris et al. (2007)
RVBP (marc, peels, seeds)	Ethanol:water 80 % (v/v) with 0.5 % 0.1 N HCl	30:1	Crushing, homogenization and centrifugation. Room temperature	Marc: 4.19 Seeds: 8.58 Peels: 3.33	Negro et al. (2003)
WVBP (seeds)	Ethanol:water 50 % (v/v)	10:1 20:1 30:1 40:1	Four PS: >0.63 mm, 0.63-0.4 mm, 0.4-0.16 mm, <0.125 mm 20 ºC to 80 ºC 120 min	25 ºC: 1.47-5.26 80 ºC: 2.39-6.68	Bucic-Kojic et al. (2007)
RVBP (seeds)	Ethanol:water 50 to 96 % (v/v)	40:1	Mean PS = 0.56mm 25 ºC, 80 ºC Shaking (200 rpm) 200 min	(50 % etanol) 7.08-12.96	Bucić-Kojić et al. (2009)
RVBP (marc)	Water with acetone, ethyl-acetate or ethanol: 50 %, 70 %, 100 % (v/v)	20:1	pH = 5 20 ºC and 60 ºC 2 h	1.73-2.02	Vatai et al. (2009)
WVBP (peels)	Ethanol	10:1 20:1 30:1	Mean PS = 0.8 mm Magnetic stirring Room temperature 9, 19 and 29 h	0.16-0.32	Casazza et al. (2012)
White grape pomace after distillation	Water	Continuous extraction (24 ml/min) (Sample 360-400 g)	Three PS: <1 mm, 3–5 mm, entire pomace 50 ºC 180 min	4.2-5.5	Sánchez et al. (2009)

Total phenol content: expressed as % (g of gallic acid equivalents -GAE- per 100 g of the raw material - dry weight).
RVBP and WVBP: red and white grape vinification by-products, respectively.

(b) The choice of the solvent or the combination of solvents, affects significantly the performance of extraction, as well as, the antioxidant power of the extract. This is mainly due to differences in polarity of the different phenolic compounds. To extract the flavanols from grape seeds, ethanol and water are the solvents most commonly used for hygienic and abundance reasons respectively, also they are more compatible with alimentary uses (EEC, 1988). An optimum ethanol and water combination is around 50 % (v/v) (Bucic-Kojic et al., 2007; Bucić-Kojić et al., 2009; Makris et al., 2007; Vatai et al., 2009).

Due to the excellent solubility of anthocyanins in water, this is the solvent commonly used to extract the pigments from grape skins. It is not unusual to add SO_2 to the water at high concentrations (e.g. 1 g/L) with the aim of avoid oxidation and enhance the extraction kinetics.

On the other hand if flavonols or resveratrol are the compounds of interest the extraction solvents should be alcohols, due to the lack of solubility of these compounds in water (Das et al., 2008; Nissiotis and Tasioula-Margari, 2002).

(c) Solvent-to-solid ratio. The higher the solvent-to-solid ratio, greater the concentration gradient between the solid and the bulk of the liquid is. This gradient is the driving force that governs the mass transfer process, and determines the final yield of the extraction. On the other hand, the decrease in the ratio leads to a reduction of the solvent purchase cost, as well as of the energy costs associated with the elimination of the solvent, when a concentrated solution or a dry-extract are the final products to obtain. Therefore, the general goal must be to use the minimum amount of solvent to obtain as much extract as possible, in order to optimize the costs of extraction. A wide range of solid-to-liquid ratios has been used in batch extractions of polyphenols from grape marcs, peels and seeds (Table 2), being values of 10:1 to 40:1 (ml solvent/g solid) the most commonly used (Bucic-Kojic et al., 2007; Bucić-Kojić et al., 2009; Casazza et al., 2012; Negro et al., 2003; Ross et al., 2011; Vatai et al., 2009).

(d) The pH of the solvent has an influence on the performance of the extraction and the quality of the extract. Acidifying the medium increases the yield and the antioxidant power of the extract. Lower pHs facilitates the hydrolysis of lignocellulosic compounds present in the pomace, releasing low molecular weight phenols trapped in the cellular structures. It is also know that polyphenols are more stable in acidic than in alkaline conditions (Amendola et al., 2010; Karvela et al., 2009; Luque-Rodríguez et al., 2006; Makris et al., 2007).

(e) Particle size reduction significantly increases the yield of phenols. Smaller sized particle have greater extraction yields, because a decrease in particle size increases the surface contact between solid and extractant (Bucic-Kojic et al., 2007; Pinelo et al., 2005; Sánchez et al., 2009). Bucic-Kojic et al. (2007) also found that in the smallest particle size trials the dispersion of the results was larger. This could be because of possible formation of clusters of particles in a variable way, which generated the variability when the extractions were performed. When reported, a wide range of particle sizes are shown to be used for the extraction of polyphenols from grape vinification by-products (table 2). For seeds, whose size reduction is easier in comparison to that of the peels, values around 0.50 mm or even lower have been reported (Bucic-Kojic et al., 2007; Bucić-Kojić et al., 2009). For the extraction from peels or from the whole marcs greater particle size are shown: from 0.8 mm (Casazza et al., 2012) to 5-5.5 mm (Pinelo et al., 2005; Sánchez et al., 2009). At commercial-scale processes, extractions on the whole marc are also common.

(g) Generally speaking a higher temperature results in a higher extraction yield due to damages on the cell membranes and the hydrolysis of lignocellulosic compounds, which results in an increased release of polyphenols and an increment in the solubility and diffusion coefficient of the polyphenols (Bucic-Kojic et al., 2007; Bucić-Kojić et al., 2009; Pinelo et al., 2005; Pinelo et al., 2005; Vatai et al., 2009). On the other hand the stability of some polyphenols mainly anthocyanins is reduced with the increment of the temperature above certain limit and changes in the color of the extracts occur (Cacace and Mazza, 2003). When seeds or the whole marcs are used as raw material, best results appear to occur at temperatures between 50 and 80 °C (Amendola et al., 2010; Bucic-Kojic et al., 2007; Bucić-Kojić et al., 2009; Pinelo et al., 2005; Pinelo et al., 2005; Sánchez et al., 2009; Vatai et al., 2009), while lower temperatures could be recommended for the extractions of polyphenols from peels. From room temperature (Casazza et al., 2012; Larrauri et al., 1997; Negro et al., 2003) to around 30-35 °C (Cacace and Mazza, 2003). However, in the commercial extraction of anthocyanins from red marcs, hot water and/or vapour are commonly used as solvents, and higher temperatures are achieved.

(h) The extraction time must also be optimized in the same way that the solvent-to-solid ratio. Not necessarily prolonged extraction times will lead to a higher phenolic content due to the occurrence oxidative reactions. Generally authors concluded that the highest amount of phenols were obtained before the longest extraction time of each work: 12.35 min of a total of 18 min of extraction (Bucic-Kojic et al., 2007), 40 min of a total of 200 min of extraction (Vatai et al., 2009); and 19 h of a total of 29 h of extraction (Casazza et al., 2012).

Optimization of the extraction could be the most important step to create an efficient and reliable extraction method. The optimization of the extraction is usually made using factorial designs analyzed by respond surface methodology (Bucic-Kojic et al., 2007; Cacace and Mazza, 2003; Casazza et al., 2012). Another possibility is to use the Peleg equation (Eq. 1) to model the extraction kinetics. Extraction curves (concentration of total polyphenols vs. time) have similar shape with the sorption curves (moisture content vs. time) therefore these curves could be model by the same equation (Bucic-Kojic et al., 2007)

$$C(t) = C_0 + \frac{t}{K_1 + K_2 * t} \tag{1}$$

Where C(t) is the concentration of total polyphenols at time t, t the extraction time, C_0 the initial concentration of total polyphenols at time t = 0, K_1 is the rate constant of Peleg and K_2 is the capacity constant of Peleg.

(i) Post-extraction treatments. At commercial level, once the liquid extract is obtained, it is subjected to additional operations in order to assure a regular and stable final product. Generally, grape seed extracts are commercialized as a dry powder. Although freeze-drying is the best solution with regard to the protection and maintenance of the functional properties of the extracts, at commercial scale spray-drying is preferred due to its higher productivity and lower investment and performance costs. Besides as a dry powder, commercial grape peels extracts could also be commercialized in the form of concentrated solutions. Vacuum evaporation is the most common choice to obtain them, although membrane separation technology (ultrafiltration) can be a suitable alternative.

Cruz et al. (2007) heated two different freeze-dried red pomace extracts at 100, 150 and 140 ºC for 120 min. They observed that the extracts had a lower susceptibility towards volatilization and higher capacity to retain antioxidant activity in comparison to two common synthetic antioxidants (BHA and BHT).

Novel Techniques to obtain Polyphenolic Extracts from Wine By-Products

Beside the traditional solid-liquid extraction method, there are novel techniques that can improve the yield and quality of the polyphenolic extracts. These novel techniques include: microwave and ultrasound-assisted extraction, enzyme-assisted extraction, pressurized-liquid, supercritical fluid extractions, subcritical water extraction, pulsed electric fields, high voltage electrical discharges (HVED) and pulsed ohmic heating (POH) assisted extractions.

Microwave-Assisted Extraction

Microwave-assisted extraction (MAE) has been reported as a novel method for extraction of bioactive compounds from different plant materials including marcs. The selected solvent should be transparent to microwave rays in order to remain at ambient temperature. However, some heating of the medium is permissible depending on the thermal stability of the aimed bioactive compounds. A microwave transparent medium can be defined as a medium that does not possess a significant dielectric constant.

The microwave rays travel freely through the microwave-transparent extraction medium and are allowed to reach the inner glandular and vascular systems of the biological material. When the biological material with favorable dielectric properties (plant material along with solvent for extraction) is reached by the microwave field, the molecules try to align with the oscillating electromagnetic field either by distortion or distribution of electron cloud within the molecule or by physical rotation of the molecular dipoles, which leads to rapid heating inside the material. The temperature keeps rising until the internal pressure exceeds the capacity of expansion of the cells walls thus creating an explosion at a cellular level. The bioactive compounds located in the cells are then free and dissolve in the extraction medium. The solid material can be removed and the extract can then be processed (Krishnaswamy et al., 2013; Lapointe et al., 1991). The main advantage of MAE over conventional extraction techniques is the better extraction efficiency in a shorter time. Another advantage is a lower solvent consumption. Reducing the extraction time diminish the loss of flavonoids due to ionization, hydrolysis, and oxidation on heat based prolonged extractions (Krishnaswamy et al., 2013; Perez-Serradilla and Luque de Castro, 2011).

Using the response surface methodology, Krishnaswamy et al. (2013) studied the effect of microwave power (100-200 W), extraction time (2-6 min), and solvent concentration (30-60 % ethanol in water) on total phenols and antioxidant capacity of seed extracts. Calculated optimal conditions to maximized variables were: 6 min with 32.6 % ethanol at 121 W. They show that too high microwave power (300W) lead to a rapid and uncontrollable sample heating. In a similar work, Li et al. (2011) concluded that ethanol concentration, liquid-to-solid ratio and time were the most important factors with regard to the extraction of polyphenols from grape seeds, being the optimal conditions 47.2 % ethanol, 45.4:1 ml solvent per gram of solid, and 4.6 min. Temperature was less relevant within the experimental conditions (40 to 60 ºC). They also concluded that microwave power was important in relation to the temperature. Inspection of the applied power profile during extraction revealed that the power was strictly modulated to maintain a constant temperature in the reaction cell, in such way that the average power was always lower than 100 W.

Ultrasound-Assisted Extraction

Ultrasound-assisted extraction (UAE) is based on the use of energy derived from ultrasound (sound waves with frequencies above 20 kHz) that cause a physical phenomenon called acoustic cavitation. This phenomenon causes the formation, growth, and violent collapse of small bubbles or voids in liquids as a result of pressure fluctuation. Collapse of these bubbles causes shock wave that passes through the solvent, enhancing the mass transfer within the system (Lieu and Le, 2010). Acoustic cavitation also generates cell walls disruption, and thus the release of bioactive compounds located within the cells. UAE also causes a fine reduction of particles size, which results in high contact surface area between solid and liquid phase (Casazza et al., 2010). UAE is particularly attractive for its simplicity and low cost of equipment (Carrera et al., 2011).

In their work with grapes Carrera et al. (2011), noted that the higher the amplitude the higher the recovery of polyphenols to a certain limit because in most cases there were not significant differences between 100 % amplitude and 50 % amplitude. However, when comparing the recoveries between 100 % and 20 % amplitude the yield at 100 % was always significant higher. The same effect was found for the duty cycle, the higher the cycle the higher the recovery.

Ghafoor et al. (2009) used the response surface methodology to optimize the solvent composition, the temperature and the extraction time in the UAE extraction of polyphenols from grape seeds. They fixed the ultrasound working conditions at 40 kHz of frequency and at 250 W of power. With some slight variations depending of the final goal, maximum total phenolic content or maximum antioxidant activity, the optimal conditions were around 52-53% ethanol in water, 56-60 ºC and 29-31 min for the three experimental factors mentioned above. Li et al. (2011) found that when comparing MAE against UAE, the former one extracted almost double of the polyphenols than the last one and the extraction was faster.

In general the main advantage of the UAE, compared to the conventional extraction methods for polyphenols from a wide variety of plant materials, are the enhancement of the extraction yield, a lower solvent consumption and a shorter extraction time (Da Porto et al., 2013). On the other hand the main drawback of this technique is that some phenolic compounds could be compromised due to their pressure and temperature stability. During the ultrasound treatment, the collapse of cavitation bubbles generates localized hot-spots that do not last long but present extremely high local temperature and pressure (Ghafoor et al., 2009; Lieu and Le, 2010).

Enzyme-Assisted Extraction

The cell wall of grape berries forms a barrier to the diffusion of phenols. It is composed of cellulose, hemicelluloses, pectin, and lignin, arranged in a complex network. The enzyme assisted extraction (EAE) mechanism of phenolic compounds from wine by-products is based on the cell-wall degrading enzymes such as cellulases, hemicellulases, pectinases, and other enzymes able to hydrolyze the bonds in plant cell-wall polysaccharides. These enzymes can weaken or break down the cell wall, leaving the polyphenolic compounds found in the intracellular materials more exposed for extraction. The use of cell wall degrading enzymes can also release the polyphenolic compounds associated within the cell wall in grapes, mainly in the lignin which is the major phenolic constituent of cell walls and it is composed of hydrophobic polymers derived from *p*-coumaryl, coniferyl, and sinapyl alcohols. Lignin is deposited together with procyanidins and hydroxycinnamic acids mainly *p*-coumaric acid and ferulic acid (Gómez-García et al., 2012; Pinelo et al., 2005).

Accessibility of the substrate is one of the most important factors affecting enzymatic activity, therefore grinding the pomace components (grape skins and seeds) is crucial to achieve an enhancement in the phenolic extraction (Maier et al., 2008). Generally the combination of enzymes that have different substrates (eg. cellulases and pectinases) results in significantly greater extraction yields (Maier et al., 2008; Meyer et al., 1998). The presence of lignin strengthens the hydrogen bonds in the surrounding cellulose−hemicellulose network creating a "shield" for the enzymatic degradation. The use of pectinases promotes the polysaccharide hydrolisis in the lignin and also acts on the esterified ferulic acid, making the cellulose−hemicellulose substrates more accessible for the action of the cellulases (Meyer et al., 1998).

EAE could be presented, as an alternative to SO_2-based extraction, mainly on grape skins. The substitution of sulfite based extractions can avoid the pseudoallergic reactions caused by foods with added sulfites (Kammerer et al., 2005).

The total phenolics extraction yields and the antioxidant capacity of the EAE in grape pomace were significantly higher than the conventional solid-liquid extractions (Kammerer et al., 2005; Meyer et al., 1998; Pinelo et al., 2005). The main drawback of this technique could be the degradation of polyphenols during long enzyme treatment, mainly the hydrolysis of glycoside anthocyanins and/or flavonols in their less stable aglycones (Maier et al., 2008; Meyer et al., 1998).

Pressurized Liquid Extraction

Pressurized liquid extraction (PLE) refers to a rapid extraction (3−20 min) in a closed and inert environment, under high pressures (3.3−20.3 MPa) and usually high temperatures (40−200 °C). The fact that the extraction solvent is pressurized allows an increase on the extraction temperature, while the solvent remains in its liquid state. These conditions improve the solubility of the polyphenols and the release kinetics from their matrices. Thus, extraction solvents that are inefficient in extracting polyphenols at low temperatures could be more efficient at elevated temperatures (Ju and Howard, 2003). Pressure is also used to increase the contact between the extracting fluid and the sample and temperature breaks the analyte−matrix bonds (Palma et al., 2001).

PLE is an improved solid-liquid extraction in which a solid sample is packed into the extraction cell and is then extracted with a suitable solvent at elevated temperature and pressure conditions. The combination of the extraction temperature, pressure, and volume of solvent used, will determine the extraction efficiency (Lou et al., 1997).

Along with the increase in extraction yields due to the rise of temperature and pressure, another advantage of using PLE is that it can be performed protected from light and under an inert atmosphere. Degradations by oxidations or by light can be avoided; therefore this method is particularly recommended for extracting compounds that react readily with oxygen or UV light but are stable to temperature, as occurs with *trans*-resveratrol (Pineiro et al., 2006). The main drawback of this technique is the loss of individual phenolic compounds that are more sensitive to temperature, such as anthocyanins. In a study of PLE of phenolic compounds from red grape skins at temperatures ranging from 20 to 140 ºC, Ju and Howard (2003) found that the highest recovery of anthocyanins occurred at 60 °C, while the highest recovery of total phenolic occurred at 120 ºC. The increased recovery of total phenolics may be explained by the greater extraction of more heat stable procyanidins and phenolic acids from grape skin.

When comparing the extraction of phenolic compounds from grape seeds and skins through different techniques, Casazza et al. (2010) found that, PLE improved the final yields up to 59 % in seeds and 511 % in peels, with regard to the conventional solid-liquid extraction method. In addition, with PLE the total polyphenols, total flavonoids and *o*-diphenols from grape skins was at least double when compared to UAE and MAE.

Supercritical Fluid Extractions

Supercritical fluid extraction (SFE) is an extraction method, based on the use of solvents in conditions above their critical point. The solvent acquires a density like a liquid, a viscosity like a gas and diffusivity values higher than typical liquids in two orders of magnitude. Carbon dioxide (CO_2) is commonly used in SFE because it is non-toxic, inexpensive and volatile. It has moderate critical conditions, therefore the gentle temperature and pressure to make CO_2 a critical fluid do not cause thermal or chemical degradations of the bioactive substances. CO_2 can be used in the food industry because it is classified as GRAS (generally recognized as safe) by the United States Food and Drug Administration (US FDA) (Oliveira et al., 2013).

The main advantage of this technique are the gentle conditions used, as well as, the absence of air and light that guarantee the conservation of bioactivities and antioxidant power of extracted compounds because degradation processes are avoided (Palma et al., 1999). Other advantages over traditional convention al extraction methods are improved selectivity, automation and environmental safety as the solvent can easily be removed from the solutes by expansion to ambient pressure (Oliveira et al., 2013).

The main disadvantage of this technique is that CO_2 is a non-polar solvent, which limits the extraction of polar compounds such as polyphenols from wine by-products and limits its solvent power (Oliveira et al., 2013). Another drawback is the high cost of the extraction, which makes it suitable only for very high value active components.

To overcome the limits of extraction of polar compounds, polar co-solvents can be used to increase the overall solvent power of the fluid phase during SFE, using an organic modifier to increase the solubility of the compounds in the supercritical fluid (Palma et al., 1999). In their invention Blasco et al. (2004) found that if the initial material is macerated in a polar solvent, containing a supercritical fluid in a proportion of less than 50 %, the extraction kinetics is faster, and the extraction yield increases. In this case the bioactive compounds are extracted by the SCF and quickly solubilized in the polar solvent; therefore the SFE does not act as a solvent but as a transport vehicle between the vegetable matrix and the solvent.

On the other hand, Palma and Taylor (1999) concluded that grape polyphenols can be quantitatively extracted with a carbon dioxide-based SFE in approximately 30 min from an inert matrix. This conclusion was made after a systematic graphical analysis concerning the extraction of eight polyphenolic compounds found in grapes, the average recovery from spiked sand ranged between 92 % and 100 %.

Another possibility is to use SFE as a pre-treatment to remove the non-polar components of grape marcs, so the polar polyphenols become more accessible. The results showed, that the pre-treatment of grape marc improved the extraction of total phenols while the amount of extracted anthocyanins was not significantly influenced (Vatai et al., 2009).

Subcritical Water Extraction

When water is used as a solvent, at high pressure and temperature, but below its critical point (22 MPa and 374 ºC), the extraction it is referred to as superheated water or subcritical water extraction (SWE).

Garcia-Marino et al. (2006) used this technique to extract catechins and proanthocyanidins from grape seeds obtained from winery by-products. They performed several assays at around 10 MPa of pressure and at both single and sequential temperatures (50, 100 and 150 ºC). Total polyphenolic contents obtained at 150 ºC and in the sequential extractions were comparable and even greater than those obtained by conventional extraction with a methanol:water (75:25) solvent. The relative composition of the extracts was significantly different depending of the extraction conditions. The extract obtained from the single extraction a 150 ºC was outstanding rich in gallic acid, and also in flavanols dimers and trimers.

SWE of red grape pomace has been optimized by Aliakbarian et al. (2012). Experimental factors were temperature (100-140 ºC) and pressure (8-15 MPa). The pressure and mainly the temperature shown to be relevant factors in the recovery of phenolics from pomace. The optimal conditions were 140 ºC and 11.6 MPa. The total polyphenolic content and the antioxidant capacity of the extracts from SWE was 19 and 3 fold, respectively, higher in comparison to the extracts obtained through conventional water extractions at atmospheric pressure and 25 ºC. When ethanol was used for conventional extraction, the differences were less remarkable, although SWE was much more efficient, reducing the extraction time close to 90 %.

Pulsed Electric Fields-Assisted Extraction

The process of the pulsed electric field assisted extraction (PEF) is based on the biological tissue damage caused by the application of external electric fields. The electroporation of cell membranes can be achieved at moderate electric fields <10 kV/cm, low specific energies <10 kJ/kg and short time and short duration 10^{-4} to10^{-2} s (Puertolas et al., 2013). The sample is placed between two electrodes in a batch or continuous treatment chamber.

The pores created in cell membranes by the PEF treatment can be reversible or irreversible depending on the intensity and duration of PEF application. Electroporation is considered to be a treatment that permeates the cell membranes but keeps the cell walls relatively intact (Boussetta et al., 2012). The important factors determining the extraction efficiency of the bioactive compounds from food products are: the intensity of the applied electric field and the input power to the substances (Takaki et al., 2011).

The main advantage of PEF is that the extraction yields can be significantly increased by a non-thermal selective extraction without losing the bioactive properties of the compounds.

All of the studies in grapes have been done in grape skins for the extraction of anthocyanins. After 60 min of extraction applying PEF, Takaki et al. (2011) obtained total phenolic contents 50 % higher than the conventional solid-liquid extraction. The results showed that many anthocyanoplasts in cells were collapsed.

Studying the extraction of anthocyanins from grape skins, Corrales et al. (2008) found that after 60 min of extraction, the antioxidant activity of the extracts using PEF increased four-fold.

High Voltage Electrical Discharges-Assisted Extraction

High voltage electric discharges assisted extraction (HVED) is a technique, in which the sample is immerse in a dielectric liquid like tap water, inside a chamber with a high voltage needle electrode and a plated grounded electrode. The sample is exposed to pulsed shockwaves typically 40–60 kV/cm during 2–5 μs (Vorobiev and Lebovka, 2010). HVED is based on the electrical breakdown in water phenomenon. With the right intensity electrical field, a streamer propagation of electrons is formed, from the high voltage needle electrode to the grounded one. The electrical breakdown is accompanied by secondary phenomena such as high-amplitude pressure shock waves, bubbles cavitation, creation of liquid turbulence, etc. Air bubbles present in water are involved into this phenomenon and accelerate the process. These secondary phenomena cause particle fragmentation and cell structure damage, accelerating the extraction of bioactive compounds from the cytoplasm of cells (Boussetta et al., 2011).

The main disadvantage of this technique could be the generation of hydroxyl radicals during water photo-dissociation, atomic hydrogen and ozone, that may damage the extracted polyphenols (Gavahian et al., 2013).

In their work with grape pomace Boussetta et al. (2011) found that HVED can increase the polyphenols yield and improve the extraction kinetics, being the optimal parameters: 80 kJ/kg, an electrodes distance gap of 5 mm and a liquid-to-solid ratio of 5. The sample was immersed in 30 % ethanol in water at 60 °C during 30 min. In a later work Boussetta et al. (2012) found that after comparing PEF and HDVE as a pretreatment on grape seeds, both can enhance the extraction kinetics. One of the advantages of using PEF is that this pre-treatment preserves the product structure and the subsequent solid–liquid separation is faster while for ground and HVED-treated samples the structure is compromised.

Pulsed Ohmic Heating-Assisted Extraction

Pulse ohmic heating (POH) is a process in which an electrical current, usually alternating current, is used to generate heat within a product. The heating occurs when electric energy is transformed into thermal energy within the material. The generated heat is directly correlated with the current induced by the voltage gradient and the electrical conductivity of the sample. The increase in temperature leads to a faster and more efficient extraction. Electrical conductivity is the most important parameter to determine the heating rate in a POH. Ohmic heating only works in water continuous systems due to the necessity of ionic mobility. POH is a methodology that combines electrical and thermal treatments resulting in a more effective extraction method of bioactive compounds using moderate electric field and moderate temperatures. The main advantage of this technology is that provides rapid and uniform heating, with the absence of a hot surface reducing thermal damage to the bioactive compounds (Gavahian et al., 2013).

El Darra et al. (2013) found that the use of POH resulted in a 50 % increase of the polyphenols extraction yield from grape pomace in water and in water–ethanol solution. Both cell permeabilization and polyphenols extraction indices increased with the electric field strength (from 200 to 800 V/cm).

Separation of Polyphenolic Extracts in Different Fractions

It is sometimes desirable to separate the crude polyphenolic extracts in different fractions in order to eliminate the non-desirable compounds of the extracts and/or to isolate the different polyphenol families for a variety of purposes in the cosmetics, food and pharmaceutical industries. Gourdin et al. (2005) achieved to separate polyphenolic extracts in three steps using a brominated polystyrene column and solvents with different eluents. The first step consisted in eliminating the non-phenolic compounds such as sugar or organic acids. The second step eluted the non-polar phenolic compounds such as condensed tannins and finally the third step eluted the polar compounds like the anthocyanins.

Another reason to fractionate the polyphenolic extracts may be to eliminate non-desirable polyphenols such as polymeric condensed tannins (degree of polymerization higher than 8) to reduce the astringency of the extracts. A crude grape seed extract (GSE) is acidified, filtered using diatomaceous earth and neutralized with a base to create compact flocculated solids to eliminate non-phenolic compounds like proteins or other polysaccharides. The clear GSE is then adsorbed on a column containing an adsorbent selected from the group consisting of styrene divinyl benzene copolymer to eliminate polymeric condensed tannins which are eluted in the first rinses of the column. To separate the polymeric condensed tannins from grape pomace extract the process is the same, with the exemption that the crude extract is treated with pectolytic enzimes (Shrikhande et al., 2003).

Palma and Taylor (1999) used different SFE conditions to obtain a fractionated GSE. Using pure CO_2 it was possible to extract the low polar compounds that resulted in yellow oil composed mainly by fatty acids (fraction A). Then CO_2/methanol (5:1) was used to extract the more polar compounds, mainly phenolics, from the same sample (fraction B). In general fraction A exhibited a stronger and wider antibacterial activity against Gram positive and negative bacteria, as well, as against human pathogens.

Size exclusion chromatograpy was applied to fractionate a crude extract obtained by conventional extraction with ethanol:water (70:30) from a white grape pomace (Torres et al., 2002). Using methanol and acetone:water as mobile phases, several flavanols and flavonols fractions of different mean degree of polymerization and percentage galloyation were generated. The fractions rich in oligomeric flavanols with a degree of polymerization between 3 and 4 and galloyated to around 30 % shown the highest radical scavenging activity and were efficient antioxidants in oil-in-water emulsions. The flavonols fractions had a low antiradical power but were also efficient in the emulsion.

Conclusion

Extraction of polyphenolic compounds from wine by-products is a delicate process. Novel techniques of extraction may enhance the extraction yields, accelerate the extraction process and even protect some of the more sensible polyphenolic compounds. Never the less, the use of these novel techniques come with an additional cost that has to be consider if these technologies are to be scale at industrial levels to make them profitable. The parameters affecting the conventional solid-liquid extractions also affect the performance of the novel extraction techniques and have to be also considered when optimizing these processes. Finally, there is still a lot of work to be done in optimizing, up-scaling and designing separation methods for the extraction of specific polyphenols families or compounds from a wide variety of plant materials including grape by-products. An optimized and specific extraction-fraction methodology in an industrial scale could represent an interesting way to generate an added value to the wine by-products.

References

Aliakbarian, B., Fathi, A., Perego, P. & Dehghani, F. (2012). Extraction of antioxidants from winery wastes using subcritical water. *J. Supercrit. Fluids*, *65*, 18-24.

Alves, N. E. G. (2012). Studies on Mechanistic Role of Natural Bioactive Compounds in the Management of Obesity An Overview. *Open Nutraceuticals Journal*, *5*, 193.

Amendola, D., De Faveri, D. M. & Spigno, G. (2010). Grape marc phenolics: Extraction kinetics, quality and stability of extracts. *J. Food Eng.*, *97*, 384-92.

Blasco, P. M., Faus, F. G., Martinez, F. J. G., Mira, F. B. E. & Subirats, H. S. (2004). Spain Patent No. EP0917876 A1. Asociacion de Investigacion de la Industria Agroalimentaria: E.P. office.

Boussetta, N., Vorobiev, E., Deloison, V., Pochez, F., Falcimaigne-Cordin, A. & Lanoiselle, J. L. (2011). Valorisation of grape pomace by the extraction of phenolic antioxidants: Application of high voltage electrical discharges. *Food Chem.*, *128*, 364-70.

Boussetta, N., Vorobiev, E., Le, L. H., Cordin-Falcimaigne, A. & Lanoiselle, J. L. (2012). Application of electrical treatments in alcoholic solvent for polyphenols extraction from grape seeds. *Lwt-Food Sci. Technol.*, *46*, 127-34.

Bucic-Kojic, A., Planinic, M., Tomas, S., Bilic, M., Velic, D. (2007). Study of solid-liquid extraction kinetics of total polyphenols from grape seeds. *J. Food Eng.*, *81*, 236-42.

Bucić-Kojić, A., Planinić, M., Tomas, S., Jakobek, L. & Šeruga, M. (2009). Influence of solvent and temperature on extraction of phenolic compounds from grape seed, antioxidant activity and colour of extract. *Int. J. Food Sci. Technol.*, *44*, 2394-401.

Cacace, J. E. & Mazza, G. (2003). Mass transfer process during extraction of phenolic compounds from milled berries. *J. Food Eng.*, *59*, 379-89.

Carrera, C., Ruiz-Rodríguez, A., Palma, M. & Barroso, C. G. (2011). Ultrasound assisted extraction of phenolic compounds from grapes. *Anal. Chim. Acta*.

Casazza, A. A., Aliakbarian, B., De Faveri, D., Fiori, L. & Perego, P. (2012). Antioxidants from winemaking wastes: A study on extraction parameters using response surface methodology. *J. Food Biochem.*, *36*, 28-37.

Casazza, A. A., Aliakbarian, B., Mantegna, S., Cravotto, G. & Perego, P. (2010). Extraction of phenolics from *Vitis vinifera* wastes using non-conventional techniques. *J. Food Eng.*, *100*, 50-55.

Castillo-Muñoz, N., Gómez-Alonso, S., García-Romero, E. & Hermosín-Gutiérrez, I. (2007). Flavonol profiles of *Vitis vinifera* red grapes and their single-cultivar wines. *J. Agric. Food. Chem.*, *55*, 992-1002.

Chira, K., Suh, J., Saucier, C. & Teissèdre, P. (2008). Les polyphénols du raisin. *Phytothérapie*, *6*, 75-82.

Corrales, M., Toepfl, S., Butz, P., Knorr, D. & Tauscher, B. (2008). Extraction of anthocyanins from grape by-products assisted by ultrasonics, high hydrostatic pressure or pulsed electric fields: A comparison. *Innov. Food Sci. Emerg. Technol.*, *9*, 85-91.

Cruz, J. M., Conde, E., Dominguez, H. & Parajo, J. C. (2007). Thermal stability of antioxidants obtained from wood and industrial wastes. *Food Chem.*, *100*, 1059-64.

Da Porto, C., Porretto, E. & Decorti, D. (2013). Comparison of ultrasound-assisted extraction with conventional extraction methods of oil and polyphenols from grape (*Vitis vinifera* L.) seeds. *Ultrason. Sonochem.*, *20*, 1076-80.

Das, S., Lin, H. S., Ho, P. C. & Ng, K. Y. (2008). The impact of aqueous solubility and dose on the pharmacokinetic profiles of resveratrol. *Pharm. Res.*, *25*, 2593-600.

Davidov-Pardo, G., Arozarena, I. & Marín-Arroyo, M. R. (2011). Stability of polyphenolic extracts from grape seeds after thermal treatments. *Eur. Food Res. Technol.*, *232*, 211-20.

Davidov-Pardo, G., Arozarena, I. & Marín-Arroyo, M. R. (2013). Grape seed extract: additive and functional ingredient. *Agro FOOD Indusrty Hi Tech*, *24*, 41-43.

Décordé, K., Teissèdre, P.-L., Sutra, T., Ventura, E., Cristol, J.-P. & Rouanet, J.-M. (2009). Chardonnay grape seed procyanidin extract supplementation prevents high-fat diet-induced obesity in hamsters by improving adipokine imbalance and oxidative stress markers. *Mol. Nutr. Food Res.*, *53*, 659-66.

Council Directive 88/344/EEC of 13 June 1988 on the approximation of the laws of the Member States on extraction solvents used in the production of foodstuffs and food ingredients (1988).

El Darra, N., Grimi, N., Vorobiev, E., Louka, N. & Maroun, R. (2013). Extraction of Polyphenols from Red Grape Pomace Assisted by Pulsed Ohmic Heating. *Food Bioprocess Technol.*, *6*, 1281-89.

Garcia-Marino, M., Rivas-Gonzalo, J.C., Ibanez, E. & Garcia-Moreno, C. (2006). Recovery of catechins and proanthocyanidins from winery by-products using subcritical water extraction. *Anal. Chim. Acta*, *563*, 44-50.

Gavahian, M., Farahnaky, A., Javidnia, K. & Majzoobi, M. (2013). A novel technology for extraction of essential oil from Myrtus communis: ohmic-assisted hydrodistillation. *J. Essent. Oil Res.*, *25*, 257-66.

Ghafoor, K., Choi, Y. H., Jeon, J. Y. & Jo, I. H. (2009). Optimization of Ultrasound-Assisted Extraction of Phenolic Compounds, Antioxidants, and Anthocyanins from Grape (*Vitis vinifera*) Seeds. *J. Agric. Food. Chem.*, *57*, 4988-94.

Gómez-García, R., Martínez-Ávila, G. G. & Aguilar, C. (2012). Enzyme-assisted extraction of antioxidative phenolics from grape (*Vitis vinifera* L.) residues. *3 Biotech*, *2*, 297-300.

Gourdin, G. T., Richheimer, S. L., Tempesta, M. S., Bailey, D. T., Nichols, R. L., Daugherty, F. J. & Freeberg, D. R. (2005). United States Patent No. US7682637 B2. Phenolics, Llc: U.S. Patents.

Hollman, P. C. H., Cassidy, A., Comte, B., Heinonen, M., Richelle, M., Richling, E., Serafini, M., Scalbert, A., Sies, H. & Vidry, S. (2011). The biological relevance of direct antioxidant effects of polyphenols for cardiovascular health in humans is not established. *J. Nutr.*, *141*, 989S-1009S.

Ju, Z. Y. & Howard, L. R. (2003). Effects of solvent and temperature on pressurized liquid extraction of anthocyanins and total phenolics from dried red grape skin. *J. Agric. Food. Chem.*, *51*, 5207-13.

Kammerer, D., Claus, A., Schieber, A. & Carle, R. (2005). A novel process for the recovery of polyphenols from grape (*Vitis vinifera* L.) pomace. *J. Food Sci.*, *70*, C157-C63.

Kar, P., Laight, D., Shaw, K. M. & Cummings, M. H. (2006). Flavonoid-rich grapeseed extracts: A new approach in high cardiovascular risk patients? *Int. J. Clin. Pract.*, *60*, 1484-92.

Karvela, E., Makris, D. P., Kalogeropoulos, N. & Karathanos, V. T. (2009). Deployment of response surface methodology to optimise recovery of grape (*Vitis vinifera*) stem polyphenols. *Talanta*, *79*, 1311-21.

Kim, S., Jeong, S., Park, W., Nam, K. C., Ahn, D. U. & Lee, S. (2006). Effect of heating conditions of grape seeds on the antioxidant activity of grape seed extracts. *Food Chem.*, *97*, 472-79.

Krishnaswamy, K., Orsat, V., Gariépy, Y. & Thangavel, K. (2013). Optimization of Microwave-Assisted Extraction of Phenolic Antioxidants from Grape Seeds (*Vitis vinifera*). *Food Bioprocess Technol.*, *6*, 441-55.

Lafka, T. I., Sinanoglou, V. & Lazos, E. S. (2007). On the extraction and antioxidant activity of phenolic compounds from winery wastes. *Food Chem.*, *104*, 1206-14.

Lapointe, J., Pare, J. R. J. & Sigouin, M. (1991). United States of America Patent No.: H.M.T.Q.I.R.O.C.A.R.B.T.M.O. Environment.

Larrauri, J. A., Ruperez, P. & Saura-Calixto, F. (1997). Effect of drying temperature on the stability of polyphenols and antioxidant activity of red grape pomace peels. *J. Agric. Food. Chem.*, *45*, 1390-93.

Li, Y., Skouroumounis, G. K., Elsey, G. M. & Taylor, D. K. (2011). Microwave-assistance provides very rapid and efficient extraction of grape seed polyphenols. *Food Chem.*, *129*, 570-76.

Lieu, L. N. & Le, V. V. M. (2010). Application of ultrasound in grape mash treatment in juice processing. *Ultrason. Sonochem.*, *17*, 273-79.

Lou, X. W., Janssen, H. G. & Cramers, C. A. (1997). Parameters affecting the accelerated solvent extraction of polymeric samples. *Anal. Chem.*, *69*, 1598-603.

Luque-Rodríguez, J. M., Pérez-Juan, P. & Luque de Castro, M. D. (2006). Extraction of polyphenols from vine shoots of *Vitis vinifera* by superheated ethanol–water mixtures. *J. Agric. Food. Chem.*, *54*, 8775-81.

Maier, T., Goeppert, A., Kammerer, D. R., Schieber, A. & Carle, R. (2008). Optimization of a process for enzyme-assisted pigment extraction from grape (*Vitis vinifera* L.) pomace. *Eur. Food Res. Technol.*, *227*, 267-75.

Makris, D. P., Boskou, G. & Andrikopoulos, N. K. (2007). Recovery of antioxidant phenolics from white vinification solid by-products employing water/ethanol mixtures. *Bioresour. Technol.*, *98*, 2963-67.

Martínez-Flóres, S., González-Gallego, J., Culebras, J. M. & Tuñón, M. J. (2002). Los flavonoides: propiedades y acciones antioxidantes. *Nutrición Hospitalaria*, *17*, 271-78.

Meyer, A. S., Jepsen, S. M. & Sorensen, N. S. (1998). Enzymatic release of antioxidants for human low-density lipoprotein from grape pomace. *J. Agric. Food. Chem.*, *46*, 2439-46.

Murcia, M. A., Vera, A. M., Martínez-Tomé, M. & Frega, N. (2003). *Sustancias antioxidantes presentes en los alimentos. Acción, dosis y su eficacia en la promoción de la salud*. Madrid: Nueva Imprenta S.A.

Negro, C., Tommasi, L. & Miceli, A. (2003). Phenolic compounds and antioxidant activity from red grape marc extracts. *Bioresour. Technol.*, *87*, 41-44.

Nissiotis, M. & Tasioula-Margari, M. (2002). Changes in antioxidant concentration of virgin olive oil during thermal oxidation. *Food Chem.*, *77*, 371-76.

Oliveira, D. A., Salvador, A. A., Smania, A. Jr., Smania, E. F. A., Maraschin, M. & Ferreira, S. R. S. (2013). Antimicrobial activity and composition profile of grape (*Vitis vinifera*) pomace extracts obtained by supercritical fluids. *J. Biotechnol.*, *164*, 423-32.

Palma, M., Pineiro, Z. & Barroso, C. G. (2001). Stability of phenolic compounds during extraction with superheated solvents. *J. Chromatogr. A*, *921*, 169-74.

Palma, M. & Taylor, L. T. (1999). Statistical design for optimization of extraction of polyphenols from an inert matrix using carbon dioxide-based fluids. *Anal. Chim. Acta*, *391*, 321-29.

Palma, M., Taylor, L. T., Varela, R. M., Cutler, S. J. & Cutler, H. G. (1999). Fractional extraction of compounds from grape seeds by supercritical fluid extraction and analysis for antimicrobial and agrochemical activities. *J. Agric. Food. Chem.*, *47*, 5044-48.

Pereira, D. M., Valentão, P., Pereira, J. A. & Andrade, P. B. (2009). Phenolics: From chemistry to biology. *Mol.*, *14*, 2202-11.

Perez-Serradilla, J. A. & Luque de Castro, M. D. (2011). Microwave-assisted extraction of phenolic compounds from wine lees and spray-drying of the extract. *Food Chem.*, *124*, 1652-59.

Pineiro, Z., Palma, M. & Barroso, C. G. (2006). Determination of trans-resveratrol in grapes by pressurised liquid extraction and fast high-performance liquid chromatography. *J. Chromatogr. A*, *1110*, 61-65.

Pinelo, M., Fabbro, P. D., Manzocco, L., Nuñez, M. J. & Nicoli, M. C. (2005). Optimization of continuous phenol extraction from *Vitis vinifera* byproducts. *Food Chem.*, *92*, 109-17.

Pinelo, M., Rubilar, M., Jerez, M., Sineiro, J. & Nunez, M. J. (2005). Effect of solvent, temperature, and solvent-to-solid ratio on the total phenolic content and antiradical activity of extracts from different components of grape pomace. *J. Agric. Food. Chem.*, *53*, 2111-17.

Puertolas, E., Cregenzan, O., Luengo, E., Alvarez, I. & Raso, J. (2013). Pulsed-electric-field-assisted extraction of anthocyanins from purple-fleshed potato. *Food Chem.*, *136*, 1330-36.

Ross, C. F., Hoye, Jr C. & Fernandez-Plotka, V. C. (2011). Influence of heating on the polyphenolic content and antioxidant activity of grape seed flour. *J. Food Sci.*, *76*, C884-C90.

Sánchez, M., Franco, D., Sineiro, J., Magariños, B. & Núñez, M. (2009). Antioxidant power, bacteriostatic activity, and characterization of white grape pomace extracts by HPLC–ESI–MS. *Eur. Food Res. Technol.*, *230*, 291-301.

Shi, J., Yu, J., Pohorly, J. E. & Kakuda, Y. (2003). Polyphenolics in grape seeds: Biochemistry and functionality. *J. Med. Food*, *6*, 291-99.

Shrikhande, A. J., Race, E. J., Wightman, J. L. D. & Sambueso, R. D. (2003). United States Patent No. US6544581 B1. Canandaigua Wine Company, Inc.: E. United States Patents.

Sies, H. (2010). Polyphenols and health: Update and perspectives. *Arch. Biochem. Biophys.*, *501*, 2-5.

Takaki, K., Hatayama, H., Koide, S. & Kawamura, Y. (2011, 19-23 June 2011). *Improvement of polyphenol extraction from grape skin by pulse electric field.* Paper presented at the Pulsed Power Conference (PPC), 2011 IEEE.

Torres, J. L., Varela, B., Garcia, M. T., Carilla, J., Matito, C., Centelles, J. J., Cascante, M., Sort, X. & Bobet, R. (2002). Valorization of grape (*Vitis vinifera*) byproducts. Antioxidant and biological properties of polyphenolic fractions differing in procyanidin composition and flavonol content. *J. Agric. Food. Chem.*, *50*, 7548-55.

Valenzuela, A. (2004). Tea consuption and health: Beneficial characteristics and properties of this ancient beverage. *Revista Chilena de Nutrición Versión Online*, *31*.

Vatai, T., Škerget, M. & Knez, Ž. (2009). Extraction of phenolic compounds from elder berry and different grape marc varieties using organic solvents and/or supercritical carbon dioxide. *J. Food Eng.*, *90*, 246-54.

Vorobiev, E. & Lebovka, N. (2010). Enhanced Extraction from Solid Foods and Biosuspensions by Pulsed Electrical Energy. *Food Eng. Rev.*, *2*, 95-108.

Yamakoshi, J., Kataoka, S., Koga, T. & Ariga, T. (1999). Proanthocyanidin-rich extract from grape seeds attenuates the development of aortic atherosclerosis in cholesterol-fed rabbits. *Atherosclerosis*, *142*, 139-49.

Yu, J. & Ahmedna, M. (2013). Functional components of grape pomace: their composition, biological properties and potential applications. *Int. J. Food Sci. Technol.*, *48*, 221-37.

In: Grapes
Editor: José S. Câmara
ISBN: 978-1-63321-402-6

Chapter 10

Grape Stalks Based Sorbents: Towards a Sustainable Alternative for the Detoxification of Divalent Heavy Metal Polluted Effluents

Carlos Escudero-Oñate[1, 2*]***, Núria Fiol***[1]***, Jordi Poch***[3]
and Isabel Villaescusa[1]

[1]Norwegian Institute for Water Research (NIVA), Oslo, Norway
[2]Chemical Engineering Department, Escola Politècnica Superior, Universitat de Girona, Girona, Spain
[3]Applied Mathematics Department, Escola Politècnica Superior, Universitat de Girona, Girona, Spain

Abstract

The increasing application and the immutable nature of heavy metals has made of this kind of pollution one of the most serious environmental problems nowadays. As they are no degradable, they can accumulate in the components of the environment where their toxicity is expressed.

Due to the special concern of heavy metal pollution in waters, there is an increasing interest towards the development of sustainable, environmentally friendly and cheap technologies for water treatment. Among the different materials that can be used for the detoxification of heavy metal polluted effluents, grape stalks (GS) has demonstrated an excellent performance. GS are the branches that connect the grape grains in the compact structure of the bunch. The material, mainly composed by cellulose and lignin, is an industrial by-product of wine elaboration industries that is generated in huge amounts in several countries worldwide. The management of these wastes, in many cases, involves an important extra cost for these companies that could be avoided if, as proposed here, it gets a value in water treatment.

---

[*] Corresponding author: Email: ces@niva.no.

In this book chapter the authors will present a novel process based on GS sorption in a continuous bed up-flow process for the detoxification of heavy metal polluted streams. The target metals of this study were Cu(II), Ni(II), Pb(II) and Cd(II), being the sorption behaviour of GS explored in both, binary and quaternary mixtures. The possible recovery of the metal in a concentrated form and regeneration of the sorbent for further sorption-desorption cycles is also presented and discussed.

Keywords: Grape stalks, heavy metals, sorption, detoxification, column

Occurrence and Toxicity of Heavy Metals

The rapid development of industries such as mining and smelting, surface finishing, energy and fuel production, fertilizers and pesticides, metallurgy, electroplating, electrolysis, leatherworking, photography, metal surface treating, aerospace and atomic energy installations has provoked that wastes containing metals are directly or indirectly discharged into the environment, having brought serious environmental pollution and threatened biolife (Bishop, 2002; Wang and Chen, 2006; Volesky, 1990). Not only anthropogenic activities have provoked the release of heavy metals into environmental systems, it worth note that some natural sources contribute also to heavy metal pollution. This is the case of some cases of running water, which can erode soil and rock, dissolving and transporting metals thousands of kilometres before being deposited.

Due to the increasing application of heavy metals and their immutable nature, this pollution has become one of the most serious environmental problems nowadays. As they are non-degradable, they can accumulate in the components of the environment where their toxicity is expressed. Heavy metals exhibit special toxic characteristics that can be summarized as follows: (1) long lasting and persistence in the environment, (2) some heavy metals even could be transformed from relevant low toxic species into more toxic forms in a certain environment (as the case of methylation of mercury to form methyl-mercury cation), (3) the bioaccumulation and bioaugmentation of heavy metals by food chain could damage normal physiological activity and endanger human life, (4) metals can only be transformed and changed in valence and species, but cannot be degraded by any methods including biotreatment, (5) toxical effects might occurs even at low concentration (Alkorta et al., 2004; Volesky, 1990).

Among the most commonly found metals in polluted environments one can find Cu(II), Ni(II), Pb(II) and Cd(II). Their sources, properties, applications and toxicological effects are briefly described next.

Copper

Copper has been known for about 10000 years and occurs in both, metallic form or in compounds with oxidation states Cu(I) and Cu(II). The metal is easily manufactured from ores in underground or open-pit mines. The most important ores contain, besides small amounts of metallic copper, Cu_2S, CuS, $CuFeS_2$, CuO and $Cu_2CO_3(OH)_2$.

Its uses are mainly derived from its excellent electrical conductivity, but this metal is also widely used in alloys, as an algaecide, in the preparation of pigments, as supplement in pastures and for the preparation of catalysers. The metal enters to the hydrological system, either groundwater or surface water, from dissolution of rock and soils, from biological cycles, atmospheric fallout and, specially, from industrial processes and waste disposal (Merian, 1991).

Copper is biologically available as Cu(I) and Cu(II) in inorganic salts and organic complexes. In high dose, this metal can cause dermatitis in contact with skin and hepatic and kidney damage, if ingestion is produced.

Nickel

Nickel is widely distributed in the environment and its bioavailability and biological effects strongly depend of the type of compounds. The metal is a silver-white, hard, malleable, ductile ferromagnetic metal that maintains a high lustre and is relatively resistant to corrosion. The oxidation states of nickel include -1, +1, +2, +3 and +4, being the Ni(II) the most stable. The main minerals containing nickel are formed mainly by oxides containing mixed nickel-iron and also sulphides as the pentlandite, $(Ni, Fe)_9S_8$.

Among its main uses it can be remarked the formation of alloys like Ni-Cr-Fe, Ni-Cu and Ni-Al. The metal is also widely used in electroplating industries, in the manufacture of Ni-Cd batteries and as catalyser forming the structure of organometallic compounds for petroleum refining (Merian, 1991).

The different dysfunctions and diseases promoted by nickel can be classified in three categories: (1) allergic processes, (2) respiratory system diseases and (3) tumoral and cancerous processes (OIT, 2001).

Lead

Metallic lead is a bluish-white, soft metal whose main oxidation state is +2. Its minerals can be found in many regions worldwide being the galene (PbS) the richest lead mineral and the most important source for the commercial production of this metal. Other minerals containing lead are the cerussite ($PbCO_3$) and anglesite ($PbSO_4$). In many cases, minerals containing lead can contain also several other toxic heavy metals.

In nature lead is a ubiquitous, non-essential element that can be found in many of the environmental media (air, soil, rocks, sediments and water) and in the components of the biosphere. Lead and its compounds may enter the environment at any point during mining, smelting, processing, use, recycling or disposal. Estimates of the dispersal lead emissions into the environment indicate that the atmosphere is the major initial recipient. Mobile and stationary sources of lead emissions tend to be concentrated in areas of high population density and near smelters and, from these emission sources, the metal moves through the atmosphere to various components of the environment. Lead is finally deposited on soil, surface waters and plants and thus is incorporated into the food chain of animals and man.

The commercial use of lead includes the lead-batteries production, alloys, glassware and ceramics. Extended uses in the past were also the preparation of lead based pigments and methyl/ethyl derivatives added as antiknock agents in gasoline. These processes were prohibited years ago in many countries (Merian, 1991).

Despite natural lead concentrations are not very high, in the last sixty years, great amounts of lead have been extracted, concentrated, used and re-emitted into the environment. Lead concentrations are now locally and regionally much higher that they used to be. Animals and humans are thus exposed to new health risks and their body, tissues and fluids may contain more lead than normal.

Once lead enters to the organism, it can be distributed in three ways: blood, soft tissues (kidney, bone-marrow, liver and brain) and in mineralized tissues (bones and teeth). The mineralized tissue contains, in adults, approximately the 95% of the total lead. Lead inhibits organism's capacity to produce haemoglobin, due to the interference with several enzymatic steps in the metabolism of hemo group. This metal is associated also to cardiotoxic effects, hypertension, inhibition of teeth and bones growing and also appearance of neurological disorders (OIT, 2001).

Cadmium

Cadmium is a relatively volatile non-essential element for plants, animals and humans. Because cadmium occurs together with zinc -from which it must be separated- cadmium production depends on the production of zinc. Cadmium is a silver-white, lustrous and ductile metal whose main oxidation state is +2. It is present in this form in rare minerals as greenockite (hexagonal CdS), hawleyte (cubic CdS), otavite ($CdCO_3$), and cadmoselite (CdSe). Cadmium mostly occurs in isomorphic form in zinc minerals such as zinc blende (ZnS) with cadmium contents ranging from 0.1 to 0.5% and smithsonite ($ZnCO_3$) with cadmium contents up to a maximum of 5%.

Cadmium has been emitted in minor amounts into the environment from the rise of industrialization, but in greatly increased quantities after World War II. Its dispersion has mainly taken place in the form of dusts and aerosols to the atmosphere, effluents into rivers and lakes and as solid from point sources (waste, slag, incineration, coal combustion and sewage sludge). The metal is very corrosion-resistant and the electrodeposited cadmium exhibits excellent properties in the protection of iron and steel. The metal has been widely employed also in rechargeable nickel-cadmium batteries, but the development of new accumulator materials and the need of lower weight in the final products are provoking its replacement in this technological application. One of its main applications is the production of cadmium soaps made with saturated and unsaturated fatty acids that play an important role as temperature and light stabilizers mainly for PVC. The metal has been also employed in nuclear reactors as a neutron absorber and in various alloys with such metals as tin, copper and aluminium (Merian, 1991).

After cadmium absorption, either digestive or by the respiratory tract, the metal is transported to the liver, where the production of low molecular weight with high sulphur content proteins begins. These proteins, named metallothioneins, avoid cadmium to exert its toxic activity when present in free form. When the amount of cadmium ingested overpass the metallothioneins production capacity of the organism, the consequence is a renal failure. Cadmium has been associated also to the appearance of tumoral and cancerous processes (OIT, 2001).

Industrial by-Products in the Detoxification of Heavy Metal Polluted Effluents: Grape Stalks Case Study

A wide variety of materials fully available have been evaluated as sorbents for the removal of heavy metals from polluted streams. According to their origin they can be grouped as natural, agricultural or industrial by-products. These materials can be used either on its raw form or after a previous treatment to enhance their sorption capacity, mechanical or mass transfer properties. Some authors have processed chemically some low-cost materials such as coconout shell, (Kirubakaran et al. 1991), pistachio shell (Abe et al. 1990), tropical wood (Maniatis and Nurmala, 1992), almond shell (Hayashi et al. 2002) and rice bran (Suzuki et al. 2007) to produce activated carbon. However, these processes are normally time consuming and expensive due to the use of chemical reagents and/or high temperatures. If the material shows a natural good sorption performance on its raw form this modification is neither feasible nor worthwhile.

The agricultural sector in Spain is one of the main economic motors and generates a vast type of different industrial vegetable by-products. This sector becomes then a huge source of potential low cost materials that could be screened, evaluated and valorised as sorbents for the removal of heavy metal polluted effluents. In this context it worths note the particular scenario of the wine industries as producer of vast amount of grape stalks (GS) agrowastes. In these industries it can be assumed that, in average, from each 100 kg of grapes material, 5 to 6 kg is GS and 94-95 is grape (Boulton et al. 2002). Despite a 5-6 % of waste material does not appear to be a large amount of residues, it has to be pointed out that a small wine industry can process about 16.000.000 kg of grape bunches on each campaign. This finally involves the production and management of about 880.000 Kg of GS residues that currently are being mainly dispersed in the fields or incinerated.

The use of grape stalks and in general of lignocellulosic materials has attracted the attention of researchers, not only for their reuse in agriculture, but also as a bioresource for the production of novel products after fractioning according to their chemical composition and properties. The use of lignocellulosic materials as a source of valuable products such as polyphenols, sugars, and lipids, is an interesting way to increase the value of this waste and to generate economic benefits (Pujol et al. 2013).

The grape stalk can reach the industry in green or ligneous state. The composition, in average, for each form is summarized next:

(i) The green contains 70 to 80% water, 1.3 to 4% tannins, 0.5 to 1.3% of tartaric and malic acid, 0.3 % of nitrogenous substances, chlorophyll, potassium bitartrate and mineral salts.
(ii) Ligneous grape stalk contains lower water content, 35 to 60%, less free acids, more potassium bitartrate and less tannin (Aleixandre, 1999).

In the next figure, a picture of GS after its separation from the grape bunch is shown:

A recent study carried out by Pujol (Pujol et al. 2013) demonstrated that grape stalks have an important amount of acidic groups whose concentration is particle size dependent. They found that the larger particle size fraction contained the highest content of acid groups, followed by medium and finest fractions. The authors stated that phenolic moieties represent more than 50% of the total acidic groups while weak acidic groups are only about 30%. The presence of these surface functional groups provides to this material an interesting potential to remove most of heavy metals, as their hydrochemical speciation is often cationic. In this situation, coulombic interactions and/or ion exchange processes are possible, leading to metal sequestration from the solution.

Cu(II), Ni(II), Pb(II) and Cd(II) Biosorption in Multimetal Systems

An overview on the few studies on multimetal biosorption reported in literature highlights that the sorbent materials come mainly from three biosources: algae, bacteria and different agrowastes. In the first group it can be found the use of algae Gelidium in the detoxification of Pb(II)/Cd(II) mixtures (Vilar et al. 2008) and Sargassum algal biosorbent in the removal of copper, zinc, cadmium (Volesky et al. 2003).

Studies on multimetal sorption by bacteria biomass have been carried out by using Pseudomonas putida to carry out metal removal from binary Pb(II)/Cu(II) mixtures (Uslu and Tanyol, 2006) and Sphaerotilus natans for the removal of Cu(II), Ni(II), Pb(II) and Cd(II) from their mixtures (Pagnanelli et al. 2003). Despite the good sorption performance demonstrated by these biological materials its use adds an unavoidable extra cost and complexity to the biosorption procedures as it involves biomass culture.

The use of agrowastes for metal removal from multimetal mixtures has been also reported. Among the different examples of use of agrowastes in the removal of heavy metals, one can find good examples in the use of grape stalk to remove Cu(II) and Ni(II) from binary mixtures (Escudero et al., 2013, Villaescusa et al. 2004) and crop milling waste for the removal of Pb(II), Cd(II), Cu(II), Ni(II) and Zn(II) from binary and ternary mixtures (Saeed et al. 2005). Some non-biological materials have been also studied as metal sorbents in multimetal mixtures. This is the case of Cu(II), Ni(II), Cd(II) and Zn(II) sorption from binary Cu/Ni, Cu/Cd and Ni/Cd mixtures by activated carbon (Seco et al. 1997), hydroxyapatite for the removal of Cd(II) Pb(II), Zn(II) and Cu(II) and soils for the removal of Cd(II), Ni(II) and Zn(II) (Antoniadis and Tsadilas, 2007).

Figure 1. Grape stalk detail.

In wastewaters containing multiple metals interactive effects appear depending on the number of metals competing for binding sites, the combination of these, the initial concentrations, the equilibrium steady state concentration of different metal ion species, limitations presented by the binding sites as well as the nature and quantity of the sorbent. In multimetal mixtures, the different nature of the metals to be adsorbed can lead thus to kinetic competition or replacement at equilibrium. This phenomenon can be explained considering a reversible reaction between metals and adsorbent sites so that, when the adsorbent in near saturation, heavy metal ions sorbed with lower affinity can be replaced by those that exhibit a higher affinity by the sorbent. When this occurs in continuous flow processes, an overconcentration in the outlet flow (overshoot) respect to the feeding solution of the metal/s uptaken with less affinity takes place (Kratochvil and Volesky, 2000).

Scope and Objective

In this chapter we present a novel process based on the sorption capacity of GS to carry out the removal of Cu(II), Ni(II) Pb(II) and Cd(II) from multi-metal solutions and in a continuous bed up-flow process. The sorption performance of the process when feeding the column with Cu(II), Ni(II), Pb(II) and Cd(II) in binary and quaternary mixtures will be evaluated. The possibility of metal recovery and sorbent bed regeneration in the quaternary scenario will be also evaluated and discussed.

Experimental

Reagents and Solutions

Grape stalks (GS) waste generated in wine production were kindly supplied by Coop. Ntra. Sra. de la Estrella (El Herrumblar, Cuenca, Spain). The material was rinsed three times with distilled water, dried in an oven at 110 °C until constant weight, cut and sieved for a particle size of 0.25–0.50 mm.

All the reagents were of analytical grade. Cu(II), Ni(II), Pb(II), Cd(II) and were purchased from Panreac (Barcelona, Spain). Solutions were prepared in MilliQ water and pH was adjusted by adding either HNO_3 or NaOH.

Apparatus and Analytical Conditions

The experimental set-up developed for the multi-metal sorption studies is presented in Figure 2. Fixed bed sorption experiments were carried out in glass columns of 10.0 cm length and 1.0 cm inner diameter (Omnifit). The columns were packed with approximately 0.5 g grape stalks waste (dry weight) previously rinsed with abundant water until no release of colored compounds was observed. By doing so, a bed height around 6.7 cm was obtained. Sorption process was carried out in the up-flow mode to avoid possible short-circuiting by clogging and channelling. The solutions were pumped at room temperature (20 ± 1 °C) upwards the grape stalks packed column by a peristaltic pump (Gilson Minipuls) at a flow rate of 30 mL·h^{-1}. By using an automatic fraction collector (Gilson FC 203B), samples of 5.5 mL were collected at different times. The pH of the collected samples was recorded.

The solutions were then acidified to avoid possible metal precipitation and analysed by FAAS. Breakthrough curves were obtained by plotting the metal concentration in the outlet stream as a function of time. The experiments were carried out in duplicate and the average results are presented.

Sorption from Multi-Metal Mixtures

Grape stalk sorption behaviour in binary equimolar Cu(II)/Ni(II), Cu(II)/Cd(II), Cu(II)/Pb(II), Ni(II)/Pb(II), Ni(II)/Cd(II), Pb(II)/Cd(II) solutions was evaluated. The metal solutions were approximately 0.2 mM concentration on each one of the metals. This concentration was also employed in sorption from quaternary mixtures.

The initial pH of the metal solutions was adjusted to 5.20 and an automatic fraction collector was programmed to carry out the sampling of the eluted stream each 30 minutes.

Metal Desorption, Recovery and Reuse of the Sorbent in Quaternary Mixtures

Continuous wastewater treatment requires a reversible sorption-desorption process in order to reuse the adsorbent and to recover the metals in a concentrated form. Four sorption and desorption cycles were carried out in a column filled with GS. Sorption experiments were carried out according to the procedure described previously. The desorption step was carried out by feeding the column containing the metals-loaded sorbent with 0.05 M HCl at the same flow rate as in sorption experiments, about 30 mL·h^{-1}.

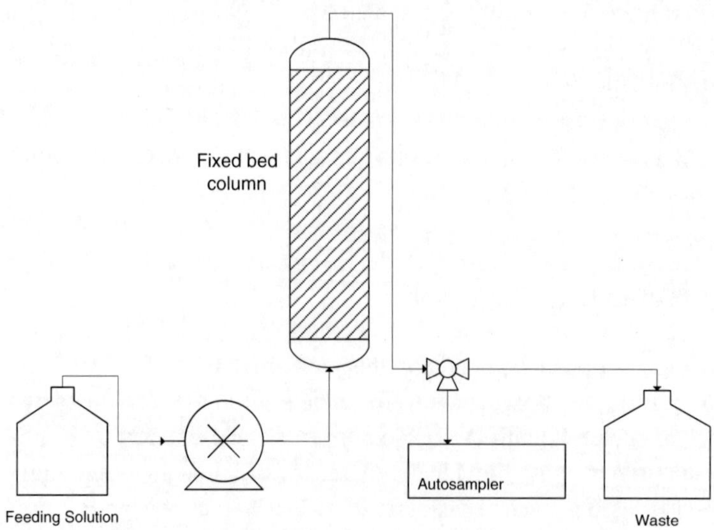

Figure 2. Experimental set-up of the continuous bed up-flow sorption process.

Due to the fast nature of the desorption process the whole volume eluted by the top of the column was collected in separated tubes during 10 minutes, without draining, obtaining fractions of about 5 mL. In these samples the pH was measured as well as Cu(II), Ni(II), Pb(II) and Cd(II) concentration. Metal desorption ratio was calculated by a mass balance.

The ability of the material to recover its sorption capacity after the acidic desorption treatment was evaluated. When acidic desorption treatment finished, the columns were fed with MilliQ water adjusted to an initial pH value of 5.2. The solution was pumped upwards the column until pH in the outlet and inlet stream became equal. Once reached this point, pumping was briefly interrupted and the MilliQ solution was replaced by the solution containing the four metals to start a new sorption cycle. According to this protocol, sorption/desorption was evaluated in four cycles.

Analysis of Samples

Samples were acidified by adding 1 μL of concentrated nitric acid. Metal concentration was analysed by flame atomic absorption spectroscopy (FAAS) using a Varian SpectrAA 220FS equipment coupled with an automatic dilutor Varian SIPS and autosampler Varian SPS3. The instrument was operated in multilamp mode and metal standard solutions of 1000 $mg \cdot L^{-1}$ were used for FAAS calibration. Measurement of pH was performed using a pHmeter PHM 250 (Meterlab).

Results and Discussion

Cu(II), Ni(II), Pb(II) and Cd(II) Sorption from Binary Mixtures

The experimental breakthrough curves of metal sorption from the binary mixtures are presented in Figure 3 and Figure 4 in terms of eluted concentration and sorbed amount respectively.

First overviews on the different plots of Figure 3 put into evidence that in all the cases, one of the metals shows the typical sigmoidal shaped breakthrough curve while the other metal shows an overconcentration respect to the feeding stream (about 0.2 mM). This phenomenon can be also observed in Figure 4, where an inflexion point appears in the normal accumulative trend of the metal that suffer overshoot and its sorbed amount decreases to reach a new equilibrium, with lower sorbed concentration.

These plots reveal that: (i) lead does not exhibit overshoot in any of the binary mixtures; (ii) copper is overshoot only when the binary mixture contains lead (iii) cadmium and nickel exhibit the highest overshoot value in the mixtures with lead and copper (iv) nickel shows a slight overshoot in the mixture with cadmium. The origin of the overshoot is based on the nature of the sorption process itself. When the column maximum capacity is approached, the species with lower affinity are pushed off by others with higher affinity and the displacement of previously sorbed species occurs (Sag and Kutsal, 2001; Kratochvil and Volesky, 1998). According to these results, an affinity ranking for the sorption of the different divalent cations onto GS can be proposed. This ranking would be, from higher to lower sorbent affinity:

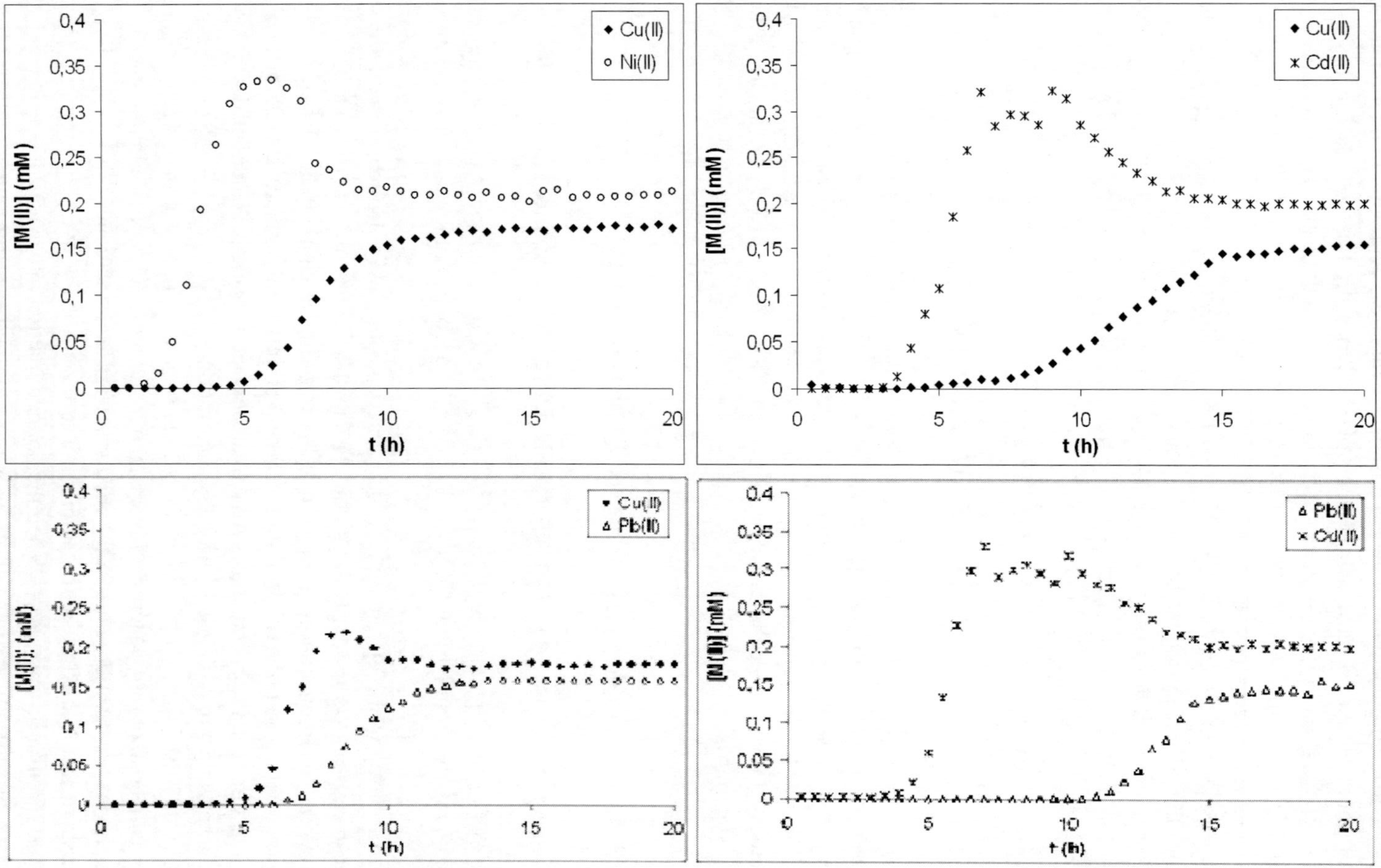

Figure 3. (Continued).

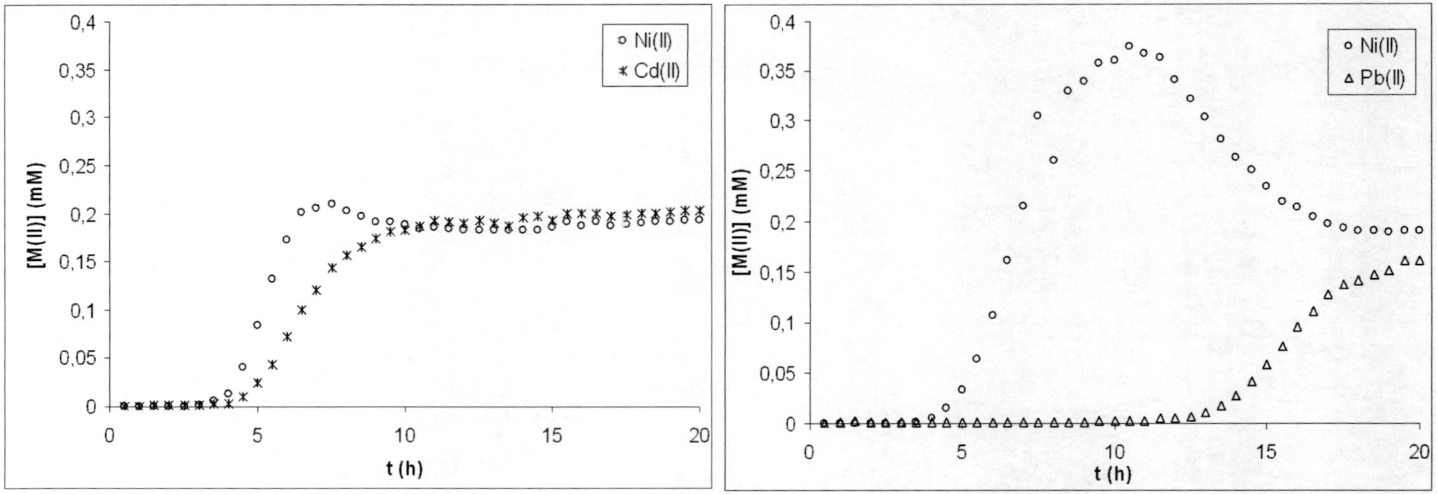

Figure 3. Experimental breakthrough curves for metal sorption onto grape stalks from binary systems. Flow rate:30 $mL \cdot h^{-1}$; metal initial concentration:0.2 mM; pH:5.0; sorbent mass:0.5 g; particle size: 0.25-0.50 mm.

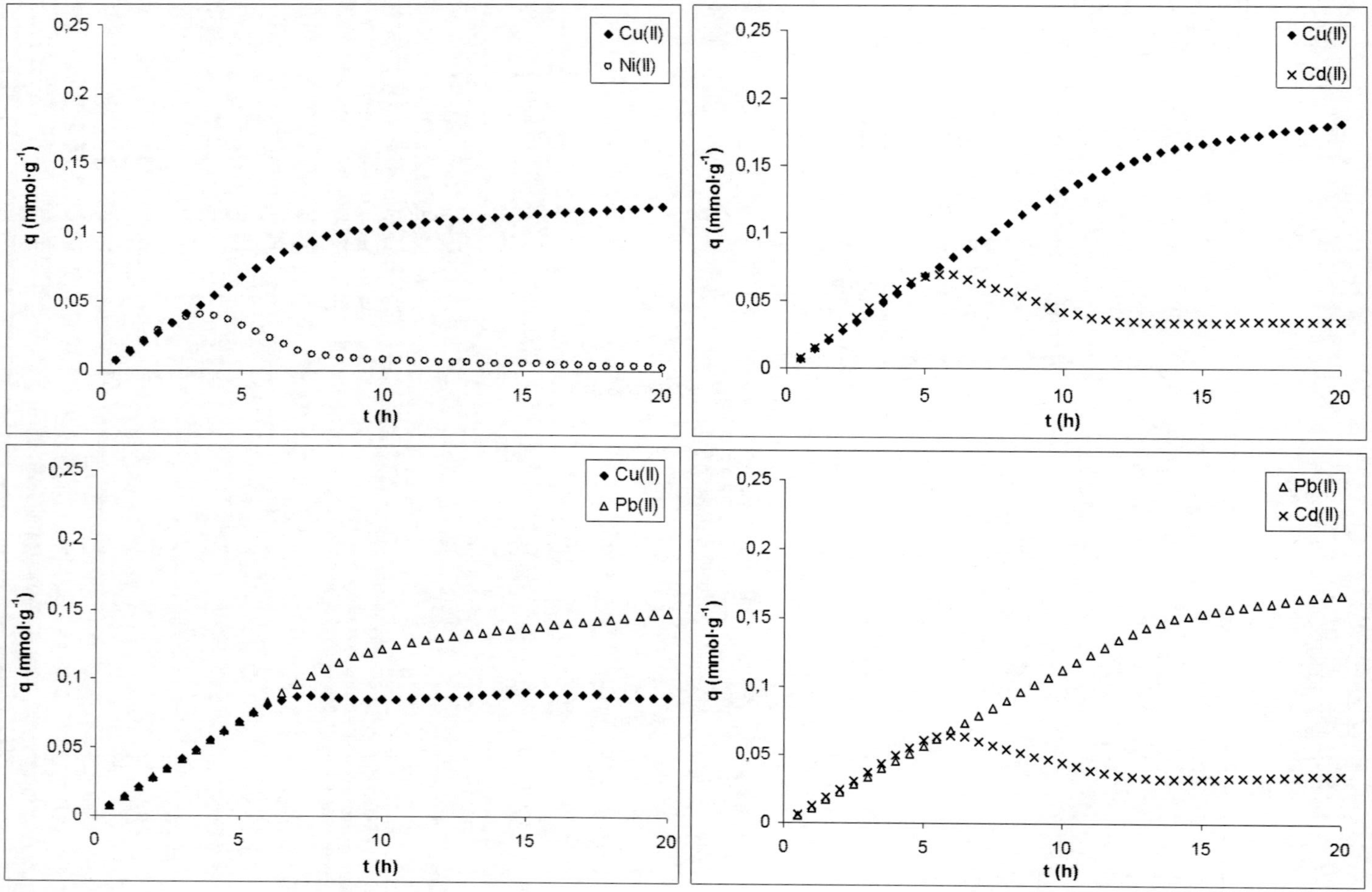

Figure 4. (Continued).

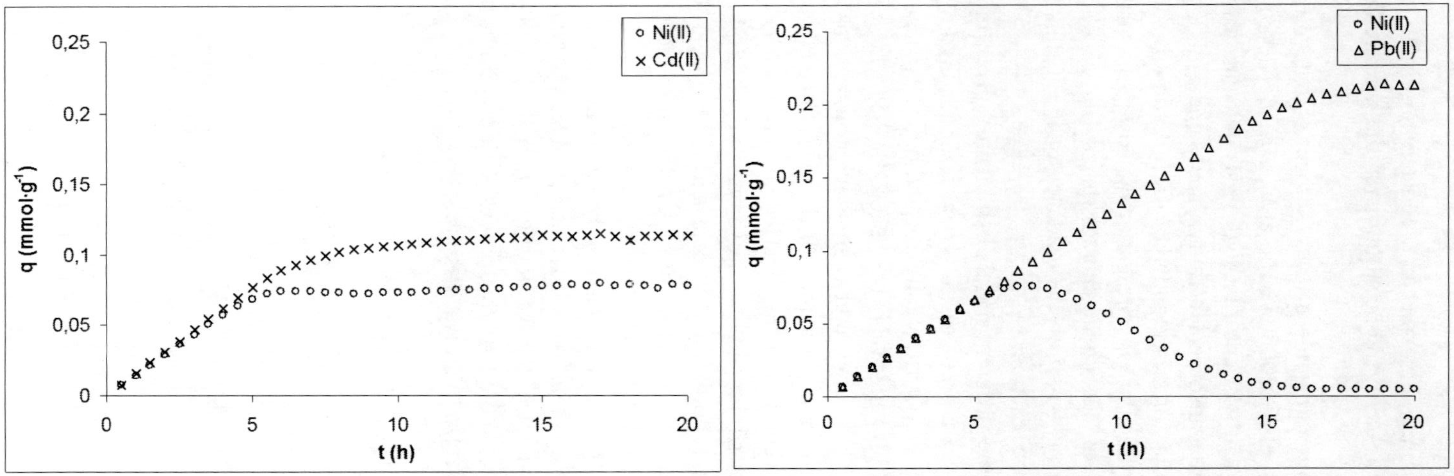

Figure 4. Sorbed metal amount onto grape stalks from binary systems as a funtion of time. Flow rate:30 $mL \cdot h^{-1}$; metal initial concentration:0.2 mM; pH:5.0; sorbent mass:0.5 g; particle size:0.25-0.50 mm.

Pb>Cu>Cd>Ni

where the metals with higher affinity are expected to replace those with lower position in the ranking provoking, in last term, an overconcentration of the weaker-bonded metal in the outlet flow of the column.

Several authors have reported that the affinity sorbent-sorbate reflects the hydrolytic properties of metals (Pagnanelli et al. 2003; Vegliò et al. 2002). These authors propose that the affinity series are in agreement with the first hydrolysis constant value so that, the more acidic the ion is, the higher the adsorbent affinity. The same authors proposed that the correlation between metal acidity property and the affinity sorbent-sorbate is even more important than the specific functional groups present on the adsorbent surface; in other words, metal speciation predominates on adsorbent characteristics (Pagnanelli et al. 2003). The explanation provided by authors is based on the tendency of the metals to form hydrolysed species $Me(OH)^+$ according to the reaction:

The sorption affinity that the different metals exhibit in a given multimetal solution is interpreted by supposing that the affinity of the sorbent for the hydrolysed metal species $Me(OH)^+$ is significantly greater than for the unhydrolysed ones (Pagnanelli et al. 2003). It is also considered that in the usual pH range (above 5) $[Me(OH)^+] << [Me^{2+}]$.

This observation is enforced by many studies using either biomass or natural materials, as it can be seen in Table 1. These results show that Pb(II) always occupies the first position in the affinity ranking, followed (when present) by Cu(II). On the other hand, Ni(II) and Cd(II) are always in the range of lower affinity and its relative order depends on the sorbent.

In the case presented in this book chapter, considering the logarithmic values of the metal's first hydrolysis constant, the next affinity ranking would be expected: $\log K_{Pb(II)} = -7.71 > \log K_{Cu(II)} = -8.00 > \log K_{Ni(II)} = -9.9 > \log K_{Cd(II)} = -10.8$ (Smith and Martell, 1976). This ranking successfully describes the observed first and second position of Pb(II) and Cu(II) respectively. However, this theory fails to describe the relative position of the pair Ni(II)-Cd(II). These results put into evidence that the description of the interaction of the different metals and the sorbent might be a more complex process that could not be merely described by the tendency of the metals to form hydrolysed species.

Table 1. Affinity series for different natural adsorbents

Sorbent	Affinity order	Reference
Microcystis and *Spirogyra*	Pb>Cu>Zn>Cd	Rai, 2000
Streptoverticillium and *Penicillium*	Pb>Cu>Zn>Cd>Ni>Co	Puranik and Paknikar, 1999
Activated sludge	Pb>Cu>Zn=Cd>Ni	Hammaini et al. 1999
Pseudomonas aeruginosa PU21	Pb>Cu>Cd	Chang and Chen, 1999
Fungal biomasses	Pb>Cu>Cd>Zn	Yin et al. 1999
Sphaerolitus natans	Pb>Cu>Zn>Cd	Pagnanelli et al. 2003
Humic umbrisols	Pb>Cr>Cu>Cd≈Ni≈Zn	Covelo et al. 2004

As previously described, in all the plots it can be observed that both metal ions accumulate progressively giving a straight line with very similar slopes until the sorbent capacity is exhausted. The slight differences observed in the slopes might be due to differences in the initial metal ion concentration. When the overshoot of one of the metals in the binary mixture occurs, the continuous tendency of this metal ion to get accumulated onto the sorbent is broken reaching a maximum accumulation (acc_{max}). From this point the concentration of this metal ion in the solid phase progressively decreases until it reaches a new equilibrium characterized by a constant accumulation value (acc_{eq}). These two data can be useful to quantify the effect of the overshoot on the total accumulation of metal in the solid phase by means of the parameter OSL. We have defined the parameter OSL (Overshoot metal Sorption Loose) according to the equation:

$$OSL = ((acc_{\max} - acc_{eq}) / acc_{\max})100 \quad (17)$$

where (acc_{max}) is the maximum accumulation reached by the metal ion before being overshoot and (acc_{eq}) is the accumulation at equilibrium.

According to OSL values, overshoots have been classified in three categories:

(i) Slight, if OSL (%) $\leq$35
(ii) Medium, if 35$\leq$OSL (%)$\leq$70
(iii) Intense, if 70$\leq$OSL (%)$\leq$100

In Table 2 it is presented a summary indicating: presence/absence (Y/N) and magnitude of overshoots (Slight: S; Medium: M; Intense: I), maximum (q_{max}) and equilibrium (q_{eq}) sorption capacities, percentage of capacity loose from the maximum to the equilibrium value for the overshot metal (OSL) and total metal sorbed amount ($q_{eq\ (M1+M2)}$).

If evaluation of the water detoxification performance on the different binary mixtures is carried out according to the European Union quality standards for human consumption, the maximum allowed concentration is dependent on the nature of the metal (Council Directive 98/83/EC). According these standards, the maximum allowed concentration values for the different metals are: Cu(II) 2 $mg \cdot L^{-1}$; Ni(II): 20 $\mu g \cdot L^{-1}$; Pb(II): 10 $\mu g \cdot L^{-1}$ and Cd(II): 5 $\mu g \cdot L^{-1}$. This relative permissiveness in the concentration level of copper, in the range of ppm, remarkably contrasts with the low allowed values for nickel, lead and cadmium, for which the European Union accepts only few ppb.

The application of this sorption technology for the different binary mixtures would be restricted thus to the time from which the concentration in the outlet flow overpass the maximum allowed concentration for one of the metals. The times of operation of the continuous bed-up flow system, according to the E.U. discharge limits for each individual metal on each binary mixture are presented in the Table 2. Underlined is indicated the maximum time of applicability that would accomplish the discharge limits for both metals in each binary mixture.

As the maximum allowed concentrations according to the E.U. criteria are metal-dependent, in the table it has been also included the time that the column can operate eluting a concentration lower than 0.1 $mg \cdot L^{-1}$ for each metal of the mixture. This concentration value has been arbitrarily fixed only for comparison sake of the sorption behaviour of the GS through the different divalent cations.

As it can be seen in the table, while in the case of the Cu(II)/Pb(II) mixture the column can operate under the specified concentration limits for both metals aproximately 5.5 hours, in the case of Pb(II)/Cd(II) mixture an unacceptable Cd(II) concentration was observed from the beginning of the process. The knowledge of the time or eluted volume at which one of the metal ions overshoots can be of great utility to prevent outlet flow contamination and determine at which point the inlet flow must be diverted to a fresh sorbent packed column.

As it had been previously discussed, the accumulated concentration in the solid phase is strongly dependent on the competitor metal in the binary mixture and the extension of metal replacement could be computed by means of the parameter previously defined as OSL. Values of OSL for Ni(II) ranging from 95.5 (almost total replacement) to approximately 0 (no replacement) were observed in the Cu(II)/Ni(II) and in the Ni(II)/Cd(II) mixtures respectively.

If the previous affinity ranking obtained from the evaluation of the overshoots is taken into account, it can be observed that the longer the distance between the metals in the ranking, the higher the capacity loose for the metal uptaken with lower affinity.

Table 2. Breakthrough parameters for divalent metal sorption in the binary mixtures

System	Metal	Overshoot/ magnitude	$t_{[M(II)]<0.1\ ppm}$ (h)	$t_{[M(II)]>EU}$ (h)	t_{qmax} (h)	q_{max} ($mmol \cdot g^{-1}$)	q_{eq} ($mmol \cdot g^{-1}$)	OSL (%)	$q_{eq\ (M1+M2)}$ ($mmol \cdot g^{-1}$)
Cu/Ni	Cu	N	4	6	--	--	0.1244	--	0.1262
	Ni	Y/Intense	1.5	1	3.5	0.0403	0.0018	95.5	
Cu/Cd	Cu	N	4.5	9	--	--	0.1866	--	0.2217
	Cd	Y/Medium	3	3	5.5	0.0696	0.0351	49.6	
Cu/Pb	Cu	Y/Slight	4	5.5	7.5	0.0884	0.0871	1.5	0.2389
	Pb	N	6	5.5	--	--	0.1527	--	
Ni/Pb	Ni	Y/Intense	3.5	3	7	0.0755	0.0047	93.8	0.2184
	Pb	N	5.5	0.5	--	--	0.2137	--	
Pb/Cd	Pb	N	9	3.5	--	--	0.1738	--	0.2110
	Cd	Y/Medium	0	0	6	0.0649	0.0372	42.7	
Ni/Cd	Ni	Y/Slight	3	3	6.5	0.0736	0.0770	0.01	0.1904
	Cd	N	3	1	--	--	0.1134	--	

* $t_{[M(II)]<0.1}$ ppm: time from which metal concentration in the outlet flow is >0.1 $mg \cdot L^{-1}$.
* $t_{[M(II)]>EU}$: time from which metal rises the maximum allowed concentration for drinking water (according to EU criteria).
* t_{qmax}: time to reach maximum capacity for the overshot metal (h).
* q_{max}: maximum metal sorbed amount for the overshot metal ($mmol \cdot g^{-1}$).
* q_{eq}: metal sorbed amount at equilibrium ($mmol \cdot g^{-1}$).
* $q_{eq\ (M1+M2)}$: total divalent metal sorbed amount ($mmol \cdot g^{-1}$).

Also the total divalent metal sorption capacity at equilibrium, $q_{eq\ (M1+M2)}$ ($mmol \cdot g^{-1}$), has been calculated for the different binary mixtures and values are presented in Table 2. As it can be observed this parameter depends also on the nature of the binary mixtures and varies from 0.1262 $mmol \cdot g^{-1}$ in the case of the Cu(II)/Ni(II) mixture to 0.2389 $mmol \cdot g^{-1}$ in the case of the Cu(II)/Pb(II). Except in the case of the Cu(II)/Ni(II) mixture, where the total capacity value is quite low, in all the other binary mixtures, independently of the metals involved, $q_{eq\,(M1+M2)}$ is close to 0.2 $mmol\cdot g^{-1}$.

Metal Sorption, Recovery and Reuse of the Sorbent in Quaternary Mixtures

In this section, copper, nickel, lead and cadmium removal from a quaternary mixture and the possibility of metal recovery by desorption in mild acidic conditions is evaluated by performing 4 consecutive sorption and desorption cycles. In the sorption experiments, when the eluted metal concentration in solution is plotted as a function of time, the characteristic breakthrough curves are obtained. In Figure 5, the experimental breakthrough curves for the different metals on the quaternary mixture in the different sorption cycles are presented. pH was also monitored and its profile has also been included.

As in the case of binary mixtures, the Figure 5 clearly indicates that some of the metal ions compete with one another for the material binding sites. It worth note that Pb(II) binding to GS is almost unaffected by the presence of Cu(II), Ni(II) and Cd(II). This result is a consequence of the highest affinity of Pb(II) through GS and is in agreement with the affinity ranking proposed in the analysis of the binary mixtures.

The figure exhibits a constant shortening of the breakthrough time from cycle to cycle for all the metals, taking place the most important shortening when the first (raw material) and second (after the first acidic desorption and re-conditioning of the bed) are compared. This is indicative of a broadening of the mass transfer sorption zone with increasing the age of the biosorbent. The reason for the shortening breakthrough time was apparently not the diminishing equilibrium uptake capacity, as it can be observed in Table 4, but rather a slight change in the overall adsorption rate of the system. This fact is indicating that while sorbent sites of the biomass are still available, they become less accessible after successive sorption/desorption cycles.

As it can be observed in Figure 5 the pH decreases from the beginning of the process and then, progressively, increases to reach final values around 4.5. While in the first sorption cycle a minimum pH of 3.5 is achieved in approximately 8 hours, in the case of the 2nd, 3rd and 4th sorption cycles (all carried out after acidic desorption and pH readjustment of the sorbent bed) more acidic pHs, about 2.7, are reached in only 1 hour after the beginning of the process. The profile of pH variation with time demonstrates that H^+ plays a very important role in the sorption mechanism after acidic regeneration of the sorbent bed. It has to be highlighted also that light metal release (K^+, Na^+, Mg^{2+} and Ca^{2+}) from the material was followed in eluted stream and a significant Ca^{2+} release was only observed in the first cycle (data not shown). In all the subsequent sorption cycles no light metal release was observed. These results point out that an ion exchange mechanism is the main responsible of metal sorption. In the raw material, sorption might be associated in some extent to a Ca^{2+} and H^+ ion exchange, according to reactions summarized as (1) and (3), while desorption proceeds by protonation of the material and subsequent release of the metal, according to reaction (2).

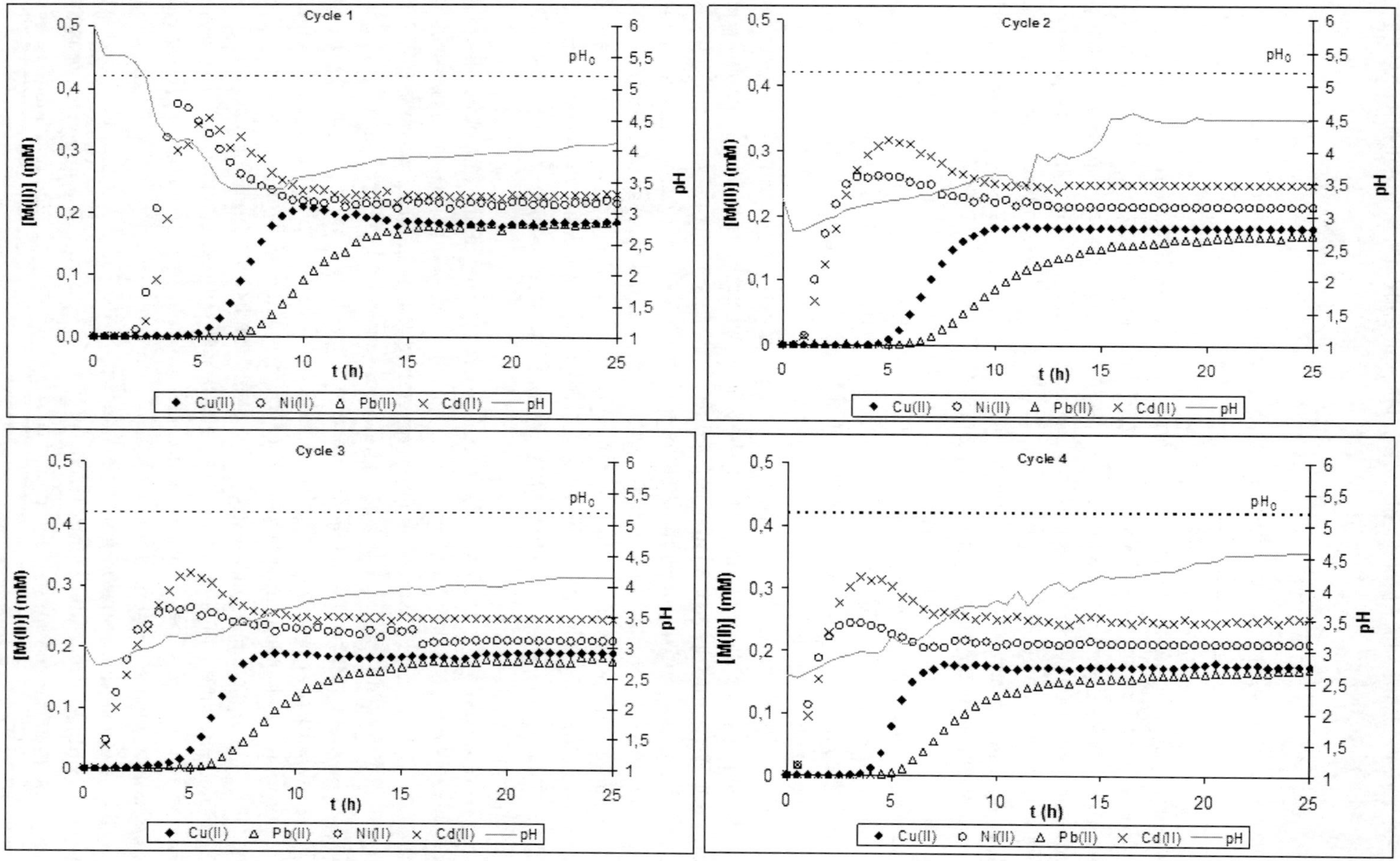

Figure 5. Cu(II), Ni(II), Pb(II) and Cd(II) concentration in solution and pH profile as a function of time for the quaternary mixture in the different sorption cycles. Solid and dotted lines represent outlet and inlet effluent pHs respectively. pH_0: 5.2; $[M(II)]_0 \approx 0.2$ mM; Flow rate≈30 $ml \cdot h^{-1}$; Sorbent mass≈0.5 g.

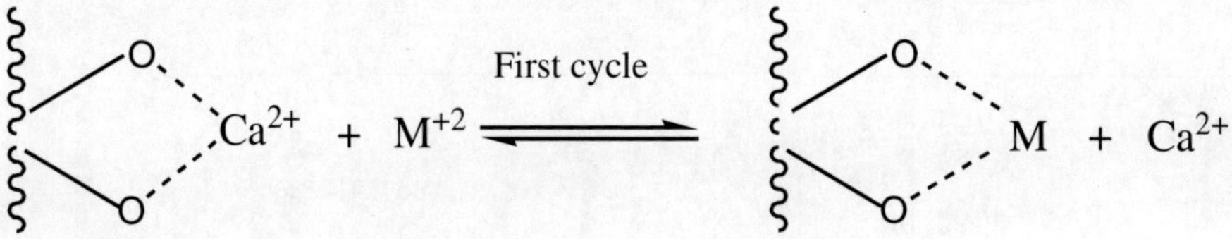

(18)

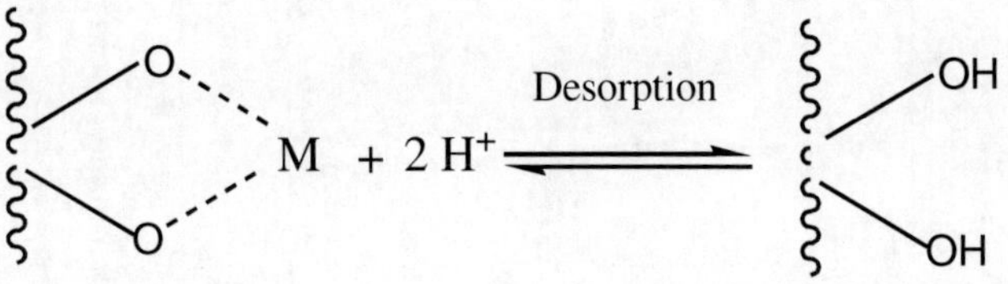

(19)

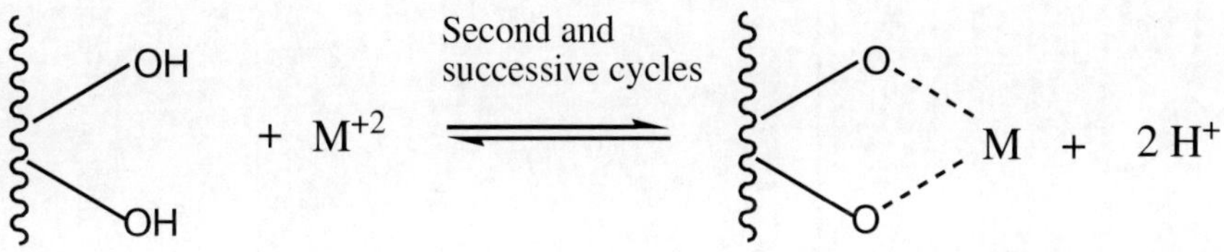

(20)

As in the case of binary metal mixtures, metal sorption from the quaternary mixture can be also discussed in terms of metal accumulation in the solid phase as a function of time. In Figure 6 Cu(II), Ni(II), Pb(II) and Cd(II) sorbed amount as a function of time for each sorption cycle is presented.

Furthermore, to evaluate the potential decay in sorption performance when submitting the material to successive sorption-desorption cycles, the sorbed amount as a function of time on each sorption cycles has been plotted separately for each one of the metals. These results are presented in Figure 7.

The effect of the acidic desorption in the equilibrium achieved in the different sorption cycles has been explored by calculating the individual metal capacity ($q_{M(II)}$ ($mmol \cdot g^{-1}$)), total metal capacity (Q_T ($mmol \cdot g^{-1}$)) and percentage of contribution of each metal to the total metal sorbed amount onto the material (R (%)). The parameter R(%) was calculated according to the equation: $R=(q_{M(II)}/Q_T) \cdot 100$. In the next table a summary of these parameters is presented.

As it can be observed in the table the sorption capacity of GS seems not to be clearly modified after the different acidic treatments used in metal recovery. In fact, the total sorbent capacity, Q_T ($mmol \cdot g^{-1}$), obtained as sum of the individual metal capacity, seems to indicate a general increasing trend. This evidence suggests the formation or activation of new sorption-active sites that could enhance the sorption capacity of the material.

The values of the parameter R(%) presented in Table 3 indicate that Cu(II) and Pb(II) are the dominant sorbed metals at equilibrium, as the sum of their individual R(%) varies from 95.5 in the 1st sorption cycle to approximately 80% in the 4th. On the other hand, Ni(II) and Cd(II) represent only a small fraction of the total sorbed metal amount. Its contribution, expressed as sum of R(%) for the two metals, varies from about 5.5 to approximately 21% in the 1st and the 4th cycle respectively. The low remaining sorbed concentration at equilibrium for Cd(II) and Ni(II) makes difficult an accurate discussion of the effect that the acidic desorption exerts in the sorption of Cd(II) and Ni(II).

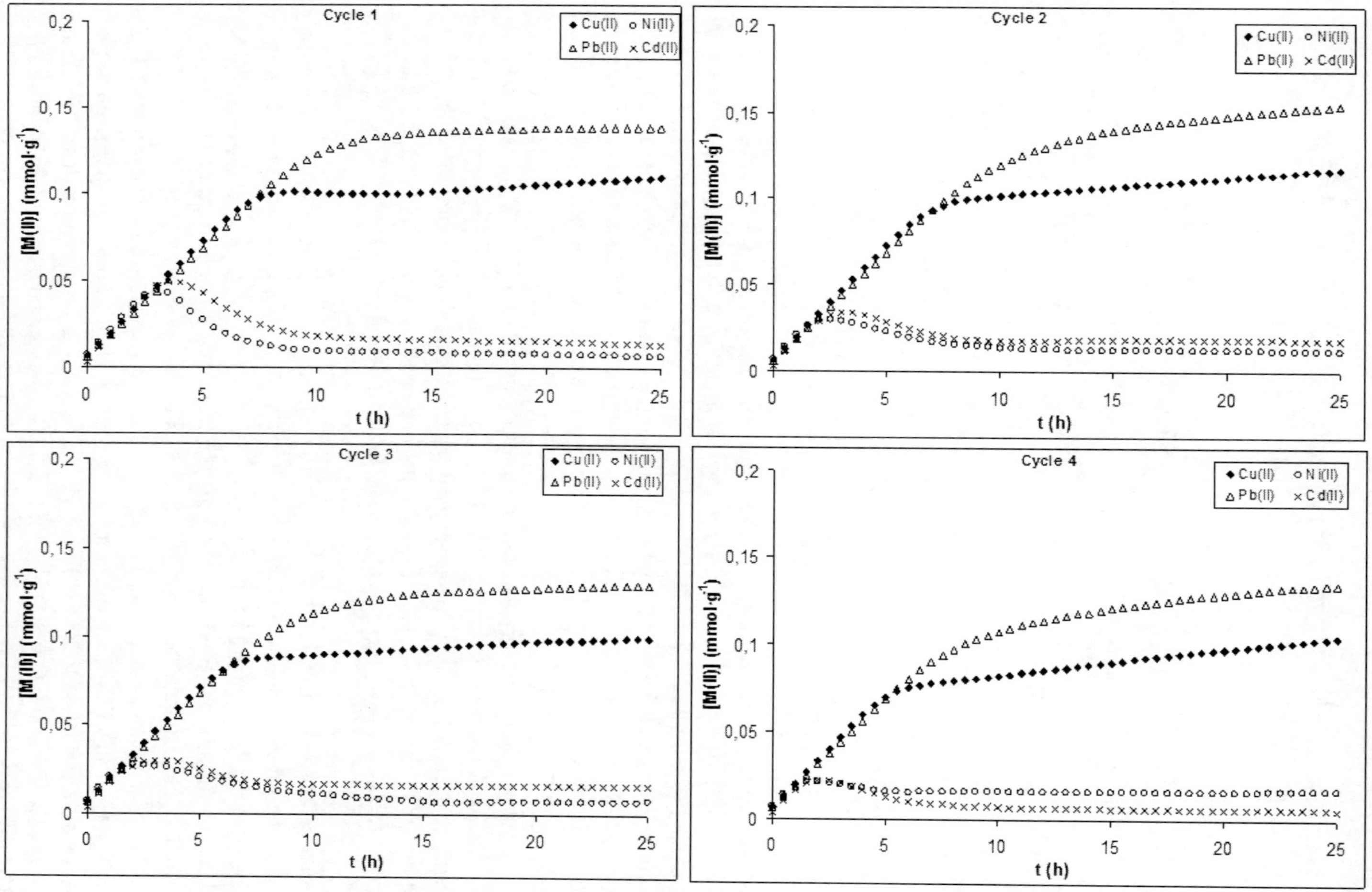

Figure 6. Capacity evolution as a function of time for the adsorption of Cu(II), Ni(II), Pb(II) and Cd(II) in different sorption cycles in the quaternary mixture. pH_0: 5.2; $[M(II)]_0 \approx 0.2$ mM; Flow rate≈30 $ml \cdot h^{-1}$; Sorbent mass≈0.5 g.

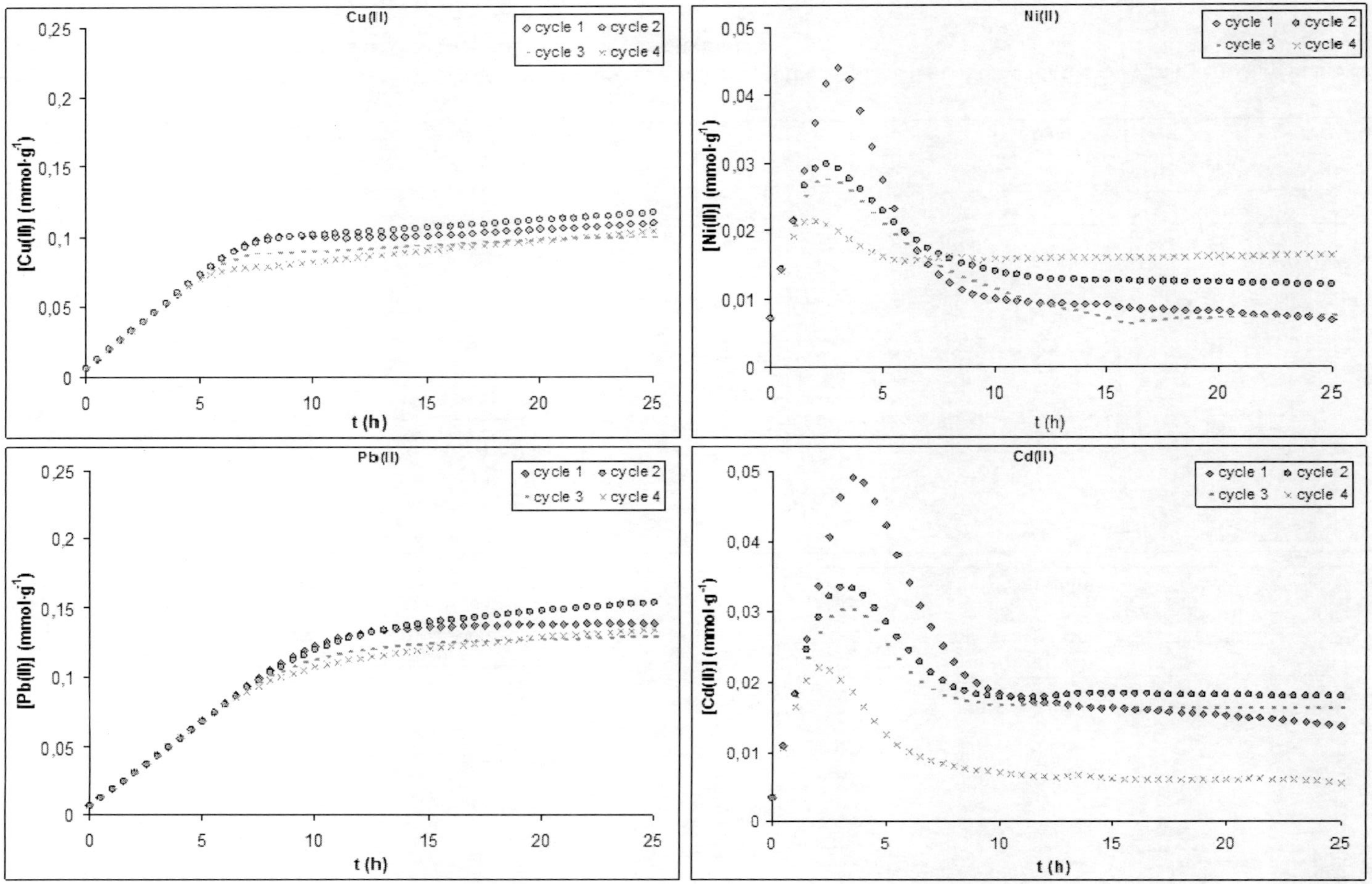

Figure 7. Individual capacity evolution as a function of time for the adsorption of Cu(II), Ni(II), Pb(II) and Cd(II) in the different sorption cycles in the quaternary mixture. pH_0: 5.2; $[M(II)]_0 \approx 0.2$ mM; Flow rate≈30 $ml \cdot h^{-1}$; Sorbent mass≈0.5 g.

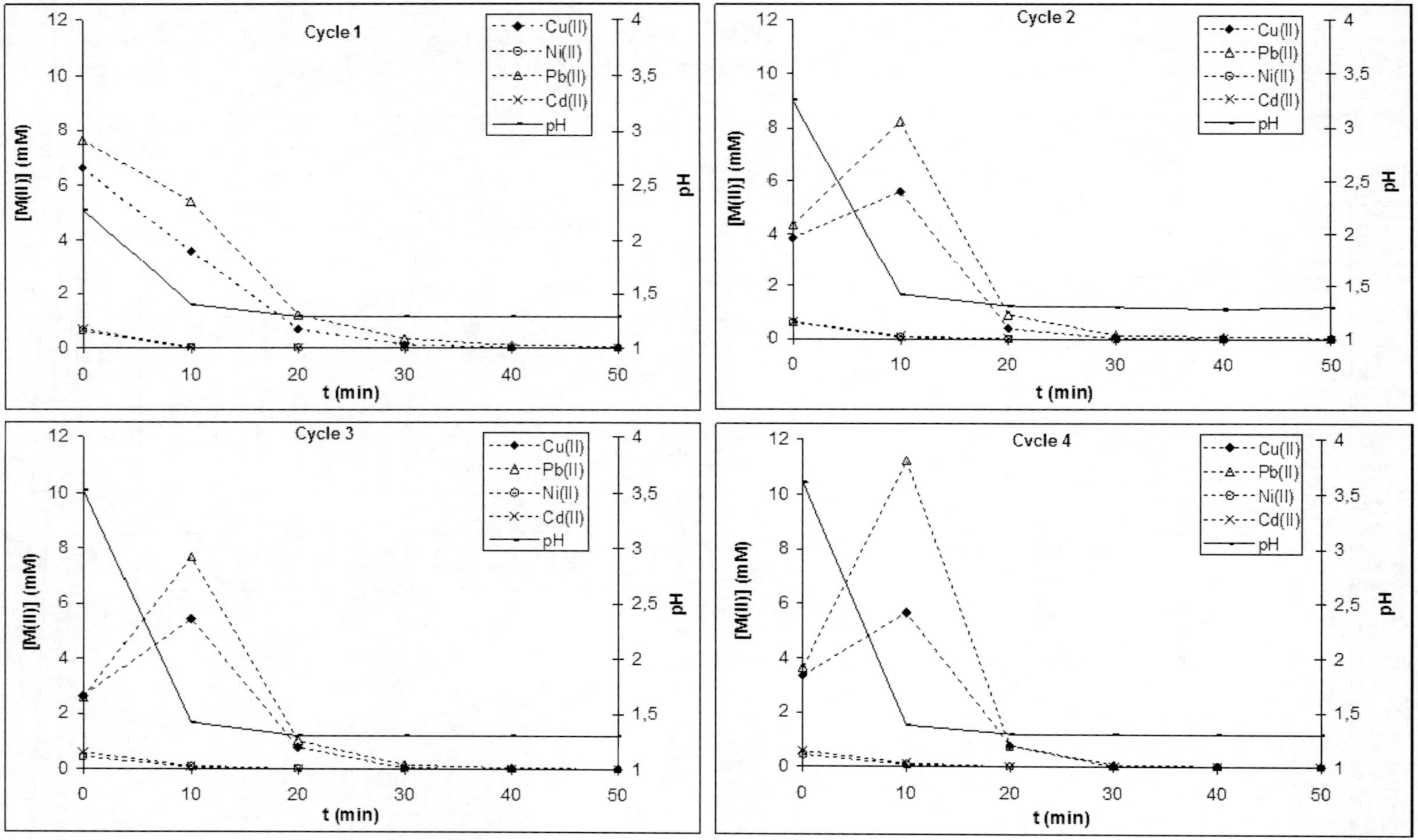

Figure 8. Cu(II), Ni(II), Pb(II) and Cd(II) concentration in solution ($mmol \cdot L^{-1}$) and pH profile as a function of time for the quaternary mixture in the different desorption cycles. Feeding HCl concentration: 0.05 M (pH0=1.30); Flow rate≈30 $ml \cdot h^{-1}$.

Table 3. Cu(II), Ni(II), Pb(II) and Cd(II), total divalent metal sorption capacity and relative percentage of each metal sorbed (R) in the different sorption cycles

Cycle	$q_{M(II)}$ (mmol·g^{-1})				Q_T (mmol·g^{-1})	R(%)			
	Cu	Ni	Pb	Cd	Cu+Ni+ Pb+Cd	Cu	Ni	Pb	Cd
1	1.11E-1	5.51E-3	1.39E-1	1.22E-2	2.64E-1	42.09	2.09	52.47	3.36
2	1.21E-1	1.18E-2	1.58E-1	1.82E-2	3.08E-1	39.11	3.82	51.17	5.91
3	1.00E-1	7.80E-3	1.29E-1	1.64E-2	2.86E-1	38.41	3.48	48.43	9.67
4	1.08E-1	1.66E-2	1.38E-1	4.34E-3	3.32E-1	35.31	15.76	44.13	4.80

* **Q_T**: Total divalent metal sorbed amount: $q_{Cu}+q_{Ni}+q_{Pb}+q_{Cd}$

* **R**: Relative percentage of Cu(II), Ni(II), Pb(II) and Cd(II) in the solid phase at equilibrium after sorption from the quaternary mixture. $R=(q_{M(II)}/Q_T)\cdot 100$

Once evaluated the performance of the material in sorption after regeneration of the sorbent bed it becomes mandatory to explore the desorptive process itself in views of a full application of this technology. In literature it can be found the use of mainly two types of desorbing agents for heavy metal cations: (i) acids (Atdor et al. 1995; Ahuja et al. 1999; Zhou et al. 1998) and (ii) complexing agents (Gong et al. 2005; Hashim et al. 2000). Despite complexing agents such as EDTA have been successfully employed as desorbing agents of different metal cations from different sorbents, the strong chelating activity of these molecules could become an important drawback for a subsequent metal recovery process from the concentrated eluate. To avoid this, an acidic desorption process was chosen for the recovery of the metals and regeneration of the sorbent bed.

When a fresh desorptive agent (that does not contain the components to be stripped) is fed through a metal laden sorbent bed in a continuous bed up flow process, it is expected to obtain a characteristic time dependent profile based on an increase of sorbate concentration in the eluate, achievement of a maximum and decrease to reach a 0 value. In our quaternary system the eluted metal concentration has been plotted as a function of time and the characteristic desorption curves have been obtained (Figure 8). Desorption results have been plotted in mM units, allowing to easily compare the behaviour of the different metals. The pH evolution of the effluent during the process was also monitored and the results have been included in the same figure.

Despite desorption seems to proceed very fast (approximately 30 minutes), the process exhibits a different profile that is metal-dependent. While for Cu(II) and Pb(II) in the 2nd, 3rd and 4th cycles the peak of maximum eluted concentration occurs for the 2nd fraction (corresponding to 10 minutes of desorption), in the case of Ni(II) and Cd(II), the highest eluted concentration is found in the 1st fraction (corresponding to the first sample eluted from the column). In the first cycle, the desorption profile observed was identical for the four metals; maximum eluted concentration observed in the first eluted fraction.

In the desorption process it worth note the important rise on the solution pH when the first volume of desorbing agent elutes after passing the bed. A pH increase from the feeding value (1.3) to values ranging from 2.25 to 3.53 from the 1st to the 4th desorption cycle respectively were detected. After this first elution, the pH rapidly decreases in the outlet flow reaching the initial 1.3 values after 20 minutes for all the cycles. This important pH increase in desorption can be explained by the strong proton consumption that occurs when metals bonded to the biomaterial are replaced by H^+.

The sorption performance and stability of the material in the different desorption cycles was evaluated by accounting the percentage of metal recovered on each desorption cycle. The recovery percentage is referred to the individual metal sorbed amount observed on each sorption cycle, $q_{M(II)}$, presented previously in Table 3. The results of metal recovery after the different desorption cycles are presented in Table 4.

If recovery percentage for the different metals is plotted in the different cycles, the next graphic is obtained.

The results obtained in the desorption study indicate that Cu(II) and Pb(II) can be quantitatively recovered from the sorbent bed in the mild acidic conditions of this study. Figure 9 puts into evidence also that the recovery percentage achieved for these two metals is not dependent on the number of sorption/desorption cycles. In the case of Ni(II), a decrease in the efficiency of the recovery process takes place when increasing the number of desorption cycles while in the case of Cd(II), the recovery seems to exhibit a general increasing trend. The different observed phenomena for Cu(II), Ni(II), Pb(II) and Cd(II) may be caused by changes on the chemistry and of the structure of the biosorbent, as well as by changes of the flow and mass transport conditions within the column. Sorption properties through a given metal may be enhanced or diminished by chemical changes of the structure derived from the acidic treatment, such as activation/deactivation or generation/destruction of specific sorption sites on lignin, which is known to play a major role in biosorption by this kind of materials (Escudero et al. 2013, Villaescusa et al. 2004).

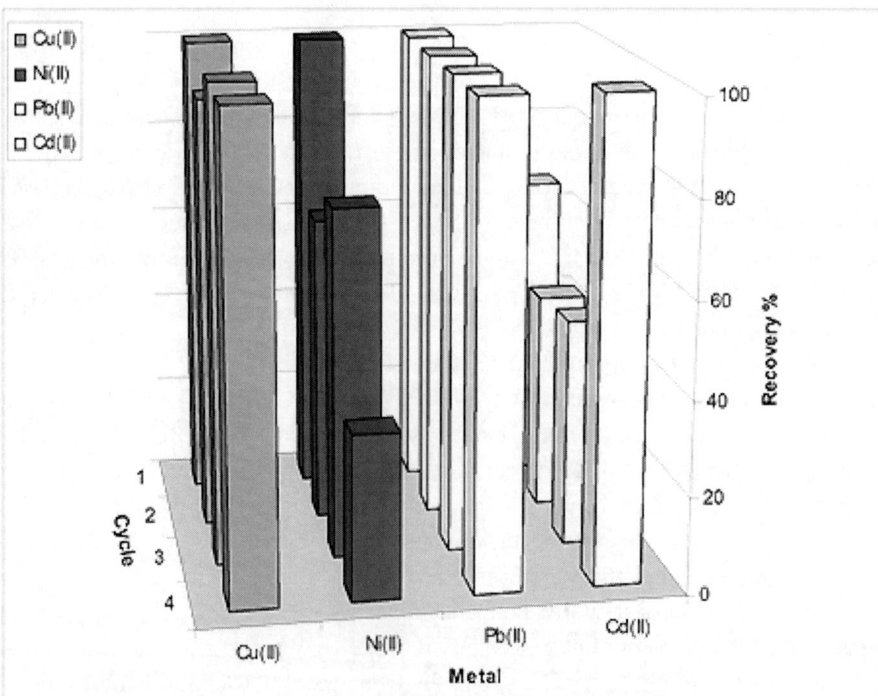

Figure 9. Recovery percentage for Cu(II), Ni(II), Pb(II) and Cd(II) in the different desorption cycles after sorbent loading with the quaternary mixture. $[M(II)]_0 \approx 0.2$ mM.

Table 4. Cu(II), Ni(II), Pb(II) and Cd(II) recovery percentage after sorption from the quaternary mixture for the different desorption cycles

Cycle	Recovery (%)			
	Cu	Ni	Pb	Cd
1	111.82	129.40	121.23	66.04
2	92.29	65.61	99.68	46.44
3	99.98	73.80	100.21	48.03
4	102.10	34.59	100.45	100.49

If GS sorption behaviour in the quaternary mixture is accepted, taking into account that only Cu(II) and Pb(II) are successfully adsorbed, it can be concluded that the reutilization of the biomaterial is possible. Desorption yields a relatively small volume of highly concentrated metal solutions making possible to eventually recover the metal by conventional electrowinning processes (Volesky, 2003).

Conclusion

In this chapter a low cost technology based on sorption onto grape stalks has been explored for the removal of divalent heavy metals in binary and quaternary mixtures in a continuous bed up-flow process. The target pollutants were Cu(II), Ni(II), Pb(II) and Cd(II) chosen the first two by their industrial recurrence and the last two because of its special high toxicity. The results revealed the existence of an intense competence between metals and allowed to obtain the next sorption affinity ranking:

$$Pb(II)>Cu(II)>Cd(II)>Ni(II)$$

The metals that exhibit higher sorption affinity are able to replace from the GS active sites to these whose affinity is lower. This replacement provokes an overconcentration respect to the feeding metal concentration (overshoot) for the metal that is being displaced.

The results presented here put into evidence that the quantitative recovery of the metal is feasible by feeding the sorbent bed with aqueous mild acidic solutions. This process allowed both: to recover the metal in a concentrated form and to reuse the sorbent bed in multiple sorption cycles.

It is worth note that the valorisation of GS wastes that otherwise would be either sent to incineration plants, field disposal or landfill is an environmentally-friendly practice and is strongly supported by government policies. The processing of sorbents, once they loss their removal performance might not pose an important problem as the heavy metal can be efficiently removed from the sorbent by desorption. The material becomes then a non-hazardous waste that could be sent to a landfill, burned in an incinerator or, after neutralization, used in composting processes or as agricultural amendment.

The evidences presented in this chapter demonstrate that GS can be used as a low cost and sustainable alternative for the removal of toxic metals from aqueous effluents.

References

Abe, I., Tatsumoto, H., Ikuta, N. & Kawafune, I. (1990). Preparation of activated carbon from pisthachio nut shell. *Chem. Exp.*, *5*, 177–180.

Ahuja, P., Gupta, R. & Saxena, R. X. (1999). Sorption and desorption of cobalt by Oscilatoria anguistissima. *Current Microbiol.*, *39*, 49–52.

Aleixandre J. L. (1999). *Vinos y bebidas alcohólicas*. Universidad Politécnica de Valencia. Valencia.

Alkorta I., Hernández-Allica Becerril J. M., Amezaga I., Albizu I. & Garbisu C. (2004). Recent findings on the phytoremediation of soils contaminated with environmentally toxic heavy metals and metalloids such as zinc, cadmium, lead, and arsenic. *Rev. Environ. Sci. Biotechnol.*, *3*, 71–90.

Antoniadis, V. & Tsadilas, C. D. (2007). Sorption of cadmium, nickel and zinc in mono- and multimetal systems. *Appl. Geochem.*, *22*, 2375-2380.

Atdor, I., Fourest, E. & Volesky, B. (1995). Desorption of cadmium from algal biosorbent. *Can. J. Chem. Eng.*, *73*, 516–522.

Bishop P. L. (2002). *Pollution prevention: fundamentals and practice*. Beijing: Tsinghua University Press.

Boulton R. B., Singleton, V. L., Bisson, L. F. & Kunkee, R. E. (2002). *Teoría y práctica de la elaboración del vino*. Acribia, S.A. Zaragoza. España.

Chang, J. S. & Chen, C. C. (1999). Biosorption of lead, copper and cadmium with continuous hollow-fiber microfiltration processes. *Sep. Sci. Technol.*, *34*, 1607-1627.

Council Directive 98/83/EC of 3 November 1998 on the quality of water intended for human consumption, *Off. J. Eur. Union*, *5*, 12.98.

Covelo, E. F., Andrade, M. L. & Vega, F. A. (2004). Heavy metal adsorption by humic umbrisols: selectivity sequences and competitive sorption kinetics. *J. Coll. Interf. Sci.*, *280*, 1-8.

Escudero, C., Poch, J. & Villaescusa, I. (2013). Modelling of breakthrough curves of single and binary mixtures of Cu(II), Cd(II), Ni(II) and Pb(II) sorption onto grape stalks waste. *Chem. Eng. J.*, *217*, 129–138.

Escudero, C. Fiol, N., Villaescusa, I. & Bollinger, J. C. (2013). Effect of chromium speciation on its sorption mechanism onto grape stalks entrapped into alginate beads. *Arabian J. Chem.* http://dx.doi.org/10.1016/j.arabjc.2013.03.011.

Gong, R., Ding, Y., Lui, H., Chem, Q. & Liu, Z. (2005). Lead biosorption and desorption by intact and pretreated Spirula maxima biomass. *Chemosphere*, *58*, 125–130.

Hammaini, A., Ballester, A., González, F., Blázquez, M. L. & Muñoz, J. A. (1999). Activated sludge as biosorbent for heavy metals. *Biohydrometallurgy and the environment toward the mining of the 21st century, Part B*. Amsterdam: Elsevier 185-192.

Hashim, M. A., Tan, H. N. & Chu, K. H. (2000). Immobilized marine algal biomass for multiple cycles of copper adsorption and desorption. *Sep. Purif. Technol.*, *19*, 39–42.

Hayashi, J., Horikawa, T., Muroyama, K. & Gomes, V. G. (2002). Activated carbon from chickpea husk by chemical activation with K_2CO_3: preparation and characterization. *Micropor. Mesopor. Mater.*, *55*, 63–68.

Kirubakaran, C. J. & Krishnaiah, Seshadri, S. K. (1991). Experimental study of the production of activated carbon from coconut shells in fluidized bed reactor. *Ind. Eng. Chem. Res.*, *27*, 2411–2416.

Kratochvil, D. & Volesky, B. (1998). Advances in the biosorption of heavy metals. *Trends Biotechnol.*, *16*, 291–300.

Kratochvil, D. & Volesky, B. (2000). Multicomponent biosorption in fixed beds. *Water Res.* *34*, 3186-3196.

Maniatis, K. & Nurmala, M. (1992). Activated carbon production from biomass. *Biomass Energy Ind. Environ.*, *274*, 1034–1308.

Merian, E. Anke, M., Stoeppler, M. & Ihnar, M. (1991). *Metals and their compounds in the environment, occurrence, analysis and biological relevance (1st ed.)*. New York. Wiley, John & Sons, Incorporated.

Organización Internacional del Trabajo (OIT). Ministerio de Trabajo y Asuntos Sociales. (2001). *Enciclopedia de Salud y Seguridad en el Trabajo*. Capítulo 63. Metales: propiedades químicas y toxicidad. 3ª Ed.

Pagnanelli, F., Esposito, A., Toro, L. & Vegliò, F. (2003). Metal speciation and pH effect on Pb, Cu, Zn and Cd biosorption onto Sphaerolitus natans: Langmuir-type empirical model. *Water Res.*, *37*, 627-633.

Pujol, D., Liu, C., Fiol, N., Olivella, M. A., Gominho, J., Villaescusa, I. & Pereira, H. (2013). Chemical characterization of different granulometric fractions of grape stalks waste. *Industrial Crops and Products.*, *50*, 494– 500

Puranik, P. R. & Paknikar, K. M. (1999). Influence of co-cations on biosorption of lead and zinc-a comparative evaluation in binary and multimetal systems. *Bioresource Technol.*, *70*, 269-76.

Rai, L. C. (2000). Metal removal from single and multimetallic systems y different biosorbent materials as evaluated by differential pulse anodic stripping voltammetry. *Process Biochem.*, *36*, 175-82.

Saeed, A., Iqbala, M. & Akhtar, M. W. (2005). Removal and recovery of lead(II) from single and multimetal (Cd, Cu, Ni, Zn) solutions by crop milling waste (black gram husk). *J. Hazard. Mater. B117*, 65–73.

Sag, Y. & Kutsal, T. (2001). Recent trends in the biosorption of heavy metals: a review. *Biotechnol. Bioprocess Eng.*, *6*, 376–385.

Seco, A., Marzal, P. & Gabaldón, C. (1997). Adsorption of heavy metals from aqueous solutions onto activated carbon in single and in binary Cu-Ni, Cu-Cd and Cu-Zn systems. *J. Chem. Technol. Biotechnol*, *68*, 23-30.

Smith, R. M. & Martell, A. E. (1976). *Critical stability constants, vol. 4*. New York: Plenum Press.

Suzuki, R. M., Andrade, A. D., Sousa, J. C. & Rollemberg, M. C. (2007). Preparation and characterization of activated carbon from rice bran. *Bioresource Technol.*, *98*, 1985–1991

Uslu, G. & Tanyol, M. (2006). Equilibrium and thermodynamic parameters of single and binary mixture biosorption of lead (II) and copper (II) ions onto Pseudomonas putida: Effect of temperature. *J. Hazard. Mater*, *B135*, 87–93.

Vegliò, F., Di Biase, A., Beolchini, F. & Pagnanelli, F. (2002). Heavy metal biosorption in binary systems: simulation in single-and two-stage UF/MF membrane reactors. *Hydrometallurgy*, *66*, 107-115.

Vilar V. J. P., Loureiro, J. M., Botelho, C. M. S. & Boaventura, R. A. R. (2008). Continuous biosorption of Pb/Cu and Pb/Cd in fixed-bed column using algae Gelidium and granulated agar extraction algal waste. *J. Hazard. Mater*, *154*, 1173-1182.

Villaescusa, I., Fiol, N., Martínez, M., Miralles, N., Poch J. & Serarols, J. (2004). Removal of copper and nickel ions from aqueous solutions by grape stalks wastes. *Water Res.*, *38*, 992–1002.

Volesky, B. (2003). *Sorption and Biosorption, first ed.* BV Sorbex, Inc., Quebec.

Volesky B. (1990). *Biosorption of heavy metals*. Florida: CRC press.

Volesky, B., Weber, J. & Park, J. M. (2003). Continuous-flow metal biosorption in a regenerable Sargassum column. *Water Res.*, *37*, 297-306.

Wang, J. & Chen, C. (2006). Biosorption of heavy metals by Saccharomyces cerevisiae: A review. *Biotechnol. Adv.*, *24*, 427-451.

Yin, P., Yu, Q., Jin, B. & Ling, Z. (1999). Biosorption removal of cadmium from aqueous solutions by using pretreated fungal biomass cultured from starch wastewater. *Water Res.*, *33*, 1960-1963.

Zhou, J. L., Huang, P. L. & Lin., R. G. (1998). Sorption and desorption of Cu and Cd by macroalgae and microalgae. *Environ. Pollut.*, *101*, 67–75.

In: Grapes
Editor: José de Sousa Câmara ISBN: 978-1-63321-402-6

Chapter 11

Thermal Stability of Major Classes of Polyphenols in Skins, Seeds and Stems of Grape Pomace

Jianmei Yu*

Food and Nutritional Sciences Program, Department of Family and Consumer Science, North Carolina A&T State University, Greensboro, NC, US

Abstract

Grape pomace (GP) is a by-product of wine industry, in which remains significant amount of polyphenols. The freshly pressed pomace is perishable and has to be dried for use at later time. Depending on the drying methods, significant loss in polyphenols may occur during drying. This chapter describes the impacts of dehydration methods such as room temperature drying and vacuum drying on the retention of polyphenols in the pomace of Cabernet Sauvignon and Muscadine Nobel grapes using freeze dried sample as control. Seeds, skins and stems were separated manually after drying and the effects of drying methods on the stability of GP polyphenols were evaluated by the retentions of total polyphenol (TP), total anthocyanins (TA) and total flavonoids (TF). Results show that seeds and stems from both grape varieties were rich in TP and TF but low in TA, while skins were rich in anthocynin. Compared to freeze drying, both room temperature and vacuum oven drying methods resulted in significant loss of TP, TA and TF in different parts of Muscadine pomace. Among different groups of polyphenols, anthocyanin was affected the most by drying methods. Flavonoids in Cabernet pomace exhibited higher heat stability than that in Muscadine pomace indicating the differences in flavonoid composition between the two grape varieties. Therefore, the polyphenol composition and stability of GP are variety dependent. The selection of drying method needs to consider what type of polyphenol to be preserved, variety of GP and the cost of drying method.

* Corresponding Author : Email: jyu@ncat.edu.

Keywords: Grape Pomace; Thermal Stability; Drying Methods; Total Polyphenol; Total Anthocyanin, Total Flavonoids

Introduction

Grapes offer a richer polyphenol profile than many other fruits. Grape pomace (GP), the residue of grape after wine making, remains significant amount of polyphenols. The health benefits of GP polyphenols have been the great interest of researchers, food manufacture and nutraceutica/pharmaceutical industry. Grape seed procyanidin extract (GSPE) modulates dyslipidemia associated with a high-fat diet in rats and repress genes controlling lipogenesis and VLDL assembling in liver (Baiges et al., 2010). Low dose (25mg per kg body weight per day) GSPE treatment of high-fat-diet (HFD) fed rats significantly reduced the adiposity index and the weight of all the white adipose tissue depots and reversed the increase in plasma phospholipids induced by the HFD feeding (Caimari et al., 2012). Chronic consumption of grape phenolics has been shown to reduce obesity development and related metabolic pathways including adipokine secretion and oxidative stress in a rat model (Décordé et al., 2009). Cancer chemopreventive and anticancer efficacy of grape seed extract and other grape-based products were summarized by Kaur et al., (2009). Grape seed extract was studied for use as an additive to inhibit the oxidation of meat products (Mielnik et al., 2006; Brannan and Mah, 2007) and bread (Peng et al., 2010).

GP is composed of grape skin, seeds and stem, and is rich in different type of extractable phenolic antioxidants (10-11% of dry weight) (Makris, et al., 2007). The grape skins, part of pomace, are proven to be rich sources of anthocyanins, hydroxycinnamic acids, flavanols, and flavonol glycosides, whereas flavanols were mainly present in the seed portion (Kammerer et al., 2004). The content of total phenolics of grape skin and grape seeds varies with grape variety, climate condition, soil type, and maturity (Fuleki 1997; Shi, 2003; McCalluma et al., 2007; Montealegre et al., 2006). Because grapes are seasonal agricultural products and the freshly pressed pomace is perishable and has to be dried or stored at freezing temperature for use at later time. Depending on the type of drying methods, significant loss in polyphenols may occur during drying.

The stability of fruit polyphenols under some thermal processing condition was studied by many researchers. Larrauri et al., (1997) reported that drying of grape seeds at 100 and 140 °C resulted in 18.6 and 32.6% reduction of extractable total polyphenols, respectively, and reduced antioxidant activity of grape seeds compared with freeze drying. It was also reported that some thermal processing methods increased the extractability of polyphenols from plant matrix. Roasting of peanuts and almonds before decoating not only increased total extractable phenolics in the skins, but also resulted in changes of phenolic composition of the skin extracts, for example, it increased procyanidin oligomers and decreased monomers (Yu et al., 2007; Garrido et al., 2008). Extrusion processing has been found to increase small molecule procyanidins in sorghum, grape seed, and grape as well as blueberry pomace while decreasing the procyanidin oligomers or polymers in the process (Khanal, 2009a; Khanal, 2009b). Dry heating of freeze dried GP in a forced air oven at 40, 60, 105, and 125 °C for 72, 48, 16, and 8 hrs respectively decreased anthocyanin and procyanidin concentrations significantly, except when heated at 40 °C for 72 hrs. The reduction of procyanidin occurred when heated at 60 °C or above with no further reduction when heating temperature increased from 105 to 125 °C, and total anthocyanin loss was highest at 125 °C (Khanal et al., 2009b and Khanal et al., 2010). Phenolic extracts of thermal processed whole and powdered grape seeds had higher *in vitro* antioxidant activity than those of untreated grape seeds (Kim et al., 2006).

However, the studies mentioned above were conducted at very high temperature. No significant loss of both procyanidin and anthocyaninin was observed when heated at 40 °C for up to 3 days (Khanal et al., 2010). Polyphenols are very sensitive to UV light (Shi et al., 2003), sundry is usually not used to dry polyphenol rich products. Freeze drying preserve the bioactive compounds but the cost is very high. Drying at mild temperature such as vacuum drying is faster and more cost effective than freeze drying, but the changes in polyphenol composition due to vacuum drying of GP has not been reported. The time needed to dry GP and the retention of polyphenol at room temperature (about 25ºC) have not been reported either. This chapter focused on the stability of GP polyphenols under mild drying condition such as room temperature drying under dim light and vacuum drying under dim light using freeze drying as control. The composition and stability of polyphenols in Cabernet Sauvignon and Muscadine Noble were compared.

Materials and Methods

Materials

Grape pomaces (GP) The pomaces of two grape varieties/cultivars were obtained from two North Carolina wineries. Fermented Cabernet Sauvignon pomace was provided by Grove Winery (Gibsonville, NC, USA), while fermented Muscadine Nobel pomace was provided by Benjamin Vineyard and Winery (Graham, NC, USA). Pomaces were picked right after pressing and were stored in refrigerator before drying.

Chemicals: Folin-Ciocalteu reagent, gallic acid and (+)-catechin were purchased from Sigma-Aldrich (St. Louis, MO, USA), other chemicals including reagent ethanol, acetonitrile (HPLC grade), acetic acid, sodium nitrite, sodium acetate, aluminum chloride and hydrochloric acid were purchased from Fisher Scientific (Atlanta, GA, USA).

Dehydration of Grape Pomace

The wet GP from each grape variety was divided into three portions. Each portion was spread as 1.0 inch thick in a tray. One portion was dried at 70ºC in an Isotemp vacuum oven (Fisher Scientific, Ashville, NC USA) for 24 hrs. Another portion was frozen at -22ºC for 48 hrs, and then dried using a FreeZone 6 Liter Freeze Dryer (Labconco, Kansas City, MO USA) for 48 hrs; the rest was dried in a well ventilated room for one week at 22ºC. The seeds, skins and stems of dry GP were manually separated and weighed. The ratios of seeds, skins and stems to total dry pomace were calculated. They are then ground into fine powders using a coffee grinder. The moisture contents of ground samples obtained from different drying methods were determined by drying the ground pomace at 80ºC for 24 hrs using the vacuum oven. Powders passed through 40 mesh sieves were packed separately in bottles and stored at 4ºC until use.

Polyphenol Extraction

Crude polyphenol was extracted using 70% ethanol/water solution (v/v). Briefly, 10.00g of dry seeds/skins/stems powder was weighed into a 250 ml amber flask, and 100 ml of extraction solvent was added to the flask. The mixture was stirred for 1 hr at room temperature under dim light, and then centrifuged at 3000g for 15 min. The resulting supernatant was collected and the residual solid was extracted one more time using the same procedure. Supernatants were combined and centrifuged again to remove any particles. The volume of extract was recorded. The final supernatants were stored at -22ºC for polyphenol analysis. All extractions and determinations were conducted in triplicate and results were expressed as mean±standard deviation.

Determination of Total Phenolic Content (TP)

The total polyphenol (TP) contents of the extracts were determined by a modified Folin-Ciocalteu method (Singleton et al, 1999). Briefly, 20 µl of properly diluted the sample solution, 1.28 ml of distilled water and 100µl Folin-Ciocalteu's reagents (Sigma-Adrich, St. Louis, MO USA) were mixed and let stand for 8 minutes at room temperature. A volume of 600µl of 10%$NaCO_3$ was added. Deionized water was used as blank to replace 20µl of sample.

The mixture was allowed to stand for 2 hrs at room temperature under dim light or at dark to develop color. The absorbance was measured at 765 nm using a Genesys™ 10 Spectrophotometer (Spectronic Unican, NY). The TP concentration was calculated according to the calibration equation developed using gallic acid as standard, and expressed as gallic acid equivalents (mg GAE/g dry pomace) according to the dilution factor, total volume of extract and weight of pomace. The linear range of the calibration curve was 50 to 1000 µg/ml (r = 0.99).

Determination of Total Anthocyanins (TA)

The TA concentration of extract was determined by AOAC method 2005.02 (AOAC, 2006). Extracts were quantitatively diluted (1:4, v/v) with pH1.0 buffer (0.25M potassium chloride) and pH 4.5 buffer (0.4M sodium acetate), respectively. The absorbance of test portion diluted with pH 1.0 buffer and pH 4.5 buffer were measured at both 520 and 700 nm within 20-50 minutes of preparation using deionized water as blank. Anthocyanin concentration was calculated using the equation below and expressed as cyanidin-3-glucoside equivalents.

$$\text{Anthocyanin (cyanidin-3-glucoside equivalents, mg/L)} = (A \times MW \times DF \times 10^3)/(\varepsilon \times L)$$

where A = (A_{520nm} – A_{700nm}) pH 1.0 – (A_{520nm} – A_{700nm}) pH 4.5; MW (molecular weight) = 449.2 g/mol for cyanidin-3-glucoside (cyd-3-glu); DF = dilution factor; L = path length in cm; ε = 26,900 molar extinction coefficient, in L x mol^{-1} x cm^{-1}, for cyd-3-glu; and 10^3 = factor for conversion from g to mg. The total anthocyanin content in dry pomace was expressed as mg/g dry pomace, according to extract volume and sample weight.

Determination of Total Flavonoid Content (TF)

The TF content was determined using a colormetric method described previously by Xu and Chang (2007)). Briefly, a dose of 0.25ml of the GP extract or (+)-catechin standard solution was mixed with 1.25 ml of distilled water in a test tube, followed by adding 0.75μl of 5% $NaNO_2$ solution. After 6 min, 150μL of a10% $AlCl_3$•$6H_2O$ solution was added and allowed to stand for another 5 min before adding 0.5 mL of 1MNaOH. The mixture was brought to 2.5 mL with distilled water and mixed well. The absorbance was measured immediately against the blank (the same mixture without the sample) at 510 nm using a Genesys™ 10 Spectrophotometer (Spectronic Unican, NY). The results were expressed as microgram of (+)-catechinn equivalents (μg CAE/g sample) using the calibration curve developed from (+)-catechin. The linear range of the calibration curve was 10-1000 μg/ml (R^2=0.9978).

HPLC Analysis of GP Extracts

A Waters HPLC system consisted of In-Line Degasser AF, 717 Autosampler, 1525 Binary HPLC Pump, and 2748 Dual Wavelength Absorbance Detector was used to monitor the changes of individual polyphenol caused by different drying methods. All supernatants were diluted using same dilution factor then filtered by PVDF syringe filters (0.2μm) before injection. The Nucleosil RP C_{18} column (250 mm x 4.6 mm, particle size 5 μm) was used as stationary phase. The samples were eluted by gradient elution mode using binary mobile phase A (2% acetic acid in DI water) and B (100% acetonitrile) at flow rate of 1.0 ml/min. In this mode B was increased from 5% to 75% in 20 min, to 100% in another 5 min, maintained at 100% for 5 min, returned to 75% in 5 min, and then to the initial condition (5%) in 5 min and remained for 5 min before next injection. Peaks eluted were detected at wavelength of 280 nm. Peaks of different samples at same retention time were overlaid for peak height comparison.

Results and Discussion

Moisture Retention of GP Dried by Different Methods

Table 1 shows that the moisture contents of GP dried by different methods differ slightly. The freeze dried samples had lowest moisture while the room temperature dried sample had highest moisture for both Muscadine and Cabernet pomace. Data shows that vacuum oven drying for one day removed more moisture from wet GP than room temperature drying for 7 days. Thererfore, vacuum oven drying is much more efficient than freeze drying and room temperature drying.

Table 1. Moisture contents of grape pomace dried by different methods

Drying method	Muscadine Pomace	Cabernet Pomace
Freeze drying (48hrs)	3.37±0.23	3.54±0.30
Room Temperature drying (1 week)	7.03±0.16	5.05±0.21
Vacuum oven drying (24 hrs)	4.79±0.14	4.36±0.12

Contribution of Seeds, Skins and Stems to the Total Extractable Polyphenols of GP

Table 2 summarizes the percentages of seeds, skins and stems in the dry GP and their contributions to the total polyphenol content of the pomace. Data were obtained from freeze dried samples.

Table 2. The contribution of seeds, skins and stems in dry GP to the total extractable polyphenols of GP (freeze dried samples)

	Muscadine Noble			Cabernet Sauvignon		
	% Dry weight	TP (mg/g)	% Contribution to TP of pomace	% DM	TP	% Contribution to TP of pomace
Seed	36.65	36.42	49.46	43.78	34.67	72.59
Skin	60.72	19.37	44.29	52.79	9.84	24.84
Stem	3.83	44.13	6.26	3.92	13.73	2.57
Total	101.10	25.98	100.01	100.00	20.91	100.01

The seeds, skins and stems contents were 36.65, 60.71 and 3.83% of the total dry weight of Muscadine Noble pomace, and 43.78, 52.79 and 3.92% of the total dry weight of Cabernet Sauvignon pomace, respectively. Although the skin contents were significantly higher than seed content, the latter contributed significantly higher percentage of polyphenols to the pomace, particularly in the Cabernet Sauvignon pomace because of the higher TP content of grape seeds. The contribution of stem to the TP of whole pomace was very limited due its small percentage in the pomace, although its TP content was the highest compared with other parts. Therefore, seeds and skins were the major contributors of GP polyphenol.

Effects of drying methods on TP, TA and TF of grape pomace

Figure 1 shows that the effects of drying methods on the stability of GP polyphenols varied significantly with grape variety. Compared to freeze drying, both room temperature drying and vacuum oven drying resulted in significant loss of TP in each part of Muscadine pomace but not Cabernet pomace. The TP in Cabernet seeds, skins and stems exhibited higher heat stability than that in Muscadine pomace.

The reasons of the observed result might include that 1) the polyphenols remained in Cabernet GP after alcohol fermentation were more stable than those in Muscadine GP, and 2) drying at higher temperature might increased the extractability of polyphenols (Kim et al., 2006) which resulted in relatively constant extractable TP content even some degraded during drying. Although the GP was dried at relatively low temperature in this chapter, the losses of TP in Cabernet and Muscadine seeds were comparable to that reported by Larrauri et al., (1997) that drying of grape seeds at 100 and 140 °C resulted in 18.6 and 32.6% reduction of extractable total polyphenols.

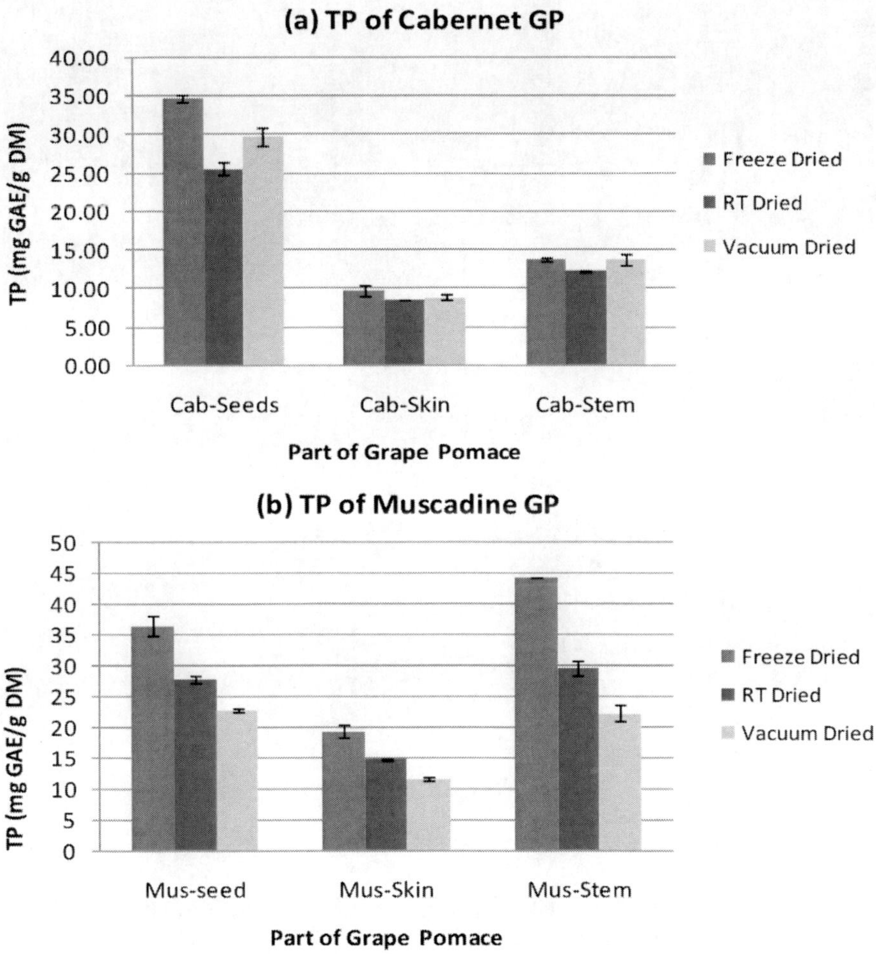

Figure 1. Effects of drying methods on total extractable polyphenol contents (TP) of GP seeds, skins and stems. (a) Cabernet Sauvignon pomace, (b) Muscadine Nobel pomace.

Comparing Figure 2a and Figure 2b, it can be seen that freeze dried Cabernet seeds and Muscadine seeds had similar total anthocyaning (TA) contents, but Muscadine skins and stems had much higher TA than Cabernet skins and stems. Drying at room temperature for 7 days and at 70ºC vacuum oven for 24 hours caused significantly loss of TA in all parts of both pomaces, and room temperature drying seems destroy more TA than vacuum drying due to longer drying time. The results are in good agreement with that reported by Khanal et al., (2009b; 2010). Therefore, the stability of GP anthocynin is very low. The TA degrades significantly within the GP matrix even at room temperature, and faster drying under reduced pressure could preserve more TA than drying at room temperature for longer time.

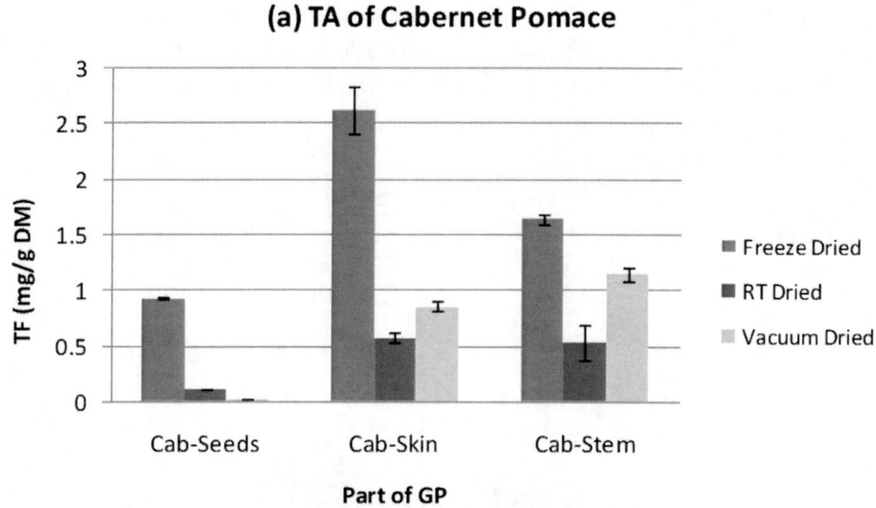

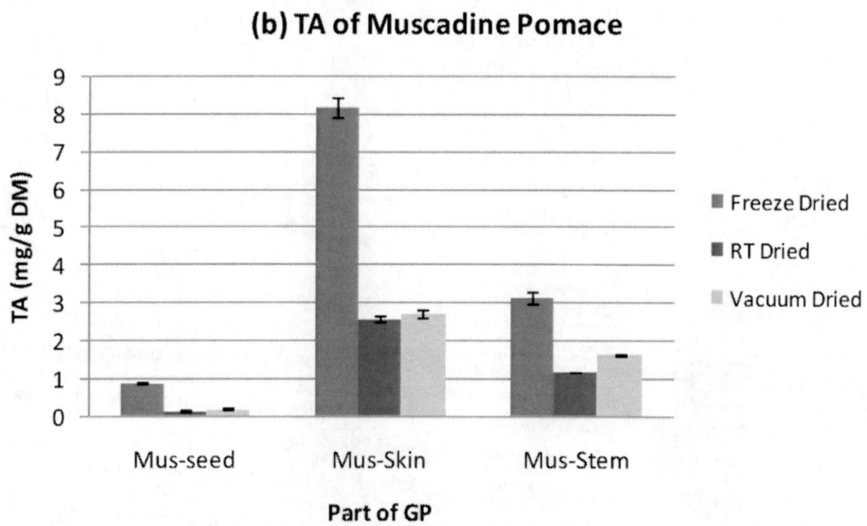

Figure 2. Effects of drying methods on total anthocyanin contents (TA) of GP seeds, skins and stems. (a) Cabernet Sauvignon pomace, (b) Muscadine Nobel pomace.

Figure 3 shows the changes of total flavonoids (TF) in GP due to different drying methods. In freeze dried samples, the TF contents of Cabernet seeds and skins were slightly higher than that of Muscadine seeds and skins, but the TF content of Cabernet stems is much lower than that of Muscadine stems. The room temperature and vacuum drying methods did not seem to cause TF loss in Cabernet seeds, but reduced TF in skins and stems significantly. This is because anthocyanins are a major polyphenols in the red/purple grape skins and stems while catechins and procyanidins are the major polyphenols in the grape seeds (Xia et al.,

2010). The TF contents in all parts of Muscadine pomace were dramatically affected by drying method. The results indicate that TF composition of Cabernet seeds is different from that Muscadine seeds. According to Mohd Zainol et al., (2009), catechin and rutin were the most stable flavonoids found in Centella asiatica. Therefore, catechins might be the major extractable flavonoids in Cabernet seeds, which contributed to the relatively higher thermal stability of TF in Cabernet seeds.

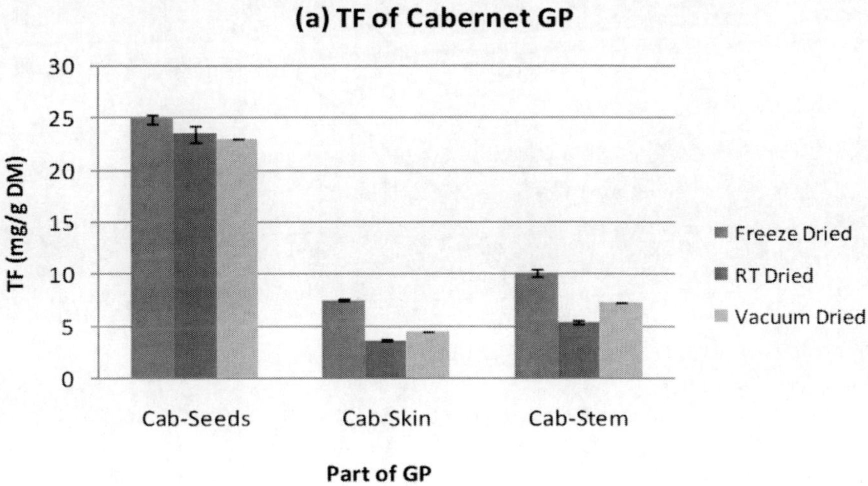

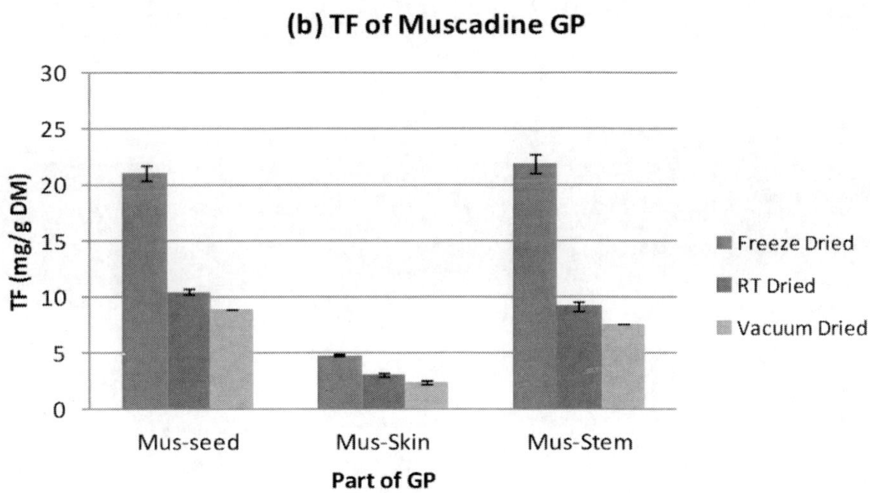

Figure 3. Effects of drying methods on total flavonoid contents (TF) of GP seeds, skins and stems. (a) Cabernet Sauvignon pomace, (b) Muscadine Nobel pomace.

Changes of Individual Polyphenols Due to Drying

Above results indicate the differences in polyphenol composition and stability between the two grape varieties. HPLC analysis was conducted to view the changes of individual polyphenols in the pomace during drying. Due to the lack of mass detector and standards to identify all individual polyphenols, major peaks separated by HPLC were numbered and their heights were qualitatively compared. Figure 4 summarized the effects of vacuum oven drying on the loss of individual polyphenols in skin, stem and seeds of the Muscadine Noble GP.

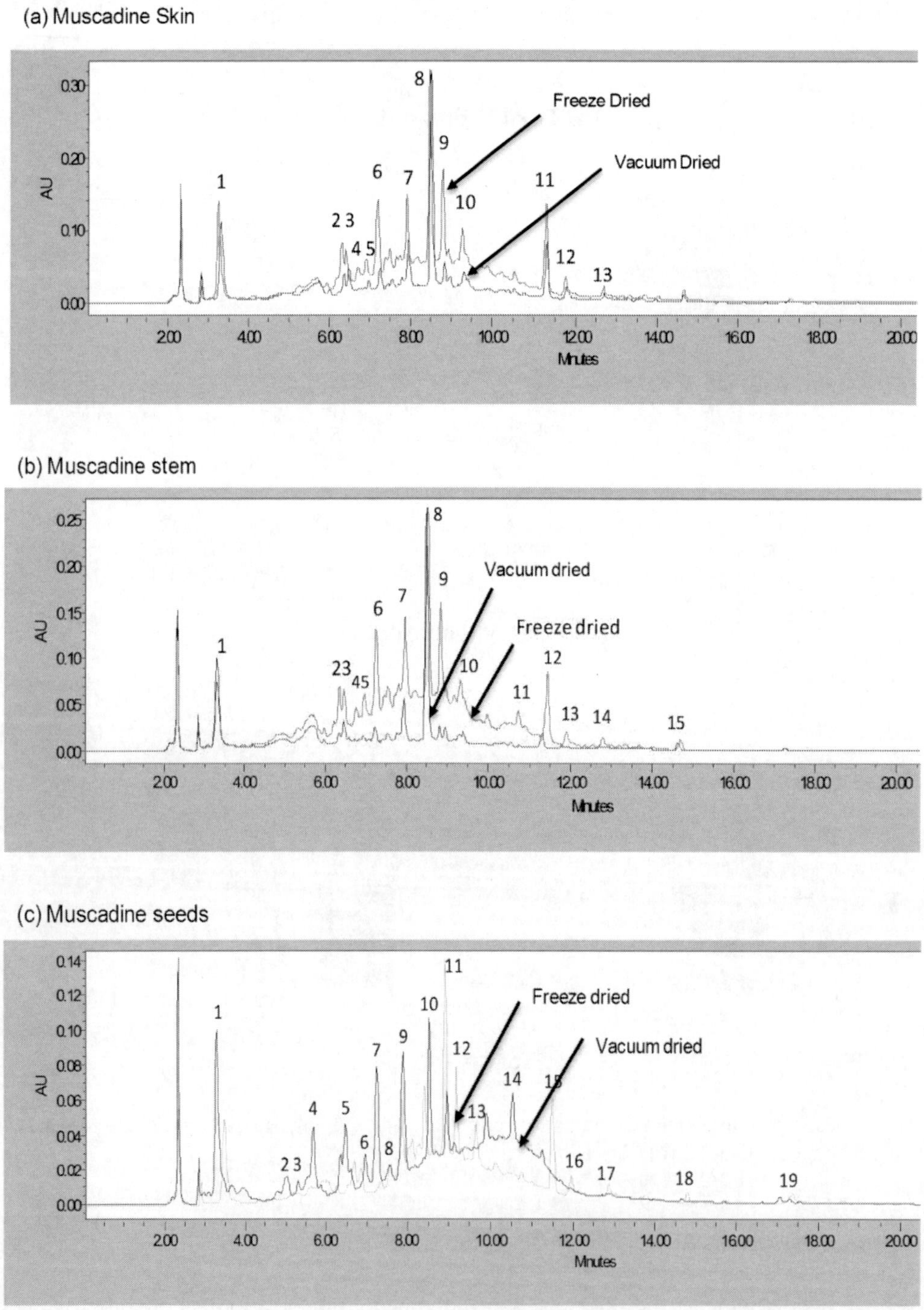

Figure 4 Effects of drying methods on individual phenolics in skin (a) stems (b) and seeds of Muscadine Nobel pomace.

Compared with freeze dried pomace, vacuum oven drying significant reduced the heights of all peaks in the extracts of skins and stems with exception of peaks 1 and 8 (Figure 4a and Figure 4b). The peaks with significantly reduced heights might represent anthocynins and anthocyanidins since they are very sensitive to heat as described above. The effects of vacuum drying on the polyphenols of grape seeds showed different pattern (Figure 4c). More peaks were resolved from seed extracts. The vacuum oven drying caused loss of peaks 11, 12, 13 and 15, but increased heights of peaks 1, 2, 3, 4, 5, 7, 9, 10 and 14 in the seed extracts of Muscadine pomace. This suggests that 1) drying at higher temperature (70ºC) results in increased extractability of some polyphenols in the plant matrix as reported by Kim et al., (2006), but also destroyed some heat sensitive polyphenols; 2) degradation of large phenolic compounds such as oligomeric- or polymeric-procyanidins formed small molecule phenolics such as catechins as reported by Khanal et al., (2009a and 2009b). The net result is the lower TP, TA and TF in vacuum dried GP than that in the freeze dried GP.

Conclusion

This chapter demonstrates that among different groups of polyphenols, anthocyanin was affected the most by drying methods. The effect of the drying method on TF depends on the variety of grape and the part of GP. While vacuum oven drying caused significant loss of TF in Muscadine pomace, it preserved more TF in Cabernet pomace, and the TF content in Cabernet seeds was only slightly affected by drying methods. Therefore, to select an appropriate drying method, it is important to consider the target polyphenols to be preserved and the variety of grape pomace. To preserve anthocyanin, freeze drying is required; to increase the extractability of some flavonoids, vacuum drying may be more suitable in terms of drying time and cost; room temperature drying does not need special equipment, but it takes significantly longer time than other methods, and it may also cause growth of mold if the drying environment is humid.

References

Baiges, I., Palmfeldt, J., Blade´ C., Gregersen, N., Arola, L., (2010). Lipogenesis is decreased by grape seed proanthocyanidins according to liver proteomics of rats fed a high fat diet. *Mol. Cell. Proteomics*. 9, 1499-1513.

Brannan, R.G., Mah, E., (2007). Grape seed extract inhibits lipid oxidation in muscle from different species during refrigerated and frozen storage and oxidation catalyzed by peroxynitrite and iron/ascorbate in a pyrogallol red model system. *Meat Sci.* 77, 540-546.

Caimari, A., del Bas, J. M., Crescenti, A., Arola, L., (2013). Low doses of grape seed procyanidins reduce adiposity and improve the plasma lipid profile in hamsters. *Int. J. Ob.* 37, 576-583.

Décordé, K., Teissèdre P.L., Sutra, T., Ventura, E., Cristol, J.P., Rouanet, J.M., (2009). Chardonnay grape seed procyanidin extract supplementation prevents high-fat diet-induced obesity in hamsters by improving adipokine imbalance and oxidative stress markers. *Mol. Nutr. Food Res.* 53, 659-666.

Fuleki, T., Silva, R.D.J.M., (1997). Catechin and procyanidin composition of seeds from grape grown in Ontario. *J. Agric. Food Chem*. 45:1157-1160.

Fuleki, T., Ricardo-Da-Silva, J. M., (2003). Effects of cultivar and processing method on the contents of catechins and procyanidins in grape juice. *J. Agric. Food Chem*. 51, 640–646.

Garrido, I., Monagas, M., Gomez-Cordoves, G., Bartolome, B., (2008). Polyphenols and antioxidant properties of almond skin, influence of industrial processing. *J. Food Sci*., 73, C106-115.

Kammerer, D., Kljusuric, J. G., Carle, R., Schieber, A., (2005). Recovery of anthocyanins from grape pomace extracts (Vitis vinifera L. cv. Cabernet Mitos) using a polymeric adsorber resin. *Eur. Food Res. Technol*. 220, 431-437.

Kaur, M., Agarwal, C., Agarwal, R., (2009). Anticancer and cancer chemopreventive potential of grape seed extract and other grape-based products. *J. Nutr*. 139, 1806S–1812S.

Khanal, R., Howard, L. R., Brownmiller, C., Prior, R. L., (2009a). Influence of extrusion processing on procyanidin composition and total anthocyanin contents of blueberry pomace. *J. Food Sci*., 74, H52–H58.

Khanal, R. C., Howard, L. R., Prior, R. L. ,(2009b). Procyanidin content of grape seed and pomace, and total anthocyanin content of grape pomace as affected by extrusion processing. *J. Food Sci*., 74, H174–H182.

Khanal, R. C., Howard, L. R., Prior, R. L. (2010). Effect of heating on the stability of grape and blueberry pomace procyanidins and total anthocyanins. *Food Res. Int*. 43, 1464-1469.

Kim, S.Y., Jeong, M., Park, W.P., Nam K.C., Ahn, D.U., Lee, S.C. (2006). Effect of heating conditions of grape seeds on the antioxidant activity of grape seed extracts. *Food Chem.,* 97, 472-479.

Larrauri J.A., Ruperez, P. Saura-Calixto, F. (1997). Effect of drying temperature on the stability of polyphenols and antioxidant activity of red grape pomace peels. *J. Agric. Food Chem*., 45, 1390-1393.

Makris, Dimitris P., Boskou G., Andrikopoulos N. K., (2007). Polyphenolic content and in vitro antioxidant characteristics of wine industry and other agri-food solid waste extracts. *J. Food Compos. Anal*. 20, 125–132.

McCalluma, J. L., Yang, R., Young, J. C., Strommer, J. N., Tsao R. (2007). Improved high performance liquid chromatographic separation of anthocyanin compounds from grapes using a novel mixed-mode ion-exchange reversed-phase column. *J. Chromatogr*. A. 1148, 38–45.

Mielnik, M.B., Olsen, E., Vogt, G., Adeline, D. Skrede, G., (2006). Grape seed extract as antioxidant in cooked, cold stored turkey meat. *LWT*, 39 191–198.

Mohd Zainol, M.K., Abdul-Hamid A., Abu Bakar, F., Pak Dek, S., (2009). Effect of different drying methods on the degradation of selected flavonoids in Centella asiatica. *Int. Food Res. J*. 16: 531-537.

Montealegre, R. R., Peces, R.R., Vozmediano, J.L.C., Gascuen, J. M., Romero, E. G., (2006). Phenolic compounds in skins and seeds of ten grape Vitis vinifera varieties grown in a warm climate. *J. Food Compos. Anal.* 19, 687–693.

Peng X., Ma, J., Cheng, K.W., Jiang, Y., Chen, F. Wang, M., (2010). The effects of grape seed extract fortification on the antioxidant activity and quality attributes of bread. *Food Chem.*, 119, 49-53.

Shi, J., Yu, J., Pohorly, J.E., Kakuda, Y., (2003). Polyphenolics in Grape Seeds—Biochemistry and Functionality. *J. Med. Food*, 6, 291 –299.

Singleton, V.L., Orthofer, R., Lamuela-Raventos, R.M., (1999). Analysis of total phenols and other oxidation substrates and antioxidants by means of Folin-Ciocalteu Reagent. *Methods in Enzymol*. 299, 152-178.

Xia, E., Deng, G., Guo Y. Li, H., (2010). Biological activities of polyphenols from grapes. *Int. J. Mol. Sci*. 11, 622-646.

Xu, B.J., Chang S.K.C., (2007). A comparative study on phenolic profiles and antioxidant activities of legumes as affected by extraction solvents. *J. Food Sci*., 72, S159-166.

Yu, J., Ahmedna, M., Goktepe, I., (2007). Peanut skin phenolics, extraction, identification, antioxidant activity and potential applications. In: Antioxidants-Measurements and Applications (F. Shahidi and Chi-Tang Ho, Editors), ACS Symposium Series #956, pp: 226-241. American Chemical Society, March, 2007.

In: Grapes
Editor: José de Sousa Câmara
ISBN: 978-1-63321-402-6

Chapter 12

Supercritical Fluid Extraction of Oil and Bioactive Compounds from Grape Residues: Experimental Optimization, Modeling and Economic Evaluation

Marcelo M. R. de Melo, Armando J. D. Silvestre and Carlos M. Silva
Department of Chemistry, CICECO, University of Aveiro,
Aveiro, Portugal

Abstract

Supercritical fluid extraction (SFE) has been a widely studied green technology to obtain oil and bioactive compounds from grape residues like seed, skin, pomace (marc). In this chapter, research advances regarding the SFE of grape residues are covered, particularly the optimization of operating conditions, modeling, scale-up and preliminary economic evaluation. Carbon dioxide is the most common solvent adopted for extraction, being generally modified with small amounts of cosolvents (e.g. ethanol) in order to enhance the solubility of the target compounds. The operating conditions found lie between 100 and 700 bar, 35°C and 80°C, 0 and 40% (v/v) of cosolvent. The existence of patents protecting the use of SFE technology upon grape residues is an acknowledgement of its suitability for food and drink industries, at least from the point of view of bulk extraction. The state-of-art concerning SFE of grape residues fully supports the commercial exploitation of this technological solution to produce oil and phenolic fractions from grape residues.

Keywords: Supercritical fluid extraction, optimization, modeling, economic evaluation, grape seed

Introduction

According to a recent report of the International Organisation of Vine and Wine (Wine 2012), the global grape production has increased 5% since 2000, reaching at the end of 2010 the rounded annual amount of 68 million tons. While nearly one third of this sum finds application on fresh consumption, the largest share is devoted to the production of wine.

Taking these records into account, the increased amount of residues from wine production assumes a growing importance to the wine industry. In this respect, some researchers have been pursuing the valorization of seeds, skins and pomace (marc) by seeking novel applications for these side materials. In this context, seeds represent 25% (wt.) of the grapes dry pomace (Rice 1976) and are known to have an oil content ranging 6-20% (wt.) (Crews et al. 2006). Furthermore, grape seeds are a source of many compounds with demonstrated bioactivity, such as catechins and procyanidins (Fuleki and da Silva 1997). These groups of phenolic compounds raise a pertinent interest from pharmaceutical industry, in view of their antioxidant, anticarcinogenic, antiviral and antimutagenic features (Palma and Taylor 1999).

The commonest technique to produce grape seed oil is by mechanical pressing, a method that is cheap but inefficient since final cakes retain significant amounts of oil (Passos et al. 2009). While organic solvents extraction is, presently, the dominant technology to overcome mechanical pressing drawbacks, supercritical fluid extraction (SFE) technology using carbon dioxide as solvent has been widely and successfully employed for the extraction of oil and valuable target compounds from biomass (Barbosa et al. 2014; de Melo et al. 2012; de Melo et al. 2013; de Melo et al. 2014a; Domingues et al. 2012a; Domingues et al. 2012b; Domingues et al. 2013; Patinha et al. 2013; Santos et al. 2012). Besides allowing undesired health and environment aspects of harmful organic solvents to be avoided, this technology provides extraction yields that are in many cases comparable to those exhibited by other extraction methods (de Melo et al. 2013; Passos et al. 2010; Rawson et al. 2012). SFE is able to provide grape seed oils with antioxidant capacities comparable to those exhibited by oils extracted with organic solvents (Passos et al. 2010).

Carbon dioxide is usually selected as solvent media for SFE by taking advantage of the moderate conditions required for operation, to the high availability and relative low cost, and to the fact that it is innocuous whatever the application scope of the final product. The critical point of CO_2 (T_c = 31.0° C and P_c = 73.7 bar (Reid et al. 1987)) is easily attainable. Owing to the existence of labile compounds in vegetable matrices in general, the low temperatures demanded by CO_2 to carry out extractions under supercritical state turns into an advantage in comparison to certain organic solvents, since it may allow undesirable thermal degradation to be avoided.

According to the foreseen potential, SFE of grape residues (particularly seeds) has been object of lab work, modeling, optimization, scale-up and economic assessment studies. This chapter documents the progresses registered in all these topics. The next section is devoted to the existing experimental work on the field on optimization of operating conditions, and it is followed by another where insights and examples of modeling work on grape SFE of grape residues are provided. In the fourth and fifth sections, scale-up studies available in literature for this technology are addressed, and a presentation of economic assessments of the SFE process with available viability results is provided. The last section presents the overall conclusions the chapter.

Optimization of Operating Conditions

As a high pressure technology whose operation involves the manipulation of the properties of carbon dioxide and the solutes involved, pressure (*P*) and temperature (*T*) play a chief role in SFE processes. In addition to those, the semi-continuous character of SFE, established by the necessary changing of raw material in the extractor, sets time (*t*) and flow rate (Q_{CO2}) also in high rank for an optimized performance of commercial SFE units. While the former four variables are mandatory for proper optimization of any SFE process, an additional factor should be mentioned, that is the cosolvent introduction for the improvement of the CO_2 solvent capacity. On a complementary basis, matrix granulometry can also influence the extraction performance in a remarkable degree, reason why the average particle diameter (d_p) of the samples is usually matter of investigation in SFE studies.

Considering the several variables involved and the significant complexity of natural matrices, the optimization of SFE processes is easier to achieve upon employment of design of experiments (DoE) and response surface methodology (RSM). As listed in Table 1, these statistical methods have been employed in SFE studies of grape matrices to disclose the most important variables affecting not only total extraction yields, but also yields or extract concentrations of specific compounds. Using an orthogonal test of three independent variables (factors) with three values (levels) for the SFE of grape seed, Cao and Ito (2003) optimized pressure, temperature and particle size values for total extraction yield (η), unsaturated fatty acids content (x_{UFA}) and saturated fatty acids content (x_{SFA}). They showed that: an intermediate particle size (20-40 mesh) is desirable for both (η) and (x_{UFA}), while bigger particles (10-20 mesh) are appropriate for (x_{SFA}).; a minimum temperature (35ºC) is preferable for any of the three responses; and maximum pressure studied (300 bar) favors both (η) and x_{UFA}, but not x_{SFA} (200 bar is preferable).

In a different work involving grapes peel, Ghafoor et al. (2010) employed an orthogonal array design (DoE) and RSM, to study the impact of pressure, temperature and ethanol content on η, total phenols concentration, total anthocyanins concentration, and extract overall antioxidant activity. The obtained results allowed these authors to conclude that 45-46ºC, 156-161 bar and 6-7% (the % basis is not referred) ethanol are preferable conditions, leading to η = 12.3% (wt.), total phenolic content of 2.2 mg_{GAE} /100 mL, an anthocyanins content of 1.2 mg/mL, and antioxidants concentration of 1.6 mg/mL. Some of their results are plotted in Figure 1 to illustrate the impact of temperature and pressure on the concentration of phenolics in grape seed oil. One of the conclusions of this essay is that the cosolvent addition (studied between 5 and 8 wt. %) is not significant for the phenolics extraction in comparison to the choice of an operating pair of pressure and temperature. Nevertheless, different conclusions on cosolvent influence were taken by Chafer et al. (2005), who analyzed the addition of ethanol in fractions up to 25% (v/v). These researchers concluded 20% is the advantageous ethanol content and also suggest performing the SFE at 250 bar and 60ºC for a high extraction of phenolic compounds.

Table 1. Literature references concerning the SFE of grape residues, and their respective features

Residue	Target Compounds	Extraction Yield (wt%)	*P* (bar)	*T* (°C)	CO_2 & Cosolvent	Other Features	Analytical technique	Ref.
bagasse	polyphenol	0-6	250-350	40	10% wt. ethanol	Economic study Soxhlet extraction Agitated bed extraction	TLC, HPLC	(Farías-Campomanes et al. 2013)
fruit/seed	phenolic		200-500	45	7.2-67.3 $kg_{CO2}\,kg^{-1}{}_{sample}$ 4% wt ethanol		UV–Vis spectrophotom.	(Fiori et al. 2009)
marc	phenolic		281-350	40-50	0-25% methanol 600 $l_{CO2}\,kg^{-1}{}_{sample}$	Trap optimization	CZE-HPLC	(Arce et al. 2001)
marc	phenolic anthocyanins		110	40	methanol	Antisolvent extraction		(Floris et al. 2010)
peel	phenolic anthocyanins	6.8-13.2	140-170	37-46	5-8% ethanol	DoE RSM		(Ghafoor et al. 2010)
pomace	anthocyanins phenolic		150-300	40	0.04-0.20% (v/v) ethanol	Fractionation	HPLC	(Vatai et al. 2009)
pomace	phenolic	2.8-28.9	80-350	25-50	0-8% ethanol	Solid-liquid extraction	HPLC	(Pinelo et al. 2007)
pomace	phytosterols	6.6-11.2	370	70	2,0-123 $kg_{CO2}\,kg^{-1}{}_{sample}$	Solvent extraction	GC-MS GC-FID HPLC	(Beveridge et al. 2005)
pomace	phenolic	4-13	300	50-60	2.1 $kg_{CO2}\,kg^{-1}{}_{sample}$ 12.5-17.5% ethanol	Soxhlet extraction Ultrasound-assisted leaching extraction	HPLC	(Oliveira et al. 2013)
pomace	resveratrol		100-400	35-55	0-5% v/v	Organic solvent extraction	HPLC	(Casas et al. 2010)
seed	phenolic		200-300	40	0-15% ethanol 0-15% methanol	Solubility study	HPLC-MS	(Murga et al. 2000)
seed		2,5-6,2	300-400	35-40	6 $l_{CO2}\,kg^{-1}{}_{sample}$ 0-10% ethanol	DoE Scale-up	HPLC–ELSD	(Cao and Ito 2003)
seed		9.1-10	655	80	24 $l_{CO2}\,kg^{-1}{}_{sample}$ 0-40% methanol	Fractionation	HPLC-UV SFC-UV HPLC-MS	(Ashraf-Khorassani and Taylor 2004)
seed		8-16	250	80	41.1 $kg_{CO2}\,kg^{-1}{}_{sample}$	species comparison	GC-MS HPLC	(Agostini et al. 2012)
seed	gallic acid protocatechuic acid p-hydroxybenzoic acid	5.7-12.3	137-167	66-75	5-8% ethanol	RSM Antiradical activity test	HPLC	(Ghafoor et al. 2012)

Residue	Target Compounds	Extraction Yield (wt%)	*P* (bar)	*T* (°C)	CO_2 & Cosolvent	Other Features	Analytical technique	Ref.
seed		13.42	350	40	12.6 $kg_{CO_2}\,kg^{-1}{}_{sample}$	Scale-up (Pilot) Economic evaluation		(Prado et al. 2012)
seed	tryacylglicerides	11.5	180-220	40-50	90 $kg_{CO_2}\,kg^{-1}{}_{sample}$	AOC-DPPH	GC-FID	(Passos et al. 2010)
seed			250-300	30-50			HPLC	(Yilmaz et al. 2011)
seed	α -tocopherol	2.0-14.4	250	40-80	25-95 $kg_{CO_2}\,kg^{-1}{}_{sample}$	Particle size study Org.Solvent extraction	HPLC	(Bravi et al. 2007)
seed		0-12	280-550	40	0.4-10 $kg_{CO_2}\,kg^{-1}{}_{sample}$	Modeling Column size study Solubility study		(Fiori 2007)
seed			550	60	4 $kg_{CO_2}\,kg^{-1}{}_{sample}$	Modeling Economic study		(Fiori 2010)
seed		0.1-7.9	60-254	30-60	20 $l_{CO_2}\,kg^{-1}{}_{sample}$	Org. solvent extraction	TIC	(Freitas et al. 2008)
seed		0-7	50-350	10-60	150 $l_{CO_2}\,kg^{-1}{}_{sample}$	Particle size study	GC	(Gómez et al. 1996)
seed		0-16	160-200	40	0-10 $kg_{CO_2}\,kg^{-1}{}_{sample}$	Enzymatic pretreatment	GC-FID	(Passos et al. 2009)
seed		0-12	180-220	40-50	0-6 $kg_{CO_2}\,kg^{-1}{}_{sample}$	Soxhlet Antioxidant capacity	GC-FID	(Passos et al. 2010)
seed	polyphenols		80-150	35-60	2% (M) methanol	Supercritical antisolvent extraction	HPLC-DAD FE-SEM	(Marqués et al. 2013)
skin	(+)-Catechin; (–)-Epicatechin Quercetin Rutin		100-300	60	10-50 $l_{CO_2}\,kg^{-1}{}_{sample}$ 5-25% v/v ethanol		HPLC	(Chafer et al. 2005)
skin	resveratrol		80-110	40	10 $l_{CO_2}\,kg^{-1}{}_{sample}$ 5-15% ethanol	Species comparison	HPLC	(Pascual-Marti et al. 2001)
pomace	oleanolic acid		250-350	40-50	5% (v/v) ethanol	Solid-liquid extraction	HPLC	(Chronopoulou et al. 2013)

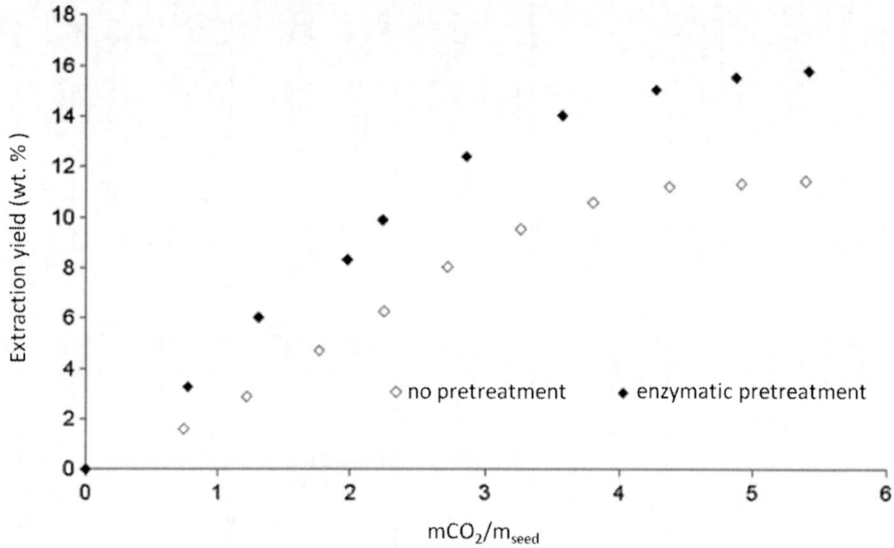

Figure 1. Impact of pressure and temperature on phenol content of grape seed oil obtained by supercritical fluid extraction with CO_2 (Ghafoor et al. 2010).

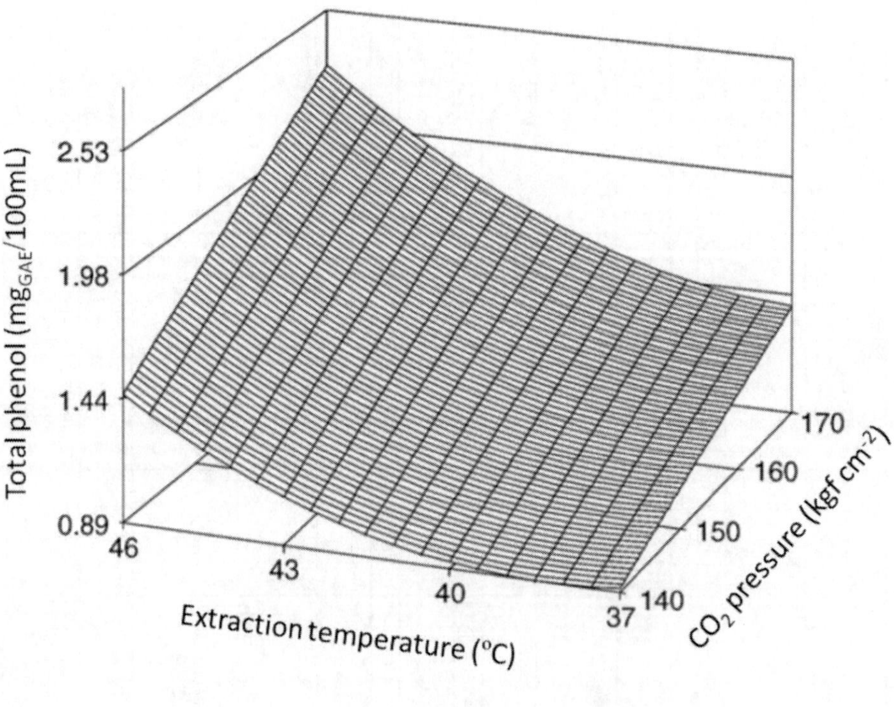

Figure 2. SFE curves of untreated and enzymatically pretreated grape seeds, using CO_2 at 40ºC and 200 bar (Passos et al. 2009).

Passos et al. (2009) introduced an original feature for the enhancement of grape seed oil yield through an enzymatic hydrolysis pretreatment step, followed by extraction with supercritical CO_2 (SC-CO_2). Extraction curves measured by these researchers are illustrated in Figure 2, where the remarkable enhancement on the pretreated run is clearly evident: the treated sample was extracted up to 16.2% (wt.) against 11.2% (wt.) in the case of an untreated matrix. As a matter of comparison, the Soxhlet extraction of this biomass with *n*-hexane provided oil yield values of 11.6% (Gómez et al. 1996; Passos et al. 2009). This means, firstly, that SFE of untreated samples is able to reach yield values very close to those of organic solvents and, secondly, that enzymatic pretreatments can improve Soxhlet reference values by nearly 40%.

More studies covering optimization of conditions are reported in Table 1, involving different grape components and also targeting specific compounds. Information regarding extraction yields obtained in each work is provided along with the corresponding operating conditions, namely pressure, temperature, cosolvent type and concentration, spent CO_2. Complementary features regarding each work are given, as well as information concerning the analytical techniques employed.

Modeling

Phenomenological modeling has been applied with advantage to SFE of grape seeds, allowing a better understanding of kinetic and equilibrium aspects of the process. One of the challenges answered by modeling comprises the knowledge of the important variables of SFE, such as convective mass transfer coefficient (k_f), intraparticle effective diffusivity (D_{eff}) and oil solubility (y^*), in order to disclose the most relevant mechanisms of the extraction and provide, at the end, an accurate analytic tool for both correlation and prediction. While diffusivity of solutes on SC-CO_2 is one of the areas where most modeling research activity has been reported in the last years (Lito et al. 2013; Magalhães et al. 2011b, 2011a, 2012; Magalhães et al. 2013c; Magalhães et al. 2013a; Magalhães et al. 2013b; Vaz et al. 2013), very complete reviews on the overall modeling may be found in the literature (Oliveira et al. 2011; Sovová and Stateva 2011).

Passos et al. measured (Passos et al. 2009, 2010) and modeled (Passos et al. 2011) the SC-CO_2 extraction curves of both enzymatically treated and untreated grape seeds, using the broken and intact cells concept (Sovová 1994, 2005).

This model takes into account physical differences in matrices so that a fraction of the biomass is composed of intact cells, while the remaining part contains cells with disrupted walls. In the case of grape residues generated in wine production processes, the biomass physical differences can be due to maceration or further grinding steps of the raw material. Through modeling of extraction curves (see Figure 3) these authors showed that saturation of SC-CO_2 was not achieved in any of the runs during the initial extraction period when maximum extraction rates were observed. The same conclusion was valid for the runs involving enzymatically treated matrices, being the higher extraction yield results explained by an increase of 35% in the fraction of broken cells and the enhancement of intraparticular diffusivity (increments ranging from 360% and 894%).

Fiori (2007) modeled extraction curves of milled grape seed with SC-CO_2 for a wider range of pressures (280-550 bar) at fixed temperature (40ºC), and also for several particle sizes. In all cases, the first period of extraction is coincident, except for the higher limiting particles of 0.93 and 0.97 mm (see Figure 4), because the available oil is sufficient to saturate the CO_2 stream. After this constant extraction rate period, each particle size determines clearly different kinetics of extraction, and gives rise to a split of the cumulative curves measured. Accordingly, Fiori (2007) computed distinct final extraction yields, i.e. a different plateau for each d_p.

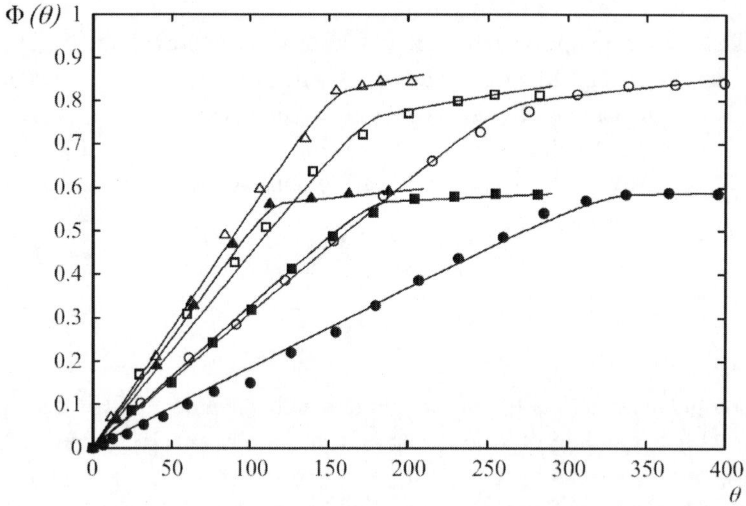

Figure 3. Supercritical extraction curves of grape seed, in a dimensionless extraction yield (Φ) *versus* time (θ) basis. Reference yield is the total (whether available or not) oil content of a seed. Experimental data: untreated seeds: ● 160 bar, 40ºC; ■ 180 bar, 40ºC; ▲ 200 bar, 40ºC; enzymatically pretreated seeds: Exp. ○ 160 bar, 40ºC; □ 180 bar, 40ºC, Δ 200 bar, 40ºC. Lines: model (Passos et al. 2011).

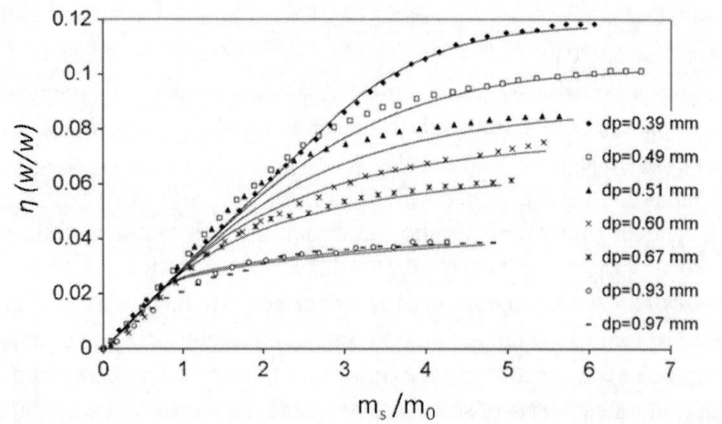

Figure 4. SFE yields of grape seed matrices with different mean particle diameters, as function of the mass of spent solvent (m_s) per mass of biomass (m_o). Experiments were performed at 550 bar and 40ºC. Data taken from Fiori (2007).

Scale-up

In view of the positive results obtained at lab scale, the SFE process of has been object of scale-up studies, for what a substantial knowledge regarding operating conditions, raw material physical properties, kinetic, and equilibrium features is recommended.

On an earlier paper, Cao and Ito (2003) presented results also for grape seed, based on SFE assays performed in extractors of 10 mL and 2 L capacity, with no noticeable scale-up criteria being referred. Disappointingly, the optimized conditions that maximize extraction yield at 10 mL (η = 5.1 wt.%) were not confirmed at 2 L scale (η = 3.1 wt.%).

Prado et al. (2012) published recently a study where SFE of grape seed is scaled-up using a 1.00 L extractor and a different one with 5.15 L capacity. Despite the quite distinct geometries of the extractors, the measured extraction curves are coincident, as depicted in Figure 5. The success of their study can be explained by the proper choice of the scale-up criterion, which in this case was to maintain constant the mass of solvent in relation to the feed mass. This criterion is suggested for situations when the process is mainly driven by the solubility of the oil in the supercritical phase. The final extraction yields between 12% and 14% (wt.) obtained both by lab and pilot scale runs are in agreement with the results obtained by other researchers at lab scale (see Table 1).

Despite the existence of works targeting the scale-up of SFE of grape seed, more results would be needed in this respect to confirm a viable transfer of scale from lab to commercial levels. Nonetheless, it is known that scale-up is many times performed by specialized companies, whose customers motivation is towards the protection of such results. This perspective is also suggested by existence of patents concerning SFE process, as it will be discussed in section 6.

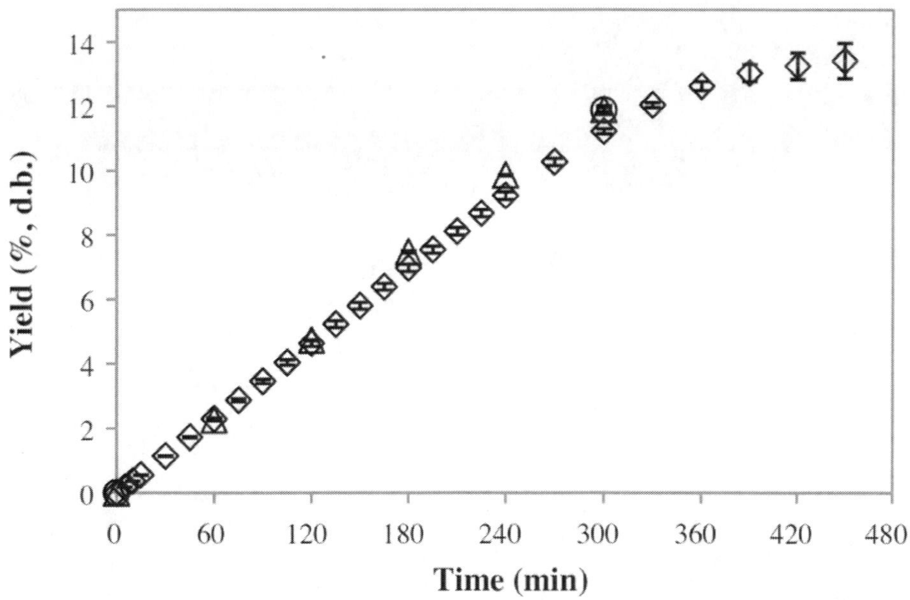

Figure 5. Supercritical fluid extraction curves of grape seed at 350 bar and 40°C, for laboratory (◊) and pilot (Δ) scales, and total collection yield at pilot scale (○) (Prado et al. 2012).

Preliminary Economic Evaluation

Despite available scale-up studies have not been performed within a desirable closeness to industrial scale, preliminary economic analyses for the application of SFE technology to grape residues are available in literature (Farías-Campomanes et al. 2013; Fiori 2010; Prado et al. 2012) aiming at the assessment of the either profitability and cost of manufacturing issues.

In view of its high pressure character, SFE unit typically demands a significant initial investment on the extraction equipment but other parcels need to be summed in view of the necessary preparation of the biomass, which may comprise drying, milling and sieving. Fiori (Fiori 2010) carried out an economic study for a putative commercial scale unit with an annual processing capacity of 3000 tons of grape seed. In his study, drying, sieving and milling costs were considered in complement to extraction step itself. Besides, a purchasing price for the grape residues has also been defined, as well as the corresponding transport costs. This study concluded 5.9 €/kg_{oil} to be the break-even production cost for a SFE unit comprising three extractors of 0.8 m^3 capacity, working in series. Since for the Italian market the commercial price of grape seed oil ranges from 10 to 30 €/kg_{oil}, the final conclusion was favorable to an investment in the SFE technology.

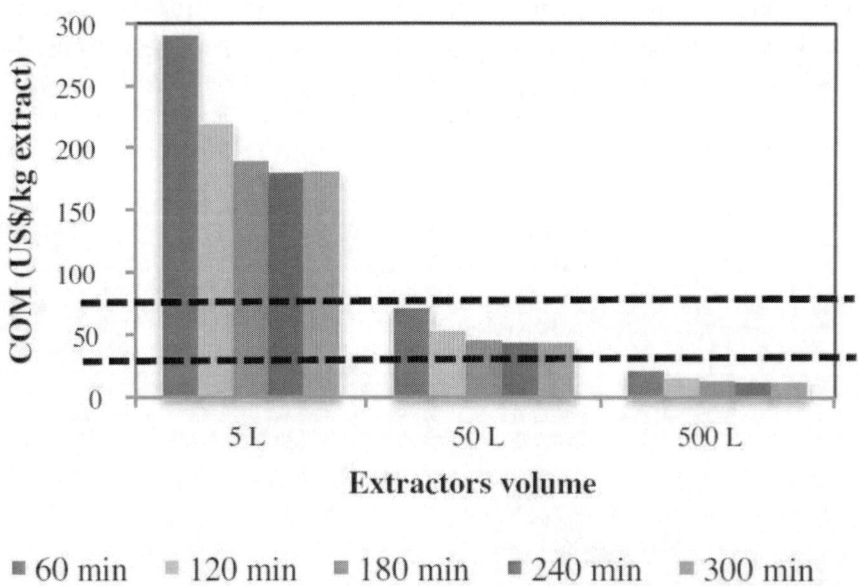

Figure 6. Cost of manufacturing (COM) estimation of grape seed oil obtained by SFE, as function of extractors volume and extraction time. The considered unit comprises two extractors in alternate working. Dotted lines represent the price of grape seed cold pressed oil in Brazil (US$ 40 kg^{-1}) and in international market (US$ 80 kg^{-1}) (Prado et al. 2012).

Prado et al., (2012) performed a cost of manufacturing (COM) estimation for SFE at three different scales (5 L, 50 L and 500 L). This type of economic calculation has been commonly applied to SFE of biomass (de Melo et al. 2014b; Pereira and Meireles 2010; Rosa and Meireles 2005). In their study, a value of US$ 2.70/ton as purchase price for grape seeds, and concluded to be economically viable to establish a SFE plant of this type in Brazil. The best results (see Figure 6) were obtained for a setting of 2 units of 500 L capacity each, where cost of manufacturing was minimal, US$ 12.08/kg_{oil}, allowing a return on investment of 1.5 years. For these calculations a selling price of US$ 40/kg_{oi}l was considered. Very recently, authors from the same group assessed the SFE of phenolics from grape bagasse with ethanol as cosolvent, using COM methodology, and concluded it is economically feasible, even for small a scale of production (Farías-Campomanes et al. 2013).

Though the comparison of technology costs in relation to conventional techniques is inevitable, it should be stressed that grape seed oil obtained by SFE can exhibit a stronger antioxidant capacity than that obtained by organic solvents, which depends mainly on the chosen operating conditions (Passos et al. 2010). If the latter aspect is matched with the fact that oil obtained by SFE has no traces of harmful organic solvents, it is expectable this oil to be set in a higher rank of quality and, thus, commercial price.

Patents

One prompt way to attest the interest and potential of the valorization of grape residues by means of supercritical fluid extraction is upon consultation of filled patents claiming rights on this subject. A search has been done and the patents described in Table 2 were found. As can be noticed from this table, intellectual rights have been claimed for very distinct processes, either for a cleaning process aiming at a special wine production (Kong et al. 2011), either as a technology for the isolation of proanthocyanidins (Hongxian 2013; Li et al. 2005; F. Wang 2007; Yang et al. 2011), and, eventually, unsaturated fatty acids (Wang 2012). An alternative approach is to obtain grape seed extracts by SFE to be used as a medium that retains constituents such as Vitamin E from other plant materials to be used in soft capsules for general population (Lin 2012).

Conclusion

Along the last decade, there has been a strong research activity on SFE of grape residues, with great emphasis on grape seeds. Taking the latter case into inspection, the necessary scientific arguments to support the consolidation of this SFE application seem to have been satisfactorily covered by research, as the various literature publications confirm.

While the available economic assessments clearly point towards the positive theoretical outcome of investing on SFE technology for valorization of grape residues, the various patents registered in the last years bear testimony of the real interest and prime opportunity that this green solution represent for the field.

Table 2. Patents comprising SFE of grape residues and purification of extracts

Year	Title	Description	Ref.
2013	Method for continuously extracting active ingredients from grape seeds	Grape seeds are treated in terms of crushing, enzymolysis and vacuum drying. SFE is performed in two steps under different conditions. The second extract is rich in proanthocyanidins and can reach 90-98% (wt.) concentration upon further separation and purification procedures.	(Hongxian 2013)
2012	Method for extracting unsaturated fatty acids from grape skin	Extraction with an organic solvent, carrying out low-temperature crystallization after dissolution to obtain an unsaturated fatty acid crude extract, and carrying out CO_2 supercritical extraction to obtain the unsaturated fatty acids.	(Wang 2012)
2012	Novel grape seed extract and preparation method thereof	Grape seed extraction by SFE and ethanol extraction, in which the oil is used as medium for constituents from other plant materials such as Vitamin E. The final product is soft capsules for human consumption.	(Lin 2012)
2011	Manufacture process of fortified grape wine	SFE of grape seed for oil extraction, so that extracted grape seeds can be further used to produce grape wine with good aesthetic quality and fortified polyphenol substances.	(Kong et al. 2011)
2011	Method for extracting grape seed oil containing oligomeric proantho cyanidins (OPC)	SFE of crushed grape seeds to obtain oil containing OPC compounds with a concentration of 2.14 mg/mL.	(Yang et al. 2011)
2007	Supercritical CO_2 extraction of grape-kernel oil and method for extracting oligoprotoanthocyanidin by solvent process	Extraction of grape-kernel oil by SFE and anthocyanidin by solvent method, wherein the raw anthocyanin obtained is subjected to chromatographic separation and microencapsulation technical treatment.	(Wang 2007)
2005	Method of extracting proto anthocyanidin by supercritical extraction grape seed oil and resin absorption	SFE of comminuted grape seed, treatment of oil in a polystyrene resin for adsorption of proanthocyanidins. Desorption with desorbent after adsorption saturation, drying the eluent under reduced pressure, finally obtaining a refined proanthocyanidin product.	(Li et al. 2005)

References

Agostini, F., et al. (2012), 'Supercritical Extraction from Vinification Residues: Fatty Acids, alpha-Tocopherol, and Phenolic Compounds in the Oil Seeds from Different Varieties of Grape', *Scientific World Journal*, 1-9.

Arce, L., et al. (2001), 'Screening of polyphenols in grape marc by on-line supercritical fluid extraction - Flow through sensor', *Analytical Letters,* 34 (9), 1461-76.

Ashraf-Khorassani, M. and Taylor, L. T. (2004), 'Sequential fractionation of grape seeds into oils, polyphenols, and procyanidins via a single system employing CO_2-based fluids', *Journal of Agricultural and Food Chemistry,* 52 (9), 2440-44.

Barbosa, H. M. A., et al. (2014), 'Optimization of the supercritical fluid coextraction of oil and diterpenes from spent coffee grounds using experimental design and response surface methodology', *The Journal of Supercritical Fluids,* 85 (0), 165-72.

Beveridge, T. H. J., et al. (2005), 'Yield and composition of grape seed oils extracted by supercritical carbon dioxide and petroleum ether: Varietal effects', *Journal of Agricultural and Food Chemistry,* 53 (5), 1799-804.

Bravi, M., et al. (2007), 'Improving the extraction of α-tocopherol-enriched oil from grape seeds by supercritical CO_2. Optimisation of the extraction conditions', *Journal of Food Engineering,* 78 (2), 488-93.

Cao, X. L. and Ito, Y. C. (2003), 'Supercritical fluid extraction of grape seed oil and subsequent separation of free fatty acids by high-speed counter-current chromatography', *Journal of Chromatography A,* 1021 (1-2), 117-24.

Casas, L., et al. (2010), 'Extraction of resveratrol from the pomace of Palomino fino grapes by supercritical carbon dioxide', *Journal of Food Engineering,* 96 (2), 304-08.

Chafer, A., et al. (2005), 'Supercritical fluid extraction and HPLC determination of relevant polyphenolic compounds in grape skin', *Journal of Separation Science,* 28 (16), 2050-56.

Chronopoulou, L., et al. (2013), 'Supercritical CO_2 extraction of oleanolic acid from grape pomace', *International Journal of Food Science and Technology,* 48 (9), 1854-60.

Crews, C., et al. (2006), 'Quantitation of the Main Constituents of Some Authentic Grape-Seed Oils of Different Origin', *Journal of Agricultural and Food Chemistry,* 54 (17), 6261-65.

de Melo, M. M. R., et al. (2012), 'Supercritical fluid extraction of triterpenic acids from *Eucalyptus globulus* bark', *Journal of Supercritical Fluids,* 70, 137-45.

de Melo, M. M. R, et al. (2013), 'Extraction and purification of triterpenoids using supercritical fluids: from lab to exploitation', *Mini-Reviews in Organic Chemistry,* DOI: 10.2174/1570193X113106660002.

de Melo, M. M. R, et al. (2014a), 'Supercritical fluid extraction of vegetable matrices: applications, trends and future perspectives of a convincing green technology', *The Journal of Supercritical Fluids,* DOI: 10.1016/j.supflu.2014.04.007.

de Melo, M. M. R, et al. (2014b), 'Supercritical fluid extraction of spent coffee grounds: Measurement of extraction curves, oil characterization and economic analysis', *The Journal of Supercritical Fluids,* 86 (0), 150-59.

Domingues, R. M. A., et al. (2012a), 'Measurement and modeling of supercritical fluid extraction curves of *Eucalyptus globulus* bark: Influence of the operating conditions upon yields and extract composition', *Journal of Supercritical Fluids,* 72, 176-85.

Domingues, R. M. A., et al. (2012b), 'Supercritical Fluid Extraction of *Eucalyptus globul*us Bark-A Promising Approach for Triterpenoid Production', *International Journal of Molecular Sciences,* 13 (6), 7648-62.

Domingues, R. M. A., et al. (2013), 'Optimization of the supercritical fluid extraction of triterpenic acids from *Eucalyptus globulus* bark using experimental design', *Journal of Supercritical Fluids,* 74, 105-14.

Freitas, L. d.S, et al. (2008), 'Extraction of Grape Seed Oil Using Compressed Carbon Dioxide and Propane: Extraction Yields and Characterization of Free Glycerol Compounds', *Journal of Agricultural and Food Chemistry,* 56 (8), 2558-64.

Farías-Campomanes, A.M., et al. (2013), 'Production of polyphenol extracts from grape bagasse using supercritical fluids: Yield, extract composition and economic evaluation', *The Journal of Supercritical Fluids,* 77 (0), 70-78.

Fiori, L. (2007), 'Grape seed oil supercritical extraction kinetic and solubility data: Critical approach and modeling', *The Journal of Supercritical Fluids,* 43 (1), 43-54.

Fiori, L., et al. (2009), 'Grape by-products: extraction of polyphenolic compounds using supercritical CO_2 and liquid organic solvent - a preliminary investigation Subproductos de la uva: extraccion de compuestos polifenolicos usando CO_2 supercritico y disolventes organicos liquidos - una investigacion preliminar', *Cyta-Journal of Food,* 7 (3), 163-71.

Fiori, L. (2010), 'Supercritical extraction of grape seed oil at industrial-scale: Plant and process design, modeling, economic feasibility', *Chemical Engineering and Processing: Process Intensification,* 49 (8), 866-72.

Floris, T., et al. (2010), 'Antioxidant compounds recovery from grape residues by a supercritical antisolvent assisted process', *The Journal of Supercritical Fluids,* 54 (2), 165-70.

Fuleki, T. and da Silva, J. M. R. (1997), 'Catechin and Procyanidin Composition of Seeds from Grape Cultivars Grown in Ontario', *Journal of Agricultural and Food Chemistry,* 45 (4), 1156-60.

Ghafoor, K., Park, J., and Choi, Y. H. (2010), 'Optimization of supercritical fluid extraction of bioactive compounds from grape (*Vitis labrusca* B.) peel by using response surface methodology', *Innovative Food Science & Emerging Technologies,* 11 (3), 485-90.

Ghafoor, K., et al. (2012), 'Supercritical Fluid Extraction of Phenolic Compounds and Antioxidants from Grape (*Vitis labrusca* B.) Seeds', *Plant Foods for Human Nutrition,* 67 (4), 407-14.

Hongxian, T. (2013), 'Method for continuously extracting active ingredients from grape seeds', (Jishou University).

Kong, W., et al. (2011), 'Manufacture process of fortified grape wine', (Shandong Yantai Agricultural School).

Li, S., et al. (2005), 'Method of extracting proto anthocyanidin by supercritical extraction grape seed oil and resin absorption', (University of Tianjin).

Lin, Y. (2012), 'Novel grape seed extract and preparation method thereof ', (Beijing Kangjian Renren Electronic Business Co Ltd.).

Lito, P.F., et al. (2013), 'Universal model for accurate calculation of tracer diffusion coefficients in gas, liquid and supercritical systems', *Journal of Chromatography A,* 1290 (0), 1-26.

Magalhães, A.L., et al.. (2011a), 'New tracer diffusion correlation for real systems over wide ranges of temperature and density', *Chemical Engineering Journal,* 166 (1), 49-72.

Magalhães, A.L., et al. (2011b), 'New models for tracer diffusion coefficients of hard sphere and real systems: Application to gases, liquids and supercritical fluids', *The Journal of Supercritical Fluids,* 55 (3), 898-923.

Magalhães, A.L., et al. (2012), 'Tracer diffusion coefficients of polar systems', *Chemical Engineering Science,* 73 (0), 151-68.

Magalhães, A.L., et al. (2013a), 'Free-volume model for the diffusion coefficients of solutes at infinite dilution in supercritical CO_2 and liquid H_2O', *The Journal of Supercritical Fluids,* 74 (0), 89-104.

Magalhães, A.L., et al. (2013b), 'Simple and accurate correlations for diffusion coefficients of solutes in liquids and supercritical fluids over wide ranges of temperature and density', *The Journal of Supercritical Fluids,* 76 (0), 94-114.

Magalhães, A.L., et al. (2013c), 'Accurate hydrodynamic models for the prediction of tracer diffusivities in supercritical carbon dioxide', *The Journal of Supercritical Fluids,* 83 (0), 15-27.

Marqués, J. L., et al. (2013), 'Supercritical antisolvent extraction of antioxidants from grape seeds after vinification', *The Journal of Supercritical Fluids,* 82 (0), 238-43.

Molero Gómez, A., et al. (1996), 'Recovery of grape seed oil by liquid and supercritical carbon dioxide extraction: a comparison with conventional solvent extraction', *The Chemical Engineering Journal and the Biochemical Engineering Journal,* 61 (3), 227-31.

Murga, R., et al. (2000), 'Extraction of natural complex phenols and tannins from grape seeds by using supercritical mixtures of carbon dioxide and alcohol', *Journal of Agricultural and Food Chemistry,* 48 (8), 3408-12.

Oliveira, D.A., et al. (2013), 'Antimicrobial activity and composition profile of grape (*Vitis vinifera*) pomace extracts obtained by supercritical fluids', *Journal of Biotechnology,* 164 (3), 423-32.

Oliveira, E.L.G., et al. (2011), 'Review of kinetic models for supercritical fluid extraction', *Chemical Engineering Research and Design,* 89 (7), 1104-17.

Palma, M. and Taylor, L. T. (1999), 'Extraction of polyphenolic compounds from grape seeds with near critical carbon dioxide', *Journal of Chromatography A,* 849 (1), 117-24.

Pascual-Marti, M. C., et al. (2001), 'Supercritical fluid extraction of resveratrol from grape skin of *Vitis vinifera* and determination by HPLC', *Talanta,* 54 (4), 735-40.

Passos, C.P., et al. (2011), 'Modelling the supercritical fluid extraction of edible oils and analysis of the effect of enzymatic pre-treatments of seed upon model parameters', *Chemical Engineering Research & Design,* 89 (7A), 1118-25.

Passos, C.P., et al. (2009), 'Enhancement of the supercritical fluid extraction of grape seed oil by using enzymatically pre-treated seed', *The Journal of Supercritical Fluids,* 48 (3), 225-29.

Passos, C.P., et al. (2010), 'Supercritical fluid extraction of grape seed (*Vitis vinifera* L.) oil. Effect of the operating conditions upon oil composition and antioxidant capacity', *Chemical Engineering Journal,* 160 (2), 634-40.

Patinha, D. J. S., et al. (2013), 'Lipophilic extractives from the bark of *Eucalyptus grandis x globulus*, a rich source of methyl morolate: Selective extraction with supercritical CO_2', *Industrial Crops and Products,* 43, 340-48.

Pereira, C.G. and Meireles, M.A.A. (2010), 'Supercritical Fluid Extraction of Bioactive Compounds: Fundamentals, Applications and Economic Perspectives', *Food and Bioprocess Technology,* 3 (3), 340-72.

Pinelo, M., et al. (2007), 'Supercritical fluid and solid-liquid extraction of phenolic antioxidants from grape pomace: a comparative study', *European Food Research and Technology,* 226 (1-2), 199-205.

Prado, J. M., et al. (2012), 'Supercritical fluid extraction of grape seed: Process scale-up, extract chemical composition and economic evaluation', *Journal of Food Engineering,* 109 (2), 249-57.

Rawson, A., et al. (2012), 'Application of Supercritical Carbon Dioxide to Fruit and Vegetables: Extraction, Processing, and Preservation', *Food Reviews International,* 28 (3), 253-76.

Reid, R.C., et al. (2001), *The properties of gases & liquids* (New York: McGraw-Hill Book Company).

Rice, A. C. (1976), 'Solid-waste generation and by-product recovery potential from winnery residues', *American Journal of Enology and Viticulture,* 27 (1), 21-26.

Rosa, P. T. V. and Meireles, M. A. A. (2005), 'Rapid estimation of the manufacturing cost of extracts obtained by supercritical fluid extraction', *Journal of Food Engineering,* 67 (1-2), 235-40.

Santos, S. A. O., et al. (2012), 'Supercritical fluid extraction of phenolic compounds from *Eucalyptus globulus* Labill bark', *Journal of Supercritical Fluids,* 71, 71-79.

Sovová, H. (1994), 'Rate of the vegetable oil extraction with supercritical CO_2—I. Modelling of extraction curves', *Chemical Engineering Science,* 49 (3), 409-14.

Sovová, H. (2005), 'Mathematical model for supercritical fluid extraction of natural products and extraction curve evaluation', *Journal of Supercritical Fluids,* 33 (1), 35-52.

Sovová, H. and Stateva, R.P. (2011), 'Supercritical fluid extraction from vegetable materials', *Reviews in Chemical Engineering,* 27 (3-4), 79-156.

Vatai, T., et al. (2009), 'Extraction of phenolic compounds from elder berry and different grape marc varieties using organic solvents and/or supercritical carbon dioxide', *Journal of Food Engineering,* 90 (2), 246-54.

Vaz, R.V., et al. (2013), 'Improved Stokes–Einstein based models for diffusivities in supercritical CO_2', *Journal of the Taiwan Institute of Chemical Engineers,* (DOI: 10.1016/j.jtice.2013.12.005).

Vaz, R.V., et al. (2013), 'Improved hydrodynamic equations for the accurate prediction of diffusivities in supercritical carbon dioxide', *Fluid Phase Equilibria,* 360, 401-15.

Wang, F. (2007), 'Supercritical CO_2 extraction of grape-kernel oil and method for extracting oligoprotoanthocyanidin by solvent process ',

Wang, M. (2012), 'Method for extracting unsaturated fatty acids from grape skin ', (Qingdao Shuaiwang Oil Chemical Ltd.).

Wine, International Organisaton of Vine and (2012), 'Statistical report on world vitiviniculture 2012'.

Yang, J. et al. (2011), 'Method for extracting grape seed oil containing oligomeric proantho cyanidins (OPC) ', in European Patent Office (ed.), (Beifang University of Nationalities).

Yilmaz, E.E., et al. (2011), 'Extraction and identification of proanthocyanidins from grape seed (*Vitis Vinifera*) using supercritical carbon dioxide', *Journal of Supercritical Fluids,* 55 (3), 924-28.

Section 2: Health Benefits of Grape Poylyphenols

In: Grapes
Editor: José de Sousa Câmara
ISBN: 978-1-63321-402-6

Chapter 13

Healthy Effects of Bioactive Metabolites from *Vitis vinifera* L. Grapes: A Review

***R. Perestrelo*[1], *C. Silva*[1], *J. Pereira*[1] *and José S. Câmara*[1,2,1*]**
[1] CQM/UMa, Centro de Química da Madeira, Universidade da Madeira, Funchal, Portugal
[2] Universidade da Madeira; Centro de Ciências Exatas e de Engenharia da Universidade da Madeira, Funchal, Portugal

Abstract

Grape is one of the main fruits produced worldwide being *Vitis vinifera* L. the most cultivated specie. Approximately 71 % of the grape production is used in winemaking, while 27 % in direct consumption as table grapes and 2 % as dried fruits. Epidemiological studies have associated the consumption of grapes, grape juice and wine with a wide variety of health-promoting effects, particularly the reduced risk of cardiovascular diseases (CVDs), type-2 diabetes, certain types of cancers, and other chronic complications. The beneficial effects of grape and grape-derived food products are believed to be related to a variety of bioactive components, namely, phenolic acids, flavonols, flavanones, flavonoids, flavon-3-ols, anthocyanins, procyanidins, sesquiterpenoids, monoterpenoids, and also C_{13} norisoprenoids. These bioactive components are present mainly in grape skins and seeds and arise from different metabolic pathways, namely acetyl-CoA and shikimic acids, via the intermediates mevalonate and aromatic amino acids, respectively. In turn, C_{13} norisoprenoids derived from carotenoid oxidation. Most of these compounds have strong antioxidant and anti-inflammatory properties and some of them were already shown to be involved in different aspects of cell homeostasis. *In vivo* study using different cancer cell lines reported that quercetin-3-glucoside, one of the major phenolic compounds of grape skins, presented notable growth inhibitory effects in colon, breast, hepatocellular and lung cancer cells. Malvidin-3-glucoside has pointed to other health benefits, such as reduction of the incidence of CVDs, through the inhibition of the oxidation of human low-density lipoproteins. There is, therefore, a great potential in grape utilization in human health because every day a new promising phytochemical is characterized in its composition and moreover the sequencing of *V. vinifera* L. genome creates new opportunities for the grape

* Corresponding Author address: Email: jsc@uma.pt.

phytochemical composition modulation. Nevertheless, more research is needed to unveil the potential of grapes composition on human health and clarify their effects in our diet.

Keywords: *Vitis vinifera* L. grapes; Biological effects; Terpenoids; Phenolic compounds

Introduction

Several epidemiological studies and clinical trials have shown that diets rich in plant-based terpenoids, phenolics, and other compounds correlate significantly with the reduction of the risk of numerous lifestyle diseases, particularly CVDs and cancer, but also rheumatoid arthritis, lung diseases and Parkinson's or Alzheimer's disease (Szajdek and Borowska 2008, Thoppil and Bishayee 2011). These studies clearly point to the plant phytochemicals, vitamin C and E, all of which presenting antioxidant properties, as the main responsibles for the observed protective effects (Szajdek and Borowska 2008). These evidences have led to a significantly increase in the awareness among consumers, academia and the medical community of the health benefits associated with a diet rich in fruits, vegetables, spices, cereals, nuts, and edible tubers or roots (Vlaykova, Dimitrova et al. 2013), which are main sources of health protective phytochemicals. These compounds are secondary plant metabolites produced using different metabolic pathways that can be divided into three main categories: terpenoids (approximately 40000), phenolics (approximately 8000), and alkaloids (approximately 12000). Phytochemicals belonging to these three categories have been linked to health protection in *in vitro*, *in vivo*, and clinical research (Azmir, Zaidul et al. 2013, Grace, Yousef et al. 2014).

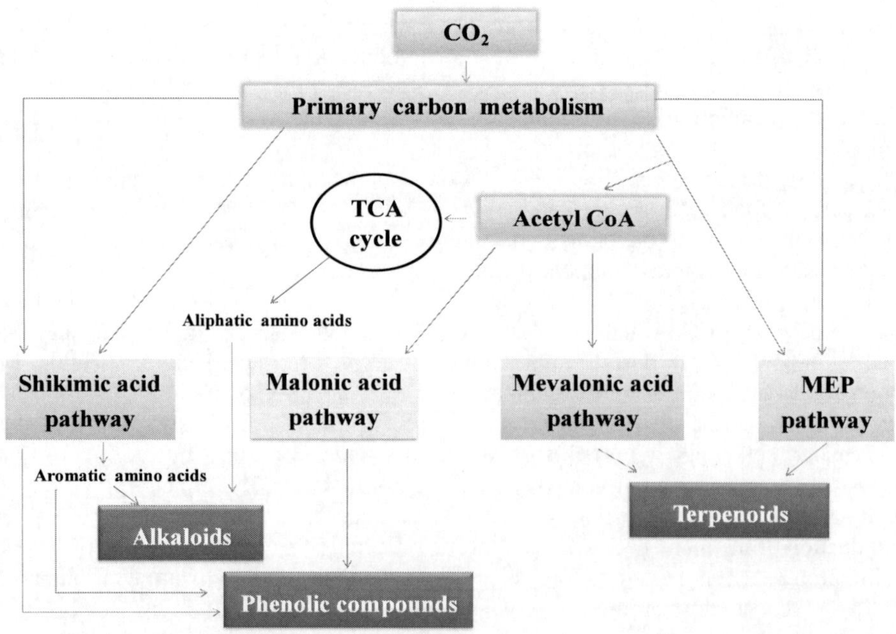

Figure 1. A simplified view of pathways for production of three major groups of plant bioactive compounds (adapted from Taiz & Zeiger (2006)).

Vitis vinifera L. Grapes Development

Grape is one of the main fruits produced worldwide being *Vitis vinifera* L. the most cultivated specie. This fruit is a rich source of phytochemicals with antioxidant properties, which are mostly present in berry skins and seeds (Perestrelo, Barros et al. 2011, Perestrelo, Lu et al. 2012). Nevertheless, they are present in vestigial amounts when compared with water (81 – 87 %) and carbohydrates (12 – 18 %), which are the main components in *V. vinifera* L. grapes. In minor proportions grapes contain proteins (0.5 – 0.6 %) and acids (0.3 – 0.4 %), significant amounts of potassium (0.1 – 0.2 %), vitamin C (0.01 – 0.02 %), vitamin A (0.001 – 0.0015 %) and also small amounts of calcium (0.01 – 0.02 %) and phosphorus (0.08 – 0.01 %). Grapes are also a major source of other nutrients like boron, a microelement very important for bone health (Çetin and Sagdic 2009, Yadav, Jain et al. 2009). From the nutritional point of view, each 100 g grapes contains ~ 78 calories of energy, ~ 0.5 g of protein, ~ 19 g of carbohydrate, ~ 0.3 g of fat (3 % calories from fat), ~ 0.18 mg of sodium, ~ 155 mg of potassium, 0.4 g of fiber, 13 mg of calcium, 9 mg of phosphorus, and 10 mg of vitamin C (Çetin and Sagdic 2009, Yadav, Jain et al. 2009).

V. vinifera L. grapes are more attractive in organoleptic, nutritional and health-promoting terms when harvested at their optimal maturity stage. During grape development several physical and biochemical changes can occur, affecting weight, sugar content, acidity, pigments development, metabolic activity, among others.

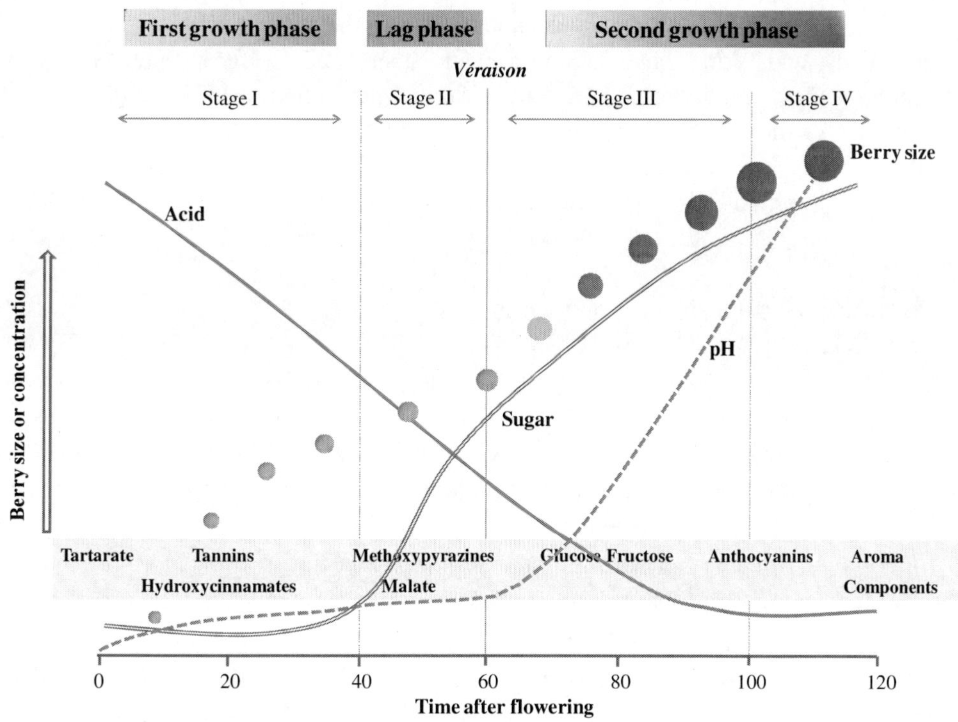

Figure 2. Schematic representation of grape berries composition change during development and ripening (adapted from Watson (2003)).

Grape berries exhibit two successive sigmoidal growth periods separated by a lag phase (Figure 2), that are divided into four stages that correspond to the berry formation (stage I and II) and berry ripening (stage III and IV) (Coombe and McCarthy 2000).

Stage I (*flowering*) begins after fruit set and lasts about 40 days. It is characterized by the increase of pericarp (flesh and skin) and seeds size, accumulation of high levels of tartaric acid and malic acid in the mesocarp (flesh), and low sugar concentration. The grape berries are green and hard. In most grape varieties, the first growth phase is followed by a lag phase (stage II) in which cell expansion slows down whereas seed maturation is completed. The duration of the first fase growth depends on the grape variety, and its end corresponds to the final grape herbaceous phase. The onset of grape ripening (*véraison*, stage III), follows the period of absence of growth, and it is characterized by an increase in the sugar content and pH and a decrease in acidity (Figure 2). Moreover, *véraison* is also characterized by a chlorophyll degradation and phenolic components accumulation, which are responsible for the color change (e.g., yellow-gold (flavonols) and pink and red (anthocyanins)). Stage IV, also known as engustment, is characterized by a full maturation and aroma and flavor components accumulation. In this last stage, the berry size reaches its maximum and sugar accumulation slows. The duration of each stage depends on the grape variety and environment and agricultural practices (Watson 2003, Conde, Silva et al. 2007, González-Barreiro, Rial-Otero et al. 2013).

A large number of reports involving different plant phytochemicals (e.g. terpenoids, phenolics, carotenoids, betalains, and vitamins) have shown that grape extracts, particularly grape seeds and skins have biological and therapeutic effects such as antioxidant, anticarcinogenic, antimicrobial, antiviral, antiaging, anti-inflammatory, antidiabetic activities as well as cardioprotective, hepatoprotective and neuroprotective effects. These results confirms why grapes are widely known by their health protective effects against several diseases (reviewed in Çetin & Sagdic (2009)).

Terpenoids and C_{13} Norisoprenoids

Terpenoids (e.g., mono, di and sesquiterpenic compounds) also referred as terpenes, and C_{13} norisoprenoids, are the two largest classes of plant metabolites, presenting a broad chemical and structural diversity. They account for about 40000 individual compounds, with several new compounds being discovered every year. They are synthesized from two five-carbon building blocks, the isoprenoid units, derived from isopentenyl (3-methyl-3-en-1-yl) pyrophosphate (Figure 3). These C_5 units are linked together in a head-to-tail manner. Based on the number of the isoprene units, terpenoids are classified as monoterpenes (C_{10}, e.g. carvone, geraniol), sesquiterpenes (C_{15}, e.g. germacrene, α-ylangene), diterpenes (C_{20}, e.g. retinol, *trans*-retinoic acid), sesterpenes (C_{25}, e.g. ceroplastol, gascardic acid), triterpenes (C_{30}, e.g. lupeol, oleanic acid), tetraterpenes (C_{40}, e.g. β-carotene, lutein) and polyterpenes. By the other hand, C_{13} norisoprenoids arise from chemical, photochemical and enzymatic degradation of carotenoids, such as β-carotene, lutein, neoxanthin, violaxanthin.

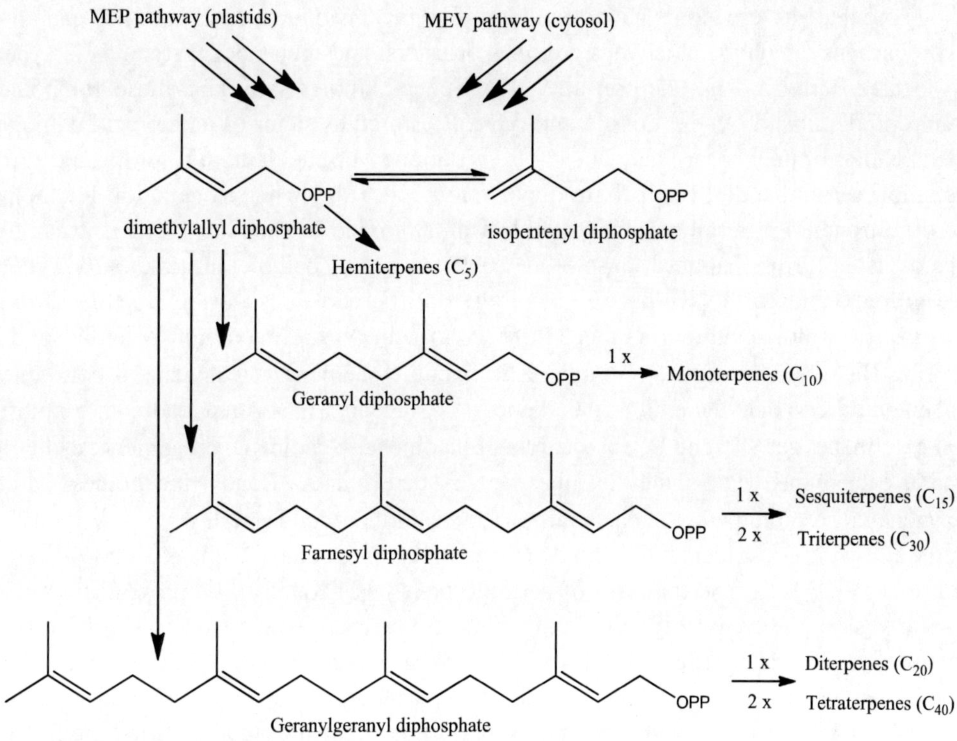

Figure 3. General scheme of plant terpenoids biosynthesis (Bohlmann & Keeling 2008).

However, this cleavage reaction is not region-specific and it leads to the formation of cyclic or linear apocarotenoids with 9, 10, 11, 13 and 15 carbon atoms (Figure 4). In grapes, the sun exposure promotes the carotenoids synthesis and/or their degradation, resulting mainly in C_{13} norisoprenoids formation.

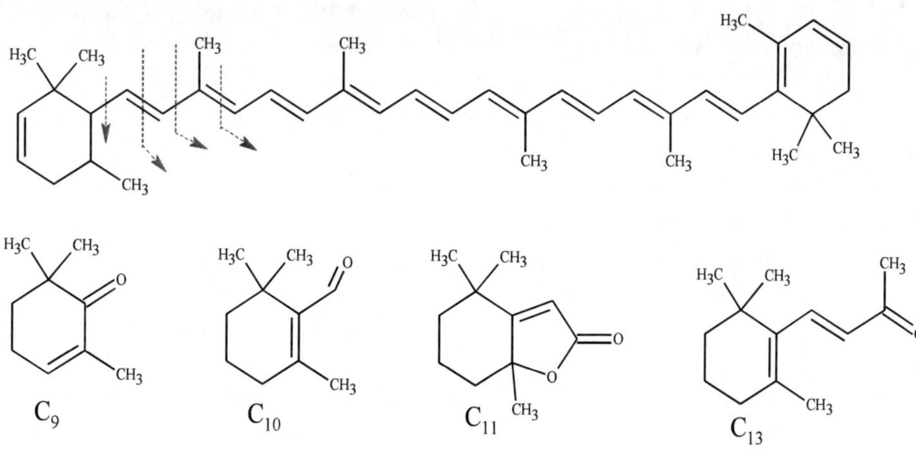

Figure 4. Formation of C9, C10, C11 and C13 norisoprenoids from carotenoids breakdown (adapted from Mendes-Pinto (2009)).

More than 40 terpenoids have been identified in *V. vinifera* L. grapes. They can appear as hydrocarbons, alcohols, aldehydes, ketones or esters, and their odor thresholds is synergistic (González-Barreiro, Rial-Otero et al. 2013). Several studies show that these compounds are found predominantly in skin tissue and typically stored as sugar or amino acid conjugates in the vacuoles of the exocarp cells. In *V. vinifera* grape Muscat Hamburg, for instance, linalool, geraniol, nerol, diendiol I and II are major free terpenoids , whereas geraniol, linalool, citral, nerol, citronellol, α-terpineol, linaloloxide I, linaloloxide II, diendiol I and II accumulate in glycosylated forms (Fenoll, Manso et al. 2009). In turn, Coelho et al. (2006, 2007) collected the grape berries of *V. vinifera* L. cv. Baga and Fernão-Pires during ripening, from half-*véraison* to a post-maturation stage, in order to achieve the maximum potential of varietal profile. Based on the results, (+)-cycloisosativene, γ-elemene, α-ylangene, β-bourbonene, β-cubenene, β-caryophyllene, 3,7-guaiadiene, (−)-isoledene, (+)-aromadendrene, α-amorphene, (−)-δ-selinene, germacrene D, epizonarene, β-cadinene, γ-cadinene, δ-cadinene, α-muurolene, and α-calacorene were some sesquiterpenoids detected in 'Baga' ripe grapes. Moreover, oxygenated sesquiterpenes, such as farnesol, nerolidol or γ-eudesmol, were described as grape berry and wine constituents (Coelho, Rocha et al. 2006, Coelho, Rocha et al. 2007, Perestrelo, Barros et al. 2011). The majority of sesquiterpenes was found to be present in grape berry exocarp (skin), whereas only minor amounts were detectable in mesocarp (flesh) at the stage of full ripeness. The terpenoids maximum content was achieved at the maturity, remaining constant until post-maturity stage. By the other hand, sixteen monoterpenoids and two C_{13} norisoprenoids were detected in Fernão-Pires grapes' and their content increase for about 3 weeks after *véraison*, and then decrease sharply. Linalool, α-terpineol and geraniol were the most representative terpenoids and their maximum content was achieved at the maturity (Coelho, Rocha et al. 2006, 2007). General, monoterpenoid content tended to increase progressively with ripening, while other varietal chemicals (sesquiterpenoids and C_{13} norisoprenoids) increased up to maturity and decreased thereafter (Coelho, Rocha et al. 2007). Regarding this, Perestrelo et al. (2011), identified 27 monoterpenoids, 27 sesquiterpenoids (plus 3 non indentified) and 10 C_{13}-norisoprenoids in the pulp and skin of the grapes Boal (white variety) and Bastardo (red variety) at maturity. This data was then used to establish the respectives global volatile signatures and was observed that both terpenoids and C_{13} norisoprenoids were significantly higher in the skin than in the pulp. (-)-β-bourbonene, α-bisabolene, and α-ylangene were the predominant components in Boal grapes, and α -caryophyllene in Bastardo grapes. Figure 5 illustrates the chemical structure of predominant terpenoids found in *V. vinifera* L. grapes with potential health benefits.

Recent studies have indicated that mechanisms underlying chemopreventive potential may be a combination of several properties, including antioxidant, anti-inflammatory, immunoenhancing, and hormone modulation effects. This synergistic action enable the modification of drug metabolizing enzymes, influence the cell cycle and differentiation, induces apoptosis, suppresses the proliferation and angiogenesis associated to the initiation and secondary modification stages of neoplastic development (Rabi and Gupta 2008). From a biological point of view, terpenes have been reported to present antimicrobial, antifungal, antiparasitic, antiviral, anti-allergenic, antispasmodic, antihyperglycemic, anti-inflammatory, and immunomodulatory properties (Table 1).

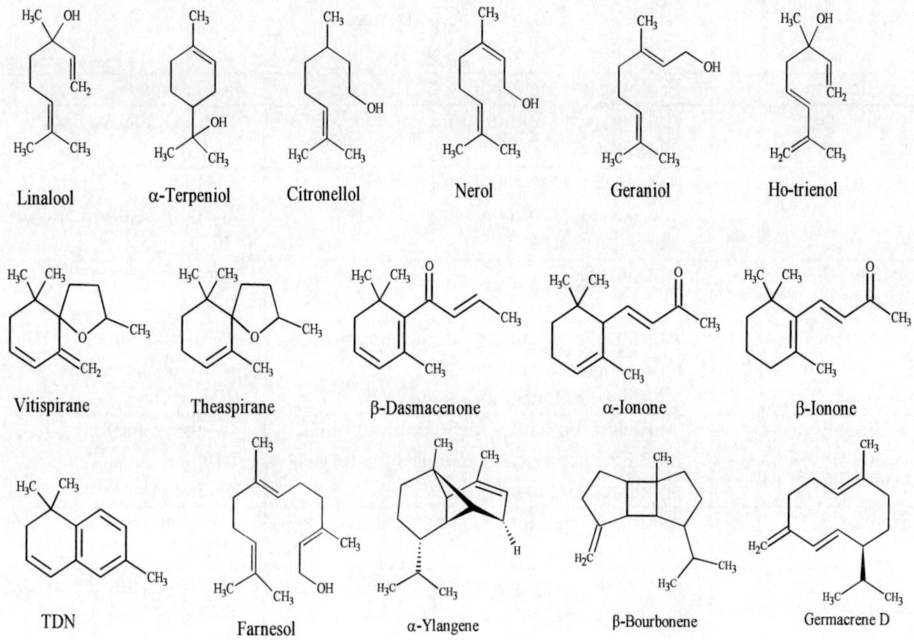

Figure 5. The most representative terpenoids found in *V. vinifera* L. grapes.

Moreover, epidemiological studies suggest that terpenes may be useful in the prevention and therapy of several cancers, including mammary, skin, lung, forestomach, colon, pancreatic and prostate carcinomas (reviewed in Thoppil & Bishayee (2011)). This is strongly supported by the observation that a number of dietary monoterpenes have antitumor activity, exhibiting not only the ability to prevent the formation or progression of cancer, but to recede existing malignant tumors (reviewed in Thoppil & Bishayee (2011)). This is the case of limonene, to which has been established chemopreventive and chemotherapeutic activities against several human cancers, as those affecting skin, liver, lung, breast, colon, and prostate (Chung, Lee et al. 2006, Kita, Miura et al. 2012, Jia, Xi et al. 2013, Miller, Lang et al. 2013) (Table 1).

Table 1. Biological effects of the terpenoids identified in *V. vinifera* L. grapes

Terpenoids	Biological effects	Reference
β-Myrcene	Analgesic, anti-inflammatory, antibiotic, antitumor	(Suhail, Wu et al. 2011)
p-Cymene	Antibiotic, anticandidal, AChE inhibitor, antiproliferative	(Razavi and Nejad-Ebrahimi 2009, Mander and Liu 2010)
Camphene	Antimicrobial, antioxidant	(Singh, Tripathi et al. 2012)
Terpinene	Anti-inflammatory, antioxidant, chemoprotective activity, antimicrobial, antiproliferation	(Ruberto and Baratta 2000, Yin, Yan et al. 2012, Rathore, Saxena et al. 2013)
Terpinolene	Antioxidant	(Ruberto and Baratta 2000)

Table 1. (Continued)

Terpenoids	Biological effects	Reference
β-Phellandrene	Antibacterial, antitumor, apoptotic, citotoxic	(Hussain, Anwar et al. 2010)
β-Pinene	Anti-inflammatory, antimicrobial, antioxidant, antineoplastic, chemoprotective.	(Mander and Liu 2010, Rathore, Saxena et al. 2013)
Terpenyl acetate	Antiproliferative	(Preethy, Padmapriya et al. 2012)
d-Limonene	Antimitogenic, antiproliferative and antimutagenic), antioxidant, antimicrobial, chemoprotective	(Miller, Lang et al. 2013, Rathore, Saxena et al. 2013)
1,8-Cineole	Antibacterial, antioxidant, antimicrobial, antibiotic, antiviral, anti-inflammatory, antineoplasic, cytotoxic	(Mander and Liu 2010, Mohamad, El-Bastawesy et al. 2011)
Camphor	Antibacterial, larvicidal, antioxidant	(Waliwitiya, Kennedy et al. 2009)
Linalool	Cytotoxic, antiproliferative, antioxidant, antimicrobial, analgesic, anti-inflammatory, antidepressant, anxiolytic	(Mander and Liu 2010, Rathore, Saxena et al. 2013)
Geraniol	Chemopreventive activity, antimutagenic	(Cardozo, de Conti et al. 2011)
Citronellol	Citotoxic, antimutagenic	(Bostancıoğlu, Kürkçüoğlu et al. 2012)
Borneol	Antimicrobial, antioxidant, antimutagenic, larvicida	(Elaissi, Rouis et al. 2012, Horváthová, Kozics et al. 2012)
Eugenol	Anticancer, antidepressant, antimicrobial, antioxidant, larvicidal	(Ruberto and Baratta 2000, Waliwitiya, Kennedy et al. 2009)
Terpineol	Antimicrobial, antioxidant, antibiotic, AChE inhibitor, antioxidant, antimalarial, cytotoxic	(Mander and Liu 2010)
Perillyl alcohol	Chemopreventive activity, antioxidant	(Mills, Chari et al. 1995, Ruberto and Baratta 2000)
Citral	Cytotoxic	(Zamanian-Azodi, Rezaie-Tavirani et al. 2011)
β-Ionone	Chemopreventive activity, antiproliferative, cytotoxic	(Rigano, Cardile et al. 2009, Cardozo, de Conti et al. 2011)
Aromadendrene	Antioxidant	(El-Ghorab, El-Massry et al. 2007)
Sesquiphellandrene	Antimicrobial, antioxidant, cytotoxic, apoptosis, anti-inflammatory	(Vukovic, Milosevic et al. 2007)
β-Elemene	Anticancer, cytotoxic	(Lu, Dang et al. 2012)
β-Bisabolene	Antifungal, antibacterial, antioxidant, antimicrobial, cytotoxic, antitumor	(Kimura, Sakamoto et al. 2012, Li, Xu et al. 2012)
β-Caryophyllene	Anticancer, antioxidant, cytoprotective, anti-inflammatory, antimalarial, antibacterial, antiproliferative, citotoxic	(Mander and Liu 2010, Chen, Lu et al. 2011)

Terpenoids	Biological effects	Reference
Germacrene	Anti-inflammatory, antibacterial, antioxidant	(Hajdari, Novak et al. 2012, Zellagui, Gherraf et al. 2012)
β-Farnesene	Antioxidant, antiproliferative	(McKay and Blumberg 2006)
α-Humulene	Antibacterial, antifungal, antioxidant, cytotoxic	(Jones, Lobo-Echeverri et al. 2007)
Farnesol	Chemopreventive activity, antioxidant	(Ruberto and Baratta 2000)
Nerolidol	Anticancer	(Kathirvel and Ravi 2012)
α-Bisabolol	Cytotoxic, antifungal, antimutagenic	(Chen, Hou et al. 2010)
Geranylgeraniol	Chemopreventive activity	(de Moura Espíndola, Mazzantini et al. 2005)

In vivo and *in vitro* studies demonstrated that the chemotherapeutic activity of *d*-limonene could be attributed to the induction of apoptosis (Lu, Zhan et al. 2004). Ji et al. (2006) found that the limonene-induced apoptotic death of human leukemia cells involves the increase in Bax protein expression, the release of cytochrome c from mitochondria and the activation of caspase. This strongly suggests that the mitochondria-mediated intrinsic death pathway may play a major role in limonene-induced death. Recent studies have demonstrated that the chemopreventive efficacy against cancer of β-ionone and *d*-limonene and some of its derivatives,(Cardozo, de Conti et al. 2011, Thoppil and Bishayee 2011, Jia, Xi et al. 2013, Miller, Lang et al. 2013) increase the anti-proliferative effects in human prostate cancer cells through the activation of the ERK pathway and induction of WAF1/p21, leading to cell cycle arrest (Chen, Lu et al. 2006). An *in vivo* study demonstrated that 0.5 % *d*-limonene in diet is responsible for a 32 % inhibition of highly prone hepatocarcinogenesis in male F-344 rats (reviewed in (Thoppil and Bishayee 2011)). By the other hand, geraniol and α-bisabolol inhibit the growth of HepG2 human hepatic carcinoma cells (Chen, Hou et al. 2010), although using different mechanisms. Geraniol, an acyclic dietary monoterpene, decreases the levels of 3-hydroxymethylglutaryl coenzyme A (HMG-CoA) reductase, which is the major rate-limiting enzyme in cholesterol biosynthesis in mammals (Polo and de Bravo 2006). Moreover, geraniol was also shown to have *in vivo* antitumor activity against murine leukemia, hepatoma and melanoma cells when administered before and after tumor cell transplantation (Yu, Hildebrandt et al. 1995, Burke, Stark et al. 1997). In turn, α-bisabolol acts in the apopotic pathway, favoring an increase in caspase-3, -8, Bax, cyt *c*., p53 and Fas, with a concomitant decrease in the Bcl-2 (Chen, Hou et al. 2010). In addition, perillyl alcohol presents chemopreventive activities against rodent mammary, pancreatic and prostatic tumors (Crowell, Ayoubi et al. 1996), and chemically induced liver cancer in rats (Mills, Chari et al. 1995). It is also very effective in preventing tumor recurrences or secondary tumors in animals treated in a chemotherapy regimen (Haag, Lindstrom et al. 1992). β-ionone, another cyclic sesquiterpene, was found to inhibit hepatic preneoplastic lesions in rats with a decrease in cell proliferation, inhibition of plasma cholesterol and amelioration of DNA damage during the initial hepatocarcinogenesis phase (de Moura Espíndola, Mazzantini et al. 2005). Among the few *in vivo* studies available, β-ionone was shown to inhibit melanoma (He, Mo et al. 1997) and breast cancer (Liu, Sun et al. 2008). In turn, the already referred geraniol inhibits growth of pancreatic adenocarcinomas (Burke, Stark et al. 1997) and both β-ionone and geraniol inhibit hepatic preneoplastic lesions when administered continuously during the initiation and selection/promotion phases of hepatocarcinogenesis (de Moura Espíndola, Mazzantini et al. 2005, Ong, Heidor et al. 2006). β-elemene, a natural sesquiterpene, exhibits broad-spectrum anti-cancer activity against many types of cancer cells, including leukemia, brain, breast, prostate, ovarian, cervical, colon, laryngeal, and lung carcinoma cells (reviewed in Lu et al. (2012)).

Concerning to the antibacterial, antimicrobial and antioxidant activity, monoterpenes hydrocarbons showed the lowest antibacterial activity, while oxygenated compounds possess a higher potential, specially phenol-type compounds as thymol and carvacrol (Ruberto and Baratta 2000, Trombetta, Castelli et al. 2005). Several studies showed that oxygenated monoterpenes, exhibit strong antimicrobial activity, particularly pronounced on whole cells, while hydrocarbon derivatives possess lower antimicrobial properties, as their low water solubility limits their diffusion through the medium (Kalemba and Kunicka 2003). Griffin et al. (2000) have shown that hydrocarbons tend to be relatively inactive regardless of their structural type, and this inactivity is closely related to their limited hydrogen bound capacity and water solubility. It has been demonstrated that α- and β-pinene are able to destroy cellular integrity, and thereby, inhibit respiration and ion transport processes (reviewed in Magwa et al. (2006)). Previous results showed that greater antimicrobial potential could be ascribed to the oxygenated terpenes (e.g. 1,8-cineole), especially phenolic compounds (Soković and van Griensven 2006, Soković, Glamočlija et al. 2010). In addition, *o*-cymene and limonene have been shown to have strong antifungal properties. Finally, *trans*-Caryophyllene and α-humulene have been shown to have antibacterial, antifungal or antioxidant properties (reviewed in Magwa et al. (2006)).

Phenolic Compounds

Phenolic components are also secondary plant metabolites, being classified in flavonoid and non-flavonoid compounds according to their carbon skeleton (Figure 6). The flavonoids are phenolics with diphenylpropanes (C6-C3-C6) skeletons that suffer different modifications in the central C-ring, originatingseveral classes, as flavonols (e.g., quercetin, kaempferol), flavones (e.g., apigenin, luteolin), flavan-3-ols (e.g., (+)-catechin, (-)-epicatechin), flavanones (e.g., taxifolin, naringenin), and anthocyanidins (e.g., cyanidin, malvidin). Non-flavonoids have C6-C1, C6-C3 or C6-C2-C6 structures, meaning that either one or three carbons are attached to the primary benzene ring (six carbons). The predominant non-flavonoids in grapes and wines are C6-C1 phenolic acids (e.g., gallic acid, protocatechuic acid), C6-C3 hydroxycinnamic acids (e.g., *p*-coumaric acid, caffeic acid) and C6-C2-C6 stilbenes (e.g., *trans*-resveratrol and its glucoside *trans*-piceid) (Xia, Deng et al. 2010, Pereira, Silva et al. 2011). Nowadays, there is a growing interest on phenolic compounds from grapes and many studies have been focused on their biological activities in human health. As a result, a myriad of protective effects have been pointed, including antioxidant, cardioprotective, anticancer and chemoprevention, anti-inflammation, antiaging and antimicrobial activities, most of them linked to the free radical scavenging ability of phenols (Ferreira and Slade 2002, Scalbert, Manach et al. 2005).

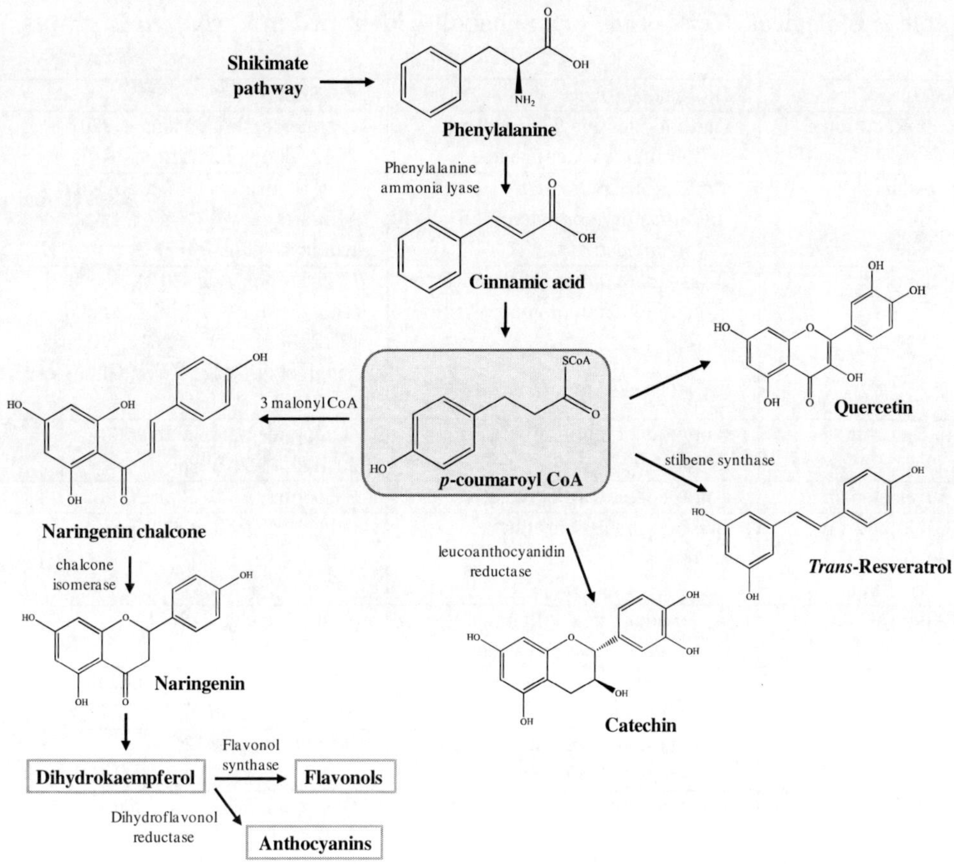

Figure 6. Overview of the biosynthesis of the main phenolics from p-coumarylCoA (adapted from (Besseau et al. 2007)].

In addition to their individual effects, phenolic compounds interact in synergistic ways and have sparing effect in which one may protect another against oxidative destruction. Table 2 summarizes the most important phenols identified in grapes and the associated biological effects.

A very popular phenolic identified in grapes was *trans*-resveratrol and several studies have been performed using *in vitro* and *in vivo* models to underlay the pathways in which this stilbene can act (Provinciali, Re et al. 2005, Juan, Alfaras et al. 2012). As a result, resveratrol has been shown to interfere with all the three stages of carcinogenesis, initiation, promotion and progression. Moreover, many other mechanisms are involved in the activity of resveratrol, at least in vitro, including modulation of transcription factors, such as the nuclear factor kappa B (NF-κB) (Leiro, Arranz et al. 2005, Rahman, Biswas et al. 2006). This is very relevant as NF-κB is an important molecular target for pharmacological intervention and drug development, regulating a wide variety of genes involved in the control of immune and inflammatory responses (including some cardiovascular pathologies such as atherosclerosis), inhibition of the cytochrome P450 and expression and augmented activity of cyclooxygenase (COX-2) enzymes (Chun, Ryu et al. 2001).

Table 2. Biological effects of the main phenolics identified in *V. vinifera* L. grapes

Phenolics	Biological effects	Reference
trans-Resveratrol	Cardioprotective, antioxidant, antiproliferative, anticancer	(Fernández-Mar, Mateos et al. 2012, Zhou and Raffoul 2012)
Piceatannol	Anticancer, antioxidative, anti-inflammatory, antimicrobial, estrogenic activity	(Vo, Madlener et al. 2010, Kita, Miura et al. 2012, Piotrowska, Kucinska et al. 2012)
Gallic acid	Antioxidant, *in vitro* anticancer effect, induction of apoptosis	(You, Moon et al. 2010, Forester, Choy et al. 2012, Liu, Li et al. 2012, Su, Lin et al. 2013, Verma, Singh et al. 2013, Yoon, Chung et al. 2013)
(+)-Catechin	Antioxidant, antiproliferative, anticancer	(Lacopini, Baldi et al. 2008, Alshatwi 2010)
(-)-Epicatechin	Antioxidant, cardioprotective	(Lacopini, Baldi et al. 2008)
Cyanidin-3-galactoside and pelargonidin	Increase in basal insulin secretion	(Pinent, Castell et al. 2008)
Malvidin	Antioxidant, anti-inflammatory, chemopreventive properties	(Fritz, Kern et al. 2006)
Naringenin	Antihyperglycemic, antioxidant, antiproliferative anticancer	(Qin, Jin et al. 2011, Annadurai, Muralidharan et al. 2012)
Myricetin	Anti-inflammatory, antioxidant, insulinomimetic effect	(Li and Ding 2012)
Rutin	Cardioprotective, analgesic and anticancer activity	(Lacopini, Baldi et al. 2008, You, Ahn et al. 2010, Lee, Shen et al. 2013)

In some cancer cell lines, *trans*-resveratrol has been shown to induce apoptosis (Aziz, Nihal et al. 2006, Faber and Chiles 2006, Riles, Erickson et al. 2006, Sareen, van Ginkel et al. 2006, Benitez, Pozo-Guisado et al. 2007). This seems to involve the activation of p53 and cyclins A, B1, and cyclin-dependent kinases cdk 1 and 2 (Cao, Fu et al. 2005). Nevertheless, most of the protective effects ascribed to resveratrol *in vitro* have not been yet demonstrated in humans due to its oral low bioavailability (Manach, Williamson et al. 2005, Juan, Alfaras et al. 2012). According to King et al. (2006), the cardioprotective effects of resveratrol may also be due, in part, to its vasorelaxation properties, decreasing LDL oxidation levels, which are an hallmark in the development of atherosclerosis. Piceatannol (3,3′,4,5′-trans-trihydroxystilbene) is another phenolic compound found in grapes and its synthesis in plants increases in response to stress conditions, e.g. UV irradiation, fungal infection, or heavy metals contamination of soil. Piceatannol is the naturally occurring hydroxylated analogue of resveratrol and, althought it is less studied than resveratrol, displays a wide spectrum of biological activities (Kita, Miura et al. 2012, Ko, Lee et al. 2012, Minakawa, Miura et al. 2012, Wu 2012, Kapoor 2013). Most studies explore its anticancer activity, based on the inactivation/activation of specific pathways that regulate cancer progression. However, some of the molecular mechanisms underlying these properties have not yet been studied in detail. Kim et al. (2008) studied the effect of piceatannol in human leukemic cells and found that this compound inhibited cell viability by inducing apoptosis through the formation of apoptotic bodies. Another study conducted by Kim et al. (2009) also showed that piceatannol induced apoptosis via the activation of the death receptor and mitochondrial-dependent pathways in androgen-insensitive DU145 prostate cancer cells. A similar result was obtained by Kita et al. (2012), demonstrating that piceatannol suppressed both the proliferation and invasion of AH109A hepatoma cells *in vitro* and *ex vivo*.

Catechin, an abundant flavanol in tea, was showed to possess anticancer activity, blocking the proliferation of MCF7 cells and inducing apoptosis, in part by modulating the expression levels of caspase-3, -8, and -9 and p53 (Alshatwi 2010). An initial study from Meyer et al. (1998) reported that catechin inhibit human LDL oxidation *in vitro* (assessed by measuring inhibition of copper-catalysed human LDL oxidation), thus reducing atherogenesis. According to Monga et al. (2013), this flavonoid inhibit lipid peroxidation both in cell culture and *in vivo* models. Malvidin, an anthocyanin widely distributed in grape skins, was recognized to interfere with the process of carcinogenesis and seems to be linked to multiple potential mechanisms of action, including inhibition of cyclooxygenase enzymes and potent antioxidant activity. Regarding this Hou et al. (2004) revealed that anthocyanins inhibit tumorigenesis by blocking the c-Jun N-terminal kinase (JNK) and the early activation of the ERK (mitogen-activated protein kinase) pathway. Many other studies have suggested that anthocyanins frequently interact with other phytochemicals to potentiate biological effects and thus the contributions from individual components are difficult to decipher (Lila 2004). Understanding the phenolics chemistry has advanced greatly in the past decade, but some questions remain, including clarification of which metabolites produce the highest biological activity and which mechanisms of action are involved. *In vivo* study using different cancer cell lines reported that quercetin-3-glucoside, one of the major phenolic compounds of grape skins, presents notable growth inhibitory effects in colon, breast, hepatocellular and lung cancer cells. Rutin, another abundant *V. vinifera* L. grapes phenolic compound, showed antiproliferative effect on five epithelial cancer cells (You, Ahn et al. 2010). Malvidin-3-glucoside has been also shown to have other health benefits, such as the reduction of the incidence of CVDs, through the inhibition of the oxidation of human low-density lipoproteins (Wang and Mazza 2002). Nevertheless, further studies must be performed in order to understand the real potential of these compounds as possible nutraceuticals, both as single compounds and as components of natural mixtures. Furthermore, it is important to study possible interactions of phenolics with other grape constituents as the synergetic reactions might potentiate the biological activities observed (Man-Ying Chan, Mattiacci et al. 2000, Maggiolini, Recchia et al. 2005, Manach, Williamson et al. 2005).

Carotenoids

Carotenoids are among the most common natural pigments, with more than 600 different compounds already characterized, being β-carotene the predominant component (Stahl and Sies 2003). In grapes berries, the presence of carotenoids is well-documented and it has been demonstrated that β-carotene and some xanthophylls (neoxanthin, flavoxanthin, and lutein) are abundant before *véraison*, decreasing dramatically in subsequent stages (Mendes-Pinto, Silva Ferreira et al. 2005). At maturity, carotenoid concentration can be as low as 0.8-2.5 mg/kg (Razungles, Babic et al. 1996). Three other xanthophylls, violaxanthin, luteoxanthin, and 5,6-epoxylutein, appear after *véraison*, when the sugar concentration reaches approximately 160 g/L. The carotenoids concentration in grape berries depends on cultivar, viticultural region, sunlight exposure, and ripening stage (Mendes-Pinto, Silva Ferreira et al. 2005, Oliveira, Barbosa et al. 2006).

Regarding to the carotenoids biological properties, they have been recognized for a long time for their antioxidant properties. More recently there is an increasing interest in carotenoids anti-cancer properties, namely their effect on the regulation of cell growth, modulation of gene expression, and, possibly, immune response (Slattery, Benson et al. 2000). *In vitro*, epidemiological and animal studies provide evidences that β-carotene, as well as other carotenoids, decreased the risk of several forms of cancer and CVDs, among others. These evidences are also supported by the correlation between β-carotene serum levels and diminished cancer risk (Mayne 1996). However, most intervention trials with β-carotene as a component of a supplement did not show any effects regarding the cancer risk (Omenn, Goodman et al. 1996, Stahl and Sies 2005). Also with respect to the prevention of CVDs, data for β-carotene are conflicting (Clarke and Armitage 2002). The compound is capable to inhibit lipid peroxidation in LDL, a process suggested to be involved in the pathogenesis of atherosclerosis, but until now there is no convincing evidence from intervention studies that β-carotene or carotenoids in general are protective against CVDs (Stahl and Sies 2005).

V. vinifera L. Grape Economical Impact

Grapes are the main fruit crop worldwide with a global production very near the 70 million tons in 2011 (FAOstat 2011). The main grape culture is obviously intended for wine production, but clearly there is a growing trend in the production of fresh table grapes, which represented around 20 % of the whole production in 2009 (Gonzalez-Centeno, Jourdes et al. 2012), being very near 25 % in 2011. This is the most recent year to which the data from Food and Agricultural Organization (FAO, (FAOstat 2011)), can be reliably compared with the one from United States (US) Department of Agriculture (indexmundi 2012). Unlike US data, FAO does not discriminate the fresh table grapes in the the whole grape production. In Figure 7 are presented the main features about grape production in terms of total grape production for wine, to be consumed as fresh table grapes in the respective internal or external markets and raisins.

Overall, the EU countries, China and US are the main players, although four different profiles can be allocated to main grape producers. The first are the winemakers (W), which are the ones that mainly produce grapes for wine production, and include the EU countries, US, Chile, Argentine, Australia and South Africa. These countries are simultaneously the major grape producers (P), although China and Turkey, that do not have a winemaking tradition, also present a substantial production of grapes not assigned as table grapes or raisins. Regarding the fresh table grapes, we can clearly distinguish countries that are consumers (C) or exporters (E), as the ones in which the grape production is mainly consumed in the internal market, or mainly allocated for exportation, respectively. Accordingly, China is presently the major fresh table grapes producer, accounting alone for almost half of the whole production (42 %). Nevertheless, this is mainly for internal consumption, and the same profile of producer for internal consumption (PC) is also the one of EU, Turkey, Brazil and India. By opposition, in smaller producers like Chile, South Africa, México and Australia, fresh grape production is mainly intended to export (E).

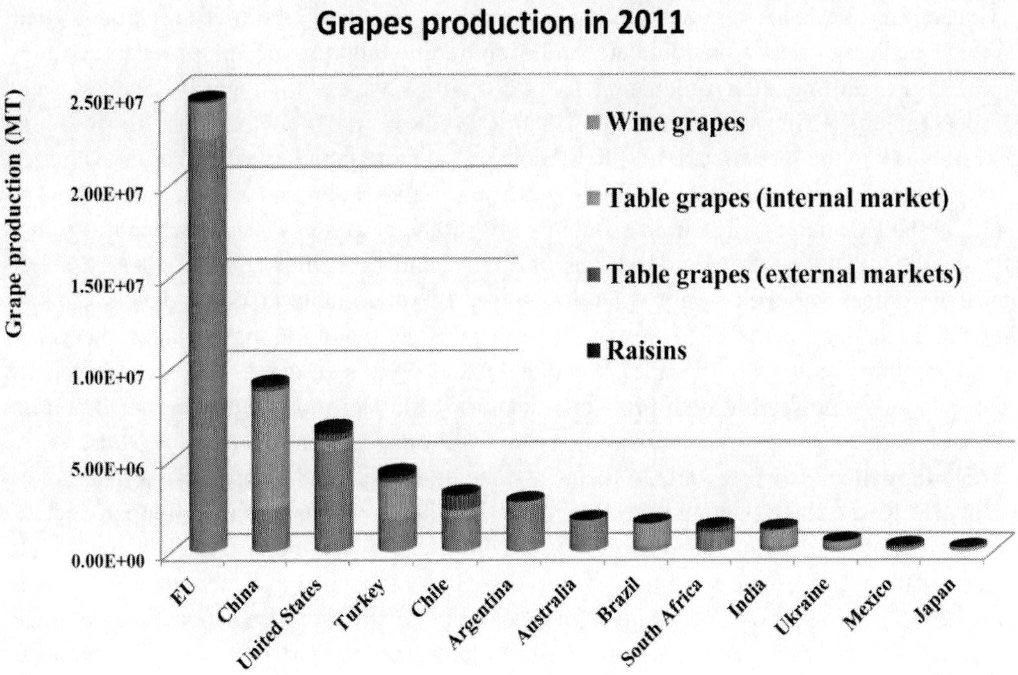

Figure 7. Some trade facts about the major players in the grape production (in mega tons –Mt) in 2011. Above each column is indicated the profile of each country: P – mainly producer, C – mainly consumer, indicating that most of the grape production is intended for internal consumption, E – mainly exporter, indicating that most of the grape production is intended for the external markets, W – winemaker, indicating that a most of the grape production is intended for wine production (data from United States Department of Agriculture (indexmundi) and FAO (FAOstat 2011)).

As for wine production, there is also a great diversity of table grape varieties around the globe, each one adapted to the local abiotic conditions and therefore presenting different organoleptic properties.

The exact number is very hard to estimate because there aren't reliable and extensive studies but is certainly above 1000 different varieties. Moreover, there are many variants and mutants of the same cultivar used in different localizations with different names and commercial brands. Nevertheless, only a few number of table grapes varieties produced for the national and international markets, correspond to almost all the worldwide production and in the last years, the table grape production is shifting for even less varieties, prevailing the seedless varieties.

In Table 3 are indicated the table grape varieties more cultivated around the globe, distributed by country and seasonality and considered in the major groups, green (or white), red and blue-black, in the seeded or seedless variants. As can be easily observed, we may found table grapes in the international markets all over the year. This is because the major producers of table grapes are localized in different latitude and longitudes, covering both hemispheres and all continents and therefore the climate conditions necessary for the table grape production can be found all over the year in different localization. Moreover, although a fragile fruit, grapes can be conditioned and refrigerated for weeks and even months since the harvesting to their consumption.

Concerning the main varieties produced nowadays, as already referred, regardless their color, the seedless varieties predominate, and even traditionally producers of seeded varieties in the EU are shifting their production for the seedless varieties Thompson Seedless and Sugarone (green), Crimpson seedless and Flame seedless (red) and Autumn Royal (blue-black). Regarding the seeded varieties, Red Globe is the more popular and one of the varieties that is produced in more latitudes, while the green varieties Italia and Victoria are still very popular in the EU markets. The high economic impact the grapes production has in the global markets and its growing demands in terms of quality and availability makes research about better table grapes varieties a very dynamic field in the main table grape producers. This is facilitated by the fact that table grapes are more responsive to cultural manipulation that other fruits. Therefore, in the last decades the main focus of this research has been driven to seedless varieties, presenting high productivities profiles under more and more harsh abiotic conditions, such as adverse climate conditions, water shortage and less productive soils.

Regarding this it is worthwhile to focus in some interesting examples. Chile's Institute of Agricultural Research (INIA) and the Biofrutals Technology Consortium developed several new mid-season harvesting seedless grapes, with firm skins, mild crunches and good postharvest lifecycles. Iniagrape-one (Uquillas 2013) was launched in the commercial circuit in April 2013 and is part of a major investment of Chile to strength its position as table grape producer. This new table grape delivers higher yields, earlier harvests and a longer post-harvest quality period (up to 60 days) which make it very attractive in commercial terms. Moreover, although there are not yet available data about this variety, its deep blue/black grape certainly confers it at least a rich composition in anthocyanins and consequent health protective effects.

Spain Grape Breeding program is also very dynamic and Table Grape Research and Technology Institute (ITUM) recently launched in the market 20 new table grape varieties (from Itum(-)one to Itum(-)twenty), comprising green, red and blue/black variants, seeded and seedless, presenting high economic potential given their high productivity and taste. United States is also a strong player in the table grape trade and has several active breeding programs aiming to develop seedless table grapes presenting better properties, as crisp textures, vibrant colors, large berries, strong stem attachment, good storage potential, high consistent yields and good flavors. This accounts for the hundreds of different table grapes varieties commercialized by US for both internal and international markets. Brazil is making an effort to increase its table grapes production, mainly the hybrid cultivar Isabel (*V. vinifera* × *V. labrusca*), which accounts for about 50 % of Brazilian grape production (Nixdorf and Hermosin-Gutierrez 2010).

In turn, South Africa has already launch in the market five Sheehan seedless table grape varieties, being three light to dark red (Allison, Krissy and Magenta), one white (Melanie) and one black (Melody). These varieties are crispy and naturally colored, highly productive, require less hand-labor and present a good storage potential.

Table 3. Table grapes seasonality around the globe

Seasonality	Production	JAN	FEB	MAR	APR	MAY	JUN	JUL	AUG	SEP	OCT	NOV	DEC
Variety													
Green grapes													
Seedless													
Thompson Seedless	AUS												
	Chile												
	China												
	EU												
	India												
	Turkey												
	SAR												
	US												
Sugraone (Superior, Menindee Seedless)	AUS												
	Brazil												
	Chile												
	China												
	EU												
	SAR												
	Turkey												
	US												

Table 3. (Continued)

Centennial Seedless	EU
	India[2]
Perlette	Chile
	EU
	India
	US
Niagara	Brazil
Autumn Seedless	India
Regal Seedless	SAR
Prime seedless	SAR
Melissa	EU
Muscatel	EU
Dominga	EU
Autumn King	US
Blanc Seedless	US
Princess	US
Sultana Seedless	Turkey

Seeded

Victoria	China	
	EU	
	SAR	
Italia	Brazil[1]	
	China	
	EU	
	SAR	
Calmeria	AUS	
	US	
Ohanes	EU	
	AUS	
Dawn	AUS	
Chasselas	EU	
Danlas	EU	
Anab-e-Shahi	India	
Dauphine	SAR	

Table 3. (Continued)

		JAN	FEB	MAR	APR	MAY	JUN	JUL	AUG	SEP	OCT	NOV	DEC
Red Grapes													
Seedless													
Crimpson seedless	AUS												
	Chile												
	China												
	EU												
	SAR												
	US												
Flame seedless	AUS												
	Chile												
	China												
	EU												
	India												
	SAR												
	US												
Ralli	AUS												
Black seedless	Chile												
Purple Seedless	China												

Apulia Rose
Muscatel
EU
Sunred Seedless
SAR
Ruby Seedless
Emperor
Holiday Seedless
US
Scarlet Royal
Sugra nineteen
Vintage Red

Seeded

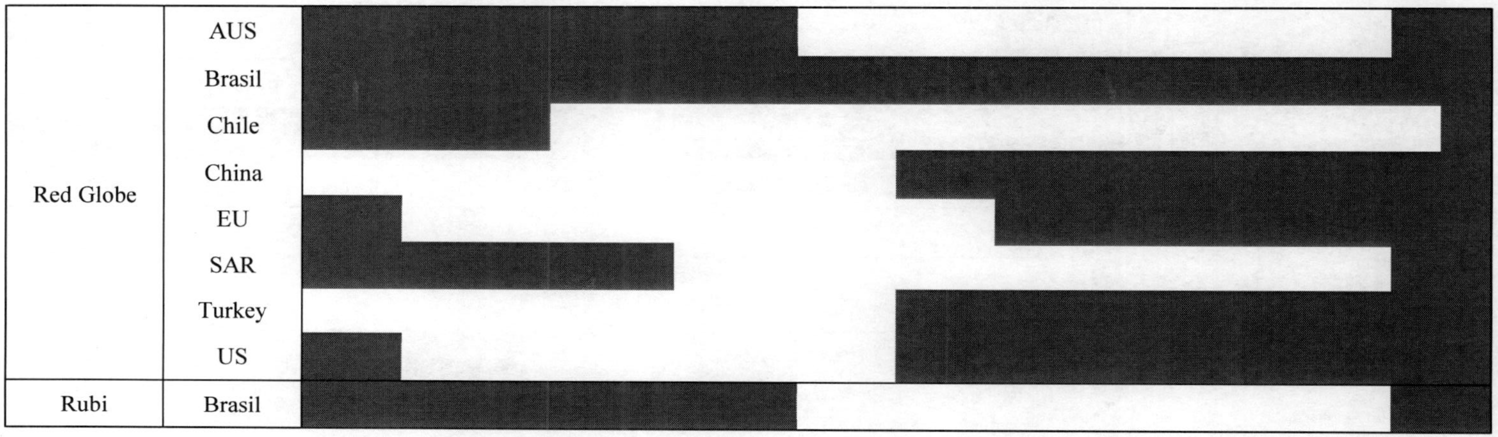
Red Globe
AUS
Brasil
Chile
China
EU
SAR
Turkey
US
Rubi
Brasil

Table 3. (Continued)

		JAN	FEB	MAR	APR	MAY	JUN	JUL	AUG	SEP	OCT	NOV	DEC
Benitaka		■	■	■	■	■							■
Christmas Rose	US									■	■	■	■
Emperatiz	US	■							■	■	■	■	■
Blue/Black Grapes													
Seedless													
Autumn Royal	AUS	■	■	■	■				■	■	■	■	■
	Chile	■	■	■									
	EU									■	■	■	
	US	■							■	■	■	■	■
Sugra thirteen	SAR	■	■									■	■
	US					■	■	■	■	■			
Big Perlon	EU							■	■	■	■	■	
Fantasy Seedless	US						■	■	■	■	■	■	■
Summer Royal	US						■	■	■	■	■		
Sharad Seedless	India	■	■	■	■								■

Seeded

Alphonse Lavallee (Ribier)
AUS
EU
SAR
Black Muscat
AUS
India
Isabella
Brasil
India
Midnight Beauty
AUS
Concord
Brasil
Black Magic
Cardinal
Lival
Muscatel
Napoleon
Palieri
Prima
Ribol
EU
Barlinka
SAR

Table 3. (Continued)

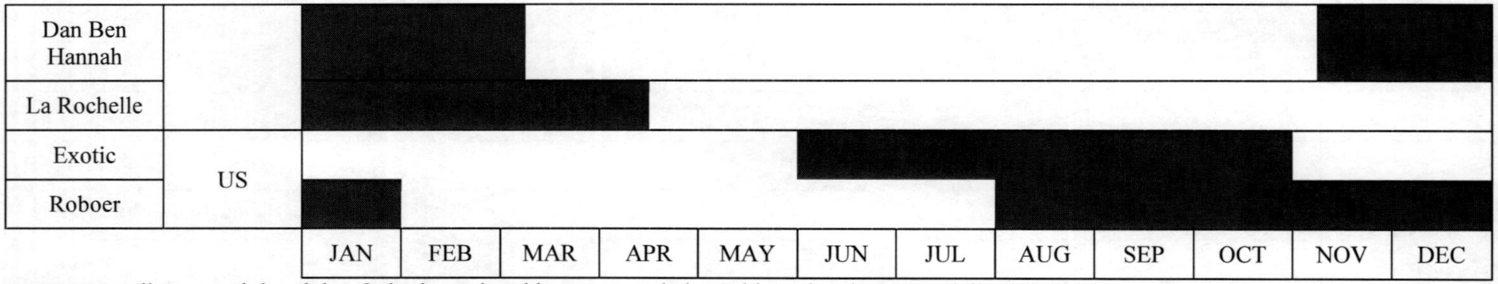

Table grapes seasonality around the globe. Only the main table grapes varieties cultivated and commercialized by the key trade players are indicated as the number of varieties is too large to be described in detail. Some varieties are locally known by other designations or trademarked brands, as Pristine® and Scarlotta Seedless® for the US Blanc Seedless and Sugranineteen varieties, respectively, or Bangalore Blue for the Isabella variety cultivated in India. AUS – Australia, US – United States, 1 – In Brazil, there are four mutants of the Italia variety widely cultivated (Italia, Ruby, Tanaka and Brasil), 2 - In India it is mainly cultivated the clonal selection of Centennial Seedless variety, Manjri Naveen, obtained in NRCG (National Research Center for Grapes, India)

Conclusion

V. vinifera L. grapes and by-products, notably wine, have been consumed for ancient times. Here the health benefits of the dietary consumption of V. vinifera L. grapes have been reviewed. The chemical composition of V. vinifera L. grapes is being gradually clarified as the interest for the different grape phytochemicals with health-protecting properties rise. However, the biological activities and the molecular mechanisms ascribed to them remain unclear. Based on the literature, terpenes such as monoterpenes, sesquiterpenes and C_{13} norisoprenoids, as well as phenolics compounds such as phenolic acids, flavonols, flavanones, flavonoids, flavon-3-ols, anthocyanins, procyanidins, have been the focus of much research pertaining to antioxidant, cardioprotective, cancer prevention, anti-inflammation, antiaging, antimicrobial activities, and many other health-protecting properties. Many promising aspects concerning daily intake of V. vinifera L. grapes have been elucidated, but much more work must be done to clearly identify in vivo health benefits from regular V. vinifera L. grapes consumption.

Acknowledgments

The authors thank the financial support of FEDER (Transnational Cooperation MAC 2007-2013 Program) through VinSaudeMAC project (MAC/1/M105) and to Portuguese Foundation for Science and Technology (FCT) through the MS Portuguese Networks (REDE/1508/RNEM/2005) and Pluriannual base funding (QUI-Madeira-674) and the postdoctoral research grant SFRH/BPD/66177/2009 given to JP.

References

Alshatwi, A. (2010). "Catechin hydrate suppresses MCF-7 proliferation through TP53/Caspase-mediated apoptosis." *Journal of Experimental & Clinical Cancer Research* 29(1): 167.

Annadurai, T., A. R. Muralidharan, T. Joseph, M. Hsu, P. Thomas and P. Geraldine (2012). "Antihyperglycemic and antioxidant effects of a flavanone, naringenin, in streptozotocin–nicotinamide-induced experimental diabetic rats." *Journal of Physiology and Biochemistry* 68(3): 307-318.

Aziz, M. H., M. Nihal, V. X. Fu, D. F. Jarrard and N. Ahmad (2006). "Resveratrol-caused apoptosis of human prostate carcinoma LNCaP cells is mediated via modulation of phosphatidylinositol 3′-kinase/Akt pathway and Bcl-2 family proteins." *Molecular Cancer Therapeutics* 5(5): 1335-1341.

Azmir, J., I. Zaidul, M. Rahman, K. Sharif, A. Mohamed, F. Sahena, M. Jahurul, K. Ghafoor, N. Norulaini and A. Omar (2013). "Techniques for extraction of bioactive compounds from plant materials: a review." *Journal of Food Engineering* 117: 426-436.

Benitez, D. A., E. Pozo-Guisado, A. Alvarez-Barrientos, P. M. Fernandez-Salguero and E. A. Castellón (2007). "Mechanisms involved in resveratrol-induced apoptosis and cell cycle arrest in prostate cancer-derived cell lines." *Journal of Andrology* 28(2): 282-293.

Besseau, S., L. Hoffmann, P. Geoffroy, C. Lapierre, B. Pollet and M. Legrand (2007). "Flavonoid accumulation in *Arabidopsis* repressed in lignin synthesis affects auxin transport and plant growth." *The Plant Cell* 19: 148-162.

Bohlmann, J. and C. I. Keeling (2008). "Terpenoid biomaterials." *The Plant Journal* 54(4): 656-669.

Bostancıoğlu, R. B., M. Kürkçüoğlu, K. H. C. Başer and A. T. Koparal (2012). "Assessment of anti-angiogenic and anti-tumoral potentials of *Origanum onites* L. essential oil." *Food and Chemical Toxicology* 50(6): 2002-2008.

Burke, Y. D., M. J. Stark, S. L. Roach, S. E. Sen and P. L. Crowell (1997). "Inhibition of pancreatic cancer growth by the dietary isoprenoids farnesol and geraniol." *Lipids* 32(2): 151-156.

Cao, Y., Z.-D. Fu, F. Wang, H.-Y. Liu and R. Han (2005). "Anti-angiogenic activity of resveratrol, a natural compound from medicinal plants." *Journal of Asian Natural Products Research* 7(3): 205-213.

Cardozo, M. T., A. de Conti, T. P. Ong, C. Scolastici, E. Purgatto, M. A. Horst, B. K. Bassoli and F. S. Moreno (2011). "Chemopreventive effects of β-ionone and geraniol during rat hepatocarcinogenesis promotion: distinct actions on cell proliferation, apoptosis, HMGCoA reductase, and RhoA." *The Journal of Nutritional Biochemistry* 22(2): 130-135.

Çetin, A. and O. Sagdic (2009). "A concise review: antioxidant effects and bioactive constituents of grape." *Erciyes Medical Journal* 31(4): 369-375.

Chen, J., M. Lu, Y. Jing and J. Dong (2006). "The synthesis of L-carvone and limonene derivatives with increased antiproliferative effect and activation of ERK pathway in prostate cancer cells." *Bioorganic & medicinal chemistry* 14(19): 6539-6547.

Chen, W., J. Hou, Y. Yin, J. Jang, Z. Zheng, H. Fan and G. Zou (2010). "α-Bisabolol induces dose-and time-dependent apoptosis in HepG2 cells via a Fas-and mitochondrial-related pathway, involves p53 and NFκB." *Biochemical Pharmacology* 80(2): 247-254.

Chen, W., Y. Lu, M. Gao, J. Wu, A. Wang and R. Shi (2011). "Anti-angiogenesis effect of essential oil from Curcuma zedoaria *in vitro* and *in vivo*." *Journal of Ethnopharmacology* 133(1): 220-226.

Chun, Y. J., S. Y. Ryu, T. C. Jeong and M. Y. Kim (2001). "Mechanism-based inhibition of human cytochrome P450 1A1 by rhapontigenin." *Drug Metabolism and Disposition* 29(4): 389-393.

Chung, B. H., H.-y. Lee, J. S. Lee and C. Y. Young (2006). "Perillyl alcohol inhibits the expression and function of the androgen receptor in human prostate cancer cells." *Cancer Letters* 236(2): 222-228.

Clarke, R. and J. Armitage (2002). "Antioxidant vitamins and risk of cardiovascular disease. Review of large-scale randomised trials." *Cardiovascular Drugs and Therapy* 16(5): 411-415.

Coelho, E., S. M. Rocha, A. S. Barros, I. Delgadillo and M. A. Coimbra (2007). "Screening of variety- and pre-fermentation-related volatile compounds during ripening of white grapes to define their evolution profile." *Analytica Chimica Acta* 597(2): 257-264.

Coelho, E., S. M. Rocha, I. Delgadillo and M. A. Coimbra (2006). "Headspace-SPME applied to varietal volatile components evolution during *Vitis vinifera* L. cv. 'Baga' ripening." *Analytica Chimica Acta* 563(1-2): 204-214.

Conde, C., P. Silva, N. Fontes, A. C. P. Dias, R. M. Tavares, M. J. Sousa, A. Agasse, S. Delrot and H. Gerós (2007). "Biochemical changes throughout grape berry development and fruit and wine quality."

Coombe, B. G. and M. G. McCarthy (2000). "Dynamics of grape berry growth and physiology of ripening." *Australian Journal of Grape and Wine Research* 6(2): 131-135.

Crowell, P. L., A. S. Ayoubi and Y. D. Burke (1996). Antitumorigenic effects of limonene and perillyl alcohol against pancreatic and breast cancer. *Dietary Phytochemicals in Cancer Prevention and Treatment*, Springer: 131-136.

de Moura Espíndola, R., R. P. Mazzantini, T. P. Ong, A. de Conti, R. Heidor and F. S. Moreno (2005). "Geranylgeraniol and β-ionone inhibit hepatic preneoplastic lesions, cell proliferation, total plasma cholesterol and DNA damage during the initial phases of hepatocarcinogenesis, but only the former inhibits NF-κB activation." *Carcinogenesis* 26(6): 1091-1099.

El-Ghorab, A., K. F. El-Massry and T. Shibamoto (2007). "Chemical composition of the volatile extract and antioxidant activities of the volatile and nonvolatile extracts of Egyptian corn silk (Zea mays L.)." *Journal of Agricultural and Food Chemistry* 55(22): 9124-9127.

Elaissi, A., Z. Rouis, S. Mabrouk, K. B. H. Salah, M. Aouni, M. L. Khouja, F. Farhat, R. Chemli and F. Harzallah-Skhiri (2012). "Correlation between chemical composition and antibacterial activity of essential oils from fifteen Eucalyptus species growing in the Korbous and Jbel Abderrahman Arboreta (North East Tunisia)." *Molecules* 17(3): 3044-3057.

Faber, A. C. and T. C. Chiles (2006). "Resveratrol induces apoptosis in transformed follicular lymphoma OCI-LY8 cells: evidence for a novel mechanism involving inhibition of BCL6 signaling." *International Journal of Oncology* 29(6): 1561.

FAOstat. (2011). "http://faostat3.fao.org/faostat-gateway/go/to/home/E."

Fenoll, J., A. Manso, P. Hellín, L. Ruiz and P. Flores (2009). "Changes in the aromatic composition of the *Vitis vinifera* grape Muscat Hamburg during ripening." *Food Chemistry* 114(2): 420-428.

Fernández-Mar, M., R. Mateos, M. García-Parrilla, B. Puertas and E. Cantos-Villar (2012). "Bioactive compounds in wine: Resveratrol, hydroxytyrosol and melatonin: A review." *Food Chemistry* 130(4): 797-813.

Ferreira, D. and D. Slade (2002). "Oligomeric proanthocyanidins: naturally occurring *O*-heterocycles." *Natural Product Reports* 19(5): 517-541.

Forester, S. C., Y. Y. Choy, A. L. Waterhouse and P. I. Oteiza (2012). "The anthocyanin metabolites gallic acid, 3-O-methylgallic acid, and 2,4,6-trihydroxybenzaldehyde decrease human colon cancer cell viability by regulating pro-oncogenic signals." *Molecular Carcinogenesis*: n/a-n/a.

Fritz, J., M. Kern, G. Pahlke, S. Vatter and D. Marko (2006). "Biological activities of malvidin, a red wine anthocyanidin." *Molecular Nutrition and Food Research* 50(4-5): 390-395.

González-Barreiro, C., R. Rial-Otero, B. Cancho-Grande and J. Simal-Gándara (2013). "Wine aroma compounds in grapes: a critical review." *Critical Reviews in Food Science and Nutrition*(just-accepted).

Gonzalez-Centeno, M. R., M. Jourdes, A. Femenia, S. Simal, C. Rossello and P. L. Teissedre (2012). "Proanthocyanidin composition and antioxidant potential of the stem winemaking byproducts from 10 different grape varieties (Vitis vinifera L.)." *Journal of Agricultural and Food Chemistry* 60(48): 11850-11858.

Grace, M. H., G. G. Yousef, S. J. Gustafson, V.-D. Truong, G. C. Yencho and M. A. Lila (2014). "Phytochemical changes in phenolics, anthocyanins, ascorbic acid, and carotenoids associated with sweetpotato storage and impacts on bioactive properties." *Food Chemistry* 145(0): 717-724.

Griffin, G. S., L. J. Markham and N. D. Leach (2000). "An agar dilution method for determination of the minimum inhibitory concentration of essential oils." *Journal of Essential Oil Research* 12(2): 249-255.

Haag, J. D., M. J. Lindstrom and M. N. Gould (1992). "Limonene-induced regression of mammary carcinomas." *Cancer Research* 52(14): 4021-4026.

Hajdari, A., J. Novak, B. Mustafa and C. Franz (2012). "Essential oil composition and antioxidant activity of *Stachys sylvatica* L. (Lamiaceae) from different wild populations in Kosovo." *Natural product research* 26(18): 1676-1681.

He, L., H. Mo, S. Hadisusilo, A. A. Qureshi and C. E. Elson (1997). "Isoprenoids suppress the growth of murine B16 melanomas in vitro and in vivo." *The Journal of Nutrition* 127(5): 668-674.

Horváthová, E., K. Kozics, A. Srančíková, Ľ. Hunáková, E. Gálová, A. Ševčovičová and D. Slameňová (2012). "Borneol administration protects primary rat hepatocytes against exogenous oxidative DNA damage." *Mutagenesis* 27(5): 581-588.

Hou, D.-X., K. Kai, J.-J. Li, S. Lin, N. Terahara, M. Wakamatsu, M. Fujii, M. R. Young and N. Colburn (2004). "Anthocyanidins inhibit activator protein 1 activity and cell transformation: structure–activity relationship and molecular mechanisms." *Carcinogenesis* 25(1): 29-36.

Hussain, A. I., F. Anwar, P. S. Nigam, M. Ashraf and A. H. Gilani (2010). "Seasonal variation in content, chemical composition and antimicrobial and cytotoxic activities of essential oils from four Mentha species." *Journal of the Science of Food and Agriculture* 90(11): 1827-1836.

indexmundi. (2012). "http://www.indexmundi.com/agriculture/?commodity=grapes."

Ji, J., L. Zhang, Y.-Y. Wu, X.-Y. Zhu, S.-Q. Lv and X.-Z. Sun (2006). "Induction of apoptosis by d-limonene is mediated by a caspase-dependent mitochondrial death pathway in human leukemia cells." *Leukemia & lymphoma* 47(12): 2617-2624.

Jia, S.-S., G.-P. Xi, M. Zhang, Y.-B. Chen, B. Lei, X.-S. Dong and Y.-M. Yang (2013). "Induction of apoptosis by d-limonene is mediated by inactivation of Akt in LS174T human colon cancer cells." *Oncology reports* 29(1): 349-354.

Jones, W. P., T. Lobo-Echeverri, Q. Mi, H.-B. Chai, D. D. Soejarto, G. A. Cordell, S. M. Swanson and A. D. Kinghorn (2007). "Cytotoxic constituents from the fruiting branches of callicarpa americana collected in southern Florida." *Journal of natural products* 70(3): 372-377.

Juan, M. E., I. Alfaras and J. M. Planas (2012). "Colorectal cancer chemoprevention by *trans*-resveratrol." *Pharmacological Research* 65(6): 584-591.

Kalemba, D. and A. Kunicka (2003). "Antibacterial and antifungal properties of essential oils." *Current Medicinal Chemistry* 10(10): 813-829.

Kapoor, S. (2013). "Tumor growth limiting effects of piceatannol." *BioMed Research International* 2013.

Kathirvel, P. and S. Ravi (2012). "Chemical composition of the essential oil from basil (Ocimum basilicum Linn.) and its in vitro cytotoxicity against HeLa and HEp-2 human cancer cell lines and NIH 3T3 mouse embryonic fibroblasts." *Natural product research* 26(12): 1112-1118.

Kim, E. J., H. Park, S. Y. Park, J.-G. Jun and J. H. Y. Park (2009). "The grape component piceatannol induces apoptosis in DU145 human prostate cancer cells via the activation of extrinsic and intrinsic pathways." *Journal of Medicinal Food* 12(5): 943-951.

Kim, Y. H., C. Park, J. O. Lee, G. Y. Kim, W. H. Lee, Y. H. Choi and C. H. Ryu (2008). "Induction of apoptosis by piceatannol in human leukemic U937 cells through down-regulation of Bcl-2 and activation of caspases." *Oncology Reports* 19(4): 961-967.

Kimura, K.-i., Y. Sakamoto, N. Fujisawa, S. Uesugi, N. Aburai, M. Kawada, S.-i. Ohba, T. Yamori, E. Tsuchiya and H. Koshino (2012). "Cleavage mechanism and anti-tumor activity of 3, 6-epidioxy-1, 10-bisaboladiene isolated from edible wild plants." *Bioorganic & medicinal chemistry* 20(12): 3887-3897.

King, R. E., J. A. Bomser and D. B. Min (2006). "Bioactivity of resveratrol." *Comprehensive Reviews in Food Science and Food Safety* 5(3): 65-70.

Kita, Y., Y. Miura and K. Yagasaki (2012). "Antiproliferative and anti-invasive effect of piceatannol, a polyphenol present in grapes and wine, against hepatoma AH109A cells." *BioMed Research International* 2012.

Ko, H. S., H.-J. Lee, S.-H. Kim and E.-O. Lee (2012). "Piceatannol suppresses breast cancer cell invasion through the inhibition of MMP-9: involvement of PI3K/AKT and NF-κB pathways." *Journal of Agricultural and Food Chemistry* 60(16): 4083-4089.

Lacopini, P., M. Baldi, P. Storchi and L. Sebastiani (2008). "Catechin, epicatechin, quercetin, rutin and resveratrol in red grape: Content, *in vitro* antioxidant activity and interactions." *Journal of Food Composition and Analysis* 21(8): 589-598.

Lee, C.-C., S.-R. Shen, Y.-J. Lai and S.-C. Wu (2013). "Rutin and quercetin, bioactive compounds from tartary buckwheat, prevent liver inflammatory injury." *Food & Function* 4(5): 794-802.

Leiro, J., J. Arranz, N. Fraiz, M. Sanmartin, E. Quezada and F. Orallo (2005). "Effect of cis-resveratrol on genes involved in nuclear factor kappa B signaling." *International Immunopharmacology* 5(2): 393-406.

Li, D., Y. Xu, C.-L. Shao, R.-Y. Yang, C.-J. Zheng, Y.-Y. Chen, X.-M. Fu, P.-Y. Qian, Z.-G. She and N. J. d. Voogd (2012). "Antibacterial bisabolane-type sesquiterpenoids from the sponge-derived fungus Aspergillus sp." *Marine drugs* 10(1): 234-241.

Li, Y. and Y. Ding (2012). "Minireview: Therapeutic potential of myricetin in diabetes mellitus." *Food Science and Human Wellness* 1(1): 19-25.

Lila, M. A. (2004). "Anthocyanins and Human Health: An In Vitro Investigative Approach." *J Biomed Biotechnol* 2004(5): 306-313.

Liu, J. R., X. R. Sun, H. W. Dong, C. H. Sun, W. G. Sun, B. Q. Chen, Y. Q. Song and B. F. Yang (2008). "β-Ionone suppresses mammary carcinogenesis, proliferative activity and induces apoptosis in the mammary gland of the Sprague-Dawley rat." *International Journal of Cancer* 122(12): 2689-2698.

Liu, Z., D. Li, L. Yu and F. Niu (2012). "Gallic acid as a cancer-selective agent induces apoptosis in pancreatic cancer cells." *Chemotherapy* 58(3): 185-194.

Lu, J.-J., Y.-Y. Dang, M. Huang, W.-S. Xu, X.-P. Chen and Y.-T. Wang (2012). "Anti-cancer properties of terpenoids isolated from *Rhizoma Curcumae* -a review." *Journal of Ethnopharmacology*.

Lu, X.-G., L.-B. Zhan, B.-A. Feng, M.-Y. Qu, L.-H. Yu and J.-H. Xie (2004). "Inhibition of growth and metastasis of human gastric cancer implanted in nude mice by d-limonene." *World Journal of Gastroenterology* 10(14): 2140-2144.

Maggiolini, M., A. Recchia, D. Bonofiglio, S. Catalano, A. Vivacqua, A. Carpino, V. Rago, R. Rossi and S. Ando (2005). "The red wine phenolics piceatannol and myricetin act as agonists for estrogen receptor α in human breast cancer cells." *Journal of Molecular Endocrinology* 35(2): 269-281.

Magwa, M. L., M. Gundidza, N. Gweru and G. Humphrey (2006). "Chemical composition and biological activities of essential oil from the leaves of Sesuvium portulacastrum." *Journal of Ethnopharmacology* 103(1): 85-89.

Man-Ying Chan, M., J. A. Mattiacci, H. S. Hwang, A. Shah and D. Fong (2000). "Synergy between ethanol and grape polyphenols, quercetin, and resveratrol, in the inhibition of the inducible nitric oxide synthase pathway." *Biochemical Pharmacology* 60(10): 1539-1548.

Manach, C., G. Williamson, C. Morand, A. Scalbert and C. Rémésy (2005). "Bioavailability and bioefficacy of polyphenols in humans. I. Review of 97 bioavailability studies." *The American journal of clinical nutrition* 81(1): 230S-242S.

Mander, L. and H.-w. Liu (2010). *Comprehensive natural products ii: chemistry and biology*, Access Online via Elsevier.

Mayne, S. T. (1996). "Beta-carotene, carotenoids, and disease prevention in humans." *The FASEB Journal* 10(7): 690-701.

McKay, D. L. and J. B. Blumberg (2006). "A review of the bioactivity and potential health benefits of chamomile tea (*Matricaria recutita* L.)." *Phytotherapy Research* 20(7): 519-530.

Mendes-Pinto, M. M. (2009). "Carotenoid breakdown products the-norisoprenoids-in wine aroma." *Arch Biochem Biophys* 483(2): 236-245.

Mendes-Pinto, M. M., A. C. Silva Ferreira, C. Caris-Veyrat and P. Guedes de Pinho (2005). "Carotenoid, chlorophyll, and chlorophyll-derived compounds in grapes and Port wines." *Journal of Agricultural and Food Chemistry* 53(26): 10034-10041.

Meyer, A. S., M. Heinonen and E. N. Frankel (1998). "Antioxidant interactions of catechin, cyanidin, caffeic acid, quercetin, and ellagic acid on human LDL oxidation." *Food Chemistry* 61(1): 71-75.

Miller, J. A., J. E. Lang, M. Ley, R. Nagle, C.-H. Hsu, P. A. Thompson, C. Cordova, A. Waer and H. S. Chow (2013). "Human breast tissue disposition and bioactivity of limonene in women with early-stage breast cancer." *Cancer Prevention Research* 6(6): 577-584.

Mills, J. J., R. S. Chari, I. J. Boyer, M. N. Gould and R. L. Jirtle (1995). "Induction of apoptosis in liver tumors by the monoterpene perillyl alcohol." *Cancer Research* 55(5): 979-983.

Minakawa, M., Y. Miura and K. Yagasaki (2012). "Piceatannol, a resveratrol derivative, promotes glucose uptake through glucose transporter 4 translocation to plasma membrane in L6 myocytes and suppresses blood glucose levels in type 2 diabetic model db/db mice." *Biochemical and Biophysical Research Communications* 422(3): 469-475.

Mohamad, R. H., A. M. El-Bastawesy, M. G. Abdel-Monem, A. M. Noor, H. A. R. Al-Mehdar, S. M. Sharawy and M. M. El-Merzabani (2011). "Antioxidant and anticarcinogenic effects of methanolic extract and volatile oil of fennel seeds (Foeniculum vulgare)." *Journal of medicinal food* 14(9): 986-1001.

Monga, J. and M. Sharma (2013). "P-0032(+)-Catechin produces antiproliferative effects in human hepatocellular carcinoma cells and inhibits tumor growth *in vivo*." *Annals of Oncology* 24(suppl 4): iv45-iv46.

Nixdorf, S. L. and I. Hermosin-Gutierrez (2010). "Brazilian red wines made from the hybrid grape cultivar Isabel: phenolic composition and antioxidant capacity." *Anal Chim Acta* 659(1-2): 208-215.

Oliveira, C., A. Barbosa, A. C. S. Ferreira, J. Guerra and P. Guedes De Pinho (2006). "Carotenoid Profile in Grapes Related to Aromatic Compounds in Wines from Douro Region." *Journal of Food Science* 71(1): S1-S7.

Omenn, G. S., G. E. Goodman, M. D. Thornquist, J. Balmes, M. R. Cullen, A. Glass, J. P. Keogh, F. L. Meyskens Jr, B. Valanis and J. H. Williams Jr (1996). "Effects of a combination of beta carotene and vitamin A on lung cancer and cardiovascular disease." *New England journal of medicine* 334(18): 1150-1155.

Ong, T. P., R. Heidor, A. de Conti, M. L. Z. Dagli and F. S. Moreno (2006). "Farnesol and geraniol chemopreventive activities during the initial phases of hepatocarcinogenesis involve similar actions on cell proliferation and DNA damage, but distinct actions on apoptosis, plasma cholesterol and HMGCoA reductase." *Carcinogenesis* 27(6): 1194-1203.

Pereira, J., C. Silva, J. Gonçalves and J. S. Câmara (2011). The Potentialities of Ultra-Performance Liquid Chromatography Combined with Photodiode Array Detection in the Analysis of Wine Metabolites with Bioactive Effects: The Case Study of Polyphenols *Wine: Types, Production and Health*. A. S. Peeters.

Perestrelo, R., A. S. Barros, S. M. Rocha and J. S. Câmara (2011). "Optimisation of solid-phase microextraction combined with gas chromatography-mass spectrometry based methodology to establish the global volatile signature in pulp and skin of *Vitis vinifera* L. grape varieties." *Talanta* 85(3): 1483-1493.

Perestrelo, R., Y. Lu, S. A. O. Santos, A. J. D. Silvestre, C. P. Neto, J. S. Câmara and S. M. Rocha (2012). "Phenolic profile of Sercial and Tinta Negra Vitis vinifera L. grape skins by HPLC–DAD–ESI-MSn." *Food Chemistry* 135(1): 94-104.

Pinent, M., A. Castell, I. Baiges, G. Montagut, L. Arola and A. Ardévol (2008). "Bioactivity of flavonoids on insulin-secreting cells." *Comprehensive Reviews in Food Science and Food Safety* 7(4): 299-308.

Piotrowska, H., M. Kucinska and M. Murias (2012). "Biological activity of piceatannol: leaving the shadow of resveratrol." *Mutation Research/Reviews in Mutation Research* 750(1): 60-82.

Polo, M. P. and M. G. de Bravo (2006). "Effect of geraniol on fatty-acid and mevalonate metabolism in the human hepatoma cell line Hep G2." *Biochemistry and cell biology* 84(1): 102-111.

Preethy, C. P., R. Padmapriya, V. S. Periasamy, A. Riyasdeen, S. Srinag, H. Krishnamurthy, A. A. Alshatwi and M. A. Akbarsha (2012). "Antiproliferative property of *n*-hexane and chloroform extracts of Anisomeles malabarica (L). R. Br. in HPV16-positive human cervical cancer cells." *Journal of pharmacology & pharmacotherapeutics* 3(1): 26.

Provinciali, M., F. Re, A. Donnini, F. Orlando, B. Bartozzi, G. Di Stasio and A. Smorlesi (2005). "Effect of resveratrol on the development of spontaneous mammary tumors in HER-2/neu transgenic mice." *International Journal of Cancer* 115(1): 36-45.

Qin, L., L. Jin, L. Lu, X. Lu, C. Zhang, F. Zhang and W. Liang (2011). "Naringenin reduces lung metastasis in a breast cancer resection model." *Protein and Cell* 2(6): 507-516.

Rabi, T. and S. Gupta (2008). "Dietary terpenoids and prostate cancer chemoprevention." *Frontiers in bioscience: a journal and virtual library* 13: 3457.

Rahman, I., S. K. Biswas and P. A. Kirkham (2006). "Regulation of inflammation and redox signaling by dietary polyphenols." *Biochemical Pharmacology* 72(11): 1439-1452.

Rathore, S., S. Saxena and B. Singh (2013). "Potential health benefits of major seed spices." *International J. Seed Spices* 3(2): 1-12.

Razavi, S. M. and S. Nejad-Ebrahimi (2009). "Chemical composition, allelopatic and cytotoxic effects of essential oils of flowering tops and leaves of Crambe orientalis L. from Iran." *Natural product research* 23(16): 1492-1498.

Razungles, A. J., I. Babic, J. C. Sapis and C. L. Bayonove (1996). "Particular behavior of epoxy xanthophylls during veraison and maturation of grape." *Journal of Agricultural and Food Chemistry* 44(12): 3821-3825.

Rigano, D., V. Cardile, C. Formisano, M. T. Maldini, S. Piacente, J. Bevilacqua, A. Russo and F. Senatore (2009). "*Genista sessilifolia* DC. and *Genista tinctoria* L. inhibit UV light and nitric oxide-induced DNA damage and human melanoma cell growth." *Chemico-biological interactions* 180(2): 211-219.

Riles, W. L., J. Erickson, S. Nayyar, M. J. Atten, B. M. Attar and O. Holian (2006). "Resveratrol engages selective apoptotic signals in gastric adenocarcinoma cells." *World Journal of Gastroenterology* 12(35): 5628.

Ruberto, G. and M. T. Baratta (2000). "Antioxidant activity of selected essential oil components in two lipid model systems." *Food Chemistry* 69(2): 167-174.

Sareen, D., P. R. van Ginkel, J. C. Takach, A. Mohiuddin, S. R. Darjatmoko, D. M. Albert and A. S. Polans (2006). "Mitochondria as the primary target of resveratrol-induced apoptosis in human retinoblastoma cells." *Investigative Ophthalmology and Visual Science* 47(9): 3708-3716.

Scalbert, A., C. Manach, C. Morand, C. Rémésy and L. Jiménez (2005). "Dietary polyphenols and the prevention of diseases." *Critical Reviews in Food Science and Nutrition* 45(4): 287-306.

Singh, B. K., M. Tripathi, B. P. Chaudhari, P. K. Pandey and P. Kakkar (2012). "Natural terpenes prevent mitochondrial dysfunction, oxidative stress and release of apoptotic proteins during nimesulide-hepatotoxicity in rats." *PLoS ONE* 7(4): e34200.

Slattery, M. L., J. Benson, K. Curtin, K.-N. Ma, D. Schaeffer and J. D. Potter (2000). "Carotenoids and colon cancer." *The American journal of clinical nutrition* 71(2): 575-582.

Soković, M., J. Glamočlija, P. D. Marin, D. Brkić and L. J. v. Griensven (2010). "Antibacterial effects of the essential oils of commonly consumed medicinal herbs using an in vitro model." *Molecules* 15(11): 7532-7546.

Soković, M. and L. J. van Griensven (2006). "Antimicrobial activity of essential oils and their components against the three major pathogens of the cultivated button mushroom, Agaricus bisporus." *European Journal of Plant Pathology* 116(3): 211-224.

Stahl, W. and H. Sies (2003). "Antioxidant activity of carotenoids." *Molecular Aspects of Medicine* 24(6): 345-351.

Stahl, W. and H. Sies (2005). "Bioactivity and protective effects of natural carotenoids." *Biochimica et Biophysica Acta (BBA)-Molecular Basis of Disease* 1740(2): 101-107.

Su, T.-R., J.-J. Lin, C.-C. Tsai, T.-K. Huang, Z.-Y. Yang, M.-O. Wu, Y.-Q. Zheng, C.-C. Su and Y.-J. Wu (2013). "Inhibition of melanogenesis by gallic acid: possible involvement of the PI3K/Akt, MEK/ERK and Wnt/β-catenin signaling pathways in B16F10 cells." *International Journal of Molecular Sciences* 14(10): 20443-20458.

Suhail, M. M., W. Wu, A. Cao, F. G. Mondalek, K.-M. Fung, P.-T. Shih, Y.-T. Fang, C. Woolley, G. Young and H.-K. Lin (2011). "Boswellia sacra essential oil induces tumor cell-specific apoptosis and suppresses tumor aggressiveness in cultured human breast cancer cells." *BMC complementary and alternative medicine* 11(1): 129.

Szajdek, A. and E. Borowska (2008). "Bioactive compounds and health-promoting properties of berry fruits: a review." *Plant Foods for Human Nutrition* 63(4): 147-156.

Taiz, L. and E. Zeiger (2006). "Secondary metabolites and plant defense." *Plant physiology* 4: 315-344.

Thoppil, R. J. and A. Bishayee (2011). "Terpenoids as potential chemopreventive and therapeutic agents in liver cancer." *World journal of hepatology* 3(9): 228.

Trombetta, D., F. Castelli, M. G. Sarpietro, V. Venuti, M. Cristani, C. Daniele, A. Saija, G. Mazzanti and G. Bisignano (2005). "Mechanisms of antibacterial action of three monoterpenes." *Antimicrobial Agents and Chemotherapy* 49(6): 2474-2478.

Uquillas, C. (2013). GRAPEVINE PLANT denominated 'Iniagrape-one', Google Patents.

Verma, S., A. Singh and A. Mishra (2013). "Gallic acid: Molecular rival of cancer." *Environmental Toxicology and Pharmacology* 35(3): 473-485.

Vlaykova, T., I. Dimitrova, I. Pavlov and T. Tacheva (2013). "Cancer prevetion-dietary anticarcinogens." *Medicine* 3(1).

Vo, N. T., S. Madlener, Z. Bago-Horvath, I. Herbacek, N. Stark, M. Gridling, P. Probst, B. Giessrigl, S. Bauer and C. Vonach (2010). "Pro-and anticarcinogenic mechanisms of piceatannol are activated dose dependently in MCF-7 breast cancer cells." *Carcinogenesis* 31(12): 2074-2081.

Vukovic, N., T. Milosevic, S. Sukdolak and S. Solujic (2007). "Antimicrobial activities of essential oil and methanol extract of Teucrium montanum." *Evidence-Based Complementary and Alternative Medicine* 4(S1): 17-20.

Waliwitiya, R., C. J. Kennedy and C. A. Lowenberger (2009). "Larvicidal and oviposition-altering activity of monoterpenoids, trans-anithole and rosemary oil to the yellow fever mosquito Aedes aegypti (Diptera: Culicidae)." *Pest management science* 65(3): 241-248.

Wang, J. and G. Mazza (2002). "Inhibitory effects of anthocyanins and other phenolic compounds on nitric oxide production in LPS/IFN-γ-activated RAW 264.7 macrophages." *J. Agric. Food Chem.* 50(4): 850-857.

Watson, B. (2003). "Evaluation of winegrape maturity." *Oregon Viticulture. Oregon State University Press, Corvallis, Oregon. USA*: 235-245.

Wu, J. M. (2012). "AKT/mTOR as novel targets of polyphenol piceatannol possibly contributing to inhibition of proliferation of cultured prostate cancer cells." *ISRN urology* 2012.

Xia, E.-Q., G.-F. Deng, Y.-J. Guo and H.-B. Li (2010). "Biological Activities of Polyphenols from Grapes." *International Journal of Molecular Sciences* 11(2): 622-646.

Yadav, M., S. Jain, A. Bhardwaj, R. Nagpal, M. Puniya, R. Tomar, V. Singh, O. Parkash, G. Prasad and F. Marotta (2009). "Biological and medicinal properties of grapes and their bioactive constituents: an update." *Journal of medicinal food* 12(3): 473-484.

Yin, Q.-h., F.-x. Yan, X.-Y. Zu, Y.-h. Wu, X.-p. Wu, M.-c. Liao, S.-w. Deng, L.-l. Yin and Y.-z. Zhuang (2012). "Anti-proliferative and pro-apoptotic effect of carvacrol on human hepatocellular carcinoma cell line HepG-2." *Cytotechnology* 64(1): 43-51.

Yoon, C.-H., S.-J. Chung, S.-W. Lee, Y.-B. Park, S.-K. Lee and M.-C. Park (2013). "Gallic acid, a natural polyphenolic acid, induces apoptosis and inhibits proinflammatory gene expressions in rheumatoid arthritis fibroblast-like synoviocytes." *Joint, Bone, Spine* 80(3): 274-279.

You, B. R., H. J. Moon, Y. H. Han and W. H. Park (2010). "Gallic acid inhibits the growth of HeLa cervical cancer cells via apoptosis and/or necrosis." *Food and Chemical Toxicology* 48(5): 1334-1340.

You, H. J., H. J. Ahn and G. E. Ji (2010). "Transformation of rutin to antiproliferative quercetin-3-glucoside by *Aspergillus niger*." *Journal of Agricultural and Food Chemistry* 58(20): 10886-10892.

Yu, S. G., L. A. Hildebrandt and C. E. Elson (1995). "Geraniol, an inhibitor of mevalonate biosynthesis, suppresses the growth of hepatomas and melanomas transplanted to rats and mice." *The Journal of Nutrition* 125(11): 2763-2767.

Zamanian-Azodi, M., M. Rezaie-Tavirani, S. Heydari-Kashal, S. Kalantari, S. Dailian and H. Zali (2011). "Proteomics analysis of MKN45 cell line before and after treatment with Lavender aqueous extract." *Gastroenterology and Hepatology from bed to bench* 5(1).

Zellagui, A., N. Gherraf and S. Rhouati (2012). "Chemical composition and antibacterial activity of the essential oils of Ferula vesceritensis Coss et Dur. leaves, endemic in Algeria." *Organic and medicinal chemistry letters* 2(1): 1-4.

Zhou, K. and J. J. Raffoul (2012). "Potential anticancer properties of grape antioxidants." *J Oncol* 2012: 803294.

In: Grapes
Editor: José de Sousa Câmara

ISBN: 978-1-63321-402-6

Chapter 14

Extraction, Characterization and Potential Health Benefits of Grape Seed Procyanidins

Joana Afonso[1], Cláudia P. Passos[2], Manuel A. Coimbra[2], Patrício Soares-da-Silva[1] and Carlos M. Silva[3]

[1]Department of Pharmacology and Therapeutics,
Faculty of Medicine, University of Porto, Porto, Portugal
[2]Department of Chemistry, QOPNA, University of Aveiro,
Campus Universitário de Santiago, Aveiro, Portugal
[3]Department of Chemistry, CICECO, University of Aveiro,
Campus Universitário de Santiago, Aveiro, Portugal

Abstract

Procyanidins are polyphenolic compounds containing flavan-3-ol units linked by C4-C8 or C4-C6 bonds (type-B structures) or doubly linked by an additional C2-C7 acetal bond (type-A structures). Gallic acid can occur as an ester at C3 position. In grape seed tissues, procyanidins represent more than 70% of the total flavan-3-ol composition.

Procyanidins have been described to possess a wide spectrum of pharmacological and therapeutic effects such as antioxidant, anti-inflammatory, and antimicrobial abilities, as well as significant cardioprotective properties and potential high blood pressure lowering benefit. The degree of polymerisation of procyanidins may influence their action intensity, which suggests previous fractionation of crude extracts in order to inspect such relation.

These issues are addressed in this chapter for the specific cases of the antioxidant activity and the inhibitory action upon Angiotensin I-converting enzyme (ACE). ACE catalyses the production of Angiotensin II (powerful vasoconstrictor) and inhibits bradykinin (powerful vasodilator). Hence, procyanidins may have potential action in lowering both systolic and diastolic blood pressure values.

Keywords: Grape; procyanidins; flavan-3-ol; degree of polymerisation; antioxidant activity; angiotensin I-converting enzyme

Introduction

Grapes (*Vitis vinifera* L.) are the world's largest fruit crop with more than 60 million tons produced annually. While 80% of the total crop is used in wine making, the pomace represents approximately 20% of the weight of grapes processed. The seeds constitute a considerable proportion of the pomace, amounting to 38–52% on a dry matter basis (Djilas et al., 2009).

The growth of the grape berry is characterized by several stages of development. The most important stage for the maturation control is named coloration or "véraison", after which the berries are chemically followed (analysed) until become ready for harvesting. During the first period of berry growth, the pericarp and seed cell number increase, and the seed approaches its full size. Then a period of little change in berry size follows, where the seed embryo develops with a concomitant hardening of the seed coat. In the second period ("véraison"), the berry turns soft and the sugars are rapidly accumulated together with anthocyanins (in the case of red varieties). At this stage, much of the seed is fully developed (Coombe 1960; Kennedy et al., 2000).

Proanthocyanidins are among the most abundant phenolic compounds in grapes (and the second most abundant natural phenolic compounds distributed in the plant kingdom), being located in stems (Souquet et al., 2000), skins (Shrikhande, 2000) and seeds (Prieur et al., 1994).

Proanthocyanidins (see Table 1), also known as condensed tannins, are oligomeric and polymeric compounds of flavan-3-ol units. If they contain exclusively (epi)catechin units, they are named procyanidins; when containing (epi)afzelechin units, they are named as propelargonidins; when containing (epi)gallocatechin units, they are named as prodelphinidins (Khanal et al., 2009).

Propelargonidins and prodelphinidins are less common in nature. The seeds (which contain exclusively procyanidins) and the outer seed coat contain the majority of both monomeric and polymeric procyanidins (two to five times more than the endosperm) (Thorngate and Singleton, 1994).

The procyanidins are linked mainly through C4-C8 bonds though C4-C6 linkages can also occur (see Figures 1.a and 1.b). Both structures are defined as type-B procyanidins, and may also be esterified through the (epi)catechin units forming 3-*O*-gallates (Escribano-Bailón et al., 1992; Fuleki and Silva, 1997; De Freitas et al., 1998; Gabetta et al., 2000; Krueger et al., 2000; Yang and Chien, 2000; Peng et al., 2001; Saucier et al., 2001; Flamini 2003; Monagas et al., 2003; Reed et al., 2005; Passos et al., 2007; Passos et al., 2010). In addition to the C4-C8 bond, the flavan-3-ol units can be doubly-linked by a C2-C7 acetal bond giving origin to type-A structures (see Figure 1.c) which result from an oxidative intramolecular reaction (Kondo et al., 2000). This last structure was reported for the first time also to occur in grape seed procyanidin extracts by Passos et al. (2007).

Depending on climatic differences and grape variety, the procyanidins composition may exhibit some discrepancies (Bordiga et al., 2011) though Harbertson et al. (2002) have shown that the differences in total seed procyanidins within several varieties are due to the number of seeds *per* berry, rather than the amount of condensed tannins *per* seed.

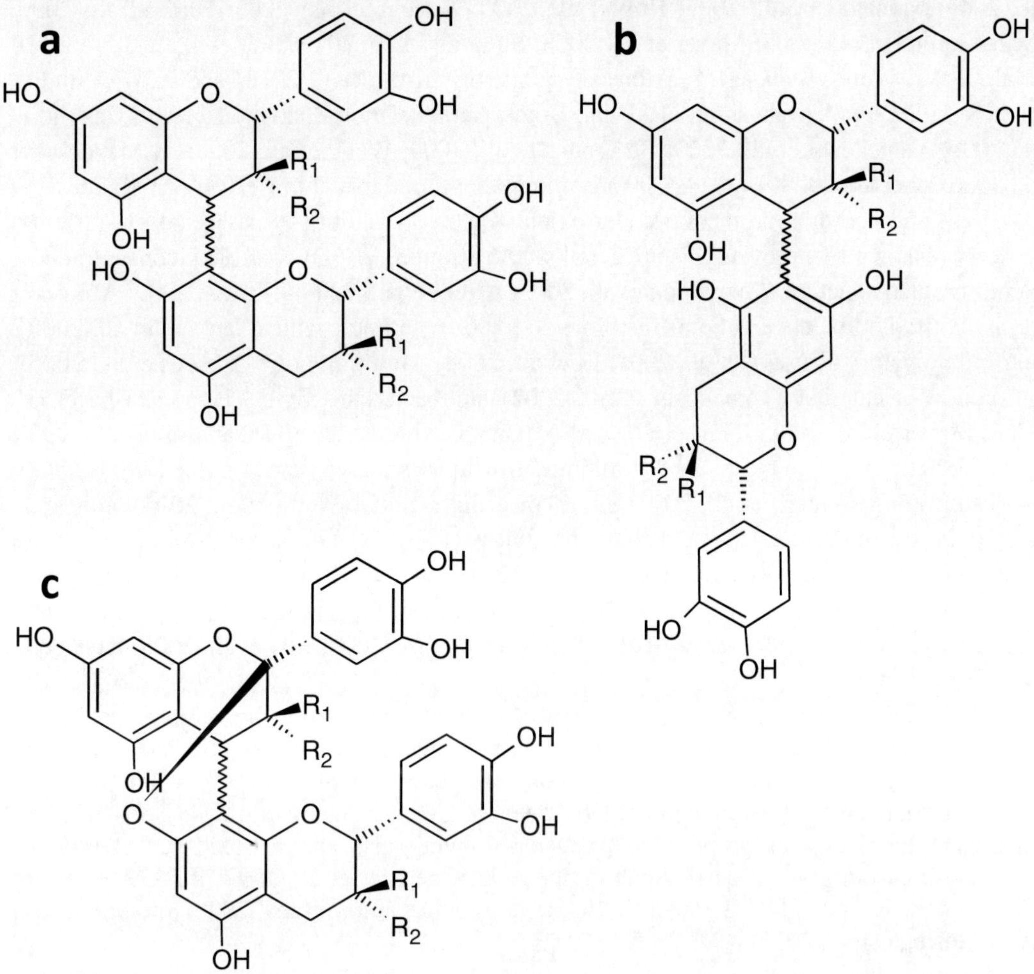

Figure 1. Structure of procyanidins dimers: (a) type-B, containing C4-C8 interflavanic linkage; (b) type-B, containing C4-C6 interflavanic linkage; (c) Type-A, containing both C4-C8 and C2-C7 interflavanic linkages

Table 1. Examples of proanthocyanidins. (G = Gallic acid residue)

Propelargonidins	*Procyanidins*	*Prodelphinidins*		
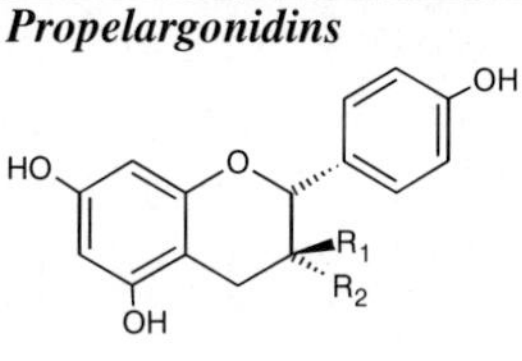	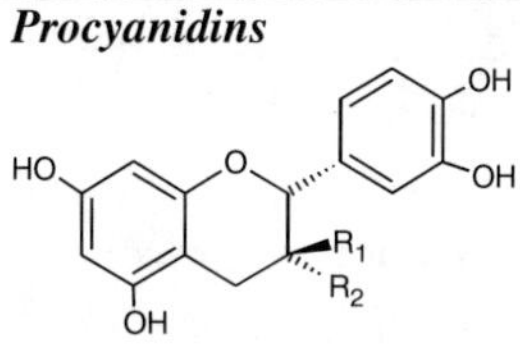	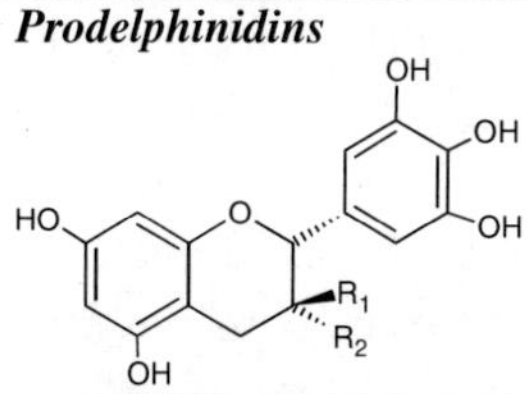	**R₁**	**R₂**
(+)-Afzelechin	(+)-Catechin	(+)-Gallocatechin	OH	H
(+)-Afzelechin-*O*-gallate	(+)-Catechin-*O*-gallate	(+)-Gallocatechin-*O*-gallate	OG	H
(-)-Epiafzelechin	(-)-Epicatechin	(-)-Epigallocatechin	H	OH
(-)-Epiafzelechin-*O*-gallate	(-)-Epicatechin-*O*-gallate	(-)-Epigallocatechin-*O*-gallate	H	OG

The therapeutic value of procyanidins has being proved using both *in vitro* models (Agarwal et al., 2000a,b; Carini et al., 2000; Catterall et al., 2000; Fitzpatrick et al., 2002; Aldini et al., 2003; Jayaprakasha et al., 2003; Moreno et al., 2003; Fan and Lou, 2004; Lu et al., 2004; Monagas et al., 2005; Roy et al., 2005; Vitseva et al., 2005; Faria et al., 2006; Actis-Goretta et al., 2008; Mayer et al., 2008; Sung and Lee, 2010; Bak et al., 2012; Vayalil et al., 2012; Wang et al., 2012; Afonso et al., 2013; Kim et al., 2013) and *in vivo* studies (Saito et al., 1998; Takahashi et al., 1998; Bomser et al., 1999; Zhao et al., 1999; Castillo et al., 2000; Yamakoshi et al., 2002; De Freitas et al., 2004; Wang et al., 2004; Al Makdessi et al., 2006; Li et al., 2008; Li et al., 2009). Among those, antiarrhythmic effect (Al Makdessi et al., 2006) and neuronal protection under oxidative stress conditions (De Freitas et al., 2004) have been shown by procyanidins after oral administration. Within controlled concentrations. cardio-protective action (Fitzpatrick et al., 2002; Aldini et al., 2003; Li et al., 2009; Afonso et al., 2013); radioprotective action (Castillo et al., 2000); antimutagenic (Catterall et al., 2000); anti-inflammatory (Vitseva et al., 2005; Li et al., 2008); antithrombotic (Vitseva et al., 2005); prevention of cataract (Yamakoshi et al., 2002); antibacterial activity (Jayaprakasha et al., 2003; Mayer et al., 2008; Yigit et al., 2009); hair growing activity (Takahashi et al., 1998); antiulcer activity (Saito et al., 1998); inhibition of lipases activity (with reduction of dietary fat absorption) (Montero et al., 2013); urate-lowering activity (Wang et al., 2004); anticaries activity (Gazzani et al., 2012); and chemopreventive action (Nandakumar et al., 2008; Zhang et al., 2009; Sung and Lee, 2010; Bak et al., 2012; Wang et al., 2012) such as prostate cancer (Agarwal et al., 2000a; Vayalil et al., 2012), breast cancer (Agarwal et al., 2000b), and skin cancer (Bomser et al., 1999; Zhao et al., 1999; Carini et al., 2000; Roy et al., 2005) have been reported.

There are evidences that the presence of galloyl groups and the oligomerization of flavanols enhance the cytotoxic effects of typical monomeric flavanols, increasing the benefits of higher molecular weight procyanidins (Fitzpatrick et al., 2002; Actis-Goretta et al., 2008). A high correlation is observed between the total content of flavan-3-ols and the antiradical efficiency of grape seed extracts, showing also the antioxidant potential of procyanidins from grape seeds (Plumb et al., 1998; Cheynier et al., 1999; Bakkalbasi et al., 2005; Monagas et al., 2005; Faria et al., 2006; Gougoulias and Mashev, 2008; Spranger et al., 2008; Bordiga et al., 2011).

The increasing number of positive health related properties has also incremented the amount of nutritional supplements based on procyanidins. However, and although there are studies that indicate a lack of toxicity of procyanidins-rich extracts (Yamakoshi et al., 2002), there are also evidences that the use of high concentrations can in some cases invert their potential and eventually induce cellular damage (Fan and Lou, 2004).

While *in vivo* studies are one step further into proving the therapeutic value of procyanidins, the majority of the studies is only *in vitro* and still needs to overcome an important milestone: the bioavailability of the compounds, particularly when higher molecular weights are involved, due to the limitations of their absorption by the organism. Prior to this step, the extraction and characterization of procyanidins are required in order to be able to draw rigorous conclusions from subsequent research.

Extraction and Characterization of Grape Seed Procyanidins

Extraction of Procyanidins

Liquid-liquid extraction is the most common methodology for the extraction of procyanidins. The most frequent solvents are methanol (Escribano-Bailón et al., 1992; Kallithraka et al., 1995; Santos-Buelga et al., 1995; Fitzpatrick et al., 2000; Monagas et al., 2003; Chung et al., 2005; Passos et al., 2007; Spranger et al., 2008), acetone (Kallithraka et al., 1995; Downey et al., 2003; Chung et al., 2005), ethyl acetate (Kallithraka et al., 1995; Chung et al., 2005; Guendez et al., 2005), ethanol (Kallithraka et al., 1995; González-Manzano et al., 2004), and aqueous mixtures of the previous solvents (Kallithraka et al., 1995; De Freitas et al., 1998; Cheynier et al., 1999; Krueger et al., 2000; Yang and Chien, 2000; Peng et al., 2001; Saucier et al., 2001; Jayaprakasha et al., 2003; Monagas et al., 2003; Chung et al., 2005; Montealegre et al., 2006; Núñez et al., 2006; Spranger et al., 2008; Khanal et al., 2009; Fanzone et al., 2011). Methanol or ethanol extracts low molecular weight fractions, including oligomeric procyanidins, whereas larger polymers are better uptaken with high contents of water (Kallithraka et al., 1995). Extracts obtained with pure water are very attractive due to the associated food grade label, especially considering their application as dietary supplements (Chedea et al., 2011). Nonetheless, they proved to be quite unstable (Chedea et al., 2011), requiring micro-encapsulating methods to extend their shelf life (Zhang et al., 2007).

Recent extraction approaches include the use of near-critical and supercritical carbon dioxide, modified or not with cosolvents like methanol (Palma and Taylor, 1999; Palma et al., 1999; Ashraf-Khorassani and Taylor, 2004) or ethanol (Palma and Taylor, 1999; Murga et al., 2000; Yilmaz et al., 2011). However, due to the low polarity of carbon dioxide, this technique is only efficient for the extraction of monomeric ((+)-catechin and (-)-epicatechin) and low molecular weight procyanidins like dimers. The application of microwave-assisted extraction using water, methanol, and water:methanol (50:50) mixtures have also been tested and proved to be an efficient approach, obtaining both monomeric and oligomeric procyanidins with good yield (Mayer et al., 2008). An ionic liquid-based ultrasonic-assisted extraction method has also been proposed for the extraction of procyanidins (Sun et al., 2013).

Procyanins Fractionation. Degree of Polymerization

The degree of polymerization (DP), which corresponds to the number of flavan-3-ol units in each molecule, is one of the most important parameters that characterize procyanidins (and proanthocyanidins in general). Considering their extraction, biological activities, and sensory effects, procyanidins often behave according to their molecular weight. For instance, the procyanidins ability to associate with proteins and polysaccharides increases with increasing DP, which emphasizes the impact of molecular weight (Guyot et al., 2000). Taking these observations into account, many studies have included fractionation steps after the initial solvent extraction and before the characterization of the resulting fractions. These preparative methods include either gel chromatography (Silva et al., 1991; Rigaud et al., 1993; Prieur et al., 1994; De Freitas et al., 1998; Sun et al., 1999; Fitzpatrick et al., 2000; Gabetta et al., 2000; Peng et al., 2001; Saucier et al., 2001; Monagas et al., 2003; Yanagida et al., 2003; Le Bourvellec et al., 2006; Mayer et al., 2008; Spranger et al., 2008), precipitations with chloroform/methanol solutions (Labarbe et al., 1999; Guyot et al., 2001; Saucier et al., 2001; Passos et al., 2010), C18 solid-phase extractions (Sun et al., 1998, 1999; Peng et al., 2001; Saucier et al., 2001; Monagas et al., 2003; Passos et al., 2007; Spranger et al., 2008), and/or combinations of the previous methods (Labarbe et al., 1999; Guyot et al., 2001; Saucier et al., 2001; Mayer et al., 2008; Passos et al., 2010). In general, these approaches allowed the separation of the initial crude extracts into several fractions with distinct mean degrees of polymerization (DPn), within more or less broad ranges according to the selected methodology: from simple fractionation into monomeric, oligomeric, and polymeric fractions (Sun et al., 1998; Monagas et al., 2003; Liu et al., 2010) to several individual fractions according to increasing DPn (Labarbe et al., 1999; Guyot et al., 2001; Passos et al., 2010; Hanlin et al., 2011).

Characterization Methods. Degree of Polymerization

The application of colorimetric methods for the characterization of procyanidins is very frequent, e.g. the use of the vanillin assay (Monagas et al., 2003; Bordiga et al., 2011; Travaglia et al., 2011; Bak et al., 2012) based on the reaction of the A-ring (Figure 2) with the phenolic aldehyde (vanillin). Without derivatization, it is only possible to identify monomers and low oligomers like dimers, trimers and eventually tetramers by reverse-phase HPLC and detection at 280 nm (Santos-Buelga et al., 1995; De Freitas and Glories, 1999; Yang and Chien, 2000; Peng et al., 2001; Montealegre et al., 2006; Travaglia et al., 2011). Therefore, most studies characterise procyanidins starting with a depolymerisation step (an acid-catalysed cleavage of the polymer in the presence of a nucleophile compound) such as thioacidolysis (Prieur et al., 1994; De Freitas et al., 1998, 1999; Labarbe et al., 1999; Kennedy et al., 2000; Fitzpatrick et al., 2002; Monagas et al., 2003; Passos et al., 2007; Spranger et al., 2008; Mattivi et al., 2009) or phloroglucinolysis (Silva et al., 1991; Downey et al., 2003; Bordiga et al., 2011; Hanlin et al., 2011; Bak et al., 2012), depending on the nucleophile compound used, combined with reverse-phase HPLC separation (Escribano-Bailón et al., 1992; Fuleki and Silva, 1997; De Freitas et al., 1998; Fitzpatrick et al., 2000; Fitzpatrick et al., 2002; Downey et al., 2003; Bakkalbasi et al., 2005; Spranger et al., 2008; Passos et al., 2010; Hanlin et al., 2011) for the quantification of the individual polymer units. This procedure allows the determination of the nature and concentration of terminal and extension units, and eventually the estimation of the mean degree of polymerization (DPn) (Labarbe et al., 1999; Kennedy et al., 2000; Guyot et al., 2001; Passos et al., 2010; Bordiga et al., 2011; Hanlin et al., 2011). While HPLC is perhaps the most frequent technique for DPn determination, other methodologies have been presented as faster and/or cheaper, or simply as alternative research methods for the same purpose. One may cite Infrared Spectroscopy (Passos et al., 2010), proton NMR (Guyot et al., 2000) and HPLC-Mass Spectrometry (HPLC-MS) (González-Manzano et al., 2006).

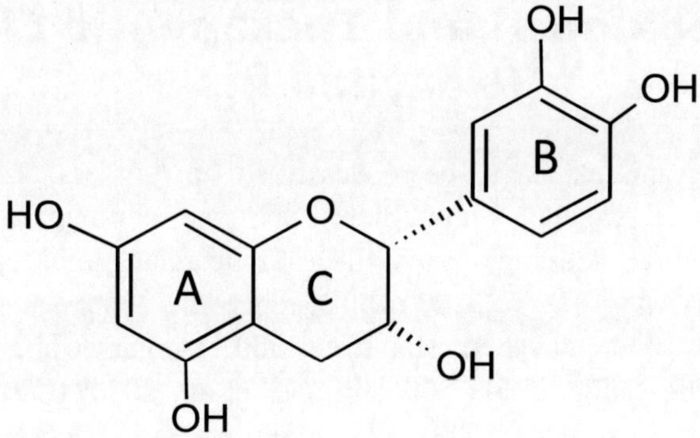

Figure 2. Example of a procyanidin monomer: (-)-epicatechin.

The individual polymer units identified in grape seed procyanidins were (+)-catechin, (-)-epicatechin and (-)-epicatechin-*O*-gallate which can occur in the midle of the chain (extension units) or at the terminal end (terminal unit). Although there are differences that may arise from the variety, the cultivar, and even from the atmospheric conditions, in most studies both (+)-catechin and (-)-epicatechin monomers appear within similar amounts (Fuleki and Silva, 1997; De Freitas and Glories, 1999), with a higher abundance of (-)-epicatechin extension units in polymeric grape seeds procyanidins (Prieur et al., 1994; Bordiga et al., 2011). Furthermore, various studies where the evolution of procyanidins in grape seeds was accompanied during several stages (from "véraison" until harvesting) showed that the percentage of terminal units of (+)-catechin and (-)-epicatechin-*O*-gallate decreases constantly, while the opposite is observed for (-)-epicatechin (Bordiga et al., 2011). In general, the DPn tends to decrease during ripening (Romeyer et al., 1986; De Freitas and Glories, 1999; Jordão et al., 2001; Ferrer-Gallego et al., 2010; Liu et al., 2010; Bordiga et al., 2011), but while monomers and procyanidins with low DP fell in proportion, procyanidins with higher DP (more than 30 units) increase correspondently (Romeyer et al., 1986; Kennedy et al., 2000; Liu et al., 2010). When compared to skin proanthocyanidins, the grape seed procyanidins have simultaneously a higher content of galloylated units and total flavanols, but a lower DPn (Sun et al., 1998; Labarbe et al., 1999; Downey et al., 2003; Monagas et al., 2003; Montealegre et al., 2006; Bordiga et al., 2011; Hanlin et al., 2011).

The use of advanced techniques based on mass spectrometry is also very common for the identification of procyanidins (De Freitas et al., 1998; Gabetta et al., 2000; Krueger et al., 2000; Yang and Chien, 2000; Peng et al., 2001; Monagas et al., 2003; Núñez et al., 2006; Passos et al., 2007; Spranger et al., 2008; Fanzone et al., 2011) both in the positive (Gabetta et al., 2000; Krueger et al., 2000; Yang and Chien, 2000; Peng et al., 2001; Passos et al., 2007) and negative modes (De Freitas et al., 1998; Yang and Chien, 2000; Monagas et al., 2003; Núñez et al., 2006; Spranger et al., 2008; Fanzone et al., 2011). While it has been possible to identify molecules with molecular weight as high as decamers with one galloyl group (Vivas et al., 2004), the quantification of procyanidins by HPLC-MS has been limited to monomers and small oligomers (i.e., dimers to tetramers) (Gabetta et al., 2000; Monagas et al., 2003; Chedea et al., 2011; Fanzone et al., 2011). Advanced equipment based on the application of two-dimensional comprehensive LC, may allow the future identification of higher molecular weight procyanidins without any depolymerisation step previous to the analyses (Montero et al., 2013).

Pharmacological and Therapeutic Effects of Procyanidins

Grape seed procyanidins have been described to possess a wide spectrum of pharmacological and therapeutic properties such as antioxidant, antiinflammatory, antibacterial and antiviral activities, as well as anticarcinogenic, cardioprotective, and neuroprotective action (Agarwal et al., 2000a,b; Bagchi et al., 2000; Rauha et al., 2000; Cui et al., 2002; Pataki et al., 2002; Jayaprakasha et al., 2003; Auger et al., 2004; Hwang et al., 2004; Feng et al., 2005; Caillet et al., 2006; Rhodes et al., 2006; Engelbrecht et al., 2007; Nassiri-Asl and Hosseinzadeh, 2009). The antioxidant effect of a grape seed proanthocyanidin extract (GSPE) protects against free radicals *in vitro*, which shown to have stronger effect than antioxidants such as ascorbic acid and tocopherols (Bagchi et al., 1997). The procyanidins structure is related with their efficient radical trapping of these compounds. The catechol unit on the aromatic B-ring (Figure 2) causes an extensive electron delocalization, which brings additional stabilization to the corresponding oxidized form of procyanidins (Rice-Evans et al., 1996). Moreover, procyanidins structures with higher molecular weight and larger number of o-dihydroxy phenolic groups provide these molecules high capacity to complex metal ions (Fe(III), Cu(II), Al(III)) and proteins. In fact, protein-procyanidin interactions disclose procyanidins ability to restrain a wide range of activities, through inhibition of a variety of eukaryotic enzymes (Porto et al., 2003). Proanthocyanidins have been shown to modulate the activity of regulatory enzymes including angiotensin-I converting enzyme (ACE), cyclooxygenase, lipooxygenase, protein kinase C and cytochrome P450 activities (Nassiri-Asl and Hosseinzadeh, 2009). ACE plays an important role in the renin-angiotensin-aldosterone system, transforming the decapeptide angiotensin-I into angiotensin-II, mainly affecting blood pressure. ACE inhibitors are therefore cardioprotective and vasculoprotective. Several authors showed procyanidins inhibition by ACE, suggesting these molecules as a future alternative to synthetic ACE inhibitors (Actis-Goretta et al., 2003, 2006; Eriz et al., 2011; Godoy et al., 2012; Afonso et al., 2013).

In summary, procyanidins may play an important role in the modulation of important metabolic pathways and contribute for novel pharmacological and therapeutic approaches. Nonetheless, bioavailability of these compounds needs yet to be completely understood. Several reports, over the last years, show that at least monomers and smaller oligomeric procyanidins are absorbed by the intestinal tract (Manach et al., 2004; Manach et al., 2005; Williamson and Manach, 2005).

Antioxidant Activity of Procyanidins

Free radicals – reactive oxygen (ROS) and reactive nitrogen (RNS) species – are produced in normal aerobic cellular metabolism. The balance of antioxidant and oxidant system is controlled by endogenous enzymes such as superoxide dismutase, glutathione peroxidase, catalase and/or non-enzymatic pathways with chemical compounds such as α-tocopherol, ascorbic acid, carotenoids, glutathione, and polyphenol compounds (Sousa et al., 2012). Nevertheless, an unbalanced antioxidant defense mechanism, due to overproduction or addition of free radicals from environment to living system, may lead to pathological cell metabolism, an onset and progression of numerous diseases. Natural compounds present in fruits and vegetables, cereals, and seeds contain an important source of antioxidants, which may be used to help human body to decrease oxidative injury. Hence, there is an increased interest in compounds exhibiting antioxidant potential effect, which can be used as specific preventive pharmaceuticals or simply as food components with protective properties to human and animal organisms (Slusarczyk et al., 2009; Amjad 2013).

The antioxidant properties of phenolics, particularly of proanthocyanidins, have been widely studied. Scavenging of free radicals (as chain breakers, as reducing agents or donors of hydrogen atoms as well as complexes of pro-oxidants metals, and quenchers of the formation of singlet oxygen), inhibition of lipid oxidation, reduction of hydroperoxide formation, and modulation of enzymatic defenses are some examples of antioxidant capacities of these molecules (Bagchi et al., 1997; Jayaprakasha et al., 2003; Yilmaz and Toledo 2004; Afonso et al., 2013).

The antioxidant activities of grape extracts have been studied by numerous researchers (Chen et al., 1996; Rice-Evans et al., 1996; Porto et al., 2003), who tried to link antioxidant ability with phenolic composition and chemical structure. Several reports refer that the availability of phenolic hydrogens can predict the antioxidant activity of proanthocyanidins, proposing phenolic hydrogens as the hydrogens donating radical scavengers and singlet oxygen quenchers. Therefore suggesting that the antioxidant capacity of such compounds is directly connected with the concentration of phenolic hydrogens (Bagchi et al., 2000; Manach et al., 2004). Nevertheless, the relationship between phenolic compounds and antioxidant capacity was inconsistent among results from different studies, which prompts the idea that, besides concentration, the antioxidant capacity is affected by other aspects. In fact, Di Majo et al. (2008) showed that Merlot and the Cabernet-Sauvignon wines were richer in polyphenols and had high antiradical activity but on the other hand wines like Syrah or Nero D'Avola had lower antioxidant power and a higher polyphenols amount. The structure of procyanidins is likely to influence radical-trapping activity as well. Porto et al. (2003) studied the importance of procyanidins structure in the protection of low-density lipoprotein oxidation, and showed that the monomers (+)-catechin and (-)-epicatechin have an identical antioxidant activity while procyanidin oligomers with higher DP exhibit higher antioxidant activities (Porto et al., 2003). In fact, polymeric procyanidins presented the highest activities, followed by oligomeric procyanidins (Spranger et al., 2008).

Beneficial effects of grape seed procyanidins, in particular their free radical scavenging capability, have received significant relevance since the antioxidant activity of grape seed polyphenols is more efficient than other well known antioxidants, such as ascorbic acid, tocopherols, and β-carotene. In fact, several reports indicate that procyanidins from grape seeds are 20 and 50 times more potent than ascorbic acid and tocopherols, respectively (Uchida et al., 1987; Shi et al., 2003). This would indicate that grape seed procyanidins might be of interest to be used as alternative antioxidants. Moreover, since phenolic compounds also protect vitamins from early oxidation, vitamins became available for their primary functions and a synergistic effect can arise between antioxidants, such as ascorbic acid and tocopherols, selenium, and carotenoids (Shi et al., 2003).

Cardiovascular diseases are linked with alterations in fatty acid metabolism and excessive lipid peroxidation of low-density lipoprotein (LDL). The resulting oxidation products are also involved in thromboxane formation and consequent increase of platelet aggregation, then to artery blockage, and lastly to thrombosis (Kushi et al., 1995; Shi et al., 2003). When there is a low level of plasma antioxidants, lipid oxidation products from LDL are likely to accumulate. The presence of grape seeds procyanidins could inhibit the oxidation of LDL and therefore reduce the risk of heart disease.

Furthermore, procyanidins of grape seeds may help to inhibit enzyme systems that are responsible for the production of free radicals and that are associated with inflammatory reactions. They are also able to inhibit ACE which is associated with cardiovascular diseases.

Cardioprotective Benefits of Grape Seed Procyanidins

The renin-angiotensin system is an important regulator of blood pressure and electrolyte homeostasis (Figure 3). Renin, an enzyme produced in the kidney, converts angiotensinogen, a globulin formed in the liver, into angiotensin I. Renin release can be activated by loss of blood volume (hemorrhage), a drop in systemic blood pressure (either directly through baroreceptors or indirectly through reduction in renal tubular fluid, as in hypotension or dehydration) or decrease in serum sodium chloride levels (Paul et al., 2006; Fyhrquist and Saijonmaa, 2008).

ACE, a glycoprotein produced mainly in the lung, converts angiotensin I into angiotensin II and also degrades bradykinin, a powerful vasodilator. Angiotensin II is a potent vasoconstrictor and neurotransmitter, which increases peripheral vascular resistance and induces sodium retention by stimulating the adrenal cortex to secrete aldosterone. Moreover, angiotensin II induces vascular smooth muscle cell migration, growth, and proliferation. Angiotensin II plays a main role in the pathogenesis of essential hypertension, congestive heart failure, myocardial infarction, and diabetic nephropathy. Hence numerous ACE inhibitors were developed as therapeutic targets (Brown and Vaughan, 1998).

Cardioprotective benefits of ACE inhibitors are well known as well as their side effects. Rash and taste disturbances are frequent with most ACE inhibitors. With captopril, these effects were found to be more prevalent, which can be attributed to its sulfhydryl moiety. Other common reactions are hypotension, cough, hyperkalemia (potassium retention), headache, dizziness, fatigue, nausea, and renal impairment (Parish and Miller, 1992). Moreover, ACE inhibitors are pharmacological drugs and their use in healthy or low-risk populations is not advisable.

Several authors refer the ability that procyanidins have to inhibit ACE activity, both *in vivo* and *in vitro* (Actis-Goretta et al., 20032006; Eriz et al., 2011; Godoy et al., 2012; Afonso et al., 2013). These findings open up the possibility that consumption of select procyanidins-rich foods may mimic synthetic ACE inhibitors and provide health benefits but without adverse effects.

Inhibitory Action upon Angiotensin I-Converting Enzyme

Several authors consider inhibition of ACE activity as a promising "therapeutic approach" for the treatment of hypertension and associated coronary diseases in humans (Loizzo et al., 2007; Eriz et al., 2011). Afonso et al. (2013) have studied the effect of phenolic compounds of grape seeds on the activity of the ACE, and analysed differences associated to diverse degrees of polymerization.

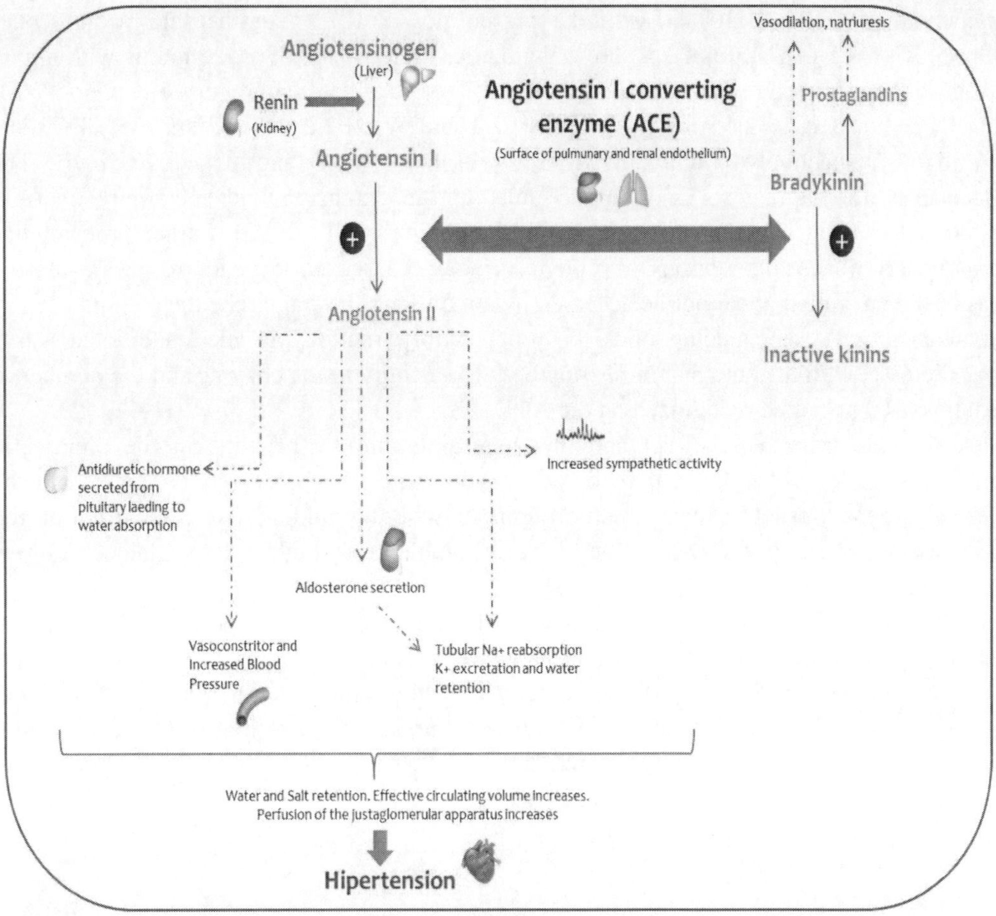

Figure 3. Role of Angiotensin-I converting enzyme (ACE).

The evaluation of ACE activity can be determined by fluorometric, spectrophotometric, or radioenzymatic methods (Geng et al., 2003; Hernández-Ledesma et al., 2003; Afonso et al., 2013). Afonso et al. (2013) have adopted a fluorometric assay. Briefly, the amount of histidyl-leucine (His-Leu) released from the N-Hippuryl-L-histidyl-L-leucine utilized in the presence of the enzyme (ACE from rabbit lung), at 37°C for 30 min, was measured adding a fluorophore: o-phtaldialdehyde. The fluorescence of the o-phtaldialdehyde condensation product of His-Leu was assessed with a spectrophotofluorometer. Four purified monomers ((+)-catechin, (-)-epicatechin, (+)-catechin-*O*-gallate and (-)-epigallocatechin), a synthetic inhibitor (captopril), and grape seed procyanidins with different degrees of polymerization have been compared. Results showed that the natural procyanidins extracts exhibit higher inhibitory potential than the monomers. IC_{50} values of (+)-catechin, (-)-epicatechin, (+)-catechin-*O*-gallate, and (–)-epigallocatechin were 11.6, 13.0, 15.5, and 18.2 mg/L, whereas the IC_{50} determined for four grape seed extracts (denoted by WM-F2.2, WM-F2.4, WM-F2.6 and WA-F2.2) were 5 to 150 times higher (0.12, 0.39, 2.65 and 2.26 mg extract/L, respectively). Furthermore, results proved that, for the same concentrations, these oligomeric procyanidin-rich extracts promote almost total ACE inhibition, whereas flavanol monomers do not.

Regarding the inhibition of ACE by grape seeds extracts, fractions with higher degree of polymerization (DPn=8, IC_{50}=0.1 mg/L) present lower IC_{50} than the intermediate ones (DPn=5, IC_{50}=3.5 mg/L). Figure 4 shows that larger molecules, i.e. procyanidins with higher average degree of polymerization, may be associated to greater inhibitory effect: for WM-F2.6, DPn=5 and IC_{50}=2.65 mg/L; for WM-F2.4 and WA-F2.2, DPn=7 and IC_{50}=0.39 and 2.26 mg/L, respectively; finally, WM-F2.2 presents DPn=8 and IC_{50}=0.12 mg/L. The influence of the number of monomeric units in the extension and specificity of ACE inhibition has been already suggested by Actis-Goretta et al. (2003). Larger procyanidins seem to exert broader interactions on protein surfaces. A larger procyanidin reaches extensive areas of the enzyme that cannot be accessed by a monomer. In that essay, the authors suggest that when more (-)-epicatechin units the procyanidin structure provides, additional active groups interact with the enzyme, increasing both the affinity to the enzyme and the number of binding sites that can be recognized on the ACE surface.

A possible synergistic effect and the increasing number of units in the oligomeric compounds could contribute in part to such results, since the heterocyclic oxygen and the hydroxyl groups present in their structural units may be the binding sites recognized on the ACE surface. Hence, in the DPn range of water soluble procyanidins, larger molecules give rise to higher inhibition.

Eriz et al. (2011) also studied the relation between the degree of polymerization of grape extracts (skin and seeds) and their ability to inhibit ACE activity. They confirmed that the size and structural composition of extracts affect ACE inhibition, since molecules with higher average degree of polymerization, and thus with larger size and higher OH availability, shown stronger inhibitory ability. It was explained that a larger structure allows these molecules to interact with more enzymatic sites.

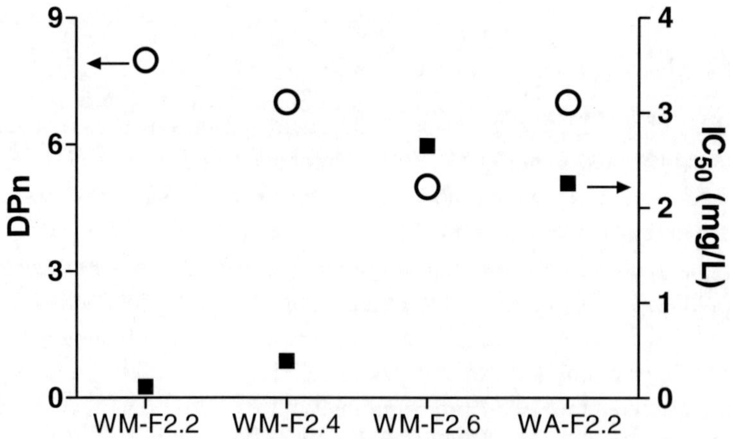

Figure 4. Inhibitory concentration 50 (IC_{50}) calculated for the inhibition of ACE activity by four white grape seed extracts (WM-F2.2, WM-F2.4, WM-F2.6, and WA-F2.2) plotted along with the average degree of polymerization of each extract.

Bioavailability of Procyanidins. Blood Pressure Studies

The amount of procyanidins consumed and their bioavailability are two important aspects to take into consideration when discussing the potential health benefits of procyanidins. The *in vivo* presence of a substance is the result of combined metabolic activities of conjugating enzymes and efflux transporters. These combined processes are present in most tissues and organs in mammals for the purposes of drug metabolism, elimination, and detoxification. Thus, it is likely that these combined processes represent a significant barrier to the oral bioavailabilities of procyanidins (Liu and Hut, 2007).

The effects of procyanidins on biological systems *in vitro* have been widely studied. Nonetheless, many of those reports do not take into consideration the influence of metabolism and absorption, and it is not certain that the same outcome can necessarily occur *in vivo*. Over the last years, new insights into procyanidins metabolism and bioavailability revealed that most classes of polyphenols, including procyanidins, appear to be satisfactorily absorbed and thus have the potential to exert biological effects. For example, after red wine ingestion, (+)-catechin can reach micromolar concentrations in the blood, which demonstrates that these compounds can cross intestinal barrier and reach the same concentrations in the bloodstream that have shown to exert effects *in vitro* (Scalbert and Williamson, 2000; Manach et al., 2005; Williamson and Manach, 2005). In fact, *in vivo* studies showed that (epi)catechins increased plasma ascorbic acid concentrations, incremented plasma antioxidant activity, decreased plasma lipid peroxidation, increased the resistance of LDL to oxidation, and increased brachial dilation (Ishikawa et al., 1997; Freese et al., 1999; Sung et al., 2000; Duffy et al., 2001a,b; Gomikawa and Ishikawa, 2002; Young et al., 2002). However, both (+)-catechin and (-)-epicatechin have short plasma half-lives (2-3 h) and their circulating forms are mainly methyl, sulfate and glucuronide conjugates of the monomers (Manach et al., 2005). Consequently, care must be taken when assuming direct extrapolations of *in vitro* results to *in vivo* effects. *In vitro* experiments must be seen as a mean to predict possible *in vivo* effects.

Oligomeric (+)-catechins and (-)-epicatechins, which are commonly dimers and trimers but can as well reach high degrees of polymerization, also demonstrate to exert potential health benefits in *in vivo* studies (Actis-Goretta et al., 2003, 2006). The main effects shown on the vascular system are endothelium-dependent blood vessel dilation and decrease of blood pressure, with positive effects on capillary fragility and permeability, and also increase microvessels diameters. Moreover oligomeric (+)-catechins also shown to contribute to the oxidative stress balance which includes substantial increase in plasma antioxidant activity; decrease of lipid peroxidation concentration; increase in acid ascorbic concentration in plasma; decrease of LDL cholesterol concentrations, and increase HDL cholesterol concentrations; decrease susceptibility of LDL to oxidation; increase concentrations of nitrosated/nitrosylated species; decrease serum thromboxane concentrations; increase plasma vitamin B6 concentrations; decrease superoxide release, and increase tocopherol concentrations (Aviram et al., 2000; Preuss et al., 2000; van der Gaag et al., 2000; Freedman et al., 2001; Hosseini et al., 2001; Heiss et al., 2003; Sano et al., 2003; Vigna et al., 2003; Sivaprakasapillai et al., 2009).

Metabolic fate of procyanidins after consumption is not yet clear. For instance, only 10 nM of procyanidins B1 was found in plasma of volunteers that had consumed 2 g of high-procyanidins grape seed extract (Sano et al., 2003). However, clinical trials indicate biological effects as described above. Possible explanations for proven *in vivo* benefits may result from effects of undisclosed metabolites of procyanidins or monomers (((+)-catechins and (-)-epicatechins) or sinergisme between procyanidins and monomers. In fact, grape seed procyanidins with higher degree of polymerization are not degraded to monomers and smaller oligomers in acidic conditions of the stomach, allowing the presence of high concentrations of these procyanidins in the intestine. The extensive microflora plays an important role in the metabolism of procyanidins absorption of intact procyanidins (with low degree of polymerization) through the colonic epithelium to blood stream and/or the breakdown to metabolites (Williamson and Manach, 2005). This can be of most interest since, as discussed above, both antioxidant and enzymatic inhibitory effects of procyanidins can depend upon the degree of polymerization and on a synergistic effect of different degrees of polymerization.

Blood Pressure Studies

Hosseini et al. (2001) showed a significant decrease in systolic blood pressure and serum thromboxane B2 levels after eight weeks of Pycnogenol supplementation (a mixture of water-soluble procyanidins, (+)-catechin, and phenolcarbonic acids) in non-smoking, mildly hypertensive patients who did not require standard pharmacological drug therapy.

A randomized controlled trial with subjects with the metabolic syndrome, conducted by Sivaprakasapillai et al. (2009), presented reduction of both systolic and diastolic blood pressures when a dose of either 300 or 150 mg/day of grape seed extracts was administered.

Quiñones et al. (2013) showed that grape seed procyanidin extract (GSPE) produced a significant decrease in systolic and diastolic blood pressure of spontaneously hypertensive rats (SHR) in a dose-dependent manner up to 375 mg/kg (maximum decrease 6 h post-administration) and did not affect blood pressure of Wistar–Kyoto rats (normotensive control rats). This research concluded that GSPE could be a potential antihypertensive ingredient for functional foods, and the improvement of oxidative stress may be associated with the benefic action on arterial blood pressure. Nevertheless, the authors suggest additional investigation to determine the real clinical value of GSPE and to clarify the mechanisms involved in its antihypertensive effect.

Gémez-Guzmán et al. (2012) reported the effect of (-)-epicatechin in rats receiving DOCA-salt. After treatment with DOCA-salt, rats presented systolic blood pressure 47 mmHg higher than in controls. When DOCA-salt-treated rats were simultaneously treated with 10 mg/kg of (-)-epicatechin, the increase of systolic blood pressure was prevented (about 70%). Control animals treated with the same amount of (-)-epicatechin showed no effect. This study prompts the view that chronic treatment with (-)-epicatechin prevents the progressive increase in systolic blood pressure and thus may reduce hypertension.

An extensive awareness of the bioavailability of procyanidins is thus essential for understanding health benefits of these compounds. New clinical trials should be designed to further understand the bioavailability of procyanidins *in vivo*. Increasing the length of *in vivo* studies and the use of validated *in vivo* markers, are significant factors to look for in future reports.

References

Actis-Goretta, L., J. I. Ottaviani, C. G. Fraga (2006). Inhibition of angiotensin converting enzyme activity by flavanol-rich foods. *J. Agric. Food Chem.*, 54(1), 229-234.

Actis-Goretta, L., J. I. Ottaviani, C. L. Keen, C. G. Fraga (2003). Inhibition of angiotensin converting enzyme (ACE) activity by flavan-3-ols and procyanidins. *FEBS Lett.,* 555(3), 597-600.

Actis-Goretta, L., L. J. Romanczyk, C. A. Rodriguez, C. Kwik-Uribe, C. L. Keen (2008). Cytotoxic effects of digalloyl dimer procyanidins in human cancer cell lines. *J Nutr Biochem.,* 19(12), 797-808.

Afonso, J., C. P. Passos, M. A. Coimbra, C. M. Silva, P. Soares-da-Silva (2013). Inhibitory effect of phenolic compounds from grape seeds (*Vitis vinifera* L.) on the activity of angiotensin I converting enzyme. *LWT-Food Sci. Technol.*, 54(1), 265-270.

Agarwal, C., Y. Sharma, R. Agarwal (2000a). Anticarcinogenic effect of a polyphenolic fraction isolated from grape seeds in human prostate carcinoma DU145 cells: Modulation of mitogenic signaling and cell-cycle regulators and induction of G1 arrest and apoptosis. *Mol. Carcinog.*, 28(3), 129-138.

Agarwal, C., Y. Sharma, J. F. Zhao, R. Agarwal (2000b). A polyphenolic fraction from grape seeds causes irreversible growth inhibition of breast carcinoma MDA-MB468 cells by inhibiting mitogen-activated protein kinases activation and inducing G(1) arrest and differentiation, 6(7), 2921-2930.

Al Makdessi, S., F. Sweidan, R. Jacob (2006). Effect of oligomer procyanidins on reperfusion arrhythmias and lactate dehydrogenase release in the isolated rat heart. *Arzneimittelforschung*, 56(5), 317-321.

Aldini, G., M. Carini, A. Piccoli, G. Rossoni, R. M. Facino (2003). Procyanidins from grape seeds protect endothelial cells from peroxynitrite damage and enhance endothelium-dependent relaxation in human artery: new evidences for cardio-protection. *Life Sci.*, 73(22), 2883-2898.

Amjad, L., Shafighi, M. (2013). Evaluation of antioxidant activity, phenolic and flavonoid content in *Punica granatum* var. Isfahan Malas flowers. *Intl. J. Agri. Crop Sci.*, 5(10), 1133-1139.

Ashraf-Khorassani, M., L. T. Taylor (2004). Sequential fractionation of grape seeds into oils, polyphenols, and procyanidins via a single system employing CO2-based fluids. J Agric Food Chem., 52(9), 2440-2444.

Auger C, G. P., Laurent-Bichon F., Porter K., Bornet A.,Caporiccio B., Cros G., Teissedre P. L., Rouanet J. M. (2004). Phenolics from commercialized grape extracts prevent early atherosclerotic lesions in hamsters by mechanisms other than antioxidant effect. *J. Agric. Food Chem.,* 52, 6.

Aviram, M., L. Dornfeld, M. Rosenblat, N. Volkova, M. Kaplan, R. Colemann, T. Hayek, D. Presser, B. Fuhrman (2000). Pomegranate juice consumption reduces oxidative stress, atherogenic modifications to LDL, and platelet aggregation: studies in humans and in atherosclerotic apolipoprotein E-deficient mice. *Am. J. Clin. Nutr.,* 71(5), 1062-1076.

Bagchi, D., M. Bagchi, S. J. Stohs, D. K. Das, S. D. Ray, C. A. Kuszynski, S. S. Joshi, H. G. Pruess (2000). Free radicals and grape seed proanthocyanidin extract: importance in human health and disease prevention. *Toxicology,* 148(2-3), 187-197.

Bagchi, D., A. Garg, R. L. Krohn, M. Bagchi, M. X. Tran, S. J. Stohs (1997). Oxygen free radical scavenging abilities of vitamins C and E, and A grape seed proanthocyanidin extract *in vitro*. *Res. Comm. Mol. Pathol. Pharmacol.*, 95(2), 179-189.

Bak, M.J., M. Jun, W.S. Jeong (2012). Procyanidins from Wild Grape (*Vitis amurensis*) Seeds Regulate ARE-Mediated Enzyme Expression via Nrf2 Coupled with p38 and PI3K/Akt Pathway in HepG2 Cells. *Int J Mol Sci.*,13(1), 801-818.

Bakkalbasi, E., O. Yemis, D. Aslanova, N. Artik (2005). Major flavan-3-ol composition and antioxidant activity of seeds from different grape cultivators grown in Turkey. *Eur Food Res Technol,* 221(6), 792-797.

Bomser, J. A., K. W. Singletary, M. A. Wallig, M. A. L. Smith (1999). Inhibition of TPA-induced tumor promotion in CD-1 mouse epidermis by a polyphenolic fraction from grape seeds. *Cancer Lett.*, 135(2), 151-157.

Bordiga, M., F. Travaglia, M. Locatelli, J. D. Coisson, M. Arlorio (2011). Characterisation of polymeric skin and seed proanthocyanidins during ripening in six *Vitis vinifera* L. cv. *Food Chem.*,127(1), 180-187.

Brown, N. J., D. E. Vaughan (1998). Angiotensin-converting enzyme inhibitors. *Circulation,* 97(14), 1411-1420.

Caillet, S., S. Salmieri, M. Lacroix (2006). Evaluation of free radical-scavenging properties of commercial grape phenol extracts by a fast colorimetric method. *Food Chem.,* 95(1), 1-8.

Carini, M., G. Aldini, E. Bombardelli, P. Morazzoni, R. M. Facino (2000). UVB-induced hemolysis of rat erythrocytes: Protective effect of procyanidins from grape seeds. *Life Sci.*, 67(15), 1799-1814.

Castillo, J., O. Benavente-García, J. Lorente, M. J. Alcaraz, A. Redondo, A. Ortuno, J. A. Del Rio (2000). Antioxidant activity and radioprotective effects against chromosomal damage induced in vivo by X-rays of flavan-3-ols (procyanidins) from grape seeds (*Vitis vinifera*): Comparative study versus other phenolic and organic compounds. *J Agri. Food Chem.,* 48(5), 1738-1745.

Catterall, F., J. M. Souquet, V. Cheynier, M. N. Clifford, C. Ioannides (2000). Modulation of the mutagenicity of food carcinogens by oligomeric and polymeric procyanidins isolated from grape seeds: synergistic genotoxicity with N-nitrosopyrrolidine, *J. Sci. Food Agric.,* 80(1), 91-101.

Chedea, V. S., C. Echim, C. Braicu, M. Andjelkovic, R. Verhe, C. Socaciu (2011). Composition in polyphenols and stability of the aqueous grape seed extract from the romanian variety "*Merlot recas*". *J. Food Biochem.*, 35(1), 92-108.

Chen, Z. Y., P. T. Chan, K. Y. Ho, K. P. Fung, J. Wang (1996). Antioxidant activity of natural flavonoids is governed by number and location of their aromatic hydroxyl groups. *Chem. Phys. Lipids*, 79(2), 157-163.

Cheynier, V., J. M. Souquet, E. Le Roux, S. Guyot, J. Rigaud (1999). Size separation of condensed tannins by normal-phase high-performance liquid chromatograph,. *Methods Enzymol.*, 299, 178-184.

Chung, H. K., C. S. Choi, W. J. Park, M. H. Kang (2005). Radical scavenging activity of grape-seed extracts prepared from different solvents. *Food Sci. Biotechnol.*,14(6), 715-721.

Coombe, B. G. (1960). Relationship of growth and development to changes in sugars, auxins, and gibberellins in fruit of seeded and seedless varieties of *Vitis vinifera*. *Plant Physiol.*, 35(2), 241-250.

Cui, J. H., G. A. Cordis, A. Tosaki, N. Maulik, D. K. Das (2002). Reduction of myocardial ischemia reperfusion injury with regular consumption of grapes. *Alcohol and Wine in Health and Disease*., 957, 302-307.

De Freitas, V., P. D. Porto, M. Assuncao, A. Cadete-Leite, J. P. Andrade, M. M. Paula-Barbosa (2004). Flavonoids from grape seeds prevent increased alcohol-induced neuronal lipofuscin formation. *Alcohol and Alcoholism*, 39(4), 303-311.

De Freitas, V. A. P., Y. Glories (1999). Concentration and compositional changes of procyanidins in grape seeds and skin of white *Vitis vinifera* varieties. *J. Sci. Food Agric.*,79(12), 1601-1606.

De Freitas, V. A. P., Y. Glories, G. Bourgeois, C. Vitry (1998). Characterisation of oligomeric and polymeric procyanidins from grape seeds by liquid secondary ion mass spectrometry. *Phytochemistry*, 49(5), 1435-1441.

De Freitas, V. A. P., Y. Glories, M. Laguerre (1998). Incidence of molecular structure in oxidation of grape seed procyanidins. *J. Agric. Food Chem*., 46(2), 376-382.

Di Majo, D., M. La Guardia, S. Giammanco, L. La Neve, M. Giammanco (2008). The antioxidant capacity of red wine in relationship with its polyphenolic constituents. *Food Chem*., 111(1), 45-49.

Djilas, S., J. Canadanovic-Brunet, G. Cetkovic (2009). By-products of fruits processing as a source of phytochemicals. *CI&CEQ*, 15(4), 191-202.

Downey, M. O., J. S. Harvey, S. P. Robinson (2003). Analysis of tannins in seeds and skins of Shiraz grapes throughout berry development. *Australian Journal of Grape and Wine Research*, 9(1), 15-27.

Duffy, S. J., J. F. Keaney, M. Holbrook, N. Gokce, P. L. Swerdloff, B. Frei, J. A. Vita (2001a). Short- and long-term black tea consumption reverses endothelial dysfunction in patients with coronary artery disease. *Circulation,* 104(2), 151-156.

Duffy, S. J., J. A. Vita, M. Holbrook, P. L. Swerdloff, J. F. Keaney (2001b). Effect of acute and chronic tea consumption on platelet aggregation in patients with coronary artery disease. *Arterioscler. Thromb. Vasc. Biol.,* 21(6), 1084-1089.

Engelbrecht, A. M., M. Mattheyse, B. Ellis, B. Loos, M. Thomas, R. Smith, S. Peters, C. Smith, K. Myburgh (2007). Proanthocyanidin from grape seeds inactivates the PI3-kinase/PKB pathway and induces apoptosis in a colon cancer cell line. *Cancer Lett.,* 258(1), 144-153.

Eriz, G., V. Sanhueza, M. Roeckel, K. Fernandez (2011). Inhibition of the angiotensin-converting enzyme by grape seed and skin proanthocyanidins extracted from *Vitis vinifera* L. *cv. Pais. LWT-Food Sci. Technol*., 44(4), 860-865.

Escribano-Bailón, T., Y. Gutiérrez-Fernandez, J. C. Rivas-Gonzalo, C. Santos-Buelga (1992). Characterization of procyanidins of *Vitis vinifera* variety Tinta del Pais grape seeds. *J. Agric. Food Chem*., 40(10), 1794-1799.

Fan, P. H., H. X. Lou (2004). Effects of polyphenols from grape seeds on oxidative damage to cellular DNA. *Mol. Cell. Biochem.,* 267(1-2), 67-74.

Fanzone, M., F. Zamora, V. Jofre, M. Assof, A. Pena-Neira (2011). Phenolic Composition of Malbec Grape Skins and Seeds from Valle de Uco (Mendoza, Argentina) during Ripening. Effect of Cluster Thinning. *J. Agric. Food Chem*., 59(11), 6120-6136.

Faria, A., C. Calhau, V. De Freitas, N. Mateus (2006). Procyanidins as antioxidants and tumor cell growth modulators. *J. Agric. Food Chem*., 54(6), 2392-2397.

Feng, Y. Z., Y. M. Liu, J. D. Fratkins, M. H. LeBlanc (2005). Grape seed extract suppresses lipid peroxidation and reduces hypoxic ischemic brain injury in neonatal rats. *Brain Res. Bull.,* 66(2), 120-127.

Ferrer-Gallego, R., M. García-Marino, J. Miguel Hernández-Hierro, J. C. Rivas-Gonzalo, M. Teresa Escribano-Bailón (2010). Statistical correlation between flavanolic composition, colour and sensorial parameters in grape seed during ripening. *Anal Chim Acta*, 660(1-2), 22-28.

Fitzpatrick, D. F., B. Bing, D. A. Maggi, R. C. Fleming, R. M. O'Malley (2002). Vasodilating procyanidins derived from grape seeds. *Annals of the New York Academy of Sciences,* 957, 78-89.

Fitzpatrick, D. F., R. C. Fleming, B. Bing, D. A. Maggi, R. M. O'Malley (2000). Isolation and characterization of endothelium-dependent vasorelaxing compounds from grape seeds. *J Agric Food Chem.*, 48(12), 6384-6390.

Flamini, R. (2003). Mass spectrometry in grape and wine chemistry. Part I: Polyphenols. *Mass Spectrom. Rev.*, 22(4), 218-250.

Freedman, J. E., C. Parker, L. Q. Li, J. A. Perlman, B. Frei, V. Ivanov, L. R. Deak, M. D. Iafrati, J. D. Folts (2001). Select flavonoids and whole juice from purple grapes inhibit platelet function and enhance nitric oxide release. *Circulation,* 103(23), 2792-2798.

Freese, R., S. Basu, E. Hietanen, J. Nair, K. Nakachi, H. Bartsch, M. Mutanen (1999). Green tea extract decreases plasma malondialdehyde concentration but does not affect other indicators of oxidative stress, nitric oxide production, or hemostatic factors during a high-linoleic acid diet in healthy females. *Eur. J. Nutr.,* 38(3), 149-157.

Fuleki, T., J. M. R. Silva (1997). Catechin and procyanidin composition of seeds from grape cultivars grown in Ontario. *J. Agric. Food Chem.*, 45(4), 1156-1160.

Fyhrquist, F., O. Saijonmaa (2008). Renin-angiotensin system revisited. *J. Intern. Med.,* 264(3), 224-236.

Gabetta, B., N. Fuzzati, A. Griffini, E. Lolla, R. Pace, T. Ruffilli, F. Peterlongo (2000). Characterization of proanthocyanidins from grape seeds. *Fitoterapia*, 71(2), 162-175.

Gazzani, G., M. Daglia, A. Papetti (2012). Food components with anticaries activity. *Curr. Opin. Biotechnol.*, 23(2), 153-159.

Geng, L. J., K. Persson, S. F. E. Nilsson (2003). Angiotensin converting enzyme (ACE) activity in porcine ocular tissue: Effects of diet and ACE inhibitors. *J. Ocular Pharmacol. Ther.*, 19(6), 589-598.

Godoy, S., M. Roeckel, K. Fernández (2012). Influence of the structure and composition of the *País* grape proanthocyanidins on the inhibition of angiotensin I-converting enzyme (ACE). *Food Chem.,* 134(1), 346-350.

Gómez-Guzmán, M., R. Jiménez, M. Sánchez, M. Jose Zarzuelo, P. Galindo, A. Maria Quintela, R. López-Sepúlveda, M. Romero, J. Tamargo, F. Vargas, F. Pérez-Vizcaíno, J. Duarte (2012). Epicatechin lowers blood pressure, restores endothelial function, and decreases oxidative stress and endothelin-1 and NADPH oxidase activity in DOCA-salt hypertension. *Free Radic. Biol. Med.,* 52(1), 70-79.

Gomikawa, S., Y. Ishikawa (2002). Effects of catechins and ground green tea drinking on the susceptibility of plasma and LDL to the oxidation in vitro and *ex vivo*. *J. Clin. Biochem. Nutr.*, 32, 55-68.

González-Manzano, S., J. C. Rivas-Gonzalo, C. Santos-Buelga (2004). Extraction of flavan-3-ols from grape seed and skin into wine using simulated maceration. *Anal. Chim. Acta*, 513(1), 283-289.

González-Manzano, S., C. Santos-Buelga, J. J. Pérez-Alonso, J. C. Rivas-Gonzalo, M. T. Escribano-Bailón (2006). Characterization of the mean degree of polymerization of proanthocyanidins in red wines using liquid chromatography-mass spectrometry (LC-MS). *J Agric Food Chem.*, 54(12), 4326-4332.

Gougoulias, N., N. Mashev (2008). Evaluation of polyphenols antioxidant activity of grape seeds (*V. vinifera*). *Oxid. Commun.,* 31(1), 88-97.

Guendez, R., S. Kallithraka, D. P. Makris, P. Kefalas (2005). An analytical survey of the polyphenols of seeds of varieties of grape (*Vitis vinifera*) cultivated in Greece: Implications for exploitation as a source of value-added phytochemicals. *Phytochem Anal.*, 16(1), 17-23.

Guyot, S., C. Le Guerneve, N. Marnet, J. F. Drilleau (2000). Methods for determining the degree of polymerization of condensed tannins: A new (1)H-NMR procedure applied to cider apple procyanidins. in: Plant Polyphenols 2: Chemistry, Biology, Pharmacology, Ecology. G. G. Gross, R. W. Hemingway and T. Yoshida, 66, 211-222.

Guyot, S., N. Marnet, J. F. Drilleau (2001). Thiolysis-HPLC characterization of apple procyanidins covering a large range of polymerization states. *J Agric. Food Chem.*, 49(1), 14-20.

Hanlin, R. L., M. A. Kelm, K. L. Wilkinson, M. O. Downey (2011). Detailed Characterization of Proanthocyanidins in Skin, Seeds, and Wine of Shiraz and Cabernet Sauvignon Wine Grapes (*Vitis vinifera*). *J Agric. Food Chem.*, 59(24), 13265-13276.

Harbertson, J. F., J. A. Kennedy, D. O. Adams (2002). Tannin in skins and seeds of Cabernet Sauvignon, Syrah, and Pinot noir berries during ripening. *Am. J. Enol. Vitic.*, 53(1), 54-59.

Heiss, C., A. Dejam, P. Kleinbongard, T. Schewe, H. Sies, M. Kelm (2003). Vascular effects of cocoa rich in flavan-3-ols. *JAMA-J. Am. Med. Assoc.*, 290(8), 1030-1031.

Hernández-Ledesma, B., P. J. Martín-Álvarez, E. Pueyo (2003). Assessment of the spectrophotometric method for determination of angiotensin-converting-enzyme activity: Influence of the inhibition type. *J. Agric. Food Chem.*, 51(15), 4175-4179.

Hosseini, S., J. Lee, R. T. Sepulveda, P. Rohdewald, R. R. Watson (2001). A randomized, double-blind, placebo-controlled, prospective, 16 week crossover study to determine the role of Pycnogenol in modifying blood pressure in mildly hypertensive patients. *Nutr. Res.*, 21(9), 1251-1260.

Hwang, I. K., K. Y. Yoo, D. S. Kim, Y. K. Jeong, J. D. Kim, H. K. Shin, S. S. Lim, I. D. Yoo, T. C. Kang, D. W. Kim, W. K. Moon, M. H. Won (2004). Neuroprotective effects of grape seed extract on neuronal injury by inhibiting DNA damage in the gerbil hippocampus after transient forebrain ischemia. *Life Sci.,* 75(16), 1989-2001.

Ishikawa, T., M. Suzukawa, T. Ito, H. Yoshida, M. Ayaori, M. Nishiwaki, A. Yonemura, Y. Hara, H. Nakamura (1997). Effect of tea flavonoid supplementation on the susceptibility of low-density lipoprotein to oxidative modification. *Am. J. Clin. Nutr.*, 66(2), 261-266.

Jayaprakasha, G. K., T. Selvi, K. K. Sakariah (2003). Antibacterial and antioxidant activities of grape (*Vitis vinifera*) seed extracts. *Food Res. Int.*, 36(2), 117-122.

Jordão, A. M., J. M. R. Silva, O. Laureano (2001). Evolution of catechins and oligomeric procyanidins during grape maturation of Castelão Francês and Touriga Francesa. *Am. J. Enol. Vitic.*, 52(3), 230-234.

Kallithraka, S., C. Garciaviguera, P. Bridle, J. Bakker (1995). Survey of solvents for the extraction of grape seed phenolics. *Phytochem. Anal.*, 6(5), 265-267.

Kennedy, J. A., M. A. Matthews, A. L. Waterhouse (2000). Changes in grape seed polyphenols during fruit ripening. *Phytochemistry*, 55(1), 77-85.

Khanal, R. C., L. R. Howard, R. L. Prior (2009). Procyanidin Composition of Selected Fruits and Fruit Byproducts Is Affected by Extraction Method and Variety. *J Agric. Food Chem.*, 57(19), 8839-8843.

Kim, Y., Y. Choi, H. Ham, H. S. Jeong, J. Lee (2013). Protective effects of oligomeric and polymeric procyanidin fractions from defatted grape seeds on tert-butyl hydroperoxide-induced oxidative damage in HepG2 cells. *Food Chem.*, 137(1-4), 136-141.

Kondo, K., M. Kurihara, K. Fukuhara, T. Tanaka, T. Suzuki, N. Miyata, M. Toyoda (2000). Conversion of procyanidin B-type (catechin dimer) to A-type: evidence for abstraction of C-2 hydrogen in catechin during radical oxidation. *Tetrahedron Lett.*, 41(4), 485-488.

Krueger, C. G., N. C. Dopke, P. M. Treichel, J. Folts, J. D. Reed (2000). Matrix-assisted laser desorption/ionization time-of-flight mass spectrometry of polygalloyl polyflavan-3-ols in grape seed extract. *J Agric Food Chem.*, 48(5), 1663-1667.

Kushi, L. H., E. B. Lenart, W. C. Willett (1995). Health implications of mediterranean diets in light of contemporary knowledge.1. Plant foods and dairy-products. *Am. J. Clin. Nutr.*, 61(6), S1407-S1415.

Labarbe, B., V. Cheynier, F. Brossaud, J. M. Souquet, M. Moutounet (1999). Quantitative fractionation of grape proanthocyanidins according to their degree of polymerization. *J Agric Food Chem.*, 47(7), 2719-2723.

Le Bourvellec, C., M. Picot, C. Renard (2006). Size-exclusion chromatography of procyanidins: Comparison between apple and grape procyanidins and application to the characterization of fractions of high degrees of polymerization. *Anal. Chim. Acta*, 563(1-2), 33-43.

Li, W., B. Xu, J. Xu, X. L. Wu (2009). Procyanidins produce significant attenuation of doxorubicin-induced cardiotoxicity via suppression of oxidative stress. *Basic Clin. Pharmacol. Toxicol.*, 104(3), 192-197.

Li, X. L., Y. Q. Cai, H. Qin, Y. J. Wu (2008). Therapeutic effect and mechanism of proanthocyanidins from grape seeds in rats with TNBS-induced ulcerative colitis. *Can. J. Physiol. Pharmacol.*, 86(12), 841-849.

Liu, Y. X., Q. H. Pan, G. L. Yan, J. J. He, C. Q. Duan (2010). Changes of flavan-3-ols with different degrees of polymerization in seeds of 'Shiraz', 'Cabernet Sauvignon' and 'Marselan' grapes after veraison. *Molecules*, 15(11), 7763-7774.

Liu, Z., M. Hut (2007). Natural polyphenol disposition via coupled metabolic pathways. *Expert Opin. Drug Metab. Toxicol.*, 3(3), 389-406.

Loizzo, M. R., R. Tundis, G. A. Statti, N. G. Passalacquai, L. Peruzzi, F. Menichini (2007). In vitro angiotensin converting enzyme inhibiting activity of *Salsola oppositifolia* Desf., *Salsola soda* L. and *Salsola tragus* L. *Nat. Prod. Res.*, 21(9), 846-851.

Lu, Y., W. Z. Zhou, Z. Chang, W. X. Chen, L. Li (2004). Procyanidins from grape seeds protect against phorbol ester-induced oxidative cellular and genotoxic damage. *Acta Pharmacol Sin.*, 25(8), 1083-1089.

Manach, C., A. Scalbert, C. Morand, C. Remesy, L. Jimenez (2004). Polyphenols: food sources and bioavailability. *Am. J. Clin. Nutr.*, 79(5), 727-747.

Manach, C., G. Williamson, C. Morand, A. Scalbert, C. Remesy (2005). Bioavailability and bioefficacy of polyphenols in humans. I. Review of 97 bioavailability studies. *Am. J. Clin. Nutr.*, 81(1), 230S-242S.

Mattivi, F., U. Vrhovsek, D. Masuero, D. Trainotti (2009). Differences in the amount and structure of extractable skin and seed tannins amongst red grape varieties. *Australian Journal of Grape and Wine Research*, 15(1), 27-35.

Mayer, R., G. Stecher, R. Wuerzner, R. C. Silva, T. Sultana, L. Trojer, I. Feuerstein, C. Krieg, G. Abel, M. Popp, O. Bobleter, G. K. Bonn (2008). Proanthocyanidins: Target compounds as antibacterial agents. *J Agric. Food Chem.*, 56(16), 6959-6966.

Monagas, M., C. Gómez-Cordovés, B. Bartolomé, O. Laureano, J. M. R. Silva (2003). Monomeric, oligomeric, and polymeric flavan-3-ol composition of wines and grapes from *Vitis vinifera* L. cv. Graciano, Tempranillo, and Cabernet Sauvignon. *J Agric. Food Chem.*, 51(22), 6475-6481.

Monagas, M., B. Hernández-Ledesma, I. Garrido, P. J. Martin-Alvarez, C. Gómez-Cordovés, B. Bartolomé (2005). Quality assessment of commercial dietary antioxidant products from *Vitis vinifera* L. grape seeds. *Nutrion and Cancer*, 53(2), 244-254.

Montealegre, R. R., R. R. Peces, J. L. C. Vozmediano, J. M. Gascueña, E. G. Romero (2006). Phenolic compounds in skins and seeds of ten grape *Vitis vinifera* varieties grown in a warm climate. *J. Food Compos. Anal.*, 19(6-7), 687-693.

Montero, L., M. Herrero, M. Prodanov, E. Ibáñez, A. Cifuentes (2013). Characterization of grape seed procyanidins by comprehensive two-dimensional hydrophilic interaction x reversed phase liquid chromatography coupled to diode array detection and tandem mass spectrometry. *Anal. Bioanal. Chem.*, 405(13), 4627-4638.

Moreno, D. A., N. Ilic, A. Poulev, D. L. Brasaemle, S. K. Fried, I. Raskin (2003). Inhibitory effects of grape seed extract on lipases. *Nutrition*, 19(10), 876-879.

Murga, R., R. Ruiz, S. Beltrán, J. L. Cabezas (2000). Extraction of natural complex phenols and tannins from grape seeds by using supercritical mixtures of carbon dioxide and alcohol. *J Agric. Food Chem.*, 48(8), 3408-3412.

Nandakumar, V., T. Singh, S. K. Katiyar (2008). Multi-targeted prevention and therapy of cancer by proanthocyanidins. *Cancer Lett.*, 269(2), 378-387.

Nassiri-Asl, M., H. Hosseinzadeh (2009). Review of the pharmacological effects of *Vitis vinifera* (Grape) and its bioactive compounds. *Phytother. Res.*, 23(9), 1197-1204.

Núñez, V., C. Gómez-Cordovés, B. Bartolomé, Y. J. Hong, A. E. Mitchell (2006). Non-galloylated and galloylated proanthocyanidin oligomers in grape seeds from *Vitis vinifera* L. cv. Graciano, Tempranillo and Cabernet Sauvignon. *J Sci Food Agric.*, 86(6), 915-921.

Palma, M., L. T. Taylor (1999). Extraction of polyphenolic compounds from grape seeds with near critical carbon dioxide. *J. Chromatogr. A*, 849(1), 117-124.

Palma, M., L. T. Taylor, R. M. Varela, S. J. Cutler, H. G. Cutler (1999). Fractional extraction of compounds from grape seeds by supercritical fluid extraction and analysis for antimicrobial and agrochemical activities. *J Agric. Food Chem.*, 47(12), 5044-5048.

Parish, R. C., L. J. Miller (1992). Adverse-effects of angiotensin converting enzyme (ACE) inhibitors - An update. *Drug Saf.,* 7(1), 14-31.

Passos, C. P., S. M. Cardoso, A. S. Barros, C. M. Silva, M. A. Coimbra (2010). Application of Fourier transform infrared spectroscopy and orthogonal projections to latent structures/partial least squares regression for estimation of procyanidins average degree of polymerisation. *Anal Chim Acta*, 661(2), 143-149.

Passos, C. P., S. M. Cardoso, M. R. M. Domingues, P. Domingues, C. M. Silva, M. A. Coimbra (2007). Evidence for galloylated type-A procyanidins in grape seeds. *Food Chem.*, 105(4), 1457-1467.

Pataki T, B. I., Kovacs P., Bagchi D., Das D. K., Tosaki A. (2002). Grape seed proanthocyanidins improved cardiac recovery during reperfusion after ischemia in isolated rat hearts. *Am. J. Clin. Nutr.*, 75, 6.

Paul, M., A. P. Mehr, R. Kreutz (2006). Physiology of local renin-angiotensin systems. *Physiol. Rev.*, 86(3), 747-803.

Peng, Z. K., Y. Hayasaka, P. G. Iland, M. Sefton, P. Hoj, E. J. Waters (2001). Quantitative analysis of polymeric procyanidins (tannins) from grape (*Vitis vinifera*) seeds by reverse phase high performance liquid chromatography. *J Agric. Food Chem.*, 49(1), 26-31.

Plumb, G. W., S. De Pascual-Teresa, C. Santos-Buelga, V. Cheynier, G. Williamson (1998). Antioxidant properties of catechins and proanthocyanidins: effect of polymerisation, galloylation and glycosylation. *Free Radic. Res.*, 29(4), 351-358.

Porto, P., J. A. N. Laranjinha, V. A. P. De Freitas (2003). Antioxidant protection of low density lipoprotein by procyanidins: structure/activity relationships. *Biochem. Pharmacol.*, 66(6), 947-954.

Preuss, H. G., D. Wallerstedt, N. Talpur, S. O. Tutuncuoglu, B. Echard, A. Myers, M. Bui, D. Bagchi (2000). Effects of niacin-bound chromium and grape seed proanthocyanidin extract on the lipid profile of hypercholesterolemic subjects: a pilot study. *J. Med.*, 31(5-6), 227-246.

Prieur, C., J. Rigaud, V. Cheynier, M. Moutounet (1994). Oligomeric and polymeric procyanidins from grape seeds. *Phytochemistry*, 36(3), 781-784.

Quiñones, M., L. Guerrero, M. Suarez, Z. Pons, A. Aleixandre, L. Arola, B. Muguerza (2013). Low-molecular procyanidin rich grape seed extract exerts antihypertensive effect in males spontaneously hypertensive rats. *Food Res. Int.*, 51(2), 587-595.

Rauha, J. P., S. Remes, M. Heinonen, A. Hopia, M. Kahkonen, T. Kujala, K. Pihlaja, H. Vuorela, P. Vuorela (2000). Antimicrobial effects of Finnish plant extracts containing flavonoids and other phenolic compounds. *Int. J. Food Microbiol.*, 56(1), 3-12.

Reed, J. D., C. G. Krueger, M. M. Vestling (2005). MALDI-TOF mass spectrometry of oligomeric food polyphenols. *Phytochemistry*, 66(18), 2248-2263.

Rhodes, P. L., J. W. Mitchell, M. W. Wilson, L. D. Melton (2006). Antilisterial activity of grape juice and grape extracts derived from *Vitis vinifera* variety Ribier. *Int. J. Food Microbiol.*, 107(3), 281-286.

Rice-Evans, C. A., N. J. Miller, G. Paganga (1996). Structure-antioxidant activity relationships of flavonoids and phenolic acids. *Free Radic. Biol. Med.*, 20(7), 933-956.

Rigaud, J., M. T. Escribano-Bailón, C. Prieur, J. M. Souquet, V. Cheynier (1993). Normal-phase high-performance liquid-chromatographic separation of procyanidins from cacao beans and grape seeds. *J. Chromatogr. A*, 654(2), 255-260.

Romeyer, F. M., J. J. Macheix, J. C. Sapis (1986). Changes and importance of oligomeric procyanidins during maturation of grape seeds. *Phytochemistry*, 25(1), 219-221.

Roy, A. M., M. S. Baliga, C. A. Elmets, S. K. Katiyar (2005). Grape seed proanthocyanidins induce apoptosis through p53, Bax, and caspase 3 pathways. *Neoplasia*, 7(1), 24-36.

Saito, M., H. Hosoyama, T. Ariga, S. Kataoka, N. Yamaji (1998). *J Agric. Food Chem.*, 46(4), 1460-1464.

Sano, A., J. Yamakoshi, S. Tokutake, K. Tobe, Y. Kubota, M. Kikuchi (2003). Procyanidin B1 is detected in human serum after intake of proanthocyanidin-rich grape seed extract. *Biosci. Biotechnol. Biochem.,* 67(5), 1140-1143.

Santos-Buelga, C., E. M. Francia-Aricha, M. T. Escribano-Bailón (1995). Comparative flavan-3-ol composition of seeds from different grape varieties. *Food Chem.*, 53(2), 197-201.

Saucier, C., M. Mirabel, F. Daviaud, A. Longieras, Y. Glories (2001). Rapid fractionation of grape seed proanthocyanidins. *J Agric. Food Chem.*, 49(12), 5732-5735.

Scalbert, A., G. Williamson (2000). Dietary intake and bioavailability of polyphenols. *J. Nutr.*, 130(8), 2073S-2085S.

Shi, J., J. Yu, J. E. Pohorly, Y. Kakuda (2003). Polyphenolics in grape seeds-biochemistry and functionality. *J. Med. Food*, 6(4), 291-299.

Shrikhande, A. J. (2000). Wine by-products with health benefits. *Food Res. Int.*, 33(6), 469-474.

Silva, J. M. R., J. Rigaud, V. Cheynier, A. Cheminat, M. Moutounet (1991). Procyanidin dimers and trimers from grape seeds. *Phytochemistry*, 30(4), 1259-1264.

Sivaprakasapillai, B., I. Edirisinghe, J. Randolph, F. Steinberg, T. Kappagoda (2009). Effect of grape seed extract on blood pressure in subjects with the metabolic syndrome. *Metab. Clin. Exp.*, 58(12), 1743-1746.

Slusarczyk, S., M. Hajnos, K. Skalicka-Wozniak, A. Matkowski (2009). Antioxidant activity of polyphenols from *Lycopus lucidus* Turcz. *Food Chem.*, 113(1), 134-138.

Souquet, J. M., B. Labarbe, C. Le Guernevé, V. Cheynier, M. Moutounet (2000). Phenolic composition of grape stems. *J Agric. Food Chem.*, 48(4), 1076-1080.

Sousa, T., J. Afonso, F. Carvalho, A. Albino-Teixeira (2012). Lipid Peroxidation and Antioxidants in Arterial Hypertension. in: Lipid Peroxidation. A. Catala. http://www.intechopen.com/books/lipid-peroxidation/lipid-peroxidation-andantioxidants-in-arterial-hypertension, InTech.

Spranger, I., B. Sun, A. M. Mateus, V. De Freitas, J. M. R. Silva (2008). Chemical characterization and antioxidant activities of oligomeric and polymeric procyanidin fractions from grape seeds. *Food Chem.*, 108(2), 519-532.

Sun, B. S., G. P. Belchior, J. M. R. Silva, M. I. Spranger (1999). Isolation and purification of dimeric and trimeric procyanidins from grape seeds. *J. Chromatogr. A*, 841(1), 115-121.

Sun, B. S., C. Leandro, J. M. R. Silva, I. Spranger (1998). Separation of grape and wine proanthocyanidins according to their degree of polymerization. *J Agric. Food Chem.*, 46(4), 1390-1396.

Sun, X., Z. Jin, L. Yang, J. Hao, Y. Zu, W. Wang, W. Liu (2013). Ultrasonic-assisted extraction of procyanidins using ionic liquid solution from *Larix gmelinii* Bark. *Journal of Chemistry*, http://dx.doi.org/10.1155/2013/541037.

Sung H., N. J., Chun S., Park H., Yang S. E., Min W. K. (2000). In vivo antioxidant effect of green tea. *Eur. J. Clin. Nutr.*, 54, 527–529.

Sung, J., J. Lee (2010). Antioxidant and antiproliferative activities of grape seeds from different cultivars. *Food Sci. Biotechnol.*, 19(2), 321-326.

Takahashi, T., T. Kamiya, Y. Yokoo (1998). Proanthocyanidins from grape seeds promote proliferation of mouse hair follicle cells *in vitro* and convert hair cycle *in vivo*. *Acta Derm Venereol.*, 78(6), 428-432.

Thorngate, J. H., V. L. Singleton (1994). Localization of procyanidins in grape seeds. *Am Soc Enol Viticulture*, 45(2), 259-262.

Travaglia, F., M. Bordiga, M. Locatelli, J. D. Coisson, M. Arlorio (2011). Polymeric proanthocyanidins in skins and seeds of 37 *Vitis vinifera* L. cultivars: a methodological comparative study. *J. Food Sci.*, 76(5), C742-C749.

Uchida, S., R. Edamatsu, M. Hiramatsu, A. Mori, G. Nonaka, I. Nishioka, M. Niwa, M. Ozaki (1987). Condensed tannins scavenge active oxygen free-radicals. *Med. Sci. Res. Biochem.,* 15(13-14), 831-832.

van der Gaag, M. S., J. B. Ubbink, P. Sillanaukee, S. Nikkari, H. F. J. Hendriks (2000). Effect of consumption of red wine, spirits, and beer on serum homocysteine. *Lancet,* 355(9214), 1522-1522.

Vayalil, P. K., A. Mittal, S. K. Katiyar (2012). Proanthocyanidins from grape seeds inhibit expression of matrix metalloproteinases in human prostate carcinoma cells, which is associated with the inhibition of activation of MAPK and NF kappa B. *Carcinogenesis.* 33(5), 1121-1121.

Vigna, G. B., F. Costantini, G. Aldini, M. Carini, A. Catapano, F. Schena, A. Tangerini, R. Zanca, E. Bombardelli, P. Morazzoni, A. Mezzetti, R. Fellin, R. M. Facino (2003). Effect of a standardized grape seed extract on low-density lipoprotein susceptibility to oxidation in heavy smokers. *Metab. Clin. Exp.*, 52(10), 1250-1257.

Vitseva, O., S. Varghese, S. Chakrabarti, J. D. Folts, J. E. Freedman (2005). Grape seed and skin extracts inhibit platelet function and release of reactive oxygen intermediates. *J. Cardiovasc. Pharmacol.*, 46(4), 445-451.

Vivas, N., M. F. Nonier, N. V. de Gaulejac, C. Absalon, A. Bertrand, M. Mirabel (2004). Differentiation of proanthocyanidin tannins from seeds, skins and stems of grapes (*Vitis vinifera*) and heartwood of Quebracho (*Schinopsis balansae*) by matrix-assisted laser desorption/ionization time-of-flight mass spectrometry and thioacidolysis/liquid chromatography/electrospray ionization mass spectrometry. *Ana. Chim. Acta*, 513(1), 247-256.

Wang, M., L. Wang, X. J. Pan, H. Zhang (2012). Monocytic differentiation of K562 cells induced by proanthocyanidins from grape seeds. *Arch Pharm Res.*, 35(1), 129-135.

Wang, Y., J. X. Zhu, L. D. Kong, C. Yang, C. H. K. Cheng, X. Zhang (2004). Administration of procyanidins from grape seeds reduces serum uric acid levels and decreases hepatic xanthine dehydrogenase/oxidase activities in oxonate-treated mice. *Basic Clin Pharmacol Toxicol.*, 94(5), 232-237.

Williamson, G., C. Manach (2005). Bioavailability and bioefficacy of polyphenols in humans. II. Review of 93 intervention studies. *Am. J. Clin. Nutr.*, 81(1), 243S-255S.

Yamakoshi, J., M. Saito, S. Kataoka, S. Tokutake (2002). Procyanidin-rich extract from grape seeds prevents cataract formation in hereditary cataractous (ICR/f) rats. *J Agric. Food Chem.*, 50(17), 4983-4988.

Yanagida, A., T. Shoji, Y. Shibusawa (2003). Separation of proanthocyanidins by degree of polymerization by means of size-exclusion chromatography and related techniques. *J Biochem. Biophys. Meth.*, 56(1-3), 311-322.

Yang, Y., M. J. Chien (2000). Characterization of grape procyanidins using high-performance liquid chromatography/mass spectrometry and matrix-assisted laser desorption/ionization time-of-flight mass spectrometry. *J Agric. Food Chem.*, 48(9), 3990-3996.

Yigit, D., N. Yigit, A. Mavi, A. Yildirim, M. Gueleryuez (2009). Antioxidant and Antimicrobial Activities of Methanol and Water Extracts of Fruits, Leaves and Seeds of *Vitis vinifera* L. cv. Karaerik. *Asian J. Chem.*, 21(1), 183-194.

Yilmaz, E. E., E. B. Ozvural, H. Vural (2011). Extraction and identification of proanthocyanidins from grape seed (*Vitis Vinifera*) using supercritical carbon dioxide. *J. Supercrit. Fluids,* 55(3), 924-928.

Yilmaz, Y., R. T. Toledo (2004). Major flavonoids in grape seeds and skins: antioxidant capacity of catechin, epicatechin, and gallic acid. *J. Agric. Food Chem.*, 52(2), 255-260.

Young, J. F., L. O. Dragsted, J. Haraldsdottir, B. Daneshvar, M. A. Kall, S. Loft, L. Nilsson, S. E. Nielsen, B. Mayer, L. H. Skibsted, T. Huynh-Ba, A. Hermetter, B. Sandstrom (2002). Green tea extract only affects markers of oxidative status postprandially: lasting antioxidant effect of flavonoid-free diet. *Br. J. Nutr.,* 87(4), 343-355.

Zhang, F. J., J. Y. Yang, Y. H. Mou, B. S. Sun, Y. F. Ping, J. M. Wang, X. W. Bian, C. F. Wu (2009). Inhibition of U-87 human glioblastoma cell proliferation and formyl peptide receptor function by oligomer procyanidins (F2) isolated from grape seeds. *Chem Biol Interact.*, 179(2-3), 419-429.

Zhang, L., D. Mou, Y. Du (2007). Procyanidins: Extraction and micro-encapsulation. *J Sci. Food Agric.*, 87(12), 2192-2197.

Zhao, J., J. N. Wang, Y. J. Chen, R. Agarwal (1999). Anti-tumor-promoting activity of a polyphenolic fraction isolated from grape seeds in the mouse skin two-stage initiation-promotion protocol and identification of procyanidin B5-3 '-gallate as the most effective antioxidant constituent. *Carcinogenesis*, 20(9), 1737-1745.

In: Grapes
Editor: José de Sousa Câmara
ISBN: 978-1-63321-402-6

Chapter 15

Cardioprotective Effects of Grapes and Phenolic Constituents

***Bolanle C. Akinwumi*[1] *and Hope D. Anderson*[1,2,*]**
[1]Faculty of Pharmacy, University of Manitoba
[2]Canadian Centre for Agri-Food Research in Health and Medicine, St. Boniface Hospital Research Centre, Winnipeg, Canada

Abstract

A grape is a plant of the genus *Vitis*, with a variety of species cultivated around the world. Grapes are eaten raw and are also used to prepare functional foods such as wine, juice or raisins. Common table grapes (*Vitis vinifera*) grow as high climbing vines up to 35 meters with fruit occurring in clusters.

Interest in the potential health benefits of grapes was revitalized upon the observation of the French paradox, in which the incidence of cardiovascular mortality was significantly reduced among the southwestern French population despite deleterious risk factors such as high intake of saturated fat. The purported salutary effects were attributed to high consumption of red wine and stimulated research into the biological activities of red wine components, apart from ethanol, that contribute to the cardioprotection. As red wine is largely derived from grapes, the components of grape have likewise garnered attention.

Phenolic constituents of grapes are considered key components that determine the quality of wine. Located in the seed and skin, grape phenolics can be classified into flavonoid and non-flavonoid compounds. The flavonoid compounds include kaempferol-3-O-glucosides, quercetin-3-O-glucosides, quercetin, myricetin, proanthocyanidin, and anthocyanin. Non-flavonoid compounds include stilbene derivatives such as resveratrol.

As there is significant public interest in grape and grape-derived products such as grape seed extract, scientific evidence is needed to support or refute the putative health benefits of these products. A number of reports have already demonstrated cardioprotective, neuroprotective, antioxidant, anti-inflammatory, anti-cancer, anti-diabetic, anti-aging, cognitive, and antimicrobial effects of phenolic grape constituents.

[*] Corresponding author: Hope D. Anderson; email: HAnderson@sbrc.ca

Finally, there is compelling epidemiological data suggesting that consumption of grapes and grape products might lower cardiovascular risk. This review focuses on the known effects of phenolic constituents of grapes on cardiovascular disease vis-à-vis vascular function, blood pressure, platelet biology, inflammation, lipid metabolism, atherosclerosis, cardiac hypertrophy and obesity. Studies reviewed in this chapter represent escalating complexity, ranging from in vitro to in vivo studies, and to human clinical trials. Potential directions in this field are discussed.

Keywords: Grapes, Phenolic constituents, Resveratrol, Cardiovascular effects

Introduction

A grape is a plant of the genus *Vitis*, with a variety of species cultivated around the world. Grape is eaten raw and is also used to prepare functional foods such as wine, juice or raisins. Common table grapes (*Vitis vinifera*) grow as high climbing vines up to 35 meters in height with the fruit occurring in clusters.

Grapes have a long history as part of the human diet and as a food known for its potential medicinal value. They contain higher amounts of polyphenols and anthocyanins than many other fruits (Briedis, Povilaityte, Kazlauskas, & Venskutonis, 2003), and these constituents are believed to confer therapeutic effects. However, interest in the salutary effects of grapes and red wine was revitalized upon the observation of the French paradox, in which the incidence of cardiovascular mortality was significantly reduced among the southwestern French population despite deleterious risk factors such as high intake of saturated fat (*The WHO MONICA Project. A worldwide monitoring system for cardiovascular diseases: cardiovascular mortality and risk factors in selected communities.*, 1989).

A WHO study, termed Worldwide Monitoring System for Cardiovascular Diseases, revealed a lower occurrence of coronary heart disease mortality in France compared to other developed countries such as the United Kingdom and the United States (St Leger, Cochrane, & Moore, 1979). This cardioprotection was observed despite the presence of similar risk factors for coronary heart disease such as high blood pressure, high body mass index, and high cholesterol. The first attempt to explain the French paradox was by Renaud in 1992 in an epidemiological review: consumption of unsaturated fat was found to positively correlate with increased CHD mortality while wine consumption was found to be negatively correlated with CHD mortality (Renaud & de Lorgeril, 1992). Since there is relatively high wine consumption among the French population, Renaud postulated that the French paradox might be due to the ability of red wine to negate the deleterious effects of dairy fat consumption (Renaud & de Lorgeril, 1992).

The purported association between red wine consumption and cardioprotection stimulated research on the biological activities of red wine components, apart from ethanol, that might be involved. As red wine is largely derived from grapes, the components of grapes have likewise garnered attention. Furthermore, there are compelling epidemiological data suggesting that consumption of grapes and grape products might lower cardiovascular risk. In this review the cardioprotective effects of grapes, grapes products and phenolic constituents will be discussed. Potential directions in this field are also highlighted.

Phenolic Constituents of Grapes

Phenolic constituents of grapes are considered key determinants of wine quality. The color of red grapes (and wine) is mainly derived from phenolic contents such as anthocyanins (Jensen, Demiray, Egebo, & Meyer, 2008), which are located in the skin. Also, the stilbenoid resveratrol is found in greater quantities in the skin of red grapes versus white grapes. Different phenolic contents have been proposed as the basis for purported greater health benefits of red wine (Opie & Lecour, 2007). A number of studies have also shown that de-alcoholized red wine exerts more protective effects on the cardiovascular system than white wine equivalents (Agewall et al., 2000; Botden et al., 2011; Karatzi et al., 2005; Serafini, Maiani, & Ferro-Luzzi, 1998).

Grape phenolics can be broadly classified into flavonoid and non-flavonoid compounds. The flavonoid compounds include kaempferol-3-O-glucosides, quercetin-3-O-glucosides, quercetin, myricetin, proanthocyanidin, and anthocyanin. Non-flavonoid compounds include stilbene derivatives such as resveratrol, pterostilbene and piceid.

The analysis and quantification of grape polyphenols is achieved by several assay methodologies such as UV-visible absorbance spectroscopy, fluorescence spectroscopy, mass spectroscopy, and high performance liquid chromatography. In any case, the method of choice depends on the specific compound to be assayed (Lorrain, Ky, Pechamat, & Teissedre, 2013). Phenolic constituents of grapes and grape products were also rapidly detected by liquid chromatography - mass spectroscopy (LC-MS) (Xu et al., 2011) wherein anthocyanins, flavonols, phenolic acids, and flavanols/proanthocyanidin were detected (Xu et al., 2011). A survey of polyphenols in grape stems detected predominantly resveratrol, and other compounds to a lesser degree including the stilbenoid piceatannol, non stilbenoid flavon-3-ols and phenolic acid glucosides (Püssa, Floren, Kuldkepp, & Raal, 2006). Among other grape phenolic constituents, resveratrol is the most widely studied with respect to its cardioprotective effects as well as its underlying mechanisms. Resveratrol is a phytoalexin (i.e. an antimicrobial compound synthesized de novo in plants to fight against fungal infection and other stress conditions) (Langcake & Pryce, 1976).

In contrast, catechin is the predominant polyphenol in grape seed (largely cultivated in Greece), while other compounds detected in lesser amounts include gallic acid, epigallocatechin, epicatechin, epicatechin gallate, procyanidin B1 and B2 (Guendez, Kallithraka, Makris, & Kefalas, 2005). The phenolic content of grapes and wines depends on standards of viticultural practices employed in the cultivation of the grapevine. Factors such as grape variety, climate, soil, use of fertilizer and spray treatment can influence the stilbene contents of grapes (Bavaresco, 2003).

Biomedical Effects of Grapes

As there is significant public interest in grapes and grape-derived products such as grape seed extract, scientific evidence is needed to support or refute the putative health benefits of these products. A number of reports have already demonstrated cardioprotective, neuroprotective (Li, Gong, Dong, & Shi, 2012; López-Miranda et al., 2012), antioxidant (Keser, Celik, & Turkoglu, 2013), anti-inflammatory (Švajger & Jeras, 2012), anti-cancer

(Aluyen et al., 2012; Whitlock & Baek, 2012), anti-diabetic (Timmers, Hesselink, & Schrauwen, 2013), anti-aging (Queen & Tollefsbol, 2010), cognitive (Harada, Zhao, Kurihara, Nakagata, & Okajima, 2011; Sharma & Gupta, 2002), and antimicrobial (Martini et al., 2009; Radovanović, Jovančićević, Radovanović, Mihajilov-Krstev, & Zvezdanović, 2012) effects of phenolic grape constituents. Here, the cardioprotective effects of grapes, grape products and phenolic constituents are reviewed.

Cardioprotective Effects of Grapes

Cardiovascular diseases (CVD) are the leading cause of death worldwide and the burden of CVD is predicted to increase over the next few decades ("Global status report on noncommunicable disaeses 2010," 2011; Mathers & Loncar, 2006). Risk of mortality due to CVD can be reduced by adjusting modifiable risk factors such as high body mass index, central obesity, dyslipidemia, high blood pressure, elevated fasting blood glucose, unhealthy diet, tobacco smoking, and physical inactivity. Modification of risk factors can be achieved through lifestyle changes, pharmacological interventions, and possibly the use of dietary supplements and functional foods.

Grapes are an example of such functional foods as they contain polyphenols that are known for their beneficial health effects. The mechanisms underlying the cardioprotective effects of grapes and wine are not yet fully understood. However, there is scientific evidence supporting effects on vascular function, blood pressure, platelet biology, inflammation, lipid metabolism, atherosclerosis, cardiac hypertrophy and obesity. Thus the overall beneficial effects derived from grapes and grape products might be a summation of their effects on various aspects of cardiovascular physiology. Studies reviewed here, summarized in Table 1, represent escalating complexity, ranging from *in vitro* to *in vivo* studies, and to human clinical trials.

Table 1. Grape products and phenolic constituents and their effects on the cardiovascular system

Grape product / constituent	Subject	Effect on cardiovascular system	Reference
Red wine	Patients with essential hypertension	Vasodilation of small resistance arteries	(Porteri et al., 2010)
Red wine polyphenolic extract	Rats	NO release and endothelium-dependent vasorelaxation in the aortic rings pre-contracted with adrenaline	(Stoclet, Kleschyov, Andriambeloson, Diebolt, & Andriantsitohaina, 1999)
Red wine polyphenolic compound (20 mg/d; 7 days)	Rats	Reduction in blood pressure	(Diebolt, Bucher, & Andriantsitohaina, 2001)
Red wine	Human endothelial cells	Increased expression of nitric oxide synthase	(Wallerath, Poleo, Li, & Förstermann, 2003)

Grape product / constituent	Subject	Effect on cardiovascular system	Reference
Red grape polyphenolic extract (600 mg)	Patients with coronary heart disease	Acutely improved endothelial dependent vasodilation	(Lekakis et al., 2005)
Grape skin extract	Rats	Reduction in blood pressure and induction of endothelium dependent vasodilation of mesenteric artery	(Soares De Moura et al., 2002)
Resveratrol	Rats	Improved vessel compliance and vasodilation	(Behbahani et al., 2010)
Freeze dried grape extract	Rats (SHR)	Improved vessel compliance and vasodilation	(Thandapilly et al., 2012)
Grape juice	Platelets (healthy humans)	Inhibition of platelet aggregation	(Pace-Asciak, Rounova, Hahn, Diamandis, & Goldberg, 1996)
Grape skin extract			(Vitseva, Varghese, Chakrabarti, Folts, & Freedman, 2005)
Grape seed extract			(Olas, Wachowicz, Stochmal, & Oleszek, 2012; Vitseva et al., 2005)
Quercetin			(Pace-Asciak, Hahn, Diamandis, Soleas, & Goldberg, 1995)
Red wine			(Pace-Asciak et al., 1995)
Resveratrol	Platelets (healthy humans, high-risk cardiac patients)		(Olas, Wachowicz, Holmsen, & Fukami, 2005; Olas, Wachowicz, Saluk-Juszczak, & Zieliński, 2002; Pace-Asciak et al., 1995; Stef, Csiszar, Lerea, Ungvari, & Veress, 2006; Zbikowska, Olas, Wachowicz, & Krajewski, 1999)
Proanthocyanidin	Different cell types	Protection against oxidative stress	(Bagchi et al., 2000; Shao, Becker, et al., 2003)
Resveratrol		Free radical scavenging	(Hwang, Kwon, Park, & Kim, 2008)
Red grape juice			(Tenore et al., 2012)
Resveratrol	Rats	Protection against ischemia reperfusion injury	(Ray et al., 1999; M. Shen, Jia, Wang, & Ma, 2006; M. Shen et al., 2012; Xiao et al., 2007)
Proanthocyanidin			(Shao et al., 2009)
Red grape juice and wine	Humans	Inhibition of plasma LDL oxidation	(Miyagi, Miwa, & Inoue, 1997)
Red grape juice	HepG2, HL-60 cell lines	Inhibition of endocytotic LDL uptake, reducing free cholesterol levels	(Dávalos et al., 2006)

Table 1. (Continued)

Grape product / constituent	Subject	Effect on cardiovascular system	Reference
Resveratrol	Human macrophages	Attenuates lipid accumulation	(Voloshyna, Hai, Littlefield, Carsons, & Reiss, 2013)
Pterostilbene	HUVEC cells, mice	Protects vascular endothelial cells against oxidized LDL-induced apoptosis	(Zhang et al., 2012)
Resveratrol	Isolated cardiomyocytes, aortic banded rats	Inhibits cardiac hypertrophy	(Chan et al., 2008; Juric, Wojciechowski, Das, & Netticadan, 2007; Thandapilly et al., 2011)
Resveratrol	Mature adipocytes	Inhibits adipogenesis, attenuates lipid accumulation and increases lipolysis in	(Baile et al., 2011)
Resveratrol (0.04%)	Mice	Increased life span of mice fed with high caloric diet	(Baur et al., 2006)
Resveratrol (150 mg/day; 30 d)	Obese humans	Mimic effects if caloric restriction	(Timmers et al., 2011)
Resveratrol	Non-human primates	Suppress body mass gain, increase metabolic rates and total energy expenditure	(Dal-Pan et al., 2011)
Lyophilized grape powder (36 g/d; 4 weeks)	Post-menopausal women	Reduced triglycerides and LDL-cholesterol levels, inflammation, and oxidative stress	(Zern et al., 2005)
Resveratrol (7 mg/d; 12 weeks)	Patients at risk for CVD; patients with stable CAD	Improved inflammatory and fibrinolytic status	(Tomé-Carneiro, Gonzálvez, Larrosa, García-Almagro, et al., 2012; Tomé-Carneiro, Gonzálvez, Larrosa, Yáñez-Gascón, et al., 2012; Tomé-Carneiro et al., 2013)

Vascular Function and Effect on Blood Pressure

Hypertension is a leading risk factor for the incidence of heart disease (Kannel, 1996). Red wine causes vasodilation (Porteri et al., 2010), induces endothelial nitric oxide release (Stoclet et al., 1999), and reduces blood pressure (Diebolt et al., 2001). These effects have been attributed to the phenolic constituents of red wine. For instance, total anthocyanin content of red wines has been correlated with their vasodilatory activities (Burns et al., 2000). Nitric oxide (NO) is a potent vasodilator (Kunes et al., 2004; Rees, Palmer, & Moncada, 1989) and as such an important pharmacological drug target. Red wine specifically from France up-regulates nitric oxide synthase (NOS) in endothelial cells (Wallerath et al., 2003). Red wine polyphenol extract also from southern France was found to increase expression of endothelial NOS (eNOS) and consequently increased release of NO in human umbilical vein endothelial cells (HUVEC) (Leikert et al., 2002). De Moura et al. proposed that the vasodilatory effects of Brazilian red wine may be due to release of NO and/or endothelial derived hyperpolarizing factor (EDHF). The role of NOS was further implicated by the ability of L-NAME, a NOS inhibitor, to attenuate the vasodilatory effect of the Brazilian red wine (de Moura et al., 2004). Red wine polyphenols improve endothelial function (Lekakis et al., 2005; Perez-Vizcaino, Duarte, & Andriantsitohaina, 2006) and thus modulate vasoconstrictive and vasodilatory factors in stressed normotensive rats, while maintaining blood pressure despite exposure to stress (Púzserová, Csizmadiová, Andriantsitohaina, & Bernátová, 2006). In addition, the red wine polyphenol, Provinol, prevented L-NAME-induced hypertension via up-regulation of NOS (Pechánová et al., 2004).

Vascular compliance is the ability of blood vessels to buffer changes in blood pressure. Resveratrol (2.5 mg/kg/day for 12 weeks) increased mesenteric small artery compliance and reduced wall stiffness in normotensive Wistar Kyoto rats, while it attenuated arterial compliance in spontaneously hypertensive rat (SHR), at least in part through inhibitory actions on pro-growth ERK signaling (Behbahani et al., 2010). In an animal model of hypertension, freeze dried grape extract (600 mg/day for 10 weeks) was also found to significantly reduce blood pressure and improve mesenteric small arterial relaxation and compliance (Thandapilly et al., 2012).

In summary, by promoting vasodilation and increasing vessel compliance, grapes and phenolic constituents may reduce vascular resistance and regulate blood pressure (López-Sepúlveda et al., 2008; Soares De Moura et al., 2002).

Platelet Biology

Platelets maintain homeostasis via formation of thrombi in order to prevent excessive blood loss when the vascular endothelium is compromised. Among others factors, activation of platelet aggregation is a major contributor to the development of atherothrombosis. In response to rupture of an unstable atherosclerotic plaque, platelets are activated, resulting in thrombus formation.(Freynhofer, Bruno, Wojta, & Huber, 2012). The detached plaque and the resulting thrombus may enter the systemic circulation and cause occlusion of blood vessels, thereby limiting blood supply to organs such as the heart (myocardial infarction) or the brain (stroke). Thus, inhibition of platelet activity is an important strategy to prevent such thrombotic events (Ling & Ovbiagele, 2009; Silva, Dusse, Vieira, & Carvalho, 2013).

Grape seed extracts (Olas et al., 2012), skin extracts (Vitseva et al., 2005), grape juice (Pace-Asciak et al., 1996), red wine (Pace-Asciak et al., 1995; Pace-Asciak et al., 1996) as well as individual phenolic grape constituents such as resveratrol (Olas et al., 2005; Olas et al., 2002; Stef et al., 2006) reportedly inhibit platelet aggregation. Moreover, the combination of grape seed and grape skin extracts exhibit greater antiplatelet action than each alone (Shanmuganayagam et al., 2002).

While the exact antiplatelet mechanisms of grape phenolic constituents are not yet well defined, numerous mechanisms have been proposed. Resveratrol (50 μg/mL) inhibits platelet aggregation induced by collagen, epinephrine, and thromboxane *in vitro* and these effects may be attributable to suppression of cyclooxygenase-1 in the arachidonic acid pathway (Zbikowska et al., 1999). Other mechanisms that may be involved include inhibition of the MAPK pathway, activation of the NO/cGMP pathway (Shen et al., 2007) and inhibition of phosphoinositide signaling (Olas et al., 2005).

Oxidative Stress

The antioxidant activity of grape polyphenols is widely established and is thought to be central to their cardioprotective effects (Göktürk Baydar, Özkan, & Yaşar, 2007; Shao, Becker, et al., 2003; Shao et al., 2009). Both red grape juice (Tenore et al., 2012) and red wine (Tenore et al., 2013) scavenge free radical markers such as DPPH (2,2-diphenyl-1-picrylhydrazyl) and also protect cardiomyocytes from doxorubicin-induced oxidant injury *in vitro*. Protection is dose dependent at lower doses (5 - 50 μg/ml), however at high concentrations (100 – 500 μg/ml), grape seed proanthocyanidin extracts exert pro-oxidant effects and lead to cardiomyocyte apoptosis (Lluís et al., 2011; Shao et al., 2006; Shao, Vanden Hoek, et al., 2003). Of particular interest is the antioxidant effect of grape seed proanthocyanin and anthocyanin (Bagchi et al., 2000). Such antioxidant and cardioprotective effects have also been demonstrated *in vivo* by grape seed extracts in animals (Busserolles et al., 2006) and by concentrated red grape juice in humans (Castilla et al., 2006).

Pre-administration of quercetin (0.033 mg/Kg/day for 4 days) protects rats against cardiac ischemia-reperfusion injury by improving diastolic and systolic cardiac function and possibly enhancing mitochondrial manganese superoxide dismutase activity (Brookes, Digerness, Parks, & Darley-Usmar, 2002). Mitochondria account for the generation of approximately 70% of total reactive oxygen species (ROS), due to continuous oxidative metabolism and the electron transport chain. Resveratrol (100 μM) also attenuates ischemia-reperfusion injury in neonatal cardiomyocytes exposed to 2 h simulated ischemia and 4 h reperfusion possibly by decreasing intracellular calcium, preventing apoptosis and enhancing activities of ROS scavenging enzymes such as superoxide dismutase (Ray et al., 1999; M. Shen et al., 2012). Other mechanisms that might confer anti-oxidant effects of resveratrol reportedly include modulation of the mitochondrial membrane permeability transition pore (mPTP) (Xi, Wang, Mueller, Norfleet, & Xu, 2009), activation of AMPK (Hwang et al., 2008), induction of nitric oxide synthase (Bradamante et al., 2003; Hattori, Otani, Maulik, & Das, 2002; M. Shen et al., 2006), and restoration of gap junction proteins such as connexin 43 and the M3 receptor (Xiao et al., 2007).

Atherosclerosis

Of the five types of lipoproteins, there are two major types that function as carriers of cholesterol in the systemic circulation: high density lipoprotein (HDL) and low density lipoproteins (LDL). High levels of HDL in the circulatory system are considered "good" while high levels of LDL are "bad." Due to their low molecular weight, a high level of LDL predisposes to lipid accumulation in the arterial wall, leading to atherogenic processes. Furthermore, LDL oxidation plays an important role in atherogenesis. Oxidized LDL promotes accumulation of inflammatory cells such as macrophages causing build-up of plaques on the vessel wall. Therefore, suppression of LDL oxidation is an important anti-atherosclerotic therapeutic target. Flavonoids from grape juice and red wine for example inhibit plasma oxidation of LDL in humans (Miyagi et al., 1997).

Red grape juice polyphenols also inhibit endocytic LDL uptake by human hepatocarcinoma HepG2 and promyelocytic HL-60 cell lines *in vitro* thus preventing intracellular processing of LDL and release of free cholesterol, possibly by preventing fusion of lysosomes and endosomes (Dávalos et al., 2006). Indeed, red grape juice was shown to up-regulate LDL receptors in a manner that is independent of antioxidant properties since other antioxidant polyphenols such as quercetin fail to affect LDL receptor expression (Dávalos et al., 2006). Furthermore, by reducing intracellular LDL in cultured HepG2 cells, red wine polyphenols purportedly activate feedback mechanisms such as up-regulation of LDL receptors and increase transcription and translation of HMG-CoA reductase, which is involved in cholesterol synthesis (Pal et al., 2003).

Anti-atherogenic effects of grape polyphenols may extend beyond modulation of lipid handling. Resveratrol modifies vascular function (Behbahani et al., 2010; Ou et al., 2006), attenuates lipid accumulation (Voloshyna et al., 2013; Zang et al., 2006),and modulates gene expression related to lipogenesis and lipolysis, whereas pterostilbene protects vascular endothelial cells against apoptosis (Zhang et al., 2012). Coupled with their inhibitory effect on platelet aggregation and lipid metabolism, grape polyphenols further protect the heart by delaying the initiation and progression of atherosclerosis.

Cardiac Hypertrophy

Cardiac hypertrophy is an increase in heart muscle cell size induced by myocardial injury or hemodynamic stress (Parker, Patterson, & Johnson, 2005). This increase in size was first thought to be compensatory leading to an increase in cardiac output (Meerson, 1961). However, it is now well known that heart enlargement is decompensatory and decreases the functionality of the heart over time (Berenji, Drazner, Rothermel, & Hill, 2005; Ho, Pinsky, Kannel, & Levy, 1993; Levy, Garrison, Savage, Kannel, & Castelli, 1990). Resveratrol prevents cardiac hypertrophy in isolated norepinephrine-treated cardiomyocytes as well as in the pressure overload model of cardiac hypertrophy (Chan et al., 2008; Juric et al., 2007; Thandapilly et al., 2011).One possible antihypertrophic mechanism is via activation of NO-AMP-activated protein kinase (AMPK) signaling (Chan et al., 2008). AMPK is an intracellular sensor of energy balance, and AMPK activation during high energy demand in cardiomyocytes triggers metabolic changes that lead to conservation of energy in the cells (Hardie, 2011). AMPK is activated by upstream phosphorylation at Thr172 by the AMPK kinase LKB1 (Beauloye, Bertrand, Horman, & Hue, 2011; Witters, Kemp, & Means, 2006). Activation of AMPK using a pharmacological activator prevents the development of hypertrophy in cultured neonatal cardiomyocytes (Chen et al., 2010).Specifically in the heart, AMPK signals increased fatty acid uptake and oxidation, increased glucose uptake, increased glycolysis and shut down energy consuming pathways such as protein synthesis (Dolinsky & Dyck, 2006).

Resveratrol induces mitochondrial biogenesis and as such improves the bioenergetics of heart muscle cells by reducing cardiac fatty acid oxidation and increasing mitochondrial protein expression (Biala et al., 2010; Rimbaud et al., 2011). Both red wine polyphenolic extracts (Al-Awwadi et al., 2004) and dealcoholized red wine (Palfi et al., 2009) prevent cardiac hypertrophy by inhibiting ROS generation. Furthermore, resveratrol protects against damage by ROS via activation of AMPK in H9c2 cardiac muscle cells (Hwang et al., 2008). Research into the protective effects of resveratrol and related polyphenols is still ongoing with the aim to exploit their ability to improve mitochondrial function, inhibit cardiac hypertrophy and prevent progression of heart injuries to heart failure.

Obesity

Obesity is characterized by excessive adipose tissue caused by increased caloric intake and or inadequate energy expenditure. As defined, body mass index greater than 30 and waist circumference greater than 88 cm and 94 cm in women and men respectively (*Obesity: preventing and managing the world epidemic*, 2000) is one of the major risk factors for the development of CVD, and weight management is part of the typically recommended life style modifications. Resveratrol inhibits adipogenesis, attenuates lipid accumulation and increases lipolysis in mature adipocytes (Baile et al., 2011). The lifespan of mice fed high caloric (60% fat) diets supplemented with 0.04% resveratrol was found to be lengthened (Baur et al., 2006). Furthermore, administration of 150 mg/day trans resveratrol over 30 days mimics the effect of caloric restriction in obese humans and thus may help to control obesity and other metabolic syndromes (Timmers et al., 2011). Similarly, resveratrol delayed age-related abnormalities albeit without significant effect on longevity in non-obese mice (Pearson et al., 2008). This indicates that resveratrol may help counter the adverse metabolic effects of obesity. In non-human primates, resveratrol was shown to suppress body mass gain, increase metabolic rates and total energy expenditure (Dal-Pan, Blanc, & Aujard, 2010; Dal-Pan et al., 2011).

Since resveratrol activates AMPK, and AMPK is involved in energy balance regulation and mitochondrial biogenesis, AMPK was proposed as a possible mechanism by which resveratrol protects against metabolic dysfunction. (Um et al., 2010). In AMPK-deficient mice, resveratrol failed to improve insulin sensitivity possibly due to non-effect on mitochondrial content and fatty acid oxidation in skeletal muscles. This caused a build-up of lipids that may inhibit insulin action (Um et al., 2010).

Furthermore, resveratrol has been shown to activate sirtuins (SIRT) (Howitz et al., 2003). SIRT1 is a NAD+-dependent protein deacetylase that, upon activation, activates peroxisome proliferator activator gamma co-activator 1 α (PGC-1α) via deacetylation and thus improves mitochondrial function and oxidative capacity. In high fat fed mice, resveratrol improved glucose tolerance by enhancing sensitivity to insulin, while protecting the mice against development of obesity (Lagouge et al., 2006). This beneficial effect of resveratrol was lost when the acetylation sites of PGC-1 α were mutated or when SIRT expression was disrupted using SIRT-deficient mice, thus revealing the role of PGC-1α-SIRT signaling (Lagouge et al., 2006).

Clinical Trials

In addition to the pre-clinical reports described above, there also exists evidence of grape-associated cardioprotection in humans. For example Lekakis et al. administered 600 mg of a red grapes polyphenolic extract (dissolved in 20 mL of water) to patients with coronary heart disease and found improved vasodilation of the brachial arteries compared to placebo (Lekakis et al., 2005). The extract contained phenolic constituents such as epicatechin, catechin, gallic acid, trans-resveratrol, rutin, epsilon-viniferin, p-coumaric acid, ferulic acid and quercetin (Lekakis et al., 2005). Furthermore, in pre and post-menopausal women, lyophilized grape powder (36 g/d for 4 weeks) reduced triglycerides and LDL-cholesterol levels, inflammation, and oxidative stress although LDL oxidation was not affected (Zern et al., 2005).

Despite the promising effects described above for grape preparations or extracts, the efficacy of specific grape polyphenols such as resveratrol is yet to be established in humans. For example, resveratrol administration (7 mg/d for 12 weeks n=15) to non-obese healthy women did not change metabolic parameters such as insulin sensitivity, mitochondrial function, AMPK signaling and inflammatory markers (Yoshino et al., 2012). The absence of effect of resveratrol on metabolic parameters in normal subjects is consistent with earlier results obtained in normal rodents. (Juan, Vinardell, & Planas, 2002; Strong et al., 2013). Thus, we may infer that resveratrol exerts its effects only in metabolic disease conditions such as obesity, type II diabetes and dyslipidemia. Effects of resveratrol administration on the primary and secondary prevention of CVD have shown some promise (Tomé-Carneiro, Gonzálvez, Larrosa, García-Almagro, et al., 2012; Tomé-Carneiro, Gonzálvez, Larrosa, Yáñez-Gascón, et al., 2012; Tomé-Carneiro et al., 2013) but small sample size (n=75) and short follow-up (1 year) limits the clinical relevance of the promising outcomes. For effects on primary prevention of CVD, resveratrol-containing grape extract (8.1 mg/d for the first 6 months and 16.2 mg/d for the next 6 months) was administered to patients with high risk for CVD in addition to other recommended preventive guidelines (Tomé-Carneiro, Gonzálvez, Larrosa, Yáñez-Gascón, et al., 2012). The resveratrol-treated patients who were also on statins showed improved inflammatory and fibrinolytic status compared to placebo and grape-extract only treated group (Tomé-Carneiro, Gonzálvez, Larrosa, Yáñez-Gascón, et al., 2012). Thus resveratrol may complement current treatment guidelines for the primary prevention of CVD. (Tomé-Carneiro, Gonzálvez, Larrosa, García-Almagro, et al., 2012; Tomé-Carneiro, Gonzálvez, Larrosa, Yáñez-Gascón, et al., 2012).

To determine the effect on secondary prevention of CVD, resveratrol-containing grape extract (8.1 mg/d for the first 6 months and 16.2 mg/d for the next 6 months) was administered to patients with stable coronary artery disease in addition to their regular medications and dietary restrictions (Tomé-Carneiro et al., 2013). Compared to the placebo and grape extract only group, there was an increase in anti-inflammatory adiponectin and a decrease in thrombogenic plasminogen activator inhibitor type 1 (PAI-1) in the group that received resveratrol (Tomé-Carneiro et al., 2013). Thus the authors concluded that resveratrol may exert its cardioprotective effects by improving anti-inflammatory response and preventing atherothrombotic signaling (Tomé-Carneiro et al., 2013).

As of 2013, a quick survey of the US clinical trial database revealed several clinical trials aimed at studying the effects of resveratrol and grape seed extracts on various conditions such as abdominal obesity, diastolic heart failure, hypertension and metabolic syndrome to name a few ("Clinicaltrials.gov," 2013). It is hoped that the results of these clinical trials will add to the evidence base supporting the therapeutic potential of grape-derived products.

Other Considerations

Resveratrol is the most widely studied grape polyphenol and is currently on the market. The possibility of drug interactions should be considered before routine use of resveratrol. For instance, administration a high dose of resveratrol (1 g/d for 4 weeks) to healthy volunteers inhibited cytochrome P450 (CYP isoforms 3A4, 2D6, 2C9) enzymes, (Chow et al., 2010) and this could influence plasma concentrations of drugs metabolized. Poor oral bioavailability is a major obstacle to achieving efficacy in humans.(Marier et al., 2002; Walle, Hsieh, DeLegge, Oatis, & Walle, 2004) Upon absorption, resveratrol is extensively metabolized to the sulfate and glucuronide conjugates.(Walle, 2011) There is also speculation that the abundance of certain metabolites such as resveratrol 3-sulfate and 3-glucuronide may contribute to additional in vivo effects.(Marier et al., 2002; Walle et al., 2004) Thus, the role of these metabolites needs to be probed further.

Toxicological data show that at low doses, resveratrol is well-tolerated in humans, although safety information upon long term administration is still lacking.(Cottart, Nivet-Antoine, Laguillier-Morizot, & Beaudeux, 2010) Furthermore, the optimum dose of resveratrol required to achieve cardioprotection remains to be unequivocally established. Evidence of cardioprotection from long-term randomized clinical trials is still limited and the clinical significance of efficacy in the management of CVD remains unclear. For instance, most of the resveratrol human studies are short term and were performed on small sample sizes.

The longest study so far, published in 2013, is by Tomé-Carneiro J. et al., lasted for 1 year and involved 75 patients with cardiovascular disease.(Tomé-Carneiro et al., 2013) Therefore additional randomized clinical studies, with larger population samples and longer follow-up periods will be important before resveratrol and other grape constituents might be used in the context of CVD.

Conclusion

In conclusion, the cardioprotective effects of grapes and their phenolic constituents are multi-faceted and it is believed that different mechanisms work in tandem to achieve an overall beneficial effect on the heart. Regular and moderate intake of red wine is recommended as part of a healthy diet (*Dietary Guidelines for Americans*, 2010). Daily intake of alcohol should not exceed 1 drink for women and 1-2 drinks for men (*Dietary Guidelines for Americans*, 2010). Consumption of grapes along with other fruit and vegetables is also recommended since they are sources of antioxidants that might protect the body against damage related to oxidative stress. Scientific evidence supporting the cardioprotective effects of grapes and grape products discussed in this review include effects on vascular function, blood pressure, platelet biology, inflammation, lipid metabolism, atherosclerosis, cardiac hypertrophy and obesity. By controlling the progression of chronic metabolic diseases, grapes products and resveratrol in particular may slow aging and extend life span.

Various resveratrol products are already being marketed for weight loss, better general health and longer life span. There is thus a need for public education about the effects of these products as they are largely available for purchase over the counter. Consumers should be made aware that these dosage forms are not yet standardized, evidence for efficacy is not well-established and there is potential for interaction with drugs and other dietary supplements.

References

Agewall, S., Wright, S., Doughty, R. N., Whalley, G. A., Duxbury, M., & Sharpe, N. (2000). Does a glass of red wine improve endothelial function? *Eur. Heart J., 21*(1), 74-78. doi: 10.1053/euhj.1999.1759.

Al-Awwadi, N. A., Bornet, A., Azay, J., Araiz, C., Delbosc, S., Cristol, J. P., . . . Teissedre, P. L. (2004). Red wine polyphenols alone or in association with ethanol prevent hypertension, cardiac hypertrophy, and production of reactive oxygen species in the insulin-resistant fructose-fed rat. *J. Agric. Food Chem., 52*(18), 5593-5597. doi: 10.1021/jf049295g.

Aluyen, J. K., Ton, Q. N., Tran, T., Yang, A. E., Gottlieb, H. B., & Bellanger, R. A. (2012). Resveratrol: potential as anticancer agent. *J. Diet Suppl., 9*(1), 45-56. doi: 10.3109/19390211.2011.650842.

Bagchi, D., Bagchi, M., Stohs, S. J., Das, D. K., Ray, S. D., Kuszynski, C. A., . . . Pruess, H. G. (2000). Free radicals and grape seed proanthocyanidin extract: importance in human health and disease prevention. *Toxicology, 148*(2-3), 187-197.

Baile, C. A., Yang, J. Y., Rayalam, S., Hartzell, D. L., Lai, C. Y., Andersen, C., & Della-Fera, M. A. (2011). Effect of resveratrol on fat mobilization. *Ann. N Y Acad. Sci., 1215*, 40-47. doi: 10.1111/j.1749-6632.2010.05845.x.

Baur, J. A., Pearson, K. J., Price, N. L., Jamieson, H. A., Lerin, C., Kalra, A., . . . Sinclair, D. A. (2006). Resveratrol improves health and survival of mice on a high-calorie diet. *Nature, 444*(7117), 337-342. doi: 10.1038/nature05354.

Bavaresco, L. (2003). Role of viticultural factors on stilbene concentrations of grapes and wine. *Drugs Exp. Clin. Res., 29*(5-6), 181-187.

Beauloye, C., Bertrand, L., Horman, S., & Hue, L. (2011). AMPK activation, a preventive therapeutic target in the transition from cardiac injury to heart failure. *Cardiovasc. Res., 90*(2), 224-233. doi: 10.1093/cvr/cvr034.

Behbahani, J., Thandapilly, S. J., Louis, X. L., Huang, Y., Shao, Z., Kopilas, M. A., . . . Anderson, H. D. (2010). Resveratrol and small artery compliance and remodeling in the spontaneously hypertensive rat. *Am. J. Hypertens, 23*(12), 1273-1278. doi: 10.1038/ajh.2010.161.

Berenji, K., Drazner, M. H., Rothermel, B. A., & Hill, J. A. (2005). Does load-induced ventricular hypertrophy progress to systolic heart failure? *Am. J. Physiol. Heart Circ. Physiol., 289*(1), H8-H16. doi: 10.1152/ajpheart.01303.2004.

Biala, A., Tauriainen, E., Siltanen, A., Shi, J., Merasto, S., Louhelainen, M., . . . Mervaala, E. (2010). Resveratrol induces mitochondrial biogenesis and ameliorates Ang II-induced cardiac remodeling in transgenic rats harboring human renin and angiotensinogen genes. *Blood Press, 19*(3), 196-205. doi: 10.3109/08037051.2010.481808.

Botden, I. P., Langendonk, J. G., Meima, M. E., Boomsma, F., Seynhaeve, A. L., ten Hagen, T. L., . . . Sijbrands, E. J. (2011). Daily red wine consumption improves vascular function by a soluble guanylyl cyclase-dependent pathway. *Am. J. Hypertens, 24*(2), 162-168. doi: 10.1038/ajh.2010.227.

Bradamante, S., Barenghi, L., Piccinini, F., Bertelli, A. A., De Jonge, R., Beemster, P., & De Jong, J. W. (2003). Resveratrol provides late-phase cardioprotection by means of a nitric oxide- and adenosine-mediated mechanism. *Eur. J. Pharmacol., 465*(1-2), 115-123.

Briedis, V., Povilaityte, V., Kazlauskas, S., & Venskutonis, P. R. (2003). [Polyphenols and anthocyanins in fruits, grapes juices and wines, and evaluation of their antioxidant activity]. *Medicina (Kaunas), 39 Suppl 2*, 104-112.

Brookes, P. S., Digerness, S. B., Parks, D. A., & Darley-Usmar, V. (2002). Mitochondrial function in response to cardiac ischemia-reperfusion after oral treatment with quercetin. *Free Radic. Biol. Med., 32*(11), 1220-1228.

Burns, J., Gardner, P. T., O'Neil, J., Crawford, S., Morecroft, I., McPhail, D. B., . . . Crozier, A. (2000). Relationship among antioxidant activity, vasodilation capacity, and phenolic content of red wines. *J. Agric. Food Chem., 48*(2), 220-230.

Busserolles, J., Gueux, E., Balasińska, B., Piriou, Y., Rock, E., Rayssiguier, Y., & Mazur, A. (2006). In vivo antioxidant activity of procyanidin-rich extracts from grape seed and pine (Pinus maritima) bark in rats. *Int. J. Vitam. Nutr. Res., 76*(1), 22-27.

Castilla, P., Echarri, R., Dávalos, A., Cerrato, F., Ortega, H., Teruel, J. L., . . . Lasunción, M. A. (2006). Concentrated red grape juice exerts antioxidant, hypolipidemic, and antiinflammatory effects in both hemodialysis patients and healthy subjects. *Am. J. Clin. Nutr., 84*(1), 252-262.

Chan, A. Y., Dolinsky, V. W., Soltys, C. L., Viollet, B., Baksh, S., Light, P. E., & Dyck, J. R. (2008). Resveratrol inhibits cardiac hypertrophy via AMP-activated protein kinase and Akt. *J. Biol. Chem., 283*(35), 24194-24201. doi: 10.1074/jbc.M802869200.

Chen, B. L., Ma, Y. D., Meng, R. S., Xiong, Z. J., Wang, H. N., Zeng, J. Y., . . . Dong, Y. G. (2010). Activation of AMPK inhibits cardiomyocyte hypertrophy by modulating of the FOXO1/MuRF1 signaling pathway in vitro. *Acta Pharmacol. Sin., 31*(7), 798-804. doi: 10.1038/aps.2010.73.

Chow, H. H., Garland, L. L., Hsu, C. H., Vining, D. R., Chew, W. M., Miller, J. A., . . . Alberts, D. S. (2010). Resveratrol modulates drug- and carcinogen-metabolizing enzymes in a healthy volunteer study. *Cancer Prev. Res. (Phila), 3*(9), 1168-1175. doi: 10.1158/1940-6207.CAPR-09-0155.

Clinicaltrials.gov. (2013). from http://clinicaltrials.gov/ct2/home

Cottart, C. H., Nivet-Antoine, V., Laguillier-Morizot, C., & Beaudeux, J. L. (2010). Resveratrol bioavailability and toxicity in humans. *Mol. Nutr. Food Res., 54*(1), 7-16. doi: 10.1002/mnfr.200900437.

Dal-Pan, A., Blanc, S., & Aujard, F. (2010). Resveratrol suppresses body mass gain in a seasonal non-human primate model of obesity. *BMC Physiol., 10*, 11. doi: 10.1186/1472-6793-10-11.

Dal-Pan, A., Terrien, J., Pifferi, F., Botalla, R., Hardy, I., Marchal, J., . . . Aujard, F. (2011). Caloric restriction or resveratrol supplementation and ageing in a non-human primate: first-year outcome of the RESTRIKAL study in Microcebus murinus. *Age (Dordr), 33*(1), 15-31. doi: 10.1007/s11357-010-9156-6.

Dávalos, A., Fernández-Hernando, C., Cerrato, F., Martínez-Botas, J., Gómez-Coronado, D., Gómez-Cordovés, C., & Lasunción, M. A. (2006). Red grape juice polyphenols alter cholesterol homeostasis and increase LDL-receptor activity in human cells in vitro. *J. Nutr., 136*(7), 1766-1773.

de Moura, R. S., Miranda, D. Z., Pinto, A. C., Sicca, R. F., Souza, M. A., Rubenich, L. M., . . . Resende, A. C. (2004). Mechanism of the endothelium-dependent vasodilation and the antihypertensive effect of Brazilian red wine. *J. Cardiovasc. Pharmacol., 44*(3), 302-309.

Diebolt, M., Bucher, B., & Andriantsitohaina, R. (2001). Wine polyphenols decrease blood pressure, improve NO vasodilatation, and induce gene expression. *Hypertension, 38*(2), 159-165.

Dietary Guidelines for Americans. (2010). Washington, DC: US government printing office Retrieved from www.dietaryguidelines.gov.

Dolinsky, V. W., & Dyck, J. R. (2006). Role of AMP-activated protein kinase in healthy and diseased hearts. *Am. J. Physiol. Heart Circ. Physiol., 291*(6), H2557-2569. doi: 10.1152/ajpheart.00329.2006.

Freynhofer, M. K., Bruno, V., Wojta, J., & Huber, K. (2012). The role of platelets in athero-thrombotic events. *Curr. Pharm Des., 18*(33), 5197-5214.

Global status report on noncommunicable disaeses 2010. (2011). Geneva: World Health Organization.

Göktürk Baydar, N., Özkan, G., & Yaşar, S. (2007). Evaluation of the antiradical and antioxidant potential of grape extracts. *Food Control, 18*(9), 1131-1136. doi: http://dx.doi.org/10.1016/j.foodcont.2006.06.011.

Guendez, R., Kallithraka, S., Makris, D. P., & Kefalas, P. (2005). An analytical survey of the polyphenols of seeds of varieties of grape (Vitis vinifera) cultivated in Greece: implications for exploitation as a source of value-added phytochemicals. *Phytochem. Anal., 16*(1), 17-23. doi: 10.1002/pca.804.

Harada, N., Zhao, J., Kurihara, H., Nakagata, N., & Okajima, K. (2011). Resveratrol improves cognitive function in mice by increasing production of insulin-like growth factor-I in the hippocampus. *J. Nutr. Biochem., 22*(12), 1150-1159. doi: 10.1016/j.jnutbio.2010.09.016.

Hardie, D. G. (2011). AMP-activated protein kinase: an energy sensor that regulates all aspects of cell function. *Genes Dev, 25*(18), 1895-1908. doi: 10.1101/gad.17420111.

Hattori, R., Otani, H., Maulik, N., & Das, D. K. (2002). Pharmacological preconditioning with resveratrol: role of nitric oxide. *Am. J. Physiol. Heart Circ. Physiol., 282*(6), H1988-1995. doi: 10.1152/ajpheart.01012.2001.

Ho, K. K., Pinsky, J. L., Kannel, W. B., & Levy, D. (1993). The epidemiology of heart failure: the Framingham Study. *J. Am. Coll. Cardiol., 22*(4 Suppl A), 6A-13A.

Howitz, K. T., Bitterman, K. J., Cohen, H. Y., Lamming, D. W., Lavu, S., Wood, J. G., . . . Sinclair, D. A. (2003). Small molecule activators of sirtuins extend Saccharomyces cerevisiae lifespan. *Nature, 425*(6954), 191-196. doi: 10.1038/nature01960.

Hwang, J. T., Kwon, D. Y., Park, O. J., & Kim, M. S. (2008). Resveratrol protects ROS-induced cell death by activating AMPK in H9c2 cardiac muscle cells. *Genes Nutr., 2*(4), 323-326. doi: 10.1007/s12263-007-0069-7.

Jensen, J. S., Demiray, S., Egebo, M., & Meyer, A. S. (2008). Prediction of wine color attributes from the phenolic profiles of red grapes (Vitis vinifera). *J. Agric. Food Chem., 56*(3), 1105-1115. doi: 10.1021/jf072541e.

Juan, M. E., Vinardell, M. P., & Planas, J. M. (2002). The daily oral administration of high doses of trans-resveratrol to rats for 28 days is not harmful. *J. Nutr., 132*(2), 257-260.

Juric, D., Wojciechowski, P., Das, D. K., & Netticadan, T. (2007). Prevention of concentric hypertrophy and diastolic impairment in aortic-banded rats treated with resveratrol. *Am. J. Physiol. Heart Circ. Physiol., 292*(5), H2138-2143. doi: 10.1152/ajpheart.00852.2006.

Kannel, W. B. (1996). Blood pressure as a cardiovascular risk factor: prevention and treatment. *JAMA, 275*(20), 1571-1576.

Karatzi, K. N., Papamichael, C. M., Karatzis, E. N., Papaioannou, T. G., Aznaouridis, K. A., Katsichti, P. P., . . . Mavrikakis, M. E. (2005). Red wine acutely induces favorable effects on wave reflections and central pressures in coronary artery disease patients. *Am. J. Hypertens, 18*(9 Pt 1), 1161-1167. doi: 10.1016/j.amjhyper.2005.03.744.

Keser, S., Celik, S., & Turkoglu, S. (2013). Total phenolic contents and free-radical scavenging activities of grape (Vitis vinifera L.) and grape products. *Int. J. Food Sci. Nutr., 64*(2), 210-216. doi: 10.3109/09637486.2012.728199.

Kunes, J., Hojná, S., Kadlecová, M., Dobesová, Z., Rauchová, H., Vokurková, M., . . . Zicha, J. (2004). Altered balance of vasoactive systems in experimental hypertension: the role of relative NO deficiency. *Physiol. Res., 53 Suppl 1*, S23-34.

Lagouge, M., Argmann, C., Gerhart-Hines, Z., Meziane, H., Lerin, C., Daussin, F., . . . Auwerx, J. (2006). Resveratrol improves mitochondrial function and protects against metabolic disease by activating SIRT1 and PGC-1alpha. *Cell, 127*(6), 1109-1122. doi: 10.1016/j.cell.2006.11.013.

Langcake, P., & Pryce, R. J. (1976). The production of resveratrol by Vitis vinifera and other members of the Vitaceae as a response to infection or injury. *Physiological Plant Pathology, 9*(1), 77-86. doi: http://dx.doi.org/10.1016/0048-4059(76)90077-1.

Leikert, J. F., Räthel, T. R., Wohlfart, P., Cheynier, V., Vollmar, A. M., & Dirsch, V. M. (2002). Red wine polyphenols enhance endothelial nitric oxide synthase expression and subsequent nitric oxide release from endothelial cells. *Circulation, 106*(13), 1614-1617.

Lekakis, J., Rallidis, L. S., Andreadou, I., Vamvakou, G., Kazantzoglou, G., Magiatis, P., . . . Kremastinos, D. T. (2005). Polyphenolic compounds from red grapes acutely improve endothelial function in patients with coronary heart disease. *Eur. J. Cardiovasc. Prev. Rehabil, 12*(6), 596-600.

Levy, D., Garrison, R. J., Savage, D. D., Kannel, W. B., & Castelli, W. P. (1990). Prognostic implications of echocardiographically determined left ventricular mass in the Framingham Heart Study. *N. Engl. J. Med, 322*(22), 1561-1566. doi: 10.1056/NEJM199005313222203.

Li, F., Gong, Q., Dong, H., & Shi, J. (2012). Resveratrol, a neuroprotective supplement for Alzheimer's disease. *Curr Pharm Des, 18*(1), 27-33.

Ling, G., & Ovbiagele, B. (2009). Oral antiplatelet therapy in the secondary prevention of atherothrombotic events. *Am. J. Cardiovasc. Drugs, 9*(3), 197-209.

Lluís, L., Muñoz, M., Nogués, M. R., Sánchez-Martos, V., Romeu, M., Giralt, M., . . . Solà, R. (2011). Toxicology evaluation of a procyanidin-rich extract from grape skins and seeds. *Food Chem. Toxicol., 49*(6), 1450-1454. doi: 10.1016/j.fct.2011.03.042.

López-Miranda, V., Soto-Montenegro, M. L., Vera, G., Herradón, E., Desco, M., & Abalo, R. (2012). [Resveratrol: a neuroprotective polyphenol in the Mediterranean diet]. *Rev. Neurol., 54*(6), 349-356.

López-Sepúlveda, R., Jiménez, R., Romero, M., Zarzuelo, M. J., Sánchez, M., Gómez-Guzmán, M., . . . Duarte, J. (2008). Wine polyphenols improve endothelial function in large vessels of female spontaneously hypertensive rats. *Hypertension, 51*(4), 1088-1095. doi: 10.1161/HYPERTENSIONAHA.107.107672.

Lorrain, B., Ky, I., Pechamat, L., & Teissedre, P. L. (2013). Evolution of analysis of polyhenols from grapes, wines, and extracts. *Molecules, 18*(1), 1076-1100. doi: 10.3390/molecules18011076.

Marier, J. F., Vachon, P., Gritsas, A., Zhang, J., Moreau, J. P., & Ducharme, M. P. (2002). Metabolism and disposition of resveratrol in rats: extent of absorption, glucuronidation, and enterohepatic recirculation evidenced by a linked-rat model. *J. Pharmacol. Exp. Ther., 302*(1), 369-373.

Martini, S., D'Addario, C., Braconi, D., Bernardini, G., Salvini, L., Bonechi, C., . . . Rossi, C. (2009). Antibacterial activity of grape extracts on cagA-positive and -negative Helicobacter pylori clinical isolates. *J. Chemother, 21*(5), 507-513.

Mathers, C. D., & Loncar, D. (2006). Projections of global mortality and burden of disease from 2002 to 2030. *PLoS Med, 3*(11), e442. doi: 10.1371/journal.pmed.0030442.

Meerson, F. Z. (1961). On the mechanism of compensatory hyperfunction and insufficiency of the heart. *Cor Vasa, 3*, 161-177.

Miyagi, Y., Miwa, K., & Inoue, H. (1997). Inhibition of human low-density lipoprotein oxidation by flavonoids in red wine and grape juice. *Am. J. Cardiol., 80*(12), 1627-1631.

Obesity: preventing and managing the world epidemic. (2000). Geneva, Switzerland: World Health Organization.

Olas, B., Wachowicz, B., Holmsen, H., & Fukami, M. H. (2005). Resveratrol inhibits polyphosphoinositide metabolism in activated platelets. *Biochim Biophys Acta, 1714*(2), 125-133. doi: 10.1016/j.bbamem.2005.06.008.

Olas, B., Wachowicz, B., Saluk-Juszczak, J., & Zieliński, T. (2002). Effect of resveratrol, a natural polyphenolic compound, on platelet activation induced by endotoxin or thrombin. *Thromb. Res., 107*(3-4), 141-145.

Olas, B., Wachowicz, B., Stochmal, A., & Oleszek, W. (2012). The polyphenol-rich extract from grape seeds inhibits platelet signaling pathways triggered by both proteolytic and non-proteolytic agonists. *Platelets, 23*(4), 282-289. doi: 10.3109/09537104.2011.618562.

Opie, L. H., & Lecour, S. (2007). The red wine hypothesis: from concepts to protective signalling molecules. *Eur. Heart J., 28*(14), 1683-1693. doi: 10.1093/eurheartj/ehm149.

Ou, H. C., Chou, F. P., Sheen, H. M., Lin, T. M., Yang, C. H., & Huey-Herng Sheu, W. (2006). Resveratrol, a polyphenolic compound in red wine, protects against oxidized LDL-induced cytotoxicity in endothelial cells. *Clin. Chim. Acta, 364*(1-2), 196-204. doi: 10.1016/j.cccn.2005.06.018.

Pace-Asciak, C. R., Hahn, S., Diamandis, E. P., Soleas, G., & Goldberg, D. M. (1995). The red wine phenolics trans-resveratrol and quercetin block human platelet aggregation and eicosanoid synthesis: implications for protection against coronary heart disease. *Clin. Chim. Acta, 235*(2), 207-219.

Pace-Asciak, C. R., Rounova, O., Hahn, S. E., Diamandis, E. P., & Goldberg, D. M. (1996). Wines and grape juices as modulators of platelet aggregation in healthy human subjects. *Clin. Chim. Acta, 246*(1-2), 163-182.

Pal, S., Ho, N., Santos, C., Dubois, P., Mamo, J., Croft, K., & Allister, E. (2003). Red wine polyphenolics increase LDL receptor expression and activity and suppress the secretion of ApoB100 from human HepG2 cells. *J. Nutr., 133*(3), 700-706.

Palfi, A., Bartha, E., Copf, L., Mark, L., Gallyas, F., Veres, B., . . . Sumegi, B. (2009). Alcohol-free red wine inhibits isoproterenol-induced cardiac remodeling in rats by the regulation of Akt1 and protein kinase C alpha/beta II. *J. Nutr. Biochem., 20*(6), 418-425. doi: 10.1016/j.jnutbio.2008.04.009.

Parker, P., Patterson, J., & Johnson, J. (2005). *Pharmacotherapy. A Pathophysiologic Approach: Heart Failure.* (Sixth ed.): The McGraw-Hill Companies, Inc.

Pearson, K. J., Baur, J. A., Lewis, K. N., Peshkin, L., Price, N. L., Labinskyy, N., . . . de Cabo, R. (2008). Resveratrol delays age-related deterioration and mimics transcriptional aspects of dietary restriction without extending life span. *Cell Metab., 8*(2), 157-168. doi: 10.1016/j.cmet.2008.06.011.

Pechánová, O., Bernátová, I., Babál, P., Martínez, M. C., Kyselá, S., Stvrtina, S., & Andriantsitohaina, R. (2004). Red wine polyphenols prevent cardiovascular alterations in L-NAME-induced hypertension. *J. Hypertens, 22*(8), 1551-1559.

Perez-Vizcaino, F., Duarte, J., & Andriantsitohaina, R. (2006). Endothelial function and cardiovascular disease: effects of quercetin and wine polyphenols. *Free Radic. Res., 40*(10), 1054-1065. doi: 10.1080/10715760600823128.

Porteri, E., Rizzoni, D., De Ciuceis, C., Boari, G. E., Platto, C., Pilu, A., . . . Agabiti Rosei, E. (2010). Vasodilator effects of red wines in subcutaneous small resistance artery of patients with essential hypertension. *Am. J. Hypertens, 23*(4), 373-378. doi: 10.1038/ajh.2009.280.

Püssa, T., Floren, J., Kuldkepp, P., & Raal, A. (2006). Survey of grapevine Vitis vinifera stem polyphenols by liquid chromatography-diode array detection-tandem mass spectrometry. *J. Agric. Food Chem., 54*(20), 7488-7494. doi: 10.1021/jf061155e.

Púzserová, A., Csizmadiová, Z., Andriantsitohaina, R., & Bernátová, I. (2006). Vascular effects of red wine polyphenols in chronic stress-exposed Wistar-Kyoto rats. *Physiol. Res., 55 Suppl 1*, S39-47.

Queen, B. L., & Tollefsbol, T. O. (2010). Polyphenols and aging. *Curr. Aging Sci., 3*(1), 34-42.

Radovanović, A. N., Jovančićević, B. S., Radovanović, B. C., Mihajilov-Krstev, T., & Zvezdanović, J. B. (2012). Antioxidant and antimicrobial potentials of Serbian red wines produced from international Vitis vinifera grape varieties. *J. Sci. Food Agric., 92*(10), 2154-2161. doi: 10.1002/jsfa.5601.

Ray, P. S., Maulik, G., Cordis, G. A., Bertelli, A. A., Bertelli, A., & Das, D. K. (1999). The red wine antioxidant resveratrol protects isolated rat hearts from ischemia reperfusion injury. *Free Radic. Biol. Med., 27*(1-2), 160-169.

Rees, D. D., Palmer, R. M., & Moncada, S. (1989). Role of endothelium-derived nitric oxide in the regulation of blood pressure. *Proc. Natl. Acad. Sci. USA, 86*(9), 3375-3378.

Renaud, S., & de Lorgeril, M. (1992). Wine, alcohol, platelets, and the French paradox for coronary heart disease. *Lancet, 339*(8808), 1523-1526.

Rimbaud, S., Ruiz, M., Piquereau, J., Mateo, P., Fortin, D., Veksler, V., . . . Ventura-Clapier, R. (2011). Resveratrol improves survival, hemodynamics and energetics in a rat model of hypertension leading to heart failure. *PLoS One, 6*(10), e26391. doi: 10.1371/journal.pone.0026391

Serafini, M., Maiani, G., & Ferro-Luzzi, A. (1998). Alcohol-free red wine enhances plasma antioxidant capacity in humans. *J. Nutr., 128*(6), 1003-1007.

Shanmuganayagam, D., Beahm, M. R., Osman, H. E., Krueger, C. G., Reed, J. D., & Folts, J. D. (2002). Grape seed and grape skin extracts elicit a greater antiplatelet effect when used in combination than when used individually in dogs and humans. *J. Nutr., 132*(12), 3592-3598.

Shao, Z. H., Becker, L. B., Vanden Hoek, T. L., Schumacker, P. T., Li, C. Q., Zhao, D., . . . Yuan, C. S. (2003). Grape seed proanthocyanidin extract attenuates oxidant injury in cardiomyocytes. *Pharmacol Res, 47*(6), 463-469.

Shao, Z. H., Hsu, C. W., Chang, W. T., Waypa, G. B., Li, J., Li, D., . . . Hoek, T. L. (2006). Cytotoxicity induced by grape seed proanthocyanidins: role of nitric oxide. *Cell Biol. Toxicol., 22*(3), 149-158. doi: 10.1007/s10565-006-0052-x.

Shao, Z. H., Vanden Hoek, T. L., Xie, J., Wojcik, K., Chan, K. C., Li, C. Q., . . . Yuan, C. S. (2003). Grape seed proanthocyanidins induce pro-oxidant toxicity in cardiomyocytes. *Cardiovasc. Toxicol., 3*(4), 331-339.

Shao, Z. H., Wojcik, K. R., Dossumbekova, A., Hsu, C., Mehendale, S. R., Li, C. Q., . . . Hoek, T. L. (2009). Grape seed proanthocyanidins protect cardiomyocytes from ischemia and reperfusion injury via Akt-NOS signaling. *J. Cell Biochem., 107*(4), 697-705. doi: 10.1002/jcb.22170.

Sharma, M., & Gupta, Y. K. (2002). Chronic treatment with trans resveratrol prevents intracerebroventricular streptozotocin induced cognitive impairment and oxidative stress in rats. *Life Sci., 71*(21), 2489-2498.

Shen, M., Jia, G. L., Wang, Y. M., & Ma, H. (2006). Cardioprotective effect of resvaratrol pretreatment on myocardial ischemia-reperfusion induced injury in rats. *Vascul Pharmacol., 45*(2), 122-126. doi: 10.1016/j.vph.2006.04.002.

Shen, M., Wu, R. X., Zhao, L., Li, J., Guo, H. T., Fan, R., . . . Pei, J. M. (2012). Resveratrol attenuates ischemia/reperfusion injury in neonatal cardiomyocytes and its underlying mechanism. *PLoS One, 7*(12), e51223. doi: 10.1371/journal.pone.0051223.

Shen, M. Y., Hsiao, G., Liu, C. L., Fong, T. H., Lin, K. H., Chou, D. S., & Sheu, J. R. (2007). Inhibitory mechanisms of resveratrol in platelet activation: pivotal roles of p38 MAPK and NO/cyclic GMP. *Br. J. Haematol., 139*(3), 475-485. doi: 10.1111/j.1365-2141.2007.06788.x.

Silva, M. V., Dusse, L. M., Vieira, L. M., & Carvalho, M. D. (2013). Platelet antiaggregants in primary and secondary prevention of atherothrombotic events. *Arq. Bras. Cardiol., 100*(6), e78-e84. doi: 10.5935/abc.20130122.

Soares De Moura, R., Costa Viana, F. S., Souza, M. A., Kovary, K., Guedes, D. C., Oliveira, E. P., . . . Gusmão Correia, M. L. (2002). Antihypertensive, vasodilator and antioxidant effects of a vinifera grape skin extract. *J. Pharm. Pharmacol., 54*(11), 1515-1520. doi: 10.1211/002235702153.

St Leger, A. S., Cochrane, A. L., & Moore, F. (1979). Factors associated with cardiac mortality in developed countries with particular reference to the consumption of wine. *Lancet, 1*(8124), 1017-1020.

Stef, G., Csiszar, A., Lerea, K., Ungvari, Z., & Veress, G. (2006). Resveratrol inhibits aggregation of platelets from high-risk cardiac patients with aspirin resistance. *J. Cardiovasc. Pharmacol., 48*(2), 1-5. doi: 10.1097/01.fjc.0000238592.67191.ab.

Stoclet, J. C., Kleschyov, A., Andriambeloson, E., Diebolt, M., & Andriantsitohaina, R. (1999). Endothelial no release caused by red wine polyphenols. *J. Physiol. Pharmacol., 50*(4), 535-540.

Strong, R., Miller, R. A., Astle, C. M., Baur, J. A., de Cabo, R., Fernandez, E., . . . Harrison, D. E. (2013). Evaluation of resveratrol, green tea extract, curcumin, oxaloacetic acid, and medium-chain triglyceride oil on life span of genetically heterogeneous mice. *J. Gerontol. A Biol. Sci. Med. Sci., 68*(1), 6-16. doi: 10.1093/gerona/gls070.

Švajger, U., & Jeras, M. (2012). Anti-inflammatory effects of resveratrol and its potential use in therapy of immune-mediated diseases. *Int. Rev. Immunol., 31*(3), 202-222. doi: 10.3109/08830185.2012.665108.

Tenore, G. C., Manfra, M., Stiuso, P., Coppola, L., Russo, M., Gomez Monterrey, I. M., & Campiglia, P. (2012). Antioxidant profile and in vitro cardiac radical-scavenging versus pro-oxidant effects of commercial red grape juices (Vitis vinifera L. cv. Aglianico N.). *J. Agric. Food Chem., 60*(38), 9680-9687. doi: 10.1021/jf301647d.

Tenore, G. C., Manfra, M., Stiuso, P., Coppola, L., Russo, M., Ritieni, A., & Campiglia, P. (2013). Polyphenolic pattern and in vitro cardioprotective properties of typical red wines from vineyards cultivated in Scafati (Salerno, Italy). *Food Chem., 140*(4), 803-809. doi: 10.1016/j.foodchem.2012.10.023.

Thandapilly, S. J., LeMaistre, J. L., Louis, X. L., Anderson, C. M., Netticadan, T., & Anderson, H. D. (2012). Vascular and cardiac effects of grape powder in the spontaneously hypertensive rat. *Am. J. Hypertens, 25*(10), 1070-1076. doi: 10.1038/ajh.2012.98.

Thandapilly, S. J., Louis, X. L., Yang, T., Stringer, D. M., Yu, L., Zhang, S., . . . Netticadan, T. (2011). Resveratrol prevents norepinephrine induced hypertrophy in adult rat cardiomyocytes, by activating NO-AMPK pathway. *Eur. J. Pharmacol., 668*(1-2), 217-224. doi: 10.1016/j.ejphar.2011.06.042.

Timmers, S., Hesselink, M. K., & Schrauwen, P. (2013). Therapeutic potential of resveratrol in obesity and type 2 diabetes: new avenues for health benefits? *Ann. N Y Acad. Sci., 1290*(1), 83-89. doi: 10.1111/nyas.12185.

Timmers, S., Konings, E., Bilet, L., Houtkooper, R. H., van de Weijer, T., Goossens, G. H., . . . Schrauwen, P. (2011). Calorie restriction-like effects of 30 days of resveratrol supplementation on energy metabolism and metabolic profile in obese humans. *Cell Metab., 14*(5), 612-622. doi: 10.1016/j.cmet.2011.10.002.

Tomé-Carneiro, J., Gonzálvez, M., Larrosa, M., García-Almagro, F. J., Avilés-Plaza, F., Parra, S., . . . Espín, J. C. (2012). Consumption of a grape extract supplement containing resveratrol decreases oxidized LDL and ApoB in patients undergoing primary prevention of cardiovascular disease: a triple-blind, 6-month follow-up, placebo-controlled, randomized trial. *Mol. Nutr. Food Res., 56*(5), 810-821. doi: 10.1002/mnfr.201100673.

Tomé-Carneiro, J., Gonzálvez, M., Larrosa, M., Yáñez-Gascón, M. J., García-Almagro, F. J., Ruiz-Ros, J. A., . . . Espín, J. C. (2012). One-year consumption of a grape nutraceutical containing resveratrol improves the inflammatory and fibrinolytic status of patients in primary prevention of cardiovascular disease. *Am. J. Cardiol., 110*(3), 356-363. doi: 10.1016/j.amjcard.2012.03.030.

Tomé-Carneiro, J., Gonzálvez, M., Larrosa, M., Yáñez-Gascón, M. J., García-Almagro, F. J., Ruiz-Ros, J. A., . . . Espín, J. C. (2013). Grape resveratrol increases serum adiponectin and downregulates inflammatory genes in peripheral blood mononuclear cells: a triple-blind, placebo-controlled, one-year clinical trial in patients with stable coronary artery disease. *Cardiovasc. Drugs Ther., 27*(1), 37-48. doi: 10.1007/s10557-012-6427-8.

Um, J. H., Park, S. J., Kang, H., Yang, S., Foretz, M., McBurney, M. W., . . . Chung, J. H. (2010). AMP-activated protein kinase-deficient mice are resistant to the metabolic effects of resveratrol. *Diabetes, 59*(3), 554-563. doi: 10.2337/db09-0482.

Vitseva, O., Varghese, S., Chakrabarti, S., Folts, J. D., & Freedman, J. E. (2005). Grape seed and skin extracts inhibit platelet function and release of reactive oxygen intermediates. *J. Cardiovasc. Pharmacol., 46*(4), 445-451.

Voloshyna, I., Hai, O., Littlefield, M. J., Carsons, S., & Reiss, A. B. (2013). Resveratrol mediates anti-atherogenic effects on cholesterol flux in human macrophages and endothelium via PPARγ and adenosine. *Eur. J. Pharmacol., 698*(1-3), 299-309. doi: 10.1016/j.ejphar.2012.08.024.

Walle, T. (2011). Bioavailability of resveratrol. *Ann. N Y Acad. Sci., 1215*, 9-15. doi: 10.1111/j.1749-6632.2010.05842.x.

Walle, T., Hsieh, F., DeLegge, M. H., Oatis, J. E., & Walle, U. K. (2004). High absorption but very low bioavailability of oral resveratrol in humans. *Drug Metab. Dispos, 32*(12), 1377-1382. doi: 10.1124/dmd.104.000885.

Wallerath, T., Poleo, D., Li, H., & Förstermann, U. (2003). Red wine increases the expression of human endothelial nitric oxide synthase: a mechanism that may contribute to its beneficial cardiovascular effects. *J. Am. Coll. Cardiol., 41*(3), 471-478.

Whitlock, N. C., & Baek, S. J. (2012). The anticancer effects of resveratrol: modulation of transcription factors. *Nutr. Cancer, 64*(4), 493-502. doi: 10.1080/01635581.2012.667862.

The WHO MONICA Project. A worldwide monitoring system for cardiovascular diseases: cardiovascular mortality and risk factors in selected communities. (1989). Geneva: World Health Organization.

Witters, L. A., Kemp, B. E., & Means, A. R. (2006). Chutes and Ladders: the search for protein kinases that act on AMPK. *Trends Biochem. Sci., 31*(1), 13-16. doi: 10.1016/j.tibs.2005.11.009.

Xi, J., Wang, H., Mueller, R. A., Norfleet, E. A., & Xu, Z. (2009). Mechanism for resveratrol-induced cardioprotection against reperfusion injury involves glycogen synthase kinase 3beta and mitochondrial permeability transition pore. *Eur. J. Pharmacol., 604*(1-3), 111-116. doi: 10.1016/j.ejphar.2008.12.024.

Xiao, J., Yue, P., Wang, Y., Zhang, Y., Huo, R., Wang, N., . . . Yang, B. F. (2007). [Resveratrol restored the structural and functional association between M3 receptor and connexin 43 gap junction proteins in ischemia-reperfusion injury of isolated rat heart]. *Yao Xue Xue Bao, 42*(1), 19-25.

Xu, Y., Simon, J. E., Welch, C., Wightman, J. D., Ferruzzi, M. G., Ho, L., . . . Wu, Q. (2011). Survey of polyphenol constituents in grapes and grape-derived products. *J. Agric. Food Chem., 59*(19), 10586-10593. doi: 10.1021/jf202438d.

Yoshino, J., Conte, C., Fontana, L., Mittendorfer, B., Imai, S., Schechtman, K. B., . . . Klein, S. (2012). Resveratrol supplementation does not improve metabolic function in nonobese women with normal glucose tolerance. *Cell Metab., 16*(5), 658-664. doi: 10.1016/j.cmet.2012.09.015.

Zang, M., Xu, S., Maitland-Toolan, K. A., Zuccollo, A., Hou, X., Jiang, B., . . . Cohen, R. A. (2006). Polyphenols stimulate AMP-activated protein kinase, lower lipids, and inhibit accelerated atherosclerosis in diabetic LDL receptor-deficient mice. *Diabetes, 55*(8), 2180-2191. doi: 10.2337/db05-1188.

Zbikowska, H. M., Olas, B., Wachowicz, B., & Krajewski, T. (1999). Response of blood platelets to resveratrol. *Platelets, 10*(4), 247-252. doi: 10.1080/09537109976103.

Zern, T. L., Wood, R. J., Greene, C., West, K. L., Liu, Y., Aggarwal, D., . . . Fernandez, M. L. (2005). Grape polyphenols exert a cardioprotective effect in pre- and postmenopausal women by lowering plasma lipids and reducing oxidative stress. *J. Nutr., 135*(8), 1911-1917.

Zhang, L., Zhou, G., Song, W., Tan, X., Guo, Y., Zhou, B., . . . Chen, L. (2012). Pterostilbene protects vascular endothelial cells against oxidized low-density lipoprotein-induced apoptosis in vitro and in vivo. *Apoptosis, 17*(1), 25-36. doi: 10.1007/s10495-011-0653-6.

In: Grapes
Editor: José de Sousa Câmara
ISBN: 978-1-63321-402-6

Chapter 16

Vitis vinifera Seed Extract Reduces Adenosine Deaminase Activity in High Glucose-Treated Human Lymphocytes

Luziane P. Bellé[1], Paula Eliete R. Bitencourt[1], Gustavo Scola[3], Mirian Salvador,[3] Claúdia Funchal[4] and Maria B. Moretto[1,2,1*]

[1]Postgraduate Program in Pharmacology Health Sciences Center, Federal University of Santa Maria (UFSM), Santa Maria, RS, Brazil
[2]Postgraduate Program in Pharmaceutical Sciences Health Sciences Center, Federal University of Santa Maria (UFSM), Santa Maria, RS, Brazil
[3]Institute of Biotechnology, University of Caxias do Sul, Caxias do Sul, RS, Brazil
[4]Postgraduate Program in Biosciences and Rehabilitation, Methodist University Center, IPA, Porto Alegre, RS, Brazil

Abstract

The consumption of grape or grape products can provide health benefits in a number of pathologies such as diabetes. Thus, the purpose of the study was to evaluate, in vitro, Adenosine deaminase (ADA) and Dipeptidyl peptidase IV, (DPP-IV) activities in lymphocytes subjected to hyperglycemic conditions and to test the effects of the aqueous grape seed extract (GSE) of *Vitis vinifera* (cv. Merlot) in this system. Blood was collected from 20-30 year old healthy subjects (n=8). The lymphocytes were isolated and exposed to different concentrations of glucose (5, 30, 50 and 100mM) and/or 300μg/mL of aqueous *Vitis vinifera* (cv. Merlot) seed extract for three hours. Thereafter, ADA and DPP-IV activities were evaluated. We observed an increase in ADA activity when lymphocytes were exposed to the highest concentration of glucose (100mM). No alteration in DPP-IV activity was observed. The grape seed extract was able to prevent the increase in ADA activity. The results indicate that aqueous seed extract of *Vitis vinifera* has hypoglycemiant properties when tested in vitro.

[1] * Corresponding Author address: Email:beatriz@smail.ufsm.br.

Keywords: Adenosine deaminase, dipeptidyl peptidase IV, grape seed extract, hyperglycemia, lymphocytes

Introduction

Adenosine deaminase (ADA; EC 3.5.4.4), an enzyme involved in purine metabolism, is present in the surface (ecto-ADA) and cytoplasm of many cell types (Dong et al. 1996). Previous studies from our research group demonstrated that ADA activity is altered in hyperglycemic conditions, in vitro (De Bona et al. 2010, 2011; Boop et al. 2009). In peripheral mononuclear blood cells, the main function of ADA is to regulate the levels of extracellular adenosine, which could be toxic to lymphocytes. Furthermore, ADA is important in the acute and protracted inflammatory responses (Dong et al. 1996).

Extracellularly, ADA interacts with membrane proteins, such as dipeptidyl peptidase IV (CD26/DPP-IV; EC 3.4.14.5), that is is a cell surface glycoprotein expressed on different cell types, including lymphocytes (Franco et al. 1997). As an exopeptidase, it cleaves dipeptides from polypeptides (e.g., chemokines, neuropeptides and hormones), and appears to play a key role in a number of aspects of lymphocyte function, including immunoregulation Gorrell et al. 2001). ADA and DPP-IV bind to lymphocytes and exert costimulatory effects mediating T cell activation Kameoka et al. 1993).

Grapes are known to be a natural source of notable bioactive compounds in particular antioxidants. Epidemiological studies have associated the consumption of grapes and their derived products such as red wine and juice with a wide variety of health beneficial effects (Pérez-Jimenez and Saura-Calixto, 2008). Several studies have demonstrated that grape extracts are effective anti-promoters of the cancer process (Singh et al. 2004; Durak et al. 2005; Asou et al. 2002), suppress lipid peroxidation in hypoxic ischemic brain (Feng et al. 2005) and is known that grape extracts have anti-hyperglycemic effects (Orhan et al. 2006; Hogan et al. 2010). Recently, Schmatz et al. (2013) report that grape juice prevented the increase of ADA activity in platelets in streptozotocin-induced diabetic rats (Schmatz et al. 2013).

Nevertheless, to our knowledge there is no data on the effects of grape extracts in lymphocytes exposed to high glucose levels. To address this issue, the objectives of this study were to evaluate, in vitro, ADA and DPP-IV activities in lymphocytes subjected to hyperglycemic conditions and to test the possible beneficial effects of the aqueous grape seed extract (GSE) of *Vitis vinifera* (cv. Merlot) in this system.

Methods

Chemicals

Adenosine was obtained from Merck (Darmstadt, Germany). Ficoll–Histopaque and Gly-Pro ρ-nitroanilide ρ-toluenesulfonate were obtained from Sigma Chemical Co. (St. Louis, MO, USA). All other chemicals were of analytical grade and were obtained from standard commercial suppliers.

Grape Seed Extract

Seeds from winery wastes of *Vitis vinifera* (cv. Merlot) were removed from the vinification tanks five days after the beginning of fermentation in January 2006. The variety was cultivated in the northeast region of Serra Gaucha, state of Rio Grande do Sul, Brazil. Voucher specimen was identified by the herbarium of the University of Caxias do Sul, Brazil (HUCS32455). The GSE and the phenolic content were performed according to Scola et al., (2010).

Isolation of Lymphocytes from Human Blood

Lymphocytes were isolated from human donors (n=8) between 20-30 years old. The research has been approved by the Human Ethics Committee of the Federal University of Santa Maria under protocol number 23081.004068-76. Blood was collected in Vacutainer tubes (BD Vacutainer, Franklin Lakes, NJ, USA) containing lithium heparin and separated on Ficoll–Histopaque density gradients, as described by Böyum et al. (1968). Cell number and viability were determined by trypan blue exclusion. Final cell suspension was performed in phosphate-buffered saline (PBS, pH 7.4) and cell concentration was adjusted to 10^6 cells/mL. Lymphocytes (3×10^5 cells/mL) were incubated for three hours with fresh glucose at the concentrations of 5, 30, 50 and 100mM and/or 300µg/mL of V*itis vinifera* seed extract. The glucose and GSE concentrations used in the present study were based on previous reports by Lunkes et al. (2008) and on previous experiments performed by our group, respectively. The concentration of 5mM was used as control because it mimics the normal physiological glycemia. Afterwards, lymphocytes were washed with PBS and used for the enzymatic assays.

ADA Activity Assay

ADA activity was estimated spectrophotometrically by the method of Giusti (1974). Lymphocytes were added to the reaction mixture containing substrate (21mM of adenosine) and the reaction was carried out for 60 min at 37°C. The reaction was stopped and the absorbance was measured at 620nm. Ammonium sulphate (75µM) was used as ammonium standard and the values were expressed in U/L.

DPP-IV Activity Assay

DPP-IV activity was determined as described by Schön et al., (1984). Lymphocytes were incubated with 0.6mM of Gly-Pro ρ-nitroanilide ρ-toluenesulfonate for 120 min at 37°C. The reaction was stopped by adding 1M of acetate buffer (pH 4.5) Absorbance was measured at 390nm and 0.5mM of ρ-nitroanilide was used as standard. The values were expressed in U/L.

Statistical Analysis

Data were analyzed by one-way analysis of variance (ANOVA) followed by the Tukey´s test. The results were expressed as mean ± Standard Error of the Mean (SEM) and the differences were considered statistically significant when p<0.05. All analyses were carried out Statistical Package for Social Sciences (SPSS) software.

Results

We observed an increase in ADA activity when lymphocytes were exposed to different concentrations of glucose. The enzymatic activity was statistically significant ($p < 0.01$) at the concentration of 100mM when compared to 5mM (physiological conditions) and this increase occurred in a dose-dependent manner (F=25.08; β= 0.0516; P=0.0376). However, GSE was able to prevent this effect in ADA activity at the highest concentration of glucose (Figure 1).

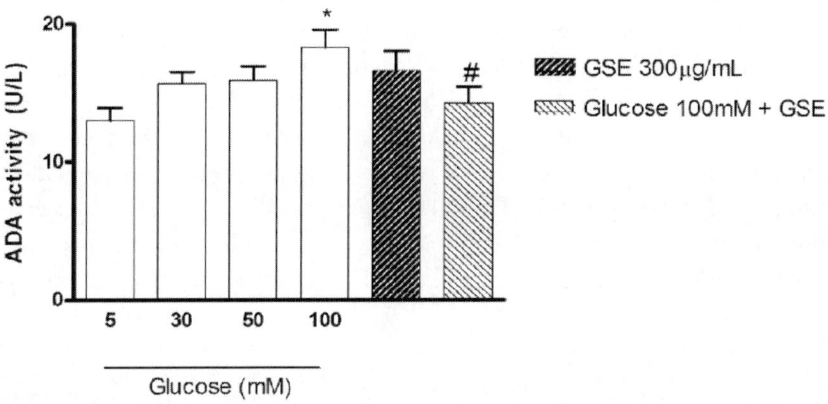

Figure 1. In vitro effect of glucose and/or GSE on ADA activity in lymphocytes. Values are expressed as mean ± sem (n=8). * p<0.01 compared with 5mm of glucose and # p<0.05 compared with 100mm of glucose.

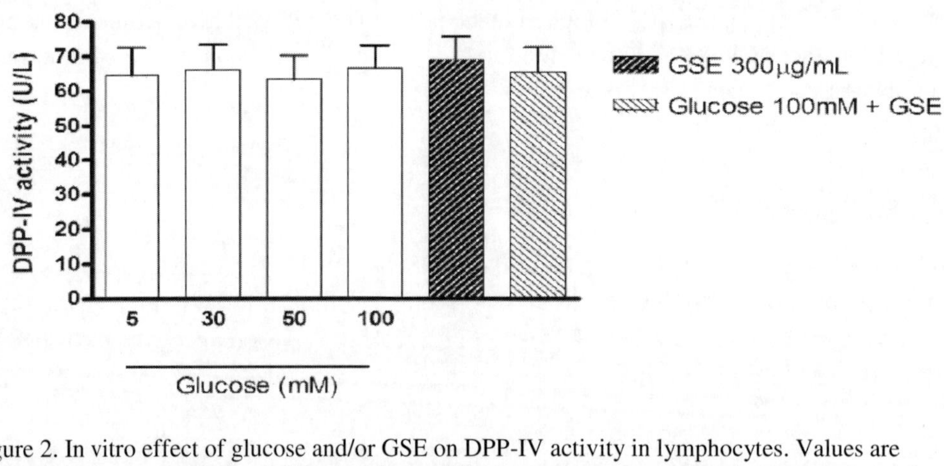

Figure 2. In vitro effect of glucose and/or GSE on DPP-IV activity in lymphocytes. Values are expressed as mean ± sem (n=8). No significant differences were seen among the groups.

On the other hand,no alteration was observed in DPP-IV activity of lymphocytes exposed to glucose and/or GSE (Figure 2).

Discussion

The results of the present investigation indicated for the first time that ADA activity increased when lymphocytes were exposed to different concentrations of glucose and that GSE was able to prevent this effect. The increase observed in ADA activity is likely to be a compensatory mechanism to degrade adenosine, which plays an important role in the regulation of many metabolic processes, including in mediating inflammatory responses and modulation of insulin action on glucose metabolism.

Recently, Kocbuch et al. (2009) observed that adenosine levels were significantly higher in human T and B lymphocytes subjected to high glucose concentrations. Moreover, high glucose levels augment their capacity to stimulate T cell proliferation and IL-2 and IL-12 secretion (Yao et al. 2006). In this context, both cytokines up-regulate ADA at the cell surface and also regulate the translocation of intracellular ADA enzyme toward the cell surface (Cordero et al. 2009).

In our study, GSE avoided the increase of lymphocyte ADA activity at the highest concentration of glucose. This effect is probably related to oxidative stress, because there is a correlation between increased ADA activity and reactive species generation (Köse et al. 2001). Furthermore, Stanković et al. (2008) observed that GSE may be effective in the prevention of oxidative damage. To our knowledge, this is the first study that demonstrated the benefic effect of aqueous GSE. Grapes contain a high concentration and a great variety of polyphenols, such as proanthocyanidins that is known to exhibit antioxidant, anti-inflammatory and immune-stimulating activities (Rice-Evans et al. 1996; Salah et al. 1995) and could be, at least in part, responsible for the effects observed in the present study.

Importantly, neither high glucose concentrations nor GSE altered lymphocyte DDP-IV activity. In a previous study developed by our group, no alteration in lymphocyte DPP-IV activity of type 2 diabetic patients was observed (Bellé et al. 2011). However, Pala et al. (2003) showed that high glucose concentrations increase endothelial DPP-IV activity.

In conclusion, high glucose concentrations in a short time are able to increase lymphocyte ADA activity, whereas DPP-IV activity remained unaltered. In addition to the antioxidant, anti-inflammatory and immune-stimulating properties of GSE, *Vitis vinifera* seed extract was able to prevent the increase in ADA activity. Thus, the aqueous GSE might act the regulation of pathological conditions caused by hyperglycemic state.

Acknowledgments

The authors are grateful to Régis Adriel Zanette for the English assistance. The first author thanks CAPES (Fundação Coordenação de Aperfeiçoamento de Pessoal de Nível Superior) for the scholarship.

References

Asou, H., Koshizuka, K., Kyo, T., Takata, N., Kamada, N., Koeffier, H.P., 2002. Resveratrol, a natural product derived from grapes, is a new inducer of differentiation in human myeloid leukemias. *Int J Hematol.* 75, 528-33.

Bellé, L.P., Bitencourt, P.E.R., De Bona, K.S., Zanette, R.A., Moresco, R.N., Moretto, M.B., 2011. Expression of CD26 and its Association with Dipeptidyl peptidase IV Activity in Lymphocytes of Type 2 Diabetes Patients. *Cell Biochem Biophys.* 61, 297–302.

Boop, A., De Bona, K.S., Bellé, L.P., Moresco, R.N., Moretto, M.B., 2009. Syzygium cumini inhibits adenosine deaminase activity and reduces glucose levels in hyperglycemic patients. *Fundam Clin Pharmacol.* 23, 501-7.

Böyum, A., 1968. Isolation of mononuclear cells and granulocytes from human blood. Isolation of mononuclear cells by one centrifugation and of granulocytes by combining centrifugation and sedimentation at 1g. *Scand J Clin Lab Invest.* 97, 77–89.

Cordero, O.J., Salgado, F.J., Nogueira, M., 2009. On the origin of serum CD26 and its altered concentration in cancer patients. *Cancer Immunol Immunother.* 58, 1723-1747.

De Bona, K.S., Bellé, L.P., Bitencourt, P.E., Bonfanti, G., Cargnelutti, L.O., Pimentel, V.C., Ruviaro, A.R., Schetinger, M.R., Emanuelli, T., Moretto, M.B., 2011. Erythrocytic enzymes and antioxidant status in people with type 2 diabetes: beneficial effect of Syzygium cumini leaf extract in vitro. *Diabetes Res Clin Pract.* 94, 84-90.

De Bona, K.S., Bellé, L.P., Sari, M.H., Thomé, G., Schetinger, M.R., Boligon, A., Athayde, M.L., Pigatto, A.S., Moretto, M.B., 2010. Syzygium cumini extract decrease adenosine deaminase, 5'nucleotidase activities and oxidative damage in platelets of diabetic patients. *Cell Physiol Biochem.* 26, 729-38.

Dong, R.P., Kameoka, J., Hege, M., Tanaka, T., Schlossman, S.F., Morimoto, C., 1996. Characterization of adenosine deaminase binding to human CD26 on T cells and its biologic role in the immune response. *J. Immunol.* 156, 1349-1355.

Durak, I., Cetin, R., Devrim, E., Ergüder, I.B., 2005. Effects of black grape extract on activities of DNA turn-over enzymes in cancerous and non cancerous human colon tissues. *Life Sci.* 76, 2995-3000.

Feng, Y., Liu, Y.M., Fratkins, J.D., LeBlanc, M.H., 2005. Grape seed extract suppresses lipid peroxidation and reduces hypoxic ischemic brain injury in neonatal rats. *Brain Res Bull.* 66, 120-7.

Franco, R., Casado, V., Ciruela, F., Saura, C., Mallol, J., Canela, E.I., et al. 1997. Cell surface adenosine deaminase much more than an ectoenzyme. *Prog Neurobiol.* 52, 283–94.

Giusti, G., 1974. Adenosine deaminase. In: Bergmeyer HV (ed.). Methods of Enzymatic Analysis, Academic Press, New York, NY, 1092–1099.

Gorrell, M.D., Gysbers, V., McCaughan, G.W., 2001. CD26: a multifunctional integral membrane and secreted protein of activated lymphocytes. *Scand J Immunol.* 54, 249-264.

Hogan, S., Zhang, L., Li, J., Sun, S., Canning, C., Zhou, K., 2010. Antioxidant rich grape pomace extract suppresses postprandial hyperglycemia in diabetic mice by specifically inhibiting alpha-glucosidase. *Nutr Metab (Lond).* 7-71.

Kameoka, J., Tanaka, T., Nojima, Y., Scholossman, S.F., Morimoto, C., 1993. Direct association of adenosine deaminase with a T cell activation antigen, CD26. *Science.* 261, 466-469.

Kocbuch, K., Sakowicz-Burkiewicz, M., Grden, M., Szutowicz, A., Pawelczyk, T., 2009. Effect of insulin and glucose on adenosine metabolizing enzymes in human B lymphocytes. *Acta Biochim Pol.* 56, 439-446.

Köse, K., Yazici, C., Aşşioğlu, O., 2001. The evaluation of lipid peroxidation and adenosine deaminase activity in patients with Behçet's disease. *Clin Biochem.* 34, 125-129.

Lunkes, G.I., Lunkes, D.S., Leal, D., Araújo Mdo, C., Corrêa, M., Becker, L., et al., 2008. Effect of high glucose levels in human platelet NTPDase and 5'-nucleotidase activities. *Diabetes Res Clin Pract.* 81, 351-357.

Orhan, N., Aslan, M., Orhan, D.D., Ergun, F., Yeşilada, E., 2006. In-vivo assessment of antidiabetic and antioxidant activities of grapevine leaves (Vitis vinifera) in diabetic rats. *J Ethnopharmacol.* 108, 280-286.

Pala, L., Mannucci, E., Pezzatini, A., Ciani, S., Sardi, J., Raimondi, L., et al., 2003. Dipeptidyl peptidase-IV expression and activity in human glomerular endothelial cells. *Biochem Biophys Res Commun.* 310, 28-31.

Pérez-Jiménez, J., Saura-Calixto, F., 2008. Grape products and cardiovascular disease risk factors. *Nutrition Research Reviews.* 21, 158–173.

Rice-Evans, C.A., Miller, N.J., Paganga, G.B., 1996. Structure-antioxidant activity relationships of flavonoids and phenolic acids. *Free Radic Biol Med.* 20, 933–956.

Salah, N., Miller, N.J., Paganga, G., Tijburg, L., Bolwell, G.P., Rice-Evans, C., 1995. Polyphenolic flavanols as scavengers of aqueous phase radicals and as chain-breaking antioxidants. *Arch Biochem Biophys.* 322, 339–346.

Schmatz, R., Mann, T.R., Spanevello, R., Machado, M.M., Zanini, D., Pimentel, V.C., et al., 2013. Moderate Red Wine and Grape Juice Consumption Modulates the Hydrolysis of the Adenine Nucleotides and Decreases Platelet Aggregation in Streptozotocin-Induced Diabetic Rats. *Cell Biochem Biophys* 65, 129–143.

Schön, E., Demuth, H.U., Barth, A., Ansorge, S., 1984. Dipeptidyl peptidase IV of human lymphocytes. Evidence for specific hydrolysis of glycylproline p-nitroanilide in T-lymphocytes. *Biochem J.* 223, 255-58.

Scola, G., Conte, D., Spada, P.W.D., Dani, C., Vanderlinde, R., Funchal, C., et al., 2010. Flavan-3-ol Compounds from Wine Wastes with in Vitro and in Vivo Antioxidant Activity. *Nutrients.* 2, 1048-1059.

Singh, R.P., Tyagi, A.K., Dhanalakshmi, S., Agarwal, R., Agarwal, C., 2004. Grape seed extract inhibits advanced human prostate tumor growth and angiogenesis and upregulates insulin-like growth factor binding protein-3. *Int J Cancer.* 108, 733-40.

Stanković, M., Tesević, V., Vajs, V., Todorović, N., Milosavljević, S., Godevac, D., 2008. Antioxidant properties of grape seed extract on human lymphocyte oxidative defence. *Planta Med.* 74, 730-735.

Yao, K., Ge, J.B., Sun, A.J., Hong, X.W., Shi, H.Y., Huang, R.C., et al., 2006. Effects and mechanism of hyperglycemia on development and maturation and immune function of human monocyte derived dendritic cells. *Zhonghua Xin Xue Guan Bing Za Zhi.* 34, 60-64

In: Grapes
Editor: José de Sousa Câmara

ISBN: 978-1-63321-402-6

Chapter 17

Phenolic Compounds of Grapes (*Vitis vinifera*): Diversity of Molecules, Content and Bioavailability

Mariane Lutz[•]
Centro de Investigación y Desarrollo de Alimentos Funcionales (CIDAF), Facultad de Farmacia, Universidad de Valparaíso, Chile

Abstract

A high number of potentially bioactive compounds has been identified in grapes, including anthocyanins, catechins, ellagic acid, lutein, lycopene, quercetin, resveratrol, many of which are potent antioxidants. Phenolic compounds are the main bioactives contained in grapes, and may exert beneficial effects by acting as direct or indirect antioxidants, anticarcinogenic, immune-modulatory, antidiabetes, antiatherogenic, and/or neuroprotective agents, among others, modulating a number of factors that are associated with health benefits.

These compounds are prevalent in different parts (e.g., fruit, stems, seeds, juice, pomace) of grapes. The reported phenolics content and antioxidant capacity of parts of various grape cultivars differ in diverse regions of the world. Phenolics in grapes are described, with emphasis on red and purple varieties containing high amounts of flavan-3-ols, hydroxycinnamates, and anthocyanins. The bioavailability of phenolics that substantiate the beneficial effects of the dietary intake of grapes and/or grape juice is discussed. Many of these bioactives or their metabolites are actually absorbed, thus able to exert their beneficial effects on the target tissues. However, current research should include the characterisation of the polyphenol metabolizing phenotypes and biomarkers of exposure should be validated.

Keywords: *Vitis vinifera*; grape cultivars; grape parts; polyphenols; bioactive compounds; bioavailability

[•] E-mail: mariane.lutz@uv.cl.

Introduction

Phytochemicals represent a major strategy for counteracting unfavourable conditions in plants. Various potentially bioactive phytochemicals have been described in *Vitis vinifera,* including secondary metabolites such as phenolics (mainly flavanols, flavonols, phenolic acids, stilbenes –resveratrol- and anthocyanins), which vary according to a series of factors related to genotype and agronomic/environmental conditions (cultivar, soil composition), climate, geographic origin, and cultivation practices or exposure to stress factors including draught, salinity, adverse climate conditions, diseases, pathogens, insects, and post-harvesting conditions (Connor et al. 2002; Proteggente et al. 2002; Xu et al. 2010). Liang et al. (2012) studied the composition and content of polyphenols in the seeds of 91 grape accessions from 17 *Vitis* species, observing tremendous variation both among and within species. In seeds, monomeric flavan-3-ols were the most abundant phenolics, followed by dimeric and trimeric flavan-3-ols, which collectively accounted for more than 96% of the total. In effect, the potentially bioactive polyphenols are prevalent in different parts of the plant (e.g. berry, stems, seeds, juice, pomace) and are characteristic of their color. Moreover, it is currently possible to correlate the phenolics profile of wines originated from the same grape variety with their geographical origin (Psomas et al. 2012). The improvement of the current analytical techniques allows the identification of a wide variety of phenolics in grapes and grape fractions. For instance, using LC gradient in different MS ionization modes and mobile phase modifiers, Xu et al. (2011) described 53 phenolic compounds in grapes, including 33 anthocyanins, 12 flavonols, 4 phenolic acids, and 4 flavanols/proanthocyanidins.

Generally, polyphenols contents are 8-fold higher in skin than in flesh and the anthocyanins are present only in the skin extracts in relevant amounts, about 50% of total phenolics (Falchi et al. 2006). The varietal difference in the color of black and red grapes results from the accumulation of anthocyanidins that are modified by the attachment of glucose moieties to form anthocyanins, and there are significant variations in the proportions of anthocyanins specific to species parentage or cultivar (Shiraishi et al. 2007). Thus, the main polyphenols found in red and purple grapes are flavan-3-ols, hydroxycinnamates, and anthocyanins, while white grapes are devoid of anthocyanins and generally contain lower quantities of phenolics than red grapes (Visloky and Fernandez, 2010). Procyanidins are monomeric, oligomeric, and polymeric polyhydroxy flavan-3-ol units: the monomeric forms are usually referred to as catechins, and the oligomeric and polymeric counterparts are called tannins. The quantity, structure, and degree of polymerization of grape proanthocyanidins differ, depending on their localization in the grape, and skins contain ca. 20-fold less monomeric, oligomeric, and polymeric flavan-3-ols compared with grape tissues (Monagas et al. 2003). The anthocyanins identified in *Vitis vinifera* spp. are the 3-O-monoglucosides and the 3-O-acylated monoglucosides of the five main anthocyanidins: delphinidin, cyanidin, petunidin, peonidin, and malvidin (Lionetto et al. 2011). Moreover, during the development of the berries chemical changes occur, and Nunan et al. (1998) observed that some fractions that give them firmness and strength increase before veraison but decrease as the berries soften.

Flavonoids are widely distributed in grapes, especially in seeds and stems, which mainly contain gallic acid, catechin, epicatechin, and procyanidin oligomers and polymers (Cantos et al. 2002; Pastrana-Bonilla et al. 2003). The skin and seeds are good sources of gallic acid, catechin, and epicatechin, while red grape skin and seeds are the main dietary source of ellagic acid, myricetin, quercetin, kaempferol, and proanthocyanidins (Hernandez-Jimenez et al. 2009; Xia et al. 2010). One of the most outstanding phenolics in *Vitis vinifera*, *labrusca*, and *muscadine* types of grapes is resveratrol (3,5,40-trans-trihydroxystilbene). The skins and seeds of these types of grapes contain 50–100 μg/g of resveratrol (Dudley et al. 2008), mainly found in red and purple grape skins, varying according to exposure to fungal infection, the agronomic conditions, and fruit processing. In fact, since fungal infections are more common in cooler climates, grapes grown in these zones have a higher concentration of resveratrol (Celotti et al. 1996), while in the equator zone grapes may contain high levels due to the high UV irradiation (Kopp, 1998). The resveratrol levels in wine are affected by the vinification process as well (eg, fermentation time when the wine is in contact with grape skins) (Perrone et al. 2007).

Phenolics in Grape Varieties

Total phenolics, anthocyanins, antioxidant capacity, and some specific phenolics were studied in grape juices made with four varieties of table grapes largely produced in Chile: blue grapes cv. Autumn Royal or Ribier exhibited higher phenolics content and antioxidant capacity than juice elaborated with red grapes cv. Red Globe or Crimson Seedless (Lutz et al. 2011). Purple grapes contain high amounts of catechin (Décordé et al. 2008), and phenolic acids are reported as the major antioxidants found in these grapes, with gallic acid prevailing over other compounds (Yilmaz and Toledo, 2004; Hogan et al. 2009). Lutz et al. (2012) determined that the predominant phenolic species in Chilean red ground grapes, juice, and skin fractions was catechin. On the other hand, Obreque-Slier et al. (2010) compared grapes skins from the varieties Cabernet Sauvignon and Carménère, observing that the second cultivar exhibited higher contents of total anthocyanins, monomeric flavan-3-ols, and total flavonoids, higher degree of polymerization, higher percentage of galloylation, higher average molecular weight of the flavanol fraction, and higher color intensity. In a study of Chilean wines cv Cabernet Sauvignon of three price categories, Cáceres et al. (2012) observed that wines from the higher price presented a higher concentration of total tannins, the differences in their concentrations were due to the different fractions of proanthocyanidins in the skins and seeds of grapes, without any differences in their proportions.

Xu et al. (2010) compared the phenolics content and antioxidant capacity of skins and seeds of 18 grape cultivars belonging to Oriental and North American *Vitis* Spp/hybrids, observing that *V. vinifera* 'Cabernet Sauvignon' exhibited the highest content of phenolics in seeds, followed by Muscadine grapes, while the lowest content appeared in the Oriental *Vitis* Spp. The Euro-Asian or Euro-American hybrids fell between their parents. The authors report that the Oriental grapes had different distributing patterns of phenolics and antioxidant properties compared with the non-Oriental grapes, and even when the total phenolics contents in the berries of two cultivars are similar, the distribution of individual phenolics compounds in seeds and skins varied greatly among them. Grape skins, in general, contain much lower concentrations of flavan-3-ols than seeds. Only the Oriental 'Purple grape' showed a higher level of flavan-3-ols in skins than in seeds, a finding similar to the report of Rodriguez Montealegre et al. (2006) in *V. vinifera* cultivars. Dani et al. (2007), Mulero et al. (2010) and

Bunea et al. (2012) observed that grapes cultivated in organic systems exhibited higher phenolics content and antioxidant activity, which may be associated with the plant exposure to environmental situations that lead to an increase of natural defense substances in the absence of pesticides.

Grape Phenolics in Waste Materials

Grape seeds make about 15% of solid waste in the wine industry and contain over 60% of total extractible grape phenolic compounds (Nawaz et al. 2006).The winery and grape juice waste mainly consists of seeds, skins, and grape stalks, which are a good source of phenolics that may exert beneficial effects (Bonilla et al. 1999; Alonso et al. 2002; Lazze et al. 2009; Khanal et al. 2009). In winery waste materials, marcs, stems and dregs (sludgy residual deposits at the bottom of fermentation vats) represent sources of antioxidants such as gallic acid, catechin, epicatechin, epigallocatechin, epicatechin gallate, epigallocatechin gallate, and procyanidins B_1 and B_2 (dimers) (Torres et al. 2002; Guendez et al. 2005a). Moreover, Guendez et al. (2005b) showed that grape seeds are an exceptional source of phenolics compounds, irrespective of the origin (white or red Greek varieties). It has been reported that press-cake or pomace, the residue remaining once the juice is extracted from grape, contains a considerable amount of procyanidins. Thus, grape pomace, that involves a major cost to the industry for its disposal, may be used to recover these phenolics, providing value-added health-promoting ingredients and may be used as functional foods ingredients, nutraceuticals or cosmeceuticals (Espin et al. 2007; Iacopini et al. 2008).

However, the elaboration processes often involves temperature elevation, making the thermal stability of the components within the product essential. Some consequences of thermal treatments on phenolics have been described, such as the epimerisation and degradation of catechins, following first-order reactions, and the rate constants of the reaction kinetics following the Arrhenius equation (Wang et al. 2006), and the polymerisation of phenolics with food processing and storage, leading to the formation of brown-colored macromolecules (Manzocco et al. 1998). Davidov-Pardo et al. (2011) observed that the thermal treatments affected the stability of grape seed extracts was dependent on the phenolics profile of each extract, being the compounds with a gallate group attached to their structure the most sensitive to heat degradation.

Grape Phenolics and Health

According to Pezzuto (2009), over 1600 compounds have been identified in grapes, including anthocyanins, catechins, ellagic acid, lutein, lycopene, quercetin, and other potent antioxidants. These potentially bioactive compounds may exert beneficial effects by acting as direct or indirect antioxidants, anticarcinogenic, immunomodulatory, antidiabetes, antiatherogenic, and/or neuroprotective agents, some also activate longevity gene SirT1, which is also responsible for the longevity mediated by calorie restriction, among other effects (Visloky and Fernandez, 2010; Pezzuto, 2009; Leifert and Abeywardena, 2008; Furiga et al. 2009; Yadav et al. 2009; Lekli et al. 2010; Huang et al. 2012; Bahadoran et al. 2013), modulating a number of factors that are associated with health benefits.

The protective effect of purple grape juice/red wine polyphenols on the vascular system is thought to include their ability to prevent oxidation of low-density lipoproteins (LDL) (Frankel et al. 1993; Stein et al. 1999), platelet aggregation and adhesion (Freedman et al. 2001), the reduction of the concentration of inflammatory biomarkers (Estruch et al. 2004; Castilla et al. 2006), reduction of myocardial ischemic reperfusion injury (Cui et al. 2002), which is associated to the presence of resveratrol (Ray et al. 1999); smooth muscle cell migration and proliferation (Iijima et al. 2002), enhanced formation of nitric oxide (NO) and endothelium-derived hyperpolarizing factor (EDHF) (Anselm et al. 2007). The ingestion of purple grape juice improves flow-mediated vasodilatation, platelet function and platelet-dependent inflammatory responses in patients with coronary artery disease (Stein et al. 1999; Chou et al. 2001), and reduces blood pressure in moderately hypertensive patients (Park et al. 2004). After being ingested, polyphenols are detected in plasma both in their soluble form and physically associated with lipoproteins and protect LDL from oxidation (Stein et al. 1999; Hayek et al. 1997; Nigdikar et al. 1998). The antioxidative properties of flavonoids have been widely described, and purple grapes and grape juice are recognized as inhibitors of the oxidation of LDL *in vitro* and *in vivo* (Olszanecki et al. 2002; O'Byrne et al. 2002). Dohadwala and Vita (2009) summarized the effects of grape phenolics on cardiovascular risk factors into four groups: those acting on hypertension, diabetes, hypercholesterolemia, and smoking, supporting the recommendation of a diet rich in fruits and vegetables. Falchi et al. (2006) observed that both he flesh and the skin of grapes are cardioprotective, exhibiting the same degree of free radicals scavenging abilities despite the fact that only the skins contain the anthocyanins and higher amounts of polyphenols, and comparable amounts of sugars and organic acids (including tartaric, malic, and shikimic acids).

In order to better understand the protective effect of purple grape juice of the endothelial function, experiments were performed to determine whether purple grape juice induces endothelium-dependent relaxation of coronary arteries, to characterize the relaxing factors involved, and the signaling pathway leading to the formation of NO. Endothelial dysfunction has been considered an early feature in the progression of atherosclerosis and an independent predictor of poor prognosis in many cardiovascular diseases (Cai and Harrison, 2000). The endothelial release of NO is relevant for vasodilation. *In vitro*, flavonoid components of purple grape juice induce endothelium-dependent vasodilation of arterial rings, a phenomenon mediated by the NO guanosine 3',5' cyclic monophosphate (NO-cGMP) pathway (Fitzpatrick et al. 1993). Ndiaye et al. (2004) observed in porcine coronary arteries that red wine phenolics induce the redox-sensitive activation of the PI3-kinase/Akt pathway in endothelial cells, causing phosphorylation of eNOS, resulting in an increased formation of NO. After a series of experiments performed *in vitro*, Anselm et al. (2007) conclude that the vasoprotective effect of polyphenols is dual: a moderate pro-oxidant action in endothelial cells to enhance vasoprotective mechanisms involving NO and EDHF, and an antioxidant action in vascular smooth muscle cells and platelets to keep them in a quiescent state. In human trials, grape juice improves endothelial function in patients with coronary heart disease (Chou et al. 2001; Mellen et al. 2010). More recently, Li et al. (2013) performed a meta-analysis of nine controlled trials describing the acute effects of grape polyphenols on endothelial function in adults. The authors concluded that endothelial function can be significantly improved in the initial 2 h after intake of grape polyphenols, as measured by flow-mediated dilation. Moreover, Karatsi et al. (2008) demonstrated that acute consumption of both red wine and green olive oil, as are usually consumed in the Mediterranean diet, led to an improvement in the postprandial endothelial function in healthy subjects.

A series of *in vitro, in vivo* and clinical studies support the beneficial effects of dietary phenolics intake on neuroprotection. In fact, grape polyphenols cross the blood-brain barrier (Passamonti et al. 2005) and may induce neuroprotective effects and influence neuronal signalling involved in memory function, exhibiting a series of neuroprotective effects, including the reverse of the course of neuronal and behavioral aging. The effects are possibly due to a multiplicity of direct and indirect mechanisms that can affect a variety of neuronal parameters, including the regulation of oxidative stress signals such as NF-kB, enhancement of neuroprotective stress shock proteins and anti-inflammatory actions through the regulation of the expression of specific inflammatory genes (IL-1β, TNFα) (Conte et al. 2003; Goyarzu et al. 2004; Shukitt-Hale et al. 2006; Shukitt-Hale et al. 2008; Krikorian et al. 2010). In order to investigate the beneficial effects of purple grape juices in reducing oxidative stress, Dani et al. (2009) used a model of aged rats (7 months old) to evaluate the antioxidant and antigenotoxic effects of organic and conventional grape juice on CCl_4-induced toxicity and observed that grape juices were capable of reducing DNA oxidative damage in the cerebral cortex, substantia nigra, and liver.

Flavonoids affect memory and learning. They seem to interact primarily with ERK and Akt pathways, leading to modulation of the transcription factor CREB as well as up-regulation of CREB gene expression (Rendeiro et al. 2012). In order to exert these effects, phenolics must reach the nervous cells. The extent of their blood–brain barrier penetration has been shown to be dependent on the lipophilicity of the compound (Youdim et al. 2003). Anthocyanins have been identified in brain regions that mediate cognition, including the medial temporal lobe and cortex (Andres-Lacueva et al. 2005), and hippocampal distribution has been associated with behavioural enhancement in animal supplementation studies (Williams et al. 2008). Among the clinical studies, the PAQUID prospective study examined cognitive performance in 1640 subjects (aged 65+ years) on four occasions over a 10-year-period and found that flavonoid-intake was associated with a significantly better cognitive performance over time (Letenneur et al. 2007).

Absorption, Metabolism and Bioavailability of Grape Phenolics

In foods, most phenolics compounds (mainly flavonoids) are found as glycosides. The rate and extent of intestinal absorption of phenolics and the nature of the metabolites circulating in the plasma are determined by the chemical structure of phenolics, while their amounts found intact in urine vary among the ingested molecules. Phenolics are poorly absorbed and rapidly metabolized into phase II conjugates, so that circulating levels of unmodified compounds are generally in the 10^{-9}-10^{-7} M range (Biasutto et al. 2010). Amongst the least absorbed polyphenols are proanthocyanidins and anthocyanins (Epriliati et al. 2012). The low bioavailability of a variety of phenolics is also limited by a small bioaccesibility in the gut due to physical-chemical interactions with indigestible constituents of the food matrix such as cell wall polysaccharides, eg dietary fiber (Palafox-Carlos et al. 2011).

On ingestion, polyphenols are hydrolyzed to aglycones by lactase phlorizin hydrolase, which is located mainly in the brush border and enterocytes (Williamson, 2004), and subsequently these diffuse into the cell (Nemeth et al. 2003). Within the enterocyte, phenolics undergo extensive phase I and II metabolism, resulting in metabolites that differ from the molecules originally found in grapes (Manach et al. 2004), many of which may enter the portal vein where they will subsequently undergo further metabolism or are effluxed. Moreover, intestinal conjugation and secretion of conjugates back to the gut lumen varies with the dose of flavonoids and saturation of the intestinal secretion of conjugates may occur (Silberberg et al. 2006). The bioavailability of various phenolics differs, and there is no direct relation between their amount in food and their bioavailability (Pandey and Rizvi, 2009).

Flavonoids are substrates for phase I (hydrolysing and oxidising) and phase II (conjugating and detoxifying) enzymes, being de-glucosylated and metabolised mainly into glucuronides, sulphates and O-methylated derivatives (Manach et al. 2004; Scalbert and Williamson, 2000), while anthocyanins do not appear to undergo extensive phase I and II metabolism of the parent anthocyanidin (Kay, 2006). Moreover, the intestinal microbiota catalyse the breakdown of phenolics by removing the conjugated moiety (formed in the small intestine) and cleave the phenolic structure (Scalbert et al. 2002; Selma et al. 2009). Barrington et al. (2009) showed that flavonols are conjugated inside the enterocyte and transported across the monolayer or effluxed back to the apical side, and observed that sulphate conjugates are effluxed in a different way compared to glucuronides, while kaempferol behaved in a different pattern. Thus, different transporters participate at the apical and the basolateral sides of the cell, affecting the absorptive process.

Goldberg et al. (2003) determined the absorptive efficiency of grape polyphenols and showed that in healthy subjects the concentration of grape polyphenols peaked in serum around 30 min after oral ingestion. Absorption of phenolics is an essential step to pass into the bloodstream and protect the body tissues against the oxidative damage. Manach et al. (2005) have found that the maximum plasma antioxidant capacity was usually reached 1–4 h after the ingestion of polyphenols. Urinary excretion of phenolics, on the other hand, provides an approach for assessing their bioavailability, although the values measured underestimate the total amount absorbed since they do not account for metabolites sequestered in body tissues and do not take into consideration other routes of elimination, including the bile (Crozier et al. 2009).

The most abundant polyphenols in grapes do not exhibit the best bioavailability profile, and in order to know the bioavailability of grape phenolics it is necessary to include their metabolites (D'Archivio et al. 2010). Taking into consideration the extent of metabolism of phenolics produced, metabolomics is a useful tool that can be used to determine a variety of metabolites that represent an approach to an individual's metabolic profile (Forester and Waterhouse, 2009). Although the studies on the bioavailability of grape and grape juice are scarce, they actually demonstrate that the potentially bioactive phenolics are bioavailable after the intake of these foods and, in spite of the low bioavailability of grape phenolics, many of these molecules (and/or their metabolites) are able to exert their beneficial effects on the target tissues (Manach et al. 2005; Stalmach et al. 2011; Lutz et al. 2014).

Since substantial quantities of the ingested phenolics pass to the colon, affecting the microbiota, the level of urinary excretion also indicates that the colonic catabolites are absorbed into the portal vein and pass through the body in the circulatory system prior to excretion (Crozier et al. 2010). Consequently, the profile of urine metabolites may be markedly different to that of plasma (Mullen et al. 2006). Colonic bioconversion of phenolics is highly variable because of large interindividual differences in the bioconversion of these compounds due to individual colonic microbiota composition, and other complex factors such as the food matrix can also modulate polyphenol–microbiota interactions. Although a limited number of bacterial species have been identified as being involved in the metabolism of polyphenols (Moco et al. 2012), the establishment of functional links between polyphenol metabolism and the gut microbiome is currently being developed through metabolomics and microbiomics (van Duynhoven et al. 2011; Bolca et al. 2013). On the other hand, phenolics modulate the gut microbial population, in a two-way phenolic-microbiota interaction that affects gut health (Cardona et al. 2013).

Conclusion

Grapes contain a variety of potentially bioactive phenolic compounds. Whereas *in vitro* studies may indicate some of the mechanisms of action of individual grape phenolics, these findings need to be supported by *in vivo* experiments, including the evaluation of their bioavailability, considering a large number of metabolites. Current research should include the characterisation of the polyphenol metabolizing phenotypes and biomarkers of exposure should be validated.

References

Alonso, A. M., Guillén, D. A., Barroso, C. G., Puertas, B., García, A., (2002). Determination of antioxidant activity of wine byproducts and its correlation with polyphenolic content. *J. Agric. Food Chem.*, 50, 5832–5836.

Andres-Lacueva, C., Shukitt-Hale, B., Galli, R. L., Jauregui, O., Lamuela-Raventos, R. M., Joseph, J. A., (2005). Anthocyanins in aged blueberry-fed rats are found centrally and may enhance memory. *Nutr. Neurosci.*, 8, 111–120.

Anselm, E., Chataigneau, M., Ndiaye, M., Chataigneau, T., Schini-Kerth, V. B., (2007). Grape juice causes endothelium-dependent relaxation via a redox-sensitive Src- and Akt-dependent activation of eNOS. *Cardiovasc. Res.*, 73, 404–413.

Bahadoran, Z., Mirmiran, P., Azizi, F., (2013). Dietary polyphenols as potential nutraceuticals in management of diabetes: a review. *J. Diab. Metab. Dis.*, 12, 43-51.

Barrington, R., Williamson, G., Bennett, R. N., Davis, B. D., Brodbelt, J. S., Kroon, P. A., (2009). Absorption, conjugation and efflux of the flavonoides, kaempferol and galangin, using the intestinal CaCo-2/TC7 cell model. *J. Functional Foods,* 1, 74-87.

Biasutto, L., Marotta, E., Garbisa, S., Zoratti, M., Paradisi, C., (2010). Determination of quercetin and resveratrol in whole blood—implications for bioavailability studies. *Molecules,* 15, 6570-6579.

Bolca, S., Van de Wiele, T., Possemiers, S., (2013). Gut metabotypes govern health effects of dietary polyphenols. *Curr. Op. Biotechnol.,* 24, 220–225.

Bonilla, F., Mayen, M., Merida, J., Medina, M., (1999). Extraction of phenolic compounds from red grape marc for use as food lipid antioxidants. *Food Chem.*, 66, 209–215.

Bunea, C. I., Pop, N., Babes, A. C., Matea, C., Dulf, F., Bunea, A., (2012). Carotenoids, total polyphenols and antioxidant activity of grapes (*Vitis vinifera*) cultivated in organic and conventional systems. *Chem. Central J.*, 6, 66-74.

Cáceres, A., Peña-Neira, A., Gálvez, A., Obreque-Slier, E, López-Solís, R, Canals, J. M., (2012). Phenolic compositions of grapes and wines from cultivar Cabernet Sauvignon produced in Chile and their relationship to commercial value. *J. Agric. Food Chem.*, 60, 8694–8702.

Cai, H., Harrison, D. G., (2000). Endothelial dysfunction in cardiovascular diseases: the role of oxidant stress. *Circ. Res.*, 87, 840–844.

Cantos, E., Espin, J. C., Tomas-Barberan, F. A., (2002). Varietal differences among the polyphenol profiles of seven table grape cultivars studied by LC-DAD-MS-MS. *J. Agric. Food Chem.,* 50, 5691–5696.

Cardona, F., Andrés-Lacueva, C., Tulipani, S., Tinahones, F. J., Queipo-Ortuño, M. I., (2013). Benefits of polyphenols on gut microbiota and implications in human health. *J. Nutr. Biochem.,* 24, 1415–1422.

Castilla, P., Echarri, R., Davalos, R., Cerrato, F., Ortega, H., Teruel, J. L., Fernandez Lucas, M., Gomez Coronado, D., Ortuño, J., Lasuncion, M. A., (2006). Concentrated red grape juice exerts antioxidant, hypolipidemic, and antiinflammatory effects in both hemodialysis patients and healthy subjects. *Am. J. Clin. Nutr.*, 84, 252–262.

Celotti, E., Ferrarini, R., Zironi, R., Conte, L. S., (1996). Resveratrol content of some wines obtained from dried Valpolicella grapes: Recioto and Amarone. *J. Chromatogr. A,* 730, 47–52.

Chou, E. J., Keevil, J. G., Aeschlimann, S., Wiebe, D. A., Folts, J. D., Stein, J. H., (2001). Effect of ingestion of purple grape juice on endothelial function in patients with coronary heart disease. *Am. J. Cardiol.,* 88, 553–555.

Connor, A. M., Luby, J. J., Tong, C. B. S., Finn, C. E., Hancock, J. F., (2002). Variation and heritability estimates for antioxidant activity total phenolic content and anthocyanin content in blueberry progenies. *J. Amer. Soc. Hortic. Sci.,* 1, 82–88.

Conte, A., Pellegrini, S., Tagliazucchi, D., (2003). Synergistic protection of PC12 cells from β-amyloid toxicity by resveratrol and catechin. *Brain Res. Bull.*, 6, 29–38.

Crozier, A., Jaganath, I. B., Clifford, M. N., (2009). Dietary phenolics: chemistry, bioavailability and effects on health. *Nat. Prod. Res.*, 26, 1001–1043.

Crozier, A., Del Rio, D., Clifford, M. N., (2010). Bioavailability of dietary flavonoids and phenolic compounds. *Mol. Asp. Medicine,* 31, 446-467.

Cui, J., Cordis, G. A., Tosaki, A., Maulik, N., Das, D. K., (2002). Reduction of myocardial ischemia reperfusion Injury with regular consumption of grapes. *Ann. N.Y. Acad. Sci.,* 957, 302–307.

Dani, C., Oliboni, L. S., Vanderlinde, R., Bonatto, D., Salvador, M., Pegas Henriques, J. A., (2007). Phenolic content and antioxidant activities of white and purple juices manufactured with organically -or conventionally- produced grapes. *Food Chem. Toxicol.,* 45,2574–2580.

Dani, C., Oliboni, L. S., Umezu, F. M., Pasquali, M. A. B., Salvador, M., Fonseca Moreira, J. C., Pegas Henriques, J. A., (2009). Antioxidant and antigenotoxic activities of purple grape juice—organic and conventional—in adult rats. *J. Med. Food,* 12, 1111–1118.

D'Archivio, M., Filesi, C., Varì, R., Scazzocchio, B., Masella, R., (2010). Bioavailability of the polyphenols: Status and controversies. *Int. J. Mol. Sci.*, 11, 1321-1342.

Davidov-Pardo, G., Arozarena, I., Marin-Arroyo, M. R., (2011). Stability of polyphenolic extracts from grape seeds after termal treatments. *Eur. Food Res. Technol.*, 232, 211–220.

Décordé, K., Teissèdre, P. L., Auger, C., Cristol, J. P., Rouanet, J. M., (2008). Phenolics from purple grape, apple, purple grape juice and apple juice prevent early atherosclerosis induced by an atherogenic diet in hamsters. *Mol. Nutr. Food Res.,* 52, 400-407.

Dohadwala, M. M., Vita, J. A., (2009). Grapes and cardiovascular disease. *J. Nutr.*, 139, 1788–1793.

Dudley, J., Das, S., Mukherjee, S., Das, D. K., (2008). Resveratrol, a unique phytoalexin present in red wine, delivers either survival signal or death signal to the ischemic myocardium depending on dose. *J. Nutr. Biochem.*, 20, 443–452.

Epriliati, I., Ginjoim, I. R., (2012). Bioavailability of Phytochemicals. In: Phytochemicals – A Global Perspective of Their Role in Nutrition and Health. In Tech; p. 401-428.

Espin, J. C., Garcia Conesa, M. T., Tomas Barberan, F. A., (2007). Nutraceuticals: Facts and fiction. *Phytochemistry,* 68, 2986–3008.

Estruch, R., Sacanella, E, Badía, E., Antúnez, E, Nicolás, J. M., Fernández-Solá, J., Rotilio, D., de Gaetano, G., Rubin, E., Urbano-Márquez, A., (2004). Different effects of red wine and gin consumption on inflammatory biomarkers of atherosclerosis: a prospective randomized crossover trial. Effects of wine on inflammatory markers. *Atherosclerosis,* 175,117–123.

Falchi, M., Bertelli, A., Lo Scalzo, R., Morassut, M., Morelli, R., Das, S., Cui, J., Das, D. K., (2006). Comparison of cardioprotective abilities between the flesh and skin of grapes. *J. Agric. Food Chem.*, 54, 6613-6622.

Fitzpatrick, D. F., Hirschfield, S. L., Coffey, R. G., (1993). Endothelium-dependent vasorelaxing activity of wine and other grape products. *Am. J. Physiol.*, 265, H774–H778.

Forester, S. C., Waterhouse, A. L., (2009). Metabolites are key to understanding health effects of wine polyphenolics. *J. Nutr.*, 138, 1824S–1831S.

Frankel, E. N., Kanner, J., German, J. B., Parks, E., Kinsella, J. E., (1993). Inhibition of oxidation of human low-density lipoprotein by phenolic substances in red wine. *Lancet,* 341, 454–457.

Freedman, J. E., Parker 3rd, C., Li, L., Perlman, J. A., Frei, B., Ivanov, V., Deak, L. R., Iafrati, M. D., Folts, J. D., (2001). Select flavonoids and whole juice from purple grapes inhibit platelet function and enhance nitric oxide release. *Circulation,* 103, 2792–2798.

Furiga, A., Lonvaud-Funel, A., Badet, C., (2009). In vitro study of antioxidant capacity and antibacterial activity on oral anaerobes of a grape seed extract. *Food Chem.*, 113, 1037–1040.

Goldberg, D. M., Yan, J., Soleas, G. J., (2003). Absorption of three wine-related polyphenols in three different matrices by healthy subjects. *Clin. Biochem.*, 36, 79–87.

Goyarzu, P., Malin, D. H., Lau, F. C., Taglialatela, G., Moon, W. D., Jennings, R., Moy, E., Moy, D., Lippold, S., Shukitt-Hale, B., Joseph, J. A., (2004). Blueberry supplemented diet: effects on object recognition memory and nuclear factor-kappa B levels in aged rats. *Nutr. Neurosci.*, 7, 75–83.

Guendez, R., Kallithraka, S., Makris, D. P., Kefalas, P., (2005a). Determination of low molecular weight polyphenolic constituents in grape (*Vitis vinifera* sp.) seed extracts: Correlation with antiradical activity. *Food Chem.*, 89, 1–9.

Guendez, R., Kallithraka, S., Makris, D. P., Kefalas, P., (2005b). An analytical survey of the polyphenols of seeds of varieties of grape (*Vitis vinifera*) cultivated in Greece: Implications for exploitation as a source of value-added phytochemicals. *Phytochem. Anal.,* 16, 17–23.

Hayek, T., Fuhrman, B., Vaya, J., Rosenblat, M., Belinky, P., Coleman, R., Elis, A., Aviram, M., (1997). Reduced progression of aterosclerosis in apolipoprotein E-deficient mice following consumption of red wine, or its polyphenols quercetin or catechin, is associated with reduced susceptibility of LDL to oxidation and aggregation. *Arterioscler. Thromb. Vasc. Biol*., 17, 2744 –2752.

Hernandez-Jimenez, A., Gomez-Plaza, E., Martinez-Cutillas, A., Kennedy, J. A., (2009). Grape skin and seed proanthocyanidins from Monastrell x Syrah grapes. *J. Agric. Food Chem.,* 57, 10798–10803.

Hogan, S., Zhang, L., Li, J., Zoecklein, B., Zhou, Z., (2009). Antioxidant properties and bioactive components of Norton (*Vitis aestivalis*) and Cabernet Franc (*Vitis vinifera*) wine grapes. *LWT - Food Sci. Technol*., 42, 1269-1274.

Huang, S., Yang, N., Liu, Y., Gao, J., Huang, T., Hu, L., Zhao, J., Li, Y., Li, C., Zhang, X., (2012). Grape seed proanthocyanidins inhibit colon cancer-induced angiogenesis through suppressing the expression of VEGF and Ang1. *Int. J. Mol. Med.,* 30, 1410-1416.

Iacopini, P., Baldi, M., Storchi, P., Sebastiani, L., (2008). Catechin, epicatechin, quercetin, rutin and resveratrol in red grape: Content, in vitro antioxidant activity and interactions. *J. Food Compos. Anal*., 21, 589– 598.

Iijima, K., Yoshizumi, M., Hashimoto, M., Akishita, M., Kozaki, K., Ako, J., Watanabe, T., Ohike, Y., Son, B., Yu, J., Nakahara, K., Ouchi, Y., (2002). Red wine polyphenols inhibit vascular smooth muscle cell migration through two distinct signaling pathways. *Circulation,* 105, 2404–2410.

Karatzi, K., Papamichael, C., Karatzis, E., Papaioannou, T. G.,Voidonikola, P. T., Vamvakou, G. D., Lekakis, J., Zampelas, A., (2008). Postprandial improvement of endothelial function by red wine and olive oil antioxidants: a synergistic effect of components of the Mediterranean diet. *J. Am. Coll. Nutr.,* 27, 448-453.

Kay, C. D., (2006). Aspects of anthocyanin absorption, metabolism and pharmacokinetics in humans. *Nutr. Res. Rev*., 19, 137–146.

Khanal, R. C., Howard, L. R., Prior, R. L., (2009). Procyanidin content of grape seed and pomace, and total anthocyanin content of grape pomace as affected by extrusion processing. *J. Food Sci*., 74, H174–H182.

Kopp, P., (1998). Resveratrol, a phytoestrogen found in red wine. A possible explanation for the conundrum of the 'French paradox'? *Eur. J. Endocrinol*., 138, 619–620.

Krikorian, R., Nash, T. A., Shidler, M. D., Shukitt-Hale, B., Joseph, J. A., (2010). Concord grape juice supplementation improves memory function in older adults with mild cognitive impairment. *Br. J. Nutr*., 103, 730–734.

Lazze, M. C., Pizzala, R., Gutierrez Pecharroman, F. J., Gaton Garnica, P., Antolin Rodriguez, J. M., Fabris, N., Bianchi, L., (2009). Grape waste extract obtained by supercritical fluid extraction contains bioactive antioxidant molecules and induces antiproliferative effects in human colon adenocarcinoma cells. *J. Med. Food,* 12, 561–568.

Leifert, W. R., Abeywardena, M. Y., (2008). Cardioprotective actions of grape polyphenols. *Nutr. Res*., 28, 729–737.

Lekli, I., Ray, D., Das, D. K., (2010). Longevity nutrients resveratrol, wines and grapes. *Genes Nutr.*, 5, 55–60.

Letenneur, L., Proust-Lima, C., Le Gouge, A., Dartigues, J. F., Barberger-Gateau, P., (2007). Flavonoid intake and cognitive decline over a 10-year period. *Am. J. Epidemiol.*, 165, 1364–1371.

Li, S. H., Tian, H. B., Zhao, H. J.,Chen, L. H., Cui, L. Q., (2013). The acute effects of grape polyphenols supplementation on endothelial function in adults: meta-analyses of controlled trials. *PLoS One,* 24; 8:e69818. doi: 10.1371/journal.pone.0069818.

Liang, Z., Yang, Y., Cheng, L., Gan-Yuan Zhong, G. Y., (2012). Characterization of polyphenolic metabolites in the seeds of Vitis germplasm. *J. Agric. Food Chem.*, 60, 1291–1299.

Lionetto, M. G., Giordano, M. E., Calisi, A., Erroi, E., De Nuccio, F., Pastrana-Bonilla, E., Akoh, C. C., Sellappan, S., Krewer, G., (2003). Phenolic content and antioxidant capacity of muscadine grapes. *J. Agric. Food Chem.*, 51, 5497–5503.

Lutz, M., Jorquera, K., Cancino, B., Ruby, R., Henríquez, C., (2011). Phenolics and antioxidant capacity of table grape (*Vitis vinifera* L.) cultivars grown in Chile. *J. Food Sci.*, 76, C1088-C1093.

Lutz, M., Cajas, Y., Henríquez, C., (2012). Phenolics content and antioxidant capacity of Chilean grapes cv. País and Cabernet Sauvignon. *J. Food CyTA,* 10, 1-7.

Lutz, M., Castro, E., García, L., Henríquez, C., (2014). Bioavailability of phenolic compounds from grape juice cv Autumn Royal. *J. Food CyTA,* 12, 48-54.

Manach, C., Scalbert, A., Morand, C., Rémésy, C., Jiménez, L., (2004). Polyphenols: food sources and bioavailability. *Am. J. Clin. Nutr.,* 79, 727–747.

Manach, C., Williamson, G., Morand, C., Scalbert, A., Remesy, C., (2005). Bioavailability and bioefficacy of polyphenols in humans. I. Review of 97 bioavailability studies. *Am. J. Clin. Nutr.*, 81, 230S–242S.

Manach, C., Mazur, A., Scalbert, A., (2005). Polyphenols and prevention of cardiovascular diseases. *Curr. Opin. Lipidol.*, 16, 77–84.

Manzocco, L., Anese, M., Nicoli, M. C., (1998). Antioxidant properties of tea extracts as affected by processing. *LWT Food Sci. Technol.*, 31, 694–698.

Mellen, P. B., Daniel, K. R., Brosnihan, K. B., Hansen, K. J., Herrington, D. M., (2010). Effect of muscadine grape seed supplementation on vascular function in subjects with or at risk for cardiovascular disease: a randomized crossover trial. *J. Am. Coll. Nutr.*, 29, 469–475.

Moco, S., Martin, F. P., Rezzi, S., (2012). Metabolomics view on gut microbiome modulation by polyphenol-rich foods. *J. Proteome Res.*, 11,4781-4790.

Monagas, M., Gomez-Cordoves, C., Bartolome, B., Laureano, O., Ricardo da Silva, J. M., (2003). Monomeric, oligomeric, and polymeric flavan-3-ol composition of wines and grapes from *Vitis vinifera* L. Cv. Graciano, Tempranillo, and Cabernet Sauvignon. *J. Agric. Food Chem.*, 51, 6475-6481.

Mulero, J., Pardo, F., Zaffrilla, P., (2010). Antioxidant activity and phenolic composition of organic and conventional grapes and wines. *J. Food Compos. Anal.*, 23,569–574.

Mullen, W., Edwards, C. A., Crozier, A., (2006). Absorption, excretion and metabolite profiling of methyl-, glucuronyl-, glucosyl- and sulpho-conjugates of quercetin in human plasma and urine after ingestion of onions. *Brit. J. Nutr.,* 96, 107-116.

Nawaz, H., Shi, J., Mittal, G. S., Kakuda, Y., (2006). Extraction of polyphenols from grape seeds and concentration by ultrafiltration. *Separ. Purif. Technol.*, 48, 176–181.

Ndiaye, M., Chataigneau, M., Lobysheva, I., Chataigneau, T., Schini-Kerth, V. B., (2004). Red wine polyphenols-induced, endothelium-dependent NO mediated relaxation is due to the redox-sensitive PI3-kinase/Akt-dependent phosphorylation of endothelial NO synthase in the isolated porcine coronary artery. *FASEB J.*, 19, 455-457.

Nemeth, K., Plumb, G. W., Berrin, J., Juge, N., Jacob, R., Naim, H. Y., Williamson, G., Swallow, D. M., Kroon, P. A., (2003). Deglycosylation by small intestinal epithelial cell b-glucosidase is a critical step in the absorption and metabolism of dietary flavonoid glycosides in humans. *Eur. J. Nutr.*, 42, 29-42.

Nigdikar, S. V., Williams, N. R., Griffin, B. A., Howard, A. N., (1998). Consumption of red wine polyphenols reduces the susceptibility of low-density lipoproteins to oxidation in vivo. *Am. J. Clin. Nutr.*, 68, 258–265.

Nunan, K. J., Sims, I. M., Bacic, A., Robinson, S. P., Fincher, G. B., (1998). Changes in cell wall composition during ripening of grape berries. *Plant Physiol.,* 118, 783–792.

Obreque-Slier, E., Peña-Neira, A., López-Solís, R., Zamora-Marín, F., Ricardo da Silva, J. M., Laureano, O., (2010). Comparative study of the phenolic composition of seeds and skins from Carménère and Cabernet Sauvignon grape varieties (*Vitis vinifera* L.) during ripening. *J. Agric. Food Chem.*, 58, 3591–3599.

O'Byrne, D. J., Devaraj, S., Grundy, S. M., Jialal, I., (2002). Comparison of the antioxidant effects of Concord grape juice flavonoids alpha-tocopherol on markers of oxidative stress in healthy adults. *Am. J. Clin. Nutr.*, 76, 1367–1374.

Olszanecki, R., Gebska, A., Kozlovski, V. I., Gryglewski, R. J., (2002). Flavonoids and nitric oxide synthase. *J. Physiol. Pharm.*, 5, 571–584.

Palafox-Carlos, H., Ayala-Zavala, J. F., González-Aguilar, G. A., (2011). The role of dietary fiber in the bioaccesibility and bioavailability of fruit and vegetable antioxidants. *J. Food Sci.*, 76, R6-R15.

Pandey, K. B., Rizvi, S. I., (2009). Plant polyphenols as dietary antioxidants in human health and disease. *Oxidative Med. Cell. Longev.*, 2, 270-278.

Park, Y. K., Kim, J. S., Kang, M. H., (2004). Concord grape juice supplementation reduces blood pressure in Korean hypertensive men: double-blind, placebo controlled intervention trial. *Biofactors,* 22, 145–147.

Passamonti, S., Vrhovsek, U., Vanzo, A., Mattivi, F., (2005). Fast access of some grape pigments to the brain. *J. Agric. Food Chem.*, 53, 7029–7034.

Pastrana-Bonilla, E., Akoh, C. C., Sellappan, S., Krewer, G., (2003). Phenolic content and antioxidant capacity of muscadine grapes. *J. Agric. Food Chem.*, 51, 5497–5503.

Perrone, G., Nicoletti, I., Pascale, M., Rossi, A. D., Girolamo, A. D., Visconti, A., (2007). Positive correlation between high levels of ochratoxin A and resveratrol-related compounds in red wines. *J. Agric. Food Chem.*, 55, 6807-6812.

Pezzuto, J. M., (2009). Grapes and human health: A perspective. *J. Agric. Food Chem.,* 56, 6777–6784.

Proteggente, A. R., Pannala, A. S., Pagana, G., Van Buren, L., Wagner, E., Wiseman, S., Van De Put, F., Dacombe, C., Rice-Evans, C. A., (2002). The antioxidant activity of regularly consumed fruit and vegetables reflects their phenolic and vitamin C composition. *Free Rad. Res.,* 36, 217–233.

Psomas, A. N., Anastasiadi, M., Haroutounian, S. A., (2012). Correlation between the polyphenolic content of Greek wines and their geographical origin. In: Grapevines: Varieties, Cultivation. Szabo, P. V. et al. (Eds), Nova Science Pub, Inc. p. 133-153.

Ray, P. S., Maulik, G., Cordis, G. A., Bertelli, A. A., Bertelli, A., Das, D. K., (1999). The red wine antioxidant resveratrol protects isolated rat hearts from ischemia reperfusion injury. *Free Rad. Biol. Med.*, 27, 160–169.

Rendeiro, C., Guerreiro, J. D., Williams, C. M., Spencer, J. P., (2012). Flavonoids as modulators of memory and learning: molecular interactions resulting in behavioural effects. *Proc. Nutr. Soc.*, 71, 246-262.

Rodriguez Montealegre, R., Romero Peces, R., Chacon Vozmediano, J. L., Martinez Gascuena, J., Garcia Romero, E., (2006). Phenolic compounds in skin and seeds of ten grape *Vitis vinifera* varieties grown in a warm climate. *J. Food Compos. Ana*l., 19, 687–693.

Scalbert, A., Williamson, G., (2000). Dietary intake and bioavailability of polyphenols. *J. Nutr.*, 130, 8S Suppl, 2073S–2085S.

Scalbert, A., Morand, C., Manach, C., Rémésy, C., (2002). Absorption and metabolism of polyphenols in the gut and impact on health. *Biomed. Pharmacother.*, 56, 276–282.

Schettino, T., (2011). Effect of the daily ingestion of a purified anthocyanin extract from grape skin on rat serum antioxidant capacity. *Physiol. Res.,* 60, 637-645.

Selma, M. V., Espín, J. C., Tomás Barberán, F. A., (2009). Interaction between phenolics and gut microbiota: role in human health. *J. Agric. Food Chem.*, 57, 6485-6491.

Shiraishi, M., Yamada, M., Mitani, N., Ueno, T., (2007). A rapid determination method for anthocyanin profiling in grape genetic resources. *J. Japan Soc. Hort. Sci.*, 76, 28–35.

Shukitt-Hale, B., Carey, A., Simon, L., Mark, D. A., Joseph, J. A., (2006). The effects of Concord grape juice on cognitive and motor deficits in aging. *Nutrition,* 22, 295–302.

Shukitt-Hale, B., Lau, F. C., Carey, A. N., Spangler, E. L., Ingram, D. K., Joseph, J. A., (2008). Blueberry polyphenols attenuate kainic acid-induced decrements in cognition and alter inflammatory gene expression in rat hippocampus. *Nutr. Neurosci.*, 11, 172–182.

Silberberg, M., Morand, C., Mathevon, T., Besson, C., Manach, C., Scalbert, A., Remesy, C., (2006). The bioavailability of polyphenols is highly governed by the capacity of the intestine and of the liver to secrete conjugated metabolites. *Eur. J. Nutr.,* 45, 88-96.

Stalmach, A., Edwards, C. A., Wightman, J. D., Crozier, A., (2011). Identification of (Poly)phenolic compounds in Concord grape juice and their metabolites in human plasma and urine after juice consumption. *J. Agric. Food Chem.,* 59, 9512-9522.

Stein, J. H., Keevil, J. G., Wiebe, D. A., Aeschlimann, S., Folts, J. D., (1999). Purple grape juice improves endothelial function and reduces the susceptibility of LDL cholesterol to oxidation in patients with coronary artery disease. *Circulation,* 100, 1050-1055.

Torres, J. L., Varela, B., Garcıa, M. T., Carilla, J., Matito, C., Centelles, J. J., Cascante, M., Sort, X., Bobet, R., (2002). Valorization of grape (*Vitis vinifera*) byproducts. Antioxidant and biological properties of polyphenolic fractions differing in procyanidin composition and flavonol content. *J. Agric. Food Chem.*, 50, 7548–7555.

van Duynhoven, J., Vaughan, E. E., Jacobs, D. M., Kemperman, R. A., van Velzen, E. J., Gross, G., Roger, L. C., Possemiers, S., Smilde, A. K., Doré, J.,Westerhuis, J. A., Van de Wiele, T., (2011). Metabolic fate of polyphenols in the human superorganism. *Proc. Natl. Acad. Sci. U.S.A,* 108 Suppl 1,4531-4538.

Visloky, L. M., Fernandez, M. L., (2010). Biomedical effects of grape products. *Nutr. Rev.,* 68, 656-670.

Wang, R., Zhou, W., Wen, R. A. H., (2006). Kinetic study of the thermal stability of tea catechins in aqueous systems using a microwave reactor. *J. Agric. Food Chem*., 54, 5924–5932.

Williams, C. M., El Moshen, M. A., Vauzour, D., Rendeiro, C., Butler, L. T., Ellis, J. A., Whiteman, M., Spencer, J. P., (2008). Blueberry-induced changes in spatial working memory correlate with changes in hippocampal CREB phosphorylation and brain-derived neurotrophic factor (BDNF) levels. *Free Radic. Biol. Med*., 45, 295–305.

Williamson, G., (2004). Common features in the pathways of absorption and metabolism of flavonoids. In: Meskin, M. S., Bidlack, W. R., Davies, A. J., Lewis, D. S. and Randolph, R. K. (Eds) Phytochemicals. Boca Raton, CRC Press.

Xu, C., Zhang, Y., Cao, L., Lu, J., (2010). Phenolic compounds and antioxidant properties of different grape cultivars grown in China. *Food Chem*., 119, 1557–1565.

Xu, Y., Simon, J. E., Welch, C., Wightman, J. D., Ferruzzi, M. G., Ho, L., Passinetti, G. M., Wu, Q., (2011). Survey of polyphenol constituents in grapes and grape-derived products. *J. Agric. Food Chem*., 59, 10586–10593.

Xia, E., Deng, G. F., Guo, Y. J., Li, H. B., (2010). Biological activities of polyphenols from grapes. *Int. J. Mol. Sci.,* 11, 622-646.

Yadav, M., Jain, S., Bhardwaj, A., Nagpal, R., Puniya, M., Tomar, R., Singh, V., Parkash, O., Prasad, G. B. K. S., Marotta, F., Yadav, H., (2009). Biological and medicinal properties of grapes and their bioactive constituents: An update. *J. Med. Food,* 12, 473–484.

Yilmaz, Y., Toledo, R. T., (2004). Major flavonoids in grape seeds and skins: antioxidant capacity of catechin, epicatechin, and gallic acid. *J. Agric. Food Chem.,* 52, 255-260.

Youdim, K. A., Dobbie, M. S., Kuhnle, G., Proteggente, A. R., Abbott, N. J., Rice-Evans, C., (2003). Interaction between flavonoids and the blood-brain barrier: *in vitro* studies. *J. Neurochem*. 85, 180–192.

In: Grapes
Editor: José de Sousa Câmara

ISBN: 978-1-63321-402-6

Chapter 18

Potential Benefits of Phenolic Compounds Extracted from Winemaking Wastes

Milla Alves Baffi[1,•], Ellen Silva Lago Vanzela[2], Eleni Gomes[2] and Roberto Da-Silva[2]
[1]Uberlandia Federal University (UFU), Uberlandia, MG, Brazil
[2]Sao Paulo State University (UNESP), Sao Jose do Rio Preto, SP, Brazil

Abstract

The grape is one of the most popular fruit crops in the world and the winemaking sector produces large amounts of under-utilized by-products (pomace), which have significant economic potential. Grape pomace (consisting mainly of skins, seeds and stems) has been evaluated as a promising source of antioxidants because of its high content of phenolic compounds (gallic acid, flavan-3-ols, flavonoids, stilbene and anthocyanins). In addition, studies have demonstrated some of the biological effects of these phenolic compounds such as anti-carcinogenic, anti-mutagenic and antimicrobial activity. As a result, there is a considerable interest in alternative approaches for the control of infections and preservation of foods using biologically active phenolic compounds extracted from winemaking wastes. However, the chemical composition and availability of these compounds in grape pomace depends on various factors: grape varieties, the contact time between the skins and the wine in fermentation and, also, environmental factors. Moreover, the yield of phenolic compounds extracted from these residues can vary from 4.0 to 65.0% depending on the extraction conditions (extraction solvent, extraction time), grape variety and grape fraction. Therefore, the recovery of these substances from this residue of the wine industry, with their potential application, could represent a significant advance in maintaining environmental balance, as well as adding value to the byproducts of winemaking. For all the mentioned reasons, the present chapter describes the importance of the recycling of grape pomace as a prospective

• E-mail: millabaffi@yahoo.com.br.

source of phenolic compounds with application in the food and pharmaceutical industries and with potential benefits to human health.

Keywords: Winemaking residues; Phenolic compounds; Antimicrobial activity

Introduction

Pathogenic microorganisms are able to bring about several diseases in human and other animals. Some bacteria produce enterotoxins and are known to cause distinct pathologies. In addition, some of them can develop resistance to antimicrobial agents, being responsible for multiple resistances in hospital and foodborne infections.

The emergence and spreading of resistant bacteria is an inevitable side effect of the use of antibiotics. In this way, a monitoring system of the prevalence and antibiotic resistance of bacteria from human, animal, food and environmental samples is necessary to understand the epidemiology of resistant strains to assure food safety and consumer health. For this reason, there is considerable interest in alternative and adjuvant approaches for the control of infections using biologically active compounds, especially antioxidants and antimicrobials from plants.

On the other hand, the use of natural methods for the control of pathogenic agents has represented a significant advance in the research associated with human health and the food and pharmacological industries. Natural products from plants, such as oils and extracts, have been studied throughout the world due to their antibacterial, antifungal and antioxidant properties and because of their potential use for the control of microorganisms which cause diseases. The application of these secondary metabolites has increased and achieved the market since they present benefits to health and low impact on the environment. Thus, phytochemical research comes up as a useful method that allows the elucidation of countless compounds with potential for application in the production of new pharmacological agents.

Natural Products As Antimicrobial Agents

The use of plants with medicinal properties is a practice well-known since the first civilizations. People from Egypt and Greece have traditionally used plants and their extracts in several procedures. In the past, the use of natural products emerged through the use of herbs by indigenous communities and it has been transferred to the colonizers, becoming widely applied in homemade medicine as teas and syrups. Nowadays, the functional properties of plant extracts with antimicrobial and antioxidant activities are the basis for their use in food preservation, pharmaceuticals, cosmetics and alternative medicine and a large number of compounds derived from plants are used in medical formulations, acting as prototypes for important synthetic therapeutic agents (Simões et al. 2004).

The use of natural products of vegetal origin for the treatment of human diseases is a millennial practice. However, in the last century, traditional medicine has been helped basically through the use of synthetic pharmacological products. Currently, around 4.000 antibiotics are known. Nevertheless, many of them are very toxic with only around 50 being non-toxic for humans and employed in drug therapy for infectious diseases. On the other hand, the increasing number of microorganisms with resistance to such substances has stimulated the research for novel antimicrobial substances.

Studies regarding the extraction of those natural compounds which are capable of avoiding microbial growth might be of special interest, mainly with respect to microorganisms associated to the deterioration of foods and disease-causing microorganisms. These natural products are generally recognized as safe and, therefore, the use of naturally occurring plant extracts may provide a potential additional barrier to inhibit the growth of hospital and food borne pathogens (Michielin et al. 2009; Oliveira et al. 2013).

The majority of these substances, when they are used for microbial control, execute their action through alterations to the bacterial cellular structure. They act specifically, changing the selective permeability of cytoplasmic membrane by denaturation of membrane proteins; causing alterations in the structure of cell wall or hindering its correct synthesis; leading modifications at protein synthesis or at the replication of bacterial DNA. In that way, they can trigger the interruption of essential processes for the maintenance of the cell and, consequently, the bacterial death.

Antibacterial actions of many plant extracts and their metabolites have been evidenced by *in vitro* studies (Mendez et al. 2012; Piccirillo et al. 2013). Oussalah et al. (2007) reported the antibacterial activity of essential oils from plants containing different monoterpenes (eugenol, timol, carvacrol and linalool) against the growth of *Escherichia coli, Salmonella typhimurium, Spaphylococcus aureus* and *Listerria monocytogenes.* SANTURIO et al. (2007) observed a strong antimicrobial activity of essential oils (EOs) from oregano against Salmonella enterica sorovars from an avian source. In a recent study, the antimicrobial activities of EOs from *Oreganum vulgare* (oregano), *Thymus vulgaris* (thyme), *Lippia graveolens* (Mexican oregano), *Cinnamomum zeylanicum* (cinnamon), *Zingiber officinale* (ginger), *Salvia officinalis* (sage), *Rosmarinus officinalis* (rosemary) and *Ocimum basilicum* (basil) were also evaluated against *Escherichia coli* strains isolated from poultry and cattle faeces (Santurio et al. 2011).

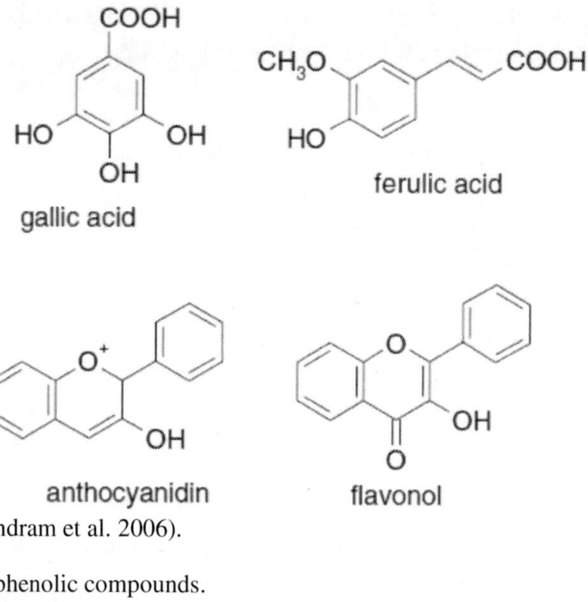

(extracted from Balasundram et al. 2006).

Figure 1. Examples of phenolic compounds.

The authors concluded that for all strains tested, regardless of their origin, the EOs that were most and least effective in antimicrobial activity were oregano and cinnamon, respectively, confirming the antimicrobial potential of some EOs.

Many of these extracts are phenolic compounds which are widely distributed in plants (Balasundram et al. 2006). Phenolics are substances which contain aromatic rings with one or more hydroxyl substitutions, including their functional groups (Figure 1). There are around 8.000 different phenolics which, according to their chemical structure, are divided into classes: phenolic acids, flavonols, estilbens and tannins in which the partially hydrophobic nature is responsible for the antimicrobial activity (Balasundram et al. 2006). Recently, the recognition of the antioxidant and antimicrobial properties of these compounds has provided a new insight into their beneficial effects for human health (Kaur and Kapoor, 2011).

Antimicrobial Activity of Red Wine

Wine is a beverage linked to human history due to its organoleptical characteristics and benefits to health. Historical records point to the medicinal use of wines by the Greeks 2000 years ago and Hipocrates (460-370 b.C.) was one of the first scholars to report the therapeutic properties of wines (Penna and Hecktheuer, 2004). This beverage is considered a food supplement due to its elevated content of several bioactive compounds, especially phenolic compounds, with antioxidant and preventive character of numerous pathologies when consumed in moderate quantities (Jackson, 2000).

The consumption of alcoholic beverages has been described as having a protective effect against bacterial infections. For example, indigenous medicinal drinks from native plants have been described in terms of their antibacterial activity and treatment of respiratory illnesses (Oliveira et al. 2007). Among them, hydroalcoholic drinks of *Aristolochia triangularis* (leaf, root and stem extracts), *Stryphnodendron adstringens* (bark extract), *Tabebuia ovellanedae* (bark extract), *Vernonia polyanthes* (root extract) were able to reduce the growth of Mycobacterium strains by two orders of magnitude in colony forming units (CFU/mL). At the same way, many studies have shown that wine, even in high dilutions, has bactericide action or can be able to reduce the viability of different genera of Gram-positive and Gram-negative bacteria, such as: *Escherichia, Listeria, Salmonella, Shigella, Staphylococcus, Streptococus, Pneumococus, Campylobacter* and *Vibrio* (Sugita-Konishi et al. 2001; Moretro and Daeschel, 2004; Gañan et al. 2009). Furthermore, the consumption of wine during meals has also been reported to have a protective effect against food-borne toxi-infections as a possible antimicrobial agent that can help the organism in the combat of external pathogenic agents mainly from bacterial food poisoning (Belido-Blasco et al. 2002; Correia et al. 2003).

Nonetheless, the exact mechanisms responsible for the antimicrobial effect of wine are not fully understood. This beverage possesses a relatively high ethanol content in addition to other antimicrobial agents like organic acids, a low pH and high polyphenols content that could affect the viability of fungi, yeasts, some viruses and countless bacteria (Just and Daeschel, 2003).

The phenolic compounds have been considered the main bactericidal or bacteriostatic substances which can inhibit the growth of some bacteria by binding to the outer membrane of the cellular wall of gram negative strains or to the microbial cytoplasmic membrane.

However, this effect depends of the phenolic compounds and of the strains tested. The antibacterial mechanisms of action of various flavonoids present in the wine could be attributed to the inhibition of nucleic acid synthesis, the inhibition of cytoplasmic membrane function or the inhibition of energy metabolism (Queipo-Ortuño et al. 2012).

One of the most important compounds found in red wines is resveratrol, which has been the focus of multiple studies (Figure 2). In vitro assays demonstrated that the resveratrol from red wines was able to inhibit the growth of different strains of *Helicobacter pylori*, with minimal inhibition concentrations between 25 and 100 μg/mL (Paulo et al. 2011). The authors not only confirmed the antimicrobial activity of resveratrol but also suggest that this compound can have a therapeutic potential as an alternative approach in the treatment of infections with antibiotics. It was also indicated that both resveratrol and the red wine itself were able to suppress the urease activity of different bacteria isolates, the repressive activity of wine being superior to the inhibitory action of resveratrol when tested separately in the same concentrations. Thus, it is likely that other components from the wine, such as ethanol, anthocyanins, tannins or organic acids can also contribute to this inhibition or can improve the effects of resveratrol (Paulo et al. 2011).

Although the majority of the benefic properties of wines have been attributed to resveratrol, other compounds, such as quercetin, cathequins, organic acids and anthocyanins have also demonstrated antimicrobial activity against various species of bacteria. The p-cumaric acid showed to be an antagonist of Gram-positive bacteria *Staphylococcus* and *Streptococcus*, while other polyphenols have been associated with the inhibition of Gram-negative bacteria, for example, *Escherichia coli, Shigella, Proteus* and *Vibrio* (Jackson, 2000). Organic acids present in wines, in conjunction with ethanol, were associated with the inhibition of the growth of *Campylobacter jejuni* while the combination of these acids with ethanol and acid pH showed strong antibacterial activity against *Escherichia coli, Listeria monocytogenes, Salmonella typhymurium* and *Staphylococcus aureus* (Boban et al. 2010).

Notwithstanding several reports regarding the protective role to human health against bacterial infections by the moderate consumption of wine, there are few studies employing *in vivo* experimental conditions. In the work of FERNANDES et al. (2007), a model stomach, containing a food matrix and a synthetic gastric fluid, was used to study the bactericidal effect of ingested wine on *Listeria innocua*.

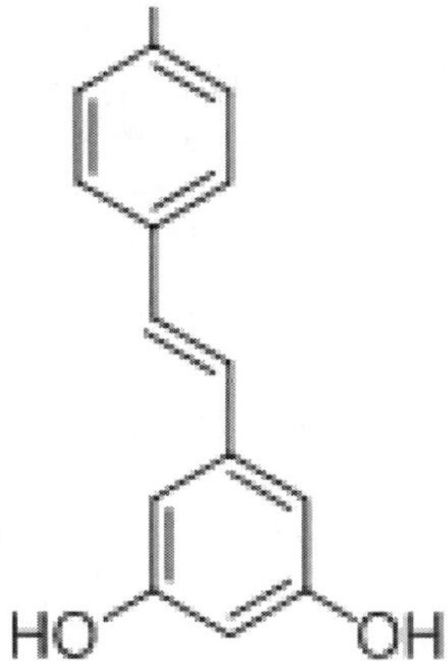

Figure 2. Structure of Trans-resveratrol.

The authors observed that volumes of wine equivalent to the ingestion of one glass led, over a period of 2 hours, to a reduction of three logarithmic cycles of the initial population. The influence of ethanol and organic acids was also investigated and the results suggested that the combination of ethanol with the organic acids act synergistically but to a lesser extent than wine itself, indicating that the ingestion of wine during a meal may reduce the quantity of *Listeria* which persists further in the alimentary tract and can help to prevent toxi-infeccions by this bacteria.

A very recent study and the first one of *in vivo* assays aimed to evaluate the possible prebiotic effect of polyphenols from red wine in the human intestinal flora (Queipo-Ortuño et al. 2012). It was observed that there were significant changes in the faecal microbiota of the participants, and that the polyphenolics from red wine demonstrated ability to prevent the growth of dangerous bacteria such as some strains of *Clostridium*. Moreover, the growth of probiotic bacteria was also verified, such as Bifidobacteria, which is implicated in the reduction of an inflammatory biomarker (reactive Protein C) and in the cholesterol concentration, bringing benefits to the host.

Studies for the evaluation of the antimicrobial activity of wine against *Bacillus cereus* vegetative cells showed that the wine causes a strong inactivation effect on vegetative cells of these bacteria (Vaz et al. 2012). In simulated gastric conditions (in the presence of food), wine diminished considerably the number of viable cells in addition to the effect of the synthetic gastric fluid. The study provided evidence that the consumption of wine during a meal may diminish the number of viable cells of *B. cereus* and reduce the impact of the germination of spores that may occur in the small intestine, thus lowering the risk of toxi-infection that may be caused by this pathogen (Vaz et al. 2012).

Grape Residues As Antimicrobial Agents

The grape is one of the fruits of major world production, with approximately 67 millions of tons per year, being Vitis vinifera the grape variety most cultivated for winemaking (Maier et al. 2009). Grapes are a rich source of different potentially bioactive phenolic compounds in high concentrations and the by-products of wine production, the skin, for example, can maintain appreciable quantities of these substances such as flavonoids (anthocyanins and flavanols), stilbenes (resveratrol), phenolic acids (derived from hydroxycinnamic and hydroxybenzoic acid) and tannins, which have antioxidants and antibacterial properties (Balasundram et al. 2006). In this sense, particular attention has been addressed to the use of grape pomace, a winery waste obtained in large amounts around the world, to produce high value products (Pinelo et al. 2007). The grape pomace is the most abundant waste in the wine industry and is composed of skins, seeds, and stems (Figure 3).

Attention to bioactive substances from grape and its products such as wine was first motivated by the "French paradox." Henceforth, some studies have revealed that phenolic compounds like gallic acid, catechin and several other phenolics are inhibitors of human low density lipoprotein (LDL) oxidation, and, therefore, can reduce the possibility of cardiovascular illnesses (Teissedre and Landrault, 2000). Several other works related grape substances to health benefits, such as epicatechin to antimutagenic and antiviral activities (Saito et al. 1998), tannic acid to a decrease in hepatic lipids (Do et al. 2011), gallic acid to the inhibition of prostate cancer (Veluri et al. 2006) and vanillic acid to antimicrobial

(Ravichandran et al. 2011), anti-mutagenic and anti-carcinogenic effects (Lirdprapamongkol et al. 2005). Esterified p-OH-benzoic acid shows antimicrobial activity and is widely used as a preservative in food, cosmetics, and pharmaceutical products (Soni et al. 2005). Gallic acid was also identified as an antibacterial component against *E. coli, S. enteritidis* and *S. aureus* (Shoko et al. 1999), confirming the importance of the reuse of grape pomace to obtain bioactive compounds.

Phenolic compounds in winemaking by-products play important roles with antioxidant, anti-inflammatory and antimicrobial effects. Moreover, grape seed extracts are usually regarded as safe for food additives and known for their antioxidant and antimicrobial activities (Rababah et al. 2004). Several types of polyphenols, including catechin and its derivates like proanthocyanidin, show antibacterial activity by the accumulation of these phenolics to the bacterial cytoplasmic membrane leading to cell death (Pyla et al. 2010). Among them, stilbenes stand out as being biologically active against many pathogens, with the species specific bacterial sensibility (Jeandet et al. 2002).

Many studies are found in the literature evaluating the *in vitro* antimicrobial activity and the composition of grape pomace extracts in several strains of bacteria and fungi (Thimothe et al. 2007; Corrales et al. 2008; Fikselova et al. 2010). Most of these studies connected the antimicrobial activity to the content of phenolic compounds like resveratrol (Chan, 2002; Filip et al. 2003; Chatterjee et al. 2005). BAYDAR et al. (2004) measured growth inhibition zones generated by the addition of phenolic extracts obtained from *Vitis vinifera* grape pomace and have considered the gallic acid as the main active compound against *E. coli.* PAPADOPOULOU et al. (2005) demonstrated the potential antimicrobial activity and the composition of alcohol-free red and white wine extracts against pathogenic strains of *S. aureus, E. coli* and *Candida albicans* and HAN (2007) showed that the grape seed extract alone inhibited the growth of *C. albicans* yeast cells, with a dose-dependent antifungal effect and a synergic effect with other antifungal (Amphorericin B) used to treat fungal infections.
ROTAVA et al. (2009) determined the *in vitro* antibacterial, antioxidant and tannin activity from grape by-products (*V. vinifera*), for their use as alternative ingredients in the poultry industry and in food conservation.

Figure 3. Grape pomace (a mixture of skins, seeds and stems) *in natura*.

The authors observed that the defatted grape seed extract (ESUD) has presented high antibacterial activity against strains of *S. aureus* and *E. coli*. BROWN et al. (2009) analyzed the inhibitory effect of grape skin and seed extracts on *Helicobacter pylori* strains and demonstrated that grape extracts and their compounds were effective in the *in vitro* inhibition of *H. pylori*, with the muscadine grape skin extract having the highest efficient.The authors also suggested that although anti-*H. pylori* activities by individual compounds have been reported, it is believed that a synergistic mode of action is more likely responsible for the extract's antimicrobial activity. SONJA et al. (2009) tested the resistance of *Campylobacter* chicken meat isolates against several selected antibiotics used in human and veterinary medicine and phenolic extracts from grape skins of different grape varieties. The authors reported that the initial screening of susceptibility of *Campylobacter* isolates to grape skin phenolic extracts was efficient and the results were very promising for further application of these extracts as antimicrobial agents for potential use in assuring safety of poultry meat and other products.

In a study carried out by KATALINIC et al. (2010), the polyphenolic composition, the antioxidative and antimicrobial properties of grape skin extracts from 14 grape varieties from Croatia were determined. The authors observed that the content of total phenols, flavonoids, catechins, flavanols and individual polyphenols ((+)-catechin, (-)-epicatechin, epicatechin gallate, procyanidin B1 and B2, quercetin glucoside, resveratrol monomers, piceid and astringin) was variety dependent. In addition, the authors also suggested that the antimicrobial activity screened against Gram-positive (*Staphylococcus aureus, Bacillus cereus*) and Gram-negative bacteria (*E. coli* O157:H7, *Salmonella Infantis, Campylobacter coli*) was confirmed against all tested organisms with minimum inhibitory concentrations (MICs) found in the range 0.014–0.59 mg of gallic acid equivalents (GAE)/ml, with lower MICs of white cultivars, especially against *Campylobacter* and *Salmonella*.

ABTAHI et al. (2011) conducted a study to measure the antibacterial activity of ethanol extract of the whole grape (*Vitis vinifera*) against several pathogenic bacteria. Aliquots of the extracts at variable concentrations were incubated with different bacterial strains, and the antimicrobial activities of the ethanolic extracts from total grapes were determined using the minimum inhibitory concentration (MIC). The authors reported that the whole grape extract significantly inhibited the growth of the tested bacterial strains and among them *S. aureus* was the most susceptible. These results suggested that total grapes extract may also be used as natural antibacterial for treatment of some diseases, especially local skin diseases (Abtahi et al. 2011).

CHENG et al. (2012) studied the effect of extraction solvent, fraction and variety on the phenolic profile, antioxidant and antimicrobial activity of extracts from wine residues of Pinot noir grapes from New Zealand and observed that 50% aqueous acetone extracts resulted in the highest antioxidant activities compared to other solvents. These extracts also presented a high antimicrobial activity against *S. aureus* which were comparable to some antibiotics.

In spite of the recent studies in different regions of the world indicating the high antimicrobial potential of grape waste extracts, in Brazil, few reports are available. Moreover, almost the totality of these studies has only used grape extracts from the *V. vinifera* variety. Recent work carried out in the state of Santa Catarina evaluated the global extraction yield, the antimicrobial activity and the composition profile of Merlot and Syrah grape pomace extracts obtained by supercritical fluids (Oliveira et al. 2013). The main components from the extracts were gallic acid, p-OH-benzoic acid, vanillic acid and epicatechin. The antibacterial and antifungal activities of these extracts were evaluated against four strains of bacteria (*S. aureus, B. cereus, E. coli and Pseudomonas aeruginosa*) and three fungi strains (*C. albicans, Candida parapsilosis, Candida krusei*) and the supercritical fluid extracts presented the highest antimicrobial effectiveness compared to the other grape pomace extracts. Merlot extracts were more active against Gram-positive bacteria. Another study performed in São Paulo state assessed the antimicrobial potential and chemical composition of grape and other agro-industrial wastes against pathogenic microorganisms of importance in foods. Among grape residues, Pinot Noir grape marc, Petit Verdot grape seed and marc and red grape fermentation lees showed antimicrobial activity against *S. aureus* and *L. monocytogenes*.

Analysis by gas chromatography - mass spectrometry (GC-MS) allowed the identification of caffeic, gallic, ferulic and p-coumaric acids, besides flavonoids quercetin, myricetin and epicatechin on wastes which exhibited antimicrobial activity (Martin et al. 2012).

All of these reported studies demonstrated that the industrial wastes from wine processing (grape pomace) represent a rich source of a variety of compounds, mainly phenolics, which can be used as new antimicrobial agents for application in the food industry as natural preservatives and also as new pharmacological agents.

Methods of Extraction of Phenolic Compounds

The quality and the composition of the extracts are strongly dependent on the extraction technique, the solvent used, the origin of the raw material, the part of the plant used (seeds, skin, leaves) and storage conditions (Louli et al. 2004). The extraction is the first stage in the recovery of phenolic compounds from the grape pomace. For that, the yield, composition and purity of the phenolics recuperated from plant materials is dependent on the chemical nature (simple and complex phenolics), sample size, extraction method (solvent, supercritical and solid-phase extractions), storage time and conditions, as well as the presence of interfering substances (Naczk and Shahidi, 2004).

The solvent type is the main factor used to determine the efficiency of an extraction process. The conventional extraction solvents which have been usually used for the recovery of polyphenols from plant materials are: acetone, ethyl acetate, ethanol and methanol, with variable efficiencies (Chavan et al. 2001; Cheng et al. 2012). Numerous works compared the application of several solvents in the extraction of polyphenols from wine wastes (Yilmaz and Toledo, 2006; Abtahi et al. 2011; Cheng et al. 2012). Among them, acetone, ethanol and methanol have been suggested as the best solvents for the extraction of polyphenols from wine residues (Kallithraka et al. 1995; Yilmaz and Toledo, 2006; Cheng et al. 2012). For the extraction of vinification solid residues, many studies have used organic/alcoholic solvent–water mixtures (Bonilla et al. 1999; Negro et al. 2003) because the presence of water can increase the permeability of cell tissue, thus improving the mass transfer as well as recovering the water soluble bioactive compounds (Alonso-Borbalan et al. 2003).

Currently, among acetone, ethanol and methanol, there is some debate regarding the best solvent extraction system. Acetone–water mixture was suggested as giving a better extraction of procyanidins from grape seeds in comparison with other extractants (Yilmaz and Toledo, 2006). Nevertheless, other studies showed that the extraction of catechins and procyanidins was more efficient when an ethanol–water mixture was employed (Jayaprakasha et al. 2002).

In contrast to these results, methanol was reported to be the best solvent for the extraction of catechins, epicatechins and epigallocatechins from grape seeds (Kallithraka et al. 1995).

A recent study was carried out to determine the effect of extraction solvent type and fraction on the biological activities (phenolic profile, antioxidant and antimicrobial activity) of extracts from wine residues to maximize their utilization as bioactive compounds (Cheng et al. 2012). The authors observed that 50% aqueous solvent was the optimum concentration for the extraction of phenolic compounds in two extractions steps (Cheng et al. 2012). They also observed that methanol/water seed extracts exhibited higher antimicrobial activity against *S. aureus* than the other extracts. Among phenolic compounds extracted, anthocyanin compounds were noted to have good MIC values (ranging between 0.0625 and 0.5 mg/mL), particularly malvidin chloride, which improved the antimicrobial properties of skin extracts.

The majority of the literature reports that the polyphenols most commonly extracted from wine residue are extracted after two extraction steps (Shi et al. 2003; Nawaz et al. 2006), at room temperature (Yilmaz and Toledo, 2006; Makris et al. 2007) and at a shorter extraction time of less than 1 hour (Yilmaz and Toledo, 2006; Cheng et al. 2012). According to BUCIC-KOJIC et al. (2007), 80–90% of total polyphenols collected within 200 min was extracted in the first 40 min of the extraction process. In addition, the extraction yield can vary from 4.0 to 65.0% (w/w) depending on the extraction solvent as well as the grape variety, fraction and if the pomace is exposed to fermentation or not (Cheng et al. 2012).

Some studies mentioned that the use of liquid-solvent-based methods to obtain the extracts presents certain limitations like: high energy costs; elevated solvent use; injurious to thermolabile substances; low selectivity; and vestiges of solvent in the solute (Pourmortazavi and Hajimirsadeghi, 2007). For this reason, alternative methods to obtain bioactive phenolic substances have been investigated, such as ultrasound-assisted leaching and supercritical fluid extraction (Oliveira et al. 2013).

The supercritical fluid extraction method (SFE) is established by the use of solvents in conditions above the critical point, resulting in a liquid-like density, gas-like viscosity and the diffusivity values within two orders of magnitude higher than that for typical liquids (Oliveira et al. 2013). This method presents various advantages over the traditional liquid-solvent-based processes including improved selectivity, automation and environmental safety since the solvent can easily be removed from the solutes (Brunner, 2005). The carbon dioxide (CO2) is normally applied in SFE because it is non-toxic, inexpensive and volatile, with moderate critical conditions (Pinelo et al. 2006). Nevertheless, CO2 is a non-polar compound, which disqualifies its use as a solvent for polar compounds and limits its solvent power. To overcome this disadvantage, polar co-solvents can be used to increase the overall solvent power during SFE (Michielin et al. 2009).

A very recent research evaluated the use of different extraction methods, such as low-pressure procedures using different solvents and supercritical fluid extraction (SFE) with CO2 and with CO2 and co-solvent, to obtain phenolic grape pomace extracts from Merlot and Syrah varietals (Oliveira et al. 2013). The extracting techniques were compared in terms of global extraction yield and antibacterial activity of the extracts, with the aim of evaluating the potential of the grape pomace to generate an alternative antimicrobial product. Among all the methods evaluated, the best antimicrobial results were obtained with SC-CO2 extracts (*S. aureus* and *B. cereus*), denoting the SFE as the most efficient method to obtain valuable antimicrobial extracts from grape pomace and contributing to increase the aggregate value of this by-product from the wine industry.

Determination of Phenolic Content

The total phenolic content (TPC) of the extracts from wine residues is generally determined spectrophotometrically by the use of the Folin–Ciocalteu colorimetric method (Singleton and Rossi, 1965). Folin-Ciocalteu (FC) analyses are used because they are appropriate, simple, and only need common equipment. Under suitable conditions, the assay includes monophenols and gives the expected reactions with the different categories of naturally-occurring phenols. Because different phenols react in variable degrees, the expression of the results as a single number—such as milligrams per liter gallic acid equivalence—is necessarily arbitrary. Also, because the reaction is independent, quantitative, and predictable, the analysis of a mixture of phenols can be recalculated in terms of any other standard (Singleton et al. 1999). The test measures all the substances which are readily oxidizable under the reaction conditions. Careful use of the assay can lead to very informative results. This analysis is an important supplement to the data from other techniques, such as high-performance liquid chromatography (HPLC), that separate a large number of individual compounds. The predictable reaction of components in a mixture makes it possible to determine a single reactant by other means and to calculate its contribution to the total FC phenol content. The reaction is based on the principle that in alkaline pHs the phenolic compounds present in the sample reduce the mixture of acids which are present in the Folin reagent to tungsten and molybdenum oxides, whose absorbances can be read using a visible light spectrophotometer. To understand the results, it is necessary to prepare a gallic acid standard calibration curve and then the results are expressed as milligrams of gallic acid equivalents per gram extract (mg GAE/g extract).

The phenolic content can also be determined by High-performance Liquid Chromatography (HPLC - formerly referred to as high-pressure liquid chromatography), , which is the chromatographic technique most commonly used to separate, identify and quantify the components in a mixture of grape pomace (Butkhupl et al. 2010; Oliveira et al. 2013).

In general, the method involves a liquid sample being passed over a solid adsorbent material packed into a column using a flow of liquid solvent. Each analyte in the sample interacts slightly differently with the adsorbent material, thus retarding the flow of the analytes. Several studies in the literature report the use of HPLC for the evaluation of the composition of phenolic compounds from derivatives of grapes and by-products of the wine industry such as anthocyanins (Ju and Howard, 2003; Corrales et al. 2008) and resveratrol (Yang et al. 2009; Oliveira et al. 2013). KAMMERER et al. (2004) characterized the phenolic compounds of 14 different samples of grape pomace using methanol as extractor solvent, identifying 13 anthocyanins, 11 phenolic acids, 13 catechins and flavonols and 2 stilbenes.

The identification and quantification of individual phenolics can also be determined by using Liquid Chromatography - Mass Spectrometry (LC–MS) to analyse the extracts from wine, using two different mobile phase systems in the separation of the compounds (Alonso et al. 1991; Cheng et al. 2012).

Another method commonly used for the quantification of phenolic compounds in order to determine the chemical composition of grape pomace extracts is the Gas Chromatography - Mass Spectrometry (GC-MS) (Markham et al. 1996; Martin et al. 2012). GC-MS is an analytical method that combines the characteristics of gas-liquid chromatography and mass spectrometry to identify different substances within a test sample. This is done by the comparison of the relative concentrations among the atomic masses in the generated spectrum. Two kinds of analysis are possible, comparative and original. The comparative analysis essentially compares the given spectrum to a spectrum library. The other method of analysis measures the peaks in relation to one another. By this technique, flavonoids, phenolic acids and derivatives are identified by comparison with data such as retention time and ionic fragmentation of authentic standards silanized and eluted under the same conditions and with the Wiley 8 library.

Antibacterial Activity Tests

The antimicrobial activity of phenolic compounds extracted from grape pomace can be evaluated by different methods. In the literature, the most well-known methods include: the agar diffusion method, macrodilution and microdilution. Simple and cheap bioassays are prioritized for a rapid screening of the extracts and fractions against various genera of bacteria (Cheng et al. 2012; Martin et al. 2012).

Firstly, a qualitative screening is carried out of the extracts in which the antibacterial activity is aimed, the agar diffusion method being the most applied. The agar diffusion technique is performed according to the Clinical and Laboratory Standards Institute-CLSI (2009). By this test, it is possible to verify if the tested extract is able to inhibit or not the growth of a micro-organism of interest.

Subsequently, the extracts which have shown antimicrobial activity in the screening method might be submitted to a quantitative analysis, such as the determination of the minimum inhibitory concentration (MIC).

This method has been the most mentioned in the literature and is based on the dilution of the extract in different concentrations in 96-well microplates, containing a standard suspension (108 UFC/mL) of the microorganism to be tested, for the determination of the minimal quantity of the extract which is able to inhibit the microbial growth. This technique is the most used due to its sensivity and minimal quantity of reagents, making a higher number of replicas possible and, thus, increasing the reliability of the results (Ostrosky et al. 2008). The lowest dilution of the extract which does not present any kind of microbial growth is considered the minimum inhibitory concentration (MIC).

Currently, there is not any standardization about the acceptable level of inhibition for natural products when they are compared to standard antibiotics. ALIGIANIS et al. (2001) proposed the classification of compounds from vegetal origin based on the MIC results, considering: strong inhibition - those extracts which present MIC up to 500 μg/mL; moderate inhibition - those with MIC of around 600 up to 1500 μg/mL; weak inhibition - the ones which showed MIC above 1600 μg/mL. Variations in the antimicrobial effect can occur due to a variety offactors such as: the technique used, the strain used in the test, the origin of the plant , the season of the harvest and the typical characteristics of each extract (Ostrosky et al. 2008).

Thus, the use of well-known standard strains is advisable, like those provided by the culture collections such as the ATCC (American Type Culture Collection) for comparisons and the assembly of more consistent results, for example: *S. aureus* (ATCC 25923) and *L. monocytogenes* (ATCC 7644), for Gram positive bacteria, and *Salmonella enteritidis* (ATCC 13076) and *E. coli* (ATCC 25922) for Gram negative microorganisms (Martin et al. 2012).

Conclusion

This chapter highlighted the importance of the reuse of winemaking wastes (grape pomace) regarding their beneficial effects to human health through their use as new natural antimicrobial agents. In addition, the relevance of the consumption of wine in moderate quantities as a food supplement is also evidenced, due to its high content of biologically active compounds. This protective role of wine and grape pomace against pathogenic microorganisms is attributed especially to the high contents of phenolic compounds (gallic acid, flavan-3-ols, flavonoids, stilbenes and anthocyanins), present in grape skins, seeds and stems.

However, the recovery of these compounds from grape pomace depends on a variety of factors, mainly from the extraction methods (extraction solvent, extraction time), grape variety and grape fraction, influencing their composition and stability. This knowledge is of great relevance since, after purchased, it would be possible to recommend alternative approaches for the control of infection and preservation of foods, by the ingestion of wine during meals as well as by the addition of food additives containing bioactive phenolic compounds extracted from grape pomace.

This research represents a promising advance for the improvement of people's quality of life, for the preservation of the environment through the recycling of winemaking residues by their application in the food and pharmaceutical industries and for giving increased value to these byproducts.

References

Abtahi, H., Ghazavi, A., Karimi, M., (2011). Antimicrobial activities of ethanol extract of black grape. *African J. Microbiol. Res.,* 5(25), 4446-4448.

Aligiannis, N., Kalpoutzakis, E., Mitaku, S., Ioanna, B. C., (2001). Composition and antimicrobial activity of the essential oils of two origanum species. *J. Agri. Food Chem.,* 49(9), 4168-417.

Alonso-Borbalan, A. M., Zorro, L., Guillen, D. A., Barroso, C. G., (2003). Study of the polyphenol content of red and white grape varieties by liquid chromatography–mass spectrometry and its relationship to antioxidant power. *J. Chromat. A.,* 1012, 31–38.

Alonso, E., Bourzeix, M., Revilla, E., (1991). Suitability of water/ethanol mixtures for the Vitis vinifera seeds contained in a winery by-product. *Seed Sci. Technol.* 19, 545–552.

Balasundram, N.; Sundram, K.; Samman, S., (2006). Phenolic compounds in plants and agri-industrial by-products: antioxidant activity, occurrence, and potential uses. *Food Chem.,* 99 (1), 191-203.

Baydar, N. G., Özkan, G., Sagdiç, O., (2004). Total phenolic contents and antibacterial activities of grape (*Vitis vinifera* L.) extracts. *Food Control,* 15, 335-339.

Belido-Blasco, J. B., Arnedo-Pena, A., Cordero-Cutillas, E., Canos-Cabedo, M., Herrero-Carot, C., Safont-Adsuara, L., (2002). The protective effect of alcoholic beverages on the occurrence of a *Salmonella* food-borne outbreak. *Epidemiol.,* 13, 228–230.

Boban, N., Tonkic, M., Budimir, D., Modun, D., Sutlovic, D., Punda-Polic, V. Boban, M., (2010). Antimicrobial effects of wine: Separating the role of polyphenols, pH, ethanol, and other wine components. *J. Food Sci.,* 75(5), M322-M326.

Bonilla, F., Mayen, M., Merida, J., Medina, M., (1999). Extraction of phenolic compounds from red grape marc for use as food lipid antioxidants. *Food Chem.,* 66, 209–215.

Brown J. C., Huang G., Haley-Zitlin V., Jiang X., (2009). Antibacterial Effects of Grape Extracts on *Helicobacter pylori. Appl. Environ. Microbiol*., 75(3), 848–852.

Brunner, G., (2005). Supercritical fluids: technology and application to food processing. *J. Food Engin.,* 67, 21-33.

Bucic-Kojic, A., Planinic, M., Tomas, S., Bilic, M., Velic, D., (2007). Study of solid–liquid extraction kinetics of total polyphenols from grape seeds. *J. Food Engin*., 81, 236–242.

Butkhupl, L., Chowtivannakul, S., Gaensakoo, R., Prathepha, P., Samappito, S., (2010). Study of the Phenolic Composition of Shiraz Red Grape Cultivar (Vitis vinfera L.) Cultivated in North-eastern Thailand and its Antioxidant and Antimicrobial Activity. *S. Afr. J. Enol. Vitic*., 31, 89-98.

Chan, M. M. Y., (2002). Antimicrobial effect of resveratrol on dermatophytes and bacterial pathogens of the skin. *Biochem.Pharmacol*., 63, 99-104.

Chatterjee, A., Bagchi, D., Yasmin, T., Stohs, S. J., (2005). Antimicrobial effects of antioxidants with and without clarithromycin on *Helicobacter pylori. Mol. Cel. Biochem*., 270, 125–130.

Chavan, U. D., Shahidi, F., Naczk, M., (2001). Extracted of condensed tannins from beach pea (*Lathyrus maritimus* L.) as affected by different solvents. *Food Chem.,* 75, 509–512.

Cheng, V. J., Bekhit, A. E. A., Mcconnell, M., Mros, S., Zhao J., (2012). Effect of extraction solvent, waste fraction and grape variety on the antimicrobial and antioxidant activities of extracts from wine residue from cool climate. *Food Chem.,* 134, 474–482.

Clinical and Laboratory Standards Institute (CLSI), (2009). Performance standards for antimicrobial disk susceptibility tests, Approved Standard M02-A10, 10th ed. Clinical and Laboratory Standards Institute, Wayne.

Corrales, M., Toepfl, S., Butz, P., Knorr, D., Tauscher, B., (2008). Extraction of anthocyanins from grape by-products assisted by ultrasonics, high hydrostatic pressure or pulsed electric fields, a comparison. *Innov. Food Sci. Emerg. Technol.,* 9, 85–91.

Correia, A., Gomes, A., Oliveira, B., Gonçalves, G., Miranda, M., Almeida, O., (2003). The protective effect of alcoholic beverages in foodborne outbreaks of *Salmonella enteritidis* PT1 in northern Portugal. *Eurosurveillance Weekly,* 7(13), 2195-2196.

Do, G. M., Kwon, E. Y., Ha, T. Y., Park, Y. B., Kim, H. J., Jeon, S. M., Lee, M. K., Choi, M. S. (2011). Tannic acid is more effective than clofibrate for the elevation of hepatic _-oxidation and the inhibition of 3-hydroxy-3-methyl-glutaryl-CoA reductase and aortic lesion formation in apo E-deficient mice. *British J. Nut.,* 106, 1855–1863.

Fernandes, J., Gomes, F., Couto, J. A., Hogg, T. (2007). The antimicrobial effect of wine on Listeria innocua in a model stomach system. *Food Control,* 18 (12), 1477-1483.

Fikselova, M., Kacaniova, M., Mellen, M., (2010). Antioxidant (antiradical) and antimicrobial (antifungal) effects of Slovak Tokaj wines. *Acta Aliment.,* 39 (3), 256–264.

Filip, V.; Plocková, M.; Šmidrkal, J.; Špicková, Z.; Melzoch, K.; Schmidt, Š., (2003). Resveratrol and its antioxidant and antimicrobial effectiveness. *Food Chem.,* 83, 585-593.

Gañan, M., Martínez-Rodríguez, A. J., Carrascosa, A. V., (2009). Antimicrobial activity of phenolic compounds of wine against Campylobacter jejuni. *Food Control,* 20, 739–742.

Han, Y., (2007). Synergic effect of grape seed extract with amphotericin B against disseminated candidiasis due to *Candida albicans. Phytomed*., 14, 733-738.

Jackson, R. S., (2000). Wine science – Principles, practice, perception. 2ª Edição, Academic Press,San Diego, Califórnia, EUA, 649p.

Jayaprakasha, G. K., Singh, R. P., Sakariah, K. K., (2001). Antioxidant activities of grape seed (Vitis vinifera) extracts on peroxidation models in vitro. *Food Chem.,* 73, 285–290.

Jeandet, P., Douillet-Breuil, A. C., Bessis, R., Debord, S., Sbaghi, M., Adrian, M., (2002). Phytoalexins from the Vitaceae: Biosynthesis, Phytoalexin Gene Expression in Transgenic Plants, Antifungal Activity, and Metabolism. *J. Agric. Food Chem.,* 50, 2731-2741.

Ju, Z. Y., Howard, L. R., (2003). Effects of Solvent and Temperature on Pressurized Liquid Extraction of Anthocyanins and Total Phenolics from Dried Red Grape Skin. *J. Agric. Food Chem.,* 51, 5207-5213.

Just, J. R., Daeschel, M.A., (2003). Antimicrobial eVects of wine on Escherichia coli O157:H7 and *Salmonella typhimurium* in a stomach model system. *J. Food Sci.,* 68, 285–290.

Kallithraka, S., Garcia-Viguera, C., Bridle, P., Bakker, J. (1995). Survey of solvents for the extraction of grape seed phenolics. *Phytochem. Anal.,* 6, 265–267.

Kammerer, D., Claus, A., Carle, R., Schieber, A., (2004). Polyphenol screening of pomace from red and white grape varieties (*Vitis vinifera* L.) by HPLC-DAD-MS/MS. *J. Agric. Food Chem*., 52, 4360-4367.

Katalinic, V., Možina, S. S., Skroza, D., Generalic, I., Abramovic, H., Miloš, M., Ljubenkov, I., Piskernik, S., Pezo, I., Terpinc, P., Boban, M., (2010). Polyphenolic profile, antioxidant properties and antimicrobial activity of grape skinextracts of 14 *Vitis vinifera* varieties grown in Dalmatia (Croatia). *Food Chem.,* 119, 715-723.

Kaur, C., Kapoor, H. C., (2001). Antioxidants in fruits and vegetables – the millennium's health. *Int. J. Food Sci. Technol*., 36(7), 703-25.

Lirdprapamongkol, K., Sakurai, H., Kawasaki, N., Choo, M.-K., Saitoh, Y., Aozuka, Y., Singhirunnusorn, P., Ruchirawat, S., Svasti, J., Saiki, I., (2005). Vanillin suppresses in vitro invasion and in vivo metastasis of mouse breast cancer cells. *European J. Pharm.Sci.,* 25 (1), 57–65.

Louli, V., Ragoussis, N., Magoulas, K., (2004). Recovery of phenolic antioxidants from wine industry by-products. *Biores. Technol*., 2, 201–208.

Makris, D. P., Boskou, G., Andrikopoulos, N. K., (2007). Recovery of antioxidant phenolics from white vinification solid by-products employing water/ethanol mixtures. *Biores. Technol*., 98, 2963–2967.

Maier, T., Schieber, A., Kammerer, D. R., Carle, R., (2009). Residues of grape (*Vitis vinifera* L.) seed oil production as a valuable source of phenolic antioxidants. *Food Chem.,* 112, 551-559.

Markham, K. R., Mitchell, K. A., Wilkins, A. L., Daldy, J. A., Lu, Y., (1996). HPLC and CGMS identification of the major organic constituents in new zeland propolis. *Phytochem*., 42, 205-211.

Martin, J. G. P., Porto, E., Corrêa, C. B., Alencar, S. M., Da Gloria, E. M.,Cabral, I. S. R., Aquino, L. M., (2012). Antimicrobial potential and chemical composition of agro-industrial wastes. *J. Nat. Prod.,* 5, 27-36.

Mendez M., Rodríguez R., Ruiza J., Morales-Adameb D., Castillo F.,Hernández-Castillo F.D., Aguilar C. N., (2012). Antibacterial activity of plant extracts obtained with alternative organicssolvents against food-borne pathogen bacteria. *Ind. Crops Prod.,* 37, 445– 450.

Michielin, E. M. Z.; Salvador, A. A.; Riehl, C. A. S.; Smânia Jr., A.; Smânia, E. F. A.; Ferreira, S. R. S., (2009). Chemical composition and antibacterial activity of *Cordia verbenacea* extracts obtained by different methods. *Biores. Technol.*, 100, 6615-6623.

Moretro, T., Daeschel, M. A., (2004). Wine is bactericidal to food borne pathogens. *J. Food Sci.,* 69, M251–M257.

Naczk, M., Shahidi, F. (2004). Extraction and analysis of phenolics in food: *Review. J. Chromat. A.,* 1054, 95–111.

Nawaz, H., Shi, J., Mittal, G. S., Kakuda, Y., (2006). Extraction of polyphenols from grape seeds and concentration by ultrafiltration. *Sep.Purif. Technol.,* 48, 176–181.

Negro, C., Tommasi, L., Miceli, A., (2003). Phenolic compounds and antioxidant activity from red grape marc extracts. *Biores. Technol.,* 87, 41–44.

Oliveira, D. A., Salvadora, A. A., Smania-Jr., A., Smania, E. F. A.,Maraschinc, M., Ferreira, S. R. S., (2013). Antimicrobial activity and composition profile of grape (*Vitis vinifera*) pomace extracts obtained by supercritical fluids. *J. Biotech.,* 164, 423– 432.

Oliveira, D. G., Prince, K. A., Higuchi, C. T., Santos, A. C. B., Lopes, L. M. X., Simões, M. J. S., Leite, C. Q. F., (2007). Antimycobacterial activity of some Brazilian indigenous medicinal drinks. *Rev. Cienc. Farm. Bas. Apl.,* 28(2), 165-169.

Ostrosky, E. A., Mizumoto M. K., Lima, M. E. L., Kaneko, T. M, Nishikawa, S. O., Freitas, B. R., (2008). Métodos para avaliação da atividade antimicrobiana e determinação da Concentração Mínima Inibitória (CMI) de plantas medicinais. *Rev. Bras. Farmacognosia,* 18(2), 301-307.

Oussalah, M., Caillet, S., Saucier, L., Lacroix, M., (2007). Inhibitory effects of selected plant essential oils on the growth of four pathogenic bacteria: *E. coli 0157:H7, Salmonella typhimurium, Staphylococcus aureus and Listerria monocytogenes. Food Control,* 18 (5), 414-420.

Papadopoulou, C., Soulti, K., Roussis, I. G., (2005). Potential antimicrobial activity of red and white wine phenolic extracts against strains of *Staphylococcus aureus, Escherichia coli* and *Candida albicans*. *Food Technol. Biotechnol*., 43, 41–46.

Paulo, L., Oleastro, M., Gallardo, E., Queiroz, J. A., Domingues, F., (2011). Anti-*Helicobacter* pylori and urease inhibitory activities of resveratrol and red wine. *Food Res. Int.,* 44(4), 964-969.

Penna, N. G., Hecktheuer, L. H. R., (2004). Vinho e Saúde: uma revisão. *Infarma*, 16(1-2), 64–67.

Piccirillo, C., Demiray, S., Silva-Ferreira, A. C., Pintado, M. E., Castro, P. M. L., (2013). Chemical composition and antibacterial properties of stem and leaf extracts from Ginja cherry plant. *Ind. Crops Prod.,* 43, 562– 569.

Pinelo, M., Ruiz-Rodríguez, A., Sineiro, J., Señoráns, F. J., Reglero, G., Núñez, M. J., (2007). Supercritical fluid and solid–liquid extraction of phenolic antioxidants from grape pomace: a comparative study. *European Food Res. Technol*., 226, 199-205.

Pourmortazavi, S. M., Hajimirsadeghi, S. H. 2007. Supercritical fluid extraction in plant essential and volatile oil analysis. *J. Chromat. A.,* 1163, 2-24, 2007.

Pyla, R., Kim, T. J., Silva, J. L., Jung, Y. S., (2010). Enhanced antimicrobial activity of starch-based film impregnated with thermally processed tannic acid, a strong antioxidant. *Int. J. Food Microbiol.,* 137(2-3): 154-160.

Queipo-Ortuño, M. I., Boto-Ordóñez, M., Murri, M., Gomez-Zumaquero, J. M., Clemente-Postigo, M., Estruch, R., Diaz, F. C., Andrés-Lacueva, C., Tinahones, F., (2012). Influence of red wine polyphenols and ethanol on the gut microbiota ecology and biochemical biomarkers. *American J. Clinical Nutrition,* 95, 1323-1334.

Rababah, T. M., Hettiarachchy, N. S., Horax, R., (2004). Total phenolics and antioxidant activities of fenugreek, green tea, black tea, grape seed, ginger, rosemary, gotu kola, and ginkgo extracts, vitamin E, and tertbutylhydroquinone. *J. Agric. Food Chem.,* 52(16), 5183-5186.

Ravichandran, M., Hettiarachchy, N. S., Ganesh, V., Ricke, S. C., Singh, S., (2011). Enhancement of antimicrobial activities of naturally occurring phenolic compounds by nanoscale delivery against Listeria monocytogenes, Escherichia coli O157:H7 and Salmonella typhimurium in broth and chicken meat system. *J. Food Safety,* 31, 462–471.

Rotava, R., Zanella, I., Da Silva, L. P., Manfron, M. P.,Ceron, C. S., Alves, S. H., Karkow, A. K., Santos, J. P. A., (2009). Atividade antibacteriana, antioxidante e tanante de subprodutos da uva. *Ciência Rural,* 39(3), 941-944.

Saito, M., Hosoyama, H., Ariga, T., Kataoka, S., Yamaji, N., (1998). Antiulcer activity of grape seed extract and procyanidins. *J. Agric. Food Chem.,* 46, 1460–1464.

Santurio, J. M., Santurio, D. F., Pozzatti, P., Moraes, C., Franchin, P. R., Alves, S. H., (2007). Atividade antimicrobiana dos óleos essenciais de orégano, tomilho e canela frente a sorovares de Salmonella entérica de origem avícola. *Ciência Rural,* 37(3), 803-808.

Santurio, D. F., Costa, M. M., Maboni, G., Cavalheiro, C. P., Sá, M. F., Pozzo, M. D., Alves, S. H., Fries, L. L. M., (2011). Atividade antimicrobiana de óleos essenciais de condimentos frente a amostras de *Escherichia coli* isoladas de aves e bovinos. *Ciência Rural,* 41(6), 1051-1056.

Shi, J., Yu, J. M., Pohorly, J. E., Young, J. C., Bryan, M., Wu, Y., (2003). Optimization of extraction of polyphenols from grape seed meal by aqueous ethanol solution. *Food Agri. Env.,* 1, 42–47.

Shoko, T., Soichi, T., Megumi, M. M., Eri, F., Jun, K., Michiko, W., (1999). Isolation and identification of an antibacterial compound from grape and its application to foods. *Nippon Nog.Kaishi,* 73, 125–128.

Simões, C. M. O., Schenkel, E. P., Gosmann, G., Mello, J. C. P., Mentz, L. A., Petrovick, P. R., (2004). Farmacognosia: da planta ao medicamento. 5. ed. Porto Alegre, RS: Ed. UFSC, 1102 p.

Singleton, V. L., Rossi, J. A., (1965). Colorimetry of total phenolics with phosphomolybdic-phosphotungstic acid reagents. *American J. Enol. Vitic.*, 16, 144-158.

Singleton, V. L., Orthofer, R., Lamuela-Raventós, R. M., Lester, P., (1999). Analysis of total phenols and other oxidation substrates and antioxidants by means of folinciocalteu reagent, Methods in enzymology. Academic Press, pp. 152-178.

Soni, M. G., Carabin, I. G., Burdock, G. A., (2005). Safety assessment of esters of phydrozybenzoic acid (parabens). *Food Chem. Toxicol.*, 43, 985–1015.

Sonja, S. M., Marija, K., Ana, K., Simona, U., Višnja, K., (2009). Prevalence and resistance against different antimicrobial compounds of *Campylobacter* spp. In/from retail poultry meat. *Tehnologija mesa.*, 50(1-2), 112-120.

Sugita-Konishi, Y., Hara-Kudo, Y., Iwamoto, T., Kondo, K., (2001). Wine has activity against entero-pathogenic bacteria in vitro but not in vivo. *Biosci. Biotech. Biochem.*, 65, 954–957.

Teissedre, P.-L., Landrault, N., (2000). Wine phenolics: contribution to dietary intake and bioavailability. *Food Res. Int.*, 33, 461–467.

Thimothe, J.; Bonsi, I. A., Padilla-Zakour, O. I., Koo, H., (2007). Chemical characterization of red wine grape (*Vitis vinifera* and *Vitis* interspecific hybrids) and Pomace Phenolic Extracts and Their Biological Activity against *Streptococcus mutans. J. Agric. Food Chem.*, 55, 10200-10207.

Vaz, M., Hogg, T., Couto, J. A., (2012). The antimicrobial effect of wine on Bacillus cereus in simulated gastro-intestinal conditions. *Food Control,* 28, 230-236.

Veluri, R., Singh, R. P., Liu, Z., Thompson, J. A., Agarwal, R., Agarwal, C., (2006). Fractionation of grape seed extract and identification of gallic acid as one of the major active constituents causing growth inhibition and apoptotic death of DU145 human prostate carcinoma cells. *Carcinogenesis,* 27 (7), 1445–1453.

Yang, J., Martinson, T. E., Liu, R. H., (2009). Phytochemical profiles and antioxidant activities of wine grapes. *Food Chem.,* 116, 332-339.

Yilmaz, Y., and Toledo, R. T., (2006). Oxygen radical absorbance capacities of grape/wine industry byproducts and effect of solvent type on extraction of grape seed polyphenols. *J. Food Comp. Anal.,* 19, 41–48.

In: Grapes
Editor: José de Sousa Câmara

ISBN: 978-1-63321-402-6

Chapter 19

Health Benefits of Grape Seed Extract

Snehal S. Joshi and Doris H. D'Souza*
Department of Food Science and Technology,
University of Tennessee-Knoxville, TN, US

Abstract

For many centuries, grapes have been consumed for their nutritive value, taste, and health benefits, and grape seeds are known for their high phenolic content. Grape seed extract (GSE), a by-product of the wine and grape processing industries is rich in bioactive compounds that include flavonoids, polyphenols, anthocyanins, proanthocyanidins, and resveratrol. GSE also contains monomeric phenolic compounds, such as catechins, epicatechin-3-O-gallate and dimeric, trimeric and tetrameric procyanidins. GSE has been reported to possess a wide gamut of beneficial properties including antimicrobial, anti-inflammatory, anti-diabetic and cardioprotective effects. The application of GSE has been studied in different food systems, environments, and in animal models to evaluate these associated-beneficial properties. GSE has been used as an effective food additive and dietary supplement with a generally recognized as safe (GRAS) status by the U.S. Food and Drug Administration. The advantages of being natural, low-in-cost, and sustainable make GSE promising for use as an appealing alternative control, treatment, or preventive option. This review intends to provide a synopsis of the currently known health benefits associated with GSE, with an emphasis on antimicrobial effects in food safety applications. The "metagenomics, proteomics, and metabolomics" era has potential to further evaluate the benefits of GSE using animal models. This "Omics" era should help to investigate the effects of GSE on the gut microbiota of animals and humans, with the overarching goal to help improve food safety and animal and human health, reduce microbial illness, and decrease metabolic disorders.

Keywords: Grape seeds, Health benefits; Anti-inflammation; Antioxidation, Anticarcinogenic, Antimicrobial effects of GSE

* Corresponding Author address. Email: ddsouza@utk.edu.

Introduction

Grapes (genus *Vitis*) are native to southern Europe and Western Asia near the Caspian Sea, and were brought to North America and Europe. The medicinal and nutritional properties of grapes have been known over centuries, where Egyptians and Greeks used the various forms of grapes (ripe, unripe, dried) for healing and treatment of various health conditions. Grape seed extract (GSE) is a natural by-product of the wine and grape juice processing industries (Shi et al., 2003). Grape seeds and grape skin have been shown to contain bioactive components such as flavonoids, polyphenols, anthocyanins, proanthocyanidins, procyanidins, and the stilbene derivative, resveratrol (Jayaprakasha et al., 2003). Grape seeds are also known to have high amounts of monomeric phenolic compounds, such as epicatechin, epicatechin-3-O-gallate, catechins, and dimeric, trimeric and tetrameric procyanidins (Yamakoshi et al., 2002).

Several pharmacological and therapeutic benefits such as antimicrobial, anti-inflammatory, antioxidative, cardioprotective, and neuroprotective effects have been reported to be associated with GSE (Nassiri-Asl et al., 2009). GSE has been approved as a generally recognized as safe (GRAS) by the U.S. Food and Drug Administration and hence is being studied extensively for its use as an antimicrobial agent in foods. ("GRAS" is Generally Recognized As Safe under sections 201(s) and 409 of the Federal Food, Drug, and Cosmetic Act) (USFDA, 2007). GSE at doses of 50 to 100 mg/ml per day are reported to not exhibit toxicity or adverse health effects in rats (Shan et al., 2011; Sivarooban et al., 2007; Yamakoshi et al., 2002). These advantages of GSE show great potential for its use as a value-added product with its low cost, sustainability, and lack of evidence of being hazardous to the environment.

This short review aims to focus on the health benefits associated with the consumption and use of GSE by the food industry. The main areas of focus include anti-inflammatory, anticarcinogenic, hepatoprotective, anti-diabetic, antibacterial, anti-parasitic, and antiviral effects with potential application in the food environment.

Role of GSE in Anti-Inflammation

Inflammation is theorized to be the underlying cause of most health issues and chronic inflammation has been linked to heart disease, hay fever, rheumatoid arthritis, obesity, and diabetes (Barnes, 2013; Eguchi et al., 2013; Mandhane et al., 2011; Schneider et al., 2013). When the anti-inflammatory and antioxidant properties of GSE were investigated in the case of septic shock caused by the intraperitoneal injection of lipopolysaccharide (LPS) derived from *Escherichia coli*, GSE at doses of 75 mg/kg/day had anti-inflammatory effects by decreasing the nitrate and nitrite (NOx) levels in the plasma, red blood cells, spleen, and liver (Pallares et al., 2013). These researchers showed that high doses of 200 mg/kg/day could down regulate the genes Il-6 and iNos (interleukin 6 and inducible nitric oxide synthase) associated with inflammatory response against the LPS endotoxin (Pallares et al., 2013). GSE at 100 mg/kg/day was also found to reduce lipid peroxides and carbonylated protein levels and also improved the antioxidant activity in plasma and hepatic tissue in diabetic rats (Chis et al., 2009). Similarly, another study showed that administration of GSE at 100 mg/kg body weight for 30 days significantly augmented the superoxide dismutase activity in brains of aged rats when compared to control aged rats (Balu et al., 2005).

In another study to investigate the effects of GSE on colitis and inflammation, mice were induced with recurrent colitis and administered GSE at 200 mg/kg per day for 7 days (Wang et al., 2010). GSE treated mice were found to have increased superoxide dismutase, glutathione peroxidase activities, and glutathione levels (all indicators of higher antioxidant capacity that helps to decrease inflammation), in colon tissues and serum of rats. These investigators also found that levels of nitric oxide levels in the colon tissues were significantly reduced from 21 μmol/L in colitis-affected serum to a normal healthy level of 6 μmol/L. Their findings suggest that GSE is efficient in regulating the inflammatory response and helps in recuperation from colonic oxidative stress by reducing nitric oxide production, that are associated with the antioxidant and anti-inflammatory properties of GSE (Wang et al., 2010).

GSE at 50 and 100 μg/mL was also studied as an anti-inflammatory agent to relieve symptoms/inflammation associated with tumor-necrosis factor (TNF-α) treatment in human umbilical vein endothelial cells (Chao et al., 2011). The study showed that GSE was found to assuage the production of 8-OHdG (oxidized nucleoside 8-hydroxydeoxyguanosine), an indicator of reactive oxygen species involved in DNA damage. Reduced adhesion of THP-1 (carcinogenic monocytic cells), expression of nuclear factor-κB inhibitors, ICAM-1 (Intercellular Adhesion Molecule 1) and COX-2, and the production of prostaglandin E2 (PGE2) were obtained, that are all induced as a result of TNF-α treatment in human umbilical vein endothelial cells, indicating that GSE helped alleviate TNF-α induced inflammation (Chao et al., 2011). Since inflammation plays a role in atherosclerosis onset, GSE might be a valuable treatment option in the prevention of atherosclerosis. GSE at 100 mg/kg when administered intraperitoneally in mice was found to reduce the formation of tartrate-resistant acid phosphatase (TRAP)-positive multinucleated cells and osteoclast activity and increased differentiation of mature osteoblasts in mice (Park et al., 2012). GSE was also found to alleviate the arthritic symptoms during the course of disease suppressing osteoclast differentiation and promoting osteoblast differentiation. This balance is important in maintaining bone homeostasis by synthesizing multiple bone matrix proteins (Park et al., 2012).

Studies have also shown that GSE can be used as a potential therapeutic option against asthma. Murine studies showed that when GSE at 100 mg/kg was administered to acute asthmatic mice, there was a decrease in the number of eosinophils, airway inflammation, hyperresponsiveness, and oxidative stress (Lee et al., 2012). Airway fibrosis was also decreased in chronic asthmatic mice, suggesting the use of dietary GSE as an adjunct therapy for asthma patients (Lee et al., 2012).

GSE and Its Effects on Atherosclerosis and Heart Disease

The increase in the consumption of high-fat and high-carbohydrate diets together with a sedentary lifestyle has led to an upsurge in the occurrence of coronary heart disease and atherosclerosis, leading to plaque build-up in the arteries made up of lipids/fat, cholesterol and calcium. Recommendations to decrease these incidences include improving dietary habits, exercise and prescription medications. Natural alternatives such as GSE have been widely studied for their cardioprotective properties, and in reducing cholesterol and plasma lipoproteins. For example, GSE was shown to alleviate the side effects such as cardiac steatosis (adipose degeneration) and lipotoxicity in rats fed on a high fat diet for 45 days at levels of 500 mg/kg/bodyweight (Charradi et al., 2011). GSE at doses of 50 and 100 mg GSE/kg body weight were found to reduce 49 and 63% foam cells, a biomarker of early stage atherosclerosis (Bagchi et al., 2003). Red grape seed extract at 200 mg/day for 8 weeks was found to reduce total cholesterol and oxidized low-density lipoprotein particles by 10.68 mg/dL and 5.47 mg/dL, respectively (Razavi et al., 2013). Thus, GSE consumption may also prove to be beneficial in reducing the risk of atherosclerosis (Razavi et al., 2013). Additionally, dry GSE at a concentration of 100 mg was found to significantly decrease the amount of serum cholesterol levels in human volunteers (Nikitina et al., 2006). GSE at 0.2 mg/g/bodyweight and 0.6 mg/g/bodyweight was found to reduce levels of plasma OX-LDL (oxidized low density lipoprotein), and ICAM-1 (Intercellular Adhesion Molecule 1) in mice, leading to decreased thickness in aortic valves of GSE-fed mice (Yu et al., 2002). Taken together, these studies show the anti-atherosclerotic and cardioprotective effects of GSE.

GSE and Its Role in Obesity and Diabetes

As mentioned above, high-fat diets and fast-food consumption pose serious health-risks to individuals prone to diabetes, heart disease, stroke, arthritis, and various forms of cancer. Diabetes is a disease characterized by high blood glucose levels, with Type 2 diabetes being the most common form. Some of the treatment options include diet changes, increased physical activity and healthy weight maintenance (American Diabetes Association, 2013).

GSE has also been explored for use in controlling obesity and preventing diabetes. Lipoprotein lipase (LPL) is an enzyme involved in the hydrolysis of triacylglycerols by releasing free fatty acids for uptake into adipocytes, involved in obesity. GSE at 1 mg/ml was found to inhibit human pancreatic lipase activity by almost 80% within 5 min of incubation in adipocytes, suggesting the benefits of GSE against obesity (Moreno et al., 2003). In addition, GSE was found to control body weight gain in mice fed on high fat diets, when complemented with GSE at 250 mg/kg body weight compared to the control mice fed on high fat diets (Park et al., 2008). The study also showed that there was significant decrease in epididymal and back fat, with down regulation of acetyl-coA carboxylase gene expression, a lipogenic gene, in GSE fed mice. These findings suggest that GSE might help in maintaining body weight, body fat and also in regulating lipid metabolism (Park et al., 2008). In another study, when albino mice were treated with GSE at 25, 50, or 100 mg/kg, T cell subsets, GITR-expressing cells, and T helper 1 cytokines, (all mediators of arthritic inflammation) were reported to be reduced (Ahmad et al., 2003). GSE was also found to up-regulate Tregs (regulatory T cells) and T helper 2 cytokine producing cells that play a role in suppressing excessive immune responses (Ahmad et al., 2013).

Another study showed that GSE at 100 mg/L reduced the expression of interleukin-6 (marker of inflammation, IL-6) and monocyte chemotactic protein (MCP-1; marker of obesity and inflammation) and augmented anti-inflammatory adipokine adiponectin levels in human adipocytes (SGBS) and macrophage-like (THP-1) cell lines (Chacon et al., 2009), indicating the benefit of GSE against obesity.

GSE was also shown to reduce postprandial plasma glucose in healthy individuals, where the plasma glucose concentrations at 15 min and 30 min after consumption of a high carbohydrate meal with 100 mg or 300 mg of GSE was significantly lower than that of a high carbohydrate meal without GSE (Sapwarobol et al., 2012). The study proposed the use of GSE a preventive measure against diabetes in otherwise healthy populations. Mice on a high fat diet were shown to be protected from nerve fiber loss compared to controls when administered 100 mg/ml GSE for 12 weeks, suggesting that GSE may provide benefits towards the treatment of diabetic neuropathy (Jin et al., 2013).

Health Benefits of GSE

GSE as an Antioxidant

GSE is known to be an efficient antioxidant and free radical scavenger (Jayaprakasha et al., 2003). GSE at 50 mg/l was shown to be an effective radical scavenger, being better than vitamins C and E (Bagchi et al., 2000). GSE components like procyanidin B4, catechin, and gallic acid at concentrations of 10-25 μmol/l were reported to be good preventive agents against cellular DNA oxidative damage, though, these compounds may induce cellular DNA damage at higher concentrations of 150 μmol/l (Fan et al., 2004). GSE at 2 mg/ml was found to inhibit the peroxidation of polyunsaturated fatty acids induced by UV–C (Bouhamidi et al., 1998). Researchers showed that rats fed a high garlic dose of 5g/kg/bodyweight, which induced anemia and a pro-oxidative state into erythrocytes, were found to be protected from the oxidative stress when supplemented with GSE at 500 mg/kg/ bodyweight (Hamlaoui et al., 2012). Their study investigated the potential of GSE as an antioxidant, where the toxic effect of high garlic dose treatment was found to be neutralized. All these finding suggest that GSE shows potential to be used as an antioxidant.

GSE as an Anticarcinogen

When grape seed polyphenols were investigated for their anti-carcinogenic effects towards 12-O-tetradecanoylphorbol-13-acetate (TPA) induced tumor reduction as a topical application, tumor incidence was inhibited by 35 and 60%, tumor multiplicity was inhibited by 61 and 83% and tumor volume was inhibited by 67 and 87% at concentrations of 0.5 and 1.5 mg/ml/mouse/application on mouse skin (Zhao et al., 1999). In another similar study, pretreatment of mouse skin with grape seed polyphenols at 5, 10, 20, and 30 mg led to a dose-dependent reduction in TPA-induced epidermal ornithine decarboxylase (an indicator of tumor promotion in mouse skin) by 27, 37, 48 and 70%, respectively (Bomser et al., 1999).

More recently in breast cancer studies, GSE at 100 μg/ml was found to increase apoptotic cell death significantly by up to 12% after 72 h incubation, when the highly metastatic MDA-MB 231 breast cancer line was used (Dinicola et al., 2014). Treatment with 25 μg/ml of GSE was found to significantly reduce (by 80%) the total number of MDA-MB231 (breast carcinoma) cells from migration (Dinicola et al., 2014). GSE at doses of 25, 50 and 100 µg/mL was reported to result in a dose-dependent growth inhibition of colorectal cancer cells (SW480 cells) after 12, 24, and 48 h treatment (Derry et al., 2012). Additionally, GSE was found to exhibit a 1.5–3-fold increase in apoptotic SW480 cells at concentrations of 25–100 µg/mL, respectively. Moreover, GSE at 25–100 µg/mL was not found to affect the normal colon epithelial cells, and did not show any growth inhibition or effect on viability (Derry et al., 2012). Taken together, these data show that GSE may prove to be useful as a potential remedy against colorectal cancer, though other chemotherapeutic drugs may be required. Another study on colon cancer showed that resveratrol potentiated the effects of GSE in inducing apoptosis in colon cancer cells and suppressed cell proliferation, suggesting that a combination therapy instead of a single drug therapy might be more efficient in treating colon cancer (Radhakrishnan et al., 2011).

Grape powder, resveratrol, and GSE when administered topically 30 min after 7,12-dimethylbenz [a] anthracene (DMBA, a skin carcinogen) exposure at concentrations of 1, 2, or 4 mg was shown to inhibit epidermal hyperplasia, proliferation, and dermal inflammation (Hanausek et al., 2011). Topical application of 1 mg resveratrol and dietary grape powder reduced DMBA-mediated Ha-ras mutations, a frequently detected oncogene, known to code for proteins that bind guanine, has GTPase activity, and plays a role in cellular membrane signaling in codon 61 by around 64% (Barbacid, 1987; Brown et al., 1990). The results from this study support the use and efficacy of GSE and its component polyphenols as potential anti-carcinogens and as supplementary anti-cancer therapeutics (Hanausek et al., 2011). GSE at 50 µg/mL was reported to significantly induce apoptotic death in non-small-cell lung cancer cell lines by causing a release of cytochrome c in the cytosol and cleavage of caspases 3 and 9 that play a role in cell death (Tyagi et al., 2013). In this study, GSE was also shown to lead to an increase in superoxide radical-associated oxidative stress and decreased intracellular reduced glutathione levels. These findings show that GSE may also be beneficial as a part of the treatment regime against lung cancer, in addition to skin, breast, and colon cancer (Tyagi et al., 2013).

Antimicrobial Properties of GSE

With regards to the antibacterial properties of GSE, the bacterial cell membrane being semipermeable allows the entry of polyphenols that can interact with the cytoplasm or cellular proteins/components, with higher ability of GSE polyphenols (that are present in undissociated forms) to enter the cell and interact with cellular components (Paulus, 1993). Thus, the benefits of the antimicrobial properties of GSE can be exploited for food or clinical applications.

Antibacterial Effects

GSE (both methanol and acetone extracts) were found to be more effective against Gram-positive bacteria than Gram-negative bacteria, when tested against several foodborne bacterial pathogens (Jayaprakasha et al., 2003).

Table 1. Effect of grape seed extract (GSE) and its components against bacteria

Type of bioactive	Concentration	Pathogen	Reduction observed	Reference
GSE	1%	*Brochothrix thermosphacta*	1.3 log CFU/ml	Margarita et al., 2009
GSE	100mg/ml	*Staphylococcus aureus*	99% inhibition	Kao et al., 2010
GSE (Gravinol-S)	50 mg /ml 78 mg /ml	*Listeria monocytogenes* *Listeria innocua ATCC 33090*	MIC	Bisha et al., 2010
2% malic acid + 3% GSE		*Salmonella Typhimurium*	2.6 to 3.3 log CFU/g	Ganesh et al., 2010
GSE	0.5%	*Listeria monocytogenes*	1.1 log CFU/g	Xi et al., 2011
GSE	20.67 µg/mL	MRSA	Complete reduction	Al-Habib et al., 2010
GSE and nisin	1% and 6400 IU	*Listeria monocytogenes*	Complete reduction	Sivarooban et al., 2007
GSE extract in Methanol:Water: Acetic acid (90:9.5:0.5)	900 ppm	*Bacillus cereus* *Bacillus subtilis* *Bacillus coagulans*	MIC	Jayaprakasha et al., 2003
	1000 ppm	*Staphylococcus aureus*		
	1250 ppm	*Escherichia coli*		
	1500 ppm	*Pseudomonas aeruginosa*		
GSE extract in Acetone:Water: Acetic acid (90:9.5:0.5)	850 ppm	*Bacillus cereus* *Bacillus subtilis* *Bacillus coagulans*		
Resveratrol and ellagic acid	6.25 to 50 g/ml	*Helicobacter pylori*	MIC	Brown et al., 2009
Gallic acid	250–280 mg/L for 10 min	*Listeria monocytogenes* ATCC 35152 *Staphylococcus aureus* ATCC 25923 *Escherichia coli* ATCC 25922	10^6 –10^7 CFU/ml to undetectable	Rhodes et al., 2006
Muscadine grape skin extract	256 to 512 g/ml	*Helicobacter pylori*	MIC	Brown et al., 2009
Polymeric phenolic fractions	1h	*Staphylococcus aureus* *Escherichia coli*	1 log reduction	Rhodes et al., 2006
Herb extracts containing grape seed	Unknown	*Listeria monocytogenes*	1.36 log CFU/gm	Shan et al., 2011
		Staphylococcus aureus	0.96 log CFU/gm	
		Salmonella enterica	1.75 log CFU/gm	
Muscadine seed extract	256 to 1,024 g/ml	*Helicobacter pylori*	MIC	Brown et al., 2009

*MIC= Minimal Inhibitory Concentration.

The minimum inhibitory concentration (MIC) with a methanol:water:acetic acid (90:9.5:0.5) grape seed extract was found to be 900 ppm for the gram-positive *Bacillus cereus*, *Bacillus subtilis* and *Bacillus coagulans* and 1000 ppm for *Staphylococcus aureus* (Jayaprakasha et al., 2003). For the gram-negative *Escherichia coli* and *Pseudomonas aeruginosa*, the MIC was found to be 1250 and 1500 ppm, respectively. When the extract contained acetone (acetone: water:acetic acid (90:9.5:0.5)) instead of methanol, differences in MIC were observed for the gram-positive bacilli, resulting in MICs of 850 ppm for *B. cereus*, *B. subtilis* and *B. coagulans*, while the other organisms showed similar MICs to the methanol:water:acetic acid (90:9.5:0.5) extract. These researchers determined that gallic acid was the bioactive component with the three hydroxyls responsible for the antibacterial activity of GSE (Jayaprakasha et al., 2003).

Muscadine grape skin extracts were also found to be effective against *Helicobacter pylori* (causative agent of ulcers) with MIC ranging from 256 to 512 g/ml, and the MIC ranged between 256 to 1,024 g/ml for muscadine seeds (Brown et al, 2009). Resveratrol and ellagic acid were also found to inhibit this ulcer causing bacteria, with an MIC range of 6.25 to 50 g/ml, while myricetin was found to have no effect (Brown et al., 2009). GSE at a concentration of 5 mg/ml was also found to inhibit the dihydrofolate reductase activity and growth of *S. aureus* at 0.5 mg/ml after 6 h, however some injured cells were recovered after 12 h (Kao et al., 2010). For GSE at 1 mg/ml, these researchers reported 99% of inhibition after 24 h, with no observed cell recovery.

In another study, commercial grape juice and skin and grape seed extract were tested against the gram-positive foodborne pathogen, *Listeria monocytogenes*, of concern to immunocompromised individuals and pregnant women. Commercial grape juice decreased *L. monocytogenes* from 10^6 –10^7 (6 to 7 log) CFU/ml to undetectable levels and *B. cereus* by 1-log within 10 min in the juice, without further reduction after 60 min (Rhodes et al., 2006). The seed and skin extracts (250–280 mg/L gallic acid) were reported to have similar anti-listerial activity to that of the commercial grape juice (Rhodes et al., 2006). Anti-bacterial activity of GSE was also tested against methicillin resistant *Staphylococcus aureus* (MRSA), which is an on-going problem in nosocomial and other related infections, where health-care providers continue to have challenges in prescribing appropriate antibiotics against these infections (Al-Habib et al., 2010). All of the 43 MRSA strains tested were found to be inactivated at 20.67 μg of GSE flavonoids (Al-Habib et al., 2010). Another study showed that GSE at 1 and 5 mg/mL was found to be effective against MRSA strains resulting in a 2.9-4.0 log CFU/ml reduction after 2 h at 37°C (Su et al., 2012). These results suggest that natural products such as GSE can be used as effective antimicrobials for multidrug resistant strains as an alternative to regular antibiotics. Red GSE at 0.5 mg/ml was found to reduce plaque formation of *Streptococcus mutans in vitro* by 40% as compared to control (Smullen et al., 2012), showing the potential of GSE to prevent dental caries. Table 1 summarizes the effect of GSE and its components against bacteria.

Antiparasitic Effects

GSE has been reported to have antiparasitic effects against *Histomona meleagridis*, a protozoan parasite responsible for histomonosis with high mortality in turkey and chicken (van der Heijden et al., 2011); *Blastocystis*, a prevalent unicellular parasite in human fecal specimens (Bart et al., 2013); *Trichomonas gallinae* that causes avian trichomonosis in various bird species, especially pigeons and doves (Jeurissen et al., 1996); and *Trichostrongylus colubriformis* a parasite of herbivorous mammals (Lattes et al., 2011).

Table 2. Effect of GSE and its components on parasites

Type of bioactive	Concentration	Parasite	Reduction	Reference
Grape seed extract	100 ppm	*Lymnea cailliaudi* snails	50% reduction (LC50)	Taher et al., 2012
Red grape seed extract	0.5 mg/ml	*Streptococcus mutans*	40% reduction in plaque formation	Smullen et al., 2013
Grape seed extract (water)	10 mg/ml	*Histomonas meleagridis*	100% reduction	Grabsteiner et al., 2008
		Tetratrichomonas Gallinarum	92.73% reduction	
		Blastocystis sp.	100% reduction	
Grape seed extract (ethanol)	10 mg/ml	*Histomonas meleagridis*	100% reduction	
		Tetratrichomonas Gallinarum	100% reduction	
		Blastocystis sp.	100% reduction	
Resveratrol	50 µM	*Encephalitozoon cuniculi*	1.5 X 10^6 spores/ml reduction	Leiro et al., 2004

*LC50 = Lethal concentration 50.

GSE at 100 ppm for 6-12 h was shown to kill 50% of exposed *Lymnea cailliaudi* snails (LC50; lethal concentration 50) with higher molluscicidal activity with increased exposure time of 12-24 h (Taher et al., 2012). However GSE showed weak effect against *Culex pipiens* larvae (responsible vectors for malaria, dengue and yellow fever) where their mortality did not increase over 50% even after increasing the GSE concentration to more than 200 ppm for 48 h (Taher et al., 2012). GSE at 5 mg/ml was also found to completely inhibit (initial number of 10^3 histomonads) the growth of *Histomona meleagridis*, *Trichomonas gallinarum* and *Blastocystis* sp. after 24 and 48 h (Grabensteiner et al., 2008). Another study tested a new product containing GSE against the avian parasite *Histomonas meleagridis*, that showed inhibition of its growth at concentration of 2.5 or 5 µl/ml after 48 h (Hauck et al., 2007). The effect of GSE and its components on parasites is shown in Table 2.

GSE at 10-20 mg/kg was reported to reduce mortality in chickens infected with *Eimeria tenella*, alleviated the oxidative stress following infection and helped improve the weight gain of chickens which is important for the poultry industry (Wang et al., 2008). Flavan-3-ols and galloyl derivatives of condensed tannins that are components of GSE were found to be completely inhibitory towards egg hatching of *Trichostrongylus colubriformis* at 1000 mg/ml, and epigallocatechin gallate at 100 mg/ml showed 100% inhibition of *T. colubriformis* larval development (Molan et al., 2003). Methanol extracts of grape seed at 300 and 600 mg were reported to be effective in killing the immature form of the leech *Limnatis nilotica* within 260 and 200 min, respectively, but was not effective towards the mature forms, suggesting that GSE should be used in combination with other treatments to boost their anti-leech effects (Gholami-Ahangaran et al., 2012).

Antiviral Activity

In addition to bacteria, GSE is known to have effects against several viruses of human health concern. Human noroviruses are recognized as the leading causes of non-bacterial gastroenteritis worldwide (Hall et al., 2013; Scallan et al., 2011). However, since these non-enveloped single-stranded RNA containing human noroviruses cannot be cultivated in the lab, cultivable surrogates such as feline calicivirus (FCV-F9) and murine norovirus (MNV-1) are used to determine inactivation by various chemical or processing technologies (Su and D'Souza., 2011). Hepatitis A virus is another non-enveloped single-stranded RNA containing foodborne virus important because of the severity of the disease that affects the liver and lasts for more than a month. GSE at 0.25, 0.5, and 1.0 mg/ml was shown to reduce FCV-F9 titers by 3.64, 4.10, and 4.61 log PFU/ml; MNV-1 by 0.82, 1.35, and 1.73 log PFU/ml; and HAV by 1.81, 2.66, and 3.20 log PFU/ml after treatment with 0.25, 0.50, and 1 mg/ml GSE at 37°C for 2 h (Su and D'Souza, 2011). Effect of GSE and its components against viruses is described on Table 3.

GSE was also shown to prevent replication of human norovirus surrogates using ELISA studies that were in agreement with the earlier studies of Su and D'Souza, 2011, where MNV-1 titers were shown to be reduced by 3 log PFU/ml when treated with 0.2 mg/ml GSE (Li et al., 2012). GSE was also evaluated for its anti-noroviral activity as a surface disinfectant by these researchers, where MNV-1 treated with 0.2 mg of GSE/ml, showed a reduction to undetectable levels in the absence of any protein load. In presence of 0.01% dried milk, reduction to undetectable levels was obtained only with 2 mg/ml GSE, while in tap water, GSE at 0.2 and 2 mg/ml was shown to cause a 1 and 2 log PFU/ml reduction in MNV-1 titers (Li et al., 2012). The effect of the symptoms of HIV-1 and AIDS is well publicized, emphasizing the need for improved control and preventive measures, where GSE may play an important role. GSE is also reported to interfere with the binding of the AIDS causing human-immunodeficiency virus (HIV-1) to its host cell receptor and thus prevents HIV-1 entry into the normal lymphocyte (Nair et al., 2002).

Table 3. Effect of GSE and its components against viruses

Type of bioactive	Concentration	Virus	Reduction	Reference
Resveratrol	10 µM	Human immunodeficiency virus – 1	20-30% reduction in replication	Heredia et al., 2000
GSE	0.2 mg/ml	MNV-1	Complete	Li et al., 2012
GSE	1 mg/ml	FCV-F9	4.61 log PFU/ml	Su and D'Souza, 2011
		MNV-1	1.73 log PFU/ml	
		HAV	3.2 log PFU/ml	
GSE	0.2 mg/ml	MNV-1	3 log PFU/ml	Li et al., 2012
Resveratrol	40 µg/mL	Influenza	Complete inhibition	Palamara et al., 2005
Resveratrol	20 uM	Polyomavirus	Complete inhibition	Berardi et al., 2009

Understanding the Mechanism of Action of GSE against Microbial Pathogens

GSE was shown to have some effect on viral adsorption of FCV-F9, MNV-1, and HAV, with less effect on replication (Su and D'Souza, 2011). These researchers showed that when the respective host cells were treated with GSE before viral infection, the titers of FCV-F9, MNV-1, and HAV were decreased by 0.8, 0.5, and 1.9 log PFU/ml, respectively. GSE treatment after the viral infection of the host cells showed a 0.1, 0.2, and 0.8 log PFU/ml reduction, respectively. The higher reduction obtained by GSE when used in pre-infection than post-infection, indicated that GSE either blocks virus binding to the host cell receptors or coats the virus to prevent attaching to the host-cell or causes clumping of viral particles.

GSE has also been shown to inhibit the replication and expression of human immunodeficiency virus type 1 (HIV-1), an enveloped virus with glycoproteins gp 120 and gp41, by blocking binding of HIV-1 to the cell receptor and preventing HIV-1 entry into the normal lymphocyte (Nair et al., 2002). The study also showed that GSE at concentrations of 0.5 to 5 mg/ml over 24 h at 37°C was found to significantly down-regulate expression of HIV co-receptors, CCR2b and CCR3 (B-chemokine receptors), in peripheral blood mononuclear cells (Nair et al., 2002).

Resveratrol (RV) is a non-flavonoid polyphenol found in grape seed and grape skin and is known to be synthesized in response to external stimuli and injury (Li et al., 2006). RV has been known to affect several intracellular signaling pathways, and strongly inhibits the *in vitro* and *in vivo* replication of influenza virus (Palamara et al., 2005). Virus replication was reportedly reduced by 20 μg/mL of RV and completely inhibited by 40 μg/mL of RV over 72 h of infection time. RV at 20 uM was also found to inhibit the replication of polyomavirus, with complete reduction of replication after 24 h of infection (Berardi et al., 2009). Polyomaviruses are small, double stranded DNA viruses known to occur in immunocompromised individuals and may cause hemorrhagic cystitis in recipients of bone marrow transplantation and progressive multifocal leukoencephalopathy (Ahsan et al., 2006).

Studies were also carried out to gain insights about the mechanism of action of GSE and its components against bacterial pathogens, and it was proposed that the antibacterial activity of tannins could be due to the inhibition of extracellular microbial enzymes (Scalbert, 1991). Inhibitory effects of GSE could also be attributed to the constituent phenolic compounds that can form phenoxyl radicals upon entering the cell, or binding to the cellular membrane and causing cellular disruption (Dorman et al., 2000). In addition, scanning electron microscopy studies carried out on MRSA cells treated with GSE showed the that the cell surface was rough and disrupted as compared to the smooth and rounded surface in control cells (Su et al., 2012), that were in agreement with the findings of Al-Habib et al., (2010). This suggests that GSE may be acting by disrupting the bacterial cell membranes causing leakage of cellular material.

Potential Application of GSE for Food Industry

Natural products that are cost-effective, environmentally friendly, and sustainable are of great interest to the food industry. GSE is one such natural product, which is being investigated as an antimicrobial surface disinfectant. Several studies have aimed to evaluate the potential of GSE for its use as a food additive or a pre-packaging wash or treatment. Kao et al., (2010) used pork slices as a "model" food to evaluate its use in the food industry. Pork slices were first immersed in 20 mg/ml, 60 mg/ml, or 100 mg/ml of GSE solutions and then inoculated with *S. aureus* suspensions, and incubated at 37°C for 8 h and 24 h, after which

>99% of inhibition was observed for pork slices treated with 100 mg/ml GSE (Kao et al., 2010). However, with such high concentrations of GSE used (due to its bitter taste), it is important to note that there may be issues with sensory acceptance of these products by consumers.

In another study by Corrales et al., (2009), GSE was introduced into pea starch films at a concentration of 1% that showed growth inhibition of *Brochothrix thermosphacta* (spoilage organism) by 1.3 log CFU/mL in pork loins during the first 4 days of incubation (Corrales et al., 2009). When GSE (commercially available Gravinol-S) was tested as an antimicrobial wash on fresh produce, MICs for *Listeria monocytogenes* and *Listeria innocua* ATCC 33090 were found to be 50 and 78 μg/ml (Bisha et al., 2010). When tested against live cells, using flow cytometry at sub-MICs and after 2 min of exposure, GSE was found to cause rapid permeabilization and clumping of *L. innocua*. GSE treatment of 1250 μg/ ml was shown to yield the highest reduction of *L. monocytogenes* by 2 log CFU/ml after 2 min on spiked Roma tomatoes. These results suggest the potential application of GSE as a surface disinfectant for fresh produce (Bisha et al., 2010).

Conventional and electrostatic spraying of malic acid (MA), tartaric acid, gallic acid and GSE against *Salmonella* Typhimurium using spinach as a model system, showed that MA alone sprayed electrostatically caused log reductions of 1.6 and 3.2 log CFU/g after 7 and 14 days, while GSE alone led to a reduction of 0 to 1.4 log CFU/g (Ganesh et al., 2010). These researchers reported that maximum reduction of 2.6 to 3.3 log CFU/g was observed for spinach leaves treated with 2% malic acid in combination with 3% GSE when sprayed electrostatically, while lower log reductions were observed (0 to 1.5 log CFU/g) when sprayed conventionally (Ganesh et al., 2010). These results showed that electrostatic combination sprays containing MA and GSE were more effective in reducing the foodborne gram-negative *S.* Typhimurium that is linked to on-going produce outbreaks.

Consumer concerns regarding the addition of nitrite in processed meats have led to the use of natural alternatives such as GSE (Xi et al., 2011). GSE at 0.5% was found to reduce *L. monocytogenes* by 1.1 log CFU/g after 9 days at 10°C in cured meats (Xi et al., 2011). When fresh pork slices that were first treated with GSE along with oregano, clove and pomegranate and then inoculated with low (10 CFU/ml) and high (10^5 CFU/mL) titers of *L. monocytogenes*, *S. aureus* and *S. enterica*, were aerobically packed and stored at 20°C, the treated samples showed reductions of 1.36, 0.96 and 1.75 log CFU/g for *L. monocytogenes*, *S. aureus*, and *S. enterica*, respectively after 9 days (Shan et al., 2011).

Irradiated meat samples when treated with a single 1% GSE treatment led to a reduction from of 6.5 to 4.5 log CFU/ml *L. monocytogenes* in 12 to 18 h but no subsequent further reduction were observed; while nisin (6400 IU/ml) treatment gave an initial reduction of 3 log CFU/ml in 12 h though the cell numbers rebounded after 18 to 24 h (Sivarooban et al., 2007). However, these researchers showed that the combined treatment with 1% GSE and nisin (6400) IU led to reduction below detectable limits (Sivarooban et al., 2007).

Application of GSE as antiviral produce wash was also studied where 0.25 to 1 mg/ml GSE was found to reduce FCV-F9 and MNV-1 titers within 5 min on spiked lettuce and peppers (Su and D'Souza, 2013). High titers (~7 log PFU/ml) of FCV-F9 on lettuce and peppers were found to be reduced by 2.71 log PFU and 3.05 with GSE at 1 mg/ml after 1 min, respectively, while MNV-1 was shown to be reduced by 0.2-0.3 log PFU on lettuce and 0.8 log PFU on peppers for low titers (~5 log PFU/ml) (Su and D'Souza, 2013).

Animal Feeding Studies and Metagenomic Approaches

A metagenome comprises of the total genetic material obtained from any particular habitat, and metagenomics is a comparatively new and emerging field that involves sequence and function based analysis of a microbiome aiming to study the microbial communities in a niche and discovering new organisms (Sleator et al., 2008). Several metagenomic studies are being carried out in diverse niches such as oceans, lakes, soil, and the animal and human gut (Sleator et al., 2008). GSE has been studied extensively for its different properties as mentioned above, however the interaction and effect of dietary GSE and its polyphenols on the gut microbiome have not been explored. Animal feeding studies are good models to study these interactions and to further study the effects of GSE. Incorporating GSE into the diet of test animals and comparing the shift in the gut microbiome before and after diet supplementation is of interest to determine effects on health. The metagenomic data obtained from such feeding studies could also provide new insights on the effects of GSE polyphenols on the original/native microbiome and beneficial gut bacteria (Frank et al., 2008; Kemperman et al., 2010; Madupu et al., 2013). Proteomics can determine the expressed protein patterns/profiles between the control and fed animals and metabolomics can provide insights into the GSE-induced pathways and anti-inflammation and antioxidant effects. These studies could also prove beneficial to assess the antimicrobial properties of GSE where its effect in animals challenged with a particular pathogen and subsequent pathogen survival and successful/unsuccessful infection can be studied. Such studies can be used as a holistic approach to gain knowledge about the use of GSE in diets for verifying its associated health benefits.

Conclusion

Overall, it is evident that GSE has several health associated benefits at concentrations where it is generally recognized as safe. However, as with consumption of any product, caution is advised when used with other antibiotics or prescription drugs that may interact adversely with GSE and discussions with the health-care provider before consumption is recommended. In addition, to better understand the antimicrobial properties and application of GSE, further research in food systems and feeding studies are needed.

References

Ahmad, S. F., Zoheir, K. M., Abdel-Hamied, H. E., Ashour, A. E., Bakheet, S. A., Attia, S. M., Abd-Allah, A. R., 2013. Grape seed proanthocyanidin extract has potent anti-arthritic effects on collagen-induced arthritis by modifying the T cell balance. *Int. Immunopharmacol.* 17, 79-87.

Ahsan, N., Shah, K. V., 2006. Polyomaviruses and human diseases. *Adv. Exp. Med. Biol.* 577, 1-18.

Al-Habib, A., Al-Saleh, E., Safer, A. M., Afzal, M., 2010. Bactericidal effect of grape seed extract on methicillin-resistant *Staphylococcus aureus* (MRSA). *J. Toxicol. Sci.* 35, 357-364.

Bagchi, D., Bagchi, M., Stohs, S. J., Das, D. K., Ray, S. D., Kuszynski, C. A., Joshi, S. S., Pruess, H. G., 2000. Free radicals and grape seed proanthocyanidin extract: importance in human health and disease prevention. *Toxicology* 148, 187-197.

Bagchi, D., Sen, C. K., Ray, S. D., Das, D. K., Bagchi, M., Preuss, H. G., Vinson, J. A., 2003. Molecular mechanisms of cardioprotection by a novel grape seed proanthocyanidin extract. *Mutat. Res.* 523-524, 87-97.

Balu, M., Sangeetha, P., Haripriya, D., Panneerselvam, C., 2005. Rejuvenation of antioxidant system in central nervous system of aged rats by grape seed extract. *Neurosci. Lett.* 383, 295-300.

Barbacid, M., 1987. ras genes. *Annu. Rev. Biochem.* 56, 779-827.

Barnes, P. J., 2013. New anti-inflammatory targets for chronic obstructive pulmonary disease. *Nat. Rev. Drug Discov.* 12, 543-559.

Bart, A., Wentink-Bonnema, E. M., Gilis, H., Verhaar, N., Wassenaar, C. J., van Vugt, M., Goorhuis, A., van Gool, T., 2013. Diagnosis and subtype analysis of *Blastocystis* sp. in 442 patients in a hospital setting in the Netherlands. *BMC Infect. Dis.* 13, 389.

Berardi, V., Ricci, F., Castelli, M., Galati, G., Risuleo, G., 2009. Resveratrol exhibits a strong cytotoxic activity in cultured cells and has an antiviral action against polyomavirus: potential clinical use. *J. Exp. Clin. Cancer Res.* 28, 96.

Bisha, B., Weinsetel, N., Brehm-Stecher, B. F., Mendonca, A., 2010. Antilisterial effects of gravinol-s grape seed extract at low levels in aqueous media and its potential application as a produce wash. *J. Food Prot.* 73, 266-273.

Bomser, J. A., Singletary, K. W., Wallig, M. A., Smith, M. A., 1999. Inhibition of TPA-induced tumor promotion in CD-1 mouse epidermis by a polyphenolic fraction from grape seeds. *Cancer Lett.* 135, 151-157.

Bouhamidi, R., Prevost, V., Nouvelot, A., 1998. High protection by grape seed proanthocyanidins (GSPC) of polyunsaturated fatty acids against UV-C induced peroxidation. CR Acad. Sci. III 321, 31-38.

Brown, J. C., Huang, G., Haley-Zitlin, V., Jiang, X., 2009. Antibacterial effects of grape extracts on *Helicobacter pylori*. *Appl. Environ. Microbiol.* 75, 848-852.

Brown, K., Buchmann, A., Balmain, A., 1990. Carcinogen-induced mutations in the mouse c-Ha-ras gene provide evidence of multiple pathways for tumor progression. *Proc. Natl. Acad. Sci. U. S. A.* 87, 538-542.

Chao, C. L., Chang, N. C., Weng, C. S., Lee, K. R., Kao, S. T., Hsu, J. C., Ho, F. M., 2011. Grape seed extract ameliorates tumor necrosis factor-alpha-induced inflammatory status of human umbilical vein endothelial cells. *Eur. J. Nutr.* 50, 401-409.

Charradi, K., Sebai, H., Elkahoui, S., Ben Hassine, F., Limam, F., Aouani, E., 2011. Grape seed extract alleviates high-fat diet-induced obesity and heart dysfunction by preventing cardiac siderosis. *Cardiovasc. Toxicol.* 11, 28-37.

Chis, I. C., Ungureanu, M. I., Marton, A., Simedrea, R., Muresan, A., Postescu, I. D., Decea, N., 2009. Antioxidant effects of a grape seed extract in a rat model of diabetes mellitus. *Diab. Vasc. Dis. Res.* 6, 200-204.

Corrales, M., Han, J. H., Tauscher, B., 2009. Antimicrobial properties of grape seed extracts and their effectiveness after incorporation into pea starch films. *Internatl. J. Food Sci. & Technology* 44, 425-433.

Derry, M., Raina, K., Agarwal, R., Agarwal, C., 2012. Differential effects of grape seed extract against human colorectal cancer cell lines: The intricate role of death receptors and mitochondria. *Cancer Lett.* Dec 23. pii: S0304-3835(12)00732-X. doi: 10.1016/j.canlet.2012.12.015.

Dinicola, S., Pasqualato, A., Cucina, A., Coluccia, P., Ferranti, F., Canipari, R., Catizone, A., Proietti, S., D'Anselmi, F., Ricci, G., Palombo, A., Bizzarri, M., 2014. Grape seed extract suppresses MDA-MB231 breast cancer cell migration and invasion. *Eur. J. Nutr.* 53(2):421-31. doi: 10.1007/s00394-013-0542-6. Epub 2013 Jun 11.

Dorman, H. J., Deans, S. G., 2000. Antimicrobial agents from plants: antibacterial activity of plant volatile oils. *J. Appl. Microbiol.* 88, 308-316.

Eguchi, K., Manabe, I., 2013. Macrophages and islet inflammation in type 2 diabetes. *Diabetes Obes. Metab.* 15 Suppl 3, 152-158.

Fan, P. H., Lou, H. X., 2004. [Isolation and structure identification of grape seed polyphenols and its effects on oxidative damage to cellular DNA]. Yao Xue Xue Bao 39, 869-875.

Frank, D. N., Pace, N. R., 2008. Gastrointestinal microbiology enters the metagenomics era. *Curr. Opin. Gastroenterol.* 24, 4-10.

Ganesh, V., Hettiarachchy, N. S., Ravichandran, M., Johnson, M. G., Griffis, C. L., Martin, E. M., Meullenet, J. F., Ricke, S. C., 2010. Electrostatic sprays of food-grade acids and plant extracts are more effective than conventional sprays in decontaminating *Salmonella* Typhimurium on spinach. *J. Food Sci.* 75, M574-579.

Gholami-Ahangaran, M., Bahmani, M., Zia-Jahromi, N., 2012. In-vitro anti-leech effects of *Vitis vinifera* L., Niclosamide and Ivermectin on mature and immature forms of leech *Limnatis nilotica*. *Global Veterinaria* 8, 229-232.

Grabensteiner, E., Liebhart, D., Arshad, N., Hess, M., 2008. Antiprotozoal activities determined in vitro and in vivo of certain plant extracts against *Histomonas meleagridis*, *Tetratrichomonas gallinarum* and *Blastocystis* sp. *Parasitol. Res.* 103, 1257-1264.

Hall AJ, Lopman BA, Payne DC, Patel MM, Gastañaduy PA, Vinjé J, et al. Norovirus disease in the United States. *Emerg. Infect. Dis.* [Internet]. 2013 Aug

Hamlaoui, S., Mokni, M., Limam, N., Zouaoui, K., Ben Rayana, M. C., Carrier, A., Limam, F., Amri, M., Marzouki, L., Aouani, E., 2012. Protective effect of grape seed and skin extract on garlic-induced erythrocyte oxidative stress. *J. Physiol. Pharmacol.* 63, 381-388.

Hanausek, M., Spears, E., Walaszek, Z., Kowalczyk, M. C., Kowalczyk, P., Wendel, C., Slaga, T. J., 2011. Inhibition of murine skin carcinogenesis by freeze-dried grape powder and other grape-derived major antioxidants. *Nutr. Cancer* 63, 28-38.

Hauck, R., Hafez, H. M., 2007. Effect of coated plant extracts on *Histomonas meleagridis* and growth of bacteria in vitro. *Avian Dis.* 51, 880-883.

Jayaprakasha, G. K., Selvi, T., Sakariah, K. K., 2003. Antibacterial and antioxidant activities of grape (Vitis vinifera) seed extracts. *Food Research International* 36, 117-122.

Jeurissen, S. H., Janse, E. M., Vermeulen, A. N., Vervelde, L., 1996. *Eimeria tenella* infections in chickens: aspects of host-parasite: interaction. *Vet. Immunol. Immunopathol.* 54, 231-238.

Jin, H. Y., Cha, Y. S., Baek, H. S., Park, T. S., 2013. Neuroprotective effects of *Vitis vinifera* extract on prediabetic mice induced by a high-fat diet. *Korean J. Intern. Med.* 28, 579-586.

Kao, T. T., Tu, H. C., Chang, W. N., Chen, B. H., Shi, Y. Y., Chang, T. C., Fu, T. F., 2010. Grape seed extract inhibits the growth and pathogenicity of *Staphylococcus aureus* by interfering with dihydrofolate reductase activity and folate-mediated one-carbon metabolism. *Int. J. Food Microbiol.* 141, 17-27.

Kemperman, R. A., Bolca, S., Roger, L. C., Vaughan, E. E., 2010. Novel approaches for analysing gut microbes and dietary polyphenols: challenges and opportunities. *Microbiology* 156, 3224-3231.

Lattes, S., Ferte, H., Delaunay, P., Depaquit, J., Vassallo, M., Vittier, M., Kokcha, S., Coulibaly, E., Marty, P., 2011. *Trichostrongylus colubriformis* nematode infections in humans, France. *Emerg. Infect. Dis.* 17, 1301-1302.

Lee, T., Kwon, H. S., Bang, B. R., Lee, Y. S., Park, M. Y., Moon, K. A., Kim, T. B., Lee, K. Y., Moon, H. B., Cho, Y. S., 2012. Grape seed proanthocyanidin extract attenuates allergic inflammation in murine models of asthma. *J. Clin. Immunol.* 32, 1292-1304.

Li, D., Baert, L., Zhang, D., Xia, M., Zhong, W., Van Coillie, E., Jiang, X., Uyttendaele, M., 2012. Effect of grape seed extract on human norovirus GII.4 and murine norovirus 1 in viral suspensions, on stainless steel discs, and in lettuce wash water. *Appl. Environ. Microbiol.* 78, 7572-7578.

Li, X., Wu, B., Wang, L., Li, S., 2006. Extractable amounts of trans-resveratrol in seed and berry skin in *Vitis* evaluated at the germplasm level. *J. Agric. Food Chem.* 54, 8804-8811.

Madupu, R., Szpakowski, S., Nelson, K. E., 2013. Microbiome in human health and disease. *Sci. Prog.* 96, 153-170.

Mandhane, S. N., Shah, J. H., Thennati, R., 2011. Allergic rhinitis: an update on disease, present treatments and future prospects. *Int. Immunopharmacol.* 11, 1646-1662.

Molan, A. L., Meagher, L. P., Spencer, P. A., Sivakumaran, S., 2003. Effect of flavan-3-ols on in vitro egg hatching, larval development and viability of infective larvae of *Trichostrongylus colubriformis*. *Int. J. Parasitol.* 33, 1691-1698.

Moreno, D. A., Ilic, N., Poulev, A., Brasaemle, D. L., Fried, S. K., Raskin, I., 2003. Inhibitory effects of grape seed extract on lipases. *Nutrition* 19, 876-879.

Nair, M. P., Kandaswami, C., Mahajan, S., Nair, H. N., Chawda, R., Shanahan, T., Schwartz, S. A., 2002. Grape seed extract proanthocyanidins downregulate HIV-1 entry coreceptors, CCR2b, CCR3 and CCR5 gene expression by normal peripheral blood mononuclear cells. *Biol. Res.* 35, 421-431.

Nassiri-Asl, M., Hosseinzadeh, H., 2009. Review of the pharmacological effects of Vitis vinifera (Grape) and its bioactive compounds. *Phytother. Res.* 23, 1197-1204.

Nikitina, N. A., Sobenin, I. A., Myasoedova, V. A., Korennaya, V. V., Mel'nichenko, A. A., Khalilov, E. M., Orekhov, A. N., 2006. Antiatherogenic effect of grape flavonoids in an ex vivo model. *Bull. Exp. Biol. Med.* 141, 712-715.

Palamara, A. T., Nencioni, L., Aquilano, K., De Chiara, G., Hernandez, L., Cozzolino, F., Ciriolo, M. R., Garaci, E., 2005. Inhibition of influenza A virus replication by resveratrol. *J. Infect. Dis.* 191, 1719-1729.

Pallares, V., Fernandez-Iglesias, A., Cedo, L., Castell-Auvi, A., Pinent, M., Ardevol, A., Salvado, M. J., Garcia-Vallve, S., Blay, M., 2013. Grape seed procyanidin extract reduces the endotoxic effects induced by lipopolysaccharide in rats. *Free Radic. Biol. Med.* 60, 107-114.

Park, J. S., Park, M. K., Oh, H. J., Woo, Y. J., Lim, M. A., Lee, J. H., Ju, J. H., Jung, Y. O., Lee, Z. H., Park, S. H., Kim, H. Y., Cho, M. L., Min, J. K., 2012. Grape-seed proanthocyanidin extract as suppressors of bone destruction in inflammatory autoimmune arthritis. *PLoS One* 7, e51377.

Park, S. H., Park, T. S., Cha, Y. S., 2008. Grape seed extract (Vitis vinifera) partially reverses high fat diet-induced obesity in C57BL/6J mice. *Nutr. Res. Pract.* 2, 227-233.

Radhakrishnan, S., Reddivari, L., Sclafani, R., Das, U. N., Vanamala, J., 2011. Resveratrol potentiates grape seed extract induced human colon cancer cell apoptosis. *Front Biosci.* (Elite Ed) 3, 1509-1523.

Razavi, S. M., Gholamin, S., Eskandari, A., Mohsenian, N., Ghorbanihaghjo, A., Delazar, A., Rashtchizadeh, N., Keshtkar-Jahromi, M., Argani, H., 2013. Red grape seed extract improves lipid profiles and decreases oxidized low-density lipoprotein in patients with mild hyperlipidemia. *J. Med. Food* 16, 255-258.

Rhodes, P. L., Mitchell, J. W., Wilson, M. W., Melton, L. D., 2006. Antilisterial activity of grape juice and grape extracts derived from Vitis vinifera variety Ribier. *Int. J. Food Microbiol.* 107, 281-286.

Sapwarobol, S., Adisakwattana, S., Changpeng, S., Ratanawachirin, W., Tanruttanawong, K., Boonyarit, W., 2012. Postprandial blood glucose response to grape seed extract in healthy participants: A pilot study. *Pharmacogn. Mag.* 8, 192-196.

Scalbert, A., 1991. Antimicrobial properties of tannins. *Phytochemistry* 30, 3875-3883.

Schneider, M., Kruger, K., 2013. Rheumatoid arthritis-early diagnosis and disease management. *Dtsch Arztebl Int.* 110, 477-484.

Shan, B., Cai, Y. Z., Brooks, J. D., Corke, H., 2011. Potential application of spice and herb extracts as natural preservatives in cheese. *J. Med. Food* 14, 284-290.

Shi, J., Yu, J., Pohorly, J. E., Kakuda, Y., 2003. Polyphenolics in grape seeds-biochemistry and functionality. *J. Med. Food* 6, 291-299.

Sivarooban, T., Hettiarachchy, N. S., Johnson, M. G., 2007. Inhibition of *Listeria monocytogenes* using nisin with grape seed extract on turkey frankfurters stored at 4 and 10 degrees C. *J. Food Prot.* 70, 1017-1020.

Sleator, R. D., Shortall, C., Hill, C., 2008. Metagenomics. *Lett. Appl. Microbiol.* 47, 361-366.

Smullen, J., Finney, M., Storey, D. M., Foster, H. A., 2012. Prevention of artificial dental plaque formation in vitro by plant extracts. *J. Appl. Microbiol.* 113, 964-973.

Su, X., D'Souza, D. H., 2011. Grape seed extract for control of human enteric viruses. *Appl. Environ. Microbiol.* 77, 3982-3987.

Su, X., D'Souza, D. H., 2013. Grape seed extract for foodborne virus reduction on produce. *Food Microbiol.* 34, 1-6.

Su, X., Howell, A. B., D'Souza, D. H., 2012. Antibacterial effects of plant-derived extracts on methicillin-resistant *Staphylococcus aureus*. *Foodborne Pathog. Dis.* 9, 573-578.

Taher, E., Mahmoud, N., Mahmoud, M., 2012. Laboratory evaluation of the effect of Egyptian native plants against some parasitic vectors. *Turkiye Parazitol. Derg.* 36, 160-165.

Tyagi, A., Raina, K., Gangar, S., Kaur, M., Agarwal, R., Agarwal, C., 2013. Differential effect of grape seed extract against human non-small-cell lung cancer cells: the role of reactive oxygen species and apoptosis induction. *Nutr. Cancer* 65 Suppl 1, 44-53.

van der Heijden, H. M., Landman, W. J., 2011. High seroprevalence of *Histomonas meleagridis* in Dutch layer chickens. *Avian Dis*. 55, 324-327.

Wang, M. L., Suo, X., Gu, J. H., Zhang, W. W., Fang, Q., Wang, X., 2008. Influence of grape seed proanthocyanidin extract in broiler chickens: effect on chicken coccidiosis and antioxidant status. *Poult. Sci*. 87, 2273-2280.

Wang, Y. H., Yang, X. L., Wang, L., Cui, M. X., Cai, Y. Q., Li, X. L., Wu, Y. J., 2010. Effects of proanthocyanidins from grape seed on treatment of recurrent ulcerative colitis in rats. *Can. J. Physiol. Pharmacol*. 88, 888-898.

Xi, Y., Sullivan, G. A., Jackson, A. L., Zhou, G. H., Sebranek, J. G., 2011. Use of natural antimicrobials to improve the control of *Listeria monocytogenes* in a cured cooked meat model system. *Meat Sci*. 88, 503-511.

Yamakoshi, J., Saito, M., Kataoka, S., Kikuchi, M., 2002. Safety evaluation of proanthocyanidin-rich extract from grape seeds. *Food Chem. Toxicol*. 40, 599-607.

Yu, H., Wang, S. E., Zhao, C., Xu, G., 2002. [Study of anti-atherosclerosic effect of grape seed extract and its mechanism]. Wei Sheng Yan Jiu 31, 263-265.

Zhao, J., Wang, J., Chen, Y., Agarwal, R., 1999. Anti-tumor-promoting activity of a polyphenolic fraction isolated from grape seeds in the mouse skin two-stage initiation-promotion protocol and identification of procyanidin B5-3'-gallate as the most effective antioxidant constituent. *Carcinogenesis* 20, 1737-1745.

Books

Paulus, W. (1993). In: Microbicides for the protection of materials –a handbook. Chapman & Hall (Eds.). London, UK, pp: 496.

Websites

U.S. Food and drug administration (http://www.accessdata.fda.gov/ scripts/fcn/gras_notices /grn000124.pdf)

American Diabetes Association (http://www.diabetes.org/living-with-diabetes/treatment-and-care/)

Index

#

21st century, 270

A

ABA, 20
absorption spectroscopy, 253
abstraction, 358
access, 67, 407
accessions, 396
accounting, 33, 115, 158, 161, 197, 268, 318
acetaldehyde, 60, 84, 85, 92, 104
acetic acid, 54, 74, 98, 228, 275, 277, 435
acetone, 230, 240, 343, 418, 419, 434, 435
acetonitrile, 139, 215, 275, 277
acetylation, 374
acidic, 6, 85, 111, 231, 249, 253, 258, 261, 263, 267, 268, 269, 352
acidity, 19, 30, 86, 110, 118, 258, 307, 308
activated carbon, 249, 250, 270, 271
active compound, 198, 417
active oxygen, 130, 362
active packaging, 165, 173, 177
active site, 66, 67, 69, 95, 96, 263, 269
acute asthma, 431
acylation, 54, 61
ADA, 387, 388, 389, 390, 391
adaptation, 9, 32, 50
additives, 208, 227
adenocarcinoma, 125, 336, 405
adenosine, x, 377, 385, 388, 389, 391, 392, 393
adenosine deaminase, x, 392, 393
adhesion, 31, 399, 431
adiponectin, 375, 384, 432
adipose, 274, 374, 432
adipose tissue, 274, 374
adiposity, 274, 283
adrenaline, 368
adsorption, 261, 264, 265, 270, 298, 438
adults, 197, 248, 399, 405, 406, 407
advancements, 168
adverse effects, 348
aerosols, 248
aerospace, 246
aesthetic, 298
Africa, 318
agar, 272, 332, 422
age, 28, 32, 197, 199, 261, 374, 382
aggregation, 213, 383, 405
aggressiveness, 337
agricultural sector, 249
agriculture, 13, 25, 44, 205, 249, 332
AIDS, 438
air temperature, 8, 229
airway inflammation, 431
albumin, 195
alcohols, 28, 186, 226, 231, 235, 310
aldehydes, 28, 310
aldosterone, 346, 348
alfalfa, 16
algae, 250, 272
Algeria, 338
alimentation, 25
alkaloids, 51, 166, 168, 306
allergic inflammation, 444
allergy, 182
almonds, 274
alpha-tocopherol, 407
alternative medicine, 337, 412
Altitude, 9
aluminium, 248
amino, 17, 30, 36, 37, 39, 67, 72, 76, 192, 305, 310
amino acid(s), 17, 30, 36, 37, 39, 67, 72, 76, 192, 305, 310
ammonia, 17
ammonium, 16, 389

amplitude, 234, 238
analgesic, 312, 316
anatomy, 224
androgen, 317, 330
anemia, 433
angiogenesis, 124, 310, 330, 393, 405
angiotensin converting enzyme, 353, 358, 359
angiotensin I-converting enzyme, 340, 356
angiotensin II, 348
ANOVA, 113, 140, 142, 143, 145, 147, 151, 155, 157, 159, 160, 390
antibacterial effects, 174
antibiotic, 311, 312, 412
antibiotic resistance, 412
anti-cancer, 109, 125, 127, 314, 318, 365, 367, 434
anticancer activity, 125, 316, 317
anticancer drug, 213
anticarcinogenic effects, 335
antidepressant, 312
antigen, 392
anti-inflammation, 314, 329, 441
anti-inflammatory agents, 226
antimicrobial activity, ix, x, 166, 170, 175, 176, 177, 178, 198, 314, 243, 301, 336, 411, 412, 413, 414, 415, 416, 417, 418, 419, 420, 422, 423, 425, 426, 427
antimicrobial properties, 177, 314, 414, 418, 420, 434, 441
Antioxidant Activity, v, 112, 128, 346, 393
antioxidant capacity, 14, 20, 58, 87, 88, 90, 123, 172, 175, 176, 177, 192, 196, 202, 205, 207, 226, 228, 229, 233, 235, 237, 297, 301, 335, 347, 355, 363, 382, 395, 397, 401, 404, 406, 407, 408, 409, 431
antioxidant properties, 5, 26, 109, 122, 123, 127, 171, 183, 184, 185, 190, 191, 198, 204, 207, 208, 212, 223, 284, 306, 307, 314, 318, 347, 373, 397, 409, 412, 425, 430
antioxidative activity, 172
antiparasitic effects, 436
antitumor, 311, 312, 313
AOC, 291
aortic valve, 432
apoptosis, 109, 123, 124, 128, 130, 131, 132, 310, 312, 313, 316, 317, 329, 330, 331, 332, 333, 334, 335, 336, 337, 338, 353, 355, 360, 370, 372, 373, 385, 434, 445
apples, 95
aqueous solutions, 271, 272
Argentina, 13, 21, 36, 40, 136, 355
arginine, 16, 37
aromatic compounds, 74, 76
aromatic rings, 51, 185, 186, 414
aromatics, 76, 78
arrest, 124, 128, 313, 329, 353
Arrhenius equation, 398
arrhythmias, 353
arsenic, 270
artery(s), 347, 353, 368, 369, 371, 374, 377, 382, 399, 407, 431
arthritis, 226, 432, 441, 445
ascorbic acid, 54, 72, 80, 86, 332, 346, 347, 351
Asia, 2, 183, 213, 430
assessment, 23, 35, 288, 296, 359, 393, 427
asthma, xii, 431, 444
astringent, 50
atherogenesis, 317, 372
atherosclerosis, 109, 182, 226, 244, 315, 316, 318, 366, 368, 373, 376, 385, 399, 404, 431
atherosclerotic plaque, 371
atmosphere, 236, 247, 248
atmospheric pressure, 237
atoms, 51, 67, 68, 69, 73
attachment, 111, 320, 396
authentication, 90, 163
authorities, ix
automation, 236, 420
auxins, 355
avian, 413, 436, 437
awareness, 28, 306, 352

B

Bacillus subtilis, 435
bacteria, 34, 36, 38, 43, 61, 66, 70, 74, 75, 80, 92, 98, 101, 170, 175, 176, 178, 198, 202, 240, 250, 412, 414, 415, 416, 417, 418, 419, 422, 426, 428, 434, 435, 436, 437, 441, 443
bacterial infection, 414, 415
bacterial pathogens, 424, 434, 439
bacterial strains, 418
bacteriocins, 166, 168
bacteriostatic, 227, 244, 414
bacterium, 168
barriers, 48, 81
base, 7, 10, 16, 44, 69, 110, 214, 237, 239, 243, 329, 375
batteries, 247, 248
beef, 172, 174, 176, 177
beer, 99, 362
Beijing, 270, 300
beneficial effect, 107, 112, 122, 127, 190, 225, 226, 227, 305, 368, 374, 376, 388, 392, 395, 398, 400, 401, 414, 423
benefits, ix, x, 34, 44, 48, 65, 81, 109, 132, 171, 182, 183, 187, 249, 274, 305, 306, 310, 317, 329, 334, 336, 342, 348, 351, 352, 361, 365, 367, 384, 387,

395, 399, 412, 414, 416, 429, 430, 432, 433, 434, 441
benzene, 62, 184, 239, 314
beverages, 199, 207, 414, 423, 424
bile, 401
bioaccumulation, 246
bioactive compounds, x, 48, 109, 140, 162, 177, 197, 212, 233, 234, 237, 238, 239, 275, 287, 300, 306, 329, 333, 359, 388, 395, 398, 414, 417, 419, 420, 429, 444
bioavailability, x, 201, 202, 203, 204, 247, 316, 334, 342, 346, 351, 352, 359, 361, 376, 378, 385, 395, 400, 401, 402, 403, 406, 407, 408, 428
biochemical processes, 48, 49
biochemistry, 9, 87, 91, 130, 178, 208, 361, 445
bioconversion, 402
biodegradation, 99
biodiversity, 13, 24, 27, 29, 33, 37, 41, 44
Biodynamic, 13, 22
biological activity(s), ix, 13, 20, 34, 51, 88, 109, 125, 201, 213, 314, 316, 317, 329, 334, 343, 365, 366, 420
Biological effects, 223, 306, 311, 312, 313, 316
biological processes, 66, 76
biological systems, 112, 351
biologically active compounds, 130, 412, 423
biomarkers, 395, 399, 402, 404, 427
biomass, 29, 34, 36, 41, 250, 258, 261, 270, 271, 272, 288, 293, 294, 296, 297
biomaterials, 330
biomolecules, 109, 183
biosphere, 247
biosynthesis, ix, 9, 13, 16, 17, 38, 78, 88, 97, 108, 112, 113, 124, 128, 152, 154, 222, 224, 309, 313, 315, 338
biosynthetic pathways, 8
biotechnology, ix
biotic, 111, 121, 212, 219, 223
black tea, 355, 427
bleaching, 76, 85
blindness, 82
blood, 98, 124, 197, 226, 248, 334, 346, 348, 351, 352, 356, 357, 361, 366, 368, 369, 370, 371, 376, 379, 382, 385, 388, 392, 399, 400, 402, 407, 409, 432, 445
blood pressure, 346, 348, 351, 352, 356, 357, 361, 366, 368, 369, 370, 371, 376, 379, 382, 399, 407
blood stream, 352
blood supply, 371
blood vessels, 226, 371
blood-brain barrier, 400, 409
bloodstream, 351, 401
body fat, 432
body mass index, 366, 368, 374
body weight, 227, 274, 431, 432
bonds, 111, 187, 235, 339, 340
bone(s), 248, 307, 431, 439, 445
bone marrow, 439
bone marrow transplant, 439
boric acid, 13
botrytis, 74
bradykinin, 339, 348
brain, 204, 248, 314, 356, 371, 388, 392, 400, 407, 409
Brazil, xiii, xiv, xv, xvi, xvii, xviii, xix, xx, 12, 14, 47, 86, 296, 297, 318, 320, 321, 322, 328, 387, 389, 411, 418
breakdown, 34, 66, 238, 309, 334, 352, 401
break-even, 296
breast cancer, 124, 125, 313, 331, 333, 334, 336, 337, 342, 425, 433, 443
breast carcinoma, 124, 353, 434
breeding, 2, 193, 224, 320
building blocks, 308
buyers, 41
by-products, 132, 152, 171, 225, 226, 227, 229, 230, 231, 235, 236, 237, 240, 241, 242, 243, 249, 300, 329, 361, 411, 416, 417, 421, 423, 424, 425

C

Ca^{2+}, 26, 261
cacao, 360
CAD, 370
cadmium, 248, 250, 253, 259, 261, 270, 272
calcium, 16, 80, 307, 431
calibration, 194, 195, 196, 253, 276, 277, 421
caloric intake, 374
caloric restriction, 370, 374
calorie, 377, 398
campaigns, 33
cancer cells, 108, 109, 124, 125, 203, 314, 317, 434
cancer progression, 316
candidates, 143, 155
candidiasis, 425
Canopy, 11, 13, 19, 129, 130, 132, 163
capillary, 196, 200, 351
carbohydrate(s), 111, 307, 431, 433
carbon, 9, 36, 42, 51, 54, 58, 59, 74, 167, 186, 192, 201, 237, 243, 244, 270, 271, 288, 289, 299, 301, 302, 308, 309, 314, 343, 359, 363, 420, 444
carbon atoms, 186, 309
carbon dioxide, 9, 167, 192, 201, 237, 243, 244, 288, 289, 299, 301, 302, 343, 359, 363, 420
carboxyl, 187
carboxylic acid, 102, 110, 196

carcinogen, 109, 124, 378, 434
carcinogenesis, 124, 126, 129, 315, 317, 333, 443
carcinoma, 128, 313, 434
cardiac muscle, 373, 379
cardiac output, 373
cardiovascular disease(s), ix, xi, 48, 65, 107, 182, 197, 305, 330, 335, 348, 366, 376, 382, 384, 385, 393, 399, 403, 404, 406
cardiovascular disorders, 181
Cardiovascular effects, 366
cardiovascular physiology, 368
cardiovascular risk, 242, 366, 379, 399
cardiovascular system, 367, 368, 369, 370
carotene, 132, 308, 317, 318, 334, 335, 347
carotenoids, 3, 21, 48, 87, 88, 130, 308, 309, 317, 318, 332, 334, 337, 346, 347
case study, 21
caspases, 333, 434
Caspian Sea, 430
catalysis, 67, 71
catalytic activity, 75
cataract, 342, 362
cation, 34, 52, 53, 77, 246
cattle, 413
CD26, 388, 392
cell biology, 335
cell culture, 94, 223, 317
cell cycle, 94, 124, 128, 203, 310, 313, 329
cell death, 109, 169, 379, 417, 433, 434
cell division, 17, 38
cell invasion, 333
cell line(s), 124, 130, 133, 305, 316, 317, 329, 333, 335, 338, 353, 355, 369, 372, 433, 434, 443
cell membranes, 226, 232, 238, 439
cell metabolism, 347
cell signaling, 172
cell size, 373
cell surface, 388, 391, 439
cellulose, 235, 245
Central Asia, 183
central nervous system, 442
central obesity, 368
cerebral cortex, 400
cervical cancer, 128, 335, 338
cervix, 125
challenges, 9, 293, 436, 444
cheese, 168, 445
chemical characteristics, 35, 36, 48, 114, 196
chemical degradation, 236
chemical interaction, 400
chemical properties, 91, 181
chemical reactions, 60, 84, 122, 228
chemical stability, 81, 183
chemical structures, 83, 182
chemokine receptor, 439
chemokines, 388
chemoprevention, 124, 314, 332, 336
chemotherapy, 124, 313
chicken, 172, 174, 177, 178, 418, 427, 436, 446
children, 182
Chile, xvi, xvii, 39, 42, 136, 165, 173, 318, 320, 321, 322, 324, 325, 326, 395, 397, 403, 406
China, 5, 12, 17, 22, 26, 32, 42, 43, 208, 211, 212, 222, 228, 318, 321, 323, 324, 325, 409
chitin, 179
chitosan, 169, 170, 177, 178, 179
chloroform, 335, 344
chlorophyll, 6, 249, 308, 334
chloroplast, 97
cholesterol, 131, 244, 313, 331, 335, 351, 366, 369, 370, 372, 374, 378, 384, 408, 416, 431
chromatograms, 139, 141, 146, 150
chromatographic technique, 421
chromatography, 93, 139, 196, 205, 206, 214, 299, 335, 343, 358, 362, 367, 419
chromium, 270, 360
chronic diseases, 187
chronic obstructive pulmonary disease, 442
circulation, 371, 372
City, 275
classes, x, 2, 8, 17, 91, 166, 181, 183, 184, 195, 308, 314, 351, 414
classification, 66, 89, 108, 129, 136, 158, 184, 203, 422
cleaning, 297
cleavage, 66, 71, 74, 76, 97, 309, 344, 434
climate(s), 1, 7, 12, 23, 25, 34, 35, 36, 37, 51, 65, 79, 90, 112, 115, 120, 127, 157, 164, 183, 199, 205, 206, 274, 284, 319, 320, 359, 367, 396, 397, 408, 424
climate change, 183
climatic factors, 7
clinical application, 434
clinical trials, 124, 306, 351, 352, 366, 368, 375, 376
clone(s), 6, 113, 114, 115
cloning, 98
clusters, 2, 9, 11, 12, 13, 17, 19, 35, 108, 145, 231, 365, 366
CMC, 168, 173
CO_2, 192, 204, 236, 239, 288, 289, 290, 291, 292, 293, 294, 298, 299, 300, 301, 302, 353, 420
coal, 248
coatings, 165, 177
cobalt, 270
cocoa, 129, 178, 357
coding, 66

codon, 434
coenzyme, 313
coffee, 197, 275, 299
cognition, 400, 408
cognitive function, 379
cognitive impairment, 383, 405
cognitive performance, 400
coherence, 229
colitis, 431
collagen, 371, 441
collateral, 82
colon, 109, 130, 305, 311, 314, 317, 331, 332, 336, 355, 392, 401, 405, 431, 434, 445
colon cancer, 109, 130, 331, 332, 336, 355, 405, 434, 445
colonisation, 27
colonization, 28
colorectal cancer, 434, 443
column, 139, 142, 147, 151, 155, 157, 159, 160, 214, 239, 246, 251, 252, 253, 258, 259, 267, 268, 272, 277, 284, 319, 421
combination therapy, 434
combustion, 248
commercial, 2, 7, 15, 18, 27, 28, 30, 33, 39, 40, 41, 44, 82, 92, 93, 102, 137, 179, 191, 198, 199, 201, 207, 228, 229, 231, 232, 247, 287, 289, 295, 296, 297, 319, 320, 354, 359, 384, 388, 403, 436
commodity, 332
community(s), 27, 32, 36, 37, 38, 39, 40, 41, 42, 306, 366, 385, 412
comparative analysis, 422
competition, 27, 35, 36, 37, 251
compilation, 130, 136
complement, 296, 375
complexity, 186, 250, 289, 366, 368
compliance, 369, 371, 377
complications, 305
compost, 13, 24
composting, 269
compound identification, 140
condensation, 59, 60, 72, 83, 84, 85, 86, 104, 349
conditioning, 261
conductivity, 239
configuration, 123
congestive heart failure, 348
Congress, 20, 128
conjugation, 123, 401, 402
conservation, 15, 41, 42, 236, 373, 417
consolidation, 297
constituents, ix, x, 48, 49, 61, 79, 81, 164, 169, 176, 178, 185, 200, 297, 298, 310, 317, 330, 332, 338, 355, 365, 366, 367, 368, 370, 371, 374, 376, 385, 400, 404, 409, 425, 428
consumers, 29, 166, 168, 173, 306, 318, 440
contact time, 194, 411
containers, 192
contamination, 38, 83, 168, 260, 316
controlled trials, 399, 406
controversies, 403
convention, 236
conventional extraction, 192, 193, 194, 226, 229, 233, 234, 237, 240, 241, 419
COOH, 62
cooking, 172
cooling, 7, 9, 208
cooling process, 208
copolymer, 239
copper, 38, 44, 66, 67, 68, 69, 70, 73, 74, 82, 95, 96, 103, 123, 171, 246, 248, 250, 253, 259, 261, 270, 271, 272, 317
coronary arteries, 399
coronary artery disease, 355, 375, 380, 384, 399, 408
coronary heart disease, 187, 202, 366, 369, 374, 380, 381, 382, 399, 403, 431
correlation(s), 109, 113, 117, 120, 125, 130, 175, 199, 205, 207, 258, 293, 300, 301, 318, 342, 356, 391, 402, 407
correlation coefficient, 113
corrosion, 247, 248
cortex, 348, 400
cosmetic(s), ix, 192, 198, 200, 221, 213, 239, 412, 417
cost, 34, 37, 38, 165, 231, 234, 236, 240, 245, 249, 250, 269, 273, 275, 283, 288, 296, 297, 302, 398, 429, 430, 439
cough, 348
covalent bond, 76
covering, 50, 119, 135, 293, 319, 357
Croatia, 418, 425
crop(s), 6, 11, 13, 14, 15, 16, 22, 24, 25, 26, 34, 35, 36, 37, 40, 41, 42, 43, 44, 57, 107, 109, 121, 136, 182, 191, 250, 271, 318, 340, 411
crop production, 41
crystal structure, 96
crystallization, 298
CT, 12, 166
Cultivar, 130, 424
cultivation, 36, 38, 65, 136, 367, 396
cultural practices, 13, 22, 118, 125
culture, 51, 79, 177, 250, 318, 422
curcumin, 383
cure, 176
customers, 295
cuticle, 50
cutin, 50
CVD, 368, 370, 374, 375, 376

cycles, 13, 246, 247, 252, 253, 261, 262, 263, 264, 265, 266, 267, 268, 269, 270, 416
cycling, 13
cyclins, 316
cyclodextrins, 96
cyclooxygenase, 315, 317, 346, 371
cysteine, 67
cystitis, 439
cytochrome(s), 76, 124, 130, 313, 315, 330, 346, 376, 434
cytokines, 226, 391, 432
cytometry, 440
cytoplasm, 67, 238, 388, 434
cytotoxicity, 333, 381

D

damages, 9, 232
damping, 7
danger, 5, 21, 163
data analysis, 113
database, 41, 99, 375
deacetylation, 374
decay, 219, 263
defects, 214
defence, 393
defense mechanisms, 67
deficiency(s), 6, 16, 19, 23, 53, 380
deficit, 7, 10, 20, 24, 25
degradation, 6, 16, 17, 34, 60, 73, 74, 76, 78, 81, 83, 84, 95, 98, 104, 105, 192, 193, 200, 235, 236, 283, 284, 308, 309, 398
degradation process, 236
degree of polymerization, 60, 84, 158, 172, 239, 240, 343, 344, 350, 352, 357, 358, 361, 362, 396, 397
dehydrate, 74
dehydration, 191, 201, 273, 348
Dekkera bruxellensis, 92
denaturation, 68, 194, 413
dendritic cell, 393
dengue, 437
Denmark, 33, 42
dental caries, 436
dental plaque, 445
Department of Agriculture, 42, 318, 319
deposits, 398
depth, 183, 191
dermatitis, 247
desorption, 246, 252, 253, 261, 263, 266, 267, 268, 269, 270, 272, 358, 362, 363
destruction, 34, 268, 445
detectable, 123, 310, 440
detection, 76, 89, 93, 190, 196, 344, 359, 382
detergents, 71, 97
detoxification, 245, 246, 250, 259, 351
developed countries, 366, 383
developmental process, 113, 127
deviation, 114, 115, 116, 228
diabetes, 65, 182, 200, 227, 305, 333, 375, 384, 387, 392, 399, 402, 430, 432, 433, 442, 443, 446
diabetic nephropathy, 348
diabetic neuropathy, 433
diabetic patients, 391, 392
diastolic blood pressure, 339, 352
dielectric constant, 233
diet, 48, 79, 107, 124, 182, 184, 185, 200, 202, 204, 227, 241, 274, 283, 306, 313, 356, 363, 366, 368, 370, 376, 377, 380, 399, 400, 404, 405, 432, 433, 441, 442, 444, 445
dietary fat, 342
dietary fiber, 400, 407
Dietary Guidelines, 376, 379
Dietary Guidelines for Americans, 376, 379
dietary habits, 432
dietary intake, 109, 209, 395, 428
diffusion, 192, 204, 232, 235, 300, 301, 314, 422
diffusivity(s), 236, 293, 301, 302, 420
dihydroxyphenylalanine, 71
dilation, 226, 351, 400
dimerization, 110
dipeptides, 388
dipeptidyl peptidase IV, 388
dipoles, 233
direct measure, 196
discharges, 226, 233, 238, 240
discrimination, 199
discs, 211, 214, 217, 218, 444
diseases, xi, 3, 11, 32, 37, 41, 48, 109, 136, 182, 190, 197, 202, 247, 306, 308, 336, 347, 348, 368, 376, 383, 396, 412, 418, 441
dislocation, 123
dispersion, 231, 248
displacement, 54, 253
disposition, 334, 358, 381
dissociation, 238
distillation, 192, 200, 205, 229, 230
distilled water, 195, 251, 276, 277
distribution, 3, 5, 29, 30, 32, 33, 34, 43, 50, 66, 98, 99, 126, 197, 233, 397, 400
diversity, ix, x, 28, 30, 31, 32, 33, 37, 38, 39, 40, 43, 44, 45, 47, 75, 135, 136, 308, 319
dizziness, 348
DNA, 33, 78, 124, 131, 132, 313, 331, 332, 335, 336, 355, 357, 392, 400, 413, 431, 433, 439, 443
DNA damage, 124, 313, 331, 332, 335, 336, 357, 431, 433

DNA strand breaks, 132
dogs, 382
DOI, 200, 202, 299, 302
donors, 78, 171, 347, 389
dopamine, 71
dosage, 25, 376
double bonds, 166, 169, 170
down-regulation, 333
draught, 78, 396
drinking water, 260
drought, 10
drug interaction, 375
drug metabolism, 351
drug therapy, 352, 413, 434
drugs, 124, 227, 333, 348, 376, 434
dry matter, 340
drying, 136, 139, 192, 205, 206, 207, 229, 232, 242, 243, 273, 274, 275, 277, 278, 279, 280, 281, 282, 283, 284, 296, 298
Drying Methods, 274
dusts, 248
dyes, 72, 182
dyslipidemia, 274, 368, 375

E

EAE, 235
East Asia, vi, 211, 212, 213
ecology, ix, 28, 32, 39, 41, 427
economic evaluation, 202, 287, 300, 302
economic losses, 37, 65
ecosystem, 28, 43
editors, 204, 223
effluent(s), 245, 248, 249, 262, 267, 269
efflux transporters, 351
egg, 437, 444
Egypt, 412
elaboration, 51, 55, 57, 79, 145, 152, 162, 245, 398
electric field, 101, 226, 233, 237, 238, 239, 241, 244, 424
electrical breakdown, 238
electrical conductivity, 239, 247
electrodes, 238
electrolysis, 246
electrolyte, 348
electromagnetic, 233
electron(s), 66, 74, 76, 77, 78, 104, 112, 123, 178, 233, 238, 346, 372
electron microscopy, 178
electron paramagnetic resonance, 104
electrophoresis, 196, 200
electroplating, 246, 247
electroporation, 238
ELISA, 438
elucidation, 88, 412
e-mail, 165
emission, 196, 247
emitters, 167
employment, 289
EMS, 140
emulsions, 171, 240
encapsulation, 363
encoding, 108
endosperm, 340
endothelial cells, 353, 368, 370, 373, 380, 381, 385, 393, 399, 431, 442
endothelial dysfunction, 355
endothelial NO synthase, 407
endothelium, 351, 353, 356, 368, 369, 371, 378, 382, 385, 399, 402, 407
energy, 11, 140, 193, 231, 234, 239, 246, 307, 373, 374, 379, 384, 415, 420
energy expenditure, 374
engineering, 208
England, 43, 100
enlargement, 18, 118, 127, 219, 373
environment(s), 6, 25, 29, 37, 38, 39, 50, 113, 132, 199, 235, 245, 246, 247, 248, 270, 271, 283, 288, 308, 346, 412, 423, 429, 430
environmental conditions, 25, 28, 396
environmental effects, 22
environmental factors, 107, 112, 411
environmental impact, 115, 129
environmental influences, 25
environmental stress, 6, 10, 25
enzymatic activity, 235, 390
enzyme inhibitors, 354
eosinophils, 431
epidemic, 374, 381
epidemiologic, 124
epidemiology, 379, 412
epidermis, 9, 50, 354, 442
epinephrine, 371
epithelial cells, 434
epithelium, 352
EPR, 73
equilibrium, 11, 251, 253, 259, 260, 261, 263, 267, 293, 295
equipment, 27, 54, 83, 138, 192, 194, 197, 234, 253, 283, 296, 345, 421
ERA, xi
erosion, 34, 37
erythrocytes, 354, 433
ESI, 87, 101, 139, 162, 163, 197, 206, 244, 335
ESR, 124
ester, 44, 61, 339, 358

estrogen, 125, 334
ethanol, 28, 74, 80, 81, 84, 192, 193, 205, 230, 231, 233, 234, 237, 238, 239, 240, 243, 275, 276, 287, 289, 290, 291, 297, 298, 334, 343, 365, 366, 377, 414, 415, 416, 418, 419, 423, 424, 425, 427, 437
ethers, 174
ethyl acetate, 138, 215, 343, 419
ethylene, 62, 167
EU, 2, 202, 260, 318, 320, 321, 322, 323, 324, 325, 326, 327
eukaryotic, 126, 346
eukaryotic cell, 126
Europe, 30, 32, 36, 109, 202, 430
European Community, 38
European Parliament, 182
European Union, 20, 40, 41, 259
evaporation, 232
evapotranspiration, 7
evidence, 38, 48, 65, 66, 94, 126, 182, 183, 199, 200, 253, 258, 263, 268, 269, 318, 331, 358, 365, 367, 368, 374, 375, 376, 416, 430, 442
evolution, 3, 49, 60, 79, 81, 89, 94, 103, 158, 199, 264, 265, 267, 330, 345
excision, 214
excitation, 196
exclusion, 240, 358, 362, 389
excretion, 401, 406
exercise, 432
experimental condition, 3, 234, 415
experimental design, 299, 300
expertise, x
exploitation, 181, 182, 183, 192, 199, 205, 287, 299, 357, 379, 405
exporter(s), 318, 319
exposure, 1, 3, 7, 8, 11, 13, 36, 67, 90, 118, 120, 124, 129, 145, 163, 164, 221, 223, 309, 318, 371, 395, 396, 397, 398, 402, 434, 437, 440
extinction, 5, 21, 163, 277
extrusion, 284, 405

F

families, 76, 85, 109, 110, 181, 185, 196, 199, 239, 240
farmers, 14
farms, 13
fasting, 368
fat, 182, 227, 241, 274, 283, 307, 366, 374, 377, 431, 432, 442, 444
fatty acids, 50, 170, 174, 191, 239, 248, 289, 297, 298, 299, 302, 432
FDA, 166, 173, 236
fermentation, 6, 27, 28, 30, 31, 35, 38, 40, 41, 42, 44, 61, 79, 80, 81, 85, 93, 101, 111, 192, 199, 201, 227, 279, 330, 389, 397, 398, 411, 419, 420
ferromagnetic, 247
ferrous ion, 83
fertility, 42
fertilization, 50
fertilizers, 1, 13, 16, 17, 19, 246
fiber, xii, 191, 270, 307, 433
fibrinolytic, 370, 375, 384
fibroblasts, 333
fibrosis, 431
films, 165, 167, 176, 177, 178, 440, 443
filters, 277
filtration, 65, 75, 79, 139
financial, 86, 173, 329
financial support, 86, 173, 329
fish, 172, 227
fitness, 202
flame, 253
flavanols, 10, 13, 15, 18, 104, 110, 112, 123, 125, 126, 127, 168, 186, 188, 195, 197, 205, 208, 226, 227, 228, 231, 237, 240, 274, 342, 345, 367, 396, 416, 418
flavonol(s), 8, 56, 57, 58, 72, 90, 91, 103, 123, 136, 137, 145, 146, 147, 148, 149, 150, 151, 152, 153, 162, 163, 189, 190, 201, 241, 244, 274, 408
flavor, 17, 20, 48, 61, 65, 76, 77, 114, 154, 172, 185, 200, 308
flavour, 30, 33, 41, 58
flight, 196, 202, 358, 362, 363
flocculation, 75
flora, 40, 42
flotation, 75
flour, 244
flowers, 11, 184, 353
fluctuations, 127
fluid, x, 54, 200, 201, 204, 226, 233, 235, 236, 243, 287, 288, 292, 295, 297, 299, 300, 301, 302, 348, 359, 405, 415, 416, 419, 420, 426, 427
fluid extract, x, 200, 201, 204, 233, 236, 243, 287, 288, 292, 295, 297, 299, 300, 301, 302, 359, 405, 419, 420, 427
fluidized bed, 271
fluorescence, 196, 201, 349, 367
folate, 444
food additive(s), 182, 192, 200, 417, 423, 429, 439
Food and Drug Administration, 236, 429, 430
food chain, 246, 247
food industry, 33, 48, 99, 168, 173, 208, 227, 236, 419, 430, 439
food intake, 125, 227
food poisoning, 414

food production, 165
food products, 170, 227, 238, 305
food safety, 177, 412, 429
food spoilage, 168
food web, 44
force, 169, 173, 226, 231
Ford, 119, 130, 163
forebrain, 357
formation, 16, 17, 44, 60, 65, 66, 68, 69, 72, 74, 79, 83, 84, 85, 92, 102, 104, 124, 131, 154, 155, 156, 172, 184, 226, 231, 234, 247, 263, 308, 309, 311, 316, 347, 355, 362, 371, 398, 399, 424, 431, 436, 437, 445
formula, 195
foundations, 25
fragility, 351
fragments, 125
France, 93, 130, 137, 182, 183, 228, 366, 370, 444
free radicals, 83, 107, 109, 112, 122, 131, 170, 171, 173, 196, 200, 226, 346, 347, 348, 399
freezing, 99, 179, 274
friction, 226
frost, 9
fructose, 75, 111, 377
fruits, 12, 14, 47, 48, 53, 54, 65, 70, 76, 79, 87, 88, 97, 99, 100, 131, 163, 182, 184, 197, 198, 207, 274, 305, 306, 307, 320, 337, 347, 355, 366, 378, 399, 416, 425
functional food, 1, 208, 352, 365, 366, 368, 398
funding, 329
fungal infection, 212, 316, 367, 397, 417
fungi, 34, 36, 37, 38, 42, 62, 65, 66, 70, 74, 80, 82, 95, 185, 414, 417, 419
Fungicide, 37
fungus, 31, 32, 74, 79, 121, 333
furan, 199
fusion, 372

G

gastroenteritis, 437
gastrointestinal tract, 203
GC-FID, 290, 291
gel, 204, 343
gene expression, 11, 17, 172, 191, 200, 318, 338, 373, 379, 400, 408, 432, 444
genes, 17, 43, 66, 78, 124, 128, 274, 315, 333, 377, 384, 400, 430, 442
genetic background, 4, 26
genetic information, 50
genetic traits, 113
genetics, 43, 183
genome, 78, 305
genotype, v, 1, 3, 33, 116, 127, 137, 161, 163, 191, 396
genus, 2, 35, 36, 37, 38, 41, 49, 61, 98, 108, 212, 365, 366, 430
geographical origin, 116, 199, 202, 396, 408
Germany, 36, 43, 137, 139, 388
germination, 416
ginger, 413, 427
gland, 333
glasses, 227
glioblastoma, 363
global markets, 320
glucose, x, 53, 62, 75, 108, 111, 125, 140, 227, 334, 368, 373, 374, 385, 387, 388, 389, 390, 391, 392, 393, 396, 432, 433, 445
glucose tolerance, 374, 385
glucosidases, 28, 30
glucoside, 3, 4, 6, 12, 15, 54, 55, 57, 58, 62, 63, 64, 86, 88, 92, 94, 113, 137, 139, 141, 142, 145, 146, 147, 149, 150, 151, 152, 156, 157, 188, 190, 192, 195, 212, 224, 276, 277, 305, 314, 317, 338, 418
glutathione, 20, 79, 80, 81, 101, 154, 346, 431, 434
glycerol, 30, 104
glycine, 34
glycogen, 385
glycolysis, 373
glycoproteins, 73, 439
glycoside, 162, 235
glycosylation, 54, 67, 110, 123, 360
God, 109, 130
grape berry, 7, 16, 17, 18, 19, 20, 26, 28, 31, 32, 37, 38, 39, 43, 50, 66, 74, 82, 87, 89, 110, 111, 115, 119, 129, 206, 222, 310, 331, 340
grape parts, 181, 395
grape phenols, 181, 183, 192, 198
Grape Pomace, vi, 241, 273, 274, 275
Grape seeds, 298, 398, 429, 430
Grape stalks, 246, 251
GRAS, 166, 173, 236, 429, 430
grass, 15, 16, 36, 37, 42, 202
gravity, 198
Greece, 41, 42, 357, 367, 379, 405, 412
Greeks, 414, 430
groundwater, 38, 247
grouping, 140
growth factor, 379, 393
GSEs, 125, 126
guanine, 434
guidelines, 375

H

H. pylori, 418

habitat(s), 212, 441
hair, 342, 362
hair follicle, 362
harmful effects, 36
harvesting, 3, 14, 137, 319, 320, 340, 345, 396
hay fever, 430
hazardous waste, 269
haze, 75, 99, 103
headache, 348
healing, 430
health condition, 430
health effects, 158, 204, 213, 368, 402, 404, 430
health risks, 166, 170, 248
heart disease, 182, 187, 202, 203, 348, 366, 369, 370, 374, 380, 381, 382, 399, 403, 430, 431, 432
heart failure, 373, 375, 377, 379, 382
heating rate, 239
heavy metals, 245, 246, 247, 249, 250, 269, 270, 271, 272, 316
height, 251, 277, 366
Helicobacter pylori, 381, 415, 418, 424, 426, 435, 436, 442
heme, 76, 77, 78
hemicellulose, 235
hemisphere, 2
hemodialysis, 208, 378, 403
hemorrhage, 348
hepatocarcinogenesis, 313, 330, 331, 335
hepatocellular carcinoma, 132, 335, 338
hepatocytes, 332
hepatoma, 313, 317, 333, 335
hepatotoxicity, 336
herbicide, 34, 35, 41, 42
heritability, 403
heterogeneity, 31, 82
hexane, 293, 335
high blood pressure, 339, 366, 368
high density lipoprotein, 372
high fat, 283, 374, 432, 433, 445
hippocampus, 357, 379, 408
histamine, 226
histidine, 68, 76
history, 167, 366, 414
HIV, 438, 439, 444
HIV-1, 438, 439, 444
homeostasis, 305, 348, 371, 378, 431
homocysteine, 362
hormone(s), ix, 78, 310 388
host, 416, 438, 439, 443
HPLC-DAD-ESI-MS, 87, 101, 139, 162, 163
HPLC-UV, 290
hue, 84
human body, 3, 347
human health, 38, 48, 65, 112, 185, 190, 191, 197, 198, 225, 226, 305, 314, 354, 377, 403, 407, 408, 412, 414, 415, 423, 429, 437, 442, 444
human leukemia cells, 313, 332
human subjects, 381
humidity, 37, 50, 74, 118
Hungary, 32
Hunter, 42
hybrid, 5, 33, 54, 66, 71, 87, 88, 89, 196, 201, 320, 335
hydrocarbons, 207, 310, 314
hydrogen, 71, 76, 77, 78, 83, 84, 104, 109, 128, 170, 171, 235, 238, 314, 347, 358
hydrogen atoms, 347
hydrogen bonds, 235
hydrogen peroxide, 71, 76, 77, 78, 83, 84, 104, 128
hydrolysis, 57, 61, 100, 102, 155, 184, 190, 213, 231, 232, 233, 235, 258, 293, 393, 432
hydroperoxides, 170, 174
hydroquinone, 75
hydroxide, 69, 131
hydroxyapatite, 250
Hydroxycinnamic acid derivatives, 155
hydroxyl, 56, 83, 110, 111, 112, 123, 132, 169, 171, 172, 183, 188, 190, 219, 238, 350, 354, 414
hydroxyl groups, 56, 111, 112, 123, 169, 171, 183, 188, 350, 354
hyperactivity, 182
hypercholesterolemia, 399
hyperglycemia, 388, 392, 393
hyperkalemia, 348
hyperlipidemia, 445
hyperplasia, 434
hypertension, 248, 348, 352, 356, 361, 368, 371, 375, 377, 380, 382, 399
hypertrophy, 366, 368, 370, 373, 376, 377, 378, 379, 384
hypodermis, 50
hypotension, 348
hypothesis, 71, 381

I

ICAM, 431, 432
ID, 201
identification, 41, 43, 54, 88, 128, 137, 139, 140, 143, 194, 209, 285, 302, 345, 363, 396, 419, 421, 425, 427, 428, 443, 446
IFN, 337
immune function, 393
immune response, 318, 392, 432
immune system, 226
immunocompromised, 436, 439

immunodeficiency, 438
immunomodulatory, 310, 398
impregnation, 175, 207
in vivo, 66, 71, 92, 94, 109, 131, 224, 227, 306, 313, 315, 317, 329, 330, 332, 335, 342, 348, 351, 352, 354, 362, 366, 368, 372, 376, 385, 399, 400, 402, 407, 415, 416, 425, 428, 439, 443
incidence, 31, 35, 48, 51, 109, 182, 305, 317, 365, 366, 370, 433
Independence, 78
independent variable, 289
India, 11, 22, 32, 33, 40, 318, 321, 322, 323, 324, 326, 327, 328
individuals, 113, 432, 433, 436, 439
inducer, 392
induction, 11, 17, 92, 124, 128, 131, 203, 221, 223, 313, 316, 353, 369, 372, 445
industrial processing, 284
industrial wastes, 241, 419, 426
industrialization, 248
industry(s), 37, 43, 132, 166, 171, 178, 181, 198, 200, 204, 205, 221, 249, 273, 274, 284, 287, 288, 398, 411, 412, 416, 417, 420, 421, 423, 425, 428, 429, 430, 437, 439
infection, 62, 74, 121, 224, 380, 416, 423, 437, 439, 441
inflammation, 109, 206, 226, 314, 329, 336, 366, 368, 370, 375, 376, 429, 430, 431, 432, 441, 443
inflammatory cells, 372
inflammatory responses, 315, 388, 391, 399
influenza, 444
infrared spectroscopy, 359
ingestion, 48, 247, 351, 399, 401, 403, 406, 408, 416, 423
ingredients, 241, 298, 300, 398, 417
inhibitor, 38, 71, 223, 311, 312, 338, 349, 371, 375
initiation, 8, 124, 171, 209, 310, 314, 315, 363, 373, 446
injury(s), 122, 200, 333, 347, 355, 356, 357, 372, 373, 377, 380, 382, 383, 385, 392, 399, 439
insects, 22, 38, 68, 70, 396
insulin, 227, 316, 335, 374, 375, 377, 379, 391, 393
insulin sensitivity, 374, 375
integration, 51
integrity, 66, 314
interface, 194
interference, 248
internal environment, 167
intervention, 48, 194, 208, 315, 318, 362, 407
intestinal flora, 416
intestinal tract, 346
intestine, 197, 213, 352, 408
intoxication, 82
intracellular calcium, 372
introns, 78
investment, 232, 296, 297, 320
ion transport, 314
ion-exchange, 284
ionization, 15, 139, 233, 358, 362, 363, 396
ions, 54, 66, 67, 82, 196, 197, 233, 271, 272
Iowa, 98
Iran, 336
iron, 23, 74, 77, 82, 103, 104, 171, 247, 248, 283
irradiation, 62, 99, 168, 201, 211, 212, 213, 214, 217, 218, 219, 220, 221, 222, 223
irrigation, 1, 10, 22, 23, 24, 25, 44
ischemia, 94, 204, 357, 360, 369, 372, 378, 382, 383, 385, 408
ischemia reperfusion injury, 369, 382, 408
ischemia-reperfusion injury, 372, 385
islands, 91, 163, 212
isolation, 93, 138, 202, 297
isomers, 5, 94, 110, 145, 156, 157, 197, 206
isoprene, 308
issues, 183, 296, 339, 430, 440
Italy, xiii, xv, 1, 12, 14, 25, 32, 43, 107, 113, 114, 115, 118, 125, 128, 130, 131, 183, 384

J

Japan, xix, xx, 211, 212, 215, 222, 224, 408

K

K^+, 261
kaempferol, 3, 8, 56, 57, 137, 145, 146, 147, 149, 150, 152, 184, 189, 204, 314, 365, 367, 397, 401, 402
ketones, 28, 310
kidney, 247, 248, 348
kill, 437
kinetic model, 301
kinetics, 67, 80, 95, 97, 221, 231, 232, 235, 237, 238, 240, 241, 294, 398, 424
Korea, 205
Kosovo, 332

L

labeling, 182
laboratory studies, 36
lactase, 401
lactate dehydrogenase, 353
lactic acid, 30, 92, 101
Lactobacillus, 61, 176

lakes, 248, 441
larvae, 437, 444
larval development, 437, 444
latency, 71
laws, 241
LC-MS, 357, 367
LDL, 130, 207, 316, 317, 318, 334, 347, 351, 353, 356, 369, 370, 372, 374, 378, 381, 384, 385, 399, 405, 408, 416, 432
leaching, 34, 181, 194, 290, 420
lead, 37, 48, 49, 60, 65, 76, 77, 80, 81, 83, 84, 85, 127, 232, 234, 247, 248, 251, 253, 259, 261, 270, 271, 346, 372, 373, 398, 421, 434
leaf disc, 211, 212, 214, 217, 218
leakage, 169, 439
learning, x, 400, 408
legume, 41, 42
lesions, 313, 331, 353
leucine, 349
leukemia, 313
leukoplakia, 203
liberation, 81, 82
lifestyle changes, 368
ligand, 76, 77, 78
light, 8, 9, 11, 13, 14, 19, 20, 36, 49, 51, 52, 53, 54, 57, 62, 112, 118, 119, 126, 127, 129, 145, 170, 193, 204, 214, 221, 236, 248, 261, 275, 276, 320, 358, 421
light conditions, 119
lignans, 183, 184
lignin, 17, 74, 76, 78, 99, 235, 245, 268, 330
linoleic acid, 191, 356
lipases, 28, 342, 359, 444
lipid metabolism, 366, 368, 373, 376, 432
lipid oxidation, 109, 166, 168, 170, 171, 172, 283, 347, 348
lipid peroxidation, 207, 317, 318, 347, 351, 356, 388, 392, 393
lipid peroxides, 430
lipids, 171, 175, 179, 230, 249, 374, 385, 416, 431
lipolysis, 227, 370, 373, 374
lipooxygenase, 346
lipoproteins, 226, 372, 399, 432
liquid chromatography, 88, 89, 90, 93, 100, 196, 200, 205, 206, 214, 243, 357, 359, 360, 362, 363, 367, 382, 421, 423
liquid phase, 234
liquids, 192, 202, 234, 236, 301, 302, 420
Listeria monocytogenes, 174, 177, 178, 415, 427, 435, 436, 440, 445, 446
lithium, 389
liver, 248, 274, 283, 311, 313, 333, 334, 337, 348, 400, 408, 430, 438
liver cancer, 313, 337
localization, 24, 60, 319, 396
longevity, 374, 398
low temperatures, 9, 235, 288
low-density lipoprotein, 103, 243, 305, 317, 347, 357, 362, 381, 385, 399, 404, 407, 432, 445
lower lip, 385
LSD, 113
lumen, 401
lung cancer, 305, 317, 335, 434, 445
lung disease, 306
lutein, 308, 317, 395, 398
lycopene, 395, 398
lymphocytes, x, 132, 387, 388, 389, 390, 391, 392, 393
lymphoma, 331, 332
lysozyme, 101

M

macroalgae, 272
macromolecules, 398
macronutrients, 23
macrophages, 337, 370, 372, 384
magnesium, 37
magnitude, 10, 89, 104, 149, 158, 221, 236, 259, 260, 414, 420
majority, 39, 50, 58, 310, 340, 342, 413, 415, 420
malaria, 437
malignant tumors, 125, 311
mammals, 38, 70, 223, 313, 351, 436
man, 203, 247
management, 1, 9, 11, 13, 15, 21, 22, 23, 24, 25, 27, 29, 30, 34, 35, 38, 40, 41, 44, 197, 245, 249, 337, 376, 402, 445
manganese, 17, 76, 372
manipulation, 289, 320
manufacturing, 48, 166, 296, 297, 302
manure, 13
marrow, 248
mass, 93, 139, 192, 196, 197, 202, 205, 206, 207, 214, 231, 234, 249, 253, 255, 257, 261, 262, 264, 265, 268, 281, 293, 294, 295, 335, 345, 355, 358, 359, 360, 362, 363, 366, 367, 368, 370, 374, 378, 380, 382, 419, 421
mass spectrometry, 93, 139, 196, 202, 205, 206, 214, 335, 345, 355, 358, 359, 360, 362, 363, 382, 419, 421
materials, 136, 165, 166, 168, 221, 233, 234, 235, 240, 245, 248, 249, 250, 258, 268, 271, 288, 297, 298, 302, 329, 398, 419, 446
matrix, 198, 203, 235, 237, 243, 274, 280, 283, 289, 293, 362, 363, 400, 402, 415, 431

matrix metalloproteinase, 203, 362
matter, 75, 289, 293
MB, 433
MCP, 432
MCP-1, 432
measurement(s), 113, 126, 195, 197
meat, 172, 174, 175, 176, 177, 178, 227, 274, 284, 418, 427, 428, 440, 446
mechanical properties, 167, 178
media, 125, 247, 288, 442
medical, xii, 130, 306, 412
medicine, ix, 335, 412, 418
Mediterranean, 6, 8, 10, 23, 49, 91, 163, 182, 183, 200, 202, 204, 205, 380, 400, 405
Mediterranean climate, 6, 8, 23
Mediterranean countries, 183
MEK, 337
melanin, 66, 67, 68, 70, 84
melanoma, 125, 313, 332, 336, 338
melatonin, 331
mellitus, 227
membrane permeability, 372
membranes, 38, 68, 238
memory, 400, 402, 404, 405, 408
memory function, 400, 405
mercury, 246
meta-analysis, 399
Metabolic, 200, 351, 408
metabolic changes, 373
metabolic disorder(s), 429
metabolic disorders, 429
metabolic pathways, xi, 10, 127, 274, 305, 306, 346, 358
metabolic syndrome, 352, 361, 374, 375
metabolism, 20, 78, 109, 118, 119, 124, 128, 130, 131, 132, 191, 197, 200, 201, 219, 223, 226, 248, 335, 346, 347, 351, 352, 372, 381, 384, 388, 391, 401, 402, 405, 407, 408, 409, 415, 444
metabolites, ix, 2, 7, 9, 28, 47, 50, 51, 107, 123, 197, 306, 308, 314, 317, 331, 337, 352, 376, 395, 396, 400, 401, 402, 406, 408, 412, 413
metabolized, 197, 376, 400
metabolizing, 310, 378, 393, 395, 402
metal ion(s), 66, 78, 103, 171, 251, 258, 259, 260, 261, 346
metal recovery, 251, 261, 263, 267, 268
metallurgy, 246
metals, 86, 109, 170, 246, 248, 251, 252, 253, 258, 259, 260, 261, 263, 267, 268, 269, 347
metastasis, 124, 125, 132, 334, 336, 425
meter, 118
methanol, 82, 138, 139, 192, 193, 195, 204, 214, 237, 239, 240, 290, 291, 337, 343, 344, 419, 420, 421, 434, 435
methodology, 195, 232, 233, 234, 239, 240, 241, 242, 289, 297, 299, 300, 335, 343, 344
methyl cellulose, 168, 173
methylation, 110, 111, 123, 246
Mg^{2+}, 261
mice, 109, 125, 227, 334, 336, 338, 353, 362, 370, 374, 377, 379, 383, 384, 385, 392, 405, 424, 431, 432, 444, 445
microbial communities, 35, 39, 441
microbiota, 29, 31, 33, 38, 39, 198, 401, 403, 408, 416, 427, 429
microclimate, 11, 118, 127, 129, 163
microorganism(s), 28, 31, 34, 36, 38, 39, 50, 167, 168, 170, 177, 178, 197, 412, 413, 419, 422, 423
Microwave assisted extraction, 204, 205
microwaves, 193, 201
middle lamella, 48, 81
migration, 348, 399, 405, 434, 443
mildew, 37, 38, 39, 40, 121, 207
milligrams, 195, 421
mitochondria, 126, 313, 443
mitogen, 125, 317, 353
mixing, 168, 195
MMP, 333
MMP-9, 333
model system, 97, 100, 103, 176, 283, 336, 425, 440, 446
modeling, 25, 92, 95, 208, 223, 287, 288, 293, 299, 300
models, 71, 96, 124, 301, 302, 315, 317, 342, 425, 429, 441, 444
modifications, 17, 54, 196, 314, 353, 374, 413
moisture, 34, 36, 37, 121, 167, 228, 229, 232, 275, 277
moisture content, 229, 232, 275, 277
mold, 74, 283
molecular mass, 68, 111
molecular oxygen, 66, 67, 69, 75
molecular structure, 2, 62, 171, 355
molecular weight, 68, 71, 81, 85, 86, 117, 128, 231, 248, 277, 342, 343, 345, 346, 372, 397, 404
molecules, 53, 58, 62, 69, 74, 75, 78, 81, 83, 84, 85, 86, 109, 111, 118, 123, 171, 182, 186, 226, 233, 267, 345, 346, 347, 350, 381, 400, 401, 405
molybdenum, 421
monolayer, 401
monomers, 4, 9, 18, 48, 58, 74, 76, 94, 123, 135, 137, 138, 139, 158, 160, 161, 188, 274, 344, 345, 346, 347, 349, 351, 352, 418
monoterpenoids, 305, 310, 337

Montenegro, 380
Moon, 316, 338, 357, 404, 444
morphology, 114, 115, 116, 119, 169, 215
mortality, 365, 366, 368, 381, 383, 385, 436, 437
motivation, 295
mRNA, 223
multivariate analysis, 202
mung bean, 74
mutagenesis, 107
mutant, 124
mutations, 434, 442
myocardial infarction, 348, 371
myocardial ischemia, 355, 383, 403
myocardium, 404

N

Na^+, 261
NAD, 374
natural compound, 184, 330, 413
natural polymers, 76
nausea, 82, 348
nebulizer, 139
necrosis, 338, 431
negative effects, 38, 182
nematode, 43, 444
neonates, 205
nerve, 433
Netherlands, 19, 442
neurodegenerative diseases, 109, 181
neurodegenerative disorders, 107, 197
neuropeptides, 388
neuroprotection, 400
neuroprotective agents, 395, 398
neurotransmitter(s), ix, 348
neutral, 196, 197
New England, 335
New Zealand, 12, 30, 32, 41, 103, 418
niacin, 360
nickel, 247, 248, 253, 259, 261, 270, 272
nicotinamide, 329
NIR, 23
nitric oxide, 226, 334, 336, 337, 356, 368, 370, 372, 377, 379, 380, 382, 383, 385, 399, 404, 407, 430, 431
nitric oxide synthase, 334, 368, 370, 372, 380, 385, 407, 430
nitrite, 275, 430, 440
nitrogen, 6, 16, 17, 19, 20, 21, 24, 30, 36, 37, 43, 50, 51, 101, 346
nitrogen compounds, 50, 51, 101
nitrogen fixation, 43
NMR, 6, 18, 19, 20, 344, 357
nodes, 12, 13, 118, 214
Non-flavonoids, 314
non-polar, 236, 237, 239, 420
norepinephrine, 373, 384
normal development, 16
North America, 4, 212, 213, 397, 430
Norway, 245
novel extraction techniques, 226, 240
Nrf2, 354
nuclear magnetic resonance, 6, 18
nucleic acid, 415
nucleic acid synthesis, 415
nucleus, 172
null, 38, 80
nutraceutical, 89, 127, 129, 221, 227, 384
nutrient, 16, 35, 36, 37
nutrient(s), ix, 13, 16, 19, 25, 34, 36, 65, 87, 307, 334, 336, 406

O

obesity, 227, 241, 274, 283, 366, 368, 374, 375, 376, 378, 384, 430, 432, 442, 445
occlusion, 371
oceans, 14, 441
OH, 38, 52, 56, 58, 59, 61, 62, 63, 69, 83, 84, 110, 112, 118, 123, 127, 172, 246, 258, 341, 350, 417, 418
Oidium, 121, 122, 128
oil, x, 108, 175, 176, 177, 191, 200, 204, 239, 240, 241, 242, 287, 288, 289, 292, 293, 294, 295, 296, 297, 298, 299, 300, 301, 302, 330, 332, 333, 334, 335, 336, 337, 383, 425, 427
oil production, 200, 204, 425
oligomerization, 342
oligomers, 9, 18, 48, 64, 185, 188, 227, 274, 344, 345, 347, 352, 359, 396
olive oil, 243, 400, 405
operations, 2, 65, 79, 200, 229, 230, 232
opportunities, 305, 444
optimization, 232, 243, 287, 288, 289, 290, 293
oral cavity, 198
oral health, 206
ores, 246
organ(s), 185, 219, 247, 351, 371
organelle(s), 66, 76, 126
Organic, 13, 19, 24, 290, 299, 338, 415
organic compounds, 83, 84, 193, 354
organic food, 29
organic matter, 34, 36
organic solvents, 244, 288, 293, 297, 302
organism, 66, 68, 248, 342, 414, 422, 440
ornithine, 433

ovarian cancer, 124
overproduction, 346
ox, 51, 169, 186, 310, 314
oxidation products, 74, 347
oxidative damage, 107, 109, 122, 197, 355, 358, 391, 392, 400, 401, 433, 443
oxidative destruction, 315
Oxidative processes, 49
oxidative reaction, 65, 79, 84, 86, 232
oxidative stress, xi, 107, 121, 129, 130, 131, 198, 241, 274, 283, 336, 342, 351, 352, 353, 356, 358, 369, 370, 375, 376, 383, 385, 391, 400, 407, 431, 433, 434, 437, 443
oxygen, 9, 66, 67, 68, 69, 73, 74, 76, 77, 78, 79, 80, 81, 82, 83, 86, 101, 103, 109, 110, 128, 131, 167, 170, 173, 196, 226, 236, 347, 350
oxygen consumption, 78, 79, 80
ozone, 92, 238

P

p53, 124, 313, 316, 317, 330, 360
Pacific, 99
palate, 60
pancreatic cancer, 330, 333
parallel, 16, 196, 199
parasite(s), 37, 436, 437, 443
parentage, 396
parents, 5, 397
partial least squares regression, 360
participants, 416, 445
partition, 93
pastures, 247
patents, 287, 295, 297
pathogenesis, 198, 318, 348
pathogens, ix, 3, 22, 42, 51, 68, 78, 166, 168, 185, 227, 240, 336, 396, 413, 417, 426
pathways, 2, 92, 124, 172, 191, 203, 306, 315, 316, 333, 346, 360, 373, 400, 409, 441, 442
PCA, 125
PCR, 29, 33, 35, 44
pea starch, 440, 443
pedicel, 49, 214
peptidase, 387, 388, 392, 393
peptide, 71, 363
peptide chain, 71
periodicity, 197
periodontal, 198, 200, 206, 207
periodontal disease, 198, 200, 206, 207
peripheral blood, 384, 439, 444
peripheral blood mononuclear cell, 384, 439, 444
permeability, 351, 385, 413, 419
permission, 218, 222
permit, 81
peroxidation, 361, 425, 433, 442
peroxide, 78, 83, 97, 131
peroxynitrite, 171, 283, 353
pesticide, 20, 41, 201, 206
pests, 37
PET, 139
petroleum, 247, 299
PGE, 227
pH, 17, 54, 55, 66, 70, 85, 111, 113, 118, 168, 170, 174, 195, 215, 225, 229, 230, 231, 251, 252, 253, 255, 257, 258, 261, 262, 266, 267, 271, 276, 277, 308, 389, 414, 415, 424
pharmaceutical(s), 20, 200, 239, 274, 288, 347, 412, 417, 423
pharmacokinetics, 405
pharmacology, 204, 335
phenol, 16, 19, 66, 69, 87, 95, 102, 110, 112, 115, 120, 181, 182, 183, 184, 193, 194, 196, 197, 198, 199, 200, 201, 207, 230, 243, 292, 314, 354, 421
phenol oxidation, 66
Phenolic acids, 110, 184
phenolic composition, ix, x, 1, 3, 4, 5, 6, 8, 10, 13, 14, 15, 18, 19, 20, 21, 23, 24, 48, 51, 65, 79, 86, 87, 88, 89, 91, 93, 101, 102, 104, 108, 109, 112, 116, 135, 136, 140, 162, 163, 199, 205, 207, 208, 225, 228, 229, 274, 335, 347, 361, 403, 406, 407
Phenolic constituents, 365, 366, 367
Phenols, vi, 94, 110, 111, 113, 181, 183, 184, 191, 192, 194, 196, 197, 198
phenomenology, 183
phenotypes, 395, 402
phenoxyl radicals, 439
phenylalanine, 17
phosphate, 16, 17, 195, 215, 389
phospholipids, 274
phosphorous, 16
phosphorus, 42, 307
phosphorylation, 125, 373, 399, 407, 409
photosynthesis, 7, 9, 11, 22
physical activity, 432
physical inactivity, 368
physical properties, 34, 295
physicochemical properties, 99
Physiological, 223, 380
physiological mechanisms, 51
physiology, ix, 9, 40, 43, 127, 331, 337
phytoremediation, 270
phytosterols, 290
PI3K, 333, 337, 354
PI3K/AKT, 333
piceatannol, 63, 211, 212, 213, 214, 217, 219, 220, 222, 223, 316, 332, 333, 334, 335, 337, 367

piceid, 62, 63, 65, 93, 94, 137, 186, 206, 211, 212, 213, 214, 215, 221, 222, 224, 314, 367, 418
pigmentation, 2, 52, 54, 197
pilot study, 360, 445
placebo, 357, 374, 375, 384, 407
planets, 14
plant growth, 330
plants, 2, 12, 14, 22, 36, 41, 47, 50, 54, 65, 66, 67, 68, 70, 71, 74, 76, 78, 95, 97, 98, 107, 111, 121, 128, 130, 138, 182, 184, 186, 197, 203, 206, 247, 248, 269, 316, 330, 333, 367, 396, 412, 413, 414, 423, 443, 445
plaque, 371, 431, 436, 437
plasma membrane, 334
plasminogen, 203, 375
platelet aggregation, 25, 131, 213, 224, 347, 353, 355, 369, 371, 373, 381, 399
platelets, 371, 379, 381, 382, 383, 385, 388, 392, 399
playing, 47, 127
pleasure, 29
polar, 236, 237, 239, 301, 420
polarity, 192, 230, 343
pollen, 16
pollen tube, 16
pollinators, 197
pollutants, 269
pollution, 226, 245, 246
polyamine(s), 39, 124
polymer(s), 18, 48, 66, 74, 84, 123, 158, 168, 171, 188, 225, 228, 235, 274, 343, 344, 345, 396
polymeric matrices, 168
polymeric products, 74
polymerization, 17, 60, 68, 84, 86, 91, 111, 123, 158, 172, 186, 239, 240, 343, 344, 349, 350, 351, 352, 357, 358, 361, 362, 396, 397
polypeptides, 166, 388
polysaccharides, 81, 102, 235, 239, 343, 400
polystyrene, 239, 298
polythene, 176
polyunsaturated fat, 166, 170, 173, 433, 442
polyunsaturated fatty acids, 166, 170, 173, 433, 442
population, 27, 29, 31, 32, 34, 37, 38, 40, 41, 43, 44, 209, 247, 297, 365, 366, 376, 402, 416
population density, 32, 247
population structure, 43
porosity, 34
portal vein, 401
Portugal, xi, xiii, xiv, xvi, xvii, xviii, xix, 43, 165, 287, 305, 339, 424
potassium, 16, 17, 19, 37, 48, 80, 177, 195, 249, 276, 307, 348
potato, 227, 244
potential benefits, x, 412
poultry, 172, 176, 413, 417, 418, 428, 437
precipitation, 81, 252
predators, 197
preparation, xii, 48, 154, 247, 270, 276, 296, 298, 300
prescription drugs, 441
preservation, 167, 411, 412, 423
preservative, 166, 417
President, xi
prevention, ix, 11, 77, 109, 124, 181, 202, 222, 270, 311, 318, 329, 334, 336, 342, 354, 359, 375, 377, 379, 380, 383, 384, 391, 406, 431, 442
primary function, 347
primate, 378
principal component analysis, 4
principles, 87, 177
probability, 50
probe, 194
probiotic, 416
Procyanidins, vi, 116, 195, 208, 339, 341, 343, 346, 348, 351, 353, 354, 356, 358, 363, 396
producers, x, 16, 30, 157, 158, 162, 183, 318, 319, 320
professionals, x
profitability, 296
prognosis, 399
progressive multifocal leukoencephalopathy, 439
pro-inflammatory, 226
project, xi, 329
proliferation, 125, 166, 310, 313, 317, 329, 330, 331, 335, 337, 348, 362, 363, 391, 399, 434
propagation, 113, 170, 171, 238
prostate cancer, 109, 125, 313, 317, 329, 330, 333, 336, 337, 342, 416
prostate carcinoma, 124, 128, 129, 311, 329, 353, 362, 428
protection, 29, 56, 81, 124, 132, 190, 204, 232, 248, 295, 306, 342, 347, 353, 360, 381, 403, 442, 446
protective role, 415, 423
protein kinase C, 346, 381
protein kinases, 353, 385
protein synthesis, 373, 413
proteins, 60, 66, 70, 72, 73, 74, 75, 76, 91, 96, 124, 126, 168, 169, 239, 248, 307, 329, 336, 343, 346, 372, 385, 388, 400, 413, 431, 434
proteolysis, 68
proteomics, 283, 429
prototypes, 412
pruning, 11, 13, 14, 118
Pseudomonas aeruginosa, 170, 258, 419, 435
public education, 376
public interest, 365, 367

pulp, 49, 50, 51, 52, 61, 107, 112, 113, 116, 191, 201, 203, 310, 335
pure water, 343
purification, 71, 96, 97, 99, 195, 196, 214, 298, 299, 361
purity, 419
PVC, 248
pyrophosphate, 308

Q

quality control, 49, 65, 206
quality improvement, 24
quality of life, 423
quality standards, 259
quantification, 204, 205, 344, 345, 367, 421
quinone(s), 67, 68, 69, 71, 72, 74, 79, 80, 83, 154, 166

R

radiation, 8, 9, 11, 13, 21, 22, 24, 51, 120
radical formation, 171
radicals, 76, 112, 122, 123, 130, 132, 170, 171, 173, 201, 226, 238, 346, 354, 362, 377, 393, 442
rainfall, 1, 7, 8, 10, 35, 127
rancid, 172
rape, 22, 26, 89, 102, 173, 213, 223, 287, 373, 428
rape seed, 173, 287
raw materials, 181
RE, 208
reactant, 421
reaction mechanism, 95, 103
reactions, 18, 49, 51, 60, 62, 65, 66, 72, 75, 76, 77, 78, 80, 83, 84, 85, 86, 98, 99, 101, 104, 109, 124, 171, 235, 261, 317, 348, 398, 421
reactive groups, 169
reactive oxygen, 107, 109, 122, 171, 200, 208, 346, 362, 372, 377, 384, 431, 445
reactive sites, 127
reactivity, 53, 96
reading, x, 195
reagents, 194, 208, 249, 251, 276, 422, 427
receptors, 373, 439, 443
recognition, 404, 414
recommendations, ix, 44
recovery, 192, 200, 203, 227, 234, 236, 237, 242, 246, 267, 268, 269, 271, 300, 302, 360, 411, 419, 423, 436
recovery process, 268
recurrence, 269
recycling, 247, 411, 423
red blood cells, 430
reducing sugars, 50, 114, 117
regenerate, 76
regeneration, 76, 246, 251, 261, 267
regions of the world, 226, 395, 418
regression, 332
relaxation, 226, 353, 371, 399, 402, 407
relevance, 6, 71, 81, 158, 242, 271, 347, 375, 423
reliability, 422
remote sensing, 23
renal failure, 248
renin, 346, 348, 360, 377
replication, 413, 438, 439
reproduction, 2
researchers, 4, 31, 65, 182, 249, 274, 288, 289, 293, 295, 347, 430, 436, 438, 440
resection, 336
residues, x, 30, 34, 39, 68, 181, 182, 183, 191, 192, 199, 200, 202, 204, 205, 206, 242, 249, 287, 288, 290, 293, 296, 297, 298, 300, 302, 411, 412, 418, 419, 420, 421, 423
resins, 50, 103
resistance, 3, 22, 124, 166, 167, 185, 219, 224, 348, 351, 368, 371, 382, 383, 412, 413, 418, 428
resolution, 197
resource availability, 15
resources, 408
respiration, 36, 38, 314
response, 10, 21, 24, 62, 121, 126, 166, 170, 173, 182, 195, 212, 213, 217, 221, 222, 223, 233, 234, 241, 242, 289, 299, 300, 316, 371, 375, 378, 380, 430, 431, 439, 445
restoration, 372
restrictions, 375
retail, 428
reticulum, 132
retinoblastoma, 336
retinol, 308
RH, 170, 171, 173
rheumatic diseases, 201
rheumatoid arthritis, 306, 338, 430
rhinitis, 444
rights, 297
rings, 2, 52, 58, 62, 110, 123, 184, 368, 399
Ripening, v, 1, 17, 163, 355
risk(s), ix, 9, 34, 107, 125, 187, 197, 201, 305, 306, 318, 330, 348, 365, 366, 368, 369, 370, 374, 375, 383, 385, 393, 406, 416, 432
risk factors, 201, 365, 366, 368, 374, 385, 393
RNA, 438
rodents, 375
ROOH, 78, 170, 171, 174

room temperature, 79, 195, 221, 232, 251, 273, 275, 276, 277, 278, 279, 280, 283, 420
root(s), 7, 62, 185, 208, 214, 219, 306, 414
routes, 401
Royal Society, 207
runoff, 34

S

safety, 36, 166, 167, 173, 236, 376, 418, 420, 429
salinity, 396
salmon, 169, 170
Salmonella, 170, 177, 413, 414, 415, 418, 422, 423, 424, 425, 426, 427, 435, 440, 443
salts, 111, 166, 247, 249
saturated fat, 289, 365, 366
saturated fatty acids, 289
saturation, 81, 251, 293, 298, 401
scaling, 240
scanning electron microscopy, 439
scavengers, 129, 171, 187, 347, 393
scholarship, 391
science, 87, 177, 337, 425
scientific knowledge, 183
scope, 288
scripts, 446
sea level, 115
seasonality, 319, 321, 328
secondary metabolism, 51
secrete, 30, 348, 408
secretion, 274, 316, 381, 391, 401
sedentary lifestyle, 431
sedimentation, 392
sediments, 193, 205, 207, 247
selectivity, 196, 236, 270, 420
selenium, 347
senescence, 121, 128
sensation, 60
sensing, 26
sensitivity, 114, 196, 374
septic shock, 430
sequencing, 33, 96, 305
serum, 131, 195, 318, 348, 351, 352, 361, 362, 384, 392, 401, 408, 431, 432
sewage, 248
shade, 12, 119
shape, 108, 232
shelf life, 76, 173, 174, 343
shock, 234, 238, 400
shock waves, 238
shoot(s), 11, 12, 13, 14, 15, 26, 119, 183, 185, 186, 188, 243
shortage, 320
showing, 48, 77, 124, 149, 184, 342, 436
side effects, 35, 348, 432
signal transduction, 128, 131
signaling pathway, 337, 378, 381, 399, 405, 439
signalling, 130, 381, 400
signals, xii, 73, 78, 227, 331, 336, 373, 400
silica, xii, 175
silk, 331
silver, 166, 247, 248
simulation, 271
skeletal muscle, 374
skeleton, 51, 62, 110, 314
skin cancer, 109, 342
skin diseases, 418
slag, 248
sludge, 205, 248, 258, 270
small intestine, 401, 416
smoking, 352, 399
smooth muscle, 226, 348, 399, 405
smooth muscle cells, 399
sodium, 71, 80, 96, 97, 174, 275, 276, 307, 348
software, 139, 390
soil type, 6, 20, 32, 274
solid phase, 195, 259, 260, 263, 267
solid surfaces, 166
solid waste, 152, 183, 204, 284, 398
solubility, 83, 192, 231, 232, 235, 237, 241, 287, 293, 295, 300, 314
solution, 42, 113, 124, 139, 194, 195, 198, 215, 231, 232, 239, 250, 251, 253, 258, 261, 262, 266, 267, 276, 277, 287, 297, 361, 427
solvents, 137, 139, 193, 225, 230, 231, 232, 235, 236, 239, 241, 243, 285, 288, 297, 343, 354, 358, 418, 419, 420, 424, 425
sorption, 232, 246, 249, 250, 251, 252, 253, 255, 258, 259, 260, 261, 262, 263, 264, 265, 267, 268, 269, 270
sorption curves, 232
sorption experiments, 251, 252, 261
sorption kinetics, 270
sorption process, 252, 253
South Africa, 18, 23, 30, 32, 92, 100, 136, 183, 318, 320
SP, 203, 411
Spain, v, xiii, xiv, xv, xvi, xvii, xviii, xix, xx, 3, 8, 27, 32, 33, 36, 41, 42, 47, 90, 91, 93, 135, 136, 137, 138, 162, 179, 181, 183, 199, 204, 209, 221, 225, 228, 240, 245, 249, 251, 320
specialists, ix, 61
specialization, xii, 66
speciation, 212, 250, 258, 270, 271
species richness, 31
spectrophotometric method, 96, 357

spectroscopy, 19, 23, 163, 367
speculation, 376
spin, 104
spleen, 430
sponge, 333
Spring, 42
squamous cell, 198
squamous cell carcinoma, 198
SS, 203
stability, 6, 34, 80, 81, 82, 88, 95, 102, 103, 123, 145, 154, 172, 175, 176, 177, 203, 207, 232, 234, 240, 241, 242, 268, 271, 273, 274, 275, 278, 280, 281, 284, 354, 398, 423
stabilization, 54, 60, 62, 75, 81, 85, 86, 94, 100, 101, 110, 158, 346
stabilizers, 248
stable complexes, 60
stable radicals, 171, 173
standard deviation, 117, 118, 120, 121, 122, 126, 276
standard error, 218, 220
standardization, 422
starch, 175, 272, 427
state(s), 2, 48, 51, 67, 68, 69, 77, 78, 107, 122, 182, 235, 246, 247, 248, 249, 251, 287, 288, 357, 389, 391, 399, 418, 433
statistics, 7
steel, 248, 444
Stilbenes, 62, 156, 185, 208
stimulation, 34
stomach, 226, 352, 415, 424, 425
storage, 25, 75, 76, 80, 82, 83, 85, 101, 102, 145, 166, 168, 172, 174, 177, 192, 221, 227, 283, 320, 332, 398, 419
stress, 1, 7, 10, 11, 16, 23, 62, 78, 109, 120, 121, 128, 131, 132, 185, 316, 367, 371, 373, 382, 396, 400, 403
stress factors, 396
stress response, 120
stroke, 371, 432
structural characteristics, 112
structural modifications, 83
structural variation, 58
style, x, 33, 374
styrene, 239
subgroups, 152
substitutes, 182
substitution(s), 53, 55, 56, 123, 172, 235, 414
substrate(s), 32, 36, 37, 66, 67, 68, 69, 71, 72, 74, 75, 76, 78, 95, 99, 154, 219, 235, 285, 389, 401, 427
sulfate(s), 16, 38, 71, 80, 96, 235, 351, 376
sulfur, 38, 44, 74, 103
sulfur dioxide, 74, 103
sulphur, 28, 38, 248
Sun, 26, 129, 178, 186, 193, 204, 208, 313, 332, 333, 343, 344, 345, 361, 363, 392, 393
supercritical fluid extraction, x, 200, 233, 243, 288, 292, 297, 299, 300, 301, 302, 359, 405, 420
supplementation, 16, 107, 241, 283, 352, 357, 378, 384, 385, 400, 405, 406, 407, 441
suppliers, 388
suppression, 227, 358, 371, 372
surface area, 14, 234
surface tension, 192
surrogates, 438
survival, 40, 44, 50, 202, 377, 382, 404, 441
susceptibility, 1, 17, 48, 101, 233, 351, 356, 357, 362, 405, 407, 408, 418, 424
suspensions, 439, 444
sustainability, 430
sweat, 61
Switzerland, 24, 39, 381
symptoms, 431, 438
synergistic effect, 170, 347, 350, 352, 405
synthesis, 6, 7, 10, 17, 18, 23, 28, 36, 50, 57, 70, 74, 78, 92, 117, 127, 212, 219, 224, 227, 309, 316, 330, 373, 381, 413
systolic blood pressure, 352

T

T cell(s), 388, 391, 392, 432, 441
Taiwan, 302
tanks, 82, 389
tannins, 5, 7, 9, 17, 18, 48, 49, 50, 58, 60, 75, 81, 82, 85, 91, 102, 111, 128, 158, 160, 166, 182, 184, 191, 195, 225, 226, 228, 239, 249, 301, 340, 341, 354, 355, 357, 359, 360, 362, 396, 397, 414, 415, 416, 424, 437, 439, 445
target, 35, 38, 122, 181, 182, 192, 193, 194, 195, 196, 246, 269, 283, 287, 288, 315, 336, 370, 372, 377, 395, 401
tartrate-resistant acid phosphatase, 431
TCC, 435
techniques, 30, 65, 81, 87, 94, 95, 181, 183, 199, 202, 225, 226, 233, 236, 240, 241, 293, 297, 345, 362, 396, 420, 421
technology(s), x, 37, 168, 178, 181, 208, 233, 239, 240, 242, 245, 259, 267, 269, 287, 288, 289, 296, 297, 299, 424, 438
teeth, 248
temporal lobe, 400
terpenes, 21, 30, 51, 308, 310, 311, 314, 329, 336
Terpenoids, 306, 308, 311, 312, 313, 337
Terroir, 20, 128
testing, 16
texture, 17, 95

Thailand, 424
therapeutic agents, 178, 337, 412
therapeutic approaches, 346
therapeutic benefits, 430
therapeutic effects, 308, 339, 366
therapeutic targets, 348
therapeutics, 434
therapy, 222, 311, 359, 380, 383, 431
thermal degradation, 104, 288
thermal energy, 239
thermal oxidation, 243
thermal stability, x, 233, 281, 398, 409
Thermal Stability, vi, 273, 274
thermal treatment, 239, 241, 398
thermodynamic parameters, 271
thermostability, 76, 98, 99
thinning, 11, 12, 13, 21, 26
thrombin, 381
thrombosis, 347
thrombus, 371
thyroid, 78, 133
thyroid cancer, 133
tides, 14
tin, 248
tissue, 8, 11, 68, 108, 109, 113, 115, 121, 237, 248, 310, 334, 356, 419, 430
TNF, 400, 431
TNF-α, 431
tobacco, 74, 368
tobacco smoking, 368
tocopherols, 191, 346, 347
tonality, 79, 84
torsion, 123
Total Anthocyanin, 274, 276
total cholesterol, 432
total energy, 370, 374
Total Flavonoids, 274
Total Polyphenol, 274
toxic effect, 182, 433
toxic metals, 269
toxicity, 34, 38, 39, 43, 80, 82, 245, 246, 269, 342, 378, 383, 400, 403, 430
TP53, 329
TPA, 354, 433, 442
trade, 202, 319, 320, 328
trade-off, 202
training, 1, 11, 14, 15, 19, 24, 26, 113, 114, 125
traits, 6, 28, 44, 86
transcription, 131, 226, 315, 373, 385, 400
transcription factors, 131, 315, 385
transformation(s), 17, 54, 65, 79, 86, 107, 332
transition metal, 84
translation, 373
translocation, 334, 391
transparent medium, 233
transpiration, 11
transplantation, 109, 313
transport, 49, 237, 268, 296, 330, 372
transport costs, 296
transportation, 66
trial, 6, 352, 375, 384, 404, 406, 407
triggers, 66, 373
triglycerides, 171, 370, 374
trypsin, 96
tumor(s), 107, 109, 124, 181, 187, 190, 209, 313, 333, 334, 335, 336, 337, 354, 356, 363, 393, 431, 433, 442, 446
tumor growth, 335, 393
tumor necrosis factor, 442
tumor progression, 124, 442
tumorigenesis, 124, 131, 317
tungsten, 421
turbulence, 238
Turkey, 318, 321, 322, 325, 354
turnover, 17, 221
type 2 diabetes, 182, 200, 384, 392, 443
tyrosine, 68, 71, 213, 223

U

ulcer, 436
ulcerative colitis, 358, 446
ultrasound, 193, 194, 200, 201, 225, 233, 234, 241, 243, 420
Ultrasound assisted extraction, 241
ultraviolet irradiation, 206, 223, 224
underlying mechanisms, 367
uniform, 239
unions, 53, 59
United Kingdom (UK), 20, 24, 177, 366, 446
United Nations, 202
United States, 109, 183, 236, 242, 244, 318, 319, 320, 328, 366, 443
unstable compounds, 48
uric acid, 362
uric acid levels, 362
urine, xi, xii, 400, 401, 406, 408
urokinase, 203
Uruguay, 14, 21
USA, xvi, 21, 23, 42, 44, 87, 88, 89, 131, 175, 177, 204, 228, 275, 276, 337, 382, 388, 389
USDA, 19
UV, 2, 8, 9, 21, 22, 24, 54, 56, 62, 63, 78, 92, 99, 139, 154, 190, 201, 208, 211, 212, 213, 214, 217, 218, 219, 220, 221, 222, 223, 236, 275, 290, 316, 336, 367, 397, 433, 442

UV irradiation, 24, 92, 213, 221, 223, 316, 397
UV light, 9, 62, 78, 154, 217, 236, 275, 336
UV radiation, 9, 56
UV-radiation, 22

V

vacuole, 52
vacuum, 273, 275, 277, 278, 279, 280, 282, 283, 298
valence, 246
Valencia, 179, 270
valorization, 197, 288, 297
variables, 3, 102, 113, 144, 148, 153, 156, 228, 234, 289, 293
variations, 4, 7, 57, 110, 114, 234, 396
vascular system, 233, 351, 399
vasodilation, 369, 370, 371, 374, 378, 399
vasodilator, 339, 348, 370, 383
vegetable oil, 201, 302
vegetables, 47, 48, 65, 70, 76, 77, 88, 95, 97, 98, 100, 121, 131, 163, 182, 184, 197, 206, 207, 227, 306, 347, 376, 399, 407, 425
vegetation, 15, 34, 35, 39, 43, 74
vein, 370, 431, 442
velocity, 81, 83
vessels, 82, 125, 380
viral infection, 65, 439
virus replication, 444
viruses, 414, 437, 438, 439, 445
viscosity, 192, 236, 420
vitamin A, 307, 335
vitamin B6, 351
vitamin C, 48, 177, 206, 306, 307, 407
vitamin E, 48, 427
vitamins, 171, 308, 330, 347, 354, 433
Vitis vinifera L. Grapes, 305, 307
VLDL, 274
volatilization, 233
vomiting, 82

W

Washington, 24, 41, 42, 90, 163, 379
waste, ix, 140, 149, 162, 171, 200, 225, 226, 247, 248, 249, 250, 251, 270, 271, 272, 302, 398, 405, 416, 418, 424
waste disposal, 247
wastewater, 252, 272
waterways, 34
wavelengths, 53, 54
weakness, 32
weight gain, 227, 437
weight loss, 376
weight management, 374
weight ratio, 22
Western blot, 126
Western countries, 125
wild grapes, x, 212, 213, 214, 215, 222, 224
wine by-products, 225, 226, 235, 236, 240
Winemaking residues, 412
wood, 76, 82, 98, 191, 219, 241, 249
working conditions, 234
working memory, 409
World Health Organization (WHO), 366, 379, 381, 385
World War I, 248
worldwide, 27, 37, 183, 199, 212, 245, 247, 305, 307, 318, 319, 366, 368, 385, 437
wound healing, 74

Y

yeast(s), 27, 28, 29, 30, 31, 32, 33, 35, 37, 38, 39, 40, 41, 42, 43, 44, 45, 74, 75, 76, 85, 92, 417
yellow fever, 337, 437
yield, 1, 6, 7, 10, 11, 13, 14, 17, 19, 22, 24, 25, 36, 37, 39, 44, 50, 114, 123, 202, 225, 229, 231, 232, 233, 234, 237, 238, 239, 289, 293, 294, 295, 343, 411, 418, 419, 420, 440

Z

zeolites, 166
zinc, 248, 250, 270, 271